The Grothendieck Festschrift
Volume I

A Collection of Articles Written in Honor of the 60th Birthday of Alexander Grothendieck

P. Cartier

L. Illusie

N.M. Katz

G. Laumon

Yu.I. Manin

K.A. Ribet

Editors

Reprint of the 1990 Edition

Birkhäuser
Boston • Basel • Berlin

Pierre Cartier
Institut des Hautes Études Scientifiques
F-91440 Bures-sur-Yvette
France

Luc Illusie
Université de Paris-Sud
Département de Mathématiques
F-91405 Orsay
France

Nicholas M. Katz
Princeton University
Department of Mathematics
Princeton, NJ 08544
U.S.A.

Gérard Laumon
Université de Paris-Sud
Département de Mathématiques
F-91405 Orsay
France

Yuri I. Manin
Max-Planck Institut für Mathematik
D-53111 Bonn
Germany

Kenneth A. Ribet
University of California
Department of Mathematics
Berkeley, CA 94720
U.S.A.

Originally published as Volume 86 in the series *Progress in Mathematics*

Cover design by Alex Gerasev.

Mathematics Subject Classification (2000): 00B15, 00B30, 01A60, 01A70, 01A75 (primary); 11G09, 11G40, 11R54, 14A15, 14A20, 14A22, 14C05, 14C17, 14C35, 14C40, 14D07, 14D15, 14D20, 14D22, 14F05, 14F20, 14F25, 14F30, 14F40, 14G10, 14G15, 14G40, 14H20, 14H40, 14H52, 14H55, 14H60, 14J20, 14J70, 14L05, 14L30, 14M17, 16E10, 16W50, 17A30, 17B66, 18G50, 18G55, 19D55, 19E20, 22E65, 32L10, 46L85, 46L87, 52B30, 55R20, 55U05, 55U15, 57N35, 57R20, 57R30, 57S20, 58A10, 58B25, 58F11, 58F18, 58G05 (secondary)

Library of Congress Control Number: 2006936966

ISBN-10: 0-8176-4566-7 e-ISBN-10: 0-8176-4574-8
ISBN-13: 978-0-8176-4566-3 e-ISBN-13: 978-0-8176-4574-8

Printed on acid-free paper.

Birkhäuser

9 8 7 6 5 4 3 2 1

www.birkhauser.com (IBT)

P. Cartier L. Illusie N.M. Katz
G. Laumon Y. Manin K.A. Ribet
Editors

The Grothendieck Festschrift

A Collection of Articles Written in Honor of the 60th Birthday of Alexander Grothendieck

Volume I

1990

Birkhäuser
Boston · Basel · Berlin

Pierre Cartier
Institut des Hautes
Etudes Mathématique
91440 Bures-sur-Yvette
France

Luc Illusie
Departement de Mathématiques
Université de Paris-Sud
Centre d'Orsay
91405 Orsay Cedex
France

Nicholas M. Katz
Department of Mathematics
Princeton University
Princeton, NJ 08544
USA

Gerard Laumon
Departement de Mathématiques
Université de Paris-Sud
Centre d'Orsay
91405 Orsay Cedex
France

Yuri Manin
Steklov Institute
Moscow
USSR

Kenneth A. Ribet
Department of Mathematics
University of California
Berkeley, CA 94720
USA

Printed on acid-free paper

ISBN 0-8176-3427-4 ISBN 0-8176-3429-0
ISBN 3-7643-3427-4 ISBN 3-7643-3429-0

Printed and bound by Edwards Brothers, Inc., Ann Arbor, Michigan.
Printed in the U.S.A.

9 8 7 6 5 4 3 2

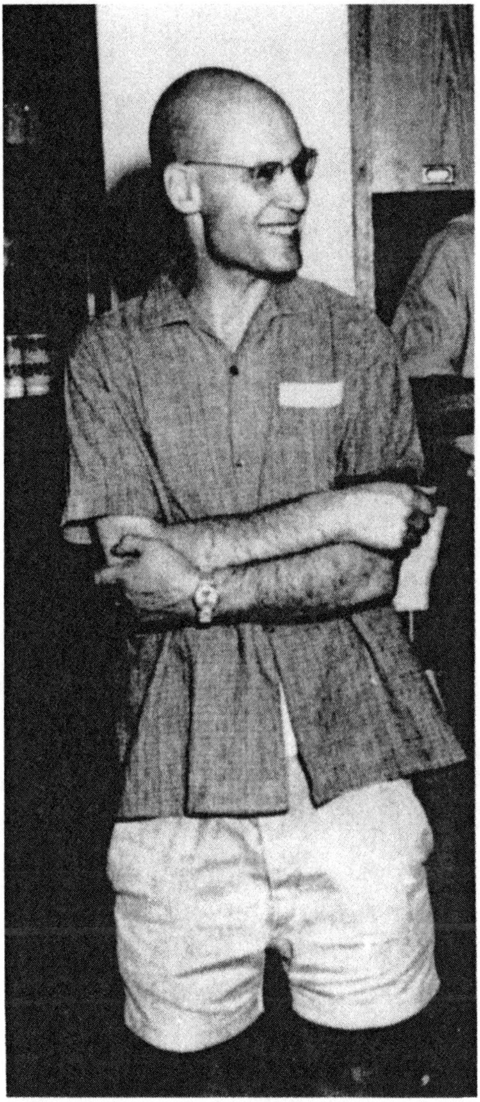

Alexander Grothendieck

Modern Birkhäuser Classics

Many of the original research and survey monographs in pure and applied mathematics published by Birkhäuser in recent decades have been groundbreaking and have come to be regarded as foundational to the subject. Through the MBC Series, a select number of these modern classics, entirely uncorrected, are being re-released in paperback (and as eBooks) to ensure that these treasures remain accessible to new generations of students, scholars, and researchers.

FOREWORD

It is difficult to grasp fully the magnitude of Alexander Grothendieck's contribution to and influence on twentieth century mathematics. He has changed the very way we think about many branches of mathematics. Many of his ideas, revolutionary when introduced, now seem so natural as to have been inevitable. Indeed, there is a whole new generation of mathematicians for whom these ideas are part of the mathematical landscape, a generation who cannot imagine that Grothendieck's ideas were ever absent.

Grothendieck was born in Berlin on March 28, 1928. After university studies at Montpellier, he spent the year 1948-49 as "auditeur libre" at the École Normale Supérieure. From 1949 to 1953 he worked in functional analysis at Nancy under J. Dieudonné and L. Schwartz; his doctoral thesis was "Produits tensoriels topologiques et espaces nucléaires" [18]. Between 1950 and 1958, he was supported by the Centre National de la Recherche Scientifique (CNRS). During this period, he spent 1953-55 at the University of São Paulo and 1955-56 at the University of Kansas.

By 1955, his interests had turned from his original field of functional analysis to topology and geometry; his Tôhoku article [28] dates from his visit to Kansas. In 1958, he outlined [32], with incredible prescience, a large part of the work that was to occupy him over the next decade. The next year he left the CNRS for a position of professor at the newly formed Institut des Hautes Etudes Scientifiques (IHES).

It is no exaggeration to speak of Grothendieck's years 1959-1970 at the IHES as a "Golden Age," during which a whole new school of mathematics flourished under Grothendieck's charismatic leadership. Grothendieck's Séminaire de Géométrie Algébrique (SGA) established the IHES as a world center of algebraic geometry, and him as its driving force. He received the Fields Medal in 1966. In looking back at this period, one marvels at the generosity with which Grothendieck shared his ideas with colleagues and students, the energy he and his collaborators devoted to meticulous redaction, the excitement with which they set out to explore a new land.

In 1970, Grothendieck left the IHES and abandoned mathematics as the exclusive focus of his energies. His active presence at the forefront of the mathematical scene has been deeply missed.

He spent 1970-72 as visiting professor at the Collège de France, and 1972-73 as visiting professor at Orsay. In 1973 he became professor at

Montpellier. From 1984 to 1988, he was on leave from Montpellier as "directeur de recherches" at the CNRS. He was awarded but declined the Crafoord Prize in 1988. He retired on October 1, 1988.

In his compelling memoir "Récoltes et Semailles: Réflexions et témoignages sur un passé de mathématicien" [56], Grothendieck discusses at length his life, his work, his relations with and his views on the mathematical community. One of Grothendieck's most striking images there is of mathematics as an edifice under construction, and of himself as a foreman-architect. He is deeply critical of what happened on the worksite after he left the job.

The mere enumeration of Grothendieck's best known contributions is overwhelming: topological tensor products and nuclear spaces, sheaf cohomology as derived functors, schemes, K-theory and Grothendieck-Riemann-Roch, the emphasis on working relative to a base, defining and constructing geometric objects via the functors they are to represent, fibred categories and descent, stacks, Grothendieck topologies (sites) and topoi, derived categories, formalisms of local and global duality (the "six operations"), étale cohomology and the cohomological interpretation of L-functions, crystalline cohomology, "standard conjectures", motives and the "yoga of weights", tensor categories and motivic Galois groups. It is difficult to imagine that they all sprang from a single mind. And yet this is only a partial picture of Grothendieck's work. There are numerous unpublished letters and manuscripts, some of which have already had an influence entirely out of proportion to their extremely limited circulation, others whose influence may only be felt in years to come. We can only hope that they will soon be published.

In contemplating Grothendieck's magnificent achievement, one is struck by the simplicity of the fundamental concepts which underlie it, and by the profound unity of thought which brought it into being. He has seen further, and shared his vision with us. By no means have all of his dreams been fully realized. Some are being actively pursued today; others may be for future generations to explore.

The articles presented here, collected on the occasion of Grothendieck's sixtieth birthday, are offered as a tribute to one of the world's greatest living mathematicians.

CONTENTS

VOLUME II

BIBLIOGRAPHIE D'ALEXANDER GROTHENDIECK

[1] *Sur la complétion du dual d'un espace vectoriel localement convexe.* C. R. Acad. Sc. Paris **230**, 605–606 (1950).

[2] *Quelques résultats relatifs à la dualité dans les espaces* $(\mathcal{F})$. C. R. Acad. Sci. Paris **230**, 1561–1563 (1950).

[3] *Critères généraux de compacité dans les espaces vectoriels localement convexes. Pathologie des espaces* $(\mathcal{LF})$. C. R. Acad. Sci. Paris **231**, 940–941 (1950).

[4] *Quelques résultats sur les espaces vectoriels topologiques.* C. R. Acad. Sci. Paris **233**, 839–841 (1951).

[5] *Sur une notion de produit tensoriel topologique d'espaces vectoriels topologiques, et une classe remarquable d'espaces vectoriels liée à cette notion.* C. R. Acad. Sci. Paris **233**, 1556-1558 (1951).

[6] *Critères de compacité dans les espaces fonctionnels généraux.* Amer. J. Math. **74**, 168–186 (1952).

[7] *Sur les applications linéaires faiblement compactes d'espaces du type* $C(K)$. Canadian J. Math. **5**, 129–173 (1953).

[8] *Sur les espaces de solutions d'une classe générale d'équations aux dérivées partielles.* J. Analyse Math. **2**, 243–280 (1953).

[9] *Sur certains espaces de fonctions holomorphes, I.* J. reine angew. Math. **192**, 35–64 (1953).

[10] *Sur certains espaces de fonctions holomorphes, II.* J. reine angew. Math. **192**, 77–95 (1953).

[11] *Quelques points de la théorie des produits tensoriels topologiques.* Segundo symposium sobre algunos problemas matemáticos que se están estudiando en Latino América, Julio 1954, 173–177. Centro de Cooperación Cientifica de la UNESCO para América Latina, Montevideo, Uruguay, 1954.

[12] *Espaces vectoriels topologiques.* Instituto de Matemática Pura e Aplicada, Universidade de São Paulo, 1954.

[13] *Résumé des résultats essentiels dans la théorie des produits tensoriels topologiques et des espaces nucléaires.* Ann. Inst. Fourier **4**, 73–112 (1952).

[14] *Sur certains sous-espaces vectoriels de* L^p. Canadian J. Math. **6**, 158–160 (1954).

[15] *Résultats nouveaux dans la théorie des opérations linéaires, I.* C. R. Acad. Sci. Paris **239**, 577–579 (1954).

[16] *Résultats nouveaux dans la théorie des opérations linéaires, II.* C. R.

Acad. Sc. Paris 239, 607–609 (1954).

[17] *Sur les espaces* $(\mathcal{F})$ *et* $(\mathcal{DF})$. Summa Brazil. Math. **3**, 57–123 (1954).

[18] *Produits tensoriels topologiques et espaces nucléaires.* Mem. Amer. Math. Soc. n° 16, 1955.

[19] *Une caractérisation vectorielle-métrique des espaces* L^1. Canad. J. Math. **7**, 552–561 (1955).

[20] *A general theory of fibre spaces with structure sheaf.* University of Kansas, 1955.

[21] Erratum au mémoire: *Produits tensoriels topologiques et espaces nucléaires.* Ann. Inst. Fourier **6**, 117–120 (1955-56).

[22] *Résumé de la théorie métrique des produits tensoriels topologiques.* Bol. Soc. Mat. São Paulo **8**, 1–79 (1956).

[23] *Théorèmes de finitude pour la cohomologie des faisceaux.* Bull. Soc. Math. France **84**, 1–7 (1956).

[24] *La théorie de Fredholm.* Bull. Soc. Math. France **84**, 319–384 (1956).

[25] *Sur la classification des fibrés holomorphes sur la sphère de Riemann.* Amer. J. Math. **79**, 121–138 (1957).

[26] *Sur certaines classes de suites dans les espaces de Banach, et le théorème de Dvoretzky-Rogers.* Bol. Soc. Mat. São Paulo **8**, 81–110, (1956).

[27] *Un résultat sur le dual d'une* C^**-algèbre.* J. Math. Pures Appl., **36**, 97–108 (1957).

[28] *Sur quelques points d'algèbre homologique.* Tôhoku Math. J. **9**, 119–221 (1957).

[29] *Algèbre homologique.* Séminaire A. Grothendieck, 1ère année, 1957, Secrétariat mathématique IHP, 11 rue Pierre et Marie Curie, 75005 Paris, 1958.

[30] *La théorie des classes de Chern.* Bull. Soc. Math. France **86**, 137–154 (1958).

[31] *Sur une note de Mattuck-Tate.* J. reine angew. Math. **200**, 208–215 (1958).

[32] *The cohomology theory of abstract algebraic varieties.* Proc. Internat. Congress Math. (Edinburgh, 1958), 103–118. Cambridge Univ. Press, New York, 1960.

[33] *The trace of certain operators.* Studia Math. **20**, 141–143 (1961).

[34] *Fondements de la géométrie algébrique* (*Extraits du Séminaire Bourbaki* 1957–62), Secrétariat mathématique IHP, 11 rue Pierre et Marie Curie, 75005 Paris, (1962).

[35] *Résidus et dualité, Prénotes pour un séminaire Hartshorne* 1963, (R. Hartshorne, Residues and Duality, Lecture Notes in Mathematics **20**, Springer-Verlag, Berlin-Heidelberg-New York, 1966.

[36] *Le groupe de Brauer*, III : *Exemples et compléments*, IHES, Mars 1966. Dix exposés sur la cohomologie des schémas, 88–188. North Holland, Amsterdam et Masson, Paris, 1968 (*).

[37] *On the de Rham cohomology of algebraic varieties.* Inst. Hautes Études Sci. Publ. Math. **29**, 95–103 (1966).

[38] *Un théorème sur les homomorphismes de schémas abéliens.* Invent. Math. **2**, 59–78 (1966).

[39] (avec Dieudonné J.) *Critères différentiels de régularité pour les localisés des algèbres analytiques.* J. Algebra **5**, 305–324 (1967).

[40] *Local cohomology.* A seminar given by A. Grothendieck, Harvard University, Fall 1961, Notes by R. Hartshorne, Lecture Notes in Mathematics **41**, Springer-Verlag, Berlin-New York, 1967.

[41] *Catégories cofibrées additives et complexe cotangent relatif.* Lecture Notes in Mathematics **79**, Springer-Verlag, Berlin-New York, 1968.

[42] *Classes de Chern et représentations linéaires des groupes discrets.* Dix exposés sur la cohomologie des schémas, 215–305. North Holland, Amsterdam; Masson, Paris, 1968.

[43] *Crystals and the de Rham cohomology of schemes*, Notes by J. Coates and O. Jussila. Dix exposés sur la cohomologie des schémas, 306–358. North Holland, Amsterdam; Masson, Paris, 1968.

[44] *Standard conjectures on algebraic cycles.* Algebraic Geometry (Internat. Colloq., Tata Inst. Fund. Res., Bombay, 1968), 193–199. Oxford Univ. Press, London, 1969.

[45] *Hodge's general conjecture is false for trivial reasons.* Topology **8**, 299–303 (1969).

[46] *Représentations linéaires et compactification profinie des groupes discrets.* (English summary) Manuscripta Math. **2**, 375–396 (1970).

[47] *The Responsibility of the Scientist Today.* Queen's Papers in Pure and Applied Mathematics, **27**, Queen's University, Kingston, Ontario 1971.

[48] (with Murre, Jacob P.) *The tame fundamental group of a formal neighbourhood of a divisor with normal crossings on a scheme.* Lecture Notes in Mathematics **208**, Springer-Verlag, Berlin-New York, 1971.

(*). Ce texte est la continuation des exposés au Séminaire Bourbaki [79] et [80].

[49] *Travaux de Heisouké Hironaka sur la résolution des singularités.* Actes du Congrès International des Mathématiciens (Nice, 1970), Tome 1, 7–9. Gauthier-Villars, Paris, 1971.

[50] *Groupes de Barsotti-Tate et cristaux.* Actes du Congrès International des Mathématiciens (Nice, 1970), Tome 1, 431–436. Gauthier-Villars, Paris, 1971.

[51] (with Seydi Hamet) *Platitude d'une adhérence schématique et lemme de Hironaka généralisé.* (English summary) Manuscripta Math. **5**, 323–339 (1971).

[52] *Topological vector spaces.* Translated from the French ([12]) by Orlando Chaljub. Notes on Mathematics and its Applications. Gordon and Breach Science Publishers, New York-London-Paris, 1973.

[53] *Groupes de Barsotti-Tate et cristaux de Dieudonné.* Séminaire de Mathématiques Supérieures. **45** (Été 1970). Les Presses de l'Université de Montréal, Montréal, Que., 1974.

[54] *A la poursuite des champs* (1983), non publié.

[55] *Esquisse d'un programme* (1984), non publié.

[56] *Récoltes et Semailles : réflexions et témoignage sur un passé de mathématicien.* Université des Sciences et Techniques du Languedoc (Montpellier) et CNRS (1985).

[EGA] : *Eléments de Géométrie Algébrique*, rédigés avec la collaboration de J. Dieudonné. Publications mathématiques de l'IHES :

[57] I. *Le langage des schémas.* **4** (1960) (seconde édition, Springer-Verlag 1971).

[58] II. *Étude globale élémentaire de quelques classes de morphismes.* **8** (1961).

[59] III. *Étude cohomologique des faisceaux cohérents. I.* **11** (1961).

[60] III. *Étude cohomologique des faisceaux cohérents. II.* **17** (1963).

[61] IV. *Étude locale des schémas et des morphismes de schémas. I.* **20** (1964)

[62] IV. *Étude locale des schémas et des morphismes de schémas. II.* **24** (1965).

[63] IV. *Étude locale des schémas et des morphismes de schémas. III.* **28** (1966).

[64] IV. *Étude locale des schémas et des morphismes de schémas. IV.* **32** (1967).

Exposés au Séminaire Bourbaki (*)

[65] *Produits tensoriels topologiques et espaces nucléaires*, 1952/53, n° 69.

[66] *La théorie de Fredholm*, 1953/54, n° 91.

[67] *Réarrangements de fonctions et inégalités de convexité dans les algèbres de von Neumann munies d'une trace*, 1954/55, n° 113.

[68] *Sur un mémoire de A. Weil : "Généralisation des fonctions abéliennes"*, 1956/57, n° 141.

[69] *Théorèmes de dualité pour les faisceaux algébriques cohérents*, 1956/57, n° 149.

[70] *Géométrie formelle et géométrie algébrique*, 1958/59, n° 182.

[71] *Technique de descente et théorèmes d'existence en géométrie algébrique, I : Généralités. Descente par morphismes fidèlement plats*, 1959/60, n° 190.

[72] *Technique de descente et théorèmes d'existence en géométrie algébrique, II : Le théorème d'existence en géométrie formelle des modules*, 1959/60, n° 195.

[73] *Techniques de construction et théorèmes d'existence en géométrie algébrique, III : Préschémas quotients*, 1960/61, n° 212.

[74] *Techniques de construction et théorèmes d'existence en géométrie algébrique, IV : Les schémas de Hilbert*, 1960/61, n° 221.

[75] *Technique de descente et théorèmes d'existence en géométrie algébrique, V : Les schémas de Picard : Théorèmes d'existence*, 1961/62, n° 232.

[76] *Technique de descente et théorèmes d'existence en géométrie algébrique, VI : Les schémas de Picard : Propriétés générales*, 1961/62, n° 236.

[77] *Fondements de la géométrie algébrique, commentaires*, 1961/62, complément.

[78] *Formule de Lefschetz et rationalité des fonctions L*, 1964/65, n° 279. (**)

[79] *Le groupe de Brauer, I : Algèbres d'Azumaya et interprétations diverses*, 1964/65, n° 290. (**)

[80] *Le groupe de Brauer, II : Théorie cohomologique*, 1965/66, n° 297. (**)

(*) publiés par W. A. Benjamin, Inc., New York, 1966.

(**) reproduit dans "Dix Exposés sur la cohomologie des schémas", North Holland, Amsterdam et Masson, Paris, 1968.

Exposés au Séminaire Chevalley (Institut Henri Poincaré, Secrétariat mathématique, 11 rue Pierre et Marie Curie, 75005 Paris).

Classification des groupes de Lie algébriques (1956/58).

[81] *Généralités sur les groupes algébriques affines. Groupes algébriques affines commutatifs.* Exp. 4.

[82] *Compléments de géométrie algébrique. Espaces de transformations.* Exp. 5.

[83] *Les théorèmes de structure fondamentaux pour les groupes algébriques affines.* Exp. 6.

[84] *Sous-groupes de Cartan, éléments réguliers. Groupes algébriques affines de dimension* 1. Exp. 7.

Anneaux de Chow et applications (1958).

[85] *Sur quelques propriétés fondamentales en théorie des intersections.* Exp. 4.

[86] *Torsion homologique et sections rationnelles.* Exp. 5.

Exposés au Séminaire Cartan 1960/61 : Familles d'espaces complexes et fondements de la géométrie analytique (W. A. Benjamin, Inc., New York, 1967).

Techniques de construction en géométrie analytique :

[87] I : *Description axiomatique de l'espace de Teichmüller et de ses variantes.* Exp. 7-8.

[88] II : *Généralités sur les espaces annelés et les espaces analytiques.* Exp. 9.

[89] III : *Produits fibrés d'espaces analytiques.* Exp. 10.

[90] IV : *Formalisme général des foncteurs représentables.* Exp. 11.

[91] V : *Fibrés vectoriels, fibrés projectifs, fibrés en drapeaux.* Exp. 12.

[92] VI : *Étude locale des morphismes ; germes d'espaces analytiques, platitude, morphismes simples.* Exp. 13.

[93] VII : *Étude locale des morphismes ; éléments de calcul infinitésimal.* Exp. 14.

[94] VIII : *Rapport sur les théorèmes de finitude de Grauert et Remmert.* Exp. 15.

[95] IX : *Quelques problèmes de modules.* Exp. 16.

[96] X : *Construction de l'espace de Teichmüller.* Exp. 17.

SGA : Séminaire de Géométrie Algébrique du Bois-Marie (*)

[97] SGA 1 *Revêtements étales et groupe fondamental*, 1960-61.
Dirigé par A. Grothendieck, Lecture Notes in Mathematics **224**, Springer-Verlag, Berlin-Heidelberg-New York, 1971.

[98] SGA 2 *Cohomologie des faisceaux cohérents et théorèmes de Lefschetz locaux et globaux*, 1961-62.
Dirigé par A. Grothendieck, North-Holland Publishing Company, Amsterdam, 1968.

[99] SGA 3 *Schémas en groupes*, 1962-64.
Dirigé par M. Demazure et A. Grothendieck.
Tome I. *Propriétés générales des schémas en groupes*, Lecture Notes in Mathematics **151**, Springer-Verlag, Berlin-Heidelberg-New York, 1970.
Tome II. *Groupes de type multiplicatif, et structure des schémas en groupes généraux*, Lecture Notes in Mathematics **152**, Springer-Verlag, Berlin-Heidelberg-New York, 1970.
Tome III. *Structure des schémas en groupes réductifs*, Lecture Notes in Mathematics **153**, Springer-Verlag, Berlin-Heidelberg-New York, 1970.

[100] SGA 4 *Théorie des topos et cohomologie étale des schémas*, 1963-64.
Dirigé par M. Artin, A. Grothendieck, J.-L. Verdier
Tome I. *Théorie des topos*, Lecture Notes in Mathematics **269**, Springer-Verlag, Berlin-Heidelberg-New York, 1972.
Tome II. Lecture Notes in Mathematics **270**, Springer-Verlag, Berlin-Heidelberg-New York, 1972.
Tome III. Lecture Notes in Mathematics **305**, Springer-Verlag, Berlin-Heidelberg-New York, 1973.

[101] SGA 5 *Cohomologie l-adique et fonctions L*, 1965-66.
Dirigé par A. Grothendieck, Lecture Notes in Mathematics **589**, Springer-Verlag, Berlin-Heidelberg-New York, 1977.

[102] SGA 6 *Théorie des intersections et théorème de Riemann-Roch*, 1966-67.
Dirigé par P. Berthelot, A. Grothendieck, L. Illusie, Lecture Notes in Mathematics **225**, Springer-Verlag, Berlin-Heidelberg-New York, 1971.

(*) Nous omettons de la liste (SGA 4 1/2, par P. Deligne, *Cohomologie étale*, Lecture Notes in Mathematics **569**, Springer-Verlag, Berlin-Heidelberg-New York, 1977), qui ne correspond à aucun séminaire du Bois-Marie.

[103] SGA 7 *Groupes de monodromie en géométrie algébrique*, 1967-69.
Tome I, Dirigé par A. Grothendieck, Lecture Notes in Mathematics **288**, Springer-Verlag, Berlin-Heidelberg-New York, 1972.
Tome II, par P. Deligne et N. Katz (*) Lecture Notes in Mathematics **340**, Springer-Verlag, Berlin-Heidelberg-New York, 1973.

(*) Bien que signée de Deligne et Katz, cette partie du séminaire a néanmoins été dirigée par Grothendieck.

De L'Analyse Fonctionnelle aux Fondements de la Géométrie Algébrique

JEAN DIEUDONNÉ

Le talent exceptionnel d'Alexander Grothendieck s'est révélé très tôt. Il avait fait ses études universitaires à Montpellier, où à cette époque l'enseignement des mathématiques à l'Université était un des plus sclérosés qu'on pût trouver en France ; ce n'est donc pas là qu'il eût pu être mis au courant des grands problèmes en suspens. Il passa l'année scolaire 1948-49 à Paris, où il suivit le premier des fameux "Séminaires H. Cartan", consacré cette année-là aux débuts de la Topologie algébrique (qui alors n'était enseignée nulle part ailleurs en France). Mais—ce qui est étonnant quand on pense à la suite de sa carrière—Grothendieck ne s'y intéressa que médiocrement ; il était plus attiré par ce qu'il avait entendu dire sur l'Analyse fonctionnelle, et sur les conseils de Cartan, il arriva à Nancy en Octobre 1949. A ce moment-là, Delsarte, Godement, Schwartz et moi-même y avions organisé un Séminaire sur les espaces vectoriels topologiques, théorie où nous travaillions tous dans diverses directions.

La théorie des espaces de Banach et de la dualité dans ces espaces, alors déjà ancienne, était bien comprise vers 1950 ; mais par contre celle des espaces localement convexes généraux ne faisait que débuter ; ceux qu'on connaissait le mieux étaient de types très particuliers, notamment les espaces de suites (Köthe), et surtout les espaces de fonctions et de distributions étudiés par Sobolev et Schwartz. Il convenait donc de chercher s'il existait des propriétés générales qui rendraient compte du comportement de ces espaces particuliers. Schwartz et moi-même avions commencé une telle étude pour les espaces de Fréchet et leurs limites inductives ; mais nous y avions rencontré toute une série de questions auxquelles nous ne savions pas répondre. Nous avons donc proposé à Grothendieck de les étudier, et le résultat dépassa rapidement nos espérances. En moins d'un an, il avait résolu tous nos problèmes, au moyen de très ingénieuses constructions ; puis, continuant sur sa lancée, il se mit à aborder de nombreuses autres questions. Quand il s'est agi en 1953 de lui décerner un doctorat, il fallut choisir entre six mémoires, dont chacun aurait fait une bonne thèse. Bien entendu, ce fut son grand mémoire sur les espaces nucléaires et les produits tensoriels topologiques qui fut retenu ; il lui assura rapidement une place de choix parmi les spécialistes internationaux d'Analyse fonctionnelle.

Il y abordait un sujet entièrement nouveau, l'étude des topologies "raisonnables" sur le produit tensoriel de deux espaces localement convexes ; seul le cas des espaces de Banach avait fait l'objet de travaux antérieurs. Dans l'étude aussi originale que profonde qu'il en fit, on reconnaît déjà sa "patte"; bien qu'à cette époque on ne parlât guère encore de catégories, c'est déjà leur esprit qui domine, dans la recherche constante de définitions "naturelles" et de propriétés "fonctorielles", qui deviendra systématique dans ses travaux ultérieurs. Mais à côté de grands théorèmes généraux, c'est à chaque instant qu'apparaît un ingénieux contre-exemple, pour délimiter exactement la portée de ces théorèmes. Sa plus remarquable découverte fut celle des espaces nucléaires, obtenue par comparaison entre deux topologies possibles sur des produits tensoriels ; cette catégorie d'espaces, jusque là totalement insoupçonnée, se révéla être la plus proche possible, par ses agréables propriétés, des espaces de dimension finie ; et Grothendieck montra que les beaux résultats connus pour les espaces de distributions (notamment le fameux "théorème des noyaux" de Schwartz) provenaient tout simplement de ce que ces espaces sont nucléaires. Depuis lors, les espaces nucléaires ont trouvé bien d'autres applications, notamment en Calcul des Probabilités. Une autre idée tout à fait nouvelle est l'étude des applications linéaires continues entre espaces localement convexes qui *se factorisent* à travers un espace $L^p(\mu)$ pour une mesure μ convenable. Grothendieck leur a consacré un mémoire spécial intitulé "Résumé de la théorie métrique des produits tensoriels topologiques", qui est devenu la source de toute une théorie consacrée à l'étude de la géométrie des espaces de Banach.

Ainsi, en moins de trois ans était créée une œuvre dont l'impact sur la théorie des espaces vectoriels topologiques ne peut, à mon avis, être comparé qu'à celui des travaux de Banach.

Mais déjà, après un cours donné à São Paulo en 1953, Grothendieck va s'en éloigner. Dans sa thèse, il avait signalé la possibilité d'appliquer ses résultats à l'algèbre homologique et à la théorie des faisceaux. Ce sont ces théories vers lesquelles il se tourne, ainsi que la Topologie algébrique, la Géométrie algébrique et la Géométrie analytique, toutes en pleine effervescence dans la décennie 1950-60, et réagissant sans cesse les unes sur les autres. Guidé en partie par Serre avec qui il entretient une abondante correspondance, Grothendieck entreprend de s'y initier dès la fin de 1954, et il y met la même rapidité déconcertante qu'il avait déjà manifestée pour ses progrès dans l'Analyse fonctionnelle ; alors qu'en 1954 il avoue ne pas se sentir à l'aise dans le maniement des suites spectrales, en 1956 il y témoigne une virtuosité qui stupéfie Serre lui-même.

En 1955, il est invité pour le premier semestre à l'Université du Kansas, où il s'occupe surtout d'espaces fibrés analytiques et algébriques, et d'algèbre homologique. Il obtient la classification des fibrés vectoriels holomorphes sur la droite projective, première étape d'une théorie encore très active aujourd'hui. En vue des applications à la Géométrie algébrique qu'il commence à envisager, il prolonge (sans le savoir tout d'abord) les idées de MacLane et de Buchsbaum sur les catégories abéliennes. Comme à son habitude, il va droit au cœur du sujet, en donnant un système d'axiomes pour les catégories abéliennes ; son résultat le plus important est la preuve que les faisceaux de modules forment une catégorie abélienne avec suffisamment d'objets injectifs, ce qui lui permet de définir la cohomologie à valeurs dans un tel faisceau sans restriction sur le type de faisceau ni sur l'espace de base. Ce travail est longtemps resté un classique pour les spécialistes de l'algèbre homologique (qui l'appelaient "Tôhoku", du nom du journal où il avait été publié).

Rentré en France en 1956, Grothendieck participe activement au mouvement mathématique de l'école française ; il s'oriente de plus en plus vers la Géométrie algébrique, et étudie de façon approfondie les mémoires de Serre (alors tout récents) sur les variétés algébriques définies sur un corps algébriquement clos, mais de caractéristique quelconque. Surtout, il amorce l'évolution qui va le conduire, en Géométrie algébrique, à passer de théorèmes "absolus" pour une variété aux théorèmes "relatifs" correspondants concernant les morphismes. Ces derniers, en Géométrie algébrique, n'avaient attiré l'attention que tardivement ; dans les ouvrages classiques, il était surtout question d'applications birationnelles, non partout définies en général ; les morphismes n'en apparaissaient que comme cas particuliers, les "applications régulières" définies partout. Dans FAC, Serre en avait donné une définition comme morphismes d'espaces localement annelés, mais sans beaucoup les utiliser. Par contre, la notion de morphisme était au cœur des travaux contemporains sur les variétés analytiques (K. Stein, Grauert et Remmert).

La notion de morphisme propre, version "relative" de la notion de "variété complète" de Weil, permit à Grothendieck de donner une version "relative" du théorème de Serre établissant que, pour une variété complète X et un $\mathcal{O}_X$-Module cohérent $\mathcal{F}$, les groupes de cohomologie $H^j(X;\mathcal{F})$ sont de dimension finie sur le corps de base k. Grothendieck prouva en 1956 que si $f : X \to Y$ est un morphisme propre, les "images supérieures" $R^q f_*(\mathcal{F})$ d'un $\mathcal{O}_X$-Module cohérent $\mathcal{F}$ sont des $\mathcal{O}_Y$-Modules cohérents.

C'est dans cette même direction qu'il obtint peu après son premier grand théorème de Géométrie algébrique, la version "relative" du théorème de Riemann-Roch-Hirzebruch. En 1953, à la suite des travaux de Cartan-Serre et de Kodaira-Spencer sur l'application de la cohomologie des faisceaux aux variétés holomorphes, Hirzebruch avait obtenu un théorème général applicable à une variété algébrique complexe M, projective, sans singularité et de dimension m ; si E est un fibré vectoriel complexe holomorphe de base M, on a la formule

$$\sum_{i=0}^{m}(-1)^i \dim H^i(\mathcal{F}) = \kappa_{2m}(ch(E) \smile td_M) \tag{1}$$

où $\mathcal{F}$ est le faisceau des germes de sections holomorphes de E, $ch(E)$ le caractère de Chern de E, et td_M la somme des polynômes de Todd $T_k(c_1, c_2, \ldots, c_k)$ en les classes de Chern du fibré tangent de M ; le second membre de (1) est la valeur $\langle u, [M]\rangle$ pour la classe fondamentale $[M]$, de la somme u des termes du cup-produit $ch(E) \smile td_M$ qui appartiennent à $H^{2m}(M; \mathbb{Q})$. Lorsque M est une courbe et E le fibré en droites associé à un diviseur de M, la formule (1) entraîne le théorème de Riemann-Roch classique.

Pour comprendre le passage de (1) à une version "relative", il est commode de se placer d'abord dans le cas où X, Y sont deux variétés projectives complexes sans singularité, de dimensions $m = \dim X$, $n = \dim Y$, et où $f : X \to Y$ est un morphisme. Le but initial est d'établir une relation entre $ch_X(E) \smile td_X$ et $ch_Y(E') \smile td_Y$, où E' est relié à E au moyen de f, d'une manière inconnue au départ ; la relation doit se réduire à (1) lorsque Y se réduit à un point.

La première idée est que, pour n'importe quel $z \in H^\bullet(X; \mathbb{Q})$, $\kappa_{2m}(z)$ doit être remplacé par $f_*(z)$, où $f_* : H^\bullet(X; \mathbb{Q}) \to H^\bullet(Y; \mathbb{Q})$ est un homomorphisme. Bien entendu, ce n'est pas "naturel" puisque $H^\bullet$ est un foncteur contravariant. Mais si l'on utilise la dualité de Poincaré pour X et Y, suivant la méthode de Hopf-Gysin, on obtient bien des homomorphismes $f_* : H^p(X; \mathbb{Q}) \to H^{p+d}(Y; \mathbb{Q})$ avec $d = 2(n - m)$. Lorsque $n = 0$, f_* n'est non nul que si $p = 2m$, et on a bien alors la relation $f_*(z) = \kappa_{2m}(z)$. Le sens à donner à E' de façon à retrouver le premier membre de (1) lorsque $n = 0$ est beaucoup plus caché. La version "relative" de $H^j(\mathcal{F})$ est l'image directe supérieure $R^j f_*(\mathcal{F})$ de Leray ; ce sont des $\mathcal{O}_Y$-Modules cohérents, nuls sauf pour un nombre fini de j. Mais par quoi remplacer la somme alternée des dimensions de $H^j(\mathcal{F})$? C'est ici qu'intervient la nouvelle idée, ce qu'on a appelé le *groupe de Grothendieck* $K(Y)$. On considère: (1) l'ensemble $C(Y)$ des classes de $\mathcal{O}_Y$-Modules cohérents pour la relation d'isomorphisme; (2) le $\mathbb{Z}$-module $\mathbb{Z}^{(C(Y))}$ des combinaisons linéaires formelles des éléments de $C(Y)$; on note $[\mathcal{F}]$ la classe d'un $\mathcal{O}_Y$-Module $\mathcal{F}$ dans $C(Y)$ et $(e_{[\mathcal{F}]})$ la base canonique indexée par $C(Y)$; (3) le sous-$\mathbb{Z}$-module $S(Y)$ engendré par les éléments $e_{[\mathcal{F}]} - e_{[\mathcal{F}']} - e_{[\mathcal{F}'']}$ pour toutes les suites exactes

$$0 \longrightarrow \mathcal{F}' \longrightarrow \mathcal{F} \longrightarrow \mathcal{F}'' \longrightarrow 0 \tag{2}$$

de $\mathcal{O}_Y$-Modules cohérents. Alors $K(Y)$ est défini par

$$K(Y) = \mathbb{Z}^{(C(Y))}/S(Y). \tag{3}$$

On reconnaît la construction usuelle d'un objet "universel" ; toute *application* $\varphi : C(Y) \to G$ dans un groupe commutatif G, telle que

$$\varphi([\mathcal{F}]) = \varphi([\mathcal{F}']) + \varphi([\mathcal{F}''']) \tag{4}$$

pour toute suite exacte (2), se factorise uniquement en

$$C(Y) \xrightarrow{e} \mathbb{Z}^{(C(Y))} \xrightarrow{\gamma_Y} K(Y) \xrightarrow{\psi} G \tag{5}$$

où $e : [\mathcal{F}] \to e_{[\mathcal{F}]}, \gamma_Y$ est l'homomorphisme canonique et ψ un *homomorphisme de groupes* .

La somme alternée

$$\sum_q (-1)^q \gamma_Y(e([R^q f_*(\mathcal{F})])) \tag{6}$$

a alors un sens et ne dépend que de la classe $[\mathcal{F}]$ dans $C(X)$. Si on l'écrit $\varphi([\mathcal{F}])$, la relation (4) est conséquence de la suite exacte de cohomologie

$$\cdots \longrightarrow R^q f_*(\mathcal{F}') \to R^q f_*(\mathcal{F}) \longrightarrow R^q f_*(\mathcal{F}'') \longrightarrow R^{q+1} f_*(\mathcal{F}') \longrightarrow \cdots$$

qui n'a qu'un nombre fini de termes $\neq 0$. L'expression (6) s'écrit donc

$$f_!(\gamma_X(e([\mathcal{F}])))$$

où $f_! : K(X) \to K(Y)$ est un homomorphisme de groupes.

Il reste à définir un "caractère de Chern"

$$ch_X : K(X) \longrightarrow H^\bullet(X; \mathbb{Q})$$

qui soit un homomorphisme. Grothendieck utilise le fait que tout $\mathcal{O}_X$-Module cohérent $\mathcal{F}$ a une résolution *finie*

$$0 \longleftarrow \mathcal{F} \longleftarrow \mathcal{L}_0 \longleftarrow \mathcal{L}_1 \longleftarrow \cdots \longleftarrow \mathcal{L}_r \longleftarrow 0,$$

où les $\mathcal{O}_X$-Modules $\mathcal{L}_j$ sont localement libres ; les classes totales de Chern $c(\mathcal{L}_j)$ sont donc définies. On définit alors la classe de Chern totale

$$c(\mathcal{F}) = c(\mathcal{L}_0)c(\mathcal{L}_1)^{-1}c(\mathcal{L}_2)\cdots c(\mathcal{L}_r)^{(-1)^r}$$

dans l'anneau $H^\bullet(X;\mathbb{Q})$, et on en déduit $ch(\mathcal{F})$ par la formule usuelle.

Le théorème de Riemann-Roch-Grothendieck peut alors s'énoncer

$$ch_Y(f_!(x)) \smile td_Y = f_*(ch_X(x) \smile td_X) \tag{7}$$

pour tout élément $x \in K(X)$.

La démonstration de Grothendieck est complètement différente de celle de Hirzebruch, et de nature purement algébrique ; le morphisme f est factorisé en $f = g \circ h$, où $h : X \to P^N \times Y$ est une immersion et $g : P^N \times Y$ est la seconde projection ; il faut naturellement prouver que $(g \circ h)! = g_! \circ h_!$. Le cas de la projection g est facile, mais pour l'immersion h il faut des arguments assez compliqués. Pour adapter la démonstration aux variétés complètes sans singularité sur un corps algébriquement clos de caractéristique quelconque, il faut remplacer $H^\bullet(X;\mathbb{Q})$ par $A(X) \otimes \mathbb{Q}$, où $A(X)$ est l'anneau des classes de cycles (anneau de Chow) et y définir des classes de Chern.

Je suis entré dans quelque détail pour montrer la richesse de l'imagination de Grothendieck. Il ne publia pas lui-même sa démonstration de (7), laissant ce soin à Borel et Serre, — qui avaient animé un séminaire à Princeton pour exposer cette preuve d'après des papiers fournis par Grothendieck —, premier exemple de ce qui allait devenir chez lui une coutume : poussé par les idées qui se pressaient en foule dans son esprit, il laissait souvent à ses collègues ou élèves le travail de leur mise au point dans tous les détails.

A peine prouvée sa version du théorème de Riemann-Roch, il allait s'embarquer dans une entreprise de tout autre envergure. Dans la conférence qu'il prononça au Congrès International d'Edimbourg en 1958, il mentionne pour la première fois le grand dessein qui va être au centre de ses préoccupations pendant dix ans : définir pour les variétés algébriques sur un corps de caractéristique $p > 0$ des groupes de cohomologie à coefficients dans un corps *de caractéristique* 0, ayant les propriétés énumérées par Weil en vue de prouver ses fameuses conjectures. Sans dévoiler encore les détails de la méthode qu'il envisageait pour y parvenir, Grothendieck conçoit qu'il lui faut pour cela procéder à une *refonte de toute la Géométrie algébrique* —semblable à celle que Weil lui-même avait dû faire dans ses "Foundations" en vue d'établir ses conjectures pour le cas particulier des courbes. Il en esquisse les traits principaux, qu'il allait développer dans les milliers de pages d'articles et de Séminaires que l'on connaît.

Depuis que Weil avait défini des "variétés abstraites" non plongées dans un espace projectif, les diverses définitions qu'on pouvait en donner reposaient sur une définition préalable des "variétés affines" que l'on "recollait" ensuite pour arriver aux variétés générales. Dans FAC, Serre avait observé qu'une "variété affine" M définie initialement comme "ensemble algébrique" dans un espace k^n (pour k algébriquement clos) correspond biunivoquement à une k-algèbre réduite sur M. Inversement, une telle algèbre A définit d'abord l'ensemble M, qu'on peut considérer comme ensemble des caractères $A \to k$, ou ensemble des idéaux maximaux de A ; puis la topologie de M, ayant pour base d'ouverts les ensembles $D(f) = \{x \in M | f(x) \neq 0\}$ pour $f \in A$; et enfin le faisceau d'anneaux locaux $\mathcal{O}_M$, tel que $\Gamma(D(f), \mathcal{O}_M)$ soit l'anneau localisé A_f pour tout $f \in A$.

Le "corps de base" k ne joue dans cette définition qu'un rôle secondaire, et Serre avait remarqué qu'il suffit de supposer que A est un anneau nœthérien dans lequel tout idéal premier est intersection d'idéaux maximaux. Kronecker avait déjà rêvé d'une géométrie algébrique qui serait "sur les entiers", et Weil avait attiré l'attention sur l'intérêt que présenterait une généralisation de ce genre dans sa conférence au Congrès International de Cambridge en 1950. Un pas vers une telle généralisation avait été fait par Chevalley et Nagata en 1955-56. Dans le Séminaire qu'il dirigea avec H. Cartan en 1955-56, Chevalley introduisit ce qu'il appelait des *schémas*, de nouveau par "recollement" de "schémas affines" ; ces derniers étaient à certain égards des généralisations des variétés affines de Serre, mais à d'autres égards ils en restreignaient la généralité. Le point de départ est encore une k-algèbre commutative A sur un corps k *quelconque*, mais A est supposé intègre. Il n'est pas question de faisceaux, et le "schéma affine" M défini par A est l'ensemble des anneaux locaux $A_{\mathfrak{p}}$ pour tous les idéaux *premiers* de A, et non seulement les idéaux maximaux. La topologie ("de Zariski") sur M a alors pour ensembles fermés les ensembles $E(\mathfrak{a})$ correspondant à tous les idéaux $\mathfrak{a}$ de A, $E(\mathfrak{a})$ étant l'ensemble des anneaux locaux $A_{\mathfrak{p}}$ pour $\mathfrak{p} \supset \mathfrak{a}$. Nagata montra aussitôt après qu'on peut étendre toutes les définitions de Chevalley en remplaçant k par un anneau de Dedekind.

Grothendieck avait suivi le Séminaire Cartan-Chevalley et connaissait le travail de Nagata sur ce qu'il appelait les "schémas arithmétiques". Il eut le mérite de voir que pour avoir une idée juste de ce que devait être la Géométrie algébrique, il fallait se débarrasser de *toutes* les restrictions qui figuraient dans ces diverses définitions. Il appela donc schéma affine le *spectre premier* $X = \text{Spec}(A)$ d'un anneau commutatif *arbitraire* A, ensemble dont les éléments sont tous les idéaux premiers de A. La topologie sur X a pour base d'ouverts les ensembles $D(f) = \{\mathfrak{p} | f \notin \mathfrak{p}\}$, pour f parcourant A. Enfin, on définit $(X, \mathcal{O}_X)$ comme espace localement annelé en prenant le faisceau d'anneaux $\mathcal{O}_X$ tel que, pour tout $f \in A$, $\Gamma(D(f), \mathcal{O}_X) = A_f$;

la fibre au point $\mathfrak{p}$ de ce faisceau est l'anneau local $A_{\mathfrak{p}}$. La définition d'un *schéma*(*) quelconque est alors la plus générale possible : c'est un espace localement annelé $(X, \mathcal{O}_X)$, admettant une base d'ouverts (dits affines) (U_α) pour lesquels $(U_\alpha, \mathcal{O}_X|U_\alpha)$ est un schéma affine. Les schémas forment une catégorie quand on prend pour morphismes les morphismes d'espaces localement annelés ; les schémas affines en forment une sous-catégorie pleine, équivalente à l'opposée de la catégorie de tous les anneaux commutatifs.

Ces définitions avaient été esquissées par Grothendieck dans sa conférence de 1958. Il y soulignait deux caractères prédominants qu'il entendait donner à sa théorie. L'un était de mettre au premier plan l'étude des morphismes plutôt que celle des schémas : un schéma de "base" S une fois fixé (et sur lequel on ne fait souvent aucune hypothèse particulière), les objets de l'étude sont les couples (X, f) formés d'un schéma X et d'un morphisme $f : X \to S$; on les appelle *S-schémas* ; un S-morphisme $(X, f) \to (Y, g)$ de S-schémas est alors un morphisme $h : X \to Y$ tel que $g \circ h = f$, de sorte que S remplit le rôle des "corps de base" classiques. Un schéma peut être considéré comme un S-schéma pour $S = \mathrm{Spec}(\mathbb{Z})$. Si S est le spectre d'un corps K (réduit à un point), les S-schémas peuvent être appelés schémas algébriques sur K.

L'autre direction privilégiée par Grothendieck était l'extension aux schémas et le développement des techniques de cohomologie des faisceaux inaugurées par Serre. Il considère les groupes de cohomologie, non seulement des faisceaux cohérents sur un schéma, mais aussi ceux des faisceaux *quasi-cohérents* utilisés par P. Cartier dans sa thèse ; sur un schéma affine $\mathrm{Spec}(A)$, ils correspondent aux A-modules quelconques.

De 1959 à 1970, Grothendieck fut membre permanent de l'Institut des Hautes Etudes Scientifiques (IHES) qui venait d'être fondé ; il y anima un Séminaire où vinrent rapidement assister de nombreux élèves, auxquels ils distribuait généreusement des idées de recherche que suscitait sa théorie, et ne cessait de prodiguer conseils et suggestions. Pendant cette décennie, il poursuivit sous deux formes la publication de ses résultats. D'une part, les notions de base de la théorie étaient exposées sous forme didactique dans les Éléments de Géométrie algébrique (EGA) publiés par livraisons successives dans les Publications mathématiques de l'IHES. Simultanément, il développait les parties plus avancées de la théorie, soit dans des exposés assez succincts au Séminaire Bourbaki, soit dans son Séminaire de l'IHES, où

(*) Initialement Grothendieck appelait ces objets des "préschémas", les schémas étant des préschémas particuliers soumis à une condition de "séparation" analogue à celle de Serre dans FAC. Plus tard, il adopta la terminologie de "schémas séparés" pour ce qu'il avait d'abord appelé des schémas.

l'aide de collègues et d'élèves lui permettait d'entrer dans plus de détails. Son influence fut immédiate et immense sur les algébristes du monde entier. Déjà en 1962, dans sa conférence au Congrès International de Stockholm, Serre pouvait constater que la théorie des schémas était dès cette époque le cadre qui semblait le mieux adapté à toute la Géométrie algébrique.

Ce succès se comprend sans peine. Même dans les parties "élémentaires" de la théorie, le point de vue des schémas donne toujours l'impression d'être celui qui convient exactement à la question, que toutes les présentations antérieures ne font qu'obscurcir ou déformer. Je me bornerai dans les exemples qui suivent aux propriétés des schémas qui ne sont pas simplement des adaptations à peu près évidentes de notions classiques.

(I) En premier lieu, l'idée de *spécialisation* d'un point x devient une notion topologique : x' est spécialisation de x si x' appartient à l'adhérence $\overline{\{x\}}$. En particulier, toute partie fermée irréductible de X est l'adhérence d'un point unique, son point *générique* ; cela permet des raisonnements de continuité (pour la topologie de Zariski) analogues à ceux des géomètres italiens pour les variétés complexes.

(II) La nouveauté peut-être la plus importante est l'existence, dans la catégorie des S-schémas, d'un "produit catégorique" $X \times_S Y$ pour deux S-schémas définis par des morphismes $f : X \to S$, $g : Y \to S$: il existe alors deux S-morphismes $p_1 : X \times_S Y \to X$, $p_2 : X \times_S Y \to Y$ tels que $f \circ p_1 = g \circ p_2$ et pour deux S-morphismes $u : T \to X, v : T \to Y$, il y a un unique S-morphisme $w : T \to X \times_S Y$ tel que $u = p_1 \circ w$ et $v = p_2 \circ w$. Pour des schémas affines $S = \mathrm{Spec}(A)$, $X = \mathrm{Spec}(B)$, $Y = \mathrm{Spec}(C)$, on a $X \times_S Y = \mathrm{Spec}(B \otimes_A C)$; pour X quelconque, $S = \mathrm{Spec}(A)$, $Y = \mathrm{Spec}(B)$, on écrit $X \otimes_A B$ en place de $X \times_S Y$. Cela donne naissance à la technique fondamentale du "changement de base" général : étant donné un morphisme $S' \to S$, on note $X_{(S')}$ le produit $X \times_S S'$ et $f_{(S')}$ le morphisme $p_2 : X_{(S')} \to S'$; on dit que le S'-schéma $X_{(S')}$ et le morphisme $f_{(S')}$ se déduisent de X et f par le changement de base $S' \to S$. Lorsque S et S' sont des spectres de corps, on retrouve la classique "extension du corps de base". Mais l'idée est applicable à une multitude d'autres situations et fournit un outil d'une extraordinaire flexibilité, dont nous nous bornerons à donner quelques exemples.

(III) Pour tout point $s \in S$, soit $\mathcal{O}_s$ l'anneau local, fibre de $\mathcal{O}_S$ en ce point, et $\kappa(s) = \mathcal{O}_s/\mathfrak{m}_s$ son corps résiduel. Pour tout S-schéma X, de morphisme structural $f : X \to S$, le schéma $X_s = X \times_S \mathrm{Spec}(\kappa(s))$ s'identifie en tant qu'espace topologique à la fibre $f^{-1}(s)$, sous-espace de X.

(IV) Un S-schéma de type fini X peut donc être considéré comme une "famille de schémas algébriques" X_s sur les corps $\kappa(s)$, indexée par S; on peut se demander s'il est possible de démontrer des propriétés de X à partir de celles des X_s ; il faut pour cela une condition qui "relie" entre eux les X_s. A cet effet, Grothendieck a beaucoup utilisé la notion de *platitude* du morphisme f : issue de l'algèbre homologique, elle avait été mise en valeur par Serre, qui avait montré son intérêt en algèbre commutative. Elle a permis à Grothendieck de définir des types de morphismes plus particuliers, les morphismes lisses (qui correspondent à l'idée de fibration locale) et les morphismes étales (qui correspondent à l'idée d'isomorphisme local) ; ils jouent un rôle considérable dans toute la théorie.

(V) Si l'on veut étudier ce qui se passe non seulement dans la fibre $f^{-1}(s)$ mais dans un voisinage de cette fibre, il suffit de remplacer $\mathrm{Spec}(\kappa(s))$ par $\mathrm{Spec}(\mathcal{O}_s)$ dans le changement de base. En effet, $\mathcal{O}_s$ peut être considéré comme limite inductive des anneaux A_λ des ouverts affines U_λ contenant s ; moyennant certaines conditions de finitude, Grothendieck a développé une technique permettant de conclure d'une propriété de $X \times_S \mathrm{Spec}(\mathcal{O}_s)$ à la même propriété pour $f^{-1}(U_\lambda)$ pour un indice λ convenable. Cette même technique de "limite inductive" permet aussi de ramener les propriétés d'un S-schéma quelconque à celles de schémas sur une $\mathbf{Z}$-algèbre de type fini, réalisant ainsi le vieux rêve de Kronecker, et ramenant beaucoup de propriétés de S-schémas au cas où S est nœthérien.

(VI) Si $S = \mathrm{Spec}(A)$ où A est un anneau local, la théorie des S-schémas a des aspects plus simples lorsque A est complet, ou hensélien, ou un anneau de valuation discrète ; les changements de base $\mathrm{Spec}(A') \to \mathrm{Spec}(A)$ où A' a l'une de ces propriétés, sont très utiles dans les démonstrations.

(VII) Comme les anneaux commutatifs qui interviennent dans la théorie des schémas sont quelconques, les anneaux locaux $\mathcal{O}_s$ pour $s \in \mathrm{Spec}(A)$ peuvent avoir des éléments nilpotents, qui jusque là avaient été soigneusement éliminés de la Géométrie algébrique. Dès 1958, Grothendieck insistait au contraire sur l'utilité de les considérer : ils donnent accès à ce qui remplace, en Géométrie algébrique "abstraite", les propriétés "infinitésimales" de la Géométrie algébrique classique. Par exemple, si A est un anneau local d'idéal maximal $\mathfrak{m}$, et $S = \mathrm{Spec}(A)$, on associe à un S-schéma X de type fini sa "réduction modulo $\mathfrak{m}$" $X_0 = X \otimes_A (A/\mathfrak{m})$, fibre de f au-dessus de l'unique point fermé $\mathfrak{m}$ de S, et qui est donc un "schéma algébrique" sur le corps $A/\mathfrak{m}$. On cherche à ramener l'étude de X à celle de X_0. Pour cela, on considère les schémas $X_n = X \otimes_A (A/\mathfrak{m}^{n+1})$, qui ont même espace topologique sous-jacent que X_0 et peuvent être considérés comme des "voisinages infinitésimaux" de X_0, intermédiaires entre X_0 et X. Lorsque X est propre sur S et que A est complet, on peut dire que la connaissance des X_n entraîne celle de X du point de vue cohomologique. De façon précise, si $\mathcal{F}$ est un $\mathcal{O}_X$-Module cohérent et $\mathcal{F}_n = \mathcal{F} \otimes_A (A/\mathfrak{m}^{n+1})$ son image sur X_n, les $H^q(X_n;\mathcal{F}_n)$ forment un système projectif, et l'on a un isomorphisme canonique

$$H^q(X;\mathcal{F}) \xrightarrow{\sim} \varprojlim H^q(X_n;\mathcal{F}_n).$$

C'est la généralisation du "théorème des fonctions holomorphes" démontré par Zariski pour $q = 0$. De cette façon, Grothendieck put donner, dès 1958, des démonstrations cohomologiques de deux théorèmes-clés de Zariski sur les variétés algébriques, le théorème de connexion et le "théorème principal", et les étendre aux schémas ; il est remarquable que sa démonstration soit une récurrence descendante sur la dimension q. L'espace X_0, muni du système projectif de faisceaux d'anneaux locaux $\mathcal{O}_{X_n}$, est un cas particulier de ce que Grothendieck appela plus tard un schéma formel ; cela correspond à l'idée de complétion en algèbre topologique.

(VIII) L'étude de certaines propriétés d'un S-schéma X est souvent plus facile pour le schéma $X_{(T)}$ déduit de X par un changement de base $T \to S$ convenable. Mais il se pose alors le problème de la "descente" : peut-on conclure d'une propriété prouvée pour $X_{(T)}$ à la même propriété pour X ? Pour les variétés algébriques sur un corps k, le problème avait déjà été considéré par Weil pour les "extensions du corps de base". Dans le premier des six exposés qu'il fit au Séminaire Bourbaki de 1959 à 1962, Grothendieck aborda le problème d'une façon générale.

L'idée originale est de considérer pour un changement de base $f : T \to S$ la suite de morphismes

$$T \times_S T \times_S T \overset{p_{12}}{\underset{p_{13}}{\overset{\xrightarrow{p_{23}}}{\rightrightarrows}}} T \times_S T \underset{p_2}{\overset{p_1}{\rightrightarrows}} T \xrightarrow{f} S$$

où les p_i et p_{ij} sont les projections canoniques. Pour tout S-schéma X, notons $p_1^*(X)$ et $p_2^*(X)$ les $(T \times_S T)$-schémas déduits de $X_{(T)}$ par les deux changements de base p_1, p_2. Il y a alors un isomorphisme canonique $\sigma : p_1^*(X) \xrightarrow{\sim} p_2^*(X)$ de $(T \times_S T)$-schémas (échange des facteurs) ; en outre, si l'on désigne par $p_{ij}^*(\sigma)$ l'isomorphisme de $(T \times_S T \times_S T)$-schémas déduit de σ par le changement de base p_{ij}, on a

$$p_{13}^*(\sigma) = p_{12}^*(\sigma) \circ p_{23}^*(\sigma). \tag{8}$$

On appelle en général *donnée de descente* pour un morphisme $f : T \to S$ et un T-schéma Y un $(T \times_S T)$-isomorphisme

$$\sigma : Y \times_T (T \times_S T)_1 \xrightarrow{\sim} Y \times_T (T \times_S T)_2$$

(où le premier changement de base est p_1 et le second p_2) vérifiant (8). Le problème est de savoir s'il existe alors un S-schéma X tel que $Y = X_{(T)}$ et que la donnée de descente corresponde à X par la construction précédente; on dit alors que cette donnée de descente est *effective*.

Il n'est pas facile de donner des critères généraux d'effectivité ; Grothendieck put toutefois montrer qu'une donnée de descente est effective si $f : T \to S$ est fidèlement plat et quasi-affine, et Y quasi-affine sur T.

(IX) Une des idées les plus originales et les plus fécondes de Grothendieck a été l'utilisation systématique de la notion de *foncteur représentable* en théorie des schémas. L'association, à un objet X d'une catégorie quelconque $\boldsymbol{C}$, du foncteur contravariant $h_X : \boldsymbol{C} \to \boldsymbol{Ens}$ défini par $h_X(Y) = \mathrm{Mor}_{\boldsymbol{C}}(Y, X)$, paraît si simple qu'on ne s'attend guère à ce qu'elle puisse être appliquée avec profit. Mais la connaissance du foncteur h_X détermine l'objet X qui le "représente" à isomorphisme près. C'est ce fait que Grothendieck a utilisé de façon très fructueuse en montrant que les propriétés des S-schémas se "traduisent" de façon beaucoup plus simple en propriétés des foncteurs correspondants. En particulier, c'est par ce biais qu'il a abordé, dans les cinq autres exposés du Séminaire Bourbaki, la construction de certains schémas, dont on ne connaissait jusque là que des cas très particuliers, comme par exemple la construction de groupes algébriques (Weil, Chow, Néron-Samuel, Matsusaka). De façon générale, Grothendieck, pour les constructions qu'il envisage, définit d'abord le foncteur qui doit correspondre au S-schéma cherché ; il s'agit alors de prouver que ce foncteur est représentable, et le S-schéma qui le représente est celui que l'on veut construire.

Pour certains de ces foncteurs, il est aisé de prouver leur représentabilité sans hypothèse sur S. Par exemple, on généralise de cette façon aux schémas la notion de grassmanniennes d'un espace vectoriel E ; les éléments de la grassmannienne $\mathrm{Grass}_n(E)$ peuvent être définis comme les espaces vectoriels quotients de E, de dimension n. Pour tout schéma S et tout $\mathcal{O}_S$-Module quasi-cohérent $\mathcal{E}$, on définit l'ensemble $\mathrm{Grass}_n(\mathcal{E})$ comme formé des $\mathcal{O}_S$-Modules quotients de $\mathcal{E}$ qui sont localement libres et de rang n. Pour tout S-schéma T, soit $\mathcal{E}_{(T)}$ le $\mathcal{O}_T$-Module image réciproque du $\mathcal{O}_S$-Module $\mathcal{E}$; on considère alors le foncteur contravariant

$$T \longmapsto \mathrm{Grass}_n(\mathcal{E}_{(T)}). \tag{9}$$

Ce foncteur est toujours représentable ; autrement dit, il existe un S-schéma $\mathbf{Grass}_n(\mathcal{E})$ tel que le foncteur (9) soit isomorphe à

$$T \longmapsto \mathrm{Mor}_S(T, \mathbf{Grass}_n(\mathcal{E})).$$

On définit de la même manière un S-schéma "fibré vectoriel" sur S, correspondant à un $\mathcal{O}_S$-Module quasi-cohérent quelconque $\mathcal{E}$ (et non seulement à un $\mathcal{O}_S$-Module localement libre).

(X) Pour d'autres constructions à l'aide de foncteurs représentables, il faut imposer des conditions restrictives. Un des exposés du Séminaire Bourbaki est consacré à la définition du "quotient" d'un schéma par une "relation d'équivalence". Pour toute catégorie $\boldsymbol{C}$, une relation d'équivalence dans un objet X de $\boldsymbol{C}$ se définit comme un triplet (R, p_1, p_2) où R est un objet de $\boldsymbol{C}$ et p_1, p_2 deux morphismes $R \underset{p_2}{\overset{p_1}{\rightrightarrows}} X$ tels que, pour tout objet T de $\boldsymbol{C}$, les applications

$$\mathrm{Mor}_{\boldsymbol{C}}(T, R) \underset{p_2'}{\overset{p_1'}{\rightrightarrows}} \mathrm{Mor}_{\boldsymbol{C}}(T, X),$$

où $p_1'(u) = pu_1$, $p_2'(u) = pu_2$, soient telles que (p_1', p_2') soit une injection dont l'image est le *graphe* d'une relation d'équivalence dans l'ensemble $\mathrm{Mor}_{\boldsymbol{C}}(T, X)$. L'objet quotient, s'il existe, est le conoyau de la paire de morphismes (p_1, p_2), autrement dit un morphisme $p : X \to Y$ tel que le diagramme

$$R \underset{p_2}{\overset{p_1}{\rightrightarrows}} X \xrightarrow{p} Y$$

soit exact. Lorsque $\boldsymbol{C} = \boldsymbol{Sch}_{/S}$, Grothendieck a montré l'existence du quotient si le morphisme p_1 est fini et localement libre (ce qui entraîne qu'il en est de même de p_2) et si en outre, pour tout $x \in X$, $p_1(p_2^{-1}(x))$ est contenu dans un ouvert affine. Des contre-exemples montrent que ce n'est pas toujours le cas, même pour une variété propre et lisse sur un corps.

(XI) Les trois derniers exposés du Séminaire Bourbaki sont relatifs à la construction de schémas qui sont devenus des outils essentiels dans toutes les recherches de Géométrie algébrique, le *schéma de Hilbert* et le *schéma de Picard* d'un S-schéma X. Le premier, qui généralise et raffine les classiques "coordonnées de Chow", représente le foncteur qui, à tout S-schéma T, associe l'ensemble des sous-T-schémas fermés de $X \times_S T$ qui sont *plats* sur T. If faut ici supposer que S est nœthérien et X projectif sur S. La construction se ramène à celle de schémas représentant chaque foncteur

$\mathrm{Quot}^P_{X/S}(T)$ qui fait correspondre à T les quotients de $\mathcal{O}_{X_{(T)}}$ ayant un polynôme donné P comme polynôme de Hilbert (d'où le nom). Des contre-exemples montrent qu'on ne peut guère affaiblir les hypothèses sur S et X.

Enfin, le schéma de Picard devrait “paramétrer” de la même manière les classes d'isomorphie de $\mathcal{O}_X$-Modules inversibles, dont l'ensemble est le groupe de Picard $\mathrm{Pic}(X)$, isomorphe à $H^1(X;\mathcal{O}_X^*)$. Mais en général le foncteur correspondant $T \mapsto \mathrm{Pic}(X \times_S T)$ n'est pas représentable, il faut le modifier de diverses manières. Un cas simple est celui où S est localement nœthérien, $f : X \to S$ projectif et plat, admet une section, et a ses fibres géométriquement intègres. On pose alors

$$P(X) = H^1(X;\mathcal{O}_X^*)/H^1(S;\mathcal{O}_S^*)$$

et le foncteur $T \mapsto P(X \times_S T)$ est représentable par un S-schéma en groupes abéliens.

Ces techniques, ainsi que celles des EGA, vont être employées dans la série des Séminaires animés ou inspirés par Grothendieck, dont chacun est consacré à un ou plusieurs grands thèmes de la Géométrie algébrique. Il ne saurait être question de résumer ces six mille pages.

Il y a peu d'exemples en mathématique d'une théorie aussi monumentale et aussi féconde, édifiée en si peu de temps et essentiellement due à un seul homme.

The presentation functor and the compactified Jacobian

ALLEN B. ALTMAN AND STEVEN L. KLEIMAN*

To A. Grothendieck
who gave us these means to treat
the Picard scheme and its compactification.

Introduction

This article continues the development, begun in [2], [4], [3], [5] and [14], of a theory of the compactified Jacobian that is modeled and based on Grothendieck's theory of the Picard scheme. Here we study the presentation functor, whose function is to bridge the gap between the compactified Jacobian J of a (relative) singular curve C and that J' of one of its partial normalizations C'. We shall place special emphasis on the case in which each geometric fiber of C has just one singularity more than C' and it is an ordinary node or cusp.

The presentation functor was introduced by Oda and Seshadri [16], who considered, by the methods of Mumford's geometric invariant theory, a possibly reducible curve C whose only singularities are ordinary nodes over an algebraically closed field. Their results about the functor were generalized in the case of an irreducible curve C by Kleppe [15], who proved the following theorem: if C is obtained by imposing a single node on a curve C' with arbitrary singularities, then the presentation functor is representable by a certain $\mathbb{P}^1$-bundle P/J' with two canonical sections, and J is obtained from P by identifying one section with a translation of the other, where the translation is induced by the difference of the two points on C lying over the node. The present authors proved that the corresponding theorem holds when a cusp is imposed: there is now only

AMS(MOS) subject classifications (1980). Primary 14H20; Secondary 14D22, 14C20.
* Partially supported by NSF grant 8801743 DMS.

one section, and it is identified with an infinitesmal translation of itself, where the translation is along a tangent vector that points out of the section and that projects in J' to a tangent vector τ, and τ is tangent to the translate of C' under the translation that carries the preimage Q' of the cusp Q to the base of τ. That result was proved by a messy direct computation. It was announced in **[13]**.

In this article, Kleppe's theorem and the corresponding theorem for a cusp are unified and generalized to families; moreover, the treatment here is more conceptual. In brief, this is what is done. The presentation functor is defined, and the associated étale sheaf is shown to be representable by a J'-scheme P. Assume now that the genus change in every fiber, δ, is 1 and that the smooth locus C'^{sm} contains the length-2, S-flat closed subscheme Q' at which $f\colon C' \to C$ is not an isomorphism. Under these hypotheses, it is proved that the presentation functor itself is representable by a certain $\mathbb{P}^1$-bundle P/J'. Furthermore, then J is obtained from P by contracting the image of a canonical embedding $\epsilon'\colon J' \times Q' \to P$; in fact, it is contracted 2–to–1 to the image of a canonical embedding $\epsilon\colon J' \to J$. More precisely, the canonical map $\kappa\colon P \to J$ is finite and surjective, it carries $P - \epsilon'(J' \times Q')$ isomorphically onto $J - \epsilon(J')$, its comorphism $\kappa^c\colon \mathcal{O}_J \to \mathcal{O}_P$ is injective, and the cokernel $\mathcal{C}ok(\kappa^c)$ is invertible on $\epsilon(J')$.

1. *Preliminaries.* Fix a connected, locally noetherian base scheme S. A *curve over* S or a projective integral S-*curve* is, by definition, a flat, projective morphism $p\colon C \to S$ all of whose fibers are geometrically integral of pure dimension 1.

Fix a projective, integral curve $p\colon C \to S$. Recall [4, p. 101] that its *compactified jacobian* $J_n := \operatorname{Pic}^n_{C/S}$ is a projective S-scheme, which represents the étale sheaf associated to the following functor on S-schemes T:

$$J_n(T) := \left\{ \begin{matrix} \text{isomorphism classes of torsion-free, rank-1 sheaves } \mathcal{I} \\ \text{on } C \times_S T, \text{ such that } \chi(I(t)) = n \text{ for all } t \in T \end{matrix} \right\},$$

where a coherent $\mathcal{O}_{C \times_S T}$-Module $\mathcal{I}$ is said to be a *torsion-free, rank-1* sheaf on $C \times_S T$ if it is T-flat and, for each $t \in T$, the fiber $\mathcal{I}(t)$ is a torsion-free, rank-1 sheaf on $C(t)$.

Let $p'\colon C' \to S$ be another projective, integral S-curve. An S-morphism $f\colon C' \to C$ is said to be *relatively birational* if the induced morphisms $f(s)\colon C'(s) \to C(s)$ on the fibers are birational for all $s \in S$. Clearly, if $T \to S$ is any morphism, then the pullback f_T is also relatively birational.

Fix, for the remainder of this article, a relatively birational morphism

$$f\colon C' \to C$$

of S-curves. Then f is projective and quasi-finite, so finite [**11**, IV$_3$, 8.11.1, p. 41]. In particular, it is affine. Since $\mathcal{O}_{C'}$ is S-flat, $f_*\mathcal{O}_{C'}$ is therefore also S-flat. Since f is relatively birational, the comorphism $f^c\colon \mathcal{O}_C \to f_*\mathcal{O}_{C'}$ is injective with S-flat cokernel whose formation commutes with base change [**11**, IV$_3$, 11.3.7, p. 135]. Put

$$\delta := \chi((f_*\mathcal{O}_{C'}/\mathcal{O}_C)(s)).$$

Then $\delta(C'/C)$ is constant as a function of s by flatness [**11**, III$_2$, 7.9.4, p. 76]. Put

$$Q := \operatorname{Supp}(f_*\mathcal{O}_{C'}/\mathcal{O}_C) \quad \text{and} \quad Q' := f^{-1}(Q).$$

Proposition 2. *If $\delta = 1$, then:*

(i) *Q is a length-1, S-flat subscheme of C;*
(ii) *There is a section $q\colon S \to C$ with image Q;*
(iii) *Q' is a length-2, S-flat subscheme of C';*
(iv) *The formation of Q and of q and of Q' commutes with base change.*

Proof. Let $\mathfrak{M}_Q$ denote the Annihilator of $f_*\mathcal{O}_{C'}/\mathcal{O}_C$; by definition, it is the Ideal of Q. Since $f_*\mathcal{O}_{C'}/\mathcal{O}_C$ has length 1 on each fiber and its formation commutes with base change, it is locally isomorphic to $\mathcal{O}_C/\mathfrak{M}_Q$. So Q is an S-flat, length-1 subscheme of C. The restriction $(p|Q)\colon Q \to S$ is therefore an isomorphism since it is flat and an isomorphism on the fibers. So $q := (p|Q)^{-1}$ is a section of p with image Q.

The Ideal of Q' is equal to the image $\mathfrak{M}_Q\mathcal{O}_{C'}$ of $\mathfrak{M}_Q \otimes_C \mathcal{O}_{C'}$ in $\mathcal{O}_{C'}$. In general, for any Ideal $\mathcal{J}$ of $\mathcal{O}_C$ we clearly have $f_*(\mathcal{J}\mathcal{O}_{C'}) = \mathcal{J}f_*\mathcal{O}_{C'}$. In the present case, $\mathfrak{M}_Q f_*\mathcal{O}_{C'}$ is contained in $\mathfrak{M}_Q$ by its very definition. So, we have

$$\mathfrak{M}_Q = f_*(\mathfrak{M}_Q\mathcal{O}_{C'}); \tag{2.1}$$

whence, $f_*(\mathcal{O}_{C'}/\mathfrak{M}_Q\mathcal{O}_{C'}) = f_*\mathcal{O}_{C'}/\mathfrak{M}_Q$. Hence, there is an exact sequence,

$$0 \longrightarrow \mathcal{O}_C/\mathfrak{M}_Q \longrightarrow f_*\mathcal{O}_{C'}/\mathfrak{M}_Q \longrightarrow f_*\mathcal{O}_{C'}/\mathcal{O}_C \longrightarrow 0.$$

It follows that Q' has length 2 and is S-flat. Finally, it is now obvious that the formations commute with base change.

Proposition 3. *Let $\mathcal{F}$ and $\mathcal{G}$ be two quasi-coherent $\mathcal{O}_{C'}$-Modules. Then:*

(i) *Any $\mathcal{O}_C$-homomorphism $u\colon f_*\mathcal{F} \to f_*\mathcal{G}$ is $f_*\mathcal{O}_{C'}$-linear.*
(ii) *The following functor is fully faithful:*

$$\begin{array}{ccc} ((\text{quasi-coherent } \mathcal{O}_{C'}\text{-Modules})) & \to & ((\text{quasi-coherent } \mathcal{O}_C\text{-Modules})) \\ \mathcal{F} & \mapsto & f_*\mathcal{F} \end{array}.$$

Proof. (i) Fix a point $x \in C$, set $s := p(x)$, and let η denote the generic point of $C'(s)$. Set $A = \mathcal{O}_{C,x}$, set $A' = (f_*\mathcal{O}_{C'})_x$, and let $\mathfrak{N}$ denote the nilradical of $\mathcal{O}_{S,s}$. Consider the following assertion:

$$\begin{array}{l}\text{For any } a' \in A' \text{ and any integer } r \geq 0, \text{ there exist} \\ b, c \in A \text{ such that } a' - \dfrac{b}{c} \text{ belongs to } (\mathfrak{N}A')^r.\end{array} \tag{3.1}$$

Here the subtraction takes place in $\mathcal{O}_\eta$, and the comorphism f^c is omitted for simplicity of notation. Once (3.1) is known, take r so that $(\mathfrak{N}A')^r = 0$. Then clearly we have

$$u_*(a'm) = u_*(\frac{b}{c}m) = \frac{b}{c}u_*(m) = a'u_*(m).$$

So u is an $(f_*\mathcal{O}_{C'})$-homomorphism.

The proof of (3.1) proceeds by induction on r. For $r = 0$, it is trivial. Now, assume $a' = b/c + n'$ with $n' \in (\mathfrak{N}A')^r$. Write $n' = \Sigma a_i' r_i$ with $r_i \in \mathfrak{N}^r$ and $a_i' \in A'$. Now, $A/\mathfrak{N}A$ and $A'/\mathfrak{N}A'$ are reduced since p and p' are flat and their fibers are reduced [**11**, IV_2, 3.3.5, p. 44]. Moreover, $\operatorname{Spec}(A') \to \operatorname{Spec}(A)$ induces an isomorphism over dense, open sets; so by the next lemma, 4, there are elements $b_i, c_i \in A$ with $a_i' = \dfrac{b_i}{c_i} + n_i$ for $n_i \in \mathfrak{N}A'$. So, $n' = \sum \dfrac{b_i}{c_i} r_i + n_i' r_i$. Since $\sum b_i r_i$ is in A and $\sum n_i' r_i$ is in $(\mathfrak{N}A')^{r+1}$, the result follows.

(ii) The second assertion follows immediately from the first and from the equivalence of categories of quasi-coherent $\mathcal{O}_{C'}$-Modules $\mathcal{H}$ and quasi-coherent $(f_*\mathcal{O}_{C'})$-Modules $f_*\mathcal{H}$ [**11**, I, 9.2.5, p. 362].

Lemma 4. *Let $A \to A'$ be an injective homomorphism of reduced noetherian rings which induces an isomorphism from a dense, open subset of* $\operatorname{Spec}(A')$ *to a dense, open subset of* $\operatorname{Spec}(A)$. *Then every element a' of A' is equal to b/c where $b, c \in A$ and c is a nonzero divisor of A.*

Proof. Since the spectra are isomorphic over open, dense subsets, the spectra have the same number of maximal (generic) points and their total fraction rings are equal. So any element $a' \in A'$ belongs to the total fraction ring of A, which is formed of quotients b/c as desired. (Note that $D^{-1}A = \prod K_i$ where $D := \{\text{nonzero divisors of } A\}$ and the K_i are the local rings of the maximal points.)

5. *The canonical embedding* $\epsilon \colon J'_n \to J_n$. Define a morphism

$$\epsilon \colon J'_n \longrightarrow J_n$$

as follows: Let T be an S-scheme, and $\mathcal{I}'$ a torsion-free, rank-1 sheaf on $C' \times_S T$. Then clearly $f_{T*}\mathcal{I}'$ is a torsion-free, rank-1 sheaf on $C \times_S T$, and $\chi(f_{T*}\mathcal{I}'(t)) = \chi(\mathcal{I}'(t))$ for all $t \in T$ because f is relatively birational, see 1. Hence, sending $\mathcal{I}'$ to $f_{T*}\mathcal{I}'$ defines a morphism $\epsilon \colon J'_n \to J_n$.

The morphism ϵ is a monomorphism by 3, ii. Since J'_n and J_n are projective S-schemes, ϵ is also proper. Hence, by [**11**, IV_3, 8.11.5, p. 42], ϵ is a closed embedding.

6. *The presentation functor.* Let $\mathcal{I}'$ be a torsion-free, rank-1 sheaf on $C' \times_S T/T$ for some S-scheme T. An injective $\mathcal{O}_{C \times_S T}$-homomorphism $h \colon \mathcal{I} \to f_{T*}\mathcal{I}'$ is said to be a *presentation of* $\mathcal{I}'$ *for* f_T if the quotient $\mathcal{N} := \operatorname{coker}(h)$ is T-flat, and if $\mathcal{N}$ has its support contained in $Q \times_S T$, and if $\operatorname{length}(\mathcal{N}(t)) = \delta$ for all $t \in T$.

Two presentations $h_1 \colon \mathcal{I}_1 \to f_{T*}\mathcal{I}'_1$ and $h \colon \mathcal{I} \to f_{T*}\mathcal{I}'$ are said to be *equivalent* if there exist an invertible sheaf $\mathcal{L}$ on T and a commutative diagram:

$$\begin{array}{ccc} \mathcal{I} & \xrightarrow{\ h\ } & f_{T*}\mathcal{I}' \\ {\scriptstyle a}\downarrow & & \downarrow{\scriptstyle u} \\ \mathcal{I}_1 \otimes_T \mathcal{L} & \xrightarrow{h_1 \otimes_T \mathcal{L}} & f_{T*}\mathcal{I}'_1 \otimes_T \mathcal{L} \end{array} \tag{6.1}$$

in which a and u are isomorphisms of $\mathcal{O}_{C \times_S T}$-Modules.

Define the *presentation functor* $\operatorname{Pres}_{f,n}$ on the category of S-schemes by

$$\begin{aligned} \operatorname{Pres}_{f,n}(T) := \big\{(\mathcal{I}', h) \bigm| & \ \mathcal{I}' \text{ represents a member of } J'_n(T), \text{ and} \\ & h \colon \mathcal{I} \to f_{T*}\mathcal{I}' \text{ is a presentation of } \mathcal{I}' \text{ for } f_T\big\}/\text{equivalence}. \end{aligned}$$

The functor is well defined. Indeed, the formations of $f_{T*}\mathcal{I}'$ and of $p_{T*}\mathcal{N}$ commute with base change since f and $p|Q$ are affine. Moreover, the pullback of $h \colon \mathcal{I} \to f_{T*}\mathcal{I}'$ is injective since $\mathcal{N}$ is T-flat.

Note that $\chi(\mathcal{I}'(t)) = \chi(f_{T*}\mathcal{I}'(t))$ for all $t \in T$; hence we have

$$\chi(\mathcal{I}'(t)) = \chi(\mathcal{I}(t)) + \delta. \tag{6.2}$$

7 *The maps π and κ.* Let T be an S-scheme. Sending a presentation $h: \mathcal{I} \to f_{T*}\mathcal{I}'$ of $\mathcal{I}'$ for f_T to the torsion-free, rank-1 sheaf $\mathcal{I}'$ defines a map (natural transformation) of functors,

$$\pi: \mathrm{Pres}_{f,n} \longrightarrow J'_n.$$

The map is well defined because, in the diagram (6.1), we have $u = f_{T*}b$ for some $\mathcal{O}_{C' \times_S T}$-isomorphism $b: \mathcal{I}' \xrightarrow{\sim} \mathcal{I}'_1 \otimes_T \mathcal{L}$ by 3, ii. Obviously, sending $h: \mathcal{I} \to f_{T*}\mathcal{I}'$ to $\mathcal{I}$ defines a map of functors,

$$\kappa: \mathrm{Pres}_{f,n} \longrightarrow J_{n-\delta}.$$

in view of (6.2).

Lemma 8. *Assume S is the spectrum of an algebraically closed field. Let $\mathcal{I}'$ and $\mathcal{I}$ be torsion-free, rank-1 sheaves on C' and C respectively. Then there are only finitely many presentations of the form $\mathcal{I} \to f_*\mathcal{I}'$.*

Proof. Clearly there is a bijection between the set of presentations $h: \mathcal{I} \to f_*\mathcal{I}'$ and the set of sub-$\mathcal{O}_C$-Modules $\mathcal{H}$ of $f_*\mathcal{I}'$ such that $\mathcal{H} \simeq \mathcal{I}$ and such that the cokernel has length δ and support in Q. Now, any two such $\mathcal{H}$ are equal off Q. In particular, they are generically isomorphic, so there is a rational function g which carries one to the other. Moreover, g must have all its zeroes and poles in Q and none can be worse than δ. Since any two rational functions with the same zeroes and poles must differ by multiplication by a constant, and since a constant will take any $\mathcal{H}$ to itself, there is a bijection between such $\mathcal{H}$ and a subset of the set of divisors $\sum n_x[x]$ on the normalization of Q having $-\delta \le n_x \le \delta$ for each point x of the normalization of Q. This latter set is clearly finite.

Proposition 9. *Assume that C' admits a Poincaré, or universal, sheaf $\mathbf{I}'$ on $C' \times_S J'_n$. Then the presentation functor $\mathrm{Pres}_{f,n}$ is represented by a projective J'_n-scheme P_n, and there is a canonical isomorphism of J'_n-schemes,*

$$P_n = \mathrm{Quot}^{\delta}_{\mathcal{F}/C \times J'_n/J'_n} \quad \text{where } \mathcal{F} := (f_{J'_n *}\mathbf{I}')|(Q \times J'_n).$$

Proof. Let $T \to J'_n$ be an S-morphism. Let $\mathcal{I}'$ be the pullback of $\mathbf{I}'$ to $C' \times T$. Since f is affine, $f_{T*}\mathcal{I}'$ is canonically isomorphic to

$(f_{J'_n *}\mathbf{I}')_T$. Hence, a T-point of Quot^δ is equal to a T-flat quotient $\mathcal{N}$ of $(f_{T*}\mathcal{I}')|(Q\times T)$ such that every $\mathcal{N}(t)$ has length δ. Let $\mathcal{I}$ denote the kernel of the composition,

$$f_{T*}\mathcal{I}' \longrightarrow (f_{T*}\mathcal{I}')|(Q\times T) \longrightarrow \mathcal{N}.$$

Then the inclusion $h\colon \mathcal{I} \to f_{T*}\mathcal{I}'$ is clearly a presentation of $\mathcal{I}'$ for f_T, and its formation clearly commutes with base change. Thus we have defined a J'_n-map of functors,

$$r\colon \mathrm{Quot}^\delta_{\mathcal{F}/C\times J'_n/J'_n} \longrightarrow \mathrm{Pres}_{f,n}.$$

To prove that the above map r is an isomorphism, construct an inverse as follows: Let $h\colon \mathcal{I} \to f_{T*}\mathcal{I}'$ be a presentation of $\mathcal{I}'$ for f_T. The cokernel map $f_{T*}\mathcal{I}' \to \mathcal{N}$ can be factored through a map $(f_{T*}\mathcal{I}')|(Q\times T) \to \mathcal{N}$ because, by definition of a presentation, $\mathcal{N}$ has support in $Q\times T$. By descent theory [**4**, 5.6(i), p. 84], there are an invertible sheaf $\mathcal{L}$ on T and an isomorphism $u\colon \mathbf{I}'_T \xrightarrow{\sim} \mathcal{I}'\otimes\mathcal{L}$. Consider the composition,

$$r'(h)\colon \mathcal{F}_T = (f_{T*}\mathbf{I}'_T)|(Q\times T) \xrightarrow{a} ((f_{T*}\mathcal{I}')|(Q\times T))\otimes\mathcal{L} \longrightarrow \mathcal{N}\otimes\mathcal{L},$$

where $a := (f_{T*}u)|(Q\times T)$. Then $r'(h)$ is an element of $\mathrm{Quot}^\delta(T)$.

Let $\mathcal{L}_1$ be another invertible sheaf on T, and let u_1 be another isomorphism $u_1\colon \mathbf{I}'_T \to \mathcal{I}'\otimes\mathcal{L}_1$. We must see that $(\mathcal{L}_1, u_1)$ defines the same element of $\mathrm{Quot}^\delta(T)$ as $(\mathcal{L}, u)$. Since $\mathcal{I}'$ is a simple sheaf [**4**, Lemma 5.4, p. 83], the functor $\mathcal{L} \mapsto \mathcal{I}'\otimes\mathcal{L}$ is fully faithful [**1**, (5), p. 119]. Therefore there is an isomorphism $w\colon \mathcal{L} \xrightarrow{\sim} \mathcal{L}_1$ such that $(\mathcal{I}'\otimes w)\circ u = u_1$ holds. Hence the diagram,

$$\begin{array}{ccccc}
\mathcal{F}_T & \longrightarrow & f_{T*}(\mathcal{I}'\otimes\mathcal{L})|(Q\times T) & \longrightarrow & \mathcal{N}\otimes\mathcal{L} \\
\| & & \downarrow{\scriptstyle f_{T*}(\mathcal{I}'\otimes w)|(Q\times T)} & & \downarrow{\scriptstyle \mathcal{N}\otimes w} \\
\mathcal{F}_T & \longrightarrow & f_{T*}(\mathcal{I}'\otimes\mathcal{L}_1)|(Q\times T) & \longrightarrow & \mathcal{N}\otimes\mathcal{L}_1
\end{array}$$

is commutative. Thus $r'(h)$ does not depend on the choice of $(\mathcal{L}, u)$. It is now clear that if h is equivalent to $h_1\colon \mathcal{I}_1 \to f_{T*}\mathcal{I}'$, then $r'(h)$ is equal to $r'(h_1)$. Hence r' is well-defined, and it is clearly an inverse to r. Thus r is an isomorphism.

Since Quot^δ is representable, the proof is therefore complete.

Theorem 10. *Assume $\delta = 1$. Assume C' admits a Poincaré sheaf $\mathbf{I}'$, and set $\mathcal{E} := (q\times 1)^*(f_{J'_n *}\mathbf{I}')$ where $q\colon S \to C$ is the section with image Q of 2. Then:*

(i) *The presentation functor $\mathrm{Pres}_{f,n}$ is represented by the J'_n-scheme $\mathbb{P}(\mathcal{E})$.*
(ii) *If $Q' := f^{-1}(Q)$ is contained in C'^{sm}, the smooth locus of C'/S, then $\mathcal{E}$ is locally free (of rank 2).*

Proof. (i) By 9 it suffices to show that $\mathbb{P}(\mathcal{E}) = \mathrm{Quot}^1_{\mathcal{F}/C\times J'_n/J'_n}$ where $\mathcal{F} := (f_{J'_n *}\mathbf{I}')|(Q \times J'_n)$. Let T be an S-scheme, and let $\mathcal{F}_T \to \mathcal{N}$ be a 1-quotient. Since $\mathcal{N}$ is flat with Hilbert polynomial 1 over T, clearly $\mathcal{N}$ must be invertible on $Q_T = T$. Thus there is a canonical isomorphism of J'_n-schemes $P_n = \mathbb{P}(\mathcal{F})$.

(ii) Consider the cartesian diagram,

$$\begin{array}{ccc} Q' \times J'_n & \xrightarrow{\ q'\ } & C' \times J'_n \\ {\scriptstyle f_Q\times 1}\downarrow & \square & \downarrow{\scriptstyle f\times J'_n} \\ J'_n & \xrightarrow{\ q_{J'_n}\ } & C \times J'_n \end{array} .$$

Now, $\mathcal{E}$ is equal to $(f_Q \times 1)_* q'^* \mathbf{I}'$ because the formation of $f_{J'_n *}\mathbf{I}'$ commutes with base change as f is affine.

Clearly f_Q is finite and flat. So, if $Q' \subset C'^{sm}$, then $\mathcal{E}$ is locally free because $q'^*\mathbf{I}'$ is invertible since a torsion-free, rank-1 sheaf is invertible at all smooth points. Moreover, its rank is 2 because f_Q is of degree 2.

11. *Remark.* The preceding results 9 and 10 were first proved in essentially this form (but over an algebraically closed field) by H. Kleppe [**15**, Proposition 7.1.2, p. 85 and Corollary 7.1.3, p. 87]. In fact, Oda and Seshadri [**16**, Proposition 12.1, p. 64] claimed to have proved a result very similar to 10 for certain reducible curves with nodes, but they gave insufficient justification. Moreover, without some modification, one must require the two reducible curves to have the same number of connected components in order for their "proof" to work [**15**, p. 23].

Theorem 12. *The étale sheaf associated to the presentation functor* $\mathrm{Pres}_{f,n}$ *is representable by a projective* J'_n*-scheme* P_n*. Moreover,* P_n *is a finite* $J_{n-\delta} \times J'_n$*-scheme.*

Proof. We consider the representability first and offer two proofs; however, the first yields the representability of the associated sheaf in the finite, flat, and finitely presentable topology only. Representability is a local matter, so it suffices in each proof to work in a neighborhood of each $s \in S$.

By [**3**, Lemma VIII, p. 33], there exist an open neighborhood S_0 of s, a finite, flat, finitely presented morphism $S' \to S_0$, and an S-morphism $S' \to C'^{sm}$, where C'^{sm} is the smooth locus of C'/S. Replace S by S_0, and write $P_{n,T}$ for $\mathrm{Pres}_{f_T,n}$, for each S-scheme T. Then since $C' \times S'/S'$ admits a section of its smooth locus, there exists a Poincaré sheaf on $C' \times J'_n \times S'$

by [**3**, (3.4, iii), p. 39], and so $P_{n,S'}$ is representable by an S'-scheme P'_n by 9. Moreover, since $\mathrm{Pres}_{f_T,n} = \mathrm{Pres}_{f,n} \times_S T$ for any S-scheme T, we have

$$\begin{aligned}\mathrm{Pres}_{f_{S'},n} \times_{\mathrm{Pres}_{f,n}} \mathrm{Pres}_{f_{S'},n} &= (\mathrm{Pres}_{f,n} \times_S S') \times_{\mathrm{Pres}_{f,n}} (\mathrm{Pres}_{f,n} \times_S S') \\ &= \mathrm{Pres}_{f,n} \times_S (S' \times S') = \mathrm{Pres}_{f_{S''},n}\end{aligned}$$

where $S'' := S' \times_S S'$. Thus $\mathrm{Pres}_{f_{S''},n}$ is representable by a projective S''-scheme P''_n and $P''_n \rightrightarrows P'_n$ is a finite, flat, finitely presented equivalence relation. Therefore the quotient P_n exists [**17**, Théorème 1. iii), p. 82], and it represents the sheaf associated to $\mathrm{Pres}_{f,n}$ in the finite, flat, and finitely presentable topology.

The second proof runs as follows. By [**11**, IV$_4$, 17.16.3, ii, p. 106], there exist an étale neighborhood $S' \to S$ of s and a section of the smooth locus of $C' \times_S S'$. So by [**3**, (3.4, iii), p. 39], there exists a Poincaré sheaf on $C' \times J'_n \times U$. So, by 9, $\mathrm{Pres}_{f_{S'},n}$ is representable by a projective $J'_n \times S'$-scheme P'_n. Similarly, if $S'' := S' \times S'$, then $\mathrm{Pres}_{f_{S''},n}$ is representable by a projective $J'_n \times S''$-scheme P''_n. Since P'_n (respectively, P''_n) is projective over $J_{n-\delta} \times J'_n \times S'$ (respectively, $J_{n-\delta} \times J'_n \times S''$) and, since the fibers are finite by 8, it is finite. So, by [**9**, VIII, 2.1, p. 202], there is a finite $J_{n-\delta} \times J'_n$-scheme P_n such that $P_n \times S' = P'_n$ and $P_n \times S' \times S' = P''_n$ hold. Since $S' \to S$ is étale, P_n represents the étale sheaf associated to $\mathrm{Pres}_{f,n}$. This proof shows moreover that P_n is a finite $J_{n-\delta} \times J'_n$-scheme.

Lemma 13. *Let $h\colon \mathcal{I} \to f_*\mathcal{I}'$ be an S-presentation. Then:*

(i) *If $\mathcal{I}$ is invertible along Q, then the canonical map $\mathcal{I} \to f_*f^*\mathcal{I}$ is an S-presentation, and it is isomorphic to $\mathcal{I} \otimes f^c$ where $f^c\colon \mathcal{O}_C \to f_*\mathcal{O}_{C'}$ is the comorphism.*

(ii) *If $\mathcal{I}$ is invertible along Q, then the adjoint $h^\sharp\colon f^*\mathcal{I} \to \mathcal{I}'$ is an isomorphism and h is canonically isomorphic to the canonical map $\mathcal{I} \to f_*f^*\mathcal{I}$.*

(iii) *$\mathcal{I}$ is invertible along Q if and only if $\mathcal{I}'$ is invertible along $Q' := f^{-1}(Q)$ and the adjoint $h^\sharp\colon f^*\mathcal{I} \to \mathcal{I}'$ is surjective.*

Proof. (i) Since $\mathcal{I}$ is invertible along Q, the pullback $f^*\mathcal{I}$ is torsion-free, rank-1 on C'. So the canonical map is injective because it is generically an isomorphism and $\mathcal{I}$ is torsion-free [**11**, IV$_3$, 11.3.7, p. 135]. Moreover, the cokernel $\mathcal{N}$ has its support contained in Q because f is an isomorphism off Q. Finally, $\chi(\mathcal{N}(s)) = \delta$ for each s because $\mathcal{I}$ is invertible along Q, which implies that $\mathcal{I}_Q \to f_{Q*}f_Q^*\mathcal{I}_Q$ is isomorphic to $\mathcal{I}_Q \otimes f^c$.

(ii) Consider the commutative diagram of S-presentations with exact rows,

$$\begin{array}{ccccccccc} 0 & \longrightarrow & \mathcal{I} & \longrightarrow & f_*f^*\mathcal{I} & \longrightarrow & \mathcal{N} & \longrightarrow & 0 \\ & & \| & & \big\downarrow{\scriptstyle f_*(h^\sharp)} & & \big\downarrow & & \\ 0 & \longrightarrow & \mathcal{I} & \xrightarrow{\ h\ } & f_*\mathcal{I}' & \longrightarrow & \mathcal{N}' & \longrightarrow & 0 \end{array}$$

Now, $f_*(h^\sharp)$ is injective on each fiber because it is generically so and the sheaves are torsion-free. Hence $\mathcal{N}(s) \to \mathcal{N}'(s)$ is injective. By (i), $\chi(\mathcal{N}(s)) = \chi(\mathcal{N}'(s))$ holds for all s. So $\mathcal{N}(s) \to \mathcal{N}'(s)$ is an isomorphism for all s, and so $\mathcal{N} \to \mathcal{N}'$ is an isomorphism. By the five lemma and 3, $h^\sharp$ is an isomorphism.

(iii) If $\mathcal{I}$ is invertible along Q, then $\mathcal{I}'$ is invertible along Q' and $h^\sharp$ is surjective by (ii). To prove the converse, note that it suffices to show $\mathcal{I}(s)$ is invertible along $Q(s)$ for any geometric point s of S by [**11**, IV_3, 11.3.10, p. 138]. Since f is affine, the formation of $h^\sharp$ commutes with base change to the fiber, and so $h^\sharp(s)$ is surjective. So, we may assume S is the spectrum of an algebraically closed field.

Set $A = \mathcal{O}_{C,Q}$, set $M = \mathcal{I}_Q$, set $A' = \mathcal{O}_{C',Q'}$, and set $M' = \mathcal{I}'_{Q'}$. The map

$$(M \otimes_A A')/\operatorname{tors}(M \otimes_A A') \to M'$$

is surjective because $h^\sharp$ is surjective. It is injective because it is generically an isomorphism and the first member is torsion-free. Now, M' is isomorphic to A' since $\mathcal{I}'$ is invertible along Q' and A' is semilocal. Replace M' by A'. Since elements of the form $m \otimes 1$ generate $M \otimes_A A'$, there is an element $m \in M$ such that the image of $m \otimes 1$ in A' generates A'. (Not every image can lie in any given maximal ideal; hence there is one that does not lie in their finite union.) Define $\mathcal{I}''$ to be the subsheaf of $\mathcal{I}$ equal to $\mathcal{I}$ away from Q and to be mA at Q. Clearly m generates $\mathcal{I}$ in a punctured neighborhood of Q (where it is nonzero). Hence $\mathcal{I}''$ is torsion-free, rank-1, invertible at Q, and equal to $\mathcal{I}$ away from Q. Hence $\mathcal{I}'' \otimes f_*\mathcal{O}_{C'} = f_*\mathcal{I}'$ because $f^*\mathcal{I}''$ is equal to $\mathcal{I}'$. Therefore tensoring the exact sequence

$$0 \to \mathcal{O}_C \to f_*\mathcal{O}_{C'} \to f_*\mathcal{O}_{C'}/\mathcal{O}_C \to 0$$

with $\mathcal{I}''$, we obtain $\chi(\mathcal{I}'')+\delta = \chi(f_*\mathcal{I}')$. Since moreover $\chi(\mathcal{I})+\delta = \chi(f_*\mathcal{I}')$ and $\mathcal{I}'' \subset \mathcal{I}$ hold, it follows that $\mathcal{I}''$ is equal to $\mathcal{I}$. Thus $\mathcal{I}$ is invertible along Q.

Theorem 14. *The natural map $\kappa\colon P_n \to J_{n-\delta}$ is an isomorphism over the open subset of $J_{n-\delta}$ parametrizing the torsion-free, rank-1 sheaves invertible along Q.*

Proof. It follows directly from 13 that κ is an isomorphism on T-points.

15. *The canonical embedding $\epsilon'\colon J'_n \times Q' \to P_n$.* Assume $\delta = 1$ and $Q' \subset C'^{sm}$. Define a closed embedding

$$\epsilon'\colon J'_n \times Q' \to P_n$$

on the level of functors as follows. Let T be an S-scheme. Let $\mathcal{I}'$ be a torsion-free, rank-1 sheaf on $C' \times_S T$, and let $q' : T \to Q'$ be an S-morphism. Let $\mathfrak{M}_{q'}$ denote the Ideal of the image of the section $q'_1 := (q', 1_T) : T \to C' \times_S T$. Then the exact sequence,

$$0 \to f_{T*}(\mathfrak{M}_{q'}\mathcal{I}') \to f_{T*}\mathcal{I}' \to f_{T*}(\mathcal{I}'|q'_1(T)) \to 0, \qquad (15.1)$$

obtained by tensoring $0 \to \mathfrak{M}_{q'} \to \mathcal{O}_{C' \times_S T} \to \mathcal{O}_{q'_1(T)} \to 0$ with $\mathcal{I}'$ and applying f_{T*}, is a presentation; indeed, the latter sequence remains exact on the fibers and $f \circ q'(T) \subset Q$ holds. Hence, sending $(\mathcal{I}', q')$ to (15.1) defines ϵ'.

The map ϵ' is a monomorphism. Indeed, $\pi \circ \epsilon' : J'_n \times Q' \to J'_n$ is equal to the first projection. Moreover, the map q' is determined by the Ideal $\mathfrak{M}_{q'}$, so by $f_{T*}\mathfrak{M}_{q'}$ as $\mathcal{I}'$ is invertible at Q' since $Q' \subset C'^{sm}$. Since $J_n \times Q'$ and P_n are both proper J_n-schemes, ϵ' is thus a proper monomorphism, hence a closed embedding [**11**, IV$_3$, 8.11.5, p. 42].

Theorem 16. *Assume $\delta = 1$ and $Q' \subset C'^{sm}$. Then:*

(i) *The following diagram is Cartesian:*

$$\begin{array}{ccc} P_n & \xrightarrow{\ \kappa\ } & J_{n-1} \\ {\scriptstyle \epsilon'}\uparrow & \square & \uparrow{\scriptstyle \epsilon} \\ J'_n \times Q' & \xrightarrow[\ \lambda\ ]{} & J'_{n-1} \end{array}$$

where $\lambda(\mathcal{I}', q') := \mathfrak{M}_{q'}\mathcal{I}'$.

(ii) *The complement $J_{n-1} - \epsilon(J'_{n-1})$ is precisely the open subset of J_{n-1} parametrizing the torsion-free, rank-1 sheaves invertible along Q.*

(iii) *The restriction of κ is an isomorphism between the complements of the images of ϵ and ϵ':*

$$\kappa|(P_n - \epsilon'(J'_n \times Q')) : (P_n - \epsilon'(J'_n \times Q')) \xrightarrow{\sim} (J_{n-1} - \epsilon(J'_{n-1})).$$

(iv) *The map κ is finite and surjective.*

Proof. (i) The diagram is obviously commutative. So there is a natural map from $J'_n \times Q'$ to $P_n \times J'_{n-1}$. It now remains to show that any T-point of $P_n \times J'_{n-1}$ has a unique lifting. Such a T-point is a presentation of the form $h : f_{T*}\mathcal{J}' \to f_{T*}\mathcal{I}'$. By 3, i, $h = f_{T*}h'$ where $h' : \mathcal{J}' \to \mathcal{I}'$. Let $\mathcal{N}$ and $\mathcal{N}'$ be the cokernels of h and h'. Then $\mathcal{N} = f_{T*}\mathcal{N}'$. Hence $\mathcal{N}'$ is T-flat and has length 1 on each fiber. Therefore, $\operatorname{Supp} \mathcal{N}'$ is the image of a section $q' : T \to Q'$, and $\mathcal{N}'$ is invertible on $\operatorname{Supp} \mathcal{N}'$. Clearly $(\mathcal{I}', q')$ is a

T-point of $J'_n \times Q'$ lifting the presentation h. Finally, the lifting is clearly unique.

(ii) It is enough to show that the two open sets have the same geometric points. So we may assume that S is the spectrum of an algebraically closed field. Let $\mathcal{I}$ be a torsion-free, rank-1 sheaf on C. We have to show that (a) the stalk $\mathcal{I}_Q$ is a module over the semilocal ring $\mathcal{O}_{Q'}$ if and only if (b) $\mathcal{I}_Q$ is not invertible over $\mathcal{O}_Q$. Obviously (a) implies (b). The converse is essentially an observation of D'Souza's, compare [8, 2.6, p. 427]. In brief, the proof is as follows. Since $Q' \subset C'^{sm}$, there is an element a in $\mathcal{I}_Q$ such that $a\mathcal{O}_{Q'} = \mathcal{I}_Q\mathcal{O}_{Q'}$. By (b), $a\mathcal{O}_Q \neq \mathcal{I}_Q$. Since $\delta = 1$ and

$$a\mathcal{O}_Q \subsetneq \mathcal{I}_Q \subseteq a\mathcal{O}_{Q'},$$

therefore $\mathcal{I}_Q = \mathcal{I}_Q\mathcal{O}_{Q'}$. Thus (a) holds.

(iii) The assertion is an immediate consequence of (i), (ii), and 14.

(iv) Because of (iii), it suffices to prove that λ is finite and surjective. To do so, observe that, if S is the spectrum of an algebraically closed field, and if $\mathcal{J}'$ is a torsion-free, rank-1 sheaf on C', then $\mathcal{J}' = \mathfrak{M}_{q'}\mathcal{I}'$ for some S-point q' of Q', because $Q' \subset C'^{sm}$; moreover, there are at most two such q' by 2, ii.

Proposition 17. *Assume $\delta = 1$ and $Q' \subset C'^{sm}$. Then, locally in the smooth topology, the map $\kappa\colon P_n \to J_{n-1}$ is isomorphic along $\epsilon(J'_{n-1})$ to the product*

$$f \times 1\colon C' \times J'_{n-1} \to C \times J'_{n-1}.$$

More precisely, there exists a smooth map $\psi\colon U \to J_{n-1}$ whose image contains $\epsilon(J'_{n-1})$, there exists a smooth map $\psi'\colon U' \to J'_{n-1}$, and there exists a Cartesian diagram,

$$\begin{array}{ccc} P_n \times_{J_{n-1}} U & \xrightarrow{\ \kappa'\ } & U \\ \downarrow & \square & \downarrow \\ C' \times U' & \xrightarrow[f\times 1]{} & C \times U' \end{array},$$

in which κ' is the projection and the two vertical maps are open embeddings. Moreover, the Cartesian diagram induced by that diagram via $Q \hookrightarrow C$ is this:

$$\begin{array}{ccc} \epsilon'(J'_n \times Q') \times_{J_{n-1}} U & \longrightarrow & \psi^{-1}\epsilon(J'_{n-1}) \\ \downarrow & \square & \downarrow \\ Q' \times U' & \xrightarrow[f\times 1]{} & Q \times U' \end{array}.$$

Proof. Let ω be the dualizing sheaf of C/S; see for example [**12**]. Fix an integer m, set

$$\mathbf{Q}^m := \mathrm{Quot}^m_{\omega/C/S},$$

and let $\mathcal{F}^m$ be the universal quotient on $C \times \mathbf{Q}^m$. Let $\mathbf{Q}^m_Q$ be the open subset parametrizing those quotients supported in the open set $C-Q$. Let U^m be the open subset of $\mathbf{Q}^1 \times \mathbf{Q}^m$ obtained by intersecting the following two open subsets: (1) $\mathbf{Q}^1 \times \mathbf{Q}^m_Q$ and (2) the complement of the closed subset

$$[p(\mathrm{Supp}(\mathcal{F}^1)) \times \mathbf{Q}^m]\bigcap[\mathbf{Q}^1 \times p(\mathrm{Supp}(\mathcal{F}^m))].$$

Consider the map $U^m \to \mathbf{Q}^{m+1}$ defined by sending the pullbacks of $\mathcal{F}^1$ and $\mathcal{F}^m$ to their sum. That map is well defined because those pullbacks have disjoint support. For a similar reason, the map is étale by the infinitesmal criterion [**11**, IV$_4$, 17.14.1, p. 98].

Let $p_a := 1 - \chi(\mathcal{O}_{C(s)})$ denote the arithmetic genus of the fibers $C(s)$. Since C is S-flat and S is connected, p_a is independent of s. Take $m := p_a - n - 1$. Then the Abel map $\mathbf{Q}^{m+1} \to J_{n-1}$ is surjective and smooth of relative dimension $-n$ provided $-n \geq p_a - 1$ [**4**, (8.4), p. 103]. Now, it is obvious that, for any integer a, twisting by $\mathcal{O}_C(a)$ defines an isomorphism $J_{n-1} \xrightarrow{\sim} J_{n-1+b}$ where $b := \deg \mathcal{O}_C(a)$. Similarly, twisting by $f^*\mathcal{O}_C(a)$ defines an isomorphism $J'_{n-1} \xrightarrow{\sim} J'_{n-1+b}$ because $b = \deg f^*\mathcal{O}_C(a)$. Hence, we may assume that $-n \geq p_a - 1$. Take $U := U^m$. Then the composition $\psi\colon U \to \mathbf{Q}^{m+1} \to J_{n-1}$ is smooth.

Let ω' be the dualizing sheaf of C'/S. It is related to ω by the following well-known formula (see for example [**12**, (9), p. 47 and (17), p. 53]):

$$f_*\omega' = \mathcal{H}om_C(f_*\mathcal{O}_{C'}, \omega).$$

Consider the following sequence of $\mathcal{O}_C$-homomorphismns whose cokernels have support in Q,

$$\mathfrak{M}_Q = f_*(\mathcal{O}_{C'}(-Q')) \longrightarrow \mathcal{O}_C \longrightarrow f_*\mathcal{O}_{C'}$$

where the equality is (2.1). Apply the functor $\mathcal{H}om_C(\bullet, \omega)$, obtaining the following dual sequence of $\mathcal{O}_C$-homomorphisms, whose cokernels have support in Q:

$$f_*\omega' \longrightarrow \omega \longrightarrow f_*(\omega'(Q')). \tag{17.1}$$

The homomorphisms are injective because ω is torsion-free.

Set $\mathbf{Q}'^m := \mathrm{Quot}^m_{\omega'(Q')/C'/S}$. Since $p'_a = p_a - 1$ (as $\delta = 1$) and since $-n \geq p_a - 1$, the Abel map $\mathbf{Q}'^m \to J'_{n-1}$ is smooth. Let U' be the open subset of $\mathbf{Q}'^m$ parametrizing those quotients supported in the open set $C' - Q'$. Finally, let $\psi' \colon U' \to J'_{n-1}$ be the restriction of the Abel map.

There is, obviously, a canonical isomorphism $U' = \mathbf{Q}^m_Q$ because the second map in (17.1) is an isomorphism off Q. Now, since $\delta = 1$ and $Q' \subset C'^{sm}$, the embedding dimension of each fiber $C(s)$ at the singular point $Q(s)$ is 2. Hence, the open subset V of C on which ω is invertible contains Q. Let W be the open subset of $\mathbf{Q}^1$ parametrizing those quotients supported in V. Then there is a canonical isomorphism $V = W$; it is defined by tensoring with ω_Q. Replace U by its intersection with $W \times \mathbf{Q}^m$. Then there is a canonical open embedding of U in $C \times U'$.

To prove that $\psi(U)$ contains $\epsilon(J'_{n-1})$, we may assume for the moment that S is the spectrum of an algebraically closed field. By 16,i,iv, any given S-point of $\epsilon(J'_{n-1})$ is representable by a sheaf of the form $f_*(\mathfrak{M}_{q'}\mathcal{I}')$ where $\mathfrak{M}_{q'}$ is the ideal of an S-point q' of Q' and $\mathcal{I}'$ is a torsion-free, rank-1 sheaf on C'.

Suppose that there is an embedding of $\mathcal{I}'$ in $\omega'(Q')$ with the following property: Let q'' be the "other" point of Q'; that is, $\mathfrak{M}_{q'}\mathfrak{M}_{q''} = \mathfrak{M}_q$ where q is the S-point at Q. If $q'' \neq q'$, then suppose $\mathcal{I}'_{q'} = \omega'(Q')_{q'}$ and $\mathcal{I}'_{q''} = \mathfrak{M}_{q''}\omega'(Q')_{q''}$. If $q'' = q'$, then suppose $\mathcal{I}'_{q'} = \mathfrak{M}_{q'}\omega'(Q')_{q'}$.

Given such an embedding, let $\mathcal{J}$ be the subsheaf of ω that is equal to $f_*\mathcal{I}'$ off Q and to ω at Q. Then, clearly, the pair $(\mathfrak{M}_q, \mathcal{J})$ defines a point of U. Moreover, the image of that point in J'_{n-1} is equal to the given point of $\epsilon(J'_{n-1})$.

Such an embedding is simply a map $u \colon \mathcal{I}' \to \omega'(q')$ that is an isomorphism at q' and q'' because $\omega'(q') = \mathfrak{M}_{q''}\omega'(Q')$. So such an embedding will exist, if

$$\dim \operatorname{Hom}(\mathcal{I}', \omega'(q')) = \dim \operatorname{Hom}(\mathcal{I}', \omega') + 1$$
$$\dim \operatorname{Hom}(\mathcal{I}', \omega'(q')) = \dim \operatorname{Hom}(\mathcal{I}', \omega'(q' - q'')) + 1.$$

because the ground field is infinite. By duality, those equations are equivalent to these:

$$\dim \mathrm{H}^1(\mathcal{I}'(-q')) = \dim \mathrm{H}^1(\mathcal{I}') + 1$$
$$\dim \mathrm{H}^1(\mathcal{I}'(-q')) = \dim \mathrm{H}^1(\mathcal{I}'(-q' + q'')) + 1.$$

By the Riemann–Roch theorem, the latter equations will hold if $\mathcal{I}'(-q')$, $\mathcal{I}'$, and $\mathcal{I}'(-q' + q'')$ have no nonzero sections. Now, if a sheaf $\mathcal{G}$ on C' has a nonzero section, then $\chi(\mathcal{G}) \geq 1 - p'_a$. However, the Euler characteristics of $\mathcal{I}'(-q')$, $\mathcal{I}'$, and $\mathcal{I}'(-q' + q'')$ are $n - 1$, n, and n. Moreover, $n + 1 \leq 2 - p_a = 1 - p'_a$. Therefore, such an embedding exists.

It remains to find the asserted open embedding of $P_n \times_{J_{n-1}} U$ in $C' \times U'$. Let V' be the open subset of C' on which ω' is invertible.

Obviously, $V' = f^{-1}V$ because $Q \subset V$ and f is an isomorphism off $f^{-1}Q$. Let W' be the open subset of $\mathbf{Q}'^{1}$ parametrizing those quotients supported in V'. Obviously, there is a canonical isomorphism $V' = W'$. Let U'' be the open subset of $V' \times U'$ parametrizing the pairs consisting of a point of V' and an m-quotient of $\omega'(Q')$ such that the support of the quotient is contained in the open set $C' - Q'$ and does not contain that point of V'. It will turn out that U'' is the image of the embedding in question.

Consider the sheaf $\mathcal{E}xt^i_C(\mathcal{O}_Q, \omega)$. On each geometric fiber, the corresponding $\mathcal{E}xt^i$ vanishes if $i \neq 1$, and it is of dimension 1 if $i = 1$ by [**6**, (4.1), p. 15]. Hence, by [**4**(1.10), p. 61], $\mathcal{E}xt^i_C(\mathcal{O}_Q, \omega)$ vanishes if $i \neq 1$; furthermore,

$$\mathcal{M} := \mathcal{E}xt^1_C(\mathcal{O}_Q, \omega) \tag{17.2}$$

is an invertible $\mathcal{O}_Q$-Module, and its formation commutes with base change. Therefore, from the functor $\mathcal{H}om_C(\bullet, \omega_C)$ and the short exact sequence,

$$0 \longrightarrow f_*(\mathcal{O}_{C'}(-Q')) \longrightarrow \mathcal{O}_C \longrightarrow \mathcal{O}_Q \longrightarrow 0,$$

we obtain the exact sequence,

$$0 \longrightarrow \omega \longrightarrow f_*(\omega'(Q')) \longrightarrow \mathcal{M} \longrightarrow 0, \tag{17.3}$$

whose formation commutes with base change.

A T-point of U'' is a pair of quotients of $\omega'(Q')_T$ of a certain type. Their sum is an $(m+1)$-quotient $\mathcal{F}$. Let $\mathcal{I}'$ be the corresponding subsheaf, and form the following diagram:

$$\begin{array}{ccccccccc}
0 & \longrightarrow & \mathcal{I} & \xrightarrow{\;h\;} & f_{T*}\mathcal{I}' & \longrightarrow & \mathcal{M}_T & \longrightarrow & 0 \\
 & & {\scriptstyle i}\downarrow & & \downarrow & & \| & & \\
0 & \longrightarrow & \omega_T & \longrightarrow & f_{T*}(\omega'(Q')_T) & \longrightarrow & \mathcal{M}_T & \longrightarrow & 0 \\
 & & \downarrow & & \downarrow & & & & \\
 & & \mathcal{F} & = & \mathcal{F} & . & & &
\end{array}$$

The map $f_{T*}\mathcal{I}' \to \mathcal{M}_T$ is surjective because it is surjective on each fiber; it is surjective on each fiber because there the images of ω_T and $f_{T*}\mathcal{I}'$ in $f_{T*}(\omega'(Q')_T)$ are unequal as ω_T is not an $f_{T*}\mathcal{O}_{C_T}$-Module. Now, the first row of the diagram is a presentation, and it is easy to see that the first column defines a T-point of U. Thus a T-point of U'' defines a T-point of $P_n \times_{J_{n-1}} U$. Moreover, the latter determines the former because the diagram is the pushout diagram of i and h, so uniquely determined by [**7**, Lemme 4.1, p. 127]. Thus, there is a natural monomorphism $U'' \to P_n \times_{J_{n-1}} U$.

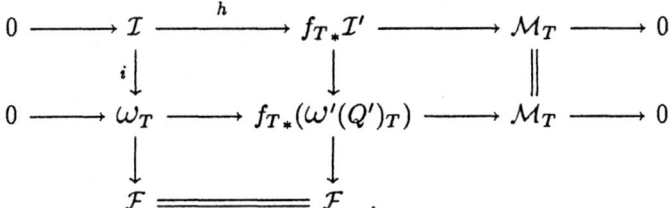

Finally, given a T-point of $P_n \times_{J_{n-1}} U$, we must lift it to T-point of U''. By descent theory, it suffices to lift it locally in the étale topology. Hence, we may assume that the given T-point of P_n is representable by a presentation $h\colon \mathcal{I} \to f_{T*}\mathcal{I}'$ whose cokernel is isomorphic to $\mathcal{O}_{Q_T}$ and that the given T-point of U is representable by an embedding $i\colon \mathcal{I} \to \omega_T$, whose cokernel $\mathcal{F}$ is an $(m+1)$-quotient of a certain type. (Initially, that is, before an étale extension, the sources of h and i need not be isomorphic.)

Form the following pushout diagram:

$$\begin{array}{ccccccccc}
0 & \longrightarrow & \mathcal{I} & \xrightarrow{h} & f_{T*}\mathcal{I}' & \longrightarrow & \mathcal{O}_{Q_T} & \longrightarrow & 0 \\
 & & {\scriptstyle i}\downarrow & & \downarrow & & \| & & \\
0 & \longrightarrow & \omega_T & \longrightarrow & \mathcal{E} & \longrightarrow & \mathcal{O}_{Q_T} & \longrightarrow & 0 \\
 & & \downarrow & & \downarrow & & & & \\
 & & \mathcal{F} & = & \mathcal{F} & . & & &
\end{array}$$

The lower exact sequence yields an element $e \in \operatorname{Ext}^1_{C_T}(\mathcal{O}_{Q_T}, \omega_T)$. That Ext^1 is equal to $\mathrm{H}^0(C, \mathcal{M})$ where $\mathcal{M}$ is given by (17.2), because the local–to–global spectral sequence degenerates. Note that the formation of e commutes with base change.

On the fibers, e is nonzero. Indeed, to check that, we may assume for the moment that T is the spectrum of an algebraically closed field and that $T = S$. Now, if $\operatorname{Supp}\mathcal{F}$ does not contain Q, then at Q the lower sequence is equal to the upper one; moreover, the upper one is not split at Q because $f_*\mathcal{I}'$ is torsion-free. Otherwise, at Q, the diagram is isomorphic to the following one:

$$\begin{array}{ccccccccc}
0 & \longrightarrow & \mathfrak{M}_{q,Q} & \longrightarrow & (f_*\mathfrak{M}_{q'})_Q & \longrightarrow & \mathcal{O}_Q & \longrightarrow & 0 \\
 & & \downarrow & & \downarrow & & \| & & \\
0 & \longrightarrow & \mathcal{O}_{C,Q} & \longrightarrow & \mathcal{E}_Q & \longrightarrow & \mathcal{O}_Q & \longrightarrow & 0 \\
 & & \downarrow & & \downarrow & & & & \\
 & & \mathcal{O}_Q & = & \mathcal{O}_Q & & & &
\end{array}$$

where q is the S-point at Q and q' is an S-point of Q'. Now, in that diagram, $(f_*\mathcal{O}'_C)_Q$ fits in place of $\mathcal{E}_Q$. However, a pushout diagram is unique [7, Lemme 4.1, p. 127]. Hence, $\mathcal{E}_Q$ is isomorphic to $(f_*\mathcal{O}'_C)_Q$, which is torsion-free. Therefore, the lower sequence is not split. Thus $e \neq 0$.

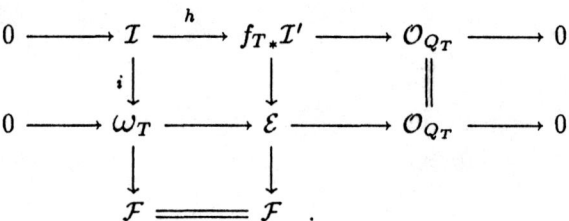

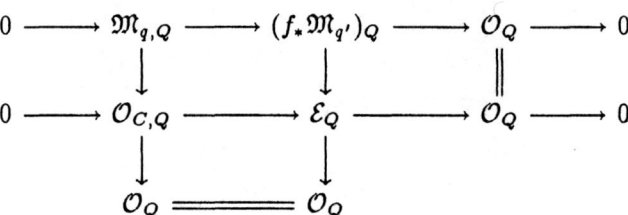

Since $\mathcal{M}$ is invertible on $\mathcal{O}_Q$ and e is nonzero on every fiber, $\mathcal{M}_T$ is free of rank 1 and e is a generator. Now, the sequence (17.3) also yields a generator. These two generators differ by multiplication by a unit on T. Therefore, we may assume that $\mathcal{E} = f_{T*}(\omega'(Q')_T)$. It is now easy to check that the second vertical sequence defines a T-point of U'' that lifts the given T-point of $P_n \times_{J_{n-1}} U$.

Finally, it is obvious that the two diagrams in the statement of the proposition are indeed Cartesian.

Theorem 18. *Assume $\delta = 1$ and $Q' \subset C'^{sm}$. Then J_{n-1} is obtained from P_n by a 2-to-1 contraction of $\epsilon'(J'_n \times Q')$ to $\epsilon(J'_{n-1})$; that is, the comorphism,*

$$\kappa^c : \mathcal{O}_{J_{n-1}} \longrightarrow \mathcal{O}_{P_n},$$

is injective, and $\mathcal{C}ok(\kappa^c)$ is an invertible sheaf on $\epsilon(J'_{n-1})$.

Proof. The assertion holds off $\epsilon(J'_{n-1})$ by 16, iii. It holds in a neighborhood of $\epsilon(J'_{n-1})$ by descent theory because of 17 and 2, i, iv.

REFERENCES

[1] A. Altman, S. Kleiman, *Algebraic systems of linearly equivalent divisor-like subschemes,* Compositio Math. **29** (1974), 113–139.

[2] A. Altman, S. Kleiman, *Compactifying the Jacobian,* Bull. Amer. Math. Soc. **82** (1976), 947–949.

[3] A. Altman, S. Kleiman, *Compactifying the Picard scheme II,* Amer. J. Math. **101** (1979), 10–41.

[4] A. Altman, S. Kleiman, *Compactifying the Picard scheme,* Adv. Math. **35** (1980), 50–112.

[5] A. Altman, A. Iarrobino, S. Kleiman, *Irreducibility of the compactified Jacobian,* in "Real and complex singularities, Oslo 1976," Proc. 9th Nordic Summer School/NAVF, P. Holm (ed.), Sijthoff and Noordhoff, 1977, pp. 1–12.

[6] H. Bass, *On the ubiquity of Gorenstein rings,* Math. Zeitschr. **82** (1963), 8–28.

[7] M. Demazure, A. Grothendieck, "Propriétés Générales des Schémas en Groupes," Lecture Notes in Math. 151, Springer Verlag, 1970.

[8] C. D'Souza, *Compactification of generalised Jacobians,* Proc. Indian Acad. Sci. **88** (1979), 419–457.

[9] A. Grothendieck, "Revêtments Étales et Groupe Fondamental," Lecture Notes in Math. 224, Springer Verlag, 1971.

[10] A. Grothendieck, with J. Dieudonné, "Éléments de Géométrie Algébrique I," Springer Verlag, 1971.

[11] A. Grothendieck, with J. Dieudonné, "Éléments de Géometrie Algébrique," Publ. Math. I.H.E.S. **11, 17, 20, 24, 28, 32**, 1961–1967.

[12] S. Kleiman, *Relative duality for quasi-coherent sheaves,* Compositio Math. **41** (1980), 39–60.

[13] S. Kleiman, *The structure of the compactified Jacobian: a review and an announcement,* in "Seminari di geometria 1982–1983," Università degli studi di Bologna, Dipartmento di matematica, 1984, pp. 81–92.

[14] S. Kleiman, H. Kleppe, *Reducibility of the compactified Jacobian,* Compositio Math. **43** (1981), 277–280.

[15] H. Kleppe, "The Picard scheme of a curve and its compactification," M.I.T. Thesis, 1981.

[16] T. Oda, C. Seshadri, *Compactifications of the generalized jacobian variety,* Trans. Amer. Math. Soc. **253** (1979), 1–90.

[17] M. Raynaud, *Passage au quotient par une relation d'équivalence plate,* in "Proceedings of a Conference on Local Fields," T. A. Springer (Ed.), Springer Verlag, 1967, pp. 78–85.

Department of Mathematics
Simon's Rock of Bard College
Great Barrington, MA 01230
and
Department of Mathematics, 2–278
M.I.T.
Cambridge, MA 02139

Some Algebras Associated to Automorphisms of Elliptic Curves

M. ARTIN, J. TATE and M. VAN DEN BERGH

1. Introduction

The main object of this paper is to relate a certain type of graded algebra, namely the regular algebras of dimension 3, to automorphisms of elliptic curves. Some of the results were announced in [V]. A graded algebra A is called *regular* if it has finite global dimension, polynomial growth, and is Gorenstein. The precise definitions are reviewed in Section 2. As was shown in [A-S], there are two basic possibilities for a regular algebra A of (global) dimension 3 which is generated in degree 1. Either A can be presented by 3 generators and 3 quadratic relations, or else by 2 generators and 2 cubic relations. Throughout this paper, A will denote an algebra so presented, over a ground field k. The number of generators will be denoted by r, and the degrees of the defining relations by s. Thus the possible values are

(1.1) $$(r,s) = \begin{cases} (3,2) \\ (2,3), \end{cases} \quad \text{and } r+s=5.$$

In ([A-S], (1.5)) it is shown that if A is regular, then there are choices of generators (x_i) and relations $(f_i = 0)$, $1 \le i \le r$, such that if we write $f_i = \sum_{j=1}^{r} m_{ij} x_j$ in the free associative algebra $T = k\langle x_1, \ldots, x_r\rangle$ generated by the x_i, then the elements $g_j = \sum x_i m_{ij}$ are also defining relations, so that there is a non-singular matrix $Q = (q_{ij}) \in G\ell_r(k)$ with scalar entries such that $\sum_{i=1}^{r} x_i m_{ij} = \sum_{l=1}^{r} q_{jl} f_l$ for $j = l, \ldots, r$. We call an algebra of this kind *standard* (see Section 2). In a standard algebra A, the linear transformation of $A_1 = \Sigma k x_i$ defined by $x_i \rightsquigarrow \sum_j x_j q_{ji}$ is canonical, independent of the choice of generators and relations. This "Q-transform" of A_1 is used in ([A-S], Section 3) to classify the standard algebras into "types," which are irreducible algebraic families; there are seven types for $r = 3$ and six types for $r = 2$. The generic algebra of each type is shown to be regular, with diagonalizable matrix Q, but in [A-S] no simple criterion is given for deciding whether a special standard algebra is regular.

In this paper we give such a criterion, and give a different sort of classification, in which the invariant Q plays no role at all. The method involves multilinearization of homogeneous elements of the algebra $T = k\langle x_1, \ldots, x_n\rangle$ (see Section 3). This T is the tensor algebra of the space A_1. To a tensor $t \in A_1^{\otimes n} = T_n$ we associate the corresponding multilinear function $\tilde{t}$ on the product of n copies of the dual space A_1^*. Since $\tilde{t}$ is multilinear, it is multihomogeneous and defines a locus of zeros in the product $(\mathbf{P}^{r-1})^n$ of n copies of the $(r-1)$-dimensional projective space of lines in A_1^*.

Let Γ be the locus of common zeros of the multilinearizations $\tilde{f}_i$ of the defining relations for A. Thus, $\Gamma \subset \mathbf{P}^2 \times \mathbf{P}^2$ if $r = 3$, and $\Gamma \subset \mathbf{P}^1 \times \mathbf{P}^1 \times \mathbf{P}^1$ if $r = 2$. Define the projections

$$\alpha\left(p^{(1)} \times \ldots \times p^{(s)}\right) = \left(p^{(1)} \times \ldots \times p^{(s-1)}\right) \quad \textit{(drop the last component)}$$

$$\beta\left(p^{(1)} \times \ldots \times p^{(s)}\right) = \left(p^{(2)} \times \ldots \times p^{(s)}\right) \quad \textit{(drop the first component).}$$

For a standard algebra A the images of Γ under these two projections are the same, both being given by the locus of zeros of $\det\widetilde{M}$, where $\widetilde{M}$ is the matrix $(\tilde{m}_{ij})$ with $m_{ij} \in T_{s-1}$ defined as above by $f_i = \Sigma m_{ij} x_j$. We call this locus $E : E = \alpha(\Gamma) = \beta(\Gamma)$, and we view Γ as the graph of a correspondence $\alpha(p) \rightsquigarrow \beta(p)$ from E to itself. If $r = 3$, then E is a cubic divisor in $\mathbf{P}^2$ or is all of $\mathbf{P}^2$ if $\det\widetilde{M}$ is identically zero. If $r = 2$, then E is a divisor of bidegree $(2,2)$ in $\mathbf{P}^1 \times \mathbf{P}^1$ or is all of $\mathbf{P}^1 \times \mathbf{P}^1$ if $\det\widetilde{M}$ is identically zero.

We call the algebra A *nondegenerate* if Γ is the graph of an automorphism $\sigma : E \to E$. For this to be so, it is necessary and sufficient that the $r \times r$ matrix $\widetilde{M}$ have rank at least $r-1$ at every point of $\mathbf{P}^2(\bar{k})$, or of $\mathbf{P}^1(\bar{k}) \times \mathbf{P}^1(\bar{k})$, where $\bar{k}$ is an algebraic closure of k, i.e., that the 9 conics in $\mathbf{P}^2$, or the 4 curves of bidegree $(1,1)$ in $\mathbf{P}^1 \times \mathbf{P}^1$ defined by the minors of $\widetilde{M}$ have no common zero. Our basic result is :

Theorem 1. *The regular algebras of global dimension* 3 *generated in degree* 1 *are exactly the nondegenerate standard algebras.*

To prove this theorem we first show (Section 5), by a brute force computation, that a degenerate standard algebra is not regular. Since regular algebras are standard, we can therefore limit our consideration to a nondegenerate standard A, and we must prove that it is regular. To such an A we associate a *triple* $\mathcal{T} = (E, \sigma, L)$ consisting of (i) a scheme $E \subset (\mathbf{P}^{r-1})^{s-1}$ which is either a divisor of the type described above, or is the whole ambient space, (ii) an automorphism σ of E and (iii) the invertible $\mathcal{O}_E$-module $L = \pi^*\mathcal{O}(1)$, where π is the inclusion of E in $\mathbf{P}^2$ if $r = 3$, or is the projection of E on the first factor $\mathbf{P}^1$ if $r = 2$. In the latter case, the projection of E on the second factor is $\pi \circ \sigma$. In both cases the map $\pi : E \longrightarrow \mathbf{P}^{r-1}$ is the morphism defined by the sections of L. Hence we can recover the subscheme $\Gamma \subset (\mathbf{P}^{r-1})^s$ from the triple $\mathcal{T} = (E, \sigma, L)$; it is the image of E under the closed immersion $p \rightsquigarrow (\pi(p) \cdot \pi(\sigma p), \dots, \pi(\sigma^s p))$. Having Γ, we can try to recover A as the quotient of the tensor algebra on $H^0(E, L) = A_1$ by the ideal generated by the tensors $f \in A_1^{\otimes S}$ whose multilinearizations $\tilde{f}$ vanish on Γ. Working this out, one is led to define a new graded algebra $B = \Sigma_n B_n$ by $B_n = H^0(E, L \otimes L^\sigma \otimes \cdots \otimes L^{\sigma^{n-1}})$, where $L^\sigma = \sigma^* L$, with multiplication defined by

$$bc = b \otimes c^{\sigma^n} \in B_{n+m},$$

for $b \in B_n$, $c \in B_m$. This $B = \mathcal{B}(\mathcal{T})$ is constructed so that there is a canonical isomorphism $A_1 \longrightarrow B_1$ which extends to a homomorphism of graded algebras $A \longrightarrow B$, and our approach to A is via B and this homomorphism. The point is that A, being defined only by generators and relations, is hard to get at, but B, whose homogeneous parts are spaces of sections of line bundles on E, is more amenable to study. In Section 6 we state (6.6) the required results about B, and show (6.8) how they not only imply that A is regular, but also yield the following theorem about the structure of A and $\mathcal{T}$.

Theorem 2. *Suppose A is regular, let $\mathcal{T} = (E, \sigma, L)$ be the triple associated to A, and let $B = \mathcal{B}(\mathcal{T})$. Then (i) If $\dim E = 2$, then $A \xrightarrow{\sim} B$. If $\dim E = 1$, then $A/gA \xrightarrow{\sim} B$, where $g \in A_{s+1}$ is a (right and left) non-zero divisor such that $Ag = gA$. (ii) Let $\lambda \in Pic\, E$ be the class of L. Then*

$$(1.2) \qquad (\sigma - 1)^2\lambda = 0 \ \textit{if } r = 3, \ \textit{and } (\sigma - 1)(\sigma^2 - 1)\lambda = 0 \ \textit{if } r = 2.$$

Section 7 is devoted to the proofs of the facts about B which are stated and used in Section 6. In case $\dim E = 2$ this is elementary. When $\dim E = 1$, i.e., E is a divisor on $\mathbf{P}^2$ or $\mathbf{P}^1 \times \mathbf{P}^1$, then it involves Riemann-Roch and duality for E and the surface. The proof is made by considering the cohomology of some relevant vector bundles on E. There are technical complications because E need not be reduced or irreducible.

In fact, it is essential for Theorem 3 below that we study (Section 7) an arbitrary triple of the type described above, without assuming it comes from a nondegenerate standard algebra A. Such triples can be viewed as coming from nondegenerate algebras on r generators with r relations of degree s of a kind more general than standard, which we call semi-standard (4.2). It turns out that the non-standard i.e., non-regular, among these give rise to triples $\mathcal{T}$ which do not satisfy (1.2). Hence, we call triples satisfying (1.2) *regular* triples. For the non-regular semi-standard algebras we have $A \xrightarrow{\sim} \mathcal{B}(\mathcal{T})$ except in one case, which we call *exceptional.* In that case, $B = A/gA$, where g is normalizing of degree $s + 1$, but is probably a zero divisor – we have not studied this case closely. In any case, the reader interested in regular algebras can ignore the discussion of non-standard algebras in Sections 5 and 6; it suffices to consider arbitrary triples in Section 7.

For regular algebras we obtain the following classification in terms of triples (6.8):

Theorem 3. *The isomorphism classes of regular graded algebras of dimension* 3 *generated in degree* 1 *are in bijective correspondence with isomorphism classes of triples $\mathcal{T}$ which are regular (*i.e., *satisfy condition* (1.2)*) and are such that, if* $\dim E = 1$, *then* $(\sigma - 1)\lambda \neq 0$ *if* $r = 3$ *and* $(\sigma^2 - 1)\lambda \neq 0$ *if* $r = 2$.

The triples excluded by the last condition are just those for which E is a divisor, but the automorphism σ can be extended from E to the whole space $\mathbf{P}^2$ or $\mathbf{P}^1 \times \mathbf{P}^1$. These triples give rise to the same A as the triples with E the whole space and σ the extension. Incidentally, as is implicit in Theorem 2, the regular algebra A is recovered from its triple by constructing $B = \mathcal{B}(\mathcal{T})$ as above, and then keeping only the relations of degree s defining B, i.e., ignoring the extra relation of degree $s + 1$ needed to define B when $\dim E = 1$.

Let us now consider some examples. The most familiar example of a regular algebra of dimension 3 with two generators is the enveloping algebra A of the Heisenberg Lie algebra. It can be presented by the cubic defining relations

$$[x,[x,y]] = [y,[x,y]] = 0. \tag{1.3}$$

Calling these f_1 and f_2 we find

$$M = \begin{bmatrix} yx - 2xy & xx \\ -yy & 2yx - xy \end{bmatrix}$$

and, using $(x_1, y_1; x_2, y_2)$ as coordinates in $\mathbf{P}^1 \times \mathbf{P}^1$,

$$\widetilde{M} = \begin{bmatrix} y_1x_2 - 2x_1y_2 & x_1x_2 \\ -y_1y_2 & 2y_1x_2 - x_1y_2 \end{bmatrix}.$$

Thus, $\det \widetilde{M} = -2(x_1y_2 - x_2y_1)^2$, so that if $\operatorname{char}(k) \neq 2$, the divisor E given by $\det \widetilde{M} = 0$ is the double diagonal $2\Delta \subset \mathbf{P}^1 \times \mathbf{P}^1$. Using the affine coordinate $t = y/x$ in $\mathbf{P}^1$, E is given by $(t_2 - t_1)^2 = 0$, i.e., by the points $(t, t + \epsilon)$ such that $\epsilon^2 = 0$, and it is easy to check that the automorphism σ is given by $\sigma(t, t+\epsilon) = (t+\epsilon, t+2\epsilon)$. Thus σ is identity on the reduced curve Δ but is not trivial on 2Δ. If char $k = 2$, then $E = \mathbf{P}^1 \times \mathbf{P}^1$ and σ is given by $(t_1 t_2) \rightsquigarrow (t_2, t_1)$.

One of the most interesting cases is that in which the automorphism σ is a translation on elliptic curve E. It turns out that the translations correspond to algebras classified as type A in ([A-S], (3.9, 11)). According to ([A-S], (10.14)), the equations defining a generic algebra of type A with $r = 3$ can be put into the form

$$\begin{aligned} f_1 &= cx^2 + bzy + ayz = 0, \\ f_2 &= azx + cy^2 + bxz = 0, \\ f_3 &= byx + axy + cz^2 = 0, \end{aligned} \tag{1.4}$$

over an algebraically closed field. The multilinear equations which correspond to (1.4) are

$$\begin{aligned} \tilde{f}_1 &= cx_1x_2 + bz_1y_2 + ay_1z_2 = 0, \\ \tilde{f}_2 &= az_1x_2 + cy_1y_2 + bx_1z_2 = 0, \\ \tilde{f}_3 &= by_1x_2 + ax_1y_2 + cz_1z_2 = 0, \end{aligned} \tag{1.5}$$

where the coordinates in $\mathbf{P}^2 \times \mathbf{P}^2$ are labeled as $(x_1, y_1, z_1; x_2, y_2, z_2)$. Let $\Gamma \subset \mathbf{P}^2 \times \mathbf{P}^2$ denote the locus of common zeros of these equations. To compute the projection $\mathrm{pr}_1\Gamma = E$, we determine the values of (x_1, y_1, z_1) for which there exist (x_2, y_2, z_2) solving (1.5). Since the equations (1.5) are linear in (x_2, y_2, z_2), the values of (x_1, y_1, z_1) in question are those such that $\det \widetilde{M} = 0$, where, dropping the subscript 1,

$$\widetilde{M} = \begin{bmatrix} cx & bz & ay \\ az & cy & bx \\ by & ax & cz \end{bmatrix}.$$

Thus E is the "level 3" cubic curve defined by the equation

$$(1.6) \qquad \left(a^3 + b^3 + c^3\right) xyz = abc\left(x^3 + y^3 + z^3\right).$$

Notice that the computation of $\mathrm{pr}_2\Gamma$ yields the same answer, in agreement with the fact that Γ is the graph of an automorphism.

The automorphism σ can also be computed from the equation (1.5). Given (x, y, z) such that $\det \widetilde{M} = 0$ we have $\sigma(x, y, z,) = (x', y', z')$ where (x', y', z') is a vector orthogonal to the row vectors of $\widetilde{M}$. Such vectors are the "outer" or "cross" products of any two of those rows, and if A is nondegenerate, i.e., if $\widetilde{M}$ has rank 2 at every point of E, then there will be a pair of rows whose outer product is not zero, so defines a point in $\mathbf{P}^2$. Taking the top two rows we find for example that in general

$$(1.7) \qquad \sigma(x, y, z) = \left(acy^2 - b^2xz, bcx^2 - a^2yz, abz^2 - c^2xy\right).$$

Let us choose the rational point $(1, -1, 0)$ on E as origin. Then $\sigma(1, -1, 0) = (a, b, c)$. It follows that σ is translation by the point (a, b, c) in the group law on E. One way to see this is to recall that the only automorphisms of a generic elliptic curve which are not translations have order 2, and this is not the case for σ, provided that (a, b, c) is generic.

Using Theorem 1 it is an easy exercise to check that, as conjectured in [A-S, 10.37i], the algebra (1.4) is regular unless either $a^3 = b^3 = c^3$, or two of the three quantities (a, b, c) are zero. (But in the case $r = 2$, 10.37i is not quite right; $A_{1,0,0}$ is regular, even linear.) If exactly one of them is zero, then E is a "triangle" consisting of the lines $x = 0$, $y = 0$, and $z = 0$, unless $a^3 + b^3 + c^3 = 0$, in which case $E = \mathbf{P}^2$. The automorphism σ either carries the lines into themselves (if $c = 0$) or permutes them cyclically, carrying the line $z = 0$ to the line containing the point (a, b, c). The non-regular case, when two of the a, b, c are zero, is the case in which the point (a, b, c) is a vertex of the triangle.

If $abc \neq 0$, then E is smooth unless $(3abc)^3 = (a^3 + b^3 + c^3)^3$, in which case, E is again a triangle, unless char $k = 3$, when E is the triple line $(x + y + z)^3 = 0$. Again, the non-regular case $a^3 = b^3 = c^3$ occurs when the point (a, b, c) is a vertex of the triangle, at least in char $\neq 3$.

In ([A-S], (6.11)), a cubic curve was introduced to describe the condition that A is a skew polynomial ring. For algebras of type A, that cubic is the locus $x^3 + y^3 + z^3 + 6uxyz = 0$, where $6u = (3a + 3b)/c$. This curve is *not* isomorphic to our curve E (1.6) for most choices of the parameters (a, b, c). Thus there are two j-invariants associated to regular algebras of dimension 3. Each of them has significance for the algebra. They agree for types E, H and S in the classification [A-S], but not for types A and B. The presence of two elliptic curves caused some confusion.

This work grew out of an attempt to understand the conjectural description of algebras of type A in [A-S]. The main thrust of these conjectures is to describe the condition that an algebra defined by equation (1.4) be a finite module over its center, in terms of the point $p = (a, b, c) \in \mathbf{P}^2$. The condition is that p must lie on one of a sequence of curves C_n defined by certain recursive relations. It has turned out that these curves are modular curves, C_n being the locus of points $p = (a, b, c) \in \mathbf{P}^2$ such that p has order n on the curve E (1.5). In other words, the algebras which are finite modules over their centers are those such that the automorphism σ has finite order. The same is true for all regular algebras of dimension 3.

It is quite easy to prove that, if σ has finite order, then the algebra $B = \mathcal{B}(E, \sigma, L)$ defined above is finite over its center. This is done in Proposition (8.5), and we make use of that fact in the last section, to prove that regular algebras of dimension 3 are left and right noetherian. However, since our description of the algebra A is indirect, our proof that A is finite over its center if σ is of finite order is considerably more complicated. It is based on a detailed study of A-modules. We will present this proof in a subsequent paper [A-T-V] (see also [V]).

The geometric approach to regular algebras A of dimension 3 which we have used, relating them to triples, works beautifully and simply for regular algebras of dimension 2, which are those of the form $A = B(\mathbf{P}^1, \sigma, \mathcal{O}(1))$, where σ is an arbitrary automorphism of $\mathbf{P}^1$. We learned only recently that the translations on elliptic curves have been used also to construct algebras in higher dimension. Already in 1983 Sklyanin [S] used an elliptic curve, translation by a point, and Theta functions θ_{ij} which are sections of a bundle of degree 4 to make an algebra with 4 generators and 6 quadratic relations which is presumably regular*, and recently Odessky and Feigin [O-F] generalized his method using sections of a bundle of degree n to obtain algebras with n generators and $\binom{n}{2}$ quadratic defining relations. The situation for $n > 3$ is more subtle than for $n = 3$, because the space of quadratic relations defining A is *not* the space of *all* quadratic relations holding in B, as it is for us in dimension 3, but is, rather, an $\binom{n}{2}$-dimensional subspace of that $(n^2 - 2n)$-dimensional space, and the description of the subspace is not at all obvious.

2. Background and notational conventions

Let k be a field. By *graded algebra* A over k, we will always mean an algebra graded in positive degrees, such that the part of degree zero is k:

(2.1) $$A = k + A_1 + A_2 + \dots .$$

*The regularity of these algebras was proved recently by P. Smith and T. Stafford.

Except for rings occurring incidentally the algebras we study will be generated by finitely many elements of degree 1. Such a ring is a quotient of the tensor algebra $T = T(A_1)$ on the vector space A_1, by an ideal I which is generated by homogeneous elements: $A \approx T/I$.

By *module* M over a graded ring A, we will always mean a graded left or right module, and homomorphisms are graded, i.e., degree-preserving, homomorphisms. To avoid confusion, we will sometimes indicate that we are dealing with a left or right module by a subscript A, viz. ${}_AM$ or M_A, depending on the case. The module M is said to be *left bounded* if $M_n = 0$ for all $n \ll 0$. A key fact about a left bounded module M is this version of the Nakayama lemma:

Proposition 2.2. *Let $I = A_1 + A_2 + \ldots$ be the augmentation ideal of A, let M be a left bounded module, and let $\varphi : N \to M$ be a homomorphism. If the composed map $N \to M/IM$ is surjective, then φ is surjective.* □

A module M is called *locally finite* if each graded piece M_n is a finite dimensional k-vector space [An]. The *Hilbert function* of a locally finite module is the function

$$n \rightsquigarrow m_n = \dim_k M_n, \tag{2.3}$$

which takes finite values for all n. Since it is a quotient of the tensor algebra, an algebra A of the form (2.1) is left bounded and locally finite, as an A-module, and so it has a Hilbert function.

The *shift* $M(r)$ of a left or right module M is the module defined by

$$M(r)_n = M_{n+r}. \tag{2.4}$$

It is an elementary exercise to prove that a left bounded module P is projective in the category of graded modules if and only if it is isomorphic to a direct sum of shifts of the module A:

$$P \approx \bigoplus_i A(-\ell_i), \tag{2.5}$$

in other words, if and only if P is free, with basis $\{e_i\}$ consisting of homogeneous elements e_i of varying degrees ℓ_i which are bounded below (see [C, Section 7, Proposition 9] or [B_1, Ch.10, Sec. 8, No. 7, Prop. 8]. Cartan's "gradué par une graduation ≥ 0" can obviously be replaced by "left bounded"). Such a module P is locally finite if and only if either the direct sum is finite or if $\ell_i \to \infty$ as $i \to \infty$. If we represent a direct sum as a row vector, then maps $f : P \to P'$ between two such projectives are given by right multiplication by row-finite matrices with homogeneous entries in A. Every left bounded module has a resolution by projective modules of this same form. In fact, it has a *minimal* resolution which is unique up to non-unique isomorphism and which is a direct summand of every resolution [C, Section 7, Definition 2]. If M is locally finite, so are the projectives in its minimal resolution.

The *projective dimension* of a module M is the minimal length of a projective resolution of M, and the algebra A is said to have *global dimension* d if every A-module M has projective dimension $\leq d$ and d is minimal with this property. A basic fact about graded algebras is that their global dimension is equal to the projective dimension of the left module ${}_Ak$. Indeed, let

$$0 \to P^d \longrightarrow \ldots \xrightarrow{f_2} P^1 \xrightarrow{f_1} P^0 \longrightarrow {}_Ak \to 0 \tag{2.6}$$

be a minimal resolution of ${}_Ak$ by left bounded, locally finite projectives of the form (2.5):

$$P^q = \bigoplus_j A\,(-\ell_{qj})\,, \quad \text{and} \quad P^d \neq 0. \tag{2.7}$$

The shape of the minimal resolution of the right module k_A is the same, as is shown by the formula

$$\operatorname{Tor}_q^A\,(k_A, {}_Ak) \approx \bigoplus k\,(-\ell_{qj})\,, \tag{2.8}$$

and in particular, its length is also d. Hence $\operatorname{Tor}_i^A(k_A, M)$ and $\operatorname{Tor}_i^A(N, {}_Ak)$ vanish for $i > d$ and for all left modules M and right modules N. If M or N has a minimal resolution, it follows that its length is $\leq d$. Thus $\operatorname{Ext}_A^i(M, L) = 0$ for all L and all $i > d$, if M is left bounded. For arbitrary modules, see [Au, Theorem 1], where the ungraded case is treated. There is no difficulty in adapting the proof to the graded case.

In general, the projectives which appear in the resolution (2.6) need not be finitely generated. However, if they are finitely generated we may consider the transposed complex of right modules, obtained by applying the functor $M \rightsquigarrow M^* = \operatorname{Hom}_A(M, A)$. Note that $A(\nu)^*$ is the right module $A(-\nu)$. An algebra of finite global dimension A will be called *Gorenstein* if

(2.9). (i) the projectives appearing in a minimal resolution of ${}_Ak$ are finitely generated, and

(ii) the transposed complex is a resolution of a right module isomorphic to k_A, shifted to some degree e:

$$0 \leftarrow k_A(e) \leftarrow P^{d*} \leftarrow \dots \leftarrow P^{1*} \leftarrow P^{0*} \leftarrow 0. \tag{2.10}$$

This last condition can also be expressed as follows:

$$\operatorname{Ext}^q_A({}_Ak, A) = 0 \text{ if } q \neq d, \text{ and } \operatorname{Ext}^d_A({}_Ak, A) \approx k_A(e). \tag{2.11}$$

Finally, a graded algebra A is called a *regular* algebra of dimension d if it satisfies these conditions:

(2.12) (i) A has global dimension d,
(ii) is Gorenstein, and
(iii) has polynomial growth,

i.e., has $\dim_k A_n \leq cn^\delta$ for some positive real numbers c, δ. For every regular algebra that we know, the minimal such δ is equal to $d-1$.

Going back to the minimal resolution (2.6) of the module ${}_Ak$, it is clear that $P^0 \approx A$. The next two terms P^1, P^2 can be described in terms of the generators and relations of A. Suppose that A is written as quotient of the free associative algebra $T = k\langle x_1, \dots, x_n\rangle$, as $A \approx T/I$, where $(x_1, \dots, x_n)$ is a minimal set of homogeneous generators. Let $\ell_{1j} = \deg x_j$. Then

$$P^1 \approx \bigoplus_{j=1}^{n} A(-\ell_{1j}),$$

and if we view the elements of P^1 as row vectors $(a_1, \dots, a_n)$, with $a_j \in A(-\ell_{1j})$, the map $P^1 \to P^0$ is right multiplication by the column vector $x = (x_1, \dots, x_n)^t$. This is rather trivial. The interpretation of P^2 in terms of defining relations is less obvious, but it is well known (cf. [W], p. 441). Let $\{f_i\}$ be a minimal set of homogeneous generators for the ideal I, and let $\ell_{2j} = \deg f_j$. Then $P^2 \approx \oplus_j A(-\ell_{2j})$, and the map $f_2 : P^2 \to P^1$ is right multiplication by the matrix M defined as follows. In the non-commutative polynomial ring, we may write the defining relations uniquely in the form

$$f_i = \sum_j m_{ij} x_j, \tag{2.13}$$

where $m_{ij} \in T_{\ell_{2i} - \ell_{1j}}$. Then M is the image in A of the matrix (m_{ij}). Thus the start of a resolution of ${}_Ak$ has the form

$$\bigoplus_i A(-\ell_{2i}) \xrightarrow{M} \bigoplus_j A(-\ell_{1j}) \xrightarrow{x} A \longrightarrow {}_Ak \longrightarrow 0, \tag{2.14}$$

x and M being defined as above. Conversely, if (2.14) is an exact sequence and (m_{ij}) is a lifting of the matrix M to the free algebra $T = k\langle x_1, \ldots, x_n \rangle$, then the elements f_i defined by (2.13) form a minimal set of defining relations for the algebra A [W, loc cit].

It is difficult to interpret the remaining terms of a minimal resolution concretely, though canonical, non-minimal resolutions are known [C-E], [An]. However, if A is a regular algebra of dimension 3, only one term in the minimal resolution remains to be found, and the Gorenstein condition together with the requirement of polynomial growth determine it completely. As is shown in ([AS], (1.6)), the minimal resolution of ${}_Ak$ for a regular algebra A of dimension 3 has the form

$$(2.15) \qquad 0 \to A(-s-1) \xrightarrow{x^t} A(-s)^r \xrightarrow{M} A(-1)^r \xrightarrow{x} A \longrightarrow {}_Ak \to 0,$$

where $(r, s) = (3, 2)$ or $(2, 3)$. Thus such an algebra has r generators and r relations of degree s, and $r+s = 5$. The Gorenstein symmetry also tells us that the entries of the row vector $g = (x^t)M$ form a minimal generating set for the ideal I. Thus

$$(2.16) \qquad g^t = \left((x^t)\, M\right)^t = QMx = Qf,$$

for some $Q \in G\ell_r(k)$.

In each case, $r = 2$ and $r = 3$, the Hilbert function of a regular algebra of dimension 3 with r generators is the same as that of a commutative polynomial algebra in three variables, the variables being of degrees $(1, 1, 1)$ if $r = 3$ and of degree $(1, 1, 2)$ if $r = 2$. Explicitly,

(2.17)
$$\begin{aligned} \dim_k A_n &= \tfrac{1}{2}\left(n^2 + 3n + 2\right) \text{ if r=3, and} \\ &= \begin{cases} \tfrac{1}{4}\left(n^2 + 4n + 4\right) & \text{if } r = 2 \text{ and } n \text{ is even,} \\ \tfrac{1}{4}\left(n^2 + 4n + 3\right) & \text{if } r = 2 \text{ and } n \text{ is odd.} \end{cases} \end{aligned}$$

We will call an algebra A a *standard algebra* if A can be presented by r generators x_j of degree 1 and r relations f_i of degree s, such that, with M defined by (2.13), $(r, s) = (3, 2)$ or $(2, 3)$ as above, and there is an element $Q \in G\ell_r(k)$ such that (2.16) holds.

For any standard algebra A, the sequence (2.15) is a complex, which, by the discussion above, is exact at the first three terms on the right. We will speak of this complex as the *potential resolution* of ${}_Ak$. This potential resolution is actually a resolution at all generic points of the variety which parametrizes such sequences, and if it is a resolution, then A is a regular algebra of dimension 3.

3. Multilinearization

Let $T = k\langle x_0, \ldots, x_n \rangle$ denote the free associative k-algebra on generators $x_0, \ldots, x_n$ of degree 1, and let $V = T_1^*$ denote the dual space of T_1. A homogeneous element $f \in T_d = T_1^{\otimes d}$ of degree d defines a linear map

$$\tilde{f} : V \otimes \ldots \otimes V = V^{\otimes d} \to k, \tag{3.1}$$

or equivalently, a multilinear form $\tilde{f} : V \times \ldots \times V = V^d \to k$. This form may be viewed as a section of the sheaf

$$\mathcal{O}(1,1,\ldots,1) = \mathrm{pr}_1^* \mathcal{O}_{\mathbf{P}}(1) \otimes \ldots \otimes pr_d^* \mathcal{O}_{\mathbf{P}}(1) \tag{3.2}$$

on the product $\mathbf{P} \times \ldots \times \mathbf{P}$ of d copies of the projective space $\mathbf{P} \approx \mathbf{P}^n$ of lines in V (or of hyperplanes in T_1). We will denote this product by $(\mathbf{P})^d$. Since $\tilde{f}$ is multilinear and hence multihomogeneous, its zeros define a locus in $(\mathbf{P})^d$.

Note that with natural conventions, we have

$$\widetilde{(fg)}(v_1, \ldots, v_p; w_1, \ldots, w_q) = \tilde{f}(v_1, \ldots, v_p)\, \tilde{g}(w_1, \ldots, w_q) \tag{3.3}$$

for $f \in T_p$, $g \in T_q$.

Let $A = T/I$ be a graded quotient ring of T, let $\tilde{I}_d$ be the space of multilinear forms $\tilde{f}$ such that $f \in I_d$. We will denote by Γ_d the scheme of zeros of $\tilde{I}_d$ in the product $(\mathbf{P})^d$:

$$\Gamma_d = \mathcal{Z}\left(\tilde{I}_d\right) \subset (\mathbf{P})^d. \tag{3.4}$$

Proposition 3.5. (i) *For any d, $\Gamma_{d+1} \subset (\mathbf{P} \times \Gamma_d) \cap (\Gamma_d \times \mathbf{P})$, with equality if $I_{d+1} = T_1 I_d + I_d T_1$, for instance if I is generated in degrees $\leq d$.*

(ii) *Let $\mathbf{P}_i$ denote the i-th factor of the product $(\mathbf{P})^d$, and for $1 \leq i < j \leq d$, let $\mathrm{pr}_{ij} = \mathrm{pr}_{ij}^{(d)}$ denote its projection to the product $\mathbf{P}_i \times \ldots \times \mathbf{P}_j = \prod_{\nu=i}^{j} \mathbf{P}_\nu$. Then $\mathrm{pr}_{ij}(\Gamma_d)$ is a closed subset of Γ_{j-i+1}.*

Proof. (i) With the obvious notation, the locus of zeros of $\widetilde{T_1}\tilde{I}_d$ in $(\mathbf{P})^{d+1}$ is $\mathbf{P} \times \Gamma_d$, and similarly, the locus of zeros of $\tilde{I}_d \widetilde{T_1}$ is $\Gamma_d \times \mathbf{P}$. Since $T_1 I_d + I_d T_1 \subset I_{d+1}$, it follows that $(\mathbf{P} \times \Gamma_d) \cap (\Gamma_d \times \mathbf{P}) \supset \Gamma_{d+1}$. If I is generated in degrees $\leq d$, then $T_1 I_d + I_d T_1 = I_{d+1}$. This implies that $(\mathbf{P} \times \Gamma_d) \cap (\Gamma_d \times \mathbf{P}) = \Gamma_{d+1}$.

(ii) The projection is closed because Γ_d is proper. Also, $\mathrm{pr}_{ij}^{-1}(\Gamma_{j-i})$ is the locus of zeros in $(\mathbf{P})^d$ of $\widetilde{T_{i-1}}\tilde{I}_{j-i+1}\widetilde{T_{d-j}}$, and $T_{i-1} I_{j-i+1}\, T_{d-j} \subset I_d$. Therefore $\mathrm{pr}_{ij}^{-1}(\Gamma_{j-i+1}) \supset \Gamma^d$, and $\Gamma_{j-i+1} \supset \mathrm{pr}_{ij}(\Gamma_d)$. □

Proposition 3.6. *Let $0 \leq i \leq d$, and let $\pi : \Gamma_{d+1} \to \Gamma_d$ denote the projection $(p_1, \ldots, p_i, q, p_{i+1}, \ldots, p_d) \rightsquigarrow (p_1, \ldots, p_d)$.*
(i) The fibres of π are linear subspaces of $\mathbf{P}$.
(ii) Let $p \in \Gamma_d$, and let L be the fibre of π at p. If $\dim L \leq 0$, then π is a closed immersion locally in a neighborhood of $p \in \Gamma_d$.

Proof. The first part follows from the fact that the defining equations of Γ_{d+1} are multilinear. This implies that the scheme-theoretic fibre L at $p \in \Gamma_d$ is defined by linear equations, and so it is a linear space. To prove the second part, note that the projection is a proper map. So if the fibre at p has dimension zero, then π is a finite map in some neighborhood of p, by Chevalley's Theorem [EGA III, 7.8]. This means that the localization of $\pi_* \mathcal{O}_{\Gamma_{d+1}}$ is a finite $\mathcal{O}_{\Gamma_{d,p}}$-module. The fibre is isomorphic to p or else is empty. Hence, by the Nakayama Lemma, the map $\mathcal{O}_{\Gamma_d} \to \mathcal{O}_{\Gamma_{d+1}}$ is surjective locally at p. □

Proposition 3.7. (i) *Assume that for some d, $\mathrm{pr}_{1d}^{(d+1)}$ defines a closed immersion from Γ_{d+1} to Γ_d, thus identifying Γ_{d+1} with a closed subscheme $E \subset \Gamma_d$. Then Γ_{d+1} defines a map $\sigma : E \to \Gamma_d$, by the rule*

$$\sigma(p_1, \ldots, p_d) = (p_2, \ldots, p_{d+1}),$$

where $(p_1, \ldots, p_{d+1})$ is the unique point of Γ_{d+1} lying over $(p_1, \ldots, p_d) \in \Gamma_d$.
(ii) *If in addition $\sigma(E) \subset E$ and if I is generated in degree $\leq d$, then $\mathrm{pr}_{1d}^{(n)} : \Gamma_n \to E$ is an isomorphism for every $n \geq d+1$.*

Proof. The first assertion follows from the identification of E with Γ_{d+1}. To prove the second one, we define an inverse function ψ as follows: If $p = (p_1, \ldots, p_d) \in E$, we set $\psi(p) = (p_1, \ldots, p_n)$, where the coordinates p_ν are chosen so that $\mathrm{pr}_{1+i,d+1}^{(n)}(p) = (p_{1+i}, \ldots, p_{d+i}) = \sigma^i(p)$ for $0 \leq i \leq n-d$. Since $\sigma^i(p) \in E$ for each i, it follows from the equality $\Gamma_{r+1} = (\Gamma_r \times \mathbf{P}) \cap (\mathbf{P} \times \Gamma_r)$ (3.5)(i) and induction that $q = \psi(p)$ is the unique point of Γ_n such that $\mathrm{pr}_{1d}^{(n)}(q) = p$ for each $p \in E$. Hence the projection $\mathrm{pr}_{1d}^{(n)}$ maps Γ_n isomorphically to E, by (3.6)(ii). □

We will now interpret the loci Γ_n associated to a graded algebra A in terms of certain A-modules which are analogous to points of projective varieties.

Definition **3.8.** A graded right A-module M will be called a *point module* if it satisfies the following conditions:
(i) M is generated in degree zero,
(ii) $M_0 = k$, and
(iii) $\dim M_i = 1$ for all $i \geq 0$.

By a *truncated point module of length* $d+1$, we mean a module M satisfying (i) and (ii), and whose Hilbert function is

$$\dim M_i = \begin{cases} 1 \text{ if } 0 \le i \le d \\ 0 \text{ otherwise} \end{cases}.$$

We may also speak of a *family* of point modules parametrized by $S = \operatorname{Spec} R$, where R is a commutative k-algebra. Such a family is, by definition a graded $R \otimes A$-module M, generated in degree zero, such that $M_0 = R$ and that M_i is locally free of rank 1 for each i. Families of truncated point modules are defined similarly.

Proposition 3.9. *With the above notation, there is a one-to-one correspondence between points* $(p_i, \dots, p_d) \in \Gamma_d$ *and truncated point modules of length* $d+1$. *More precisely,* Γ_d *represents the functor of flat families of such modules: If* R *is a commutative* k*-algebra and* $S = \operatorname{Spec} R$*, then points of* Γ_d *with coordinates in* R *correspond to isomorphism classes of families of truncated point modules, parametrized by* S.

Proof. Let M be a family of truncated point modules. Since point modules have been rigidified by the requirement that $M_0 = R$, they are compatible with descent. So we may localize R, to reduce to the case that each graded piece M_j, $1 \le j \le d$, is a free R-module of rank 1. We choose a basis m_j for each M_j. For $1 \le j \le d$ and $0 \le i \le n$, we write out the products of the bases m_j by the generators x_i of A:

$$m_{j-1} x_i = m_j a_{ij}, \tag{3.10}$$

for some $a_{ij} \in R$. In this way, we obtain a set of d points

$$a_j = (a_{0j}, \dots, a_{nj}) \in \mathbf{P} \tag{3.11}$$

with coordinates in R, i.e., an R-valued point of $(\mathbf{P})^d$. This point has the property that if $f(x)$ is any polynomial in $R \otimes T$ of degree d, then the formula

$$m_0 f(x) = m_d \tilde{f}(a_1, \dots, a_d)$$

holds, where M is given the structure of T-module via the map $T \to A$. If $f(x) \in I_d$, then $f(x)$ represents zero in A, hence $\tilde{f}(a_1, \dots, a_d) = 0$. So by definition of the locus Γ_d, $(a_1, \dots, a_d) \in \Gamma_d(R)$. This procedure is reversible: Starting with a point of $\Gamma_d(R)$, we localize R so that the point can be represented by a d-tuple of homogeneous coordinates $(a_1, \dots, a_d)$, with $a_{ij} \in R$. We let M_j denote a free module of rank one with basis m_j, and we define a structure of module on the direct sum $M = M_0 \oplus \dots \oplus M_d$

by the rule (3.10). It is easily checked that this definition produces a truncated point module, and that the two functors thus defined are quasi-inverses of each other. □

If $A = T$ is the free associative algebra itself, there are no equations, so $\Gamma_d = \Gamma_d(T)$ is just the product $(\mathbf{P})^d$. In this case, the universal family of truncated point modules corresponding to the functor we have described is a sheaf $M = M_0 \oplus \ldots \oplus M_d$ on $(\mathbf{P})^d$, where

$$(3.12) \qquad M_j = \mathrm{pr}_1^* \mathcal{O}_{\mathbf{P}}(1) \otimes \ldots \otimes \mathrm{pr}_j^* \mathcal{O}_{\mathbf{P}}(1) = \mathcal{O}(1, \ldots, 1, 0, \ldots, 0).$$

Since $T_1 = H^0(\mathbf{P}, \mathcal{O}(1))$, multiplication of tensors provides us with a map $M_j \otimes_k T_1 \to M_{j+1}$, which defines the T-module structure on the sheaf M.

For an arbitrary quotient algebra A of T, the universal truncated point module is obtained by restricting the sheaf M defined above to $\Gamma_r(A)$.

Note that the functorial description of the schemes Γ_d shows that they are intrinsic to the algebra A, and do not depend on its presentation as a quotient of T.

Corollary 3.13. *Let Γ denote the inverse limit of the sets Γ_d. Then the points of Γ are in one-to-one correspondence with point modules.*

We may also obtain an object which represents the functor of point modules analogous to (3.9), by considering Γ as a pro-scheme instead of taking the inverse limit. Or, Γ has the structure of an infinite-dimensional scheme, namely the subscheme of the infinite product $\mathbf{P} \times \mathbf{P} \times \ldots$ defined by the relations $\tilde{f} = 0$ for all $f \in I$.

Note that if one of the projections $\mathrm{pr}_{1d} : \Gamma_{d+1} \to \Gamma_d$ is an isomorphism, then the inverse system $\{\Gamma_i\}$ is constant for $i \geq d$, by Proposition (3.7)(ii), and hence $\Gamma \approx \Gamma_d$. This occurs in many cases, such as with $d = s = 5 - r$, when A is regular algebra of dimension 3.

As a simple example, we may consider the commutative ring $A = k[x_0, \ldots, x_n]$. Here the defining equations are $x_i x_j = x_j x_i$, and the associated bilinear equations define the diagonal $\Gamma_2 = \Delta \subset \mathbf{P} \times \mathbf{P}$. Proposition (3.7) applies, σ is the identity map $\mathbf{P} \to \mathbf{P}$, and $\Gamma_d \approx \mathbf{P}$ for all d. By the above Corollary, point modules are in bijective correspondence with points of the projective space $\mathbf{P}$.

As another example, we may consider the regular algebras of global dimension 2 which are generated in degree 1. They are defined by a single quadratic relation in two variables of the form

$$(3.14) \qquad (cx + dy)x - (ax + by)y = 0,$$

with $ad - bc \neq 0$. For these algebras, Γ_2 is the graph of an automorphism of $\mathbf{P}^1$, namely

$$(3.15) \qquad \sigma(x, y) = (ax + by, cx + dy),$$

and two such algebras are isomorphic if and only if the conjugacy classes of the corresponding automorphisms of $\mathbf{P}^1$ are the same. Proposition (3.7) shows that for these algebras as well, point modules are in bijective correspondence with points of $\mathbf{P}^1$. On the other hand, the relation $yx = 0$ defines an algebra which is of global dimension 2 but is not Gorenstein, hence is not a regular algebra. For this algebra, $\Gamma_2 = (p \times \mathbf{P}^1) \cup (\mathbf{P}^1 \times q)$, where $p = (1,0)$ and $q = (0,1)$, is the graph of a degenerate correspondence on $\mathbf{P}^1$. In this case, the schemes Γ_n are all non-isomorphic.

Question **3.16**. Assume that A is noetherian and finitely presented. Is it true that the inverse system $\{\Gamma_d\}$ is constant for large d?

3.17. *The algebra B*: The most naive attempt to recover an algebra A from its multilinearization correspondences Γ_d yields an algebra B and a canonical homomorphism $A \to B$, which we will now describe. In fact, we can associate canonically a graded algebra B to any sequence of subschemes $\{Z_d \subset (\mathbf{P})^d\}$ having the property

$$\mathrm{pr}_{1,d-1}(Z_d) \subset Z_{d-1} \text{ and } \mathrm{pr}_{2,d}(Z_d) \subset Z_{d-1} \text{ for all } d. \tag{3.18}$$

It follows from this condition that $\mathrm{pr}_{ij}(Z_d) \subset Z_{j-i+1}$ for all $1 \leq i < j \leq d$, as in Proposition (3.5)(ii). Denote by L_d the restriction of the invertible sheaf $\mathcal{O}(1,1,\dots,1) = \mathrm{pr}_1^*\mathcal{O}_{\mathbf{P}}(1) \otimes \dots \otimes \mathrm{pr}_d^*\mathcal{O}_{\mathbf{P}}(1)$ to Z_d:

$$L_d = \mathcal{O}(1,1,\dots,1) \otimes_{\mathcal{O}_{(\mathbf{P})^d}} \mathcal{O}_{Z_d}, \tag{3.19}$$

and let $B_d = H^0(Z_d, L_d)$. Since $\mathrm{pr}_{1,i}(Z_{i+j}) \subset Z_i$ and $\mathrm{pr}_{i+1,i+j}(Z_{i+j}) \subset Z_j$, we have $Z_i \times Z_j \supset Z_{i+j}$. Therefore we can define a multiplication map $B_i \times B_j \to B_{i+j}$ by applying H^0 to the obvious isomorphism $\mathrm{pr}_{1,i}^*(L_i) \otimes_{\mathcal{O}_{\Gamma_d}} \mathrm{pr}_{i+1,i+j}^*(L_j) \to L_{i+j}$. This multiplication makes $B = \oplus B_d$ into a graded associative algebra. We will refer to B as the algebra *associated* to the sequence $\{Z_d\}$. We remark that B need not be generated in degree 1, and if it is, then the sequence $\{Z_d\}$ may be properly contained in the sequence $\{\Gamma_d = \Gamma_d(B)\}$.

If $Z_d = (\mathbf{P})^d$ for all d, the associated algebra B is the free algebra T. But if we start with an arbitrary algebra $A = T/I$ then the algebra B associated to the sequence $\{Z_d = \Gamma_d(A)\}$ need not be isomorphic to A, though the two algebras are always related by a homomorphism:

Proposition 3.20. *Let $A = T/I$ be a quotient of the free ring, and let $\Gamma_d = \Gamma_d(A)$. Let B be the algebra associated to the sequence $\{\Gamma_d\}$. There is a canonical homomorphism,*

$$\varphi : A \to B,$$

which is bijective in degree 1.

Proof. The functorial maps

$$H^0\left((\mathbf{P})^d, \mathcal{O}(1,1,\dots,1)\right) \to H^0\left(\Gamma_d, L_d\right)$$

define a homomorphism from the free algebra T to B, which carries an element $f \in T_d$ to the restriction to Γ_d of the corresponding section $\tilde{f}$ of $\mathcal{O}(1,\dots,1)$. Since Γ_d is the scheme of zeros of $\tilde{I}_d = \{\tilde{f} | f \in I_d\}$, the ideal I is in the kernel of the homomorphism $T \to B$. Therefore this homomorphism factors through $T/I = A$. In degree 1, Γ_1 is the projective space of hyperplanes in $T_1/I_1 = A_1$, and our homomorphism is the canonical isomorphism $A_1 \xrightarrow{\sim} H^0(\Gamma_1, \mathcal{O}(1)) = B_1$.

4. Multilinearization of semi-standard algebras

Throughout this section, A will denote an algebra which is presented in the form $A = T/I$, where T is a non-commutative polynomial ring with r generators of degree 1 and I is an ideal generated by r linearly independent relations of degree s. We assume moreover that $(r, s) = (3, 2)$ or $(2, 3)$ as in the end of Section 2.

Let $f = (f_1, \dots, f_r)^t$ be a column vector of defining relations, i.e., of generators for I. As in Section 2, there are uniquely defined matrices M, N with entries in the tensor algebra T such that

$$f = Mx \quad \text{and} \quad f^t = x^t N. \tag{4.1}$$

Let $\mathbf{P} = \mathbf{P}^{r-1}$. We consider the sequence of subschemes $\Gamma_d \subset (\mathbf{P})^d$ defined in the previous section as the zeros of the multilinearizations $\tilde{I}_d$. It follows from Proposition (3.5) that the whole sequence of schemes is determined by $\Gamma = \Gamma_s$, the locus of zeros of the multilinearized relations $\tilde{f}_i$ in $(\mathbf{P})^s$.

Let E be the scheme-theoretic image of Γ via the projection $\mathrm{pr}_{1,s-1}$. We denote the restriction of this projection to Γ by $\pi_1 : \Gamma \to E$, and similarly, we denote by $\pi_2 : \Gamma \to E'$ the map from Γ to its image under the map $\mathrm{pr}_{2,s}$. By definition, E is the locus of points $p = (p_1, \dots, p_{s-1}) \in (\mathbf{P})^{s-1}$ for which there exists $p_s \in \mathbf{P}$ such that $(p_1, \dots, p_{s-1}, p_s) \in \Gamma$, i.e., for which the system of linear equations $\widetilde{M}(p_1, \dots, p_{s-1})x = 0$ has a non-trivial solution $x = p_s$. So E is the locus of zeros in $(\mathbf{P})^{s-1}$ of the multihomogeneous polynomial $\det \widetilde{M} = 0$. Similarly, E' is the locus $\det \widetilde{N} = 0$. Thus E is either all of $(\mathbf{P})^{s-1}$ if $\det \widetilde{M} = 0$, or is a Cartier divisor in $(\mathbf{P})^{s-1}$, and similarly for E'.

We will call an algebra A with r generators and r relations of degree s as above *semi-standard* if the schemes E and E' are equal, or equivalently, if

$$\det \widetilde{M} = c \det \widetilde{N} \tag{4.2}$$

for some $c \in k^*$. The constant c is independent of change of basis in x and f, provided that $\det \widetilde{M}$ is not identically zero.

Suppose that our algebra A is semi-standard, so that $E = E'$. We may view Γ as the graph of a correspondence on E via the closed immersion $(\pi_1, \pi_2) : \Gamma \to E \times E$, and we call A *nondegenerate* if Γ is the graph of an automorphism σ of E, and *degenerate* otherwise. This terminology conflicts slightly with the terminology of [A-S, Section 2], where degenerate means not regular. However, we will show in the next section that regular algebras are nondegenerate, and ultimately, for the algebras considered in [A-S, Tables (3.9, 11)], that the nondegenerate ones are regular.

Proposition 4.3. *Let A be a semi-standard algebra. With the above notation, the scheme E is either a divisor in $(\mathbf{P})^{s-1}$, or else $E = (\mathbf{P})^{s-1}$. More precisely, we have the following cases.*

Case $r = 3$: Either

(a) *E is a divisor of degree 3 in $\mathbf{P}^2$, and Γ is the graph of an automorphism σ of E, or*
(b) *$E = \mathbf{P}^2$, and Γ is the graph of an automorphism σ of $\mathbf{P}^2$, or*
(c) *the scheme Γ contains a subset of the form $p \times \ell$, where ℓ is a line and p is a point in $\mathbf{P}^2$.*

Case $r = 2$: Either

(a) *E is a divisor in $\mathbf{P}^1 \times \mathbf{P}^1$ of bidegree (2,2), and Γ defines an automorphism σ of E, of the form $\sigma(p,q) = (q, f(p,q))$,*
(b) *$E = \mathbf{P}^1 \times \mathbf{P}^1$, and σ defines an automorphism of $\mathbf{P}^1 \times \mathbf{P}^1$ which has the form $\sigma(p,q) = (q, \tau(p))$, where τ is an automorphism of $\mathbf{P}^1$, or*
(c) *the scheme Γ contains a subset of the form $p \times q \times \mathbf{P}^1$, where p, q are points of $\mathbf{P}^1$.*

In both cases, A is degenerate if and only if (c) holds.

Proof. The main differences between the cases $r = 2$ and $r = 3$ are notational, so since there is a small additional point when $r = 2$, we will omit the proof of the case $r = 3$.

We have seen in Proposition (3.6) that A is degenerate unless the projections $\pi_i : \Gamma \to E$ are isomorphisms. Assume that they are. Then $(\pi_1, \pi_2)\Gamma$ is the graph of the automorphism $\sigma = \pi_2 {\pi_1}^{-1}$ of E, and E, being the scheme of zeros of the "biquadratic" form $\det \widetilde{M}$, is either all of $(\mathbf{P})^2 \approx \mathbf{P}^1 \times \mathbf{P}^1$, if $\det \widetilde{M} = 0$, or is a divisor of bidegree $(2,2)$.

A priori, the automorphism σ has the form $\sigma(p,q) = (q, f(p,q))$ for some map $f : E \to \mathbf{P}^1$ (see (3.7)(i)). So to complete the proof, we need to show that if $E = \mathbf{P}^1 \times \mathbf{P}^1$, then $f(p,q)$ is independent of q. Now since the automorphism σ has the above form, the map $p \rightsquigarrow f(p,q)$ defines an automorphism of $\mathbf{P}^1$ for each fixed point q. Varying q, we obtain a map $\mathbf{P}^1 \to PG\ell_2$. This map is constant because $\mathbf{P}^1$ is proper and $PG\ell_2$ is affine. Thus $f(p,q)$ is independent of q, as claimed. The fact that τ is an automorphism of $\mathbf{P}^1$ follows from the fact that σ is an automorphism.

It remains to discuss the asymmetry in possibility (c) in case A is degenerate. This is taken care of by the following lemma.

Lemma 4.4. *A semi-standard algebra A is nondegenerate if and only if one of the projections $\pi_i : \Gamma \to E$ is an isomorphism, or equivalently, if either one of the matrices $\widetilde{M}$, $\widetilde{N}$ has rank at least $r-1$ at every point.*

Proof. As we have already remarked, it follows from Proposition (3.6) that π_1 is an isomorphism if and only if $\widetilde{M}$ has rank at least $r-1$ at every point. Now if π_1 is an isomorphism then Γ is the graph of a surjective, dominant map $\sigma : E \to E$. Then since E is either a curve or $\mathbf{P}^2$ or $\mathbf{P}^1 \times \mathbf{P}^1$, the fibre of σ can not contain a line, hence σ is an isomorphism. If E is a curve, this follows from the fact that E contains finitely many irreducible components. If $E = \mathbf{P}^2$ or $\mathbf{P}^1 \times \mathbf{P}^1$ it follows from the fact that these surfaces do not contain curves which can be contracted to a point. The same reasoning works if π_2 is an isomorphism. $\square$

The generic situation for a semi-standard algebra is Case (a). Case (b) also occurs; for example, when $r = 3$, it includes the commutative polynomial ring. As we have already remarked, we will show in Theorem (5.1) that regular algebras are nondegenerate, hence that Case (c) does not occur for them.

Proposition (4.3) tells us that we can associate to every nondegenerate semi-standard algebra A the triple $\mathcal{T}(A) = (E, \sigma, L)$, where $E = \pi_1(\Gamma)$, σ is the automorphism of E defined by Γ, and L is the invertible sheaf $pr_1^*\mathcal{O}(1)$. It will be convenient to consider triples given abstractly as well. So we make the following definition:

Definition 4.5. A *triple* $\mathcal{T}$ is a set (E, σ, L), where E is a scheme, σ is an automorphism of E, and L is an invertible sheaf on E whose global sections define a morphism $\pi : E \to \mathbf{P} \approx \mathbf{P}^{r-1}$, and where one of the following holds:

Case $r = 3$:

(a) E is a divisor of degree 3 in $\mathbf{P}^2$, and L is the restriction of $\mathcal{O}_{\mathbf{P}^2}(1)$,
(b) $E = \mathbf{P}^2$, and $L = \mathcal{O}(1)$,

Case $r = 2$:

(a) E is a divisor of bidegree $(2,2)$ in $\mathbf{P}^1 \times \mathbf{P}^1$, σ has the form $\sigma(p,q) = (q, f(p,q))$, and $L = pr_1^*(\mathcal{O}_{\mathbf{P}^1}(1))$.

(b) $E = \mathbf{P}^1 \times \mathbf{P}^1$, σ has the form $\sigma(p,q) = (q, \tau(p))$, where τ is an automorphism of $\mathbf{P}^1$, and $L = pr^*(\mathcal{O}_{\mathbf{P}^1}(1))$.

We will say that a triple of type (a) is *elliptic*. Similarly, we will say that a nondegenerate semi-standard algebra A is *elliptic* if the associated triple $\mathcal{T}$ is elliptic, i.e., if we are in case (a) of (4.3). We have chosen this terminology because smooth divisors E of degree 3 in $\mathbf{P}^2$ or of bidegree $(2,2)$ in $\mathbf{P}^1 \times \mathbf{P}^1$ are elliptic curves. However, we do not require our divisor E to be smooth.

Triples of type (b) will be called *linear*.

4.6. *Standard algebras and regular triples.* Standard algebras are semi-standard. This follows from equation (2.16), which expresses the property that the entries of the row-vector $x^t M$ generate the defining ideal: $x^t M = f^t Q^t$ for some $Q \in Gl_r(k)$. Combining this with (4.1), we find for a regular algebra that

$$M = NQ^t, \quad Q \in Gl_r(k). \tag{4.7}$$

This condition obviously implies (4.2), with $c = \det Q$.

Next, let $\mathcal{T}(A) = (E, \sigma, L)$ be the triple associated to a regular algebra, and let λ denote the class of L in Pic E. It will turn out that the operation of the automorphism σ on the class λ satisfies the following relation:

$$\begin{cases} (\sigma - 1)^2 \lambda = 0 & \text{if } r = 3 \\ (\sigma - 1)(\sigma^2 - 1)\lambda = 0 & \text{if } r = 2 \end{cases}. \tag{4.8}$$

Recall that a triple $\mathcal{T}$ is said to be *regular* if it satisfies (4.8). These definitions of the words semi-standard, standard and regular form local terminology, introduced in this paper for convenience. Note that regularity of a triple has nothing to do with whether E is a regular scheme.

Note that all linear triples are regular. A linear triple $\mathcal{T}$ satisfies the stronger condition

$$\begin{cases} \sigma\lambda = \lambda & \text{if } r = 3 \\ \sigma^2\lambda = \lambda & \text{if } r = 2 \end{cases}, \tag{4.8$'$}$$

because Aut $\mathbf{P}$ acts trivially on Pic $\mathbf{P} = \mathbb{Z}$. In fact, this condition characterizes linear triples.

Our main results, Theorem (6.8) and Proposition (6.7) assert that the regular algebras of dimension 3 are precisely the nondegenerate standard algebras.

4.9. *Exceptional triples and exceptional algebras.* Triples of the following types play a special role:

(4.10) r=3: E is the union of a line and a conic, and σ interchanges these two components;
r=2: E is the union of three components of bidegrees $(1,0)$, $(0,1)$, and $(1,1)$ respectively, and σ permutees these components cyclically.

We will call a triple satisfying (4.10) *exceptional*, and similarly, we will call a semi-standard algebra A *exceptional* if it is nondegenerate and the associated triple is exceptional.

The following proposition may be helpful for intuition, though it is not needed elsewhere.

Proposition 4.11. *Exceptional triples are not regular.*

Proof. Suppose that $r = 3$, and let $E = C_1 + C_2$, where C_1 is a line and C_2 is a conic. Denote by a subscript i the restriction of a sheaf to C_i. Then $\deg L_1 = \deg (\sigma^2 L)_1 = 1$, while $\deg (\sigma L)_1 = 2$. Thus $\deg ((\sigma - 1)^2 L)_1 = -2$, which shows that $(\sigma - 1)^2 L$ is not even zero numerically. The proof is similar for the case $r = 2$. □

Remarks 4.12. (i) It is not difficult to check that the only triples for which the condition (4.8) does not hold at least *numerically* are the exceptional triples.

(ii) It is possible to specialize a translation on a plane cubic curve to an automorphism which interchanges a line and a conic, but nevertheless the above proposition and Theorem (6.8ii) show that regular algebras are not exceptional.

For each type of regular algebra of dimension 3 in the classification of [A-S, (3.9, 11)], the generic algebras are elliptic. The divisors E and automorphisms σ which they determine are described below, when the ground field is algebraically closed. If E is a smooth curve, the description involves the tacit choice of a suitable base point p_0 to make E into an abelian variety. In each case, the automorphism σ and line bundle L are related by the condition that $\mathcal{T}$ be regular. The relation (4.8) follows from the theorem of the square for type A, for which σ is a translation, and so it does not impose any condition on L for these algebras. But it is not automatic for some of the other types.

4.13. *The divisors E and automorphism σ associated to generic regular algebras of dimension* 3.

Case $r = 3$:

type A : E is a smooth curve, and σ is a translation by a point of E.

type B : E is a smooth curve, σ is multiplication by -1, and $L \approx \mathcal{O}_E(2p_0 + q)$, where q is a point of order 4 of E.

type E : E is a smooth curve, $j(E) = 0$, σ is an automorphism of order 3, multiplication by a cube root of unity, and $L \approx \mathcal{O}_E(2(p_0) + (q))$, where q is a point of order 3 on E not fixed by σ.

type H: E is a smooth curve, $j(E) = 12^3$, σ is an automorphism of order 4, multiplication by i, and $L \approx \mathcal{O}_E(2(p_0) + (q))$, where q is a point of order 2 of E which is not fixed by σ.

type S_1 : E is a triangle, and σ stabilizes the three components.

type S_1' : E is the union of a line and a conic meeting the line in two points, and σ stabilizes the components.

type S_2 : E is a triangle, and σ interchanges two of its sides.

Case $r = 2$:

type A : E is a smooth curve, and σ is a translation by a point of E.

type E: E is a smooth curve, $j(E) = 0$, σ is a complex multiplication order 3, and $L \approx \mathcal{O}_E(p_0 + q)$, where q is a point of order 3 of which is not fixed by σ.

type H: E is a smooth curve, $j(E) = 12^3$, σ is a complex multiplication of order 4, and $L \approx \mathcal{O}_E(p_0 + q)$, where q is a point of order 4 such that $2q$ is not fixed by σ.

type S_1 : E is the union of two curves of bidegree $(1,1)$, and σ stabilizes the components.

type S_2 : $E = C_1 + C_1 + C_1' + C_1'$, where the bidegrees of C_i, C_i' are $(1,0), (0,1)$ respectively, and σ interchanges C_i and C_i'.

type S_2' : $E = C_1 + C_1' + C_2$ is a union of curves of bidegree $(1,0), (0,1), (1,1)$ respectively, and σ interchanges C_1 and C_1'.

5. Discussion of degenerate cases

We retain the notation of Section 4. In this section we show that regular algebras are nondegenerate, by showing that the degenerate standard algebras are the ones listed in [A-S, Lemma (4.2)].

Theorem 5.1. (1) *Regular algebras are neither degenerate nor exceptional.*
(2) *Let A be a semi-standard algebra. Consider the following assertions:*
(i) *There is a basis for the defining ideal of A having one of the forms of* [A-S, (4.2)], *viz:*

$$\begin{aligned} &\textit{Case } r=3: && \text{(a)}\,(f_1,f_2,f_3) = (x^2,{}^*,{}^*)\,,\\ & && \text{(b)}\,(f_1,f_2,f_3) = (yz+cx^2, zx+dy^2, xy)\,.\\ &\textit{Case } r=2: && \text{(a)}\,(f_1,f_2) \quad = (x^3,{}^*)\,,\\ & && \text{(b)}\,(f_1,f_2) \quad = (yxy+cx^3, xyx)\,, \end{aligned}$$

(ii) *There are non-zero elements $u \in T_{s-1}$ and $v \in T_1$ whose product uv is in I_s.*

(iii) *A is degenerate or exceptional.*

The following implications hold: (i) $\Rightarrow$ (ii) $\Rightarrow$ (iii). *Moreover if A is either a degenerate standard algebra or an exceptional algebra, then* (i) *holds.*

Proof. (1): We know that regular algebras are standard. It is proved in [A-S (4.2)] that the standard algebras of the form (2i) are not regular, so the fact that neither degenerate nor exceptional algebras are regular follows from the last assertion of the theorem.

(2): It is trivial that (i) implies (ii). We prove that (ii) implies (iii) by showing that if (ii) holds and A is nondegenerate, then A is exceptional. So we assume, with our usual notation, that Γ is the graph of an automorphism σ of E. Suppose first that $r=3$. Then by hypothesis, I_2 contains an element uv which is the product of two linear forms. We may take $uv = f_1$ as one of the three defining equations f_1, f_2, f_3. The locus of zeros of $\tilde{u}\tilde{v}$ in $\mathbf{P}\times\mathbf{P}$ $(\mathbf{P}=\mathbf{P}^2)$ has the form $Z = (\ell_u\times\mathbf{P})\cup(\mathbf{P}\times\ell_v)$, where ℓ_u and ℓ_v are the zeros of u and v respectively. Since $\Gamma\subset Z$, we may decompose it accordingly: $\Gamma=\Gamma_u\cup\Gamma_v$, where $\Gamma_u=\{\tilde{f}_2=\tilde{f}_3=0\}\cap(\ell_u\times\mathbf{P})$ and $\Gamma_v=\{\tilde{f}_2=\tilde{f}_3=0\}\cap(\mathbf{P}\times\ell_v)$. Since Γ is the graph of an automorphism, the triple $\mathcal{T}=\mathcal{T}(A)$ must be elliptic, and the divisor E is reducible. Thus Γ_u and Γ_v are schemes of dimension 1 in $\mathbf{P}\times\mathbf{P}$, and Γ_u is the graph of an isomorphism from ℓ_u to $C_u=\mathrm{pr}_2\Gamma_u$. Similarly, Γ_v is the graph of an isomorphism from $C_v=pr_1\Gamma_v$ to ℓ_v, and we have $E=\ell_u+C_v=\ell_v+C_u$. Since E is of degree three, C_u and C_v are conics. Hence $\ell_u=\ell_v$, $C_u=C_v$, and the triple is exceptional.

Suppose that $r=2$. Then there is a defining equation $f_1=uv$, where u is quadratic and v is linear, and one other cubic defining relation f_2. As before, we have a decomposition $\Gamma=\Gamma_u\cup\Gamma_v$. Setting $\mathbf{P}=\mathbf{P}^1$, we have $\Gamma_v = \{\tilde{f}_2 = 0\} \cap (\mathbf{P} \times \mathbf{P} \times p_v)$, where p_v is the zero of ℓ_v in $\mathbf{P}$ and $\Gamma_u = \{\tilde{f}_2 = 0\} \cap (D \times \mathbf{P})$, where D is the locus of zeros of $\tilde{u}$, a divisor of bidegree $(1,1)$ in $\mathbf{P} \times \mathbf{P}$. Then Γ_v is a curve of tridegree $(1,1,0)$ in $\mathbf{P} \times \mathbf{P} \times \mathbf{P}$, and Γ_u has tridegree $(1,1,2)$. Since σ is an automorphism, Γ_v is the graph of an isomorphism $\mathrm{pr}_{12}\Gamma_v = C_0 \to C_1 = \mathrm{pr}_{23}\Gamma_v$, where C_0 and C_1 are components of E. So $E = C_0 \cup C_1 \cup C_2$, where C_2 has degree $(0,1)$. Then $\mathrm{pr}_{12}\Gamma_u = C_1 + C_2$, and $\mathrm{pr}_{23}\Gamma_u = C_0 + C_2$. A consideration of the degrees shows that σ permutes the three components cyclically, hence that A is exceptional. This completes the proof that (ii) implies (iii).

The following lemma shows that exceptional algebras have defining equations of the form (2i)(a).

Lemma 5.2. *Let A be an exceptional algebra* (4.9). *There exists a non-zero element $x \in A_1$ such that $x^s = 0$ in A_s.*

This lemma will be proved in the course of proof of Proposition (6.7ii).

We now come to the main part of the proof, the verification that every degenerate standard algebra has defining equations of one of the forms (2i). This is done by an explicit computation.

We first treat the case $r = 3$. Let A be a standard algebra which defines a degenerate correspondence. As we have seen (4.3), this implies that Γ contains a subset of the form $p \times \tilde{\ell}$. Adjusting coordinates (x,y,z) in $\mathbf{P}^2$ appropriately, we may assume that the pair (p,ℓ) is in one of two standard positions, according as p is contained in ℓ or not:

Case 1: $p = (1,0,0), \ell = \{(0,y,z)\}$,

Case 2: $p = (1,0,0), \ell = \{(x,y,0\}$.

We write the defining equations in the form

(5.3)

$$\begin{aligned} f_1 &= a_1x^2 + a_2xy + a_3xz + a_4yx + a_5y^2 + a_6yz + a_7zx + a_8zy + a_9z^2, \\ f_2 &= b_1x^2 + b_2xy + b_3xz + b_4yx + b_5y^2 + b_6yz + b_7zx + b_8zy + b_9z^2, \\ f_3 &= c_1x^2 + c_2xy + c_3xz + c_4yx + c_5y^2 + c_6yz + c_7zx + c_8zy + c_9z^2. \end{aligned}$$

Bilinearizing and substituting $(1,0,0)$ for the first variable yields

$$\tilde{f}_1((1,0,0),(x,y,z)) = a_1x + a_2y + a_3z,$$

etc. Therefore the coefficients a_i, b_i, c_i vanish if $i = 2,3$ in Case 1, and if $i = 1,2$ in Case 2. We now analyze $(x,y,z)M = (g_1,g_2,g_3)$, which by hypothesis is also a set of generators for the ideal (f_1,f_2,f_3) (cf. 2.16).

Case 1: The generators f_i and g_i have the form:

$$\begin{aligned}
f_1 &= a_1x^2 && + a_4yx + a_5y^2 + a_6yz + a_7zx + a_8zy + a_9z^2,\\
f_2 &= b_1x^2 && + b_4yx + b_5y^2 + b_6yz + b_7zx + b_8zy + b_9z^2,\\
f_3 &= c_1x^2 && + c_4yx + c_5y^2 + c_6yz + c_7zx + c_8zy + c_9z^2,\\
g_1 &= a_1x^2 + a_4xy + a_7xz && + b_1yx + b_4y^2 + b_7yz + c_1zx + c_4zy + c_7z^2,\\
g_2 &= a_5xy + a_8xz && + b_5y^2 + b_8yz + c_5zy + c_8z^2,\\
g_3 &= a_6xy + a_9xz && + b_6y^2 + b_9yz + c_6zy + c_9z^2.
\end{aligned}$$

Here the xy and xz coefficients of g_i must vanish because they are not present in f_i. Hence $a_i = 0$ for all $i \geq 1$, and $f_1 = a_1x^2$. This algebra has the form (a).

Case 2: In this case, the generators have the form:

$$\begin{aligned}
f_1 &= && a_3xz + a_4yx + a_5y^2 + a_6yz + a_7zx + a_8zy + a_9z^2,\\
f_2 &= && b_3xz + b_4yx + b_5y^2 + b_6yz + b_7zx + b_8zy + b_9z^2,\\
f_3 &= && c_3xz + c_4yx + c_5y^2 + c_6yz + c_7zx + c_8zy + c_9z^2,\\
g_1 &= && a_4xy + a_7xz + b_4y^2 + b_7yz + c_4zy + c_7z^2,\\
g_2 &= && a_5xy + a_8xz + b_5y^2 + b_8yz + c_5zy + c_8z^2,\\
g_3 &= a_3x^2 && + a_6xy + a_9xz + b_3yx + b_6y^2 + b_9yz + c_3zx + c_6zy + c_9z^2.
\end{aligned}$$

As in the previous case, it follows that $a_3 = a_4 = a_5 = a_6 = 0$. This reduces f_i and g_i to

$$\begin{aligned}
f_1 &= a_7zx + a_8zy + a_9z^2,\\
f_2 &= b_3xz + b_4yx + b_5y^2 + b_6yz + b_7zx + b_8zy + b_9z^2,\\
f_3 &= c_3xz + c_4yx + c_5y^2 + c_6yz + c_7zx + c_8zy + c_9z^2,\\
g_1 &= a_7xz + b_4y^2 + b_7yz + c_4zy + c_7z^2,\\
g_2 &= a_8xz + b_5y^2 + b_8yz + c_5zy + c_8z^2,\\
g_3 &= a_9xz + b_3yx + b_6y^2 + b_9yz \quad c_3zx + c_6zy + c_9z^2.
\end{aligned}$$

Subcase (2a): $a_7 \neq 0$. Then it follows that $c_3 \neq 0$ as well, and that if we express f_1 as linear combination of the g_i, the coefficient of g_3 is not zero. Therefore $b_3 = 0$, which implies that $b_4 = c_4 = 0$ too.

We are still allowed a change of coordinates which preserves the flag $p \subset \ell$, i.e., one of the form

$$\begin{aligned} x &\rightsquigarrow d_{11}x+d_{12}y+d_{13}z, \\ y &\rightsquigarrow \qquad\quad d_{22}y+d_{23}z, \\ z &\rightsquigarrow \qquad\qquad\qquad d_{33}z. \end{aligned}$$

Using the substitution for x, we can eliminate the coefficients a_8, a_9. If also $b_5 = 0$, then we conclude in turn that $b_6 = c_5 = c_6 = b_7 = b_8 = b_9 = 0$, hence that $f_2 = 0$, which shows that A can be defined by 2 relations, contradicting our assumption that A is standard. Assume that $b_5 \neq 0$. Then we can use this coefficient to eliminate b_6 by a substitution $y \rightsquigarrow d_{22}y + d_{23}z$. Since f_1 is a linear combination of g_2, g_3, the vectors (b_5, b_8, c_5, c_8) and (b_6, b_9, c_6, c_9) are proportional. Since $b_6 = 0$, it follows that $b_9 = c_6 = c_9 = 0$, which forces $c_7 = c_8 = b_8 = c_5 = 0$. Hence $g_2 = b_5 y^2$, and the algebra has the form (a).

Subcase (2b): $a_7 = 0$. If also $a_8 = 0$, the algebra has the form (a). If not, we eliminate a_9 by a change of variable y. Then if $b_3 = 0$, it follows that $c_3 \neq 0$, but $b_4 = c_4 = 0$. In this case, f_1 is a linear combination of g_1 and g_3. This forces $c_3 = 0$, which means that $f_3 = 0$. If $b_3 \neq 0$, we can eliminate b_6, b_9 by a change of variable x. When this is done, f_1 is a multiple of g_1, hence $b_4 = b_7 = c_7 = 0$, and $c_3 = 0$. Then f_i is in the span of $\{g_1, g_i\}$, which implies that $b_8 = c_8 = c_5 = c_6 = 0$. This algebra has the form (b).

We now consider the case $r = 2$. Here $\Gamma \subset \mathbf{P}^1 \times \mathbf{P}^1 \times \mathbf{P}^1$, and (4.3) we can assume that the projection pr_{12} is not an immersion at $p = (p_1, p_2) \in \mathbf{P}^1 \times \mathbf{P}^1$, hence that the fibre is all of $\mathbf{P}^1$. We write the equations as

(5.4)

$$\begin{aligned} f_1 &= a_1x^3 + a_2x^2y + a_3xyx + a_4xy^2 + a_5yx^2 + a_6yxy + a_7y^2x + a_8y^3, \\ f_2 &= b_1x^3 + b_2x^2y + b_3xyx + b_4xy^2 + b_5yx^2 + b_6yxy + b_7y^2x + b_8y^3, \end{aligned}$$

so that

$$\begin{aligned} g_1 &= a_1x^3 + a_3x^2y + a_5xyx + a_7xy^2 + b_1yx^2 + b_3yxy + b_5y^2x + b_7y^3, \\ g_2 &= a_2x^3 + a_4x^2y + a_6xyx + a_8xy^2 + b_2yx^2 + b_4yxy + b_6y^2x + b_8y^3. \end{aligned}$$

Adjusting coordinates, we have again two cases, according as $p_1 \neq p_2$ or $p_1 = p_2$:

Case 1: $p = ((1,0)(0,1))$. Substituting p into $\tilde{f}_1$ gives $\tilde{f}_1(1,0;0,1;x,y) = a_3x + a_4y$. Thus $a_3 = a_4 = 0$, and similarly, $b_3 = b_4 = 0$. It follows in order that $a_2 = b_2 = a_6 = b_6 = a_5 = a_7 = a_8 = 0$,, hence that $f_1 = a_1x^3$, which is of the form (a).

Case 2: $p = ((1,0)(1,0))$. Substituting p into $\tilde{f}_1$ gives $a_1x + a_2y$. Thus $a_1 = a_2 = 0$ and $b_1 = b_2 = 0$. It follows that $a_3 = a_4 = a_5 = b_5 = 0$, which reduces the relations to

$$\begin{aligned}
f_1 &= a_6yxy + a_7y^2x + a_8y^3,\\
f_2 &= b_3xyx + b_4xy^2 + b_6yxy + b_7y^2x + b_8y^3,\\
g_1 &= a_7xy^2 + b_3yxy + b_7y^3,\\
g_2 &= a_6xyx + a_8xy^2 + b_4yxy + b_6y^2x + b_8y^3.
\end{aligned}$$

Subcase (2a): $A_6 \neq 0$, hence $b_3 \neq 0$. In this case, f_1, is a multiple of g_1, which implies that $a_7 = 0$. We can make a change of variable fixing p, i.e., of the form $x \rightsquigarrow d_1x + d_2y$. Using such a change, we can eliminate a_8. Then it follows in order that $b_4 = b_7 = b_6 = 0$. This algebra has the form (b).

Subcase (2b): $a_6 = 0$, hence $b_3 = 0$. If $a_7 = 0$, then A has the form (a). If not we eliminate a_8 by a change of variable. Then f_1 is a multiple of g_2, hence $b_4 = 0$, and this contradicts $a_7 \neq 0$. This completes the proof of Theorem 5.1. □

6. The algebras A and B defined by a triple

We have seen how multilinearization determines a triple $\mathcal{T}(A) = (E, \sigma, L)$ from a nondegenerate semi-standard algebra A. In this section we start with an arbitrary triple $\mathcal{T}$ and construct from it two algebras $\mathcal{A}(\mathcal{T})$ and $\mathcal{B}(\mathcal{T})$, and a homomorphism $\beta = \beta(\mathcal{T}) : \mathcal{A}(\mathcal{T}) \to \mathcal{B}(\mathcal{T})$. This construction has the property that if A is a nondegenerate semi-standard algebra and if $\mathcal{T} = \mathcal{T}(A)$, then there is a canonical isomorphism $A \to \mathcal{A}(\mathcal{T})$ which allows us to identify the homomorphism β with the map $A \to B$ constructed at the end of Section 3.

We will prove (6.8ii) that if $\mathcal{T}$ is a regular triple, then $\mathcal{A}(\mathcal{T})$ is a regular algebra of dimension 3, by using the homomorphism $\beta(\mathcal{T})$ together with the properties of $\mathcal{B}(\mathcal{T})$ stated in Theorem 6.6 and proved in the Section 7. We also show (6.7iii) that $\mathcal{T}(A)$ is regular if A is nondegenerate standard. The fact that A is isomorphic to $\mathcal{A}(\mathcal{T}(A))$ then shows that a nondegenerate standard algebra is regular.

We also show (6.8iii) that if $\mathcal{T}$ is regular then we recover $\mathcal{T} = \mathcal{T}(A)$ from the algebra $A = \mathcal{A}(\mathcal{T})$, unless $\mathcal{T}$ is elliptic and satisfies (4.8′). In that case, the automorphism σ of E extends uniquely to the ambient space $X = \mathbf{P}^2$ or $\mathbf{P}^1 \times \mathbf{P}^1$, and the triple attached to $\mathcal{A}(\mathcal{T})$ is the linear one obtained from that extension of σ.

Construction of the algebras $\mathcal{A}(\mathcal{T})$ and $\mathcal{B}(\mathcal{T})$. Let $\mathcal{T} = (E, \sigma, L)$ be a triple and $\pi : E \to \mathbf{P}^{r-1}$ ($r = 2$ or 3) be the morphism determined by the global sections of L. With $\mathcal{T}$ we associate a graded ring $B = \mathcal{B}(\mathcal{T})$ as follows: For each integer $n \geq 0$, set

$$L_n = L \otimes L^\sigma \otimes \ldots \otimes L^{\sigma^{n-1}}. \tag{6.2}$$

Here $L^\sigma = \sigma^* L$ as before, $L_0 = \mathcal{O}_E$, and tensor products are taken over $\mathcal{O}_E$. As a graded vector space, $\mathcal{B}(\mathcal{T}) = B = \Sigma B_n$ is defined by the rule

$$B_n = H^0(E, L_n). \tag{6.3}$$

For every pair of integers $m, n \geq 0$, there is a canonical isomorphism

$$L_m \otimes L_n^{\sigma^m} \to L_{m+n}, \tag{6.4}$$

and hence a map

$$\mu_{m,n} : H^0(E, L_m) \otimes_k H^0(E, L_n) \to H^0(E, L_{m+n})$$

which defines the multiplication in B : If $u \in H^0(E, L_m)$ and $v \in H^0(E, L_n)$, we define the product to be

$$uv = \mu_{m,n}\left(u \otimes v^{\sigma^m}\right) \in B_{m+n}, \tag{6.5}$$

where $v^\sigma = v \circ \sigma$. In this way we obtain an associative algebra $B = \mathcal{B}(\mathcal{T})$ from the triple $\mathcal{T}$.

To define $\mathcal{A}(\mathcal{T})$, let $T = \Sigma T_n$ be the tensor algebra over k on $T_1 = H^0(\mathbf{P}, \mathcal{O}_{\mathbf{P}}(1))$. The isomorphism

$$\pi^* : T_1 \xrightarrow{\sim} B_1$$

induces a homomorphism $T \to B$. Let $J = \Sigma J_n$ be its kernel, and let I be the two-sided ideal of T generated by J_s, where $s = 5 - r$. We then put $\mathcal{A}(\mathcal{T}) = T/I$. The composition of the natural homomorphisms

$$\mathcal{A}(\mathcal{T}) = T/I \to T/J \to B = \mathcal{B}(\mathcal{T})$$

gives us a canonical homomorphism $\mathcal{B}(\mathcal{T}) : \mathcal{A}(\mathcal{T}) \to \mathcal{B}(\mathcal{T})$ which is bijective in degree 1. Note that in fact $A = \mathcal{A}(\mathcal{T})$ depends only on $B = \mathcal{B}(\mathcal{T})$. Indeed, A is the quotient of the tensor algebra on B_1 by the relations of degree s defining B. Key properties of B are listed in the following theorem.

Theorem 6.6. (Properties of $\mathcal{B}(\mathcal{T})$). *Let $\mathcal{T}$ be a triple, let $B = \mathcal{B}(\mathcal{T})$, and call J the kernel of the map $T \to B$ defined above. Let $r = \dim B_1$ and $s = 5 - r$. Then*

(i) *$B = k[B_1]$, and the right and left socles of B are zero (i.e., if $b \in B$ and if either $B_1 b = 0$ or $bB_1 = 0$, then $b = 0$).*

(ii) *If $\mathcal{T}$ is elliptic, then $\dim_k B_n = rn$ for $n > 0$. If $\mathcal{T}$ is linear, then the Hilbert function of B is the same as that of a regular algebra of dimension 3 with r generators* (cf. (2.17)).

(iii) *$J_n = 0$ for $n < s$, and $\dim_k J_s = r$.*

(iv) *For $n \geq 2$, $J_{s+n} = T_1 J_{s+n-1} + J_s T_n = T_n J_s + J_{s+n-1} T_1$.*

(v) *Let $U = J_{s+1}/(T_1 J_s + J_s T_1)$ and $W = T_1 J_s \cap J_s T_1$. Then $\dim_k U = 1$ if $\mathcal{T}$ is elliptic and is either regular or exceptional, and $\dim_k U = 0$ otherwise. Moreover,*

$$\dim_k W = \begin{cases} \dim_k U & \text{if } \mathcal{T} \text{ is elliptic} \\ 1 & \text{if } \mathcal{T} \text{ is linear.} \end{cases}$$

(vi) *Suppose that $\mathcal{T}$ is regular, so that in particular, $\dim_k W = 1$. Then W is a nondegenerate subspace of $T_1 J_s$ and of $J_s T_1$, in the following sense: If w is a non-zero element of W, and if we write $w = \Sigma x_i f_i = \Sigma g_i x_i$, where $\{x_i\}$ is a basis for T_1, then both $\{f_i\}$ and $\{g_i\}$ are bases for J_s.*

A proof of this theorem will be given in the next section. It involves a study of various aspects of the cohomology of certain sheaves on E. For the rest of this section, we assume the theorem and derive the promised consequences for the algebra $\mathcal{A}(\mathcal{T}) = T/I$.

Proposition 6.7. *Let $\mathcal{T} = \mathcal{T}(\mathcal{A}') = (E, \sigma, L)$ be the triple associated to a nondegenerate semi-standard algebra A' as in Section 4, and let $A = \mathcal{A}(\mathcal{T})$. Then*

(i) *The algebras A' and A are canonically isomorphic.*

(ii) *If the algebra A' is exceptional, it is not standard.*

(iii) *If A' is standard, then $\mathcal{T}$ is regular.*

Proof. (i) Both A and A' are quotients of the tensor algebra T on $A'_1 = H^0(\mathbf{P}, \mathcal{O}(1))$ by ideals $I \supset I'$ generated in degree s. By (6.6iii), $\dim I_s = r$, and by definition of a semi-standard algebra, $\dim I'_s = r$ too (Section 4). Hence $I = I'$.

(ii) Suppose A' is exceptional. We will show that there exists a non-zero element $x \in A'_1$ such that $x^s = 0$ in A'_s. The non-standardness of A' follows, because a standard algebra with such an element x is degenerate, as is shown in [A-S, (4.2)]. Let $B = \mathcal{B}(\mathcal{T})$. If $r = 3$, E is the union of a line C_1 and a conic $C_2 = C_1^\sigma$, and we take x to be a non-zero element of $T_1 = H^0(\mathbf{P}, \mathcal{O}(1)) = H^0(E, L) = B_1$ which vanishes on C_1. Then

$x \otimes x^\sigma \in H^0(E, L \otimes L^\sigma)$ vanishes on all of E, so $x^2 = 0$ in $B_2 = T_2/J_2 = T_2/I_2 = A_2 = A'_2$. Similarly, if $r = 2$, then E is a union $C \cup \sigma C \cup \sigma^2 C$, where C is a "line" of the form $p \times \mathbf{P}$, and we take x to be a non-zero element of T_1 such that $x(p) = 0$. Then $x \otimes x^\sigma \otimes x^{\sigma^2}$ vanishes on all of E, so $x^3 = 0$ in $B_3 = T_3/J_3 = T_3/I_3 = A_3 = A'_3$.
(iii) Recall that if $A' = T/I$ is a standard algebra, there is a non-zero element $w = \Sigma x_i m_{ij} x_j \in T_1 I_s \cap I_s T_1$. Since $J_s = I_s$, it follows from (6.6v) that $\mathcal{T}$ is either regular or exceptional (recall that linear triples are regular). But by (ii), a standard algebra is not exceptional. □

Theorem 6.8. *Let $\mathcal{T}$ be a triple. Let $A = \mathcal{A}(\mathcal{T})$ and $B = \mathcal{B}(\mathcal{T})$. Then*
(i) *The canonical homomorphism $A \to B$ is always surjective. It is an isomorphism unless $\mathcal{T}$ is elliptic and either regular or exceptional, in which case its kernel has the form $gA = Ag$, where g is a non-zero normalizing element of degree $s+1$.*
(ii) *If $\mathcal{T}$ is regular, then A is a regular algebra of global dimension 3, and in the elliptic case the element g of (i) is left and right regular.*
(iii) *Let $\mathcal{T}' = (E', \sigma', L')$ be the triple $\mathcal{T}(A)$ attached to A. Then $\mathcal{T} = \mathcal{T}'$ unless $\mathcal{T}$ is elliptic and satisfies the condition (4.8′). In that case $\mathcal{T}'$ is linear, E' is the ambient space $\mathbf{P}^2$ or $\mathbf{P}^1 \times \mathbf{P}^1$ in which E is embedded, and σ' is the unique extension of σ from E to an automorphism of E'.*

Proof. (i) We have $B = k[B_1]$ by (6.6i). Since $A_1 \approx B_1$, the homomorphism $A \to B$ is surjective. Denote its kernel by $K = J/I$, where I is the ideal of T generated by J_s. By (6.6iii), $K_n = 0$ for $n \le s$, and by (6.6iv), $K_n = T_1 K_{n-1} = K_{n-1} T_1$ for $n \ge s+2$. It follows by induction on n that K is generated as left ideal and as right ideal by $K_{s+1} = U$, and now (i) follows from (6.6v), on taking for g a non-zero element of U in case $\dim U = 1$. In that case, let Z and Z' be the right and left annihilators of U in A. For each integer n, we have exact sequences

$$(6.9) \qquad \begin{aligned} 0 \to Z_{n-s-1} \to A_{n-s-1} \xrightarrow{\lambda(g)} A_n \to B_n \to 0 \\ 0 \to Z'_{n-s-1} \to A_{n-s-1} \xrightarrow{\rho(g)} A_n \to B_n \to 0, \end{aligned}$$

where $\lambda(g)$ and $\rho(g)$ denote left and right multiplication by g respectively. Comparing these two sequences, we see that $\dim Z_n = \dim Z'_n$ for all n. Hence, to prove the statement about g in (ii), it suffices to show that $Z = 0$ in case $\mathcal{T}$ is elliptic and regular.
(ii) We now suppose that $\mathcal{T}$ is regular and we prove that A is regular, and if $\mathcal{T}$ is elliptic, that $Z = 0$. By (6.6v), $\dim W = 1$. Let w be a non-zero element of W and write $w = \Sigma x_i m_{ij} x_j$, where (x_i) is a basis for $A_1 = T_1$, with $m_{ij} \in A_{s-1} = T_{s-1}$. By (6.6vi), A is a standard algebra because the

elements $f_i = \Sigma_{ij} x_j$ span J_s, and so do the $g_j = \Sigma x_i m_{ij}$. The standard potential resolution of the left A-module k (2.15) is the complex whose degree n part is

$$(6.10) \qquad 0 \to A_{n-s-1} \xrightarrow{x^t} A^r_{n-s} \xrightarrow{M} A^r_{n-1} \xrightarrow{x} A_n \to k_n \to 0$$

(where $k_n = 0$ if $n \neq 0$, $k_0 = k$, x is the column vector (x_i) of length r, and $M = (m_{ij})$. To say that A is a regular algebra means that this sequence is exact for each n. Since the elements f_i span I_s, it is exact to the right of M. Let P_{n-s-1} and Q_{n-s} be the homology groups of the complex (6.10) at A_{n-s-1} and A^r_{n-s} respectively. Thus $P_n = \{f \in A_n | fA_1 = 0\}$ is the degree n part of the left socle of A.

If a graded vector space is denoted by a capital letter, say X, then we will use the lower case letter to denote its dimension, viz. $x_n = \dim_k X_n$. With this notation, we need to show that $p_n = q_n = 0$ for all n, and also $z_n = 0$ in case T is elliptic. We also set

$$(6.11) \qquad \begin{aligned} e_n &= \text{Hilbert function of a regular algebra} \\ &\quad\ \text{of dimension 3 on } r \text{ generators, and} \\ d_n &= e_n - a_n. \end{aligned}$$

From (6.10), the relation

$$a_{n-s-1} - ra_{n-s} + ra_{n-1} - a_n + \delta_{n,0} = p_{n-s-1} - q_{n-s}$$

holds for all $n \in \mathbb{Z}$, where δ is the Kronecker delta, and since (6.10) is exact for regular algebras, we also have

$$e_{n-s-1} - re_{n-s} + re_{n-1} - e_n + \delta_{n,0} = 0$$

for all n.

Subtracting these two equations, we find

$$(6.12) \qquad -d_{n-s-1} + rd_{n-s} - rd_{n-1} + d_n = p_{n-s-1} - q_{n-s}.$$

If T is linear, we have $b_n = e_n$ by (6.6ii), and since we have seen that $A \approx B$ in this case, we have $a_n = e_n$, i.e., $d_n = 0$ for all n. By (6.6i), A has no socle, so $p_n = 0$ for all n. Then (6.12) shows that $q_n = 0$ for all n as well, hence that A is regular.

Suppose that T is elliptic. In this case we use induction on n to prove simultaneously that p_n, q_{n+1}, and z_n are zero, and thereby finish the proof of (ii). The key point is the following lemma:

Lemma (6.13). *Let j be an integer. If $z_{j-s} = 0$ and $p_{j-s-1} = 0$, then $p_j = 0$.*

Proof. Let $u \in P_j$. Then since B has trivial socle (6.6i), the image of u in B is zero. By (6.9), there is an element $h \in A_{j-s-1}$ such that $u = gh$. Then $ghA_1 = uA_1 = 0$, hence $hA_1 \in Z_{j-s}$. If $Z_{j-s} = 0$, then $h \in P_{j-s-1}$, and if $P_{j-s-1} = 0$ as well, then $h = 0$, and $u = gh = 0$, as was to be shown. □

By (6.6ii), we have $b_n = rn$ if $n \geq 0$, and by (6.6i), $b_0 = 1$, and $b_n = 0$ for $n < 0$. From this fact, or by examining a particular regular algebra of dimension 3, it is easy to check that $b_n = e_n - e_{n-s-1}$. From (6.9), it follows that

$$z_{n-s-1} - a_{n-s-1} + a_n = e_n - e_{n-s-1},$$

i.e., that $z_{n-s-1} = d_n - d_{n-s-1}$ for all n. Together with (6.12), this gives

$$x_{n-s-1} + rd_{n-s} - rd_{n-1} = p_{n-s-1} - q_{n-s} \tag{6.14}$$

for all n.

We now show that $d_n = p_n = q_n = z_n = 0$ for all n by induction on n. For each integer m, let S_m denote the statement

$$S_m : \begin{cases} d_i = 0 & \text{for } i \leq m-1 \\ p_i = 0 & \text{for } i \leq m-1-s \\ q_i = 0 & \text{for } i \leq m-s \\ z_i = 0 & \text{for } i \leq m-1-s \end{cases}.$$

For $m \leq 0$, S_m is trivially true. We show that S_m implies S_{m+1}. Suppose that S_m is true. By (6.12), it follows that $d_i = 0$ for $i \leq m$. By (6.14), with n replaced by $m+1$

$$z_{m-s} = p_{m-s} - q_{m+1-s}. \tag{6.15}$$

Now the lemma, with $j = m - s$, shows that $p_{m-s} = 0$, and consequently, by (6.15), that $z_{m-s} = 0$ and $q_{m+1-s} = 0$. Thus S_{m+1} is true. This completes the induction step and, therewith the proof of (ii).

(iii) It is tautological that $\mathcal{T} \subset \mathcal{T}'$, in the sense that $E \subset E'$ and that σ and L are the restrictions to E of σ' and L'. Thus $E = E'$, and hence $\mathcal{T} = \mathcal{T}'$, unless E and E' have different dimensions, i.e., unless $\mathcal{T}$ is elliptic and $\mathcal{T}'$ is linear. If $\mathcal{T}'$ is linear, then $\mathcal{T}'$ satisfies (4.8′), and so does $\mathcal{T}$ since $\mathcal{T} \subset \mathcal{T}'$. Conversely, suppose $\mathcal{T}$ is elliptic and satisfies (4.8′). Let $\pi : E \to \mathbf{P}^{r-1}$ be the map given by sections of L. If $r = 3$, then $L \approx L^\sigma$ and this means there is an automorphism σ' of $\mathbf{P}^2$ such that $\pi\sigma = \sigma'\pi$, and Γ' is the graph of σ', so $E' = \mathbf{P}^2$. If $r = 2$, then $L \approx L^{\sigma^2}$, which means there is an automorphism τ of $\mathbf{P}^1$ such that $\pi\sigma^2 = \tau\pi$. Thus σ extends to the automorphism $\sigma' : (p,q) \rightsquigarrow (q,\tau p)$ of $\mathbf{P}^1 \times \mathbf{P}^1$, Γ' is the locus of points $(p,q,\tau p)$, and $E' = \mathbf{P}^1 \times \mathbf{P}^1$. □

7. Proof of Theorem 6.6

Let $\mathcal{T} = (E,\sigma,L)$ be a triple, $B = \mathcal{B}(\mathcal{T})$ and let $r = \dim B_1$ and $s = 5 - r$ as usual. To prove Theorem 6.6, we will treat the cases $\mathcal{T}$ linear and $\mathcal{T}$ elliptic separately, but we begin with some remarks which apply in both cases.

Lemma 7.1. *To prove* (6.6i) *for* $\mathcal{T}$ *it suffices to show that*
(i) $H^0(E,\mathcal{O}_E) = k$, *and that*
(ii) *For each* $n > 0$, *the multiplication map*

$$H^0(E,L) \otimes H^0(E,L_n^\sigma) \to H^0(E,L \otimes L_n^\sigma) = H^0(E,L_{n+1})$$

is surjective.

Proof. These statements mean that $B_0 = k$ and that $B_1 \otimes B_n \to B_{n+1}$ is surjective. They imply that $B = k[B_1]$ by induction on the degree. The rest of (6.6i), that B has trivial socle, follows from the fact that, by definition of a triple, L is invertible and generated by its sections. Hence if a section of any sheaf is killed by multiplication with every element of $B_1 = H^0(E,L)$, it is zero. □

Lemma 7.2. *The ideal* J *is generated by* J_s *if and only if* (6.6iii) *and* (6.6iv) *hold and the space* U *in* (6.6v) *is* 0.

This is evident and is left to the reader. □

Lemma 7.3. *The statement about* $\dim_k W$ *in* (6.6iv) *follows from the preceding parts of* (6.6).

Proof. We have

$$\begin{aligned}\dim J_{s+1} &= \dim U + \dim(T_1 J_s + J_s T_1)\\ &= \dim U + (T_1 J_s) + \dim(J_s T_1) - \dim W\\ &= \dim U + 2r^2 - \dim W.\end{aligned}$$

Combining this with $\dim J_n = r^n - \dim B_n$ gives the result in all cases, as we leave to the reader to check. □

Suppose that $\mathcal{T}$ is a linear triple. Then E is either $\mathbf{P}^2$ or $\mathbf{P}^1 \times \mathbf{P}^1$. Up to isomorphism, the invertible sheaves on E are just $\mathcal{O}(n)$ or $\mathcal{O}(m,n)$

according to the case, and the effect of σ on them is given by $\mathcal{O}(n)^\sigma = \mathcal{O}(n)$ in the first case and $\mathcal{O}(m,n)^\sigma = \mathcal{O}(n,m)$ in the second one. In the first case, $L = \mathcal{O}(1)$, hence $L_n \approx \mathcal{O}(n)$ for all $n \geq 0$. In the second case, $L = \mathcal{O}(1,0)$, hence $L_{2m} \approx \mathcal{O}(m,m)$ and $L_{2m+1} \approx \mathcal{O}(m+1,m)$ for all $m \geq 0$. Statements (6.6i) and (ii) follow easily from these facts and from Lemma (7.1).

Choose an isomorphism $\varphi : L \to L^\sigma$ if $E = \mathbf{P}^2$, or an isomorphism $\varphi : L \to L^{\sigma^2}$ if $E = \mathbf{P}^1 \times \mathbf{P}^1$, and let τ be the automorphism of $B_1 = H^0(E,L)$, such that, for $x \in B_1$, $\varphi(x) = (x^\tau)^\sigma$ or $(x^\tau)^{\sigma^2}$, according to the case. Let $\{x_i\}$, $1 \leq i \leq r$, be a basis for $B_1 = T_1$.

Proposition 7.4. *If $r = 3$, J is generated by the three elements*

$$x_i x_j^\tau - x_j x_i^\tau, \quad 1 \leq i < j \leq 3. \tag{7.4'}$$

If $r = 2$, J is generated by the two elements

$$x_1 x_j x_2^\tau - x_2 x_j x_1^\tau, \quad j = 1,2. \tag{7.4''}$$

By lemmas (7.2) and (7.3), parts (iii-v) of Theorem 6.6 follow from this proposition, and part (vi) does too, using, in the two cases, the formulas

$$w = x_1 x_2^\tau x_3^{\tau^2} - x_1 x_3^\tau x_2^{\tau^2} + x_2 x_3^\tau x_1^{\tau^2} - x_2 x_1^\tau x_3^{\tau^2} + x_3 x_1^\tau x_2^{\tau^2} - x_3 x_2^\tau x_1^{\tau^2}$$

and

$$w = x_1^2 (x_2^\tau)^2 - x_1 x_2 x_2^\tau x_1^\tau - x_2 x_1 x_1^\tau x_2^\tau - x_2^2 (x_1^\tau)^2,$$

as we leave to the reader to check.

Proof of 7.4. If $r = 3$, the isomorphism

$$L \otimes L \xrightarrow{1 \otimes \varphi} L \otimes L^\sigma$$

carries the relation $xy = yx$ for $x, y \in H^0(E,L) = B_1$ into the relation $x(y^\tau)^\sigma = y(x^\tau)^\sigma$, which shows, by the definition (6.5) of multiplication in B, that we have $xy^\tau = yx^\tau$ in B for any $x, y \in B_1$. If $r = 2$, the isomorphism

$$L \otimes L^\sigma \otimes L \xrightarrow{1 \otimes 1 \otimes \varphi} L \otimes L^\sigma \otimes L^{\sigma^2}$$

shows similarly that the relation

$$xyz^\tau = zyx^\tau$$

holds for any $x, y, z \in B_1$. Thus the elements (7.4′) and (7.4″) are in J_s.

Let $J' \subset J$ be the ideal generated by the relations (7.4′) or (7.4″), as the case may be, and let $B' = T/J'$.

Using those relations, it is easy to check that B'_n is spanned by the monomials

$$x_{i_1} x_{i_2}^{\tau} \dots x_{i_n}^{\tau^{n-1}}, \quad \text{with} \quad i_1 \leq i_2 \leq \dots \leq i_n$$

in the first case, and by the monomials of degree n of the form

$$x_{i_1} x_{j_1} x_{i_2}^{\tau} x_{j_2}^{\tau} x_{i_3}^{\tau^2} x_{i_3}^{\tau^2} \dots, \quad \text{with} \quad \begin{cases} i_1 \leq i_2 \leq \dots \\ j_1 \leq j_2 \leq \dots \end{cases}$$

in the second case. The number of such monomials is $\dim B_n$ in each case. Hence $B' = B$, $J' = J$, and incidentally, these monomials are linearly independent. This completes the proof of Theorem 6.6 in the case that $\mathcal{T}$ is linear.

For the rest of this section, we assume that our triple is elliptic. Thus if $r = 3$, then E is a divisor of degree 3 in $X = \mathbf{P}^2$, and if $r = 2$, then E is a divisor of bidegree $(2,2)$ in $X = \mathbf{P}^1 \times \mathbf{P}^1$. In both cases the canonical sheaf ω_E is isomorphic to $\mathcal{O}_E$, and E has arithmetic genus 1. This follows from the adjunction formula ([A-K Ch. I, 2.5], [M p.81, Theorem 3])).

We write the divisor E in the form $E = \Sigma r_i C_i$, where C_i are the irreducible components of its support. If M is an invertible sheaf on E, we denote by $d_i(M)$ the degree of its restriction to C_i. By *degree* $\deg M$ of M we mean its total degree: $\deg M = \Sigma r_i d_i(M)$.

Lemma 7.5. *Suppose that $\mathcal{T}$ is not exceptional, i.e., does not satisfy (4.9). Then for all i we have $d_i(L^\sigma) = d_i(L)$ if $r = 3$, and $d_i(L^{\sigma^2}) = d_i(L)$ if $r = 2$.*

Proof. The assertion is trivial if E has only one component. If $r = 3$, there are at most three components, and if there are three, then they are all lines. In that case L has the same degree on each component, and so every automorphism preserves the degrees d_i. There remains the case of two components interchanged by σ, which is the exceptional configuration (4.10).

If $r = 2$, then we need to show that σ^2 preserves the degree $d_i(L)$, hence we need only consider cases in which the set of components has a σ-orbit of order at least 3. Now E has at most four components, and if there are 4, then they consist of two pairs of rulings – a "quadrilateral." In that case a permutation of order 4 exists, but it satisfies the relation $d_i(L^{\sigma^2}) = d_i(L)$. The remaining case of three components is the exceptional configuration (4.10). □

If M is a locally free sheaf on E, we will use the standard notation

$$(7.6) \qquad \begin{gathered} H^i(M) = H^i(E, M), \;\; h^i(M) = \dim_k H^i(M), \\ \det M = \Lambda^n M, \;\; \text{if } n = \text{rank } M, \\ \deg M = \deg(\det M), \;\; M^* = \mathrm{Hom}_{\mathcal{O}_E}(M, \mathcal{O}_E). \end{gathered}$$

The Riemann-Roch theorem and duality give

$$(7.7) \qquad \chi(M) := h^0(M) - h^1(M) = \deg M, \;\; \text{and} \;\; h^1(M) = h^0(M^*).$$

(see [M] p. 79 Ths. 1,2 or [A-K]).

In order to fix attention on cases in which this formula can be used to compute h^0 and h^1 separately in terms of the degree, we introduce some more local terminology:

Definition 7.8: An invertible sheaf M on E is *tame* if either $h^0(M) = 0$ or $h^1(M) = 0$, or $M \approx \mathcal{O}_E$.

If E is reduced and irreducible, e.g., a smooth elliptic curve, then all invertible sheaves are tame, because in that case an invertible M of degree ≤ 0 has no non-zero section unless $M \approx \mathcal{O}_E$, and dually, if $\deg M \geq 0$, then $h^1(M) = 0$ unless $M \approx \mathcal{O}_E$. However, on a reducible divisor E, not all invertible sheaves are so nice. For example, suppose that $E = C_1 + C_2$ consists of two curves C_i of genus zero, meeting with multiplicity $(C_1 \cdot C_2) = 2$. If the first degree d_1 of an invertible sheaf M on E is at least 2, then M will have non-zero sections which vanish on C_2 no matter how negative the total degree $d_1 + d_2$ is. In order to overcome this difficulty, we make a digression.

We recall the definition of numerically connected divisors [R]: Let D be a positive divisor on a smooth surface X. A *decomposition* of D is a pair of strictly positive divisors A, B such that $D = A + B$. The divisor D is said to be *numerically connected* if $(A \cdot B) > 0$ for every such decomposition. Ramanujam proves the following facts concerning numerically connected divisors:

Proposition 7.9. (i) *Suppose that the divisor D moves in an algebraic family without fixed points on X, and that $(D^2) > 0$. Then D is numerically connected.*

(ii) *Let D be a numerically connected divisor. Then $H^0(D, \mathcal{O}_D) = k$.*

Ramanujam's method of proof of (ii) can be used to show the following:

Proposition 7.10. *Let M be an invertible sheaf on a numerically connected divisor D, with the following properties:*

(i) $\deg(M) \leq 0$, *and*

(ii) *for every decomposition* $D = A + B$, $\deg(M_B) < (A \cdot B)$, *where* M_B *denotes the restriction of* M *to* B.

Then either $M \approx \mathcal{O}_D$, *or else* $H^0(D, M) = 0$.

Proof. Let α be a non-zero section of M. Assume first that α does not vanish identically on any component C of D. Then $\deg(M_C) \geq 0$ for every C. Since the total degree of M is non-positive, it follows that $\deg(M_C) = 0$ for every C. Then since M_C has a non-zero section, it is the trivial sheaf and α does not vanish anywhere on C. This is true for each component, so α is a non-vanishing section on D, and defines an isomorphism $\mathcal{O}_D \xrightarrow{\sim} M$.

We want to rule out the possibility that α vanishes identically on some component of D. Assume that it does, and let A be the largest divisor $< D$ on which α vanishes identically. Let $B = D - A$, and consider the exact sequence

$$0 \to M_B(-A) \to M \to M_A \to 0.$$

By definition of A, α is a section of $M_B(-A)$. Since the degree of that sheaf is $\deg(M_B) - (A \cdot B) < 0$, it follows that α vanishes on a component of B, contradicting maximality. □

We now return to our elliptic triple $\mathcal{T} = (E, \sigma, L)$.

Proposition 7.11. *The divisor* E *is numerically connected, and* $H^0(E, \mathcal{O}_E) = k$.

This is an immediate consequence of Proposition 7.9, because every positive divisor on $\mathbf{P}^2$ or $\mathbf{P}^1 \times \mathbf{P}^1$ moves in the family of divisors of the same degree or bidegree, which is without fixed points, and because $(E^2) = r^s > 0$. □

Proposition 7.12. *An invertible sheaf* M *on* E *is tame if it satisfies any one of the following conditions:*

(i) *The hypotheses of* (7.10) *hold with* D *replaced by* E.
(ii) M^* *is tame.*
(iii) $d_i(M) \leq 0$ *for all* i *or* $d_i(M) \geq 0$ *for all* i.
(iv) M *is generated by its sections.*

Proof. Proposition 7.10 shows that M is tame in case (i). Case (ii) follows from the definition of tame and duality. If $d_i(M) \leq 0$ for all i, then $\deg L_B \leq 0 < (A \cdot B)$ for all decompositions $E = A + B$, so this is a subcase of (i). The case $d_i(M) \geq 0$ for all i follows by duality, and it includes the case (iv). □

Proposition 7.13. *An invertible sheaf M on E such that $\deg M \geq 2$ and that $d_i(M) \geq 0$ for all i is generated by its sections.*

Proof. We may suppose k algebraically closed. Let $d = \deg M \geq 2$. For each i, we will show that M is generated by its sections at every point of C_i.

Case 1: $d_i = d$. Then the irreducible component $C = C_i$ has multiplicity 1 in E, and the smooth points of C are smooth points of E. Let p be such a point. Then the sheaf $M(-(d-1)p)$ is of degree 1, and is tame by (7.12iii), so it has a non-zero section x. The divisor of zeros of x is of degree 1, hence it consists of a single smooth point q. So x generates M everywhere except at p and q. Since $M(-p)$ is tame of degree $d-1$, it has fewer sections than M, so the sections of M generate M at p. The same argument shows that they generate M at q.

Case 2: $d_i < d$. Let $C = C_i$ and $B = E - C$. Then $C \approx \mathbf{P}^1$, because it is reduced and irreducible of genus 0. Moreover, B is numerically connected. If $r = 3$, this is obvious. If $r = 2$, it follows from the fact that B does not have bidegree $(2,0)$ or $(0,2)$ because, being reduced and irreducible, C does not. Consider the exact sequence

$$0 \to M_B(-C) \to M \to M_C \to 0.$$

The sheaf M_C is of degree $d_i \geq 0$ on C, and is therefore generated by its sections. To prove that M is generated by its sections at points of C, it suffices to show that the map $H^0(M) \to H^0(M_C)$ is surjective. To do this we show that $h^1(M_B(-C)) = 0$ by duality on B. Since $\omega_E \approx \mathcal{O}_E$, we have $\omega_B \approx \mathcal{O}_B(-C)$, and therefore

$$\begin{aligned} h^1\left(M_B(-C)\right) = h^0\left(\omega_B \otimes (M_B(-C))^*\right) &= h^0\left(\mathcal{O}_B(-C) \otimes M_B^*(C)\right) \\ &= h^0\left(M_B^*\right) = 0 \end{aligned}$$

by (7.10), because M_B^* is an invertible sheaf on the numerically connected divisor B whose degree on each irreducible component is ≤ 0, and whose total degree is $d_i - d < 0$. □

Notation 7.14. Let M be a locally free sheaf which is generated by its sections. We denote by [M] the canonical exact sequence

$$[\mathrm{M}] \qquad 0 \to M'' \to M' \xrightarrow{\varphi} m \to 0,$$

where $M' = \mathcal{O}_E \otimes_k H^0(M)$, and M'' is the kernel of the canonical map $\varphi : M' \to M$, which is surjective because M is generated by its sections. Thus M' is a free $\mathcal{O}_E$-module of rank $h^0(M)$, M'' is locally free, and since $H^0(\mathcal{O}_E) = k$,

$$H^0(M'') = 0. \tag{7.15}$$

Moreover,

(7.16)
$$\det M'' = (\det M)^*, \quad \deg M'' = -\deg M, \quad \text{and} \quad \chi(M'') = -\chi(M).$$

Proposition 7.17. *Let M and N be locally free sheaves on E such that M is generated by its sections and that $h^1(N) = 0$. Let*

$$\mu_{M,N} : H^0(M) \otimes H^0(N) \to H^0(M \otimes N)$$

be the multiplication map. Then $\ker \mu_{M,N} \approx H^0(M'' \otimes N)$, $\operatorname{coker} \mu_{M,N} \approx H^1(M'' \otimes N)$ and $H^1(M \otimes N) = 0$.

This follows immediately from the exact sequence of cohomology of the exact sequence $[M] \otimes N$, which is

$$0 \to H^0(M'' \otimes N) \to H^0(M) \otimes H^0(N) \xrightarrow{\mu_{M,N}} H^0(M \otimes N)$$
$$\to H^1(M'' \otimes N) \to 0 \to H^1(M \otimes N) \to 0.$$

□

Corollary. *Let M and N be invertible sheaves on E such that M is generated by its sections and N is tame of degree ≥ 0. Then $M \otimes N$ is tame of degree ≥ 0.*

Proof. If N is isomorphic to $\mathcal{O}_E$, this follows from (7.12iv). If not, then $H^1(N) = 0$, and hence $H^1(M \otimes N) = 0$ by the proposition. □

Lemma 7.18. *The sheaf $L^* \otimes L_n^\sigma$ is tame for all $n \geq 1$.*

Proof. We have

$$L^* \otimes L_n^\sigma = (L^* \otimes L^\sigma) \otimes L^{\sigma^2} \otimes \ldots \otimes L^{\sigma^n}.$$

Since each L^{σ^i} is generated by sections and since $\deg(L^* \otimes L^\sigma) = 0$, the above corollary and induction reduces us to showing that $L^* \otimes L^\sigma$ is tame. We verify the condition of Proposition 7.10. Let $E = A + B$ be a decomposition of the divisor E. If $r = 3$, we may assume that $\deg A = \deg L_A = 1$, and $\deg B = \deg L_B = 2$. There are two possibilities: Either $\deg L_A^\sigma = 1$ and $\deg L_B^\sigma = 2$, or else $\deg L_A^\sigma = 2$ and $\deg L_B^\sigma = 1$. In each case, $\deg(L^* \otimes L^\sigma)_A$ and $\deg(L^* \otimes L^\sigma)_B$ are less than $(A \cdot B) = 2$, as required.

Assume that $r = 2$, and say that the bidegree of B is (i, j). Then the bidegree of A is $(2 - i, 2 - j)$, and $\deg L_B = i$, $\deg L_B^\sigma = j$. In each case, $\deg(L^* \otimes L^\sigma)_B = i - j < (A \cdot B) = 2(i + j - ij)$.

We can now prove parts (i-iii) of Theorem 6.6. Part (ii) is easy. We have

$$\dim B_n = h^0(L_n) = \deg L_n = rn,$$

for $n \geq 1$, because L_n is the product of n sheaves L^{σ^i}, each of which is of degree r and is generated by its sections. Hence L_n is generated by its sections, so it is a tame sheaf of degree rn.

Part (iii) of (6.6) follows from parts (i) and (ii), which imply that $\dim J_n = r^n - rn$ for all $n \geq 1$.

Part (i) is, by (7.1), a consequence of the fact that $H^0(L) \approx k$ (7.11), Proposition 7.17 (with $M = L$, $N = L_n^\sigma$) and

Lemma 7.19. *$H^1(L'' \otimes L_n^\sigma) = 0$ for $n \geq 1$.*

Proof. If $r = 2$, then L'' is of rank 1, so $L'' \approx \det L'' \approx L^*$, and $L'' \otimes L_n^\sigma \approx L^* \otimes L_n^\sigma$. This sheaf is tame by Lemma 7.18, and of degree $2(n-1) \geq 0$. Moreover, when the degree is zero, the sheaf in question is $L^* \otimes L^\sigma$, which is not isomorphic to $\mathcal{O}_E$. Indeed, L is not isomorphic to L^σ, because $\pi_L : E \to \mathbf{P}^1$ is of degree 2, while $(\pi_L, \pi_{L^\sigma}) : E \to \mathbf{P}^1 \times \mathbf{P}^1$ is a closed immersion.

If $r = 3$, then L'' is of rank 2. The pairing $L'' \otimes L'' \to \det L''$ shows in this case that

$$(L'')^* \approx L'' \otimes (\det L'')^* \approx L'' \otimes L. \tag{7.20}$$

We use Lemma 7.18 and put $M = L_n^\sigma$ in the following lemma to get our result. ☐

Lemma 7.21. *Assume that $r = 3$. Let M be an invertible sheaf on E of degree at least 3 such that $L^* \otimes M$ is tame. Then $H^1(L'' \otimes M) = 0$.*

Proof. Dualizing the sequence $[L]$ and tensoring it with $L^* \otimes M$ using (7.20) gives the exact sequence

$$0 \to (L^*)^{\otimes 2} \otimes M \to (L^* \otimes M)^3 \to L'' \otimes M \to 0.$$

If L and M are not isomorphic, then $H^1(L^* \otimes M) = 0$ because $\deg(L^* \otimes M) \geq 0$, and hence $H^1(L'' \otimes M) = 0$. To show that $H^1(L'' \otimes L) = 0$ too, we use (7.20), duality, and the fact that $H^0(L'') = 0$ (7.15). ☐

In order to complete the proof of Theorem 6.6 in the case $r = 3$, we need one more result about generation by sections.

Proposition 7.22. *Assume that $r = 3$. Let M be an invertible sheaf of degree 3 whose sections define an embedding of E as a divisor of degree 3 in $\mathbf{P}^2$. Then $L'' \otimes M$ is generated by its sections.*

Proof. Let $N = L'' \otimes M$. This sheaf is of rank 2, and its determinant

$$\det N = \wedge^2 N = \wedge^2(L'' \otimes M) \approx \wedge^2(L'') \otimes M^{\otimes 2} \approx L^* \otimes M^{\otimes 2}$$

is of degree 3. Moreover, $d_i(\wedge^2 N) \geq 0$ for each i. For, if E has more than one irreducible component, then the equation $3 = \Sigma d_i(L)$, together with the fact that $d_i(L) > 0$ for each i, imply that $1 \leq d_i(L) \leq 2$ for each i. The same holds for M, so that $2d_i(M) - d_i(L) \geq 0$ for each i, as claimed. Thus by (7.13), $\wedge^2 N$ is generated by its sections.

The same argument we used in the proof of Lemma 7.18 to prove that $L^* \otimes L^\sigma$ is tame when $r = 3$ shows that $L^* \otimes M$ is tame. Therefore $h^1(N) = 0$ by the preceding lemma, and consequently $h^0(N) = \deg N = 3$. Thus each of the three vector spaces

$$X = H^0(L), \quad Y = H^0(M), \quad \text{and} \quad Z = H^0(N)$$

has dimension 3.

The sheaf exact sequence $[L] \otimes M$ gives an injection $i : N \to X \otimes_k M$. Applying the functor $\wedge^2$, we have another sheaf map

$$\text{(7.23)} \qquad \wedge^2 N \xrightarrow{j} \wedge(X \otimes_k M) = (\wedge_k^2 X) \otimes_k M^{\otimes 2}.$$

Applying H^0 to these maps, we obtain an injection

$$\text{(7.24)} \qquad Z \xrightarrow{i_*} X \otimes Y,$$

which identifies Z with $\ker \mu_{L,M}$ (cf. (7.17)), and a commutative diagram

$$\begin{array}{ccc} \wedge_k^2 Z & \xrightarrow{\wedge^2 i_*} & \wedge_k^2 (X \otimes_k Y) \\ \downarrow \varphi & & \downarrow \psi \\ H^0(\wedge_k^2 N) & \xrightarrow{j_*} & (\wedge_k^2 X) \otimes \mathrm{Symm}_2(Y), \end{array}$$

in which the upper horizontal arrow is injective. The vertical arrows are the canonical maps $\wedge^2 H^0 \to H^0 \wedge^2$. At the bottom right we have used the isomorphism $\mathrm{Symm}_2(Y) = H^0(M^{\otimes 2})$, which follows from our hypothesis that the sections of M embed E as a plane cubic in $\mathbb{P}^2 = \mathbb{P}(Y)$, the projective space of planes in Y; the restriction map

$$\mathrm{Symm}_k(Y) = \bigoplus_{n \geq 0} H^0(\mathbb{P}^2, \mathcal{O}_{\mathbb{P}^2}(n)) \to \bigoplus_{n \geq 0} H^0(E, M^{\otimes n})$$

is surjective, and its kernel is generated by a form of degree 3. So it is an isomorphism in degrees ≤ 2.

We have seen in the first paragraph of this proof that $\wedge^2 N$ is generated by its sections. In order to conclude that N itself is generated by sections, it suffices to prove that the map φ is surjective. For if that is so, then for every point $p \in E$ there exist elements $z, z' \in Z$ such that $\varphi(z \wedge z')(p) = z(p) \wedge z'(p)$ is not zero, and such z and z' generate N at p.

To show φ is surjective, it suffices to show φ injective, because φ is a linear transformation from one three-dimensional vector space to another.

It is an elementary fact that every element of $\wedge^2 Z$ is a decomposable tensor, of the form $z{\wedge}z'$. (In general, all elements of the $(n-1)$-st exterior power of an n-dimensional vector space are decomposable [B1], Ch. III, Sect. 11, No. 13, Cor. of Prop. 15.) We have therefore only to show that for $z, z' \in Z$, $\varphi(z \wedge z') = 0$ implies $z \wedge z' = 0$.

Let $\{x_i\}$, $i = 1, 2, 3$, be a basis for X, and write

$$i_* z = \sum_i x_i \otimes y_i, \quad i_* z' = \sum_i x_i \otimes y'_i,$$

with $y_i, y'_i \in Y$. Then if $\varphi(z{\wedge}z') = 0$, we have

$$\begin{aligned} 0 = j_*\varphi(z \wedge z') &= \psi\,(\wedge^2 i_*)\,(z \wedge z') = \psi\,(i_* z \wedge i_* z') \\ &= \psi((\sum_i x_i \otimes y_i) \wedge (\sum_i x_i \otimes y'_i)) = \sum_{i,j} (x_i \wedge x_j) \otimes y_i y'_j \\ &= \sum_{1 \le i,j \le 3} (x_i \wedge x_j)\,(y_i y'_j - y_j y'_i)\,. \end{aligned}$$

Since the elements $x_i \wedge x_j$ for $i < j$ form a basis for $\wedge^2 X$, it follows that $y_i y'_j - y_j y'_i = 0$ in $\mathrm{Symm}_2 Y$ for all i, j, and from this we wish to conclude that $z \wedge z' = 0$.

We use unique factorization in the polynomial algebra Symm Y. If $y_i = 0$ for all i, then $i_* z = 0$, hence $z = 0$. So renumbering if necessary, we can suppose $y_1 \neq 0$. We have

$$y_1 y'_j = y_j y'_1 \tag{7.26}$$

for $j = 1, 2, 3$, and the y's are linear forms, with $y_1 \neq 0$. Hence either y_1 divides y'_1, in which case $y'_1 = c y_1$ for some $c \in k$, or else y_1 divides y_j for each j, in which case there are constants $c_j \in k$ such that $y_j = c_j y_1$. Substituting in (7.26) and canceling y_1, we conclude that for all j,

$$y'_j = \begin{cases} c y_j & \text{in the first case,} \\ c_j y'_1 & \text{in the second case} \end{cases}.$$

In the first case the conclusion is $i_* z' = c i_* z$, which means $z' = cz$ and $z{\wedge}z' = 0$, as desired. In the second case, putting $x = \Sigma c_i x_i$, we have

$$i_* z = x \otimes y_1 \quad \text{and} \quad i_* z' = x \otimes y_1'.$$

We now use the fact that $i_* Z$ is the kernel of the multiplication map $\mu_{L,M}$, which implies that

$$x y_1 = 0 = x y_1'$$

in $H^0(L \otimes M)$. Since $c_1 = 1 \neq 0$, the section x is not identically zero, and, since E has no embedded components, it follows that the support of x is not finite. Hence the linear forms y_1 and y_1' have an infinite set of common zeros on E. This means that the lines $y_1 = 0$ and $y_1' = 0$ in $\mathbf{P}(Y) = \mathbf{P}^2$ have an infinite number of common points, which implies that $y_1' = c' y_1$ for some $c' \in k$. Then $i_* z' = c' i_* z$, $z' = c' z$, and $z \wedge z' = 0$, as before. This completes the proof of Lemma 7.22. □

The lemma we have just proved is essentially equivalent to the following corollary, which may be of some interest in its own right.

Corollary. *Let E be scheme, and let L and M be invertible sheaves on E whose spaces of sections $X = H^0(E, L)$ and $Y = H^0(E, M)$ are of dimension 3 and give embeddings $\varphi_L : E \to \mathbf{P}(X)$ and $\varphi_M : E \to \mathbf{P}(Y)$ of E as a cubic divisor in $\mathbf{P}^2$. Let $Z \subset X \otimes Y$ be the kernel of the multiplication map $\mu_{L,M} : X \otimes Y \to H^0(L \otimes M)$, and let $\Gamma \subset \mathbf{P}(X) \times \mathbf{P}(Y)$ be the locus of zeros of Z. Then $\Gamma \cap (\mathbf{P}(X) \times \varphi_M(E))$ is the graph of the isomorphism $\theta : \varphi_L(E) \to \varphi_M(E)$ such that $\varphi_M = \theta \circ \varphi_L$.*

Proof. With notation as in the proof just completed, let $(z^{(\nu)})$, $1 \leq \nu \leq 3$, be a basis for Z, and let $i_* z^{(\nu)} = \sum_j x_j y_j^{(\nu)}$. The lemma we have just proved amounts to the fact that the sections $\varphi(z^{(\mu)} \wedge z^{(\nu)})$ have no common zeros on E. The map j (7.23) is an injection of sheaves and splits locally, because the same is true of $i : N \to X \otimes_k M$. It follows that the sections

$$j_* \varphi\left(z^{(\mu)} \wedge z^{(\nu)}\right) = \sum_{i<j} (x_i \wedge x_j) \otimes \left(y_i^{(\mu)} y_j^{(\nu)} - y_j^{(\mu)} y_i^{(\nu)}\right)$$

have no common zero, i.e., that the 3×3 matrix $(y_i^{(\mu)})$ has rank ≥ 2 at every point of E. Thus there is at most one solution $p \in \mathbf{P}(X)$ to the equations

$$\sum_j x_j(p) y_j^{(\nu)}(q) = 0, \qquad \nu = 1, 2, 3,$$

for each point $q \in \varphi_M(E)$. And of course there is one solution, namely $p = \theta^{-1}(q)$. □

We now return to the proof of Theorem 6.6, to establish the remaining properties (6.6iv)-(6.6vi) of the algebra $B = \mathcal{B}(T)$. For $\ell, m, n \geq 0$, let

$$K_{l,m} = \ker (B_l \otimes B_m \to B_{l+m}) \subset B_l \otimes B_m,$$

and

$$K_{l,m,n} = \ker (B_l \otimes B_m \otimes B_n \to B_{l+m+n}) \subset B_{l+m+n}.$$

Let θ_n denote the canonical map $T_n \to B_n$. The following elementary lemma has nothing to do with our special situation. It holds for any graded algebra B such that $B = k[B_1]$.

Lemma 7.27. *For each triple of positive integers ℓ, m, n, the map*

$$\theta_{l,m,n} = \theta_l \otimes \theta_m \otimes \theta_n : T_{l+m+n} \to B_l \otimes B_m \otimes B_n$$

induces an isomorphism

(∗)
$$J_{l+m+n} / (T_l \otimes J_{m+n} + J_{l+m} \otimes T_n) \xrightarrow{\sim} K_{l,m,n} / (B_l \otimes K_{m,n} + K_{l,m} \otimes B_n).$$

Proof. The map $\theta = \theta_{l,m,n}$ is surjective because $B = k[B_1]$. We have immediately from the definitions

$$\begin{aligned} \theta^{-1}(K_{l,m,n}) &= J_{l+m+n}, \\ \theta^{-1}(B_l \otimes K_{m,n}) &= T_l \otimes J_{m+n} + J_l \otimes T_{m+n}, \\ \theta^{-1}(K_{l,m} \otimes B_n) &= T_{l+m} \otimes J_n + J_{l+m} \otimes T_n. \end{aligned}$$

Since $J_l \otimes T_{m+n} \subset J_{l+m} \otimes T_n$ and $T_{l+m} \otimes J_n \subset T_l \otimes J_{m+n}$, it follows that the inverse images, under θ, of the top and bottom of the quotient on the right of (∗) are, respectively, the top and bottom groups on the left. □

For $m \geq 1$ and $n \geq 1$, let $M = L_m^\sigma$, and $N = L_n^{\sigma^{n+1}}$, so that $L \otimes M = L_{m+1}$, $M \otimes N = L_{m+n}^\sigma$, and $L \otimes M \otimes N = L_{1+m+n}$.

Lemma 7.28. *For every $m, n \geq 1$, there is a commutative diagram*

$$\begin{array}{ccc} K_{1,m} \otimes B_n & \xrightarrow{\alpha} & K_{1,m,n} / (B_1 \otimes K_{m,n}) \\ \downarrow & & \downarrow \\ H^0(L'' \otimes M) \otimes H^0(N) & \xrightarrow{\mu} & H^0(L'' \otimes M \otimes N) \end{array}$$

in which the vertical arrows are isomorphisms, α is the map induced by the inclusion $K_{1,m} \otimes B_n \subset K_{1,m,n}$, and $\mu = \mu_{L'' \otimes M, N}$ is the multiplication map.

Proof. Applying the functor H^0 to the diagram

$$\begin{array}{c}[L]\otimes M\otimes N'\\ \big\downarrow\\ [L]\otimes M\otimes N\end{array}$$

and making such identifications as $B_m = H^0(M)$ via σ so that $\mu_{L,M}$ becomes the multiplication map $B_1 \otimes B_m \to B_{1+m}$, etc., we obtain the exact commutative diagram

$$\begin{array}{ccccccc} 0 \longrightarrow H^0(L'' \otimes M)\otimes H^0(N) & \xrightarrow{\alpha} & B_1\otimes B_m\otimes B_n & \xrightarrow{\mu\otimes 1} & B_{1+m}\otimes B_n & \longrightarrow 0 \\ \big\downarrow{\scriptstyle \mu_{L''\otimes M,N}} & & \big\downarrow{\scriptstyle 1\otimes\mu} & & \big\downarrow{\scriptstyle \mu} & \\ H^0(L''\otimes M\otimes N) & \longrightarrow & B_1\otimes B_{m+n} & \xrightarrow{\mu} & B_{1+m+n} & \longrightarrow 0 \end{array}$$

in which the μ's without subscripts are multiplication in B. The top row shows that $H^0(L''\otimes M)\otimes H^0(N) = K_{1,m}\otimes B_n$ (cf. (7.17)). The diagram shows that the image of $H^0(L''\otimes M\otimes N)$ in $B_1\otimes B_{m+n}$ is the same as the image of the subspace $K_{1,m,n}$ of $B_1\otimes B_m\otimes B_n$, and this proves the lemma since the kernel of the vertical map of $B_1\otimes B_m\otimes B_n$ is $B_1\otimes K_{m,n}$. □

From now on we set $m = s-1$, and we suppose $n \geq 1$. Combining the last two lemmas, we obtain

Lemma 7.29. *For $n \geq 1$, $J_{s+n}/(T_1 J_{s+n-1} + J_s T_n)$ is isomorphic to the cokernel of the multiplication map*

$$\mu_{L''\otimes M,N} : H^0(L''\otimes M)\otimes H^0(N) \to H^0(L''\otimes M\otimes N),$$

where $M = L^{\sigma}_{s-1}$ and $N = L^{\sigma^s}_n$.

Lemma 7.30. *The sheaf $L''\otimes M$ is generated by its sections. It has rank $r-1$, its determinant is $L^*\otimes M^{\otimes(r-1)}$ and its degree is r. Moreover, $h^0(L''\otimes M) = r$ and $h^1(L''\otimes M) = 0$.*

Proof. The fact that $h^1 = 0$ is a consequence of (7.19). The rest is straightforward except the generation by sections. This follows from (7.22) if $r=3$, because then $L''\otimes M = L''\otimes L^{\sigma}$. If $r=2$, then $L''\otimes M \approx L^*\otimes L^{\sigma}\otimes L^{\sigma^2}$, and it follows from (7.13) in that case, provided that we show $d_i(L''\otimes M) \geq 0$ for all i. This is true. If the triple is not exceptional, (7.6) shows that $d_i(L''\otimes M) = 0$ for each i. If the triple is exceptional and the components are ordered so that the multidegree of L is $(d_1, d_2, d_3) = (1,0,1)$, it is easy to compute that the multidegree of $L^*\otimes L^{\sigma}\otimes L^{\sigma^2}$ is $(0,2,0)$. □

Since $L''\otimes M$ is generated by sections, we can analyze the multiplication map μ in (7.29) via (7.17) and the cohomology sequence of the sheaf exact sequence $[L''\otimes M]\otimes N$, instead of the sheaf exact sequence $(L''\otimes M)\otimes[N]$. The advantage of this is that $(L''\otimes M)''$ is of rank 1, whereas N'' is not, in general. Thus $(L''\otimes M)''$ is equal to its determinant, which is $L\otimes(M^*)^{\otimes r-1}$. By (7.17), $J_{s+n}/(T_1J_{s+n-1}+J_sT_n)$, which is the cokernel of the map μ in (7.29), is isomorphic to $H^1(L\otimes(M^*)^{\otimes r-1}\otimes N)$, i.e., to $H^1(F\otimes L_{n-1}^{\sigma^{s+1}})$, where

$$F=\begin{cases} L\otimes((L^*)^{\sigma})^{\otimes^2}\otimes L^{\sigma^2} & \text{if } r=3,\\ L\otimes(L^*)^{\sigma}\otimes(L^*)^{\sigma^2}\otimes L^{\sigma^3} & \text{if } r=2.\end{cases} \tag{7.31}$$

Suppose that E is not exceptional. Then, by (7.6), we have $d_i(F)=0$ for all i. Consequently, the sheaf $F\otimes L_{n-1}^{\sigma^{s+1}}$ is tame by (7.12iii)) for all $n\geq 1$; and since the total degree of that sheaf is $(n-1)r$, which is >0 for $n\geq 2$, statement (6.6iv) follows.

Remark 7.32. It is enough to prove one half of (6.6iv), as we have just done. The other half follows on replacing σ by σ^{-1} in our triple, because the algebra $\mathcal{B}(E,\sigma^{-1},L)$ is the opposite of $\mathcal{B}(E,\sigma,L)$. This is immediate from the definitions: Call the algebras B' and B. Then the isomorphisms

$$\begin{aligned}\theta_n: B'_n = H^0\left(L\otimes L^{\sigma^{-1}}\otimes\ldots\otimes L^{\sigma^{-n+1}}\right)\\ \xrightarrow{\sigma^{n-1}} H^0\left(L^{\sigma^{n-1}}\otimes\ldots\otimes L^{\sigma}\otimes L\right)\\ \xrightarrow{\tau} H^0\left(L\otimes L^{\sigma^1}\otimes\ldots\otimes L^{\sigma^{n-1}}\right)=B_n,\end{aligned}$$

where τ reverses the order of the tensors, satisfy the relation

$$\theta_{m+n}(fg)=\theta_m(g)\theta_n(f).$$

Taking $n=1$, we get (6.6v) in the non-exceptional case, because $F\approx\mathcal{O}_E$ if and only if $\mathcal{T}$ is regular. (Recall that, by (7.3), the assertion for W in (6.6v) follows from the assertion for U, once (6.6i)-(6.6iv) are proven.)

If E is exceptional, then F is not tame, but we can check (6.6iv) and (6.6v) by explicit computation.

Assume that E is exceptional and $r=3$. Then $E=C_1+C_2$, each C_i is isomorphic to $\mathbf{P}^1$, and $(C_1\cdot C_2)=2$. Let G be an invertible sheaf on E. If $d_1(G)\leq -1$, then $h^0(G)=\text{Max }(0,d_2(G)-1)$, because sections of G are sections of $G_{C_2}(-C_1)$, a sheaf of degree $d_2(G)-2$ on a curve of genus zero. If we order the components so that L has multidegree $(d_1,d_2)=(1,2)$, then F has multidegree $(-2,2)$, so $h^0(F)=h^1(F)=1$. This proves (6.6v) in that case. For $n \geq 2$, the sheaf $F \otimes L_{n-1}^{\sigma^3}$ has degrees $d_1, d_2 \geq 0$, so it is tame and since its total degree is positive, $h^1 = 0$. This proves (6.6iv).

Assume that E is exceptional and that $r = 2$. Then $E = C_1 + C_2 + C_3$ with components $C_i \approx \mathbf{P}^1$, and $(C_i \cdot C_j) = 1$ for all i, j. Let G be an invertible sheaf on E. If $d_1(G) \leq -1$ and $d_2(G) \leq 0$, then $h^0(G) = \mathrm{Max}\,(0, d_3(G) - 1)$. (Restricting to C_1, then to C_2, one sees that sections of G are the same as sections of $G_{C_3}(-C_1 - C_2)$.) Ordering the C_i so that L has multidegree $(1, 1, 0)$, we check that F has multidegree $(1, 1, -2)$, so the discussion above applies to F^*, giving $h^0(F^*) = h^1(F) = 1$ as before. The sheaf $F \otimes L^{\sigma^4}$ has multidegree $(2, -1, 1)$ (or $(1, -1, 2)$) and again, after changing C_2 and C_3, the discussion above applies to F^* and gives $0 = h^0(F^*) = h^1(F)$. For $n \geq 3$, the sheaf $F \otimes L_{n-1}^{\sigma^4}$ has $d_i \geq 0$ for each i, so it is tame, of total degree > 0, and $h^1 = 0$.

There remains (6.6vi). We assume $\mathcal{T}$ is regular and must show that W is a nondegenerate subspace of $J_s T_1$ and of $T_1 J_s$. By Remark 7.32, it suffices to treat only the inclusion $W \subset J_s T_1$. In the notation explained just before (7.27), we have $J_s = K_{1,s-1}$. Indeed if $r = 3$, $s = 2$, the two spaces coincide by definition. If $r = 2$, $s = 3$, then

$$J_3 = \ker(B_1 \otimes B_1 \otimes B_1 \to B_3) = \ker(B_1 \otimes B_2 \to B_3) = K_{1,2}$$

because $B_1 \otimes B_1 \to B_2$ is an isomorphism in this case. Similarly, $J_s = K_{s-1,1}$. Hence, taking $m = s - 1$ and $n = 1$ in Lemma 7.28, we find that W, which by definition is the kernel of the map α in that lemma, fits into an exact sequence

$$0 \to W \xrightarrow{i} J_s \otimes T_1 \xrightarrow{\mu} H^0(L'' \otimes M \otimes N). \tag{7.33}$$

This is the sequence obtained by applying H^0 to the sheaf exact sequence $[L'' \otimes M] \otimes N$, which is of the form

$$0 \to F \xrightarrow{i} J_s \otimes N \xrightarrow{j} L'' \otimes M \otimes N \to 0, \tag{7.34}$$

and $F \approx \mathcal{O}_E$ because our triple is assumed to be regular.

Let w be a basis for W and let $\{x_i\}$, $1 \leq i \leq r$, be a basis for T_1. Letting

$$i(w) = \sum_{j=1}^{r} g_j x_j,$$

we wish to show that the g_j span J_s. To do this, we let $\wedge : J_s \to k$ be a linear form vanishing on the g_j, and we proceed to show that $\wedge = 0$. Consider the homomorphism of $\mathcal{O}_E$-modules

$$\lambda \otimes 1 : J_s \otimes_k N \to k \otimes_k N = N.$$

By construction, the composed map $(\lambda \otimes 1) \circ i$ kills $W = H^0(F)$, and hence it kills F, because $F \approx \mathcal{O}_E$. Thus, by the exactness of (7.33), the sheaf map $\lambda \otimes 1$ is of the form $\gamma \circ j$ for some $\gamma : L'' \otimes M \otimes N \to N$. But $\mathrm{Hom}(L'' \otimes M \otimes N, N) = \mathrm{Hom}(L'' \otimes M, \mathcal{O}_E) = \mathrm{Hom}((L'' \otimes M)^*) = 0$ by duality and (7.30). Therefore $\lambda \otimes 1 = 0$ and this implies $\lambda = 0$, as was to be shown. This completes the proof of Theorem 6.6. $\square$.

8. Proof that regular algebras of dimension 3 are noetherian

The object of this section is to prove the following theorem:

Theorem 8.1. *Every nondegenerate semi-standard algebra, and in particular, every regular algebra of dimension 3, is left and right noetherian.*

Suppose that A is such an algebra. By (6.7i) we have $A \approx \mathcal{A}(\mathcal{T})$, where $\mathcal{T}$ is the triple derived from A, and hence, by (6.8i), A contains a normalizing element g, possibly zero, of degree $s+1$, such that $A/gA \approx B = \mathcal{B}(\mathcal{T})$. Thus the following lemma reduces Theorem 8.1 to Theorem 8.3, which is stated below.

Lemma 8.2. *Let A be a graded k-algebra, and let $g \in A$ be a homogeneous normalizing element of positive degree. If $B = A/gA$ is left or right noetherian, so is A.*

Proof. Left multiplication by g on a left A-module M:

$$\ell_g : M \to M$$

need not be A-linear, but nevertheless the kernel and image of ℓ_g are submodules of M, and ℓ_g induces a bijection between the A-submodules of $M/\ker \ell_g$ and the A-submodules of $\ell_g M$.

Assume that A is not left noetherian. Then ([B], Ch. III, Sect. 2, no. 10, Lemme 1) A has a *graded* left ideal which is not finitely generated. Using Zorn's Lemma, we may select a graded left ideal L which is not finitely generated and is maximal with that property. Then $\bar{A} = A/L$ is a noetherian A-module. We consider left multiplication by g on the short exact sequence

$$0 \to L \to A \to \bar{A} \to 0.$$

The snake lemma provides us with an exact sequence

$$K \xrightarrow{\delta} L/gL \xrightarrow{\epsilon} A/gA = B,$$

where $K = \ker(\ell_g : \bar{A} \to \bar{A})$. The map ϵ is A-linear, and so im ϵ is a finitely generated submodule of the noetherian A-module B. Also, K is a finitely generated module, because it is a submodule of the noetherian module $\bar{A}$.

The map δ is defined as follows: Let $\bar{y} \in K$ be represented by $y \in A$. Then $\delta(y)$ is represented by the element gy of L. If $x_1, \ldots, x_n \in A$ represent generators $\bar{x}_i$ of K, and if $\bar{y} = \Sigma a_i \bar{x}_i$, then $\delta(\bar{y})$ is represented by $g(\Sigma a_i x_i)$, and this element can be rewritten in the form $\Sigma a_i' g x_i$. This shows that the elements $gx_i = \delta(\bar{x}_i)$ generate $\delta(K) = \ker \epsilon$, and hence that L/gL is finitely generated. Since g is of degree > 0, it follows that L is finitely generated. (If M is a finitely generated graded submodule of L such that $L = M + gL$, then, by induction on n, $L = M + g^n L$ for all $n > 0$. This implies $L_n = M_n$ for all n, hence $L = M$.) □

Theorem 8.3. *Let $\mathcal{T}$ be a triple, and $B = \mathcal{B}(\mathcal{T})$. Then B is (left and right) noetherian.*

Let us call a ring A *essentially finite* if it is a finite module over its center $Z(A)$.

Lemma 8.4. *Let A be a finitely generated graded algebra over k. Suppose A is essentially finite. Then its center $Z(A)$ is also a finitely generated graded k- algebra, and A is noetherian.*

For the proof, see [A-T]. The relevant theorem is stated there for commutative rings, but only the fact that the subring is central in the big one is used in the proof. □

Lemma 8.5. *Let $\mathcal{T} = (E, \sigma, L)$ be a triple and let $n > 0$ be an integer such that $\sigma^n = 1$. Let $B = \mathcal{B}(\mathcal{T})$, and denote by $B\langle n \rangle$ the graded ring $\sum_\nu B_{\nu n}$. Then*

(i) *B is essentially finite.*

(ii) *There is an automorphism τ of $B\langle n \rangle$, such that $(x_1 \ldots x_n)^\tau = x_n x_1 \ldots x_{n-1}$ for all $x_i \in B_1$.*

(iii) *For $b \in B_i$ and $f \in B\langle n \rangle$, we have $bf = f^{\tau^i} b$. In particular, $B\langle n \rangle$ is commutative and the subring of invariants*

$$C = \{f \in B\langle n \rangle | f^\tau = f\}$$

is in the center of B.

Proof. Let i and j be integers such that $i + j \equiv 0$ (modulo n). Then

$$fgh = hgf, \tag{8.6}$$

for all $f, h \in B_i$ and $g \in B_j$, because

$$fgh = f \otimes g^{\sigma^i} \otimes h^{\sigma^{i+j}} = f \otimes g^{\sigma^i} \otimes h = h \otimes g^{\sigma^i} \otimes f = h \otimes g^{\sigma^i} \otimes f^{\sigma^{i+j}} = hgf.$$

For $v \in B_{\nu n}$, write $v = \Sigma_\alpha g_\alpha y_\alpha$ with $g_\alpha \in B_{\nu n-1}$, $y_\alpha \in B_1$, and put $v^\tau = \Sigma_\alpha y_\alpha g_\alpha$. Then, by (8.6),

$$xv = v^\tau x, \tag{8.7}$$

for all $x \in B_1$. Since the socle of B is zero, (8.7) shows that the element v^τ depends only on v, and is independent of the choice of the expression $\Sigma g_\alpha y_\alpha$. The map $v \rightsquigarrow v^\tau$ is clearly k-linear, and it is also multiplicative, since $xv_1v_2 = v_2^\tau x v_2 = v_1^\tau v_2^\tau x$, and $(v_1v_2)^\tau$ is the *unique* element such that $xv_1v_2 = (v_1v_2)^\tau x$ for all $x \in B_1$. This proves (ii), and (iii) follows immediately, by induction on i, from (8.7).

To prove (i), we note that since $B = k[B_1]$, we have $B\langle n\rangle = k[B_n]$, and so the commutative ring $B\langle n\rangle$ is a finitely generated k-algebra. Since τ is of finite order, C is finitely generated and $B\langle n\rangle$ is a finite C-module ([B], Ch. V, Sect.1, No. 9, Th. 2). Also, B is a finite left $B\langle n\rangle$-module generated by $B_0 + B_1 + \ldots + B_{n-1}$. Hence B is a finite C-module. Since C is noetherian, so is B. In fact, any ring D between C and B is finitely generated over k, and B is a finite left D-module. This applies in particular to the center $D = Z(B)$. □

To prove Theorem 8.3, we will use the fact that a triple consists of algebro-geometric data, and consequently that there is a reduction of $\mathcal{T}$ (modulo p) to a finite field. To be precise, we define a family $\mathcal{T}_R = (E_R, \sigma_R, L_R)$ of triples, parametrized by a commutative ring R, to be a flat and proper family of schemes E_R over R, an automorphism σ_R of this R-scheme, and an invertible sheaf L_R on E_R, such that the conditions of (4.5) hold for each geometric fibre of E_R over R.

Lemma 8.8. *Let $\mathcal{T}_R$ be a family of triples. Let $(L_R)_n$ be defined as in (6.2). Then $H^0(E_R, (L_R)_n)$ is a locally free R-module for every $n \geq 0$.*

Proof. This follows from ([EGA] III, 7.8, or [H], p. 288, Cor. 12.9), because $H^1(E_k, (L_k)_n) = 0$ for every geometric point $R \to k$ of Spec$R$ (7.12). □

Because of this lemma, the construction of the algebra $\mathcal{B}(\mathcal{T})$ from a triple $\mathcal{T}$ carries over to a family $\mathcal{T}_R$, and provides us with a flat graded R-algebra B_R with the property that for every ring homomorphism $R \to S$, the algebra $S \otimes_R B_R$ is the one associated to the triple $\mathcal{T}_S = S \otimes_R \mathcal{T}_R$. Now if $\mathcal{T}_K$ is a triple defined over a field K, then there is a subring R of K which is finitely generated over $\mathbb{Z}$, and a family of triples $\mathcal{T}_R$ parametrized by Spec R, such that $K \otimes_R \mathcal{T}_R \approx \mathcal{T}_K$. Moreover, being a non-zero ring of finite type over $\mathbb{Z}$, R contains a maximal ideal M, and the residue field $R/M = k'$ is finite. We may ([B], Ch. 6, Sect. 1, No. 2, Th. 2) choose a valuation ring R of K which dominates R/M and with algebraic residue field extension. Then Lemma 8.9 below reduces us to the case that K is algebraic over the prime field $\mathbf{F}_p$ for some prime integer p. In that case, we will show that B is noetherian by proving (Lemma 8.10) that B satisfies the hypothesis of Lemma 8.5 and is therefore a finite module over its center.

Lemma 8.9. *Let R be a valuation ring with fraction field K and residue field k, and let A_R be a graded R-algebra with the property that $(A_R)_n$ is a free R-module of finite rank for each n. Let $A_K = A_R \otimes_R K$, and $A_k = A_R \otimes_R k$. If A_k is noetherian, so is A_K, and if R is a discrete valuation ring, then A_R is noetherian too.*

Proof. As in the proof of Lemma 8.2, it suffices to prove *graded* ideals are finitely generated. Let L be a graded left ideal of A_K. Then for each n, $L_n \cap (A_R)_n$ is a direct summand of $(A_R)_n$, because the quotient

$$(A_R)_n \,/\, (L_n \cap (A_R)_n) \approx (L_n + (A_R)_n) \,/ L_n$$

is torsion-free as R-module, being a submodule of the K-vector space $(A_K)_n/L_n$, and is finitely generated as R-module because it is a homomorphic image of $(A_R)_n$. Such an R-module is free [B$_2$, Ch. 6, Sect. 3, No. 6, Lemme 1]. Thus

$$L_R := L \cap A_R$$

is a direct summand of A_R, as R-module. This implies that $L_k := L_R \otimes_R k$ is a direct summand of A_k. So the map $L_k \to A_k$ is injective, and maps L_k isomorphically to a left ideal of A_k. Therefore by our hypothesis on A_k, L_k is a finitely generated module over A_k. Let S be a finite set of homogeneous elements of L_R whose residues generate L_k. The ordinary Nakayama Lemma, applied to each degree separately, shows that L_R is generated over A_R by S, and therefore that L is generated over A_K by S as well. This shows that A_K is noetherian.

Assume that R is a discrete valuation ring with maximal ideal p, and let L be a left ideal of A_R. Let $L' = A_R \cap L_K$, where $L_K = K \otimes_R L$. It was shown above that L' is finitely generated, from which it follows that $p^r L' \subset L$ for suitable r. The successive quotients of the filtration $p^\nu L' \cap L$ are submodules of the finitely generated A_k-modules $p^\nu L'/p^{\nu+1}L'$, and hence are finitely generated. Also, $p^r L'$ is finitely generated. It follows that L is finitely generated. Thus A_R is noetherian. □

Now the proof of Theorem 8.3 is completed by

Lemma 8.10. *An automorphism σ of a one-dimensional projective scheme E defined over a finite field K_0 has finite order.*

Proof. Let $L = \mathcal{O}_E(1)$. Replacing σ by a power, we may assume that it stabilizes each irreducible component of E. Then the divisor class

$[L^\sigma] - [L]$ is in the connected component of Pic E, which has finitely many K_0-rational points. So replacing σ by a power again, we may assume that $[L^\sigma] = [L]$. When this is so, σ extends to an automorphism of the ambient projective space, and so it is an element of projective general linear group over K_0, which is finite. □

REFERENCES

[An] D. Anick, *On the homology of associative algebras*, Trans. Amer. Math. Soc. **296** (1986), 641-659.

[A-K] A. Altman and S. Kleiman, *Introduction to Grothendieck Duality Theory*, Lecture Notes in Math. **146** (1970).

[A-T] E. Artin and J. Tate, *A note on finite ring extensions*, J. Math. Soc. Japan **3** (1951), 74-77.

[A-S] M. Artin and W. Schelter, *Graded algebras of global dimension* 3, Advances in Math **66** (1987), 171-216.

[A-T-V] M. Artin, M., J. Tate, and M. Van Den Bergh, *Modules over regular algebras of dimension* 3, (in preparation).

[Au] M. Auslander, *On the dimension of modules and algebras* III, Nagoya Math. J. **9** (1955), 67-77.

[B_1] N. Bourbaki, *Algèbre, Eléments de Mathématiques*, Hermann, Paris 1960-65.

[B_2] N. Bourbaki, *Algèbre Commutative, Eléments de Mathématiques*, Hermann, Paris 1960-65.

[C] H. Cartan, *Homologie et cohomologie d'une algèbre graduée*, Séminaire Cartan, 11e année, 57-58, exposé 15.

[C-E] H. Cartan and S. Eilenberg, *Homological Algebra*, Princeton University Press, Princeton 1956.

[EGA] A. Grothendieck and J. Dieudonné, *Eléments de géométrie algébrique*, Pub. Math Inst. Hautes Études Sci. 1960-67.

[H] R. Hartshorne, *Algebraic Geometry*, Springer-Verlag, New York 1977.

[O-F] **Одесский, А. Б.** and **Б. Л. Фейгнн, Алгебры Склянин, ассоциированные с зллиптической кривой**, (manuscript).

[N-O] C. Nastacescu and R. Van Oystaeyen, *Graded Ring Theory*, North-Holland, Amsterdam 1982.

[R] C. P. Ramanujam, *Remarks on the Kodaira vanishing theorem*, J. Indian Math. Soc. **36** (1972), 41-51.

[S] **Склянин, Е. К., О некоторых алгебраических структурах, связанных с уравнением янга-Бакстера.**, II **Представления квантовой алгебры, – Функцион анализ 17** (1983), 34-38.

[W] Wall, C. T. C., *Generators and relation for the Steenrod algebras*, Annals of Math **72** (1960), 429-444.

[V] Van Den Bergh, M., *Regular algebras of dimension* 3, Séminaire Dubreil-Malliavin 1986, Lecture Notes in Math. **1296**, Springer-Verlag, Berlin 1987, 228-234.

Michael Artin
Dept. of Mathematics
MIT
Cambridge, MA 02139

John Tate
Dept. of Mathematics
Harvard University
Cambridge, MA 02138

M. Van den Bergh
Dept. Wisk. en Inform.
UIA
Wilrijk, Belgium

Cohomology of a Moduli Space of Vector Bundles

V. BALAJI* and C. S. SESHADRI

Dedicated to A. Grothendieck on his 60th birthday

Introduction

Let X be a smooth projective curve of genus g, defined over the field of complex numbers. Let M_0 be the moduli space of semi-stable vector bundles V of rank two and trivial determinant (cf. [8]).

In [7] a canonical desingularisation $p : N \to M_0$, has been defined. Recall that p is an isomorphism over M_0^s, where M_0^s is the open subscheme of M_0 consisting of stable vector bundles and further $\dim_{\mathbb{C}} N = 3g - 3$. In this paper, we determine the ranks of the cohomology groups $H^k(N)$ for k up to one-third of the real dimension of N. Hence, by duality, the ranks of $H^k(N)$, for $k \geq (2/3) \dim_{\mathbb{R}} N$, are also determined (see Section 4, Remark 34 for precise details).

We have a canonical family of quadratic forms $\{Q_x\}_{x \in N}$ on a 3 dimensional vector space parametrised by N. We define closed subschemes $\{N_i\}$, $i = 1, 2, 3$ by the condition

$$N_i = \{x \in N \mid \text{ rank of } \quad Q_x \leq 3 - i\}$$

(cf. [2]). One sees that $N_1 - N_2$, $N_2 - N_3$ are all smooth subschemes of N. Besides $N_1 = p^{-1}(M_0 - M_0^s)$, and N_2 is the inverse image by p of the nodal points of $M_0 - M_0^s$, which is the Kummer variety associated to the Jacobian of X. In [2], $N_1 - N_2$ has been explicitly determined, as well as its normal bundle in N. Using this information some low cohomology

*NBHM Research Scholar

groups of N have been determined in [2] e.g. $H^3(N)$. By using the same method, one could compute more cohomology groups of N if one could determine explicitly the subschemes $N_2 - N_3$ and N_3 as well as their normal bundles in N, the basic principle being that this information leads to the splitting of the Thom-Gysin sequence into short exact sequences (see also Atiyah-Bott [1] pp.537). This indeed is the main theme of this paper. Then we are able to compute the ranks of the cohomology groups $H^k(N)$, $k \leq 1/3 \dim_{\mathbb{R}} N$. In this range, for $g \geq 4$, we have

$$H^k(N) = \{ H^{k-12}(N_3) \oplus H^{k-6}(N_2 - N_3) \oplus \\ H^{k-2}(N_1 - N_2) \oplus H^k(N - N_1) \} .$$

Note that $N - N_1 = M_0^s$.

Since the subschemes N_3, $N_2 - N_3$ and $N_1 - N_2$ are explicitly determined, all their cohomology groups could be obtained; however, the above decomposition holds only in the above range. Moreover, we determine the Betti numbers of M_0^s only in the above range (see Section 3 for precise details). This last computation has been done earlier by Kirwan [4]. We give an independent proof for this using the "Hecke Correspondence" between the moduli spaces of degree one and degree zero (cf. [6], [2]), and the computation of the Betti numbers of the moduli spaces of vector bundles of rank two and degree one (cf. [3], [1]).

Preliminaries

In this section, we shall recall briefly some of the facts in [2], which will be used extensively in the present work. Proofs of these statements can be found in [2] and [7]. *We state at the outset that for us the ground field is the field $\mathbb{C}$ of complex numbers.*

Let X be a smooth projective curve of genus g over $\mathbb{C}$. Let M_0 be the moduli space of semi-stable vector bundles of rank 2 and degree 0 with trivial determinant. Let M_0^s be the open subset of stable points in M_0 and N the canonical desingularisation model for M_0, constructed in [7]. (Note that this space was denoted by N_0 in [2]). If PV_4 is the space of parabolic semi-stable bundles V of rank 4 and degree 0 (cf. [7]), then N is in fact, the set of isomorphism classes of bundles V in PV_4 such that $\det V$ is trivial and End V is a specialisation of $M(2)$ - the (2×2)-matrix algebra.

In [2], a generalized conic bundle is defined on the smooth projective variety N, which in turn gives rise to a stratification of N as

$$N \supset N_1 \supset N_2 \supset N_3$$

where, if C_q^+ is the set of even-degree elements in the Clifford algebra C_q corresponding to a quadratic form q on a 3-dimensional vector space, then

$$\begin{aligned} N - N_1 &= \{V \in N \mid \text{End}\,(V) \simeq C_q^+, \quad \text{rank} \quad (q) = 3\} \\ N_1 - N_2 &= \{V \in N \mid \text{End}\,(V) \simeq C_q^+, \quad \text{rank} \quad (q) = 2\} \\ N_2 - N_3 &= \{V \in N \mid \text{End}\,(V) \simeq C_q^+, \quad \text{rank} \quad (q) = 1\} \quad \text{and} \\ N_3 &= \{V \in N \mid \text{End}\,(V) \simeq C_q^+, \quad \text{rank} \quad (q) = 0\}. \end{aligned}$$

Then we can summarise the main theorem of [2] as follows: As in [2], we denote the space $N - N_2$ by Z and $N_1 - N_2$ by Y, and therefore $Z - Y = N - N_1$ is isomorphic to M_0^s. Then we have

Theorem (cf. [2]). (i) *The smooth divisor Y of Z is a $\mathbf{P}^{g-2} \times \mathbf{P}^{g-2}$ bundle over $K - K_0$, K being the Kummer variety of* dim g *associated to the Jacobian of X, and K_0 its nodal points.*

(ii) *Let n_Y be the normal bundle of Y in Z and $e_Y \in H^2(Y, \mathbf{R})$ be its Euler class. Let h be the Gysin homomorphism for the pair $(Z, Z - Y)$ namely,*

$$\begin{aligned} H^{k-2}(Y, \mathbf{R}) &\xrightarrow{h} H^k(Z, \mathbf{R}) \\ a &\longmapsto a \wedge e_Y. \end{aligned}$$

Then h is injective for $k \leq 2g-4$. Hence, the Thom-Gysin sequence splits to give

$$H^k(Z, \mathbf{R}) \simeq H^{k-2}(Y, \mathbf{R}) \oplus H^k(Z - Y, \mathbf{R}) \quad k \leq 2g - 4.$$

1. Determination of the spaces $N_2 - N_3$ and N_3

1. *Reduction of structure group.* Let us denote by R' the space $N_2 - N_3$. If the quadratic form q of the previous section is given in terms of the variables x_1, x_2 and x_3, then we can, for simplicity, assume that points in R' have an endomorphism algebra to be C_q^+, $q = x_1^2$, or more precisely, if $a \in R'$ and V_a is the corresponding vector bundle, then

$$\text{End}\,(V_a) \simeq C_q^+, \quad q = X_1^2.$$

Now it is not hard to see that C_q^+, $q = x_1^2$, can be identified with the exterior algebra on a 2-dimensional vector space, which we call by A. Therefore, C_q^+, $q = x_1^2$, is given by the following defining relations: If 1, y_1, y_2, y_3 be the elements in a basis of A, then we have the relations

$$\begin{aligned} y_1^2 = y_2^2 = y_3^2 = y_1y_3 = y_2y_3 = 0 \\ y_1y_2 = -y_2y_1 = y_3. \end{aligned} \tag{i}$$

Hence, the left regular representation of A is given as follows:

$$\begin{aligned} \text{Let } a &= \lambda_0 + \lambda_1 y_1 + \lambda_2 y_2 + \lambda_3 y_3 \in A, \\ a \cdot 1 &= \lambda_0 \cdot 1 + \lambda_1 \cdot y_1 + \lambda_2 \cdot y_2 + \lambda_3 \cdot y_3 \\ a \cdot y_1 &= 0 \cdot 1 + \lambda_0 \cdot y_1 + 0 \cdot y_2 + (-\lambda_2) y_3 \cdot \\ a \cdot y_2 &= 0 \cdot 1 + 0 \cdot y_1 + \lambda_0 y_2 + \lambda_1 y_3 \cdot \\ a \cdot y_3 &= 0 \cdot 1 + 0 \cdot y_1 + 0 \cdot y_2 \cdot + \lambda_0 y_3. \end{aligned}$$

From this, we see that the transpose of the left regular representation of A is given by

$$(\lambda_0 + \lambda_1 y_2 + \lambda_2 \cdot y_2 + \lambda_3 y_3) \mapsto \begin{bmatrix} \lambda_0 & \lambda_1 & \lambda_2 & \lambda_3 \\ 0 & \lambda_0 & 0 & -\lambda_2 \\ 0 & 0 & \lambda_0 & \lambda_1 \\ 0 & 0 & 0 & \lambda_0 \end{bmatrix}.$$

Since these matrices form a sub-algebra of $M(4)$, the (4×4) matrix algebra, it follows that the transpose of the left regular representation of A is isomorphic to this subalgebra of $M(4)$.

Let A_ℓ denote the left A-module whose underlying vector space is A and the A-action being left multiplication. Let $A_\ell^\vee$ be the dual of A_ℓ. Then $A_\ell^\vee$ acquires a canonical algebra structure by compositions of transposes of left multiplications. The algebra relations for $A_\ell^\vee$ are also defined by (i) so that we have

$$A_\ell^\vee \simeq A \simeq A^0 \simeq A_r^\vee$$

where A^0 is the opposite algebra of A and $A_r^\vee$ the dual of the right regular representation of A.

If $V \in R'$, i.e., V is a vector bundle on X defined by an element of R' then the associated graded gr V is $\simeq \overset{4}{\oplus} L$, where L is a line bundle with $L^2 \simeq \mathcal{O}_x$ (of [7], [2], [8]). So, let $R \subset R'$ be defined as

$$R = \left\{ V \in R' \mid \text{gr } V = \overset{4}{\oplus} \mathcal{O}_x \right\}$$

so that R' is the disjoint union of 2^{2g} copies of R (2^{2g} being the cardinality of $K_0 = \{L \in K \mid L^2 \simeq \mathcal{O}_x\}$).

Proposition 1. *Let H be the subgroup of* GL(4) *given by*

$$H = \begin{bmatrix} 1 & \lambda_1 & \lambda_2 & \lambda_3 \\ 0 & 1 & 0 & -\lambda_2 \\ 0 & 0 & 1 & \lambda_1 \\ 0 & 0 & 0 & 1 \end{bmatrix}.$$

Then the structure group for the principal bundle of $V \in R$ can be reduced to H. Further, for $V \in R$ with transition functions in H, End V can be canonically identified with the algebra A_1 of the form

$$A_1 = \left\{ \begin{bmatrix} a & b & c & d \\ 0 & a & 0 & c \\ 0 & 0 & a & -b \\ 0 & 0 & 0 & a \end{bmatrix} \right\}$$

which is the transpose of the right regular representation of A (w.r.t. the basis $1, y_1, y_2, y_3$ of A).

Proof. Let $C(A)^*$ be the group of units of the commutant of A in End V_x, $x \in X$. Since by Proposition 3 of [7] the A modules V_x and $A_r^\vee$ are isomorphic, $C(A)$ is simply the commutant of A in End $(A_r^\vee)$. Now, the commutant of A_r is A_ℓ, therefore, we see that $C(A) \simeq A_\ell^\vee$ and hence $C(A)$ is given by the algebra of matrices in (i) above.

Again by Proposition 3 of [7], for $V \in R'$, the structure group for the principal bundle of V can be reduced to $H' = C(A)^*$. Now, let H be the unipotent subgroup of $C(A)^*$. Then by the relations (i) which define $C(A)$, we see that H is given by the subgroup of GL(4)

$$H = \begin{bmatrix} 1 & \lambda_1 & \lambda_2 & \lambda_3 \\ 0 & 1 & 0 & -\lambda_2 \\ 0 & 0 & 1 & \lambda_1 \\ 0 & 0 & 0 & 1 \end{bmatrix}.$$

As we have noted already, if $V \in R'$, the corresponding structure group can be reduced to H', and

$$H' = \{\lambda_0 h \mid \lambda_0 \in \mathbb{C},\ h \in H\}.$$

Thus, for a suitable open covering $\{U_i\}$ of X, the transition functions of V in H' can be given by $\{a_{ij}\phi_{ij}\}$, where ϕ_{ij} take values in H and a_{ij} are scalar functions. Now, if gr $V = \overset{4}{\oplus} L$, then the $\{a_{ij}\}$ are, indeed, the transition functions of the line bundle L. Therefore, we conclude that if $V \in W$ (i.e., gr $V = \overset{4}{\oplus} \mathcal{O}_x$), the corresponding structure group can be reduced to H. The proof of the second assertion is obvious.

2. *Isomorphism classes of H-bundles.* Let $x_1, x_2 \in X$ be two distinct points and consider the open cover $\{U_i\}_{i=1,2}$, $U_i = X - x_i$, $i = 1, 2$. Then,

if $V \in R$, $i = 1,2$, $V\,|\,_{U_i}$ is trivial (indeed, on an affine smooth curve, if V is a vector bundle, then $V \simeq I_{n-1} \oplus L$, with $L \simeq \det V$; hence if $\det V$ is trivial, such a bundle itself becomes trivial). Therefore, $V \in R$ can be given in terms of a single transition function, say $\{\theta_{12}\}$, i.e., just a matrix of the form

$$\begin{bmatrix} 1 & \lambda & \mu & \nu \\ 0 & 1 & 0 & -\mu \\ 0 & 0 & 1 & \lambda \\ 0 & 0 & 0 & 1 \end{bmatrix}$$

with λ, μ, ν functions on

$$U_{12} = U_1 \cap U_2.$$

We see that

$$Z^1\left(\{U_i\}, \mathcal{O}_x\right) = Z^1(X, \mathcal{O}), \quad \text{(cocycles with values in } \mathcal{O}).$$

Thus $\{\lambda, \mu, \nu\} \in \overset{3}{\oplus} Z^1(X, \mathcal{O})$, and any element in $\overset{3}{\oplus} Z^1(X, \mathcal{O})$ determines an element in R.

Let P be a linear subspace of coset representatives in $Z^1(X, \mathcal{O})$, such that the natural map

$$P \to H^1(X, \mathcal{O})$$

is an *isomorphism*.

Proposition 2. *Let $V \in R$ be represented by $\{\lambda, \mu, \nu\} \in \overset{3}{\oplus} P$. Then the λ, μ in P are linearly independent and, conversely, given $\{\lambda, \mu, \nu\} \in \overset{3}{\oplus} P$, with λ, μ linearly independent, the transition function*

$$\begin{bmatrix} 1 & \lambda & \mu & \nu \\ 0 & 1 & 0 & -\mu \\ 0 & 0 & 1 & \lambda \\ 0 & 0 & 0 & 1 \end{bmatrix}$$

defines an element in R.

Proof. Indeed, the only condition that needs to be checked for a transition function

$$\begin{bmatrix} 1 & \lambda & \mu & \nu \\ 0 & 1 & 0 & -\mu \\ 0 & 0 & 1 & \lambda \\ 0 & 0 & 0 & 1 \end{bmatrix}$$

to define an element in R, is the condition in Proposition 1 of [7], which in fact, characterises elements in PV_4 itself. More precisely by Proposition 1 of [7], it suffices to prove the following assertion:

If $\{\lambda,\mu,\nu\} \in \overset{3}{\oplus} P$, defining $V \in R$, are such that λ,μ are *linearly dependent*, then there exists a sub-bundle for V of the form $\overset{r}{\oplus} L$, $r \geq 2$, L a line bundle of degree 0, and conversely.

So, suppose $\lambda,\mu, \in P$ are linearly dependent. Then, there exist non-vanishing constant functions a, b such that

$$a\lambda + b\mu = 0.$$

Let m be the element in GL(4), given by

$$\begin{bmatrix} 1 & 0 & 0 & 0 \\ a & a & 0 & 0 \\ b & b & 1 & 0 \\ 0 & 0 & 0 & 1 \end{bmatrix}$$

and let

$$\nu = \begin{bmatrix} 1 & \lambda & \mu & \nu \\ 0 & 1 & 0 & -\mu \\ 0 & 0 & 1 & \lambda \\ 0 & 0 & 0 & 1 \end{bmatrix}$$

be the transition function for V in U_{12}. Consider the matrix $m^{-1} \cdot \nu \cdot m$.

Then by using the condition $a\lambda + b\mu = 0$, we get

$$m^{-1}\nu m = \left[\begin{array}{cc:cc} 1 & 0 & & * \\ 0 & 1 & & \\ \hdashline 0 & 0 & 1 & \lambda \\ 0 & 0 & 0 & 1 \end{array}\right],$$

i.e., by change of coordinates one can get $(\mathcal{O}_x \oplus \mathcal{O}_x)$ itself to be a sub-bundle of V. The converse is rather easy to see.

Let $Q = \{(\lambda,\mu) \in P \times P \,|\, \lambda,\mu$ are linearly independent$\}$. Then by Proposition 2 above, we have a canonical map

$$\xi : P \times Q \to R.$$

Proposition 3. *This map ξ is a surjection. In fact, if $\mathcal{H}$ is the set of isomorphism classes of H-bundles, with $\lambda,\mu \in P$, linearly independent, then the induced map*

$$\xi' : P \times Q \to \mathcal{H}.$$

is itself surjective.

Proof. Indeed, we have a canonical homomorphism

$$H \to \mathbf{G}_a \times \mathbf{G}_a, \; (\lambda,\mu,\nu) \mapsto (\lambda,\mu)$$

which gives a map

$$\psi : \{\text{Isom classes of } H\text{-bundles}\} \to \{\text{Isom classes of } \mathbb{G}_a \times \mathbb{G}_a \text{ bundles}\}.$$

Since the image space can be identified with $H^1(X, \mathcal{O}) \times H^1(X, \mathcal{O})$, can be identified with a map:

$$\psi : \mathcal{H} \to H^1(X, \mathcal{O}) \times H^1(X, \mathcal{O}).$$

It is clear that ψ is surjective. Simply lift the transition function in U_{12} for $\mathbb{G}_a \times \mathbb{G}_a$ into that in H by taking

$$\begin{bmatrix} 1 & \lambda & \mu & \nu \\ 0 & 1 & 0 & -\mu \\ 0 & 0 & 1 & \lambda \\ 0 & 0 & 0 & 1 \end{bmatrix} \quad (\lambda,\ \mu \in \mathbb{G}_a).$$

Hence, to prove the proposition, it is enough to show that, given a fibre of ψ, it lies in the image of ξ'. To prove this, we can suppose that $(\lambda, \mu) \in Q$ and $\nu \in Z^1(X, \mathcal{O})$, and it suffices to find a $\nu' \in P$ such that the bundle defined by $\{\lambda, \mu, \nu\}$ is H-"equivalent" to the one defined by $\{\lambda, \mu, \nu'\}$.

Explicitly writing the condition for two H-bundles given by $\{\lambda, \mu, \nu\}$ and $\{\lambda, \mu, \nu'\}$, $\lambda, \mu \in Q$, $\nu, \nu' \in Z^1(X, \mathcal{O})$ to be isomorphic as H-bundles, we get $a, b, c, \in H^0(U_1, \mathcal{O})$ and $a', b', c' \in H^0(U_2, \mathcal{O})$ such that

$$\begin{bmatrix} 1 & a & b & c \\ 0 & 1 & 0 & -b \\ 0 & 0 & 1 & a \\ 0 & 0 & 0 & 1 \end{bmatrix} \begin{bmatrix} 1 & \lambda & \mu & \nu \\ 0 & 1 & 0 & -\mu \\ 0 & 0 & 1 & \lambda \\ 0 & 0 & 0 & 1 \end{bmatrix} = \begin{bmatrix} 1 & \lambda & \mu & \nu' \\ 0 & 1 & 0 & -\mu \\ 0 & 0 & 1 & \lambda \\ 0 & 0 & 0 & 1 \end{bmatrix} \begin{bmatrix} 1 & a' & b' & c' \\ 0 & 1 & 0 & -b' \\ 0 & 0 & 1 & a' \\ 0 & 0 & 0 & 1 \end{bmatrix}$$

This implies,

$$\lambda + a = \lambda + a'$$
$$\mu + b = \mu + b'.$$

By the choice of λ, μ in P (cf. definition before Proposition 2), we get $a = a' =$ constant and $b = b' =$ constant.

Further,

$$-a\mu + b\lambda + \nu + c = -b\lambda + a\mu + \nu' + c'$$

i.e., $(\nu - \nu') = (c' - c) + 2(a\mu - b\lambda)$. Thus, $\{\lambda, \mu, \nu\}$ and $\{\lambda, \mu, \nu'\}$ give isomorphic H-bundles $\Leftrightarrow$ there exist constants, a, b such that $(\nu - \nu') + 2(-a\mu + b\lambda)$, which is a function on U_{12}, is a coboundary. Thus, if $(\nu - \nu')$ is a coboundary, then taking $a = b = 0$, we get

$$\{\lambda, \mu, \nu\} \sim \{\lambda, \mu, \nu'\} \text{ as } H\text{-bundles.}$$

But, given $\nu \in Z^1(X, \mathcal{O})$, one can obviously choose a $\nu' \in P$ so that $(\nu - \nu')$ is a coboundary, and therefore the proposition follows.

3. *Identification* of $N_2 - N_3 = R'$. Consider the set

$$M(3,g)^* = \{(3 \times g) - \text{matrices},$$

with the last two rows linearly independent}. Then clearly, $P \times Q$ can be identified with $M(3,g)^*$ since $g = \dim H^1(X, \mathcal{O})$.

Proposition 4. *We have an isomorphism*

$$R \simeq G \backslash M(3,g)^*$$

where the group

$$G \text{ is } = \left\{ \begin{bmatrix} a & * & * \\ 0 & & C \end{bmatrix} \in GL(3), \mid \det C = a \right\}.$$

Proof. Let $\{\lambda, \mu, \nu\}$ and $\{\lambda', \mu', \nu'\}$ in $P \times Q$ be two triples. These determine isomorphic bundles in R iff there are functions $(a_{ij}) \in \mathrm{GL}(4, \mathcal{O}_{U_1})$ and $(b_{ij}) \in \mathrm{GL}(4, \mathcal{O}_{U_2})$ such that

$$\begin{bmatrix} a_{11} & \cdot & \cdot & \cdot & a_{14} \\ \cdot & \cdot & \cdot & \cdot & \cdot \\ \cdot & \cdot & \cdot & \cdot & \cdot \\ a_{41} & \cdot & \cdot & \cdot & a_{44} \end{bmatrix} \begin{bmatrix} 1 & \lambda & \mu & \nu \\ 0 & 1 & 0 & -\mu \\ 0 & 0 & 1 & \lambda \\ 0 & 0 & 0 & 1 \end{bmatrix}$$

$$= \begin{bmatrix} 1 & \lambda' & \mu' & \nu' \\ 0 & 1 & 0 & -\mu' \\ 0 & 0 & 1 & \lambda' \\ 0 & 0 & 0 & 1 \end{bmatrix} \begin{bmatrix} b_1 & \cdot & \cdot & \cdot & b_{14} \\ \cdot & \cdot & \cdot & \cdot & \cdot \\ \cdot & \cdot & \cdot & \cdot & \cdot \\ \cdot & \cdot & \cdot & \cdot & \cdot \\ b_{41} & \cdot & \cdot & \cdot & b_{44} \end{bmatrix}.$$

Expanding this matrix expression, we can draw the following conclusions:

(i) Equating the last row, we get

(a) $a_{41} = b_{41}$ in $U_{12} \Rightarrow a_{41}$ is a globally defined holomorphic function on X and hence $a_{41} = b_{41} =$ Constant.

(b) $a_{41}\lambda + a_{42} = b_{42} \Rightarrow a_{41} \cdot \lambda$ is a coboundary: but by hypothesis $(\lambda, \mu) \in Q$, implying $a_{41} = 0$ and $a_{42} = b_{42} =$ Constant. Similarly $a_{43} = b_{43} =$ Constant.

(c) Since $a_{41} = 0$, we have $-a_{42}\mu + a_{43}\lambda = b_{44} - a_{44}$. Now since a_{42}, a_{43} are constants and $a_{42}\mu + a_{43}\lambda$ is a coboundary, the linear independence of $(\lambda, \mu) \in Q$ implies that

$$a_{42} = a_{43} = 0, \quad \text{i.e.,} \quad a_{44} = b_{44} = \text{ Constant}$$
$$a_{42} = b_{42} = 0, \qquad a_{43} = b_{43} = 0.$$

Thus

(ii) Equating the first column, we get
$a_{31} = b_{31} + b_{41}\lambda' \Rightarrow b_{41}\lambda'$ is a coboundary
hence $b_{41} = 0$ and, $a_{31} = b_{31} =$ Constant.

(iii) Equating the third row, we get
$a_{31}\lambda + a_{32} = b_{32}$ (since $b_{42} = 0$)
$a_{31} = 0$, and $a_{32} = b_{32} =$ Constant. Similarly
$a_{33} = b_{33} =$ Constant, $a_{31} = b_{31} = 0$.

(iv) Equating the second row, we get
$a_{21} = b_{21} =$ Constant
$a_{21}\lambda + a_{22} = b_{22} \Rightarrow a_{21} = b_{21} = 0$
$a_{22} = b_{22} =$ Constant. Similarly $a_{23} = b_{23} =$ Constant.

(v) Equating the first row, we get $a_{11} = b_{11} =$ Constant
(i) - (v) after simplification, imply that
$a_{11}\lambda = a_{22}\lambda' + a_{32}\mu' + (b_{12} - a_{12})$

and three similar expressions. Thus, if "$\equiv$" denotes equality of elements in $H^1(X, \mathcal{O})$, we have

$$\begin{aligned} a_{11}\lambda &\equiv a_{22}\lambda' + a_{32}\mu' \\ a_{11}\mu &\equiv a_{23}\lambda' + a_{33}\mu' \\ a_{44}\lambda' &\equiv a_{33}\lambda - a_{32}\mu \\ a_{44}\mu' &\equiv -a_{23}\lambda + a_{22}\mu \end{aligned} \tag{**}$$

We can also suppose that $\lambda, \mu, \nu, \lambda', \mu', \nu' \in P$. Then the relations '$\equiv$' in (**) become equalities and we get

$$\begin{aligned} b_{12} = a_{12} = \text{ Constant}, \quad b_{13} = a_{13} = \text{ Constant} \\ a_{34} = b_{34} = \text{ Constant}, \quad a_{24} = b_{24} = \text{ Constant}. \end{aligned} \tag{§}$$

Also we have

$a_{11}\nu - a_{12}\mu + a_{13}\lambda = \lambda' a_{24} + \mu' a_{34} + \nu' a_{44} + (b_{14} - a_{14})$, i.e., $a_{11}\nu - a_{12}\mu + a_{13}\lambda \equiv \lambda' a_{24} + \mu' a_{34} + \nu' a_{44}$ using (ˆ§) and letting $\lambda, \mu, \nu, \lambda', \mu', \nu' \in P$, we thus have $a_{11} - a_{12}\mu + a_{13}\lambda = \lambda' a_{24} + \mu' a_{34} + \nu' a_{44}$.

Therefore, for $\{\lambda, \mu, \nu\}$, $\{\lambda', \mu', \nu'\} \in Q \times P$, the conditions determined above give a matrix

$$\begin{bmatrix} a_{11} & a_{12} & a_{13} & * \\ 0 & a_{22} & a_{23} & a_{24} \\ 0 & a_{32} & a_{33} & a_{34} \\ 0 & 0 & 0 & a_{44} \end{bmatrix}$$

(with all its entries except (*) being constants), such that,

$$\begin{aligned}
\lambda a_{11} &= \lambda' a_{22} + \mu' a_{32}, \ \lambda' a_{44} = \lambda a_{33} - \mu a_{32} \\
\mu a_{11} &= \lambda' a_{23} + \mu' a_{33}, \ \mu' a_{44} = -\lambda a_{23} + \mu a_{22} \\
a_{11}\nu - a_{12}\mu + a_{13}\lambda &= a_{24}\lambda' + a_{34}\mu' + a_{44}\nu'.
\end{aligned} \tag{$\dagger$}$$

Substituting for λ', μ' from ($\dagger$), we get

$$\nu' a_{44} = a_{11}\nu + (\text{Const.})\lambda + (\text{Const.})\mu$$

i.e.,

$$\nu' = \left(\frac{a_{11}}{a_{44}}\right) \cdot \nu + (\text{Const.})\lambda + (\text{Const.})\mu$$

and

$$\lambda' = \left(\frac{a_{33}}{a_{44}}\right)\lambda - \left(\frac{a_{32}}{a_{44}}\right)\mu, \ \mu' = \left(\frac{-a_{23}}{a_{44}}\right)\lambda + \left(\frac{a_{22}}{a_{44}}\right)\mu.$$

Similarly we get λ, μ in terms of λ', and μ'. These yield,

$$\frac{1}{a_{44}} \cdot \begin{bmatrix} a_{33} & -a_{32} \\ -a_{23} & a_{22} \end{bmatrix} \frac{1}{a_{11}} \cdot \begin{bmatrix} a_{22} & a_{32} \\ a_{23} & a_{33} \end{bmatrix} = I.$$

Let $B = \begin{bmatrix} a_{22} & a_{23} \\ a_{32} & a_{33} \end{bmatrix}$, then $\det B = a_{11} \cdot a_{44}$, i.e.,

$$\frac{\det B}{a_{44}^2} = \frac{a_{11}}{a_{44}}.$$

Thus

$$\begin{vmatrix} \nu' \\ \lambda' \\ \mu' \end{vmatrix} = \begin{bmatrix} \frac{a_{11}}{a_{44}} & * & * \\ 0 & & C \\ 0 & & \end{bmatrix} \begin{vmatrix} \nu \\ \lambda \\ \mu \end{vmatrix} = \begin{bmatrix} a & * & * \\ 0 & C & \end{bmatrix} \begin{vmatrix} \nu \\ \lambda \\ \mu \end{vmatrix}$$

$a = \frac{a_{11}}{a_{44}}$, $C = \frac{(1)}{a_{44}} \cdot B$ and $\det B = a$. Summarising, we get $(\lambda, \mu, \nu) \sim (\lambda', \mu', \nu') \Leftrightarrow$ there exists a constant matrix of the form

$$\begin{bmatrix} a & * & * \\ 0 & C & \end{bmatrix}$$

with $\det C = a$, such that

$$\begin{bmatrix} a & * & * \\ 0 & C & \end{bmatrix} \begin{vmatrix} \nu \\ \lambda \\ \mu \end{vmatrix} = \begin{vmatrix} \nu' \\ \lambda' \\ \mu' \end{vmatrix} \tag{$***$}$$

Since $P \times Q$ is the same as $M(3, g)^*$, we see that $(***)$ yields the proposition.

Proposition 5. *$G \backslash M(3,g)$ (and therefore R) is a vector bundle of rank $(g-2)$ over $\mathrm{Gr}(2,g)$ - the Grassmanian of 2 dimensional subspaces of a g-dimensional vector space.*

Proof. Let M_{12} be the subset of $M(3,g)^*$ of the form

$$\begin{bmatrix} 0 & 0 & * & * & \cdot & \cdot & \cdot \\ 1 & 0 & * & * & \cdot & \cdot & \cdot \\ 0 & 1 & \underbrace{* & * & \cdot & \cdot & \cdot}_{g-2} \end{bmatrix}$$

Deleting the first row of every element of M_{12}, we obtain a subset M'_{12} of the space of $(2 \times g)$-matrices. We have $M_{12} = M'_{12} \times$ (affine space of dimension $(g-2)$). Let U_{12} be the G-stable open subset of $M(3,g)^*$ such that the minor formed by the first and second columns and the second and third rows is non-singular. We can similarly define M_{ij}, M'_{ij}, U_{ij} corresponding to the ith and j^{th} columns of $M(3,g)^*$. Then U_{ij} is G-stable and M_{ij} can be identified with $G \backslash U_{ij}$. We can cover $M(3,g)^*$ by the U_{ij} and one knows that M'_{ij} form an open covering of $\mathrm{Gr}(2,g)$. From these considerations the proposition follows easily.

4. *Identification of N_3.* We shall follow the same procedure as for $N_2 - N_3$. Let $S = \{V \in N_3 \mid \mathrm{gr}\, V = \overset{4}{\oplus} \mathcal{O}_X\}$. Then N_3 is the disjoint union of 2^{2g} copies of S.

Proposition 6. *There is an isomorphism*

$$S \simeq Gr(3,g)$$

where $Gr(3,g)$ is the Grassmanian of 3 dimensional subspaces in a g-dimensional vector space.

Proof. Consider the algebra C_q^+, where q is the zero quadratic form. Then, it is not hard to see that C_q^+, $q=0$, is isomorphic to $\mathbb{C}[u,v,w]/m^2$, $m = (u,v,w)$, i.e., it is a commutative algebra. Denoting this commutative algebra by D, we see that for $V \in N_3$, End $(V) \simeq D$.

The defining relations for D in terms of 1, u, v, w are $u^2 = v^2 = w^2 = uv = vw = uw = 0$. Thus the transpose of the left regular representation of D as a subalgebra of $M(4)$, the (4×4) matrix algebra, takes the form

$$\begin{bmatrix} \lambda_0 & \lambda_1 & \lambda_2 & \lambda_3 \\ 0 & \lambda_0 & 0 & 0 \\ 0 & 0 & \lambda_0 & 0 \\ 0 & 0 & 0 & \lambda_0 \end{bmatrix}.$$

Keeping in view Proposition 3 of [7] and following Proposition 1, it is easy to see that the structure group for the principal bundle of $V \in S$ can be reduced to the subgroup of GL(4) given by

$$F = \begin{bmatrix} 1 & \lambda & \mu & \nu \\ 0 & 1 & 0 & 0 \\ 0 & 0 & 1 & 0 \\ 0 & 0 & 0 & 1 \end{bmatrix}$$

Proposition 6 follows verbatim except that, in this case the F-bundles which give elements of S are given by triples $\{\lambda, \mu, \nu\} \in \overset{3}{\oplus} P$ with all three of λ, μ, ν *linearly independent.* Thus if $Q' = \{(\lambda, \mu, \nu) \in \overset{3}{\oplus} P \mid \lambda, \mu, \nu$ linearly independent$\}$ then Q' can be identified with $M(3,g)^0 - (3 \times g)$ matrices of rank 3.

If $\{\lambda, \mu, \nu\}$ and $\{\lambda', \mu', \nu'\}$ be two triples in Q', imitating Proposition 4, one sees that the condition for them to represent isomorphic GL(4)-bundles (i.e., they define the same point of S) is *equivalent* to the existence of a constant matrix $M \in \mathrm{GL}(3)$ such that

$$M \begin{vmatrix} \lambda \\ \mu \\ \nu \end{vmatrix} = \begin{vmatrix} \lambda' \\ \mu' \\ \nu' \end{vmatrix}$$

which implies $S \simeq \mathrm{GL}(3)\backslash Q' \simeq \mathrm{GL}(3)\backslash M(3,g)^0\backslash \simeq \mathrm{Gr}(3,g)$.

2. The Normal bundle of R and S (relative to N)

5. *Preliminary Remarks.* These remarks are concerning the "specialisation" condition considered in [7] (cf. [8] also). We shall keep the same notations and conventions in what follows below:

Let Φ be the space of quadratic forms in 3 - variables or equivalently the space of (3×3) - symmetric matrices. Then there is a canonical action of GL(3) on Φ, namely

$$A \circ X := AXA^t, \quad A \in \mathrm{GL}(3), \quad X \in \Phi.$$

For $\phi \in \Phi$, let C_ϕ^+ be the subset of even degree elements in the Clifford algebra C_ϕ corresponding to ϕ and let J be the 3-dimensional vector space C_ϕ^+ mod (e_0), e_0 being the identity element of C_ϕ^+. Then the Lie algebra structure induced on C_ϕ^+ by the associative algebra structure goes down to a Lie algebra structure on J. This in turn gives rise to a skew-symmetric bilinear form

$$\delta_\phi : J \times J \to J$$

or equivalently, a linear map

$$\delta_\phi : \overset{2}{\wedge} J \to J.$$

We have a canonical GL(3) action on Hom $(\overset{2}{\wedge}J, J)$ and there is a non-degenerate GL(3)-equivalent pairing

$$J \times \overset{2}{\wedge}J \to \overset{3}{\wedge}J \simeq \mathbf{C}$$
$$(x, y) \mapsto (x \wedge y)$$

where on $\mathbf{C}$, GL(3) acts through the "determinant." Identify now GL(3) with Aut J, and denote by $J(-1)$, the "twist" of the GL(3) module J by the character $(\det)^{-1}$. This gives a canonical GL(3)- equivariant pairing

$$J(-1) \times \overset{2}{\wedge}J \to \mathbf{C}$$

the GL(3)-action, now on $\mathbf{C}$ being the trivial action. Hence we have

$$(\overset{2}{\wedge}J)^{\vee} \xrightarrow{\sim} J(-1)$$

where $(\overset{2}{\wedge}J)^{\vee}$ is the dual of $\overset{2}{\wedge}J$ which implies that

$$\text{Hom}\,(\overset{2}{\wedge}J, J) \simeq J \otimes J(-1)$$

as GL(3)-modules. Moreover the association $\phi \mapsto \delta_\phi$ gives a bijection onto the symmetric elements in $J \otimes J(-1)$ or equivalently an element of $S^2(J)(-1)$. (cf. [7], [8].)

Consider now the moduli space N. The functor defining it being representable, we have a defining vector bundle E on $X \times N$ of rank 4. Let $f : X \times N \to N$, be the canonical projection and End E the vector bundle associated to the sheaf of endomorphisms of E. Set

$$B = f_*(\text{End } E).$$

Then $B = \{B_n\}_{n \in N}$ is a sheaf of $\mathcal{O}_N$ algebras and defines the canonical family of "specialisations" of $M(2)$, the (2×2) matrix algebra, parametrised by N (cf. [7], [2]). The underlying $\mathcal{O}_N$ module of B is a vector bundle of rank four. Let e_0 be the identity section of B and $\mathsf{J} = B/\mathcal{O}_N \cdot e_0$. Then there is a natural Lie algebra structure on J and moreover, the "specialisation condition" actually says that this Lie algebra structure on J is given by a symmetric section of $\mathsf{J} \otimes \mathsf{J}(-1)$ or equivalently a section of $S^2(\mathsf{J})(-1)$. (This can be seen using the earlier discussion in the paragraph together with Section 2 of [7].) We call this the *canonical section* ψ_N or simply ψ of $S^2(\mathsf{J})(-1)$.

We now make the following.

Definition. Let Sym(3) denote the space of (3×3) - symmetric matrices. Then we call the action of GL(3) on Sym(3) given by,

$$A \circ X := (\det A)^{-1} \cdot AXA^t, \quad A \in \text{GL}(3), \quad X \in \text{Sym}(3)$$

the "Canonical twisted action" and shall denote this GL(3) module by Sym(3)(−1).

6. *More Remarks.* (a) Let T be an algebraic variety and let $F = \{F_t\}_{t\in T}$ be a family of vector bundles on X parametrised by T, so that if $p : X \times T \to T$ is the natural projection, then $p_*(\text{End } F)$ satisfies the "specialisation condition" (cf. (5)). Denote $p_*(\text{End } F)$ by A_T. Then we have by (5) a vector bundle J_T of rank 3 on T together with a *canonical global section*

$$\psi_T \in S^2(\mathsf{J}_T)(-1).$$

We observe that this section is "functorial in T," that is to say, if $f : S \to T$ is a morphism, then the universal property of N would show that $\mathsf{J}_S = f^*(\mathsf{J}_T)$, $\psi_S = f^*(\psi_T)$, etc. In particular for the subvariety $R \subset N$ we see that the restriction of J to R is J_R, and the restriction of the canonical section ψ to R gives the canonical section ψ_R on R.

(b) Consider the vector bundle J of rank 3 on N. Then $S^2(\mathsf{J})(-1)$ is the associated bundle with fibre the GL(3)-module Sym(3)(−1). Denote by S_1 the subvariety of Sym(3) consisting of rank 1 matrices and let $\mathbb{S}_1$ be the sub-bundle space of $S^2(\mathsf{J})(-1)$ with fibre type, the GL(3)-subvariety S_1 of Sym(3). (S_1 is stable for the twisted action also). Then we note the following:

(∗): By our construction, for $R \subset N$, the restriction of the canonical section ψ to R, which as we noted in (a) is the canonical section ψ_R on R, actually lies in $\mathbb{S}_1\,|_R$.

(∗∗): Moreover, if we consider the point $s \in S$ given by $s = \begin{bmatrix} 1 & 0 & 0 \\ 0 & 0 & 0 \\ 0 & 0 & 0 \end{bmatrix}$

then we claim that the isotropy subgroup of GL(3) at s for the "canonical twisted action," is the group G (of Section 1) and thus $S_1 \simeq \mathrm{GL}(3)/G$.

Let $B \in \mathrm{GL}(3)$ be such that

$$(\det B)^{-1} \cdot B \cdot s \cdot B^t = s, \quad \text{i.e.,}$$

$$B \cdot \begin{bmatrix} 1 & 0 & 0 \\ 0 & 0 & 0 \\ 0 & 0 & 0 \end{bmatrix} \cdot B^t = \begin{bmatrix} \det B & 0 & 0 \\ 0 & 0 & 0 \\ 0 & 0 & 0 \end{bmatrix}.$$

We have

$$B = \begin{bmatrix} b_{11} & b_{12} & b_{13} \\ b_{21} & b_{22} & b_{23} \\ b_{31} & b_{32} & b_{33} \end{bmatrix}, \quad \text{then } B \cdot s \cdot B^t = \begin{bmatrix} b_{11}^2 & b_{11}b_{21} & b_{11}b_{31} \\ b_{11}b_{21} & b_{21}^2 & b_{21}b_{31} \\ b_{11}b_{31} & b_{21}b_{31} & b_{31}^2 \end{bmatrix}.$$

We then deduce

$$b_{11}^2 = \det B,\ \ b_{21} = b_{31} = 0$$

so that B is of the form

$$B = \begin{bmatrix} b_{11} & b_{12} & b_{13} \\ 0 & b_{22} & b_{23} \\ 0 & b_{32} & b_{33} \end{bmatrix},\ \ \det B = b_{11}^2$$

which implies that $B \in G$.

From (**) it follows that $S_1 \simeq \mathrm{GL}(3)/G$. Then from (*) and (**) we deduce that ψ_R induces a reduction of the structure group of the bundle J_R to G, which is therefore canonical.

(c) Consider the map $q : M(3,g)^* \to R$ of Proposition 4, which gives an isomorphism $R \simeq G\backslash M(3,g)^*$. Then we have a canonical family of bundles $F = \{F_t\}_{t\in M(3,g)^*}$ on X, parametrised by $M(3,g)^*$, namely by identifying as in Section 1, a triple $\{\lambda, \mu, \nu\}$ with an element of $M(3,g)^*$ The group G operates naturally on F. Therefore, F "descends" to a bundle F_1 on X parametrised by R. Indeed, by the discussion in Section 1, F_1 is actually the defining bundle on $X \times R$. If

$$p : X \times M(3,g)^* \to M(3,g)^*$$

be the canonical projection, then by the construction of the family F, it is not difficult to see that the direct image $p_*(\mathrm{End}\ F)$ of the vector bundle End F is the *trivial* family of algebras $M(3,g)^* \times A_1$, where of course $A_1 \simeq C_q^+$, $q = x_f^2$. One way of seeing this is that, if A_1 is the subalgebra of $M(4)$ defined by

$$A_1 = \left\{ \begin{bmatrix} a & b & c & d \\ 0 & a & 0 & c \\ 0 & 0 & a & -b \\ 0 & 0 & 0 & a \end{bmatrix} \right\}$$

then A_1 commutes with the elements of the group H (see Proposition 1) in which lie the values of the transition functions of all the vector bundles of the family defined by F (see Section 1). We see then that if V is any vector bundle on X in the family F, then End V is canonically isomorphic to A_1, which proves the required assertion.

Let $x,\ x' \in M(3,g)^*$ be represented by $x = \{\lambda, \mu, \nu\}$ and $x' = \{\lambda', \mu', \nu'\}$ and let V_x and $V_{x'}$ be the vector bundles on X given by the associated transition functions as in Section 1. Suppose that $x' = h \cdot x$, $h \in G$, so that h defines an isomorphism of V_x onto $V_{x'}$. We see then that h induces an automorphism of the algebra A and that this induces a *canonical* action of G as a group of automorphisms of the algebra A_1.

Now the canonical action of G on the family F induces an action of G on $p_*(\mathrm{End}\ F) \simeq M(3,g)^* \times A_1$. It is clear that this action of G

on $M(3,g)^* \times A_1$ is the diagonal action (we take the canonical action of G on A_1 as above). It follows easily that the family of algebras $B = p_*(\text{End } F)$ on X parametrised by R is the bundle with fibre the G-module A_1, associated to the principal G-fibre space

$$M(3,g)^* \to R \simeq G\backslash M(3,g)^*.$$

It is clear that the corresponding rank 3 vector bundle $B/\mathcal{O}_R e_0$ (e_0 the unit section of B) is isomorphic to $\mathbf{J}_R$ (see (6) (a) above). Let $A_1' = A_1/(\text{id})$. Then we see that $\mathbf{J}_R$ is the bundle associated to the principal bundle $M(3,g)^* \to R$, with fibre the G-module A_1'. Let s_0 be the point in $S^2(A_1')(-1) \simeq \text{Sym}(3)(-1)$ represented by the canonical algebra structure on A_1'. Now G is the isotropy group at s_0, since G is the group of automorphisms of A_1 (of course, we should have $s_0 = s$ the element in $\text{Sym}(3)(-1)$, defined above when we do the correct identification). The point s_0 is a G-subvariety of $S^2(A_1')(-1)$ and if s_0, $S^2(A_1')(-1)$, denote the bundles, asssociated to the principal G-bundle $\text{M}(3,g)^* \to R$, we see that s_0 gives a section ϕ of $S^2(A_1')(-1)$. Through the identification $S^2(A_1')(-1) \simeq S^2(\mathbf{J}_R)(-1)$ we see that ϕ becomes the canonical section ψ_R of $S^2(\mathbf{J}_R)(-1)$ (see (6)(a)). The transition functions of the principal G-bundle $M(3,g)^* \to R$, which are in the group G, obviously leave invariant the section Φ. This means that the reduction to the structure group G defined by Φ is isomorphic to the principal G-bundle $M(3,g)^* \to R$. Thus we have proved:

Proposition 7. *The canonical reduction of the structure group of $\mathbf{J}_R$ to G, induced by the section ψ_R, is (isomorphic to) the principal G-bundle $M(3,g)^* \to R$.*

7. *Identification of the normal bundle of R (relative to N).* To compute the normal bundle of R relative to N, we require the determination of the *mini-versal deformation space* of the moduli space of quadratic forms in three variables, at a quadratic form of rank one. To be more precise, let F be the functor, on the category $\mathcal{C}$ of *pointed analytic spaces or schemes*, (T,t_0) over $\mathbf{C}$, t_0 being a $\mathbf{C}$-valued point (if $T = \text{Spec } A$, A local, t_0 is the point corresponding to the maximal ideal), defined as follows.

If $T = \text{Spec } A$,

$F(T,t_0) = \{$equivalence classes of quadratic forms or (3×3) symmetric matrices Q with values in A such that $Q(t_0)$ is of rank one, say $s = \begin{bmatrix} 1 & 0 & 0 \\ 0 & 0 & 0 \\ 0 & 0 & 0 \end{bmatrix}$ without loss of generality$\}$.

Here the equivalence $Q \sim Q'$ means:

$$Q' = B \cdot Q \cdot B^t, \; B \in \text{GL}(3,A).$$

We make a similar definition of F when $\mathcal{C}$ is the category of pointed analytic spaces. Consider now the following morphism.

$$v: \ \mathrm{Sym}(3) \to \mathrm{Sym}(2)$$

defined in a neighbourhood of the quadratic form $s = \begin{bmatrix} 1 & 0 & 0 \\ 0 & 0 & 0 \\ 0 & 0 & 0 \end{bmatrix}$, as follows:

$$v \cdot \begin{bmatrix} a & h & g \\ h & b & f \\ g & f & c \end{bmatrix} = \frac{1}{a} \cdot \begin{bmatrix} b - h^2/a & f - gh/a \\ f - gh/a & c - g^2/a \end{bmatrix}.$$

We have the following computation:

$$\text{Let } B = (1/\sqrt{a}) \cdot \begin{bmatrix} 1 & 0 & 0 \\ -\frac{g}{a} & 1 & 0 \\ -\frac{h}{a} & 0 & 1 \end{bmatrix}, \qquad \text{then}$$

$$\text{if } \ q = \begin{bmatrix} a & h & g \\ h & b & f \\ g & f & c \end{bmatrix}, \ B \cdot q \cdot B^t = \begin{bmatrix} \frac{1}{a} & 0 & 0 \\ 0 & \multicolumn{2}{c}{v(q)} \\ 0 & & \end{bmatrix}.$$

Suppose then that *units in A have square roots* (e.g., A is a complete local $\mathbb{C}$-algebra) then we have the equivalence

$$(*) \quad q \sim \begin{bmatrix} \frac{1}{a} & 0 & 0 \\ 0 & \multicolumn{2}{c}{v(q)} \\ 0 & & \end{bmatrix}.$$

Lemma 8. *The mini-versal deformation space for the functor F can be identified (and the precise manner of this identification is given in the proof below) with the complete local ring associated to the closed point* (0) *of* Sym(2) *(when $\mathcal{C}$ is the category of pointed analytic spaces, the mini-versal deformation is the germ of* Sym(2) *at* (0)*).*

Proof. We have the canonical family of quadratic forms in three variables parametrised by Sym(3) and obviously the complete local ring of Sym(3) at s is a *versal deformation* space for the functor F. Say $T = \mathrm{Spec}A$, A – Artin local. Then this versal property gives a morphism $\phi : T \to \mathrm{Sym}(3)$, (the point t_0 maps to s). Because of the equivalence assertion $(*)$ above, we see easily that a neighbourhood of (0) in Sym(2) parametrises a family of symmetric matrices of rank 3 such that at (0) the symmetric matrix is s; besides the morphism

$$v \cdot \phi : T \to \mathrm{Sym}(2)$$

gives the versal property of this family parametrised by a neighbourhood of Sym(2). That this family is a mini-versal (for the functor F) is easily checked, say by a simple dimension count.

Remark 9. By the series of remarks in (5), we have, over an open subset U of N, where $\mathbf{J}$ is trivial, the section $\psi \in S^2(\mathbf{J})(-1)$ given by $\psi_U : U \to \mathrm{Sym}(3)$ (or more precisely, $\mathrm{Sym}(3)(-1)$).

We note that, $\psi_U(U \cap R) \subset S_1$ (symmetric matrices of rank 1) so that ψ_U defines an analytic family Q of quadratic forms in (three variables), such that at $p \in U \cap R$, $Q(p)$ is of rank one. It is easily seen that given $p \in U \cap R$, there is a (Euclidean) neighbourhood V in which,

$$Q \sim Q' \text{ with } Q(p) = s = \begin{bmatrix} 1 & 0 & 0 \\ 0 & 0 & 0 \\ 0 & 0 & 0 \end{bmatrix}, \quad p \in V \cap R.$$

Hence we can suitably modify the open subsets U in the *Euclidean topology* so that ψ maps $U \cap R$ to s.

Lemma 10. *Let $\{U_i\}$ be an open cover for N (by Euclidean neighbourhoods) such that the morphisms*

$$\psi_i = \psi_{U_i} : U_i \to \mathrm{Sym}(3)$$

maps $U_i \cap R$ to the point $s \in S_1$. Consider the diagram

$$\begin{array}{ccc} U_i & \xrightarrow{\psi_i} & \mathrm{Sym}(3) \\ \| & & \downarrow v \\ U_i & \xrightarrow[\phi_i]{} & \mathrm{Sym}(2). \end{array}$$

Then the ϕ_i's are smooth.

Proof. Let us recall very briefly, one of the main results of [7]. There it is shown that the morphism from the functor defining the moduli space N to the functor defining family of algebras which are "specialisations" of $M(2)$ (or equivalently, rank 3 quadratic forms) is *formally smooth.*

Now, Lemma 8 tells us that Sym(2) is the local moduli space for the functor F. Hence it follows that the ϕ_i's are smooth.

Remark 11. In [7], attention was concentrated at proving the smoothness at the point in N corresponding to the commutative algebra A_0 (in a sense the most difficult case), whose mini-versal deformation space was proved to be Sym(3).

Remark 12. By Remark 9, we have an open cover $\{U_i\}$ of N by Euclidean open subsets and morphisms

$$\psi_i : U_i \to \mathrm{Sym}(3)$$

so that ψ_i maps $U_i \cap R$ to the point $s = \begin{bmatrix} 1 & 0 & 0 \\ 0 & 0 & 0 \\ 0 & 0 & 0 \end{bmatrix}$. Further, if $\{\Theta_{ij}\}$ are the transition functions of the GL(3)-bundle $\mathbf{J}$,

$$\psi_i = \Theta_{ij} \cdot \psi_i \quad \text{in} \quad U_{ij} = U_i \cap U_j$$
$$\text{i.e., } \psi_i(n) = \Theta_{ij}(n) \cdot \psi_j(n), \quad n \in U_{ij}$$

where on the right hand side the "dot" denotes the canonical twisted action of GL(3) on Sym(3). As already noted in (6)(b), the isotropy subgroup at $s \in \mathrm{Sym}(3)$ is the group $G = \begin{bmatrix} \alpha & * \\ 0 & C \end{bmatrix} \det C = \alpha$, and by Proposition 7 it follows that, for $r \in R$, $\Theta_{ij}(r) \in G$.

Lemma 13. *Let $r \in R \cap U_i$ and $s \in S_1$, as above. If n_r be the normal space of $R \cap U_i$ in U_i at r relative to U_i and similarly n_s the normal space of S_1 at s relative to* Sym(3)*, then the differential $(d\psi_i)(r)$ of $\psi_i : U \to \mathrm{Sym}(3)$ induces an isomorphism $n_r \simeq n_s$.*

Proof. By Lemma 10, $\psi_i(U_i \cap R) = s \in S_1$. Hence $(d\psi_i)(r)$ maps the tangent space to R at r into the tangent space to S_1 at s, so that $(d\psi_i)(r)$ induces a map $(\overline{d\psi_i})(r) : n_r \to n_s$. By similar considerations applied to dv and $d\phi_i$ (ϕ_i as in Lemma 10), we get the following commutative diagram

$$\begin{array}{ccc} n_r & \xrightarrow{(\overline{d\psi_i})(r)} & n_s \\ & \overline{d\psi_i}(r) \searrow & \downarrow \overline{dv}(s) \\ & & n_0 \end{array}$$

where n_0 is the normal space at (0) relative to Sym(2) i.e., in this case the tangent space to Sym(2) at (0). Here since ϕ_i is smooth, $(d\phi_i)(r)$ is surjective and since $\dim n_r = \dim n_0 = 3$, it turns out to be an isomorphism. It follows that $(d\psi_i)(r)$ maps n_r *injectively* into n_s. Again since $\dim n_r = \dim n_s = 3$, it follows that $(d\psi_i)(r)$ is an isomorphism of n_r onto n_s.

Lemma 14. *Let L be the subspace of* Sym(3) *of the form* $\begin{bmatrix} * & * & * \\ * & 0 & 0 \\ * & 0 & 0 \end{bmatrix}$. *Then L can be canonically identified as a G-module with the tangent space to S_1 at s. Hence the normal space n_s to S_1 at s (relative to* Sym(3)*) can be identified as a G-module with* Sym(3)$/L$. *We observe that*

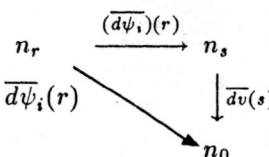

$$\mathrm{Sym}(3)/L \simeq \begin{bmatrix} 0 & 0 & 0 \\ 0 & * & * \\ 0 & * & * \end{bmatrix} \simeq \begin{bmatrix} 0 & 0 & 0 \\ 0 & \mathrm{Sym}(2) & \\ 0 & & \end{bmatrix}$$

as linear spaces.

Proof. The only thing we need to show is that $L \simeq T_s(S_1)$. Now, Sym(3) being an affine space, T_s (Sym(3)) $\simeq$ Sym(3). We claim that $T_s(S_1) = \{A \in \mathrm{Sym}(3) \,|\, A \cdot \ker \cdot s \subseteq \ \mathrm{Im}\ s\}$. In fact this is true for any point in S_1. To see this, let $t \in S_1$ be a matrix of rank 1. Then we have $\ker t$ to be of dimension 2. Let v_1, v_2 be a basis of $\ker t$, then we have the relation $tv_1 = 0$, $tv_2 = 0$. Therefore, if $B \in \mathrm{Sym}(3)$ is a tangent vector at t, then we have by differentiating the relation,

$$\begin{array}{lll} B \text{ tangent vector} & \} \Rightarrow (t + \epsilon B)(v_1 + \epsilon w_1) = 0, & \\ \text{to } S_1 \text{ at } s & \} \quad (t + \epsilon B)(v_2 + \epsilon w_2) = 0 & \end{array} \qquad (\text{with } \epsilon^2 = 0)$$

for some $w_1, w_2 \in \mathrm{Sym}(3)$.

$\Leftrightarrow Bv_1 = -tw_1,\ Bv_2 = -tw_2$, for some $w_1,\ w_2 \in \mathrm{Sym}(3)$, $B \ker t \subseteq \mathrm{Im}\, t$.

Now, $s = \begin{bmatrix} 1 & 0 & 0 \\ 0 & 0 & 0 \\ 0 & 0 & 0 \end{bmatrix}$. Therefore, if $B \cdot \ker s \subset \ \mathrm{Im}\ t$, we see that B is of the type $\begin{bmatrix} a & h & g \\ h & 0 & 0 \\ g & 0 & 0 \end{bmatrix}$, which proves the lemma.

Remark 15. Recall that the (twisted) G-module structure on Sym(3) is given by

$$A \cdot X := (\det A)^{-1} \cdot A \cdot X \cdot A^t,\ A \in G,\ X \in \mathrm{Sym}(3).$$

By Lemma 14, the space of matrices of the form $\begin{bmatrix} 0 & 0 & 0 \\ 0 & \mathrm{Sym}(2) & \\ 0 & & \end{bmatrix}$ acquires a G-module structure by its identification with n_s. It is now easy to see that this action is given as follows:

$$A' = \begin{bmatrix} \alpha & * \\ 0 & B \end{bmatrix}, X = \begin{bmatrix} 0 & 0 & 0 \\ 0 & Y & \\ 0 & & \end{bmatrix}, Y \in \mathrm{Sym}(2).$$

Then $A' \cdot X = (\det B)^{-2} \cdot B \cdot Y \cdot B^t$.

Let ξ be the homomorphism $\xi : G \to \ \mathrm{GL}(2)$ defined by

$$A = \begin{bmatrix} \alpha & 0 & 0 \\ 0 & B & \\ 0 & & \end{bmatrix} \stackrel{\xi}{\longmapsto} B.$$

Through ξ, the GL(2)-module Sym(2)(−2) acquires a G-module structure, and we see that the G-module n_s is isomorphic to the G-module Sym(2)(−2). Hence n_s acquires now a canonical GL(2)-module structure which is just Sym(2)(−2).

Proposition 16. *Let $\{U_i\}$ be the covering of N with Euclidean neighbourhoods as in Remark 12, and let $\{\Theta_{ij}\}$ be the transition functions of the GL(3)-bundle $\mathbf{J}$ with respect to $\{U_i\}$. Then the normal bundle n_R of the subvariety R relative to N is defined by the restriction of the transition functions Θ_{ij} to R. More precisely, the normal bundle of $U_i \cap R$ in U_i for all i, is the trivial bundle with fibre type n_S and these glue up to give the normal bundle of R in N as a G-bundle via the transition functions Θ_{ij}. (Note that for $r \in R$, $\Theta_{ij}(r) \in G$.)*

Proof. (∗) Consider the diagram

$$\begin{array}{ccc} U_i \cap U_j = U_{ij} & \xrightarrow{\psi_i} & \mathrm{Sym}(3) \\ \psi_j & \searrow & \\ & & \mathrm{Sym}(3) \end{array}$$

By Remark 12, for points $n \in U_{ij}$, $\psi_i(n) = \Theta_{ij}(n) \cdot \psi_j(n)$, and $\Theta_{ij}(r) \in G$, for $r \in R \cap U_{ij}$. Written explicitly, this gives us

$$\psi_i(n) = \det \Theta_{ij}(n)^{-1} \cdot \Theta_{ij}(n) \cdot \psi_j(n) \cdot \Theta_{ij}(n)^t. \qquad (*)$$

Taking differentials and inducing the maps to the respective normal spaces, *we claim* that the following diagram commutes:

$$\begin{array}{ccc} n_r(R \cap U_{ij}) & \xrightarrow{(\overline{d\psi_i})(r)} & n_s \\ \overline{d\psi_i}(r) & \searrow & \downarrow \Theta_{ij}(r) \\ & & n_s \end{array} \qquad (**)$$

by which we mean

$$\frac{1}{\det \Theta_{ij}(r)} \cdot \Theta_{ij}(r)(d\psi_j) \cdot \Theta_{ij}(r)^t = (d\psi_i)(r) \qquad (***)$$
$$r \in R \cap U_{ij} \text{ (modulo the tangent space to } S_1 \text{ at } s)$$

To prove this claim, we take differentials in the relation (∗) to get for $r \in R \cap U_{ij}$

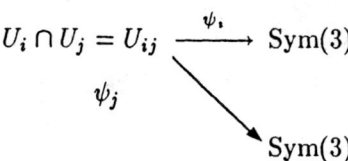

$$\begin{aligned}(d\psi_i)(r) =&(d(\det\Theta_{ij})(r)\cdot\Theta_{ij}(r)\cdot\psi_j(r)\cdot\Theta_{ij}(r))\\ =&\{\det\Theta_{ij}(r)\Theta_{ij}(r)(d\psi_j)(r)\Theta_{ij}(r)\}\\ &+\{d(\det\Theta_{ij}(r)^{-1})\cdot\Theta_{ij}(r)\cdot\psi_j(r)\cdot\Theta_{ij}(r)^t\}\\ &+\{\det\Theta_{ij}(r)^{-1}\cdot(d\Theta_{ij})(r)\cdot\psi_j(r)\Theta_{ij}(r)^t\}\\ &+\{\det\Theta_{ij}(r)\cdot\psi_j(r)\cdot(d\Theta_{ij})(r)^t\}.\end{aligned}$$

We will show that the terms in the last two brackets actually lie in $T_s(S_1)$ and hence the relation $(***)$ would follow. Firstly, $\psi_j(r) = s = \begin{bmatrix}1&0&0\\0&0&0\\0&0&0\end{bmatrix}$ and $\Theta_{ij}(r)\in G$.

We check easily that

$$(d(\det\Theta_{ij})(r)^{-1})\cdot\Theta_{ij}(r)\cdot\psi_j(r)\cdot\Theta_{ij}(r)^t = \begin{bmatrix}\gamma&0&0\\0&0&0\\0&0&0\end{bmatrix}$$

so that by Lemma 14, this lies in $T_s(S_1)$. Secondly, let $\Theta_{ij}(r) = \begin{bmatrix}\alpha & * \quad *\\ 0 & C\end{bmatrix}$ then we remark that $(d\Theta_{ij})(r)^t = (d\Theta_{ij}^t)(r)$. Therefore, if $(d\Theta_{ij})(r) = (a_{ij})_{3\times3}$ with $a_{ij} = a_{ji}$, then

$$\begin{aligned}&(\det\Theta_{ij}(r))(d\Theta_{ij})(r)\cdot\psi_j(r)\cdot\Theta_{ij}(r)^t+\Theta_{ij}(r)\cdot\psi_j(r)(d\Theta_{ij})(r)^t)\\ &=\frac{1}{\alpha^2}\alpha\begin{bmatrix}a_{11}&0&0\\a_{12}&0&0\\a_{13}&0&0\end{bmatrix}+\alpha\begin{bmatrix}a_{11}&a_{12}&a_{13}\\0&0&0\\0&0&0\end{bmatrix}\\ &=\frac{1}{\alpha}\begin{bmatrix}2a_{11}&a_{12}&a_{13}\\a_{21}&0&0\\a_{13}&0&0\end{bmatrix}\end{aligned}$$

which again by Lemma 12, lies in $T_s(S_1)$ and this proves the claim. Now it is clear from $(**)$, that the transition functions Θ_{ij} restricted to R indeed define n_R in N.

Remark 17. The crucial claim $(**)$ in the proof of Proposition 16 can be formulated more generally as follows:

Let X be a complex manifold and Y a submanifold. Let G be a complex Lie group and V a finite dimensional G-module. Let H = Isotropy subgroup at $v\in V$ and W the G-orbit of v. Let $\psi : X\to V$ and $\Theta : X\to G$ be holomorphic maps such that $\psi(Y) = v$ and $\Theta(Y)\subset W$. Let $\Theta\cdot\psi : X\to V$, be the map

$$(\Theta \cdot \psi)(x) = \Theta(x) \cdot \psi(x).$$

Let $y \in Y$. Then if $\xi \in T_y(X)$, we have

$$[(d(\Theta \cdot \psi)(y))(\xi)] \equiv \Theta(y) \cdot [((d\psi)(y))(\xi)] (\text{mod } T_v(W)).$$

This assertion is easily proved.

Theorem 18. *Let $q : M(3,g)^* \to G\backslash M(3,g)^* \simeq R$ be the principal G-bundle on R (cf. Proposition 4). Then the normal bundle n_R of R in N is* isomorphic *to the vector bundle associated to q with fibre n_s under the G-action described in Remark* 15.

Proof. We observe that the G-valued transition functions $\{\Theta_{ij}(r)\}$, $r \in R$ for $\mathbf{J}_R$ in Proposition 16 define the canonical reduction to the group G (as they leave invariant this section ψ_R). Now Theorem 18 follows from Proposition 16 and Proposition 7.

Remark 19. Recall that (see (3)) $M(3,g)^* = P \times Q$, where P is the space of $(1 \times g)$-matrices and Q is the space of $(2 \times g)$-matrices of rank 2. Hence we have

$$R \simeq G\backslash M(3,g)^* \simeq G\backslash P \times Q.$$

Therefore by Theorem 18, for the normal bundle n_R of R relative to N, we have

$$n_R \simeq G\backslash(P \times Q \times n_s)$$

where we take the canonical action of G on n_s.

Proposition 20. *Consider the morphism $\eta : R \to \mathrm{GL}(2)\backslash Q \cong \mathrm{Gr}(2,g)$ of Proposition* 5. *Then n_R is the pullback by η of the bundle associated to the principal* GL(2)*-bundle $Q \to \mathrm{Gr}(2,g)$ with fibre the* GL(2)*-module* Sym(2)(−2).

Proof. Let p be the canonical projection

$$p : (P \times Q) \times n_s \to Q \times n_s$$

We have the diagonal action of G on $P \times Q \times n_s$ and the diagonal GL(2)-action on $Q \times n_s$. (Recall that the GL(2)- module structure on $n_s \simeq$ Sym(2)(−2), see Remark 15). Then p is equivariant for these actions via the homomorphism $\xi : G \to \mathrm{GL}(2)$ (see Remark 15). Hence we get a commutative diagram of vector bundles

$$\begin{array}{ccc} G\backslash(P \times Q \times n_s) & \longrightarrow & \mathrm{GL}(2)\backslash(Q \times n_s) \\ \downarrow & & \\ R & \longrightarrow & \mathrm{Gr}(2,g) \simeq \mathrm{GL}(2)\backslash Q \end{array}$$

where the maps are obvious. By Remark 18, the normal bundle $n_R \simeq G\backslash(P \times Q \times n_s)$. The commutativity of this diagram shows that n_R is the pullback of the required vector bundle on $\mathrm{Gr}(2, g)$.

8. *Normal bundle of N_3 (relative to N).* By (4), $S \simeq \mathrm{Gr}(3, g)$ and N_3 is the disjoint union of 2^{2g} copies of S.

Theorem 21. *The normal bundle n_S of S on N is isomorphic to the bundle* $\mathrm{Sym}^2(\gamma)(-1)$, *$\gamma$ being the tautological rank 3 bundle on* $\mathrm{Gr}(3, g)$.

Proof. Once again, as in (4), we shall repeat the procedure as for R. Note that in this case the sections $\psi : U \to \mathrm{Sym}(3)$ are smooth since $\mathrm{Sym}(3)$ is the local deformation space for points in N_3 and ψ_U are such that ψ_U restricted to $U \cap S$ is mapped onto the zero matrix $z \in \mathrm{Sym}(3)$. Therefore the normal space n_z at z is isomorphic to $\mathrm{Sym}(3)$ itself. However,the canonical $\mathrm{GL}(3)$ action on n_z is once again a twisted one, i.e.,

$$A \cdot X = (\det A)^{-1} \cdot A \cdot X \cdot A^t, \quad X \in n_z \simeq \mathrm{Sym}(3).$$

Thus, as a $\mathrm{GL}(3)$-module n_z is isomorphic to $\mathrm{Sym}(3)(-1)$. Following arguments as in (7) we get the normal bundle n_S to be the vector bundle associated to the principal $\mathrm{GL}(3)$-bundle

$$p : M(3, g)^0 \to S \simeq \mathrm{Gr}(3, g) \quad (\text{cf.}(4))$$

with fibre n_z under the twisted $\mathrm{GL}(3)$ action on n_z. This implies that $n_S \simeq \mathrm{Sym}^2(\gamma)(-1)$.

3. Betti numbers of M_0^s

8. *Hecke Correspondences.* For the computation of the Betti numbers of M_0^s, we shall recall briefly the Hecke Correspondence of [6] (for details, of [2], Propositions 3.2, 3.3 and 3.6).

Let $\widetilde{M_0}$ be the moduli space of parabolic stable bundles of rank 2 and trivial determinant, and $M_{-1,x}$ the moduli space of stable bundles of rank 2 and determinant L_x. Then we have a diagram

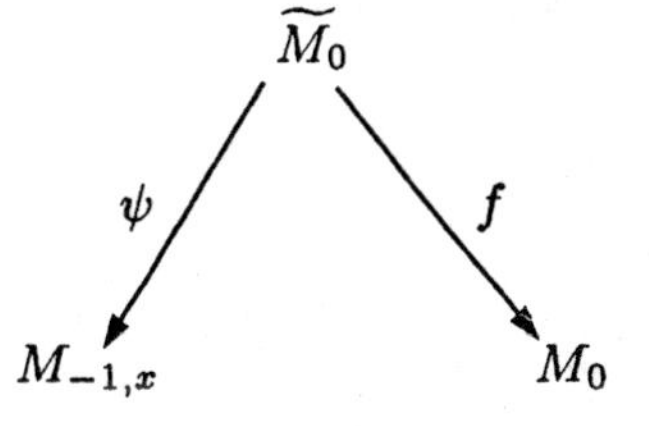

which gives a correspondence between $M_{-1,x}$ and M_0. Then we have the following properties:
H.(1) The map ψ is a $\mathbf{P}^1$-bundle, locally trivial in the Zariski topology and hence the cohomology groups of $\widetilde{M}_0$ are the same as those $\mathbf{P}^1 \times M_{-1,x}$ (cf. Propositions 3.2 and 3.6(b) of [2]).

The cohomology groups of $M_{-1,x}$ can be obtained from Atiyah-Bott [1] and hence those of $\widetilde{M}_0$.
H.(2) Let $K \subset M_0$ be the Kummer variety, consisting of semi-stable points which are not stable. Then $M_0^s = M_0 - K$ (cf. [5]) and f is a $\mathbf{P}^1$-bundle over M_0^s. Moreover the complex codimension of $f^{-1}(K)$ in $\widetilde{M}_0$ is $(g-1)$ (cf. Propositions 3.3, 3.6 of [2]).

9. *Cohomology computation.* For any pair (X, Y) in the complex projective space, $H^*(X, Y)$ will denote cohomology with coefficients in $\mathbf{R}$, in the usual topology. We shall mainly be dealing with cohomology (or homology) groups of the type

(i) $H^r(X)$, X a quasi-projective variety.
(ii) $H^r(X, Y)$, X a projective variety, Y a closed subvariety.
(iii) $H^r(X, X - Y)$ under same conditions as in (ii).

Hence by Spanier [9], the singular cohomology groups coincide with the $\overline{H}$ groups of 6.1 of [9].

Lemma 22. *Let $D = f^{-1}(K) \subset \widetilde{M}_0$. Then we have an isomorphism*

$$H_k(\widetilde{M}_0 - D) \xrightarrow{\sim} H_k(\widetilde{M}_0) \quad \text{for } k < 2g-3.$$

Proof. Consider the pair $(\widetilde{M}_0, \widetilde{M}_0 - D)$ which falls under the type (iii) above. Writing the homology exact sequence for this pair, we have

$$\to H_{k+1}(\widetilde{M}_0, \widetilde{M}_0 - D) \to H_k(\widetilde{M}_0 - D) \to H_k(\widetilde{M}_0) \to H_k(\widetilde{M}_0, \widetilde{M}_0 - D) \to$$

Now, $\widetilde{M}_0$ is a smooth projective variety, therefore by Alexander duality theorem (6.2.16 of [9]), we get

$$H_k(\widetilde{M}_0, \widetilde{M}_0 - D) \xrightarrow{\sim} H^{\lambda-k}(D), \qquad \lambda = \dim_{\mathbf{R}} \widetilde{M}_0. \tag{$*$}$$

If $\lambda - k > \dim_{\mathbf{R}} D$, then $k <$ $\operatorname{codim}_{\mathbf{R}} D$, and by (8) H.(2) $\operatorname{codim}_{\mathbf{R}} D = 2(g-1)$. Also if $\lambda - k > \dim_{\mathbf{R}} D, H^{\lambda-k}(D) = 0$, hence we get using $(*)$

$$H_k(\widetilde{M}_0, \widetilde{M}_0 - D) = 0 \quad \text{for } k < 2g-2.$$

Hence the exact sequence of the pair gives

$$H_k(\widetilde{M}_0 - D) \simeq H_k(\widetilde{M}_0) \quad \text{for } k < 2g - 3.$$

Lemma 23. *The Leray-Hirsch theorem for real cohomology groups holds for the*

$$\mathbf{P}^1 - \textit{fibration} \qquad f : \widetilde{M}_0 - D \to M_0^s \quad \textit{and hence}$$
$$H^*(\widetilde{M}_0 - D) \simeq H^*(M_0^s) \otimes H^*(\mathbf{P}^1).$$

Proof. The following form of the Leray-Hirsch theorem will suit our purposes.
Leray-Hirsch: Let $E \to B$ be a fibre bundle of smooth manifolds with fibre F. Suppose that the cohomology of f is finite dimensional. If there are global cohomology classes $e_1, \ldots, e_r$ on E which when restricted to each fibre freely generate the cohomology of the fibre, then $H^*(E)$ is a free-module over $H^*(B)$ with basis $e_1, \ldots, e_r$ or

$$H^*(E) \simeq H^*(B) \otimes H^*(F).$$

Since f is a projective morphism, we can consider the relatively ample line bundle on $\widetilde{M}_0 - D$, (or in this case, we could just restrict the ample line bundle on the projective variety $\widetilde{M}_0$ to $\widetilde{M}_0 - D$). This when restricted to the fibres will give a power of the hyperplane bundle. Since our cohomology groups have coefficients in $\mathbf{R}$ and since $H^*(\mathbf{P}^1)$ is generated by the hyperplane line bundle, it is clear that the conditions for the Leray-Hirsch theorem are satisfied by f and the lemma follows.

Remark 24. From the above Lemma 23 and $H^*(\mathbf{P}^1)$ we obtain the following relation between the Betti numbers of M_0^s in terms of those of $\widetilde{M}_0 - D$:

$$B_k(\widetilde{M}_0 - D) = B_k(M_0^s) + B_{k-2}(M_0^s).$$

Also, Lemma 22 gives

$$B_k(\widetilde{M}_0) = B_k(\widetilde{M}_0 - D), \ k < 2g - 3.$$

As already remarked in (8) H.(1), the Betti numbers of $\widetilde{M}_0$ are the same as those of $\mathbf{P}^1 \times M_{-1,x}$ and can therefore be obtained using [1].

By [1], pp.593, the Poincaré polynomial of $M_{-1,x}$ is given by

$$P_t(M_{-1,x}) = \frac{(1+t^3)^{2g} - t^{2g}(1+t)^{2g}}{(1-t^2)(1-t^4)}$$

and therefore

$$P_t(\widetilde{M_0}) = \frac{(1+t^3)^{2g} - t^{2g}(1+t)^{2g}}{(1-t^2)^2}.$$

Using this we can compute recursively, the Betti numbers of M_0^s as far as $k < 2g-3$.

4. Betti numbers of N

10. *The main theorem.*

Lemma 25. *Let V be the vector bundle on $\mathrm{Gr}(2,g)$ given by $V = \mathrm{Sym.}^2(\gamma)\otimes L$, where γ is the tautological rank two bundle on $\mathrm{Gr}(2,g)$ and $L^{-1} = (\det\gamma)^2$. Then the Euler class $e(V)$ (i.e., the top Chern class $C_3(V)$) is non zero.*

Proof. Let C_1 and C_2 denote the Chern classes of γ. Now the "splitting principle" states that "to prove a polynomial identity in the Chern classes of complex vector bundles, it suffices to prove it under the assumption that the bundles are direct sum of line bundles".

So let $\gamma = L_1 \oplus L_2$, then $\mathrm{Sym.}^2(\gamma) = (L_1\otimes L_1)\oplus(L_1\otimes L_2)\oplus(L_2\otimes L_2)$. Let $\alpha_i = C_i(L_i)$, $i = 1,2$, be the "Chern roots" of and $C_t(\gamma) = (1+\alpha_1\cdot t)(1+\alpha_2\cdot t)$ its Chern polynomial. Then $C_1 = \alpha_1+\alpha_2$ and $C_2 = \alpha_1\alpha_2$.

From the decomposition of $\mathrm{Sym.}^2(\gamma)$, it is clear that the Chern roots of $\mathrm{Sym.}^2(\gamma)$ are $C_1(L_1^2)$, $C_1(L_1\otimes L_2)$ and $C_1(L_2^2)$, i.e., $2\alpha_1$, $\alpha_1+\alpha_2$ and $2\alpha_2$ respectively.

Thus the top Chern class $C_3(\mathrm{Sym.}^2(\gamma))$, obtained from the Chern polynomial is the third symmetric function in its Chern roots, i.e.,

$$C_3(\mathrm{Sym.}^2(\gamma)) = 2\alpha_1\cdot 2\alpha_2(\alpha_1+\alpha_2) = 4C_1C_2.$$

Similarly, $C_3(\mathrm{Sym.}^2(\gamma)\otimes L)$ is given by

$$C_3(\mathrm{Sym.}^2(\gamma)) + C_2(\mathrm{Sym.}^2(\gamma))\cdot C_1(L) + C_1(\mathrm{Sym.}^2(\gamma))\cdot C_1(L)^2 + C_1(L)^3 \ldots (*).$$

Since $L = (\det\gamma)^{-2}$, $C_1(L) = -2C_1$

$$\begin{aligned} C_2(\mathrm{Sym.}^2(\gamma)) &= 2\alpha_1(\alpha_1+\alpha_2) + 2\alpha_2(\alpha_1+\alpha_2) + 4\alpha_1\alpha_2 \\ &= (\alpha_1+\alpha_2)^2 + 4\alpha_1\alpha_2 = 2C_1^2 + 4C_2. \\ C_1(\mathrm{Sym.}^2(\gamma)) &= 3(\alpha_1+\alpha_2) = 3C_1. \end{aligned}$$

Fitting these into (*) we get

$$C_3(V) = -4C_1C_2$$

is a non-zero polynomial in C_1 and C_2.

Lemma 26. *Consider the Thom-Gysin homomorphism induced by V on* $\mathrm{Gr}(2,g)$, *i.e.,*

$$H^{p-6}(\mathrm{Gr}(2,g)) \xrightarrow{\cup C_3(V)} H^p(\mathrm{Gr}(2,g)).$$

Then this is injective for $p \leq 2g-2$.

Proof. Consider the cohomology ring of $\mathrm{Gr}(2,g)$. Recall that γ sits canonically as a sub-bundle of the trivial rank g bundles E on $\mathrm{Gr}(2,g)$. Let β be the "orthogonal complement" of γ in E, or equivalently the quotient bundle E/γ. Let $d_1, \dots, d_{g-2}$ be the Chern classes of β. Then $H^*(\mathrm{Gr}(2,g))$ is a polynomial ring in C_1 and C_2 with the relation

$$C(\gamma) \cdot C(\beta) = 1, \ C - \text{total Chern class} \tag{*}$$

(*) then gives an inductive method to obtain $d_1, \dots, d_{g-2}$ in terms of C_1 and C_2. Proceeding further, we have the following relations:

Set $d_{g-1} = d_g = 0$ since rank $\beta = g-2$.

Now,

$$\begin{aligned} d_{g-1} &= d_{g-2} \cdot C_1 + d_{g-3} \cdot C_2 \\ d_g &= d_{g-1} \cdot C_2. \end{aligned} \tag{**}$$

And (**) clearly gives the only non-trivial generating relations among C_1 and C_2, namely, $d_{g-1} = 0$ in $H^{2g-2}(\mathrm{Gr}(2,g))$ and $d_g = 0$ in $H^{2g-1}(\mathrm{Gr}(2,g))$, observe that, d_{g-1} is a class in $H^{2g-2}(\mathrm{Gr}(2,g))$. Thus for $k < 2g-2$, the Thom-Gysin homomorphism

$$H^{k-6}(\mathrm{Gr}(2,g)) \xrightarrow{\cup C_3(V)} H^k(\mathrm{Gr}(2,g)) \tag{A}$$

is injective, as till $H^{2g-2}(\mathrm{Gr}(2,g))$, there are no relations in C_1 and C_2.

In fact, more is true by Lemma 25. This gives $e(V) = C_3(V) = -4C_1C_2$ and inductively, from (*) it can be seen that d_{g-1} has a term C_1^{g-1} in it. Also $d_{g-1} = 0$, is the only relation in $H^{2g-2}(\mathrm{Gr}(2,g))$. Therefore, since the polynomial d_{g-1} is not divisible by C_1C_2, it follows that (A) holds for $k = 2g-2$ as well.

Lemma 27. *Let* $E \xrightarrow{\phi} \mathrm{Gr}(2,g)$, *be a complex vector bundle over* $\mathrm{Gr}(2,g)$ *and let $W = \phi^*(V)$, V as in Lemma 25. Then the Thom-Gysin homomorphism on E, induced by W is an injection, for $k \leq 2g-2$.*

Proof. Immediate, since ϕ induces an isomorphism of the cohomology rings $H^*(E) \xrightarrow{\phi^*} H^*(\mathrm{Gr}(2,g))$.

Theorem 28. *Consider the closed submanifold $N_2 - N_3$ of $N - N_3$ of complex codimension 3 (notations as in the Preliminary Section). Let $e(n) = e$ be the Euler class of the normal bundle $n = n_{N_2 - N_3}$. Then the Thom-Gysin homomorphism*

$$H^{k-6}(N_2 - N_3) \xrightarrow{\cup e} H^k(N_2 - N_3)$$

is an injection, for $k \leq 2g - 2$.

Proof. Lemma 27, together with Proposition 20 and Proposition 5 proves the theorem. Note that $N_2 - N_3$ is a *disjoint union* of 2^{2g} copies of R.

Corollary 29. *The Thom-Gysin homomorphism for the pair $(N_2 - N_3, N - N_3)$ is also an injection for $k \leq 2g - 2$.*

Proof. This follows immediately from Theorem 28 and the following commutative diagram.

$$\begin{array}{ccc} H^{k-6}(N_2 - N_3) & \xrightarrow{\text{Gysin}} & H^k(N - N_3) \\ & \searrow \qquad \swarrow \text{Res.} & \\ & H^k(N_2 - N_3) & \end{array}$$

Proposition 30. *Let γ_1 be the tautological rank 3 bundle on* $\mathrm{Gr}(3, g)$ *and let $V_1 = \mathrm{Sym.}^2(\gamma_1) \otimes L_1$, where*

$$L_1 = (\det \gamma_1)^{-1}.$$

Then the Thom-Gysin homomorphism, induced by V_1 (a 6 − plane, complex bundle), i.e.,

$$H^{k-12}(\mathrm{Gr}(3, g)) \xrightarrow{\cup C_6(V_1)} H^k(\mathrm{Gr}(3, g))$$

is injective, for $k \leq 2g - 4$.

Proof. Firstly we have to compute explicitly the top Chern class $C_6(V_1)$. To this end, we proceed as in Lemma 25, using the splitting principle. So let $\gamma_1 = F_1 \oplus F_2 \oplus F_3$, F_i line bundles. Then,

$$\mathrm{Sym.}^2(\gamma_1) = \sum_{i=1}^{3} F_i \otimes F_i \oplus \sum_{i \neq j} F_i \otimes F_j. \qquad \text{(i)}$$

Therefore, if $\alpha_i = C_1(F_i)$, then the Chern polynomial is given by

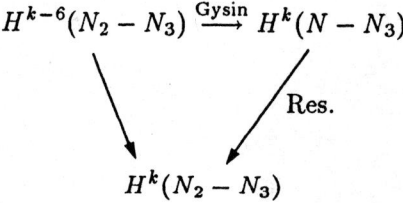

$$\begin{aligned} C_t(\gamma_1) &= (1+d_1t)(1+d_2t)(1+d_3t) \\ C_1 = C_1(\gamma_1) &= \alpha_1+\alpha_2+\alpha_3 \\ C_2 = C_2(\gamma_1) &= \alpha_1\alpha_2+\alpha_1\alpha_3+\alpha_2\alpha_3 \\ C_3 = C_3(\gamma_1) &= \alpha_1\alpha_2\alpha_3. \end{aligned} \tag{ii}$$

The Chern roots of $\mathrm{Sym.}^2(\gamma_1)$ from (i) are

$$2\alpha_1,\ 2\alpha_2,\ 2\alpha_3,\ \alpha_1+\alpha_2,\ \alpha_1+\alpha_3,\ \alpha_2+\alpha_3.$$

Now, the various Chern classes of $\mathrm{Sym.}^2(\gamma_1)$ are the symmetric function in the Chern roots and we give them below, in terms of C_1, C_2 and C_3 using (ii):

$$\begin{aligned} C_1(\mathrm{Sym.}^2(\gamma_1)) &= 4C_1 \\ C_2(\mathrm{Sym.}^2(\gamma_1)) &= 5C_2+5C_1^2 \\ C_3(\mathrm{Sym.}^2(\gamma_1)) &= 6C_3+11C_1C_2+2C_1^3 \\ C_4(\mathrm{Sym.}^2(\gamma_1)) &= 6C_1^2C_2+12C_1C_3+4C_2^2 \\ C_5(\mathrm{Sym.}^2(\gamma_1)) &= 4C_1C_2^2+8C_1^2C_3 \\ C_6(\mathrm{Sym.}^2(\gamma_1)) &= 8C_1C_2C_3-16C_3^2. \end{aligned}$$

$$\begin{aligned} &\therefore\ C_6(\mathrm{Sym.}^2(\gamma_1)\otimes L_1) \\ &= C_1(L)^6+C_1(L)^5C_1(\mathrm{Sym.}^2(\gamma_1))+\cdots+C_6(\mathrm{Sym.}^2(\gamma_1)) \\ &= 8C_1C_2C_3-16C_3^2-2C_1^3C_3, \text{ noting that } C_1(L_1)=-C_1. \end{aligned} \tag{B}$$

$$\therefore C_3(V_1) = 8C_1C_2C_3-16C_3^2-2C_1^3C_3$$

implying that $C_3(V_1)\neq 0$ in $H^*(\mathrm{Gr}(3,g))$.

Now, we consider $H^*(\mathrm{Gr}(3,g))$. Let β_1, be the orthogonal complement of γ_1, γ_1 considered as a rank 3 sub-bundle of the trivial rank of bundle E or equivalently, E/γ_1.

Let $d_1,\dots,d_{g-3}$ be the Chern classes of β_1. Then $H^*(\mathrm{Gr}(3,g))$ is a polynomial ring in C_1, C_2 and C_3 with the relation

$$C(\gamma_1)\cdot C(\beta_1)=1,$$

i.e.,

$$(1+C_1+C_2+C_3)\cdot(1+d_1+\cdots+d_{g-3})=1. \tag{C}$$

From (C) we can inductively obtain $d_1,\dots,d_{g-3}$ in terms of C_1, C_2 and C_3. Proceeding further, we have the following relations

$$d_{g-2} = d_{g-1} = d_g = 0$$

since rank $\beta_1 = g - 3$.
Now,

$$\begin{aligned} d_{g-2} &= d_{g-3}C_1 + d_{g-4}C_2 + d_{g-5}C_3 \\ d_{g-1} &= d_{g-3}C_2 + d_{d-4}C_3 \\ d_g &= d_{g-3}C_3. \end{aligned} \tag{D}$$

And (D), clearly gives the only non-trivial generating relations in C_1, C_2 and C_3. Observe that, d_{g-2} is a class in $H^{2g-4}(\mathrm{Gr}(3,g))$ and thus, for $k < 2g - 4$, the Thom-Gysin homomorphism

$$H^{k-12}(\mathrm{Gr}(3,g)) \xrightarrow{\cup C_6(V_1)} H^k(\mathrm{Gr}(3,g)) \tag{E}$$

is injective, since there are no relations in $H^k(\mathrm{Gr}(3,g))$, $k < 2g-4$. In fact more is true. For consider (B), which gives $C_6(V_1) = 8C_1C_2C_3 - 16C_3 - 2C_1^2 \cdot C_3$, and inductively it can be seen that d_{g-2} has the term C_1^{g-2} in it. Also the equation $d_{g-2} = 0$ is the only relation in $H^{2g-4}(\mathrm{Gr}(3,g))$. Since as a polynomial in C_1, C_2 and C_3, d_{g-2} is not divisible by $C_6(V_1)$, it implies that (E) holds for $k \leq 2g - 4$, as well.

Theorem 31. *Consider the closed submanifold N_3 of N, of complex codimension 6 (notations as in the Preliminary Section). Then the Thom-Gysin homomorphism induced by $e(N_3)$, i.e.,*

$$H^{k-12}(N_3) \xrightarrow{\cup e(N_3)} H^k(N_3)$$

is injective, for $k \leq 2g - 4$.

Proof. This follows from Proposition 30, Theorem 21 and Proposition 6. (Note again that N_3 is a disjoint union of 2^{2g} copies of S).

Corollary 32. *The Thom-Gysin homomorphism for (N_3, N) is also an injection for $k \leq 2g - 4$.*

Proof. Immediate as in Corollary 29.

Remark 33. Keeping the discussion of the Preliminary Section in mind, consider the stratification

$$N \supset N_1 \supset N_2 \supset N_3.$$

Then we have the following relations in cohomology

(i) Corollary 3.13 of [2] gives

$$H^k(Z) = H^{k-2}(Y) \oplus H^k(Z - Y), \qquad k \leq 2g - 4$$

where $Z = N - N_2$, $Y = N_1 - N_2$, $Z - Y \simeq M_0^s$.

(ii) Corollary 29 gives

$$H^k(N_2 - N_3) = H^{k-6}(N_2 - N_3) \oplus H^k(Z), \qquad k \leq 2g - 2.$$

(iii) Corollary 32 gives

$$H^k(N) = H^{k-12}(N_3) \oplus H^k(N - N_3), \qquad k \leq 2g - 4.$$

(i), (ii) and (iii) gives

$$\begin{aligned} H^k(N) = H^{k-12}(N_3) \oplus H^{k-6}(N_2 - N_3) \oplus H^{k-2}(y) \\ \oplus H^k(Z - Y), \quad \text{for } k \leq 2g - 4. \end{aligned} \tag{*}$$

Remark 34. The cohomology of the individual summands in (*) above are easily obtained as follows.

C.(1) By (4), N_3 is the disjoint union of 2^{2g} copies of S and by Proposition 6, $S \simeq \mathrm{Gr}(3, g)$, hence

$$B_{k-12}(N_3) = 2^{2g} \cdot B_{k-12}(\mathrm{Gr}(3, g)) \ \ (\text{where } B_k^k(T) = \dim_{\mathbb{R}} H^k(T, \mathbb{R})).$$

C.(2) By (1), Section 1, $N_2 - N_3 = R'$ is the disjoint union of 2^{2g} copies of R and by Proposition 5, R is a vector bundle over $\mathrm{Gr}(2, g)$, hence

$$B_{k-6}(N_2 - N_3) = 2^{2g} \cdot B_{k-6}(\mathrm{Gr}(2, g)).$$

C.(3) By Proposition 3.4. of [2],

$$H^*(Y, \mathbb{R}) \simeq H^*(K - K_0, \mathbb{R}) \otimes H^*(\mathbb{P}^{g-2} \times \mathbb{P}^{g-2}, \mathbb{R}).$$

This, together with Spanier [10], provides us with $B_{k-2}(Y)$ in terms of $B_j(K - K_0)$ and $B_i(\mathbb{P}^{g-2} \times \mathbb{P}^{g-2})$.

C.(4) Identifying $Z - Y$ with M_0^s together with Remark 24, gives us

$$B_k(Z - Y).$$

C.(1), C.(2), C.(3) and C.(4) provide us with the ingredients for $B_k(N)$, $k \leq 2g - 4$.

REFERENCES

[1] M. F. Atiyah and R. Bott, *The Yang-Mills equations on a Riemann surface*, Phil. Trans. R. Soc. Lond. **308** (1982), 523-621.

[2] V. Balaji, *Cohomology of certain moduli spaces of vector bundles*, to appear in Proc. Indian Acad. Sci. (Math. Sci.).

[3] G. Harder and M. S. Narasimhan, *On the cohomology groups of moduli spaces of vector bundles over curves*, Math. Annlan **212** (1975), 215-248.

[4] F. Kirwan, *On the homology of the compactifications of moduli spaces of vector bundles over a Riemann surface*, Proc. Lond. Math. Soc. **53** (1986), 237-267.

[5] M. S. Narasimhan and S. Ramanan, *Moduli of vector bundles on a compact Riemann surface*, Ann. of Math. **89** (1969), 14-51.

[6] M. S. Narasimhan and S. Ramanan, *Geometry of Hecke Cycles - I*, C. P. Ramanujam, A Tribute, T. I. F. R. Bombay, 1978.

[7] C. S. Seshadri, *Desingularisation of moduli varieties of vector bundles on curves*, Int. Symp. on Algebraic Geometry, Kyoto, 1977, 155-184.

[8] C. S. Seshadri, *Fibrés vectorials sur les courbes algébriques*, Astérisque **96** (1982).

[9] E. H. Spanier, *Algebraic Topology*, Springer Verlag, 1987.

[10] E. H. Spanier, *The homology of Kummer manifolds*, Proc. Amer. Math. Soc. **7** (1956), 155-160.

Institute of Mathematical Sciences
Tharamani
Madras 113, India

Sur les hypersurfaces dont les sections hyperplanes sont à module constant

ARNAUD BEAUVILLE

Introduction

Cette note a pour origine une question de L. Illusie (motivée par [I]) : quelles sont les hypersurfaces dont toutes les sections hyperplanes lisses sont isomorphes entre elles? La réponse est très simple : outre les hyperquadriques, ce sont (à un isomorphisme projectif près) *les hypersurfaces d'équation* $\sum T_i^{q+1} = 0$, *où q est une puissance de la caractéristique.*

Ce résultat est basé sur une étude infinitésimale. Au §1, nous examinons à quelle condition toutes les déformations au premier ordre d'une section hyperplane donnée sont triviales; nous en déduisons le théorème aux §2 et 3. On a besoin au §1 d'une propriété de *l'anneau jacobien* de l'hypersurface, et plus précisément de son socle; cette propriété est très simple lorsque le degré de l'hypersurface est premier à la caractéristique, mais elle m'a donné du fil à retordre dans le cas contraire. J'en ai profité pour décrire au §4 la structure de cet anneau, avec un peu plus de détails qu'il n'est nécessaire pour établir le théorème. D. Eisenbud et C. Huneke montrent dans l'appendice qu'on peut déduire ces résultats de théorèmes plus généraux d'algèbre commutative.

Je tiens à remercier D. Eisenbud de l'intérêt qu'il a manifesté pour ce travail, et C. Voisin à qui je suis redevable de l'essentiel du §3.

1. Déformations d'une section hyperplane

Soit k un corps algébriquement clos, de caractéristique $p \geq 0$. Soit X une hypersurface lisse de degré d dans l'espace projectif $\mathbf{P}_k^n (n \geq 2)$, d'équation $F(T_0, \ldots, T_n) = 0$. Nous noterons R l'anneau jacobien de X : c'est par définition l'anneau gradué quotient de $k[T_0, \ldots, T_n]$ par *l'idéal jacobien* $(F'_{T_0}, \ldots, F'_{T_n}, F)$.

Commençons par une remarque préliminaire. Considérons la suite exacte normale

$$0 \to T_X \to T_{\mathbf{P}|X} \to \mathcal{O}_X(d) \to 0.$$

Par produit tensoriel avec $\mathcal{O}_X(j)$, pour $j \in \mathbf{Z}$, on en déduit un homomorphisme de liaison $\partial_j : H^0(X, \mathcal{O}_X(d+j)) \to H^1(X, T_X(j))$.

Lemme 1. *L'homomorphisme ∂_j induit un homomorphisme de R_{d+j} dans $H^1(X, T_X(j))$, qui est injectif sauf si $n = 2$ et $j = d - 3$.*

La suite exacte de cohomologie associée à la suite normale s'écrit

$$H^0(X, T_{\mathbf{P}}(j)_{|X}) \to H^0(X, \mathcal{O}_X(d+j) \to H^1(X, T_X(j));$$

ainsi ∂_j annule l'image de $\pi : H^0(X, T_{\mathbf{P}}(j)_{|X}) \longrightarrow H^0(X, \mathcal{O}_X(d+j))$, donc aussi celle de $\pi' : H^0(\mathbf{P}^n, T_{\mathbf{P}}(j) \to H^0(X, \mathcal{O}_X(d+j))$. Les éléments de $H^0(\mathbf{P}^n, T_{\mathbf{P}}(j))$ sont de la forme $\sum A_i(T)\partial/\partial T_i$, avec deg $(A_i) = j+1$; l'homomorphisme π' associe à un tel champ de vecteurs le polynôme $\sum A_i(T)F'_{T_i}$. La première assertion du lemme en résulte.

A l'aide de la suite exacte

$$0 \to \mathcal{O}_X(j) \to \mathcal{O}_X^{n+1}(j+1) \to T_{\mathbf{P}}(j)_{|X} \to 0,$$

on montre sans difficultés que l'homomorphisme $H^0(\mathbf{P}^n, T_{\mathbf{P}}(j)) \to H^0(X, T_{\mathbf{P}}(j)_{|X})$ est surjectif sauf pour $n = 2$ et $j = d - 3$. Ce cas étant exclu, on en déduit qu'on a Ker $\partial_j =$ Im π', d'où le lemme.

Rappelons que si $m = (a_0, \ldots, a_n)$ est un point de $\mathbf{P}^n$, le *diviseur polaire* de m dans X est le diviseur découpé sur X par l'hypersurface $\sum a_i F'_{T_i} = 0$; son support est l'ensemble des points x de X tels que l'hyperplan tangent à X en x passe par m.

Proposition 1. *On exclut le cas $p = d =$ dim $(X) = 3$. Soit H un hyperplan de $\mathbf{P}^n$, tel que la section hyperplane $Y = X \cap H$ de X soit lisse. Pour que toute déformation au premier ordre de Y dans X soit triviale en tant que déformation (non plongée) de Y, il faut et il suffit qu'il existe un point m de $\mathbf{P}^n$, non situé sur H, dont le diviseur polaire contienne Y.*

Démonstration. On peut supposer que H est défini par $T_0 = 0$. Toute déformation au premier ordre de Y dans X est découpée par l'hyperplan $T_0 = \epsilon L$ sur $k[\epsilon]$ ($\epsilon^2 = 0$), où L est une forme linéaire en $T_1, \ldots, T_n$; une telle déformation est isomorphe à l'hypersurface dans $\mathbf{P}^{n-1}_{k[\epsilon]}$ d'équation

$$F(\epsilon L, T_1, \ldots, T_n) = 0,$$

soit encore $F(0, T_1, \ldots, T_n) + \epsilon F'_{T_0}(0, T_1, \ldots, T_n)L(T_1, \ldots, T_n) = 0$.

Posons $G(T_1, \ldots, T_n) = F(0, T_1, \ldots, T_n)$. Toute déformation de Y dans $\mathbf{P}^{n-1}$ est donnée par une équation de la forme $G(T_1, \ldots, T_n) + \epsilon\Gamma(T_1, \ldots, T_n) = 0$, où Γ est un polynôme homogène de degré d; pour que cette déformation soit triviale en tant que déformation non plongée de Y, il faut et il suffit que Γ appartienne au noyau de l'homomorphisme de liaison $\partial_0 : H^0(Y, \mathcal{O}_Y(d)) \to H^1(Y, T_Y)$ associé à la suite exacte normale

$$0 \to T_Y \to T_{\mathbf{P}|Y} \to \mathcal{O}_Y(d) \to 0.$$

Excluons d'abord le cas $d = n = 3$. D'après le lemme 1, la condition ci-dessus signifie encore que Γ s'annule dans l'anneau jacobien R de Y.

Revenant à la situation de la proposition, on voit que l'hypothèse que toute déformation au premier ordre de Y est triviale équivaut à dire que pour tout i le polynôme $T_i F'_{T_0}(0, T_1, \ldots, T_n)$ appartient à l'idéal jacobien de Y, ou encore que la classe de $F'_{T_0}(0, T_1, \ldots, T_n)$ dans R_{d-1} est annulée par l'idéal maximal de R. L'ensemble des éléments de R possédant cette propriété est un idéal homogène, le *socle* de R; nous verrons au §3 qu'il est engendré par deux éléments homogènes s et t, de degrés respectifs $d\left[\frac{n+1}{2}\right] - n$ et $n(d-2)$, avec $s = 0$ si d est premier à p. Ces deux nombres sont $\geq d$ (les cas $d = n = 3$ et $p = d = n - 1 = 3$ étant exclus), donc la classe de $F'_{T_0}(0, T_1, \ldots, T_n)$ dans R_{d-1} appartient au socle de R si et seulement si elle est nulle. Cela signifie qu'il existe des scalaires $\alpha_1, \ldots, \alpha_n$ tels qu'on ait

$$F'_{T_0}(0, T_1, \ldots, T_n) = \sum \alpha_i G'_{T_i}(T_1, \ldots, T_n) = \sum \alpha_i F'_{T_i}(0, T_1, \ldots, T_n),$$

ce qui revient à dire que le diviseur polaire dans X du point $(-1, \alpha_1, \ldots, \alpha_n)$ contient la section hyperplane Y.

Reste le cas $d = n = 3$ (de sorte que Y est une cubique plane). D'après la démonstration qui précède, la trivialité des déformations au premier ordre de Y se traduit par la condition $\partial_0(T_i F'_{T_0}(0, T_1, \ldots, T_n)) = 0$ pour tout i. Or l'application naturelle

$$H^1(Y, T_Y(-1)) \to \operatorname{Hom}\,(H^0(Y, \mathcal{O}_Y(1)), H^1(Y, T_Y))$$

est injective (elle s'identifie à l'isomorphisme $H^1(Y, \mathcal{O}_Y(-1)) \to H^0(Y, \mathcal{O}_Y(1))^*$ donné par la dualité de Serre). La condition ci-dessus équivaut donc à $\partial_{-1}(F'_{T_0}(0, T_1, \ldots, T_n)) = 0$, ce qui signifie d'après le lemme 1 que la classe de $F'_{T_0}(0, T_1, \ldots, T_n)$ dans R_{d-1} est nulle. On termine alors la démonstration comme dans le cas général.

2. Le résultat principal

Théorème. *Soit X une hypersurface lisse de degré $d \geq 3$ dans l'espace projectif $\mathbf{P}^n_k$ $(n \geq 3)$, d'équation $F(T_0, \ldots, T_n) = 0$, sur un corps* *algébriquement clos k de caractéristique $p \geq 0$. Les conditions suivantes sont équivalentes :*

(i) *Les sections hyperplanes lisses de X sont isomorphes entre elles.*

(ii) *Pour toute section hyperplane lisse Y de X, toute déformation au premier ordre de Y dans X est triviale en tant que déformation (non plongée) de Y.*

(iii) *Pour tout point m de $\mathbf{P}^n$, le diviseur polaire de m dans X est multiple d'une section hyperplane.*

(iv) *Le nombre $d-1$ est une puissance de p, et les dérivées partielles $F'_{T_0}, \ldots, F'_{T_n}$ sont des puissances de formes linéaires.*

(v) *Le nombre $d-1$ est une puissance de p, et X est isomorphe à l'hypersurface de Fermat $\sum T_i^d = 0$.*

(i) $\Rightarrow$ (ii) : Notons U l'ouvert de l'espace dual $(\mathbf{P}^n)^*$ formé des hyperplans transverses à X, et $\mathcal{M}$ l'espace des modules (grossier) des hypersurfaces de degré d dans $\mathbf{P}^{n-1}$; l'application qui associe à un hyperplan H de U la classe d'isomorphisme de $H \cap X$ est un morphisme $h : U \to \mathcal{M}$. L'hypothèse (i) signifie que h est constant. Soit $H \in U$, et soit $Y = H \cap X$; l'homomorphisme $\widehat{h}^* : \widehat{\mathcal{O}}_{\mathcal{M},Y} \to \widehat{\mathcal{O}}_{U,H}$ induit sur les anneaux locaux complétés annule l'idéal maximal. Or cet homomorphisme se factorise en $\widehat{\mathcal{O}}_{\mathcal{M},Y} \xrightarrow{u} R \xrightarrow{v} \widehat{\mathcal{O}}_{U,H}$, où R est la k-algèbre locale complète qui représente les déformations formelles de Y. Comme u fait de R une $\widehat{\mathcal{O}}_{\mathcal{M},Y}$-algèbre finie, on conclut que v annule l'idéal maximal de R, donc que l'application $T_{U,H} \to T_R$ induite sur les espaces tangents est nulle, ce qui n'est autre que la condition (ii).

(ii) $\Rightarrow$ (iii): Nous excluons ici le cas $p = d = \dim(X) = 3$, qui est traité au paragraphe suivant. Compte tenu de la proposition 1, l'hypothèse (ii) implique que toute section hyperplane de X est composante d'un diviseur polaire. Un argument de dimension montre alors qu'un diviseur polaire quelconque contient une section hyperplane de X. Or le système linéaire des diviseurs polaires est sans point base (un tel point serait un point singulier de X), et définit un morphisme fini de X dans l'espace projectif dual $(\mathbf{P}^n)^*$; le théorème de Bertini [Z] entraîne donc qu'un diviseur polaire assez général dans X est de la forme $p^e D$, où D est intègre. Par suite D est une section hyperplane de X, ce qui implique (iii).

L'argument ci-dessus montre que $d-1$ est une puissance de p si la condition (iii) est satisfaite, d'où l'implication (iii) $\Rightarrow$ (iv).

(iv) $\Rightarrow$ (v) : Sous l'hypothèse (iv), il existe des formes linéaires

$L_0, \dots, L_n$ sur $\mathbf{P}^n$ telles qu'on ait

$$F(T_0, \dots, T_n) = T_0 L_0^q + \dots + T_n L_n^q,$$

où $q = d - 1$ est une puissance de p.

Il est commode d'utiliser la notation matricielle: nous écrirons

$$T = \begin{pmatrix} T_0 \\ \vdots \\ T_n \end{pmatrix}, \ L = \begin{pmatrix} L_0 \\ \vdots \\ L_n \end{pmatrix}.$$

Il existe une matrice $A \in M_{n+1}(k)$ telle qu'on ait $L = AT$. La lissité de X signifie que les formes $L_0, \dots, L_n$ sont linéairement indépendantes, c'est-à-dire que la matrice A est *inversible*.

Pour toute matrice $B = (b_{ij})$ sur k, convenons de noter $B^{(q)}$ la matrice (b_{ij}^q); le polynôme F s'écrit alors ${}^tTA^{(q)}T^{(q)}$. Une variante du théorème de Lang (cf. [S], théorème 1.10) affirme que si φ est un automorphisme d'un groupe algébrique G sur k, tout élément de G peut s'écrire sous la forme $\varphi(g)^{-1}g^{(q)}$ pour un élément $g \in G$ convenable. Appliquant ceci à l'automorphisme $g \mapsto {}^tg^{-1}$ de $GL_{n+1}(k)$, on en déduit que la matrice $A^{(q)}$ peut s'écrire ${}^tB\,B^{(q)}$, donc que le polynôme F s'écrit ${}^tU\,U^{(q)}$, avec $U = BT$. La matrice B fournit donc un isomorphisme linéaire de X sur l'hypersurface de Fermat, d'où (v).

Il est clair que (v) $\Rightarrow$ (iv).

(iv) $\Rightarrow$ (i) : Si X vérifie la condition (iv), toute section hyperplane lisse de X la vérifie également, et donc est isomorphe à l'hypersurface de Fermat en vertu de l'implication (iv) $\Rightarrow$ (v) démontrée ci-dessus. CQFD

Remarque. Un raisonnement analogue à celui de la démonstration de (iv) $\Rightarrow$ (v) montre que les sections hyperplanes singulières de l'hypersurface de Fermat de degré $d = p^e + 1$ dans $\mathbf{P}^n$ sont isomorphes à l'hypersurface $\sum_{i=1}^{n-1} T_i^d = 0$ dans $\mathbf{P}^{n-1}$.

3. Le cas de l'hypersurface cubique dans $\mathbf{P}^4$

La proposition 1, et par conséquent la démonstration de l'implication (i) $\Rightarrow$ (ii) du théorème, ne s'appliquent pas à l'hypersurface cubique dans $\mathbf{P}^4$ en caractéristique 3. Nous allons régler ce problème par une méthode *ad hoc*, qui utilise les propriétés particulières de ces hypersurfaces.

Soit donc X une hypersurface cubique lisse dans $\mathbf{P}_k^4$, où la caractéristique de k est supposée $\neq 2$.

Lemme 2. *Il existe un hyperplan H dans $\mathbf{P}^4$ tel que la surface $H \cap X$ ait comme seule singularité un point double ordinaire.*

Dans la terminologie de [K], cela signifie que le plongement de X dans $\mathbf{P}^4$ est un *plongement de Lefschetz*; il suffit d'ailleurs de prouver qu'il existe un hyperplan H de X tel que la surface $H \cap X$ admette un point double ordinaire (loc. cit., 3.3 et 3.5).

Soit x un point de X. Dans un système de coordonnées affines $(t_1, \ldots, t_4)$ centré en x, l'équation de X s'écrit

$$F(t_1, \ldots, t_4) = h(t_1, \ldots, t_4) + q(t_1, \ldots, t_4) + g(t_1, \ldots, t_4),$$

avec deg $(h) = 1$, deg $(q) = 2$, deg $(g) = 3$.

L'hyperplan H_x d'équation $h = 0$ est l'hyperplan tangent en x à X, le cône quadratique Q_x dans H_x défini par $h = q = 0$ est le cône osculateur en x à X, et la variété G_x d'équation $h = q = g = 0$ est réunion des droites de X passant par x. Si x est assez général, G_x est formée de 6 droites distinctes [M]; il en résulte que le cône Q_x est de rang ≥ 2. Il s'agit de prouver que ce rang est 3 (pour x assez général).

Soit l une droite générale de X. D'après [M], la famille des droites rencontrant X est paramétrée par une courbe lisse connexe C_l. Si l'on avait $rg(Q_x) = 2$ pour x assez général dans X, l'ensemble G_x serait réunion de 3 droites coplanaires l_1, l_2, l_3 et de deux droites l'_1, l'_2 coplanaires avec l, ce qui contredirait l'irréductibilité de C_l.

Lemme 3. *Soit S une surface cubique dans $\mathbf{P}^3$, lisse ou admettant un seul point double ordinaire, et soit $\mathcal{J}(S)$ l'idéal jacobien de S. On a* dim $\mathcal{J}_3(S) = 16$.

Soit $F = 0$ une équation de S. L'idéal $\mathcal{J}_3(S)$ est engendré par les éléments F et $T_i F'_{T_j}$, pour $0 \leq i, j \leq 3$; il s'agit de prouver qu'il n'existe pas d'autre relation entre ces polynômes que la relation d'Euler. Si S est lisse et si le corps k est de caractéristique $\neq 3$, cette assertion est bien connu : elle résulte facilement de ce que la suite $(F'_{T_0}, \ldots, F'_{T_n})$ est régulière. La proposition 2 (c) ci-dessous entraîne alors le résultat en caractéristique 3 (pour S lisse).

On peut donc supposer que S admet un point double ordinaire, par exemple le point $s = (1, 0, 0, 0)$. L'équation de S est alors de la forme

$$F(T_0, \ldots, T_3) = T_0 Q(T_1, T_2, T_3) + G(T_1, T_2, T_3),$$

où Q est une forme quadratique non dégénérée, et G une forme cubique.

Soient α une constante et $L_0, \dots, L_3$ des formes linéaires, telles qu'on ait

$$\sum L_i F'_{T_i} = \alpha F.$$

En considérant le coefficient de T_0^2 dans cette relation, on voit que les formes L_1, L_2, L_3 s'annulent en s; en ajoutant à cette relation un multiple de la relation d'Euler, on peut de plus supposer qu'il en est de même de L_0. En considérant le coefficient de T_0, puis le coefficient constant, on obtient

$$\sum_{i=1}^{3} L_i Q'_{T_i} = \alpha Q \quad ; \quad L_0 Q + \sum_{i=1}^{3} L_i G'_{T_i} = \alpha G.$$

Ces relations entraînent que le champ de vecteurs $\sum_{i=1}^{3} L_i \frac{\partial}{\partial T_i}$ sur $\mathbf{P}^2$ est tangent aux courbes $Q = 0$ et $G = 0$ en leurs points d'intersection. Or la lissité de S en dehors de s implique que ces deux courbes se coupent en 6 points distincts, en lesquels elles sont transverses; on conclut que les formes $L_i (1 \leq i \leq 3)$ s'annulent en ces 6 points, donc sont identiquement nulles. Les relations ci-dessus entraînent finalement $\alpha = 0$ et $L_0 = 0$, ce qui achève de prouver le lemme.

Démonstration du théorème (fin) : Il reste à vérifier qu'une hypersurface cubique de dimension 3 en caractéristique $\neq 2$ ne peut vérifier la condition (ii). Pour $\lambda \in k$, notons Y_λ la surface découpée sur X par l'hyperplan $T_0 = \lambda T_1$. En vertu du lemme 2, on peut supposer que Y_0 admet comme seule singularité un point double ordinaire s_0, et que Y_λ est lisse lorsque λ parcourt un ouvert V de la droite affine.

Reprenant la démonstration de la proposition 1 (§1), on voit que l'hypothèse (ii) entraîne que pour tout i et pour tout $\lambda \in V$, le polynôme $T_i F'_{T_0}(0, T_1, \dots, T_4)$ s'annule dans l'anneau jacobien $R(Y_\lambda)$. Il résulte du lemme 3 que les espaces $R_3(Y_\lambda)$ s'organisent en un fibré vectoriel au-dessus de $V \cup \{0\}$; par suite les polynômes $T_i F'_{T_0}(0, T_1, \dots, T_4)$ s'annulent encore dans $R_3(Y_0)$. On en déduit que F'_{T_0} s'annule en s_0, donc que X est singulière en s_0, ce qui contredit l'hypothèse. CQFD

4. L'anneau jacobien en caractéristique p

Soit X une hypersurface lisse de degré d dans $\mathbf{P}^n$, définie par une équation $F(T_0, \dots, T_n) = 0$. Rappelons qu'on appelle *idéal jacobien* de X l'idéal homogène $\mathcal{J}_X = (F'_{T_0}, \dots, F'_{T_n}, F)$ de $k[T_0, \dots, T_n]$, et *anneau jacobien* de X l'anneau gradué quotient $R = k[T_0, \dots, T_n]/\mathcal{J}_X$. Dire que X est lisse signifie que R est un anneau artinien.

Le *socle* de R est l'idéal homogène de R formé des éléments annulés par l'idéal maximal. Il est de dimension un (sur k) si et seulement si R est un anneau de Gorenstein; dans ce cas il coïncide avec le composant homogène de plus haut degré de R. Soit τ ce degré; si $i + j = \tau$, la multiplication $R_i \otimes R_j \to R_\tau$ est une dualité parfaite.

Si la caractéristique p de k ne divise pas d, l'anneau R est intersection complète (grâce à la relation d'Euler, l'idéal $\mathcal{J}_X$ est défini par la suite $(F'_{T_0}, \ldots, F'_{T_n})$, qui est régulière puisque la sous-variété de k^{n+1} qu'elle définit est réduite à zéro). Alors R est un anneau artinien de Gorenstein, son socle est le composant homogène non nul de degré maximum de R, et ce degré est égal à $(n+1)(d-2)$. De plus la dimension des composants homogènes R_i de R ne dépend ni de p, ni de F, mais seulement des entiers d et n.

Nous supposerons désormais que p divise d.

Proposition 2. (a) *Le socle de R est de dimension 2. Il est engendré par des éléments homogènes s de degré $\sigma = ([\frac{n}{2}]+1)d - n - 1$ et t de degré $\tau = (n+1)(d-2)$.*

(b) *L'anneau quotient $R' = R/ks$ est un anneau de Gorenstein.*

(c) *Notons $R^{(0)}$ l'anneau jacobien d'une hypersurface lisse de degré d dans $\mathbf{P}^n_{\mathbf{C}}$. On a*

$$\dim_k R_i = \dim_{\mathbf{C}} R_i^{(0)},$$

sauf si n est pair et $i = \sigma$; dans ce cas on a $\dim_k R_\sigma - 1 = \dim_k R'_\sigma = \dim_{\mathbf{C}} R_\sigma^{(0)}$.

D. Eisenbud et C. Huneke savent déduire cette proposition d'un énoncé nettement plus général d'algèbre commutative; leur démonstration est esquissée en appendice. Je vais indiquer ici une démonstration géométrique de la proposition 2. Le cas $d = 3$, $n = 2$ demande quelques modifications, que je vais laisser au lecteur pour ne pas trop alourdir le texte. J'exclus donc ce cas dans la suite.

Démonstration : (1) La suite exacte

$$0 \to \mathcal{O}_X(-d) \to \Omega^1_{\mathbf{P}|X} \to \Omega^1_X \to 0$$

fournit pour $1 \le q \le n-1$ une suite exacte

$$(S_q) \qquad 0 \to \Omega_X^{n-q-1}(-d) \to \Omega^{n-q}_{\mathbf{P}|X} \to \Omega_X^{n-q} \to 0,$$

à laquelle est associée une classe d'extension $\xi_q \in \mathrm{Ext}^1(\Omega_X^{n-q}, \Omega_X^{n-q-1}(-d))$.

Considérons en particulier la suite (S_1). Après produit tensoriel par $\mathcal{O}_X(j+n+1-d)$, elle s'identifie via les isomorphismes canoniques $T_X \cong \Omega_X^{n-2} \otimes K_X^{-1}$ et $T_{\mathbf{P}} \cong \Omega_{\mathbf{P}}^{n-1} \otimes K_{\mathbf{P}}^{-1}$ à la suite exacte normale

$$0 \to T_X(j-d) \to T_{\mathbf{P}}(j-d)_{|X} \to \mathcal{O}_X(j) \to 0.$$

En vertu du lemme 1 (§1), l'homomorphisme

$$\cup\xi_1 : H^0(X, \mathcal{O}_X(j)) \to H^1(X, \Omega_X^{n-2}(j+n+1-2d))$$

induit un homomorphisme (noté de la même manière)

$$R_j \to H^1(X, \Omega_X^{n-2}(j+n+1-2d)).$$

Considérons la suite d'homomorphismes

$$R_j \xrightarrow{\cup\xi_1} H^1(X, \Omega_X^{n-2}(j+n+1-2d)) \xrightarrow{\cup\xi_2} H^2(X, \Omega_X^{n-3}(j+n+1-3d)) \to \dots$$
$$\xrightarrow{\cup\xi_{n-2}} H^{n-2}(X, \Omega_X^1(j+n+1-(n-1)d)).$$

(2) Nous allons voir que sauf dans des cas exceptionnels, les homomorphismes $\cup\xi_i$ sont *bijectifs*.

Rappelons d'abord le lemme d'annulation de Bott: pour $1 \leq i \leq n-1$, les espaces $H^i(\mathbf{P}, \Omega_{\mathbf{P}}^q(k))$ sont nuls à l'exception des $H^q(\mathbf{P}, \Omega_{\mathbf{P}}^q)$, qui sont de dimension 1 (voir [D] pour une démonstration valable en toute caractéristique).

On déduit alors de la suite exacte

$$0 \to \Omega_{\mathbf{P}}^q(k-d) \to \Omega_{\mathbf{P}}^q(k) \to \Omega_{\mathbf{P}}^q(k)_{|X} \to 0$$

que les espaces $H^i(X, \Omega_{\mathbf{P}}^q(k)_{|X})$ sont nuls pour $1 \leq i \leq n-2$, sauf $H^{q-1}(X, \Omega_{\mathbf{P}}^q(d)_{|X})$ et $H^q(X, \Omega_{\mathbf{P}}^q{}_{|X})$ qui sont de dimension 1. En considérant la suite exacte de cohomologie associée à la suite (S_i), on voit que *l'homomorphisme* $\cup\xi_i$ *est bijectif en dehors des cas suivants :* $n = 2i$ *et* $j = \sigma$ *ou* $\tau - \sigma$; $n = 2i \pm 1$ *et* $j = \sigma$ (l'injectivité de $\cup\xi_1$ provient du lemme 1 du §1).

(3) Pour $1 \leq q \leq n-1$, notons $\xi^{(q)}$ la classe d'extension $\xi_q \cdot \xi_{q-1} \cdot \dots \cdot \xi_1$; elle vit dans $\mathrm{Ext}^q(\Omega_X^{n-1}, \Omega_X^{n-q-1}(-qd))$, que l'on peut identifier à $H^q(X, \Omega_X^{n-q-1}(n+1-(q+1)d))$. Soient i, j des entiers distincts de σ, tels que $i + j = \tau$. Posons $k = i+n+1-2d = -j-n-1+(n-1)d$. Considérons le diagramme commutatif

$$\begin{array}{ccc} R_i \otimes R_j & \xrightarrow{(\cup\xi_1)\otimes(\cup\xi^{(n-2)})} & H^1(X,\Omega_X^{n-2}(k)) \otimes H^{n-2}(X,\Omega_X^1(-k)) \\ {\scriptstyle m}\downarrow & & \downarrow{\scriptstyle \mu} \\ R_\tau & \xrightarrow{\cup(\xi_1\,\xi^{(n-2)})} & H^{n-1}(X,\Omega_X^{n-1}). \end{array}$$

La flèche μ, définie par le cup-produit, induit la dualité de Serre; d'après (2), les homomorphismes $\cup\xi_1$ et $\cup\xi^{(n-2)}$ sont bijectifs. Prenant d'abord $i = 0$, on obtient que R_τ s'identifie au dual de R_0, donc est de dimension 1, et que la flèche $\cup(\xi_1 \cdot \xi^{(n-2)}) : R_\tau \to H^{n-1}(X,\Omega_X^{n-1})$ est bijective. On voit alors que *pour tout i distinct de σ ou $\tau-\sigma$, la multiplication $R_i \otimes R_{\tau-i} \to R_\tau$ définit une dualité parfaite.*

(4) Reste à considérer le cas $i = \sigma$. Supposons d'abord n pair, et posons $n = 2\nu$. L'homomorphisme $\cup\xi^{(\nu-1)} : R_\sigma \to H_\nu(X,\Omega_X^{\nu-1}(d))$ est bijectif d'après (2). On déduit de (S_ν) la suite exacte

$$H^{\nu-1}(X,\Omega_X^{\nu-1}) \xrightarrow{\alpha} H^{\nu-1}(X,\Omega_{\mathbf{P}}^\nu(d)_{|X}) \to H^{\nu-1}(X,\Omega_X^\nu(d)) \xrightarrow{\cup\xi_\nu} H^\nu(X,\Omega_X^{\nu-1}) \to 0.$$

Prouvons que la flèche α est nulle. Par dualité de Serre, sa transposée s'identifie à l'homomorphisme de restriction $H^\nu(X,\Omega^\nu_{\mathbf{P}|X}) \to H^\nu(X,\Omega_X^\nu)$. Notons $h_{\mathbf{P}}$ la classe dans $H^1(\mathbf{P},\Omega_{\mathbf{P}}^1)$ d'une section hyperplane, et $h_X \in H^1(X,\Omega_X^1)$ sa restriction à X; l'élément $h_{\mathbf{P}}^q$ engendre $H^q(\mathbf{P},\Omega_{\mathbf{P}}^q)$ pour tout q, et sa restriction h_X^q engendre $H^q(X,\Omega_X^q)$ pour $q < \nu$ [D]. Comme le degré de X est divisible par p, on a $h_X^\nu \cdot h_X^{\nu-1} = 0$ dans $H^{n-1}(X,\Omega_X^{n-1})$, d'où $h_X^\nu = 0$, ce qui prouve que la flèche α est nulle.

Par suite l'homomorphisme $\cup\xi^{(\nu)} : R_\sigma \to H^\nu(X,\Omega_X^{\nu-1})$ est surjectif, et son noyau est engendré par un élément non nul s de R_σ; d'autre part l'homomorphisme $\cup\xi^{(\nu-1)} : R_{\tau-\sigma} \to H^{\nu-1}(X,\Omega_X^\nu)$ est bijectif d'après (2). On déduit alors du diagramme

$$\begin{array}{ccc} R_\sigma\otimes R_{\tau-\sigma} & \to & H^\nu(X,\Omega_X^{\nu-1})\otimes H^{\nu-1}(X,\Omega_X^\nu) \\ \downarrow & & \downarrow \\ R_\tau & \longrightarrow & H^{n-1}(X,\Omega_X^{n-1}) \end{array}$$

qu'on a $s \cdot R_{\tau-\sigma} = 0$, et que *la multiplication $R_\sigma \otimes R_{\tau-\sigma} \to R_\tau$ induit une dualité parfaite entre R_σ/ks* et $R_{\tau-\sigma}$. Compte tenu de (3), cela entraîne les assertions (a) et (b) dans le cas où n est pair.

(5) Traitons le cas n impair : posons $n = 2\nu + 1$. D'après (2), l'homomorphisme $\cup\xi^{(\nu-1)} : R_\sigma \to H^{\nu-1}(X, \Omega_X^{\nu+1}(d))$ est bijectif. La suite (S_ν) fournit une suite exacte

$$0 \to H^{\nu-1}(X, \Omega_X^{\nu-1}(d)) \xrightarrow{\cup\xi_\nu} H^\nu(X, \Omega_X^\nu) \xrightarrow{\beta} H^\nu(X, \Omega_{\mathbf{P}}^{\nu+1}(d)_{|X}).$$

Par dualité de Serre, la transposée de β s'identifie de nouveau à l'application de restriction $H^\nu(X, \Omega^\nu_{\mathbf{P}|X}) \to H^\nu(X, \Omega_X^\nu)$. Avec les notations de (4), on conclut donc que *l'homomorphisme* $\cup\xi^{(\nu)}$ *identifie* R_σ *à l'orthogonal de* h_X^ν *dans* $H^\nu(X, \Omega_X^\nu)$. Considérons le diagramme commutatif

$$\begin{array}{ccc} R_\sigma \otimes R_\sigma & \longrightarrow & H^\nu(X, \Omega_X^\nu) \otimes H^\nu(X, \Omega_X^\nu) \\ \downarrow & & \downarrow \\ R_\tau & \longrightarrow & H^{n-1}(X, \Omega_X^{n-1}) \end{array}$$

L'élément h_X^ν de $H^\nu(X, \Omega_X^\nu)$ est non nul [D], de carré nul puisque p divise d. Si s désigne l'élément correspondant de R_σ, il en résulte que la forme bilinéaire (symétrique) $m : R_\sigma \otimes R_\sigma \to R_\tau$ a pour noyau ks, et par suite induit une dualité parfaite $(R_\sigma/ks) \otimes (R_\sigma/ks) \to R_\tau$. Compte tenu de (3), cela entraîne les assertions (a) et (b) dans ce cas.

(6) Il reste à prouver l'assertion (c). Soit W l'anneau des vecteurs de Witt sur k, et soit $\mathcal{F}(T_0, \ldots, T_n)$ un polynôme homogène de degré d dont la réduction (mod p) est égale à F. Notons $\mathcal{X}$ l'hypersurface d'équation $\mathcal{F} = 0$ dans $\mathbf{P}^n_W$, et $\mathcal{R}$ la W-algèbre $W[T_0, \ldots, T_n]/(\mathcal{F}'_{T_0}, \ldots, \mathcal{F}'_{T_n}, \mathcal{F})$ (observons que $\mathcal{X}$ est nécessairement lisse sur W).

Il résulte de ce qui précède et du lemme de Nakayama que $\mathcal{R}_i$ est nul pour $i > \tau$, et que le W-module $\mathcal{R}_\tau$ est libre de rang un. Montrons que $\mathcal{R}_i$ est sans torsion (donc libre) pour $i \neq \sigma$: dans le cas contraire il contient un élément de torsion x dont la classe $\bar{x}$ dans R_i n'est pas nulle; on a nécessairement $x \cdot \mathcal{R}_{\tau-i} = 0$ dans $\mathcal{R}_\tau$, donc $\bar{x} \cdot R_{\tau-i} = 0$, ce qui contredit (3) ou (4).

Si n est impair, on déduit de (5) que $\mathcal{R}_\sigma$ s'identifie à un sous-module du W-module $H^\nu(\mathcal{X}, \Omega^\nu)$; comme celui-ci est libre [D], il en est de même de $\mathcal{R}_\sigma$.

Enfin si n est pair, on déduit du lemme de Nakayama une suite exacte

$$0 \to \mathcal{T} \to \mathcal{R}_\sigma \to (\mathcal{R}_{\tau-\sigma})^* \to 0,$$

où $\mathcal{T}$ est un W-module monogène de torsion. Le W-module $\mathcal{R}_\sigma/\mathcal{T}$ est libre, et sa réduction (mod p) s'identifie à R_σ/ks.

L'assertion (c) résulte immédiatement de ces remarques. CQFD

APPENDICE

Ideals with a regular sequence as syzygy

DAVID EISENBUD and CRAIG HUNEKE

We sketch an alternate approach to Proposition 2, reducing it to results of Huneke and Ulrich [H-U] and Kustin [Ku] (results similar to those of Kustin were also obtained by M. Stillman). In [H-U] the authors work over a ring containing a field, but the results are general, and are done explicitly without this hypothesis in [Ku].

Assume that R is a local Nœtherian ring, that $x_1, \ldots, x_n$ is a regular sequence in R and that $f_1, \ldots, f_n$ are elements of R satisfying the relation

$$(*) \qquad x_1 f_1 + \ldots + x_n f_n = 0.$$

We further set $I = (f_1, \ldots, f_n)$ and suppose that the grade of I is $n-1$, the largest possible value.

If f is a form in $k[x_1, \ldots, x_n]$ defining a nonsingular hypersurface, and if char(k) divides the degree of f, then Euler's relation shows that these hypotheses are satisfied by the partial derivatives of f in the localization of $k[x_1, \ldots, x_n]$.

Theorem. *If grade$(I) = n-1$, then*

(i) *if n is odd, R/I is perfect of Cohen-Macaulay type* 2.

(ii) *if n is even, there exists an element $f \notin I$ such that*

$$I : (x_1, \ldots, x_n) = (I, f),$$

and $R/(I, f)$ is perfect of Cohen-Macaulay type 1.

Proof : The most interesting point is the identity of the element f : the relation (*) shows that the vector (f_i) is a linear combination of the syzygies of the x_i. Since the x_i form a regular sequence, their syzygies are given by the first map of the Koszul complex $k : \wedge^2 R^n \to \wedge^1 R^n$, so there exists a skew-symmetric matrix A such that $(f_i) = A(x_j)$. The element f is then the Pfaffian of A.

The result follows by specialization from the generic case, which is treated in [H-U], 5.8, 5.9 and 5.12, and in [Ku] . QED

Corollary. *If R is regular, $x_1, \ldots, x_n$ generate the maximal ideal, and g is an element of R such that $ht(I, g) = n$, then the socle of $R/(I, g)$ is two-dimensional.*

Proof : If n is odd, the corollary follows at once from (i). If n is even, it follows from (ii) because g must be a nonzero divisor mod(I, f).

Graded free resolutions for the generic forms of the ideals I and (I, f) as in the Theorem can be found in [Ku], Theorem 6.3. By local duality, this gives the degrees of the socle elements in the corollary (alternatively, one can use linkage, as was done in [H-U]). Applying this to the case of partial derivatives of the equation of a nonsingular hypersurface, one recovers the degree results of Beauville.

BIBLIOGRAPHIE

[D] P. Deligne, *Cohomologie des intersections complètes*, SGA 7 II, exp. 11, Lecture Notes **340**, Springer-Verlag, Berlin-Heidelberg-New York (1973).

[H-U] C. Huneke, B. Ulrich, *Divisor class groups and deformations*, Amer. J. of Math. **107** (1985), 1265-1303.

[I] L. Illusie, *Ordinarité des intersections complètes générales*, ce volume.

[K] N. Katz, *Pinceaux de Lefschetz: théorème d'existence*, SGA 7 II, exp. 17, Lecture Notes **340**, Springer-Verlag, Berlin-Heidelberg-New York (1973).

[Ku] A. Kustin, *The minimal free resolutions of the Huneke-Ulrich deviation two Gorenstein ideals*, J. of Algebra **100** (1986), 265-304.

[M] J. P. Murre, *Algebraic equivalence modulo rational equivalence on a cubic threefold*, Compositio math. **25** (1972), 161-206.

[S] R. Steinberg, *Endomorphisms of linear algebraic groups*, Mem. Amer. Math. Soc. **80** (1968).

[Z] O. Zariski, *Introduction to the problem of minimal models in the theory of algebraic surfaces*, Publ. Math. Soc. Japan 4 (1958), 1-89.

Arnaud Beauville
Université Paris-Sud
Orsay

Aomoto Dilogarithms, Mixed Hodge Structures and Motivic Cohomology of Pairs of Triangles on the Plane

A. A. BEILINSON, A. B. GONCHAROV
V. V. SCHECHTMAN, and A. N. VARCHENKO

à Alexandre Grothendieck pour son 60ᵉ anniversaire

It is known that a group of linear combinations of polytopes in $\mathbb{R}^3$ considered up to movements with respect to cutting of polytopes may be embedded into $\mathbb{R} \otimes \mathbb{R}/2\pi\mathbb{Z} \oplus \mathbb{R}$; this embedding assigns to a polytope its Dehn invariant and volume [C]. The study of motivic cohomology of a projective plane with two distinguished families of projective lines leads to an analogous problem: to describe a group of linear combinations of pairs of triangles on a plane considered up to the action of PGL(3), with respect to a cutting of any triangle of a pair. It turns out that this group is isomorphic up to 12–torsion to $B_2 \oplus S^2 B_1$, where $S^2 B_1$ is the symmetric square of the multiplicative group of a ground field, and B_2 — the Bloch group of this field. This is the first main result of the paper (see Theorems 2.12, 3.8 and 3.6.2).

The second main result is the description of the motivic cohomology of a pair of straight line configurations on a plane (2.17). Section 2 is devoted to motivic cohomology. In the first section a certain fragment of the linear algebra of mixed Hodge structures is contained and serves as the motivation for the constructions of Section 2. Many proofs in this paper are omitted or sketched. The complete proofs will be published later.

We are grateful to Don Zagier and Spencer Bloch for useful discussions. The first and third authors wish to thank the Mathematics Department of M. I. T. where this work was finished for its hospitality, and especially Anne Richard for the careful typing of this manuscript.

The idea of motives belongs to Alexandre Grothendieck (cf. [Ma]). It is a pleasure for us to dedicate our paper to this great mathematician.

Notation. If A is an abelian group then $A_{\mathbb{Q}}$ denotes $A \otimes_{\mathbb{Z}} \mathbb{Q}$.

1. Aomoto Dilogarithms and the Hodge-Tate Algebra

1.1. *Aomoto Dilogarithms.* Call a *triangle* on a complex projective plane $\mathbf{P}^2(\mathbf{C})$ a triple of straight lines $L = (L_0, L_1, L_2)$, its *vertices*– points $L_{ij} = L_i \cap L_j$. With a triangle, the differential form $\omega_L = d\log(z_1/z_0) \wedge d\log(z_2/z_0)$ is connected where $z_i = 0$ is a homogeneous equation of L_i. With a pair of triangles in sufficiently general position, an integral

$$a(L, M) = \int_{\Delta_M} \omega_L \tag{1.1.1}$$

is connected, where an oriented two-chain Δ_M is a curved triangle with sides on lines $\{M_j\}$, and vertices in $\{M_{ij}\}$, which does not intersect L and is oriented as in Figure 1.1. The integral depends on the choice of Δ_M,

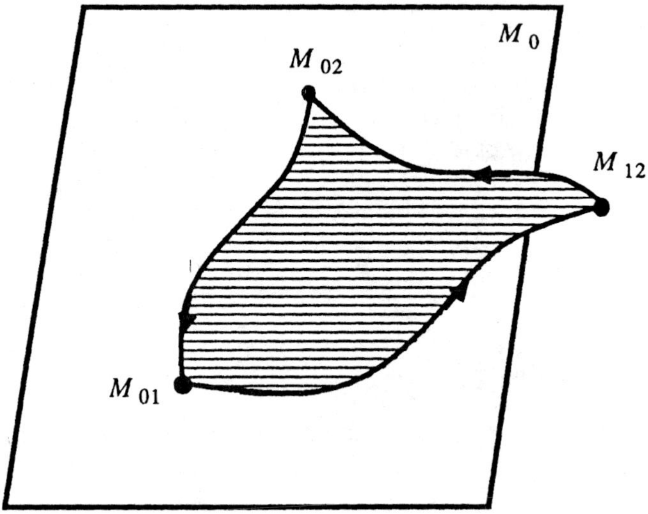

Figure 1.1. Δ_M.

but does not change under its continuous deformation. We call it an *Aomoto dilogarithm* (cf.[A]). It has the following properties.

1.1.2. *Skew symmetry.* After a renumbering of the family L or M an integral is multiplied by the sign of the corresponding permutation.

1.1.3. *Additivity.*

(a) *Additivity with respect to a chain.* Suppose that a straight line M_3 goes through M_{12}. Put $M' = (M_0, M_1, M_3)$, $M'' = (M_0, M_3, M_2)$. Then

$$a(L,M) = a(L,M') + a(L,M''),$$

if $\Delta_M = \Delta_{M'} + \Delta_{M''}$, see Figure 1.2.

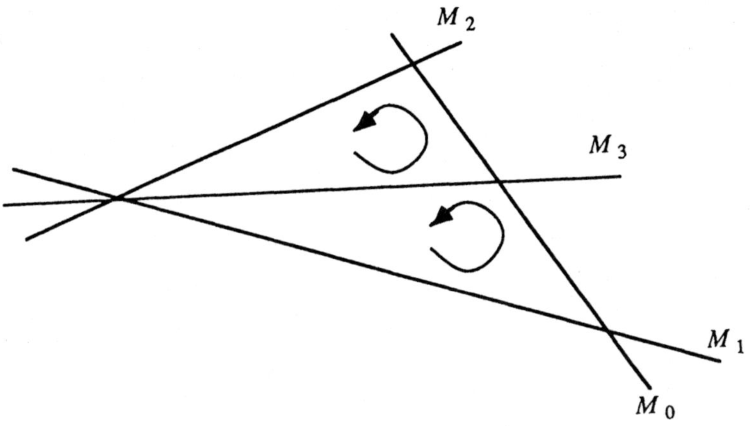

Figure 1.2.

(b) *Additivity with respect to a differential form.* If a straight line L_3 goes through L_{12}, then

$$a(L,M) = a(L',M) + a(L'',M),$$

(in the notations analogous to (a)), since $\omega_L = \omega_{L'} + \omega_{L''}$.

1.1.4. *Projective invariance.* For every $g \in \mathrm{PGL}(3,\mathbb{C})$

$$a(gL, gM) = a(L,M),$$

if $\Delta_{gM} = g\Delta_M$.

Aomoto polylogarithm of order n $a_n(L,M)$ is defined in an analogous way. It is associated with two families of hyperplanes("n-simplices") in

$\mathbf{P}^n(\mathbf{C})$: $L = (L_0, L_1, \dots, L_n)$ and $M = (M_0, \dots, M_n)$, and is equal to an integral of the n-form ω_L over an n-simplex Δ_M, which are defined analogously to the case $n = 2$. Aomoto polylogarithms have the properties analogous to (1.1.2)-(1.1.4).

For example, an Aomoto polylogarithm of order 1 is defined by two pairs of points $L = (L_0, L_1), M = (M_0, M_1)$ on $\mathbf{P}^1$, and is equal to an integral of $\omega_L = d\log(z_1/z_0)$ over a path going from M_1 to M_0. Let $r(L_0, L_1, M_0, M_1) \in \mathbf{C}^*$ be a cross ratio, i.e., the number to which M_0 goes under the action of the projective transformation which takes (L_0, L_1, M_1) to $(\infty, 0, 1)$. Then $a_1(L, M) = \log r(L_0, L_1, M_0, M_1)$.

Aomoto polylogarithms of different orders are connected with each other: a product of Aomoto polylogarithms of orders p and q may be expressed as a sum of Aomoto polylogarithms of order $p + q$. For example, consider a configuration of straight lines on $\mathbf{P}^2$ showed in Figure 1.3. Then after a suitable choice of 2-chains,

$$a_1(\ell, m)a_1(\ell', m') = a_2(L; M_0', M_1, M_2) + a_2(L; M, M_1', M_0)$$

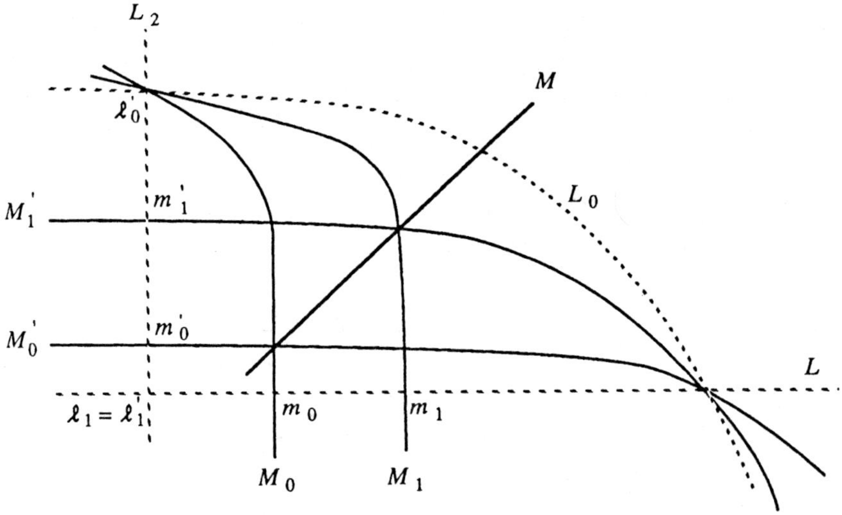

Figure 1.3. Product of Aomoto 1-logarithms.

1.2. *The Classical Euler Dilogarithm* [L] is defined for $t \in \mathbf{R}$, $0 < t < 1$, by a formula

$$Li_2(t) = \sum_{n=0}^{\infty} t^n/n^2 \tag{1.2.1}$$

The Rogers dilogarithm [R], is

$$L_2(t) = Li_2(t) + \frac{1}{2}\log(t)\log(I - t) \tag{1.2.2}$$

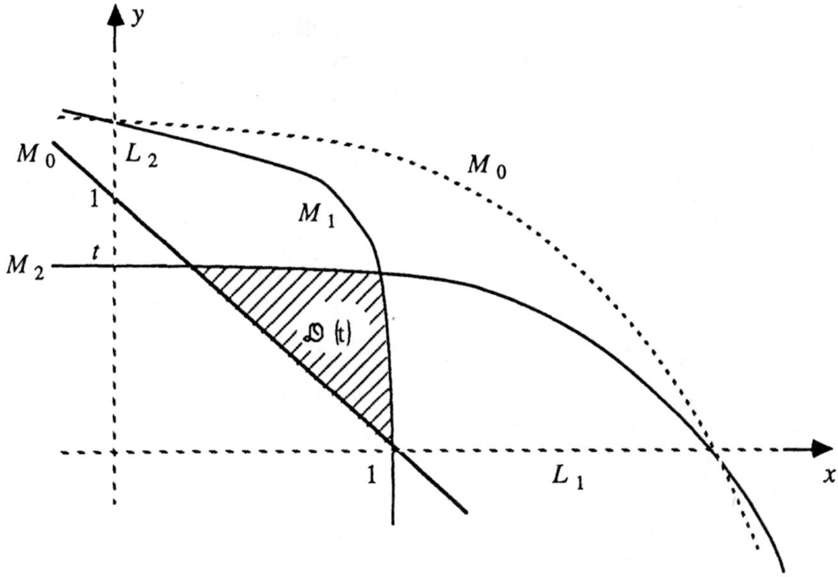

Figure 1.4. The Euler dilogarithm.

The functions Li_2, L_2 may be continued to a multivalued analytic function on $\mathbf{P}^1 - \{0, 1, \infty\}$. The following remarkable functional equation holds: for every 5-tuple of different points $\{z_0, z_1, z_2, z_3, z_4\} \subset \mathbf{P}^1(\mathbf{C})$ we have

$$\sum_{i=0}^{4}(-1)^i L_2(r_i) = \frac{\pi^2}{6} = -\frac{(2\pi i)^2}{24} \tag{1.2.3}$$

where $r_i = r(z_0, \ldots, \hat{z}_i, \ldots, z_4)$, $\frac{\pi^2}{6} = Li_2(1)$ (cf. [L] 1.27).*

---

*Precisely, one has $L_2(x) - L_2(y) + L_2(\frac{y}{x}) - L_2\frac{(1-x^{-1})}{(1-y^{-1})} + L_2\frac{(1-x)}{(1-y)} = \frac{\pi^2}{6}$ for $0 < y < x < 1$.

The Euler dilogarithm is a particular case of the Aomoto dilogarithm. Namely, consider a pair of triangles shown in Figure 1.4. Then

$$Li_2(t) = a_2(L; M) = \int_{\mathcal{D}(t)} \frac{dx}{x} \wedge \frac{dy}{y} \tag{1.2.4}$$

1.3. *Mixed Hodge Structures.* Consider the second complex cohomology group of a pair $(\mathbf{P}^2(\mathbf{C})\backslash L, M\backslash M \cap L)$. The space has a canonical mixed Hodge structure [D], which we denote as $H(L, M)$. The weights of $H(L, M)$ are 0,2,4 and the corresponding graded quotients are isomorphic to $\mathbf{Z}(0), \mathbf{Z}(-1)^k, \mathbf{Z}(-2)$, where $\mathbf{Z}(j)$ is the Tate structure of weight $-2j$, $k = 0, 1, 2, 3$, or 4 (it depends on the position of triangles). $H(L, M)$ has the following *frame*: in $\mathrm{gr}_4^W H$ a vector is distinguished – the class of the form $\omega_L/(2\pi i)^2$, and in $\mathrm{gr}_0^W H$ a covector is distinguished – the one that comes from the class of the chain Δ_M.

Consider a mixed Hodge structure with weight quotients isomorphic to $\mathbf{Z}(0), \mathbf{Z}(-1)^k, \mathbf{Z}(-2)$. We say that H is *framed* if the isomorphisms $\mathbf{Z}(-1) \xrightarrow[\sim]{} \mathrm{gr}_2^W(H), \mathrm{gr}_0^W(H) \xrightarrow[\sim]{} \mathbf{Z}(0)$ are fixed.

Consider the set of all framed mixed Hodge structures with weight quotients $\mathbf{Z}(0), \mathbf{Z}(-1)^k \cdot \mathbf{Z}(-2)$. Introduce on this set the equivalence relation: namely, the coarsest one, for which H_1 is equivalent to H_2 if there is a morphism of mixed Hodge structures $H_1 \to H_2$ compatible with frames. Denote by $\mathcal{H}_2$ the set of equivalence classes.

One may introduce on $\mathcal{H}_2$ a structure of an abelian group. The construction is analogous to the definition of the Baer sum on Ext groups. Namely, for $[H_1], [H_2] \in \mathcal{H}_2$ let $H \subset H_1 \oplus H_2$ be the sub-Hodge structure consisting of vectors projecting to the diagonal defined by the frames in $\mathrm{gr}_4^W(H_1 \oplus H_2)$.

The frames in $\mathrm{gr}_0^W H_i$ define the maps $W_0 H = W_0 H_1 \oplus W_0 H_2 \to \mathbf{Z}(0) \oplus \mathbf{Z}(0)$. Consider the composition of this map with the sum map $\mathbf{Z}(0) \oplus \mathbf{Z}(0) \to \mathbf{Z}(0)$. Let ker be the kernel of this composition. Define $[H_1]+[H_2]$ to be the class of the structure $H/\ker$. A frame in $W_0(H/\ker)$ will be induced by the above composition, and in $\mathrm{gr}_4^W(H/\ker)$ will be the sum of frames in $\mathrm{gr}_0^W H_i$.

Let $-[H]$ be the class of H in which the frame in $\mathrm{gr}_4^W H$ is multiplied by -1. (Note that if we multiply by -1 the frame covector on $\mathrm{gr}_0^W H$, we get the structure lying in the same class.)

It turns out that classes $[H(L, M)] \in \mathcal{H}_2$ have the properties analogous to (1.1.2)-(1.1.4).

1.3.1. *Skew symmetry*, with respect to a renumbering of a family L or M.

1.3.2. *Additivity.*

$$[H(L,M)] = [H(L,M')] + [H(L,M'')];$$
$$[H(L,M)] = [H(L',M)] + [H(L'',M)].$$

1.3.3. *Projective invariance.*

$$[H(L,M)] = [H(gL,gM)].$$

Here we used the notations of 1.1.

In fact, skew-symmetry and projective invariance are obvious. Let us prove the additivity with respect to L. Consider the mixed Hodge structure H_1 in the second cohomology of a pair $(\mathbf{P}^2(\mathbf{C}) \setminus \bigcup_{j=0}^{3} L_j, M \setminus (M \cap \bigcup_{j=0}^{3} L_j))$. One has restriction maps

$$\pi : H(L';M) \oplus H(L'';M) \to H_1; \rho : H(L,M) \to H_1.$$

Consider the framed mixed Hodge structure $H/\ker$ constructed above, which represents the sum $[H(L';M)] + H(L'';M)]$ in $\mathcal{H}_2$. It is easy to see that π induces the map $\sigma : M/\ker \to H_1$, and the images of frame vectors in $\mathrm{gr}_4^W H/\ker$ and $\mathrm{gr}_4^W H(L;M)$ in $\mathrm{gr}_4^W H_1$ coincide.

On the other hand there exists a covector on $\mathrm{gr}_0^W H_1$ which goes to the frame covectors in $\mathrm{gr}_0^W(H/\ker)$ and $\mathrm{gr}_0^W H_1$. This easily implies that $[H/\ker] = [H(L;M)]$. Additivity with respect to M is similarly proved.

Let $L = (L_0, L_1, \dots, L_n)$ and $M(M_0, M_1, \dots, M_n)$ be two families of hyperplanes in general position in $\mathbf{P}^n(\mathbf{C})$. Denote by $H(L;M)$ the mixed Hodge structure in the n-th cohomology of the pair $(\mathbf{P}^n(\mathbf{C}) - \bigcup_i L_i, M - (M \cap U_i L_i))$. Its weight quotients are isomorphic to $\mathbf{Z}(0), \mathbf{Z}(-1)^{k_1}, \dots, \mathbf{Z}(-n+1)^{k_{n-1}}, \mathbf{Z}(-n)$. Moreover, $H(L,M)$ has canonical frames in $(\mathrm{gr}_0^W)^*$ and gr_{2n}^W. By analogy with $\mathcal{H}_2$, construct the group of equivalence classes of framed mixed Hodge structures of the above kind. It turns out that classes $H(L,M)$ in $\mathcal{H}_n$ has properties of skew-symmetry, additivity and projective invariance, analogous to (1.3.1)-(1.3.3).

The group $\mathcal{H}_1$ is isomorphic to $\mathbf{C}^*$. In fact, let H be a representative form $[H] \in \mathcal{H}_1$. H is determined by a position of a one-dimensional Hodge subspace F^1 with respect to a one-dimensional weight subspace W_0 in the two-dimensional H. Let $e_1 \in W_0$ be the vector dual to the frame covector, $e_2 \in H$ — an integer vector projecting to the frame in $\mathrm{gr}_2^W H$. e_2 is defined up to the addition of $ke_1, k \in \mathbf{Z}$. Let $v = \lambda e_1 + e_2$ be a vector from F^1. Then the number

$$s(H) = \exp(2\pi i\lambda)$$

does not depend on the choice of e_2. This defines the isomorphism $s : \mathcal{H}_1 \xrightarrow{\sim} \mathbf{C}^*$. We have

$$s(H(L,M)) = \exp(a_1(L,M)).$$

Similarly, the Aomoto dilogarithm is one of the elements of the period matrix of a pair of triangles on the plane.

1.3.4. *Variants of the definition of groups $\mathcal{H}_n$.* Let us call a *Hodge-Tate structure* a mixed Hodge structure without a torsion with weight factors isomorphic to $\mathbf{Z}(i)^k$. Say that H is *n-framed* if it is supplied with two maps $\mathbf{Z}(-n) \to H$ and $H \to \mathbf{Z}(0)$. Introduce an equivalence relation on the set of all n-framed Hodge-Tate structures: namely, the coarsest one, for which $H_1 \sim H_2$ if there is a map $H_1 \to H_2$ compatible with frames. Let $\mathcal{H}'_n$ be the set of equivalence classes. Similarly to 1.3, one may introduce on $\mathcal{H}'_n$ a structure of an abelian group.

Define a map $f : \mathcal{H}'_n \to \mathcal{H}_n$. Let H be a framed Hodge-Tate structure. Consider the Hodge-Tate structure $H_1 = W_{2n}H/W_{-1}H$. The frame of H defines maps $\mathbf{Z}(-n) \to \mathrm{gr}^W_{2n} H_1$, $\mathrm{gr}^W_0 H_1 \to \mathbf{Z}(0)$. As in 1.3, (by pull back and push forward), these maps allow to construct from H_1 a framed Hodge-Tate structure $\tilde{H}_1$ with one-dimensional weight quotients $\mathrm{gr}^W_0 \tilde{H}_1, \mathrm{gr}^W_{2n} \tilde{H}_1$. Put $f[H] = [\tilde{H}_1]$. One easily shows that f is an isomorphism.

Certainly, we may consider $\mathbf{Q}$–Hodge structures instead of $\mathbf{Z}$ ones, the corresponding group is just $\mathcal{H}_{\mathbf{Q}} = \mathcal{H} \otimes \mathbf{Q}$. It has also the following description. Call a n-framed $\mathbf{Q}$–Hodge-Tate structure minimal if it has no non-trivial subquotient compatible with framing. It is easy to see that a non-zero minimal n-framed H has $\mathrm{gr}^W_i(H) = 0$ for $i < 0$ and $i > 2n$, and the frame defines isomorphisms $\mathrm{gr}^W_0(H) = \mathbf{Q}(0)$, $\mathrm{gr}^W_{2n}(H) = \mathbf{Q}(-n)$.

Let $\mathcal{H}''_{\mathbf{Q}n}$ be the set of isomorphism classes of n- minimal framed Hodge-Tate structures. One has the evident map $g : \mathcal{H}''_{\mathbf{Q}n} \to \mathcal{H}'_{\mathbf{Q}n}$. One can show, that every n-framed Hodge-Tate structure has the unique minimal subquotient. This implies that g is an isomorphism.

1.4. *Multiplication and comultiplication of framed mixed Hodge structures.* Different groups $\mathcal{H}_n$ are connected with each other by the following maps of multiplication and comultiplication.

The multiplication

$$\mu : \mathcal{H}_k \otimes \mathcal{H}_\ell \to \mathcal{H}_{k+\ell}$$

is induced by the tensor product of Hodge structures. It is commutative.

Define the comultiplication

$$\nu = \bigoplus_{k+\ell=n} \nu_{k\ell} : \mathcal{H}_n \to \bigoplus_{k+\ell=n} \mathcal{H}_k \otimes \mathcal{H}_\ell .$$

Let $[H] \in \mathcal{H}_n$, $R \subset \mathrm{gr}^W_{2k} H$ be a lattice. Define a homomorphism $\varphi : R \to \mathcal{H}_k$. Namely, for $x \in R$, $\varphi(x)$ is the class of the sub-Hodge structure of H, consisting of vectors in $W_{2k}H$ whose projection is $\mathrm{gr}^W_{2k} H$ and is proportional to x. As a frame in the upper weight factor we choose x.

Dually, let R^* be an integer lattice in the dual space $(\mathrm{gr}^W_{2k} H)^*$. Define the homomorphism $\psi : R^* \to \mathcal{H}_{n-k}$. Namely, for $x \in R^*$ $\psi(x)$ is the class of factor-structure $(H/W_{2k-1}H)/\ker x$; as a frame in the lower weight factor we choose the covector x.

We have $R \otimes R^* \cong \mathrm{End}\, R$; let $1 \in R \otimes R^*$ be the identity operator. Put

$$\nu_{k,n-k}([H]) = (\varphi \otimes \psi)\ (1).$$

In other words, if $\{e_j\}$, $\{e^j\}$ are dual bases in R, R^*, then

$$\nu_{k,n-k}([H]) = \sum_j \varphi(e_j) \otimes \psi(e^j). \tag{1.4.1}$$

μ and ν are compatible:

$$\nu(a \cdot b) = \nu(a) \cdot \nu(b), \tag{1.4.2}$$

where $a \cdot b = \mu(a,b)$. This follows without difficulty from definitions.

Consider the graded abelian group $\mathcal{H} = \bigoplus_{n=0}^{\infty} \mathcal{H}_n$, where $\mathcal{H}_0 = \mathbf{Z}$. By the above, $\mathcal{H}$ has a structure of a graded Hopf algebra with the commutative multiplication μ and the comultiplication ν. (For the definition and properties of Hopf algebras (see [Q1]).)

We will call $\mathcal{H}$ a Hodge-Tate algebra; let $\mathcal{H}_{\mathbb{Q}} = \mathcal{H} \otimes \mathbb{Q}$ be the corresponding Q-algebra.

Lemma 1.4.3. *The Hodge-Tate algebra $\mathcal{H}_{\mathbb{Q}}$ is a cofree graded commutative Hopf algebra with the space of generators in degree n $(n > 0)$ equal to $\mathbb{C}/(2\pi i)^n\mathbb{Q}$.*

The lemma follows from 1.6 (next page) plus the vanishing of Ext^2 for mixed Hodge structures [B]. (One has $\mathbb{C}/(2\pi i)^n\mathbb{Q} = \mathrm{Ext}^1(\mathbb{Q}(0), \mathbb{Q}(n))$, $n > 0$.) The dual space $\mathcal{H}^*_{\mathbb{Q}}$ is a cocommutative Hopf algebra (in a topological sense); it is a universal enveloping algebra of a Lie algebra L – a free pronilpotent graded Lie algebra $L_{\mathcal{H}}$ with generators in degree $-n$ equal to $(\mathbb{C}/(2\pi i)^n\mathbb{Q})^* = \mathrm{Hom}_{\mathbb{Q}}(\mathbb{C}/(2\pi i)^n\mathbb{Q},\ \mathbb{Q})$. Note that there is no canonical way to embed $L_{\mathcal{H}}/[L_{\mathcal{H}}, L_{\mathcal{H}}]$ into $L_{\mathcal{H}}$.

1.5. *Poincaré duality.* Let $[H] \in \mathcal{H}_n$, H^* be the mixed Hodge structure dual to H. The frame in H induces the one in H^*; we have $[H^* \otimes \mathbf{Z}(-n)] \in \mathcal{H}_n$. Define the involution $i : \mathcal{H}_n \to \mathcal{H}_n$ by a formula $i[H] = [H^* \otimes \mathbf{Z}(-n)]$.

Theorem 1.5.1. (a). *i is an antipode, i.e., is an algebra automorphism. $(i(x \cdot y) = i(x) \cdot i(y))$ and i is a coalgebra anti-automorphism $(\nu(i(x)) = t(i(\nu(x)))$ where*

$$t(a \otimes b) = b \otimes a).$$

(b). *$i([H(L,M)]) = (-1)^n [H(M,L)]$, where $[H(L,M)]$ is the class of the mixed Hodge associated to a pair of families of hyperplanes in $\mathbf{P}^n(\mathbf{C})$ (see 1.3).*

(a) follows easily from the definitions and (b) from the Poincaré duality.

i induces the involution on $\mathcal{H}^*_{\mathbf{Q}}$ (to be denoted by the same letter) which is a coalgebra automorphism and an algebra anti-automorphism.

Corollary 1.5.2. *On primitive elements of $\mathcal{H}^*_{\mathbf{Q}}$ i acts as the multiplication by -1.*

Corollary 1.5.3. *For every $x \in \mathcal{H}_{\mathbf{Q},2}$*

$$i(x) = -x + \mu\nu_{11}(x).$$

1.6. *Mixed Hodge structures and comodules over Hodge-Tate algebra.*

Lemma. *The category of $\mathbf{Q}$-mixed Hodge-Tate structures is canonically equivalent to the category of finite dimensional graded $\mathcal{H}_{\mathbf{Q}}$-comodules.*

Namely, the equivalence assigns to a Hodge structure H the graded comodule $M(H)$, $M(H)_n = \mathrm{gr}^W_{2n}(H)$ with $\mathcal{H}_{\mathbf{Q}}$- action $M(H) \otimes M(H)^* \to \mathcal{H}_{\mathbf{Q}}$ given by the formula $x_m \otimes y_n \mapsto$ class of Hodge structure H framed by x_m, y_n.

For details see [BMS].

1.7. *Motivic cohomology of a pair of straight line configurations on the plane.* Recently some attempts were made to construct the motivic cohomology of algebraic varieties which has to be the arithmetico-algebraic analogue of the singular cohomology (cf. [BMS]). In the present paper we propose the definition of the motivic cohomology of a pair $(\mathbf{P}^2 \setminus L, M \setminus (L \cap M))$ where L, M are straight line configurations on the projective plane.

The construction of motivic cohomology follows the example of mixed Hodge structures. Namely, as was shown, to give a mixed Hodge structure with weight quotients of a certain kind is the same as to give a graded comodule over the Hodge-Tate algebra. Similarly, we define the motivic cohomology of a pair $(\mathbf{P}^2 \setminus L, L \setminus (L \cap M))$ as a graded comodule $H^*_{\mathcal{M}}(L,M) = H^*_{\mathcal{M}}(L,M)_0 \oplus H^*_{\mathcal{M}}(L,M)_1 \oplus H^*_{\mathcal{M}}(L,M)_2$ over a suitable graded (commutative) Hopf algebra $A_{\mathbf{Q}}$.

Groups A_n are defined in 2.1, proceeding from the properties of skew symmetry, additivity and projective invariance (1.3.1)-(1.3.3). From our definitions it follows easily that A_1 is isomorphic to the multiplicative group of a ground field. The main result of the paper is the description of A_2.

In a module $H^*_{\mathcal{M}}(L,M)$ operators from $A^*_{k,\mathbf{Q}}$ decrease the grading by k, hence for its description it is sufficient to know only the components A_0, A_1, A_2, and the operations of multiplication and comultiplication connecting them. This is what is done in this paper (see 2.2, 2.10 and 2.12). General definitions of multiplication and comultiplication will be given in the next publication. The module $H^*_{\mathcal{M}}(L,M)$ is described in 2.17.

2. Algebra of pairs of simplices.

2.1. *The groups A_n.* Consider the n–dimensional projective space $\mathbf{P}^n = \mathbf{P}^n(k)$ over a field k. Call *an n–simplex* a family of $n+1$ hyperplanes $L = (L_0, \dots, L_n)$. Call an n–simplex *non-degenerate* if the hyperplanes are in general position. Call *a face* of an n–simplex any non-empty intersection of hyperplanes from L. Call a pair of n–simplices (L,M) *admissible*, if L and M have no common faces.

Define the group $A_n = A_n(k)$ as the group with generators $(L;M)$, where (L,M) runs through all admissible pairs of simplices, and the following relations.

2.1.1. If one of the simplices L or M is degenerate, then $(L;M) = 0$.

2.1.2. *Skew symmetry.* For every permutation $\sigma : \{0,1,\dots,n\} \to \{0,1,\dots,n\}$

$$(\sigma L; M) = (L; \sigma M) = (-1)^{|\sigma|}(L;M),$$

where $\sigma L = (L_{\sigma(0)}, \dots, L_{\sigma(n)})$, $\mid \sigma \mid$ is the parity of σ.

2.1.3. *Additivity. Additivity in L.* For every family of hyperplanes $(L_0, \dots, L_{n+1})$ and an n–simplex M

$$\sum_{j=0}^{n+1} (-1)^j \ (\hat{L}^j; M) = 0,$$

where $\hat{L}^j = (L_0, \dots, \hat{L}_j, \dots, L_{n+1})$.

Additivity in M. For every family $(M_0, \ldots, M_{n+1})$ and an n-simplex L

$$\sum_{j=0}^{n+1}(-1)^j \ (L; \hat{M}^j) = 0.$$

2.1.4. *Projective invariance*. For every $g \in PGL_{n+1}(k)$

$$(gL; gM) = (L, M).$$

Put $A_0 = \mathbf{Z}$.

Remark 2.1.5. Let $k = \mathbf{C}$. Associate with every admissible pair (L, M) the mixed Hodge structure $H(L; M)$. According to the properties (1.3.1)-(1.3.3) this defines the homomorphism $H : A_n \to \mathcal{H}_n$.

One may consider the group A_n as a formal analogue of $\mathcal{H}_n$, and homomorphisms $A_n \to \mathcal{H}_n$ as formal analogues of Aomoto polylogarithms.

Define the canonical isomorphism $r : A_1 \xrightarrow{\sim} k^*$, where k^* is the multiplicative group of k. Namely, associate with a quadruple $(L_0, L_1; M_0, M_1)$ of different points on $\mathbf{P}^1$ their cross-ratio $r(L_0, L_1; M_0, M_1) \in k^*$ (see (1.1)). The zero of A_1 send to $1 \in k^*$. It is easy to see that the skew-symmetry relation is sent to the relation

$$r(L_1, L_0; M_0, M_1) = r(L_0, L_1; M_1, M_0) = r(L_0, L_1; M_0, M_1)^{-1},$$

and the additivity relations – to the relations

$$\begin{aligned} r(L_0, L_1; M_0, M_1) r(L_1, L_2; M_0, M_1) &= r(L_0, L_2; M_0, M_1), \\ r(L_0, L_1; M_0, M_1) r(L_0, L_1; M_1, M_2) &= r(L_0, L_1; M_0, M_2). \end{aligned}$$

2.1.6. *Poincaré involution*. Let (L, M) be an admissible pair of n-simplices in $\mathbf{P}^n$. Put $P(L, M) = (M, L)$. It is easy to see that by this formula the involution $P_n : A_n \to A_n$ is correctly defined. This involution is formal analogue of Poincaré duality (see 1.5).

It follows from properties of the cross-ratio, that $P_1 : A_1 \to A_1$ is the identity. For the description of $P_2 : A_2 \to A_2$, see 1.3.

2.2. *Multiplication* $\mu : A_1 \otimes A_1 \to A_2$. Let $(\ell_0, \ell_1; m_0, m_1)$, $(\ell'_0, \ell'_1; m'_0, m'_1)$ be two quadruples of points on $\mathbf{P}^1$, such that $(\ell_0 \neq \ell_1,\ \ell'_0 \neq \ell'_1$. Consider the configuration of straight lines in $\mathbf{P}^2$ shown in Figure 2.1.

In this configuration the quadruple of lines $(L_0, L_2; M_0, M_1)$ cuts the quadruple of points $(\ell; m)$ on L_1, and $(L_0, L_1; M'_0, M'_1)$ cuts (ℓ', m') on L_2.

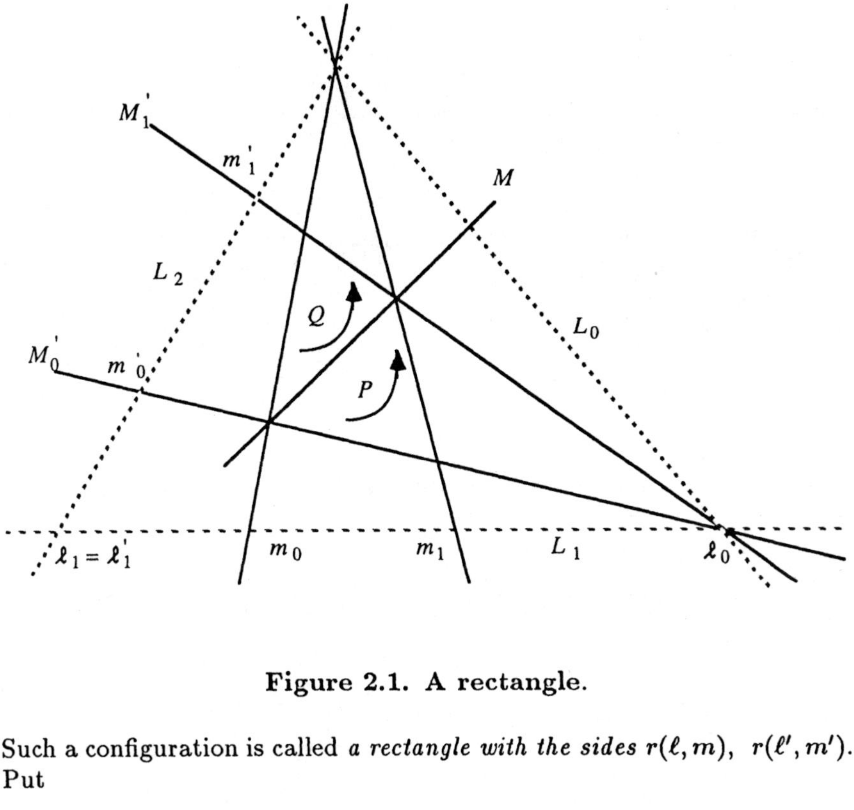

Figure 2.1. A rectangle.

Such a configuration is called *a rectangle with the sides* $r(\ell, m)$, $r(\ell', m')$. Put

$$\mu = ((\ell; m) \otimes (\ell'; m')) = (L; P) + (L; Q) \tag{2.2.1}$$

where $L = (L_0, L_1, L_2)$, $P = (M_0', M_1, M_2)$, $Q = (M, M_1', M_0)$ (cf. Figure 1.3).

Proposition 2.2.2. (a) *The previous construction correctly defines the homomorphism* $\mu : A_1 \otimes A_1 \to A_2$. *We call it multiplication.*

(b) *The multiplication is commutative, i.e.,* $\mu(a \otimes b) = \mu(b \otimes a)$.

We shall also denote $\mu(a \otimes b)$ simply by $a \cdot b$.

Remark. The homomorphism $H : \mathcal{H}_n \to A_n(\mathbb{C})$ defined in 2.1.5 is compatible with the multiplication:

$$[H(L;P)] + [H(L;Q)] = [H(\ell;m)] \cdot [H(\ell';m')] \tag{2.2.3}$$

This follows from the Künneth formula. (2.2.3) is the motivation of the definition of the multiplication.

The higher-dimensional construction analogous to Figure 2.1 is the following. Let $(L';M') \subset \mathbf{P}^{n'}$, $(L'',M'') \subset \mathbf{P}^{n''}$ be two admissible pairs of non-degenerate simplices, $L = (L_0, L_1, \ldots, L_n)$ a non-degenerate simplex in $\mathbf{P}^n$, $n = n' + n''$. Identify $\mathbf{P}^n \backslash L_0$ with $(\mathbf{P}^{n'} \backslash L_0') \times (\mathbf{P}^{n''} \backslash L_0'')$. Then the simplices M', M'' define the prism Π in $\mathbf{P}^n$. A cutting of Π into simplices, $\Pi = \cup \Delta_j$ defines the element $\Sigma(L;\Delta_j) \in A_n$. In such a way the multiplication $A_{n'} \otimes A_{n''} \to A_n$ is defined.

2.3. *The comodule of a pair of configurations.* In the next three n°'s we associate with a pair of configurations of straight lines (L,M) on a plane the graded group $G(L,M) = G_0(L;M) \oplus G_1(L,M) \oplus G_2(L;M)$ and construct homomorphisms

$$\varphi_{ij}(L,M) : G_i(L,M)^*_{\mathbb{Q}} \otimes G_j(L,M)_{\mathbb{Q}} \to A_{j-i,\mathbb{Q}},$$

$0 \leq i \leq j \leq 2$.

At first, for a pair (L,M) the graded complex $C^*(L,M) = \oplus C^*(L,M)_{2j}$ will be defined. This complex is determined by the combinatorics of the positions of straight lines. The complex corresponding to a pair (L,M) is canonically dual to the one corresponding to a pair (M,L). The zero-th cohomology has the grading $0,2,4$: $H^0C^*(L,M) = G_0(L,M) \oplus G_1(L,M) \oplus G_2(L,M)$, where $G_i(L,M)$ has the grading $2i$. Groups G_i are formal analogues of weight factors $\mathrm{gr}^W_{2i} H(L,M)$ of the mixed Hodge structure $H(L,M)$.

Groups G_0 and G_2 are free and satisfy the relations

$$G_i(L,M)^* \cong G_{2-i}(M,L), \quad i = 0,2. \tag{2.3.1}$$

We do not know whether in general $G_1(L,M)$ is free. At any rate, after tensoring by $\mathbb{Q}$, one has

$$G_1(L,M)^*_{\mathbb{Q}} \cong G_1(M,L)_{\mathbb{Q}}. \tag{2.3.2}$$

A vector x in $G_2(L,M)$ is represented by a linear combination $\Sigma n_i \Delta_i$ of triangles from L, and a covector y on $G_0(L,M)$ — by a linear combination $\Sigma m_j \Lambda_j$ of triangles from M. Hence to a pair (x,y) corresponds the element $\Sigma_{i,j}\; n_i m_j (\Delta_i, \Lambda_j)$ from A_2. This gives the map φ_{02}. The map

φ_{01}, φ_{12} are generalizations of the cross-ratio. Maps φ_{ii} are canonical pairings of a vector and a covector.

2.4. *Complex* $C^*(L;M)$. Let $L = (L_0, \dots, L_p)$, $M = (M_0, \dots, M_q)$ be a pair of straight line configurations in $\mathbf{P}^2$. Assume that configurations L, M have no common lines or common vertices.

Call a point x in $\mathbf{P}^2$ *an* L*−vertex* if at least two different lines from L intersect at x; *an* M*−vertex* if at least two different lines from M intersect at x, and *an* LM*−vertex* if at x intersect at least one line from L and at least one line from M. A one point may have several types. Call *an* L*−flag* a pair (ℓ, L_i), where ℓ is an L−vertex which lies on L_i. M−flags are defined analogously.

Let

$$C^*(L;M) = C^*(L;M)_0 \oplus C^*(L,M)_2 \oplus C^*(L,M)_4,$$

where

$$C^*(L;M)_0 : 0 \to C_0^{-2} \xrightarrow{d_0^{-2}} C_0^{-1} \xrightarrow{d_0^{-1}} C_0^0 \to 0$$

$$C^*(L;M)_2 : 0 \to C_2^{-1} \xrightarrow{d_2^{-1}} C_2^0 \xrightarrow{D_2^0} C_2^1 \to 0$$

$$C^*(L;M)_4 : 0 \to C_4^0 \xrightarrow{d_4^0} C_4^1 \xrightarrow{d_4^1} C_4^2 \to 0$$

By definition, $C_i^j = C^j(L;M)_i$ for $(i,j) \neq (0,0)$ and $(4,0)$ is a free abelian group with the following sets of generators:

$$C_0^{-2} : \{(\mathbf{P}^2)\},\ C_0^{-1} : \{(M_i) \mid i = 0, \dots, q\};$$
$$C_2^{-1} : \{(L_i) \mid i = 0, \dots, P\},$$
$$C_2^{-1} : \{(\mathbf{P}^2),\ (N) \mid N \text{ runs over all } LM - \text{vertices}\};$$
$$C_4^1 : \{(L_i) \mid i = 0, \dots, P\},\ C_4^2 - \{(\mathbf{P}^2)\}$$
$$C_2^1 : \{(M_i) \mid i = 0, \dots, q\}.$$

Let us define C_0^0 and C_4^0. Let D_L be the free abelian group with generators $\{(\ell, L_i)\}$,where (ℓ, L_i) runs over all L−flags. By definition, $C_4^0 \subset D_L$ is the subgroup consisting of linear combinations $\sum a_{(\ell,L_i)}(\ell, L_i)$ such that for every L−vertex $\ell \sum_{\ell \in L_i} a_{(L_i,\ell)} = 0$. Analogously, let D_M be the free abelian group with generators $\{(m, M_j)\}$ where (m, M_j) runs over all M−flags. For an M−vertex m put $x_m = \sum_{m \in M_j} (m, M_j) \in D_M$. Let $B_M \subset D_M$ be the subgroup generated by all x_m. By definition, $C_0^0 = D_M/B_M$.

Define differentials:

$$d_0^{-2}((\mathbf{P}^2)) = \sum_{i=0}^{q}(M_i),\ d_0^{-1}((M_i)) = \sum_{m\in M_i}(m, M_i);$$
$$d_2^{-1}((L_i)) = (\mathbf{P}^2) + \sum(N),$$

where N runs through all L–vertices which lie on L_i;

$$d_2^0((\mathbf{P}^2)) = -\sum_{i=0}^{q}(M_i),\ d_2^0((N)) = \sum_{N\in M_j}(M_i);$$
$$d_4^0((\ell, L)) = (L),\ d_4^1((L_i)) = (\mathbf{P}^2)$$

d_4^0 defines the homomorphism $D_L \to C_4^1$ and its restriction to C_4^0 is the differential.

The symbolic picture of C^* is shown in Figure 2.2. Here continuous lines mean lines from L, and dotted ones – from M, $\square$ means the plane $\mathbf{P}^2$.

It is easy to see that $C^*(L; M)$ is a complex, i.e., $d^2 = 0$. Note that $C_0^*(L; M)$ depends only on the configuration M, and $C_4^*(L; M)$ – only on L.

Functoriality. Let $L \subset L'$. Consider the complexes $C^*(L; M)$, $C^*(L'; M)$. Generators of the first complex are contained among the generators of the second one. This defines the inclusion map $i : C^*(L; M) \to C^*(L'; M)$.

Proposition 2.4.1. *i is the morphism of complexes.*

Let $M \subset M'$. Generators of the complex $C^*(L; M)$ are contained among the generators of $C^*(L; M')$. Define the projection map

$$j : C^*(L; M') \to C^*(L; M)$$

by sending a generator of $C^*(L; M')$ to a generator of the same name in $C^*(L; M)$ if the last one exists, and otherwise to zero.

Proposition 2.4.2. *j is a morphism of complexes.*

Propositions 2.4.1 and 2.4.2 are proved by direct verification.

2.4.3. *Duality*. The complex $C^*(L; M)$ is dual to $C^*(M; L)$. More exactly,

$$C^*(L; M)_i \cong C^*(M; L)_{4-i}^*,\ i = 0, 2, 4.$$

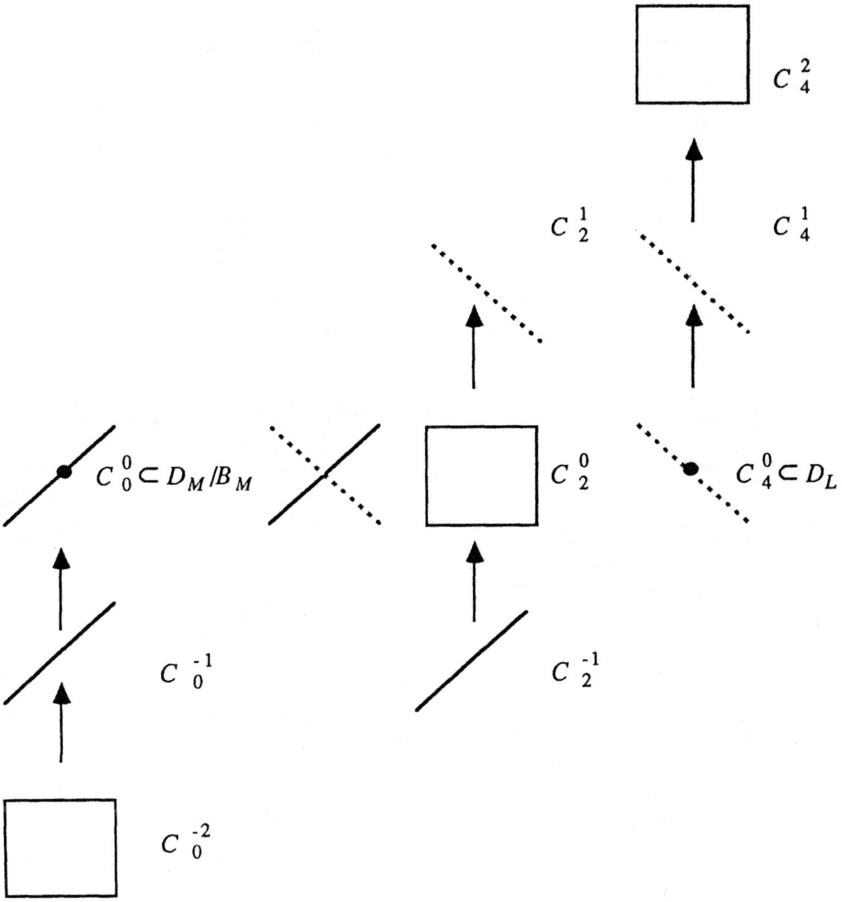

Figure 2.2. $C^*(L;M)$.

To describe this isomorphism, define the scalar product on generators in $C^*(L;M)$ and $C^*(L;M)$ and $C^*(M;L)$. Namely, the scalar product of generators of the same name is equal to 1, with the one exception: $<(\mathbb{P}^2),\ (\mathbb{P}^2)>=-1$, where the first argument lies in $C^0(L;M)_2$, and the second in $C^0(M;L)_2$. The scalar product of generators with different names is equal to zero.

For example, $<(M_i),(L_j)>=0$; $<(M_i),(M_j)>=\delta_{ij}$, where the left argument lies in $C^{-1}(L;M)_0$, and the right in $C^1(M;L)_4$, etc.

Put $G_i(L;M) = H^0C^*(L;M)_{2i}$, so $H^0 = G_0 \oplus G_1 \oplus G_2$.

Proposition 2.4.4. (a) *If $p \neq 0$, then $H^i(C^*(L,M)_4) = 0$ for $i \neq 0$. If $q \neq 0$ then $H^i(C^*(L,M)_0) = 0$ for $i \neq 0$.*

(b) *$G_0(L,M)$ and $G_2(L,M)$ are free abelian groups.*

(c) *If $p = q = 3$ then $G_1(L,M)$ is free.*

Corollary 2.4.5. (a) *For $i = 0, 2$ and any p, q*

$$G_i(L,M)^* \cong G_{2-i}(M,L)$$

(b) *For $p = q = 3$*

$$G_1(L,M)^* \cong G_1(M,L)$$

(c) *For any p, q and $0 \leq i \leq 2$*

$$G_i(L,M)^*_Q \cong G_{2-i}(M,L)_Q.$$

Let us describe generators and relations in G_2. Let $0 \leq i < j \leq p$. Denote by (L_{ij}) the chain $(L_i \cap L_j,\ L_i) - (L_i \cap L_j,\ L_j) \in C_4^0(L;M)$. For $0 \leq i < j < k \leq p$, put $L_{ijk} = (L_{ij}) - (L_{ik}) + (L_{jk})$. L_{ijk} is a cocycle, i.e., it defines the element $(L_{ijk}) \in G_2(L,M)$.

Proposition 2.4.6. (a) *Elements (L_{ijk}) satisfy the following relations*

(1) *$(L_{ijk}) = 0$ if the triangle (L_i, L_j, L_k) is degenerate;*

(2) *for $0 \leq i < j < k < \ell \leq p$*

$$(L_{ijk}) - (L_{ij\ell}) + (L_{ik\ell}) - (L_{jk\ell}) = 0.$$

(b) *Elements (L_{ijk}) and relations (1) and (2) give the complete system of generators and relations of the group $G_2(L;M)$.*

So $G_2(L;M)$ is the group of linear combinations of triangles (L_{ijk}) with relations (1) and (2).

Proposition 2.4.7. *Let (L,M) be an admissible pair of non-degenerate triangles. Then $rk\ G_i(L;M) = 1$ for $i = 0, 2$, and $rk\ G_1(L;M) = 4 - k_1 - k_2$ where k_1 is the number of vertices of the triangle L which lie on the sides of M, and k_2 is the number of vertices of M which lie on the sides of L.*

2.5. *The Map $\varphi_{02} : G_0(L;M)^* \otimes G_2(L;M) \to A_2$.* According to (2.4.6) an element $y \in G_2(L,M)$ is a linear combination $\sum n_i\Delta_i$ of triangles of the configuration L up to relations (1) and (2) (2.4.6). According to (2.4.5), (2.4.6), and element $x \in G_0(L;M)^*$ is a linear combination $\sum m_j\Lambda_j$ of triangles of M, up to analogous relations. Let $\varphi_{02}(x \otimes y)$ be equal to $\sum n_i m_j(\Delta_i, \Lambda_j) \in A_2$. The correctness of this definition follows from the relations (1) and (2), (2.4.6).

2.6. *The Map* $\varphi_{01} : G_0(L;M)^* \otimes G_1(L;M) \to A_1$ (cf. 1.4).

2.6.1. *Cross-ratio.* Let f be a non-zero meromorphic function on $\mathbf{P}^1$, $a = \sum n_i a_i$ a divisor on $\mathbf{P}^1$ whose support does not contain zeros and poles of f. Let $f(a) = \prod_i f(a_i)^{n_i} \in k^*$.

Let a, b be two divisors of degree zero on $\mathbf{P}^1$. Let

$$r(a;b) = f_a(b) \in k^*, \tag{2.6.1.1}$$

where f_a is any function with $\operatorname{div} f_a = a$. This number does not depend on the choice of f_a since $\deg b = 0$. If $a = a_1 - a_0$, $b = b_0 - b_1$, then

$$r(a;b) = r(a_0, a_1; b_0, b_1)$$

in the sense of 1.1.4.

2.6.2. *The cross-ratio of a cycle and a cocycle.* A cocycle in $C_2^0(L;M)$ is a linear combination $z = \sum a_N(N) - n(\mathbf{P}^2)$, N being LM–vertices such that for every line $M_i \in M$ $\sum_{N \in M_i} a_N = n$. Call the number n *the degree* of the cocycle. The cocycle z defines on every line M_i the divisor $z_i = \sum_{N \in L_i} a_N(N)$ of degree n. A coboundary $d_2^{-1}(L_i)$ of a line $L_i \in L$ is equal to $\sum_{N \in M_i}(N)$. $G_0(L;M)^*$ is a group of linear combinations of triangles $(M_i, M_j, M_k) \subset M$, $0 \leq i < j < k \leq q$, up to relations (1),(2),(2.4.6).

Let $\Delta = (M_i, M_j, M_k)$ $i < j < k$ be a non-degenerate triangle, $z \in C_2^0(L;M)$ a cocycle. Say that z is *fortunate with respect to the side* M_i if the support of z_i does not contain vertices of Δ. Say that z is *fortunate with respect to* Δ if it is fortunate with respect to all its sides.

Suppose that z is fortunate with respect to Δ. Choose homogeneous linear equations of the sides of $\Delta : f_i = 0$, $f_j = 0$, $f_k = 0$. Let

$$\varphi(\Delta, z) = (-1)^{\deg z} \frac{f_j(z_i) f_k(z_j) f_i(z_k)}{f_k(z_i) f_i(z_j) f_j(z_k)} \in k^* \tag{2.6.2.1}$$

It is clear that this number does not depend on the choice of equations f_i, f_j, f_k.

Proposition 2.6.3. *Let z be a fortunate cocycle of degree zero. Then*

$$\varphi(\Delta, z) = r(a_i, z_i)\ r(a_j, z_j)\ r(a_k, z_k),$$

where a_i is the divisor $(M_j \cap M_i) - (M_k \cap M_i)$ on the line M_i, a_j, a_k the analogous divisors on M_j, M_k; r the cross-ratio (2.6.1.1).

Proposition 2.6.4. (a) *If a fortunate cocycle is a coboundary, then* $\varphi(\Delta; z) = 1$.

(b) *If cocycles z_1, z_2 are fortunate then* $\varphi(\Delta, z_1+z_2) = \varphi(\Delta, z_1)\varphi(\Delta, z_2)$.

Proposition 2.6.5. *Every element $[z] \in G_1(L; M)$ has a fortunate representative.*

Let $[z] \in G_1(L; M)$, and z be its fortunate representative. Put

$$\varphi(\Delta, [z]) = \varphi(\Delta, z). \tag{2.6.6}$$

By the previous propositions, (2.6.6) correctly defines the homomorphism

$$\varphi(\Delta, \cdot) : G_1(L, M) \to k^*.$$

Set $\varphi(\Delta, \cdot) = 1$ if Δ is degenerate.

Proposition 2.6.7. *Let $0 \leq i_1 < i_2 < i_3 < i_4 \leq q$. Denote by Δ_j the triangle $(M_{i_1}, \dots, \hat{M}_{i_j}, \dots, M_{i_4})$. Then for every $z \in G_1(L; M)$ we have*

$$\prod_{j=1}^{4} \varphi(\Delta_j, z)^{(-1)^j} = 1.$$

By (2.4.6) and (2.6.7) φ correctly defines the homomorphism

$$G_0(L; M)^* \otimes G_1(L; M) \to k^*.$$

By (2.1) the group k^* is canonically isomorphic to $A_1(k)$, so φ defines the map

$$\varphi_{01} = \varphi_{01}(L; M) : G_0(L; M)^* \otimes G_1(L; M) \to A_1. \tag{2.6.8}$$

2.7. *Homomorphisms* $\varphi_{12} : G_1(L, M)^*_{\mathbf{Q}} \otimes G_2(L; M)_{\mathbf{Q}} \to A_{1,\mathbf{Q}}$. By (2.4.5) $G_1(L; M)^*_{\mathbf{Q}} \otimes G_2(L; M)_{\mathbf{Q}} \cong G_0(M; L)^*_{\mathbf{Q}} \otimes G_1(M; L)_{\mathbf{Q}}$. Let by definition $\varphi_{12} = \varphi_{12}(L; M)$ be equal to the composition of this isomorphism and a map $\varphi_{01}(M; L)$. Let $p = q = 3$. Then by (2.4.4) and (2.4.5) the same construction defines the map

$$\varphi_{12} : G_1(L, M)^* \otimes G_2(L, M) \to A_1.$$

2.8. *Functoriality.*

Proposition. (a) *Let $i_k : G_k(L;M) \to G_k(L';M)$, $k = 0,1,2$. be the natural homomorphisms induced by an nclusion of configurations $L \subset L'$ see* (2.4). *Then for every $x \in G_k(L',M)^*$, $y \in G_\ell(L;M)$*

$$\varphi_{k\ell}(i_k^* x \otimes y) = \varphi_{k\ell}(x \otimes i_\ell y).$$

(b) *Let $j_k : G_k(L;M') \to G_k(L;M)$ be the natural homomorphisms induced by an inclusion of configurations $M \subset M'$. Then for every $x \in G_k(L;M)^*$, $y \in G_\ell(L,M')$*

$$\varphi_{k\ell}(j_k^* x \otimes y) = \varphi_{k\ell}(x \otimes j_\ell y).$$

Remark 2.9. One can show that the cross-ratio homomorphism φ_{01} is uniquely determined by the properties (2.6.3),(2.6.4),(2.6.7) and (2.8).

2.10. *Comultiplication.* $\nu_{11} : A_2 \to A_1 \otimes A_1$ is defined similarly to the construction of comultiplication in $\mathcal{H}$ (see (1.4)). Let (L,M) be an admissible pair of non-degenerate triangles on $\mathbb{P}^2$. Consider the free graded group $G_0(L;M) \oplus G_1(L;M) \oplus G_2(L;M)$. The triangle M defines the covector $e \in G_0(L,M)^*$; the triangle L defines the vector $g \in G_2(L;M)$ (see (2.4)). Choose a base $f_1, \dots, f_k$ in $G_1(L;M)$; let $\{f^1, \dots, f^k\} \subset G(L;M)^*$ be the dual base. Put

$$\nu_{11}(L;M) = \sum_{j=1}^{k} \varphi_{01}(e \otimes f_j) \otimes \varphi_{12}(f^j \otimes g) \in A_1 \otimes A_1. \qquad (2.10.1)$$

Proposition 2.10.2. (2.10.1) *correctly defines the homomorphism ν_{11} : $A_2 \to A_1 \otimes A_1$.*

Proposition 2.11. *Let (L,M) be an admissible pair of triangles. Then*

$$\nu_{11}(M,L) = t\nu_{11}(L,M)$$

where $t : A_1 \otimes A_1 \to A_1 \otimes A_1$ is the transposition, $t(a \otimes b) = b \otimes a$.

2.12. *Compatibility of the multiplication and comultiplication.*

Proposition. *For every $a, b \in A_1$*

$$\nu_{11}(a \cdot b) = a \otimes b + b \otimes a.$$

Recall that $a \cdot b = \mu(a \otimes b)$ is the rectangle with the sides a, b (see Figure 2.1).

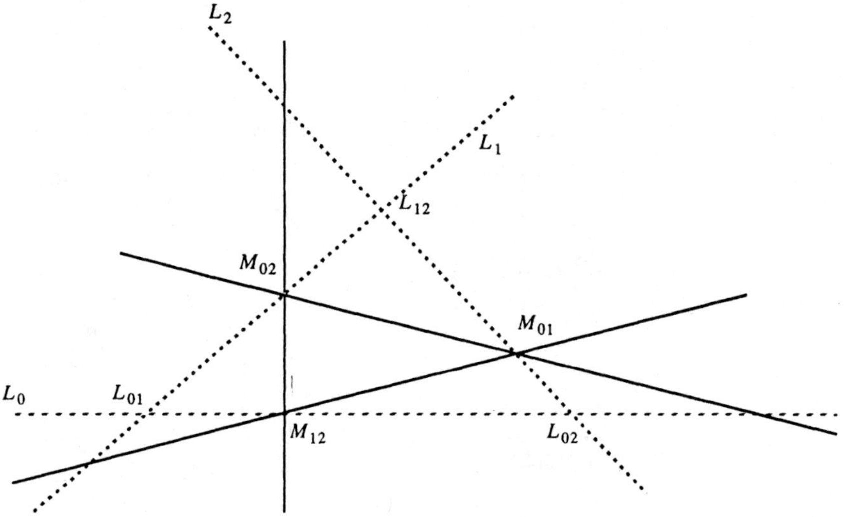

Figure 2.3. $K(a)$.

Example 2.13. Consider a configuration shown in Figure 2.3. Its projective equivalence class is defined by a number $a \in k^*$, the cross-ratio $r(L_0 \cap M_0, M_{12}; L_{01}, L_{02})$ on the line L_0. Denote this configuration $K(a)$.

Proposition. $\nu_{11}(K(a)) = a \otimes a + (-1) \otimes a$.

Let us prove that $\nu_{11}(K(a)) = a \otimes a +$ some element annihilated by a power of 2. Consider the dual configuration $PK(a)$ (see Figure 2.4a). It is easy to see that $2PK(a)$ is "a square" with the side a (see Figure 2.4b). From (2.12) follows that $\nu_{11}(2PK(a)) = 2a \otimes a$. Applying 2.11, we get the desired result.

Example 2.14. Consider a dilogarithmic configuration $\mathcal{D}(a)$ shown on Figure 1.4.

Proposition. $\nu_{11}(\mathcal{D}(a)) = -a \otimes (1-a)$ *for* $a \neq 1$. $\nu_{11}(\mathcal{D}(1)) = 0$.

Note that $3\mathcal{D}(1) = K(-1)$ (cf. Figure 2.5).

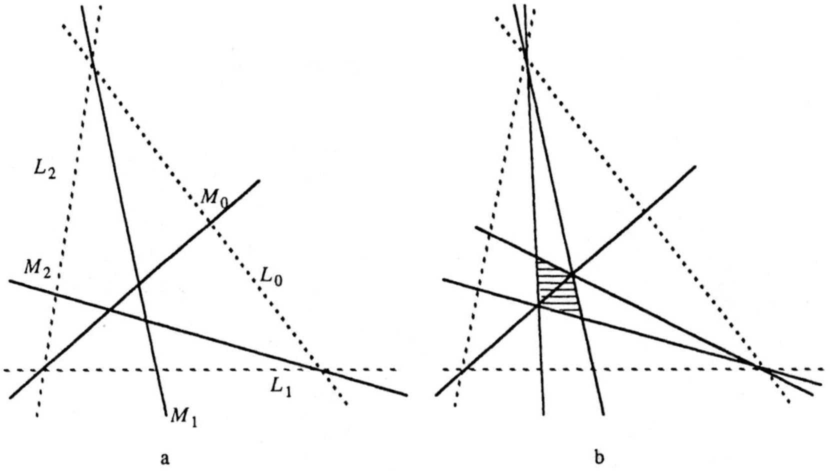

Figure 2.4.

2.15. *The Poincaré-Birkhoff-Witt decomposition for A_2.* Let $\sqcap = \sqcap(k) \subset A_2(k)$ be the subgroup generated by all rectangles, i.e., $\sqcap = \mu(A_1 \otimes A_1)$ where μ is the multiplication.

Let $S^2A_1 \subset A_1 \otimes A_1$ be the subgroup generated by tensors $a \otimes b + b \otimes a$. By 2.12 $\nu_{11}(\sqcap) \subset S^2A_1$.

Main theorem 1. (a) *The kernel and cokernel of*

$$\nu_{11}\,|_{\sqcap} : \sqcap \to S^2A_1$$

are 2*–torsion.*

(b) *The kernel and cokernel of the map*

$$\xi : A_2 \to \sqcap \oplus A_2/\sqcap,$$

where $\xi(x) = \mu\nu_{1,1}(x) \oplus [x]$, *are* 2*–torsion.*

Proof. It is easy to see that $\mu\nu_{11} : A_2 \to A_2$ restricted to $\sqcap$ is equal to the multiplication by 2. Similarly, $\nu_{11}\mu$ restricted to S^2A_1 is equal to multiplication by 2. This implies (a).

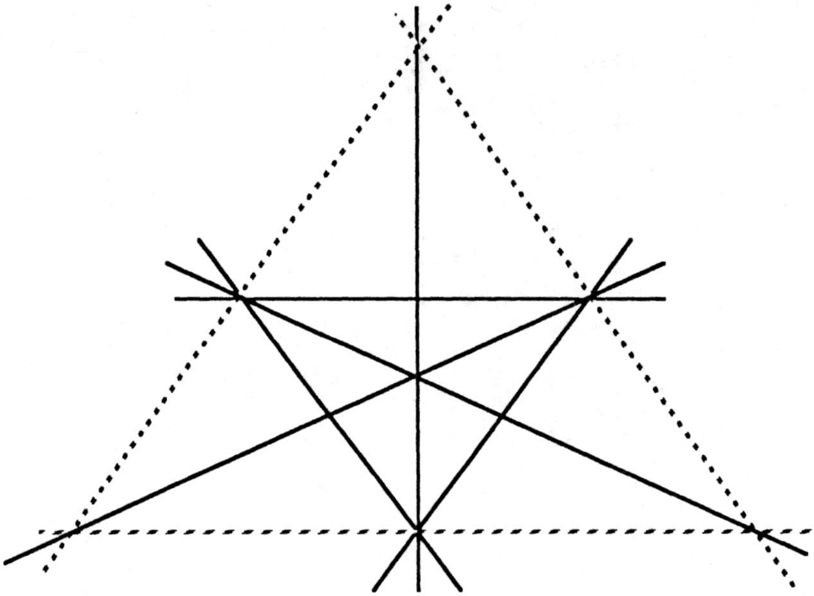

Figure 2.5. $3\mathcal{D}(1) = K(-1)$.

Define the map $\eta : \sqcap \oplus A_2/\sqcap \longrightarrow A_2$ by the formula $\pi \oplus [x] \longmapsto \pi + 2x = \mu\nu_{11}x$, where x is some representative of $[x]$. It is easy to see that η is correctly defined and $\xi\eta$ and $\eta\xi$ are multiplications by 2. This implies b.

QED

2.16. *The Hopf algebra A.* Let us give a sketch of the construction of comultiplication maps

$$\nu_{k\ell} : A_{k\ell} \rightarrow A_k \otimes A_\ell \tag{2.16.1}$$

generalizing the construction of ν_{11}. Put $n = k + \ell$. To every admissible pair of configurations of hyperplanes $(L; M)$ in $\mathbf{P}^n$ one can associate a certain complex $C^*(L; M) = \bigoplus_{i=0}^{n} C^*(L; M)_{2i}$. It satisfies the properties analogous to the ones formulated in (2.4) for $n = 2$. Put $G_i(L; M) = H^0(C^*(L; M)_{2i})$. We have $G_i(L; M)_{\mathbf{Q}} \cong G_{n-i}(M; L)^*_{\mathbf{Q}}$. One can define maps

$$\varphi_{k\ell} : G_k(L; M)^*_{\mathbf{Q}} \otimes G_\ell(L; M)_{\mathbf{Q}} \rightarrow A_{\ell-k,\mathbf{Q}}. \tag{2.16.2}$$

Groups $G_0(L, M)$ and $G_n(L, M)$ are free. $G_n(L, M)$ depends only on the configuration L and is the group of $n-$simplices of L up to additivity relations (see (2.4.6)). (This is a "Steinberg module of the configuration L"). $G_0(L, M) = G_n(M, L)^*$. It follows that if (L, M) is an admissible pair of non-degenerate simplices then in $G_n(L,M)$ a vector is distinguished, and on $G_0(L;M)$ – a covector. Then, acting as in 2.10 we get maps $\varphi_{k\ell}$ (2.1.6.1).

If n–simplices L, M are in general position, then $\nu_{k\ell}(L,M)$ is defined by the formula written down in [BMS], p. 708 (in[BMS] $\nu_{k\ell}$ is denoted by $\mu_{k,n-k}$)[1].

Together with the multiplication maps whose construction is sketched in at the end of 2.2, this comultiplication defines on $A = \bigoplus_{n\geq 0} A_n$ the structure of a commutative Hopf algebra. In particular, the multiplication and comultiplication are compatible: $\nu(a \cdot b) = \nu(a) \cdot \nu(b)$. Proposition 2.12 is just a particular case of this compatibility in grading (1,1).

Precise constructions and proofs will be published later.

Consider the dual Hopf algebra $A^*_{\mathbb{Q}}$. It is a cocommutative Hopf algebra and by [Q1] is an enveloping algebra of the graded Lie algebra $L = L(k) = \bigoplus_{n\geq 0} L_n$, where

$$L_n = (A_n / \sum_{\substack{k+\ell=n \\ k,\ \ell>0}} A_k \cdot A_\ell)^*_{\mathbb{Q}}.$$

In particular, $L_0 = \mathbb{Q}$, $L_1 = A^*_{1,\mathbb{Q}}$, $L_2 = (A_2/\sqcap)^*_{\mathbb{Q}}$. The last group will be calculated in Section 3. In (3.16) the conjectures about the connection between the cohomology of $L.(k)$ and algebraic K–theory of k will be formulated.

2.17. *Motivic cohomology of a pair of configurations.* Let (L,M) be an admissible pair of configurations of straight lines on $\mathbf{P}^2$. Consider the pair $(\mathbf{P}^2 \backslash L, M \backslash (M \cap L))$. By definition, the motivic cohomology of this pair (with coefficients in $\mathbb{Q}(0)$) is called the following graded $A^*_{\mathbb{Q}}$–module $H^*_{\mathcal{M}}(L,M) = \bigoplus_{a=0}^{4} H^a_{\mathcal{M}}(L,M)$. As a $\mathbb{Q}$–vector space

$$H^*_{\mathcal{M}}(L,M) = H^*(C^*(L,M)_{\mathbb{Q}}).$$

Every space $H^a_{\mathcal{M}}(L,M)$ is graded by weights:

$$H^a_{\mathcal{M}}(L,M) = \bigoplus_j H^a(L,M)_j$$

where

$$H^a(L,M)_j = H^a(C^*(L,M)_{2j,\mathbb{Q}}).$$

It is easy to see that for $a \neq 0$, H^a has at most one non-trivial weight component. By definition,

[1]There is a misprint in [BMS] in this place: in rows 16, 17 one has to replace j_k by j_{n-k}, and j_{n-k} by j_k.

$$H^0_{\mathcal{M}}(L,M) = G_0(L,M)_{\mathbb{Q}} \oplus G_1(L,M)_{\mathbb{Q}} \oplus G_2(L,M)_{\mathbb{Q}}.$$

Define the action

$$H^a_{\mathcal{M}}(L,M)_j \otimes A^*_{j-i,\mathbb{Q}} \to H^a_{\mathcal{M}}(L,M)_i.$$

For $a \neq 0$ this action is zero for $j \neq i$ and the identity for $i = j$. For $a = 0$ these maps may be non-zero only for $0 \le i \le j \le 2$ and are induced by maps φ_{ij} constructed in (2.5)-2.7).

2.18. *Poincaré duality in motivic cohomology.* Let $M = M_0 \oplus M_1 \oplus M_2$ be a graded $A^*_{\mathbb{Q}}$-module. Define *the Poincaré dual* $A^*_{\mathbb{Q}}$-module $M^v = M^v_0 \oplus M^v_1 \oplus M^v_2$ as follows. As vector spaces

$$M^v_i = (M_{2-i})^*.$$

The $A^*_{\mathbb{Q}}$-action

$$M^v_j \otimes A^*_{j-i} \to M^v_i$$

is induced by the composition

$$M_{2-i} \otimes A^*_{j-i,\mathbb{Q}} \to M_{2-i} \otimes A^*_{j-i,\mathbb{Q}} \to M_{2-j}$$

where the first arrow sends $m \otimes a$ into $m \otimes Pa$, P being the Poincaré involution, and the second one is defined by the $A^*_{\mathbb{Q}}$-module structure on M.

Proposition. *For a pair of (L,M) of configurations on the plane, an $A^*_{\mathbb{Q}}$-module $H^a_{\mathcal{M}}(M,L)$ is Poincaré dual to the $A^*_{\mathbb{Q}}$-module $H^{-a}_{\mathcal{M}}(L,M)$.*

This follows easily from the definitions.

3. The Bloch group and calculation of $A_2/\sqcap$.

3.1. *The Bloch group.* Define the Bloch group $B_2 = B_2(k)$ to be an abelian group whose generators are symbols (x_1, x_2, x_3, x_4) where $\{x_1, \dots, x_4\}$ runs through all quadruples of points in $\mathbf{P}^1(k)$, subject to the following relations

$$(x_1, x_2, x_3, x_4) = 0 \text{ if } x_i = x_j \text{ for some } i \neq j. \qquad (3.1.1)$$

3.1.2. *Skew symmetry.* For every permutation $\sigma : \{1,2,3,4\} \to \{1,2,3,4\}$

$$(x_{\sigma(1)}, x_{\sigma(2)}, x_{\sigma(3)}, x_{\sigma(4)}) = (-1)^{|\sigma|}(x_1, x_2, x_3, x_4)$$

where $|\sigma|$ is the parity of σ.

3.1.3. *Additivity*. For every 5–uple $\{x_1, x_2, x_3, x_4, x_5\}$

$$\sum_{i=1}^{5} (-1)^i \ (x_1, \dots, \hat{x}_i, \dots, x_5) = 0.$$

3.1.4. *Projective invariance*. For every $g \in PGL_2(k)$

$$(gx_1, gx_2, gx_3, gx_4) = (x_1, x_2, x_3, x_4).$$

3.1.5. *Variant*. Define the group $B_2'(k)$ as an abelian group with generators $[t]$, $t \in k$, and relations

(i) $[0] = [1] = 0$;

(ii) $[t] + [t^{-1}] = 0$, $[t] + [1-t] = 0$ for every $t \in k^*$;

(iii) For every t, $s \neq 1$

$$[t] - [s] + [s/t] - \left[\frac{1-t^{-1}}{1-s^{-1}}\right] + \left[\frac{1-t}{1-s}\right] = 0.$$

Define the map $r : B_2 \to B_2'$, associating to a quadruple (x_1, x_2, x_3, x_4) the cross-ratio $r(x_2, x_1, x_4, x_3)$. One easily sees that r is an isomorphism. We shall identify B_2 and B_2' by means of r.

Remark 3.1.6. Put $B_1(k) = A_1(k) = k^*$. Note that $B_1(k)$ is isomorphic to the group with generators $[t]$, $t \in k^*$, and relations

$$[t] - [s] + [s/t] = 0$$

for all $t, s \in k^*$.

3.1.7. *Traditional Bloch group*. Define the group $B\ell_2 = B\ell_2(k)$ as an abelian group with generators (x_1, x_2, x_3, x_4), where $\{x_i\}$ runs through all quadruples of *pairwise distinct* points in $\mathbf{P}^1(k)$, subject to relations (3.1.4) and (3.1.5) for pairwise distinct $\{x_1, \dots, x_5\}$. The group $B\ell_2$ was studied by S. Bloch [Bℓ].

We have an evident epimorphism $f : B\ell_2 \to B_2$. In [DS] and by A. Suslin it is shown that the following relations hold in $B\ell_2$:

$$6((x_1, x_2, x_3, x_4) + (x_1, x_4, x_3, x_2)) = 0,$$
$$2((x_1, x_2, x_3, x_4) + (x_1, x_2, x_4, x_3)) = 0$$

Corollary. *The kernel of $f : B\ell_2 \to B_2$ is 6−torsion.*

3.2. *The Bloch complex.* Consider the complex

$$B\ell(2) : 0 \to B\ell_2 \xrightarrow{s} B_1 \wedge B_1 \to 0. \qquad (3.2.1)$$

Here $B_1 \wedge B_1 = B_1 \otimes B_1 / S^2 B_1$, where $S^2 B_1$ is the subgroup generated by elements $a \otimes b + b \otimes a \cdot a$, $b \in B_1$. By definition $s(t) = t \wedge (1-t)$ where $a \wedge b$ denotes the image of $a \otimes b$ under projection $B_1 \otimes B_1 \to B_1 \wedge B_1$. The direct calculation shows that s is correctly defined. Call the complex (3.2.1) the traditional Bloch complex.

One has the following remarkable.

Theorem 3.2.2. (a) (Matsumoto [Mi]). *There exists a canonical isomorphism Coker $s \cong K_2(k)$.*

(b) (Suslin [S]). *There exists a canonical epimorphism $K_3(k)^{\text{ind}} \to \ker s$, whose kernel is $\text{Tor}_{\mathbb{Z}}(k^*, k^*)$, and hence is a torsion group.*

Here $K_i(k)$ are the Quillen groups of k [Q2]; $K_3(k)^{\text{ind}} = K_3(k)/K_1(k)^3$ – the indecomposable part of K_3. Multiplication maps $K_i \otimes K_j \to K_{i+j}$ are defined in [Lo].

Consider the complex

$$B(2) : 0 \to B_2 \xrightarrow{\tilde{s}} B_1 \tilde{\wedge} B_1 \to 0. \qquad (3.2.3)$$

Here $B_1 \tilde{\wedge} B_1 = B_1 \otimes B_1 / \tilde{S}^2 B_1$, where $\tilde{S}^2 B_1 = \{x \in B_1 \otimes B_1 | 2x \in S^2 B_1\}$, $\tilde{s}(t) = t \wedge (1-t)$.

One has an evident epimorphism $g : B_1 \wedge B_1 \to B_1 \tilde{\wedge} B_1$ whose kernel is annihilated by 2. The diagram

$$\begin{array}{ccc} B\ell_2 & \xrightarrow{s} & B_1 \wedge B_1 \\ \downarrow{\scriptstyle f} & & \downarrow{\scriptstyle g} \\ B_2 & \xrightarrow{\tilde{s}} & B_1 \tilde{\wedge} B_1 \end{array}$$

is commutative. Hence, g and f induce maps

$$g_c : \text{Coker}\, s \to \text{Coker}\, \tilde{s}$$

and

$$f_k : \ker s \to \ker \tilde{s}.$$

Proposition 3.2.4. (a) *g_c is an epimorphism whose kernel is annihilated by 2.*

(b) *Coker f_k is annihilated by 2.*

(c) *If char $k \neq 2$ then $\ker f_k$ is annihilated by 6.*
In fact, a,b follow easily from definitions, and c follows from 3.1.7.

Corollary 3.2.5. (a) *There exists a canonical epimorphism $K_2(k) \to$ Coker $\tilde{s}$ whose kernel is annihilated by* 2.

(b) *There exists a canonical homomorphism $K_3^{\rm ind}(k) \to \ker \tilde{s}$, whose cokernel is annihilated by* 2. *Its kernel is a torsion group.*

In the next two n°'s a homomorphism $A_2 \to B_1$ will be defined.

3.3. *Flags and triangles*. Call *a flag* in $\mathbf{P}^2$ a pair (m, M), where $M \subset \mathbf{P}^2$ is a straight line, and $m \in M$ a point. Denote by F a free abelian group whose generators are all flags in $\mathbf{P}^2$.

Associated to every triple $M = (M_0, M_1, M_2)$ of distinct straight lines in $\mathbf{P}^2$ the following linear combination of flags

$$\psi(M) = \sum_{\sigma \in S_3} (-1)^{|\sigma|} (M_{\sigma(0)\sigma(1)}, M_{\sigma(1)}),$$

where $M_{ij} = M_i \cap M_j$. Note that $\psi(M) = 0$ if all M_i intersect at one point.

Say that a flag $f = (m, M)$ is *admissible with respect to a triangle* $L = (L_0, L_1, L_2)$ if M does not coincide with a side of L. Define a homomorphism $\rho_L : F \to B_2$ by setting $\rho_L(f)$ to be equal to the class of the quadruple $(M \cap L_0, M \cap L_1, M \cap L_2, m)$ on the line M if a flag f is admissible with respect to L, and otherwise to zero.

Let (L, M) be an admissible pair of triangles. Put

$$\eta(L; M) = \rho_L(\psi(M))$$

if L has no coinciding sides, and

$$\eta(L, M) = 0$$

otherwise.

Proposition. *The above construction correctly defines a homomorphism*

$$\eta : A_2 \to B_2.$$

We call the homomorphism η the flag, or the six-term one. The proof will be done in the next n°.

3.4. *Nine-term homomorphism*. Let us give another construction of the flag homomorphism. Let (L, M) be an admissible pair of triangles, and

M has no coinciding sides. For $0 \leq i,j \leq 2$ consider the ordered pairs $\{i_1, i_2\} = \{0,1,2\} - \{i\}$, $i_1 < i_2$; $\{j_1, j_2\} = \{0,1,2\} - \{j\}$, $j_1 < j_2$. Put

$$\eta_{ij}(L,M) = (L_{i_1} \cap M_j, \ L_{i_2} \cap M_j, M_{j_1 j}, M_{j_2 j}) \in B_2.$$

Put

$$\eta'(L,M) = \sum_{i,j=0}^{2} (-1)^{i+j+1} \eta_{ij}(L,M). \tag{3.4.1}$$

Proposition. (a) *The above construction correctly defines a homomorphism $\eta' : A_2 \to B_2$.*
(b) *We have $\eta = \eta'$.*

Remark. Consider cross-ratios of quadruples from (3.4.1). These numbers take part in the first nine terms of the formula for $\mu_2\langle a_1, a_2, b_1, b_2\rangle$ on p. 709 from [BMS]. The geometric interpretation of these numbers as such cross-ratios was given by D. Zagier in 1987.[2]

Proof of Propositions 3.3 *and* 3.4. Let (L,M) be an admissible pair of triangles. The expression $\eta(L,M) - \eta'(L,M)$ has 15 terms. Every 5 from these terms corresponding to quadruples which lie on a fixed side M_i, represent a five-term relation (3.13) in B_2. Hence, $\eta = \eta'$. Let us prove the correctness of η. It is evident that η is additive and skew-symmetric with respect to M. On the other hand, it is not difficult to show that η' is additive and skew-symmetric by L. QED

3.5. *The subgroup of half-squares* $\sqcap \subset A_2$. Call *a half-square* with the side t the admissible pair of triangles shown in Figure 3.1.

Denote by $\tilde{\sqcap} \subset A_2$ the subgroup generated by half-squares.

Proposition 3.5.1. $\tilde{\Pi} \supset \Pi$ *The quotient group* $\tilde{\sqcap}/\sqcap$ *is* $2-$*torsion.*

3.6. *Dilogarithmic map.* Associate with a number $t \in k^*$ the dilogarithmic configuration

$$\delta(t) = (L, \mathcal{D}(t))$$

shown in Figure 1.4. If $x = (x_1, \ldots, x_4)$ is a quadruple of points in $\mathbf{P}^1$, put $\delta(x) = \delta(r(x))$ where r is the cross-ratio, if x_i are distinct and $\delta(x) = 0$ otherwise.

---

[2]One has a misprint in [BMS] in this formula. Namely, the summand $a_1 b_1^{-1} \otimes (1 - a_1 b_1^{-1})$ must have the sign $+$.

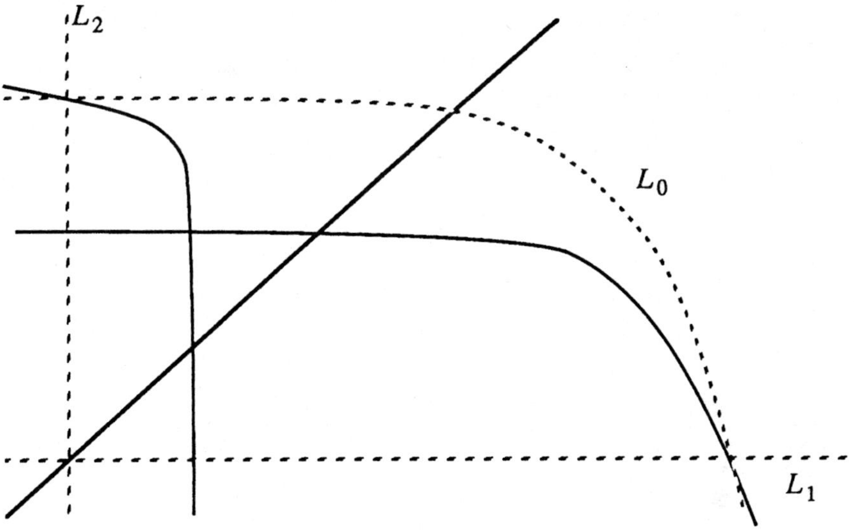

Figure 3.1.

Proposition 3.6.1. $12\delta(1) = 0$.

This follows easily from the equality $3\delta(1) = K(-1)$ (cf. 2.14). Consider the subgroup $\Sigma \subset A_2$ generated by $\tilde{\sqcap}$ and $\delta(1)$.

Corollary 3.6.2. *The quotient* $\Sigma/\sqcap$ *is annihilated by* 12.

Proposition 3.7. (a) $\eta(\Sigma) = 0$, *i.e.,* η *correctly defines a map*

$$\eta : A_2/\Sigma \rightarrow B_2$$

(b) δ *correctly defines the map*

$$\delta : B_2 \rightarrow A_2/\Sigma.$$

The first part follows immediately from definitions. The second one will be proven in 3.10.

3.8. Main Theorem 2. η *and* δ *are mutually inverse isomorphisms*

$$A_2/\Sigma \rightleftarrows B_2.$$

The proof of 3.8 and 3.7 (b) is based on the following geometric construction.

3.9. *Dilogarithmic decomposition.* Let (L, M) be an admissible pair of triangles; $\eta(L, M)$ – a linear combination of 6 (or less) quadruples constructed in 3.2; $\delta\eta(L;M)$ - the corresponding linear combination of dilogarithmic configurations.

Lemma 3.9.1. *The difference* $(L,M) - \delta\eta(L,M)$ *considered as an element of* A_2 *is the sum of (two or less) rectangles and, may be,* $[\delta(1)]$ *– the class of* $\delta(1)$.

The proof in the case of triangles in general position is shown in Figure 3.2.

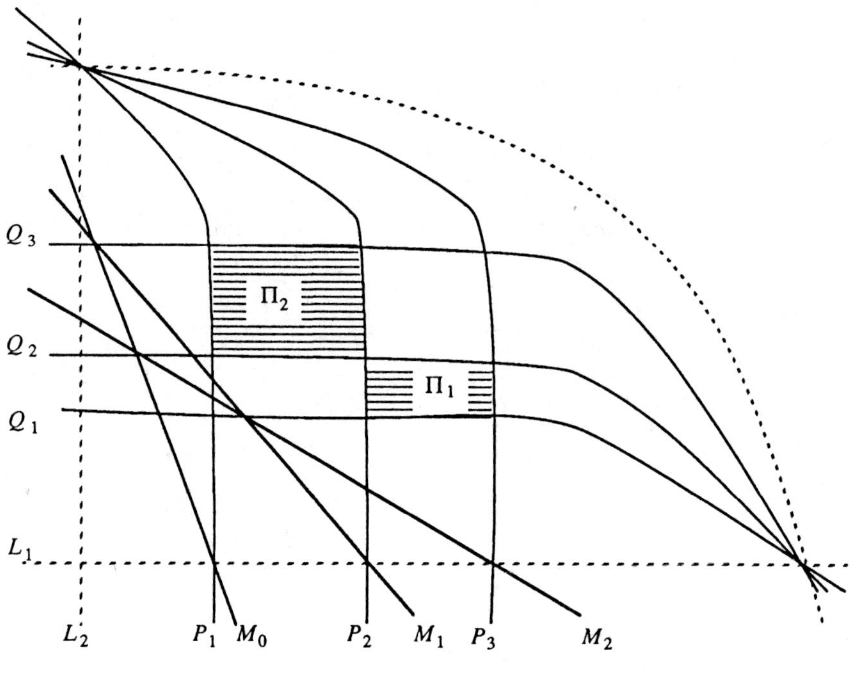

Figure 3.2. The dilogarithmic decomposition.

Namely, we have

$$(M_0,M_1,M_2) = (M_0,Q_3,P_1) - (M_1,Q_3,P_2) - (M_0,Q_3,P_1) + (M_2,Q_2,P_3) - (M_2,Q_1,P_3) + (M_1,Q_1,P_2) - \sqcap_2 + \sqcap_1,$$

The difference $(L,M) - \delta\eta(L,M)$ is equal to $(L,\sqcap_1) - (L,\sqcap_2)$. The degenerated cases are treated similarly. The element $[\delta(1)]$ arises in such a difference, when the triangle M has a vertex on L_2 and one of its sides which go through that vertex, contains L_{01}. For example, this the case, when $(L,M) = \delta(1)$.

This lemma means that every configuration may be canonically decomposed into the sum of (6 in general position) dilogarithmic ones, some rectangles, and may be $[\delta(1)]$.

Corollary 3.9.2. *An Aomoto dilogarithm may be represented as a sum of Euler dilogarithms, products of logarithms, and* $n \cdot \frac{\pi^2}{6}$, $n \in \mathbb{Z}$.

3.10. *Five term relation in* A_2/Σ. Let us show that δ sends the five-term relation (3.1.2) in B_2 to zero in A_2/Σ. Consider the dilogarithmic configuration $\delta(t)$ and cut the coordinate triangle as is shown in Figure 3.3. Put $L = (L_2L_0L_1)$, $L' = (L_2L_3L_1)$, $L'' = (L_3L_2L_0)$.

We have

$$(L,M) = (L',M) + (L'',M). \tag{3.10.1}$$

Decompose configurations (L',M) and (L'',M) into the sum of dilogarithmic ones. Using the construction (3.9), one easily sees in both configurations the triangle M is decomposed into the sum of two dilogarithmic ones. Substituting these decompositions into (3.10.1) we get the five-term relation in A_2/Σ which is the image of the five- term relation in B_2 corresponding to the points $(x_1,\dots,x_5)$ on the line M_0 (see Figure 3.3).

This proves that the dilogarithmic map δ sends the five-term relation into zero in A_2/Σ. Similarly, one can prove that skew symmetry relations (3.1.2) are sent to zero.

This implies 3.7 (b).

3.11. **Proof of Theorem 3.8.** It is easy to see that for $px \neq 1$ $\eta\delta(x) = 1[x]$. On the other hand, by 3.9 $\delta\eta = \mathrm{id}_{A_2/\Sigma}$. QED

3.12. *Rogers Function.* Define the map

$$L : B_2 \longrightarrow A_2$$

by the formula

$$L([t]) = 12\delta(t) + 6\mu(t \otimes (1-t)).$$

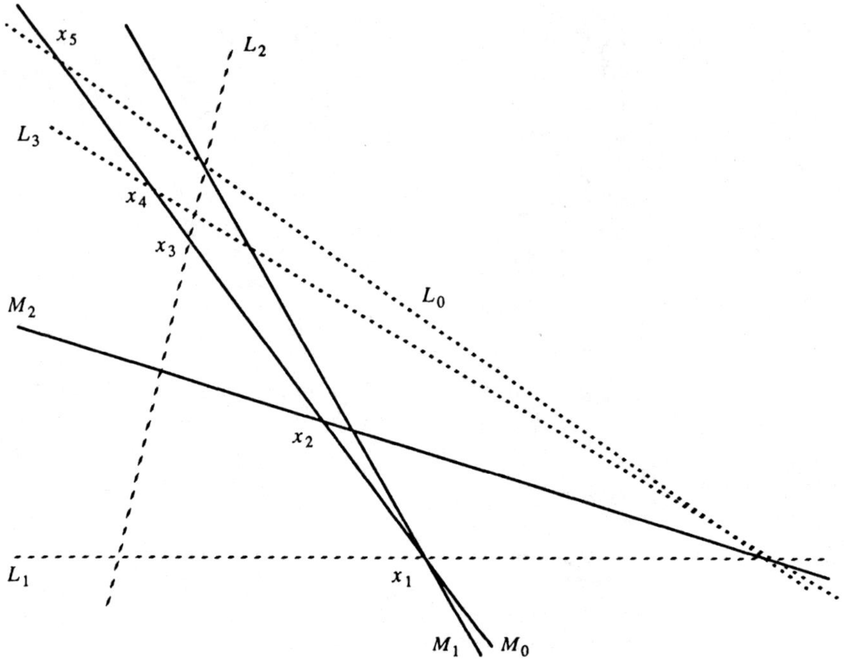

Figure 3.3. The five-term relation in A_2/Σ.

From (3.7) and (2.14 (a)) it follows that L is correctly defined. It is the formal analogue of the Rogers function (1.2.2), multiplied by 12.

From the proof of 3.5 it follows that the homomorphism

$$(i, L) : \sqcap \oplus B_2 \rightarrow A_2 \tag{3.12.1}$$

where $i : \sqcap \rightarrow A_2$ is $6x$ natural inclusion and has 12—torsion kernel and cokernel.

3.13. *Poincaré Duality.*

Theorem 3.13.1. *Let $P_2 : A_2 \rightarrow A_2$ be the Poincaré involution, (2.1.6). Then for every $x \in A_2$*

$$P_2(x) = -x + \mu\nu_{11}(x).$$

Corollary 3.13.2. *If $x \in \sqcap$ then $P_2(x) = x$.*

If $x \in L(B_2)$ then $P_2(x) = -x$.

3.14. *Adjunction Involution.* Introduce on the group A_2 the other involution. Namely, let $\mathbf{P}^{2^*}$ be the dual projective plane. Its points are straight lines in $\mathbf{P}^2$; and straight lines in $\mathbf{P}^{2^*}$ are in one-to-one correspondence with points in $\mathbf{P}^2$.

Associate with a non-degenerate triangle $L = (L_0, L_1, L_2)$ in $\mathbf{P}^2$ a non-degenerate triangle $L^* = (L_0^*, L_1^*, L_2^*)$ in $\mathbf{P}^{2^*}$ such that its sides correspond to vertices (L_{12}, L_{02}, L_{01}) of L.

It is not difficult to see, that associating to a pair of non-degenerate triangles (L, M) in $\mathbf{P}^2$ a pair (L^*, M^*) in $\mathbf{P}^{2^*}$ correctly defines an involution

$$Q_2 : A_2 \to A_2$$

(A similar construction defines involutions $Q_n : A_n \to A_n$).

Theorem. $Q_2 = P_2$.

Note that the configuration $P_2Q_2(L, M) = (M^*, L^*)$ is not, in general, projectively equivalent to (L, M), but defines the same element of A_2.

3.15. Consider the diagram of complexes

$$\begin{array}{ccccccccc}
0 & & & & 0 & & 0 & & \\
\downarrow & & & & \downarrow & & \downarrow & & \\
C(2) & & 0 & \longrightarrow & \Sigma & \xrightarrow{\nu} & \tilde{S}^2 A_1 & \longrightarrow & 0 \\
{\scriptstyle i}\downarrow & & & & \downarrow & & \downarrow & & \\
A(2) & & 0 & \longrightarrow & A_2 & \xrightarrow{\nu} & A_1 \otimes A_1 & \longrightarrow & 0 \\
\downarrow & & & & \downarrow & & \downarrow & & \\
B(2) & & 0 & \longrightarrow & B_2 & \longrightarrow & B_1 \tilde{\wedge} B_1 & \longrightarrow & 0 \\
\downarrow & & & & \downarrow & & \downarrow & & \\
0 & & & & 0 & & 0 & &
\end{array} \tag{3.15.1}$$

We will consider complexes $A(2)$,$B(2)$,$C(2)$, to be placed in degrees 1,2.

Note that according to 2.14(a) and 3.6.2 the cohomology of $C(2)$ are 2−torsion. By definition, the right column is exact. By 3.8 the left column is also exact.

Define the map

$$B_1\tilde{\Lambda}B_1 \to B_1 \otimes B_1 = A_1 \otimes A_1$$

by the formula $a\tilde{\Lambda}b \longmapsto 6(a\otimes b - b\otimes a)$. Together with the Rogers map $L : B_2 \to A_2$ this defines the morphism of complexes

$$L(2) : B(2) \to A(2),$$

hence a map

$$(i, L(2)) : C(2) \oplus B(2) \to A(2). \qquad (3.15.2)$$

Theorem 3.15.3. (a) *diagram* (3.15.1) *is commutative.*
(b) *The kernel and cokernel of the map* (3.15.2) *are* 12*−torsion.*

Corollary 3.15.4. *There exist canonical isomorphisms*
(a) $H^2(A(2)) \otimes \mathbf{Z}[\frac{1}{2}] \cong K_2(k) \otimes \mathbf{Z}[\frac{1}{2}]$
(b) $H^1(A(2))_{\mathbf{Q}} \cong K_3(k)^{\text{ind}}_{\mathbf{Q}}$.

3.16. *Cohomology of $A^*_{\mathbf{Q}}(k)$ and algebraic K-theory of k.* Consider Hopf algebras $A_{\mathbf{Q}}$ and $A^*_{\mathbf{Q}}$ introduced in 2.15. Recall that $A^*_{\mathbf{Q}}$ is the enveloping algebra $U(L)$ of a certain pronilpotent graded Lie algebra $L = L(k)$.

For every $n \geq 1$ define the complex

$$A(n)_{\mathbf{Q}} : 0 \to A(n)^1_{\mathbf{Q}} \to \cdots \to A(n)^n_{\mathbf{Q}} \to 0,$$

placed in degrees from 1 to n, as follows. Put

$$A(n)^i_{\mathbf{Q}} = \Sigma \ A_{p_1,\mathbf{Q}} \otimes \cdots \otimes A_{p_i,\mathbf{Q}},$$

where the summing ranges over all families $(p_1, \dots, p_i)$ such that $\sum_{j=1}^{i} p_j = n$, and $p_j \geq 1$ for all j.

The differential

$$d^i : A(n)^i_{\mathbf{Q}} \to A(n)^{i+1}_{\mathbf{Q}}$$

is, by definition, equal to the alternating sum $\sum_{j=1}^{i-1}(-1)^{j+1} \ d^i_j$, where

$$d^i_j(a_1 \otimes \ldots \otimes a_j) = \sum_{\substack{p'_j+p''_j=p_j \\ p'_j,p''_j\geq 1}} a_1 \otimes \ldots \otimes a_{j-1} \otimes \nu_{p'_j p''_j}(a_j) \otimes a_{j+1} \otimes \ldots \otimes a_i,$$

$a_1 \otimes \ldots \otimes a_i \in A_{p_1,\mathbf{Q}} \otimes \ldots \otimes A_{p_i,\mathbf{Q}}$. The equality $d^2 = 0$ is the direct consequence of the associativity of ν. Put $A(0) = A_0$ placed in degree 0.

Let $\mathbf{Q}(j)$ be the graded$A^*_{\mathbf{Q}}$−module with $\mathbf{Q}(j)_j = \mathbf{Q}$, $\mathbf{Q}(j)_i = 0$ for $i \neq j$. One has canonical isomorphisms

$$H^i(A(n)_{\mathbf{Q}}) = \mathrm{Ext}^i_{A^*_{\mathbf{Q}}}(\mathbf{Q}(0),\ \mathbf{Q}(n))$$

see [Mac].

The complexes $A(n)_{\mathbf{Q}}$ contain quasi-isomorphic subcomplexes $B(n)_{\mathbf{Q}}$ such that $\underset{n}{\oplus} B(n)_{\mathbf{Q}}$ is the standard complex of cochains of $L : C^*(L) = \Lambda_{\mathbf{Q}} L^*$.

For $n = 1$, $A(1)_{\mathbf{Q}} = B(1)_{\mathbf{Q}}$ reduces to one term equal to $A_{1,\mathbf{Q}} = B_{1,\mathbf{Q}}$. For $n = 2$ we get complexes $B(2)_{\mathbf{Q}}$ considered above.

Recall that *Milnor K-groups* $K_i^M(k)$, $i \geq 1$ are defined to be

$$K_i^M(k) = (k^*)^{\otimes i}/R$$

where $(k^*)^{\otimes i} = \underbrace{k^* \otimes k^* \otimes \ldots \otimes k^*}_{i \text{ times}}$, and R is the subgroup generated by elements of the form $b_1 \otimes \ldots \otimes b_p \otimes a \otimes (1-a) \otimes b_{p+1} \otimes \ldots \otimes b_{i-2}$, $b_j \in k^*$, $a \in k^* - \{1\}$.

From 3.15.4 (a) follows

Proposition 3.16.1. *One has canonical isomorphisms*

$$K_n^M(k) \otimes \mathbf{Z}\left[\frac{1}{2}\right] \cong H^n(A(n) \otimes \mathbf{Z}\left[\frac{1}{2}\right]),$$

for all $n \geq 1$.

Now consider Quillen $K-$*groups* (k), $i \geq 0$. *Note that* $K_0(k) = \mathbf{Z}$, $K_i(k) = K_i^M(k)$ *for* $i = 1, 2$.

3.16.2. *Conjecture.* For all $i \geq 0$, there exist canonical isomorphisms

$$K_i(k)_{\mathbf{Q}} \cong \bigoplus_{n=[1/2]+1}^{i} H^{2n-i}(A(n)_{\mathbf{Q}}).$$

These isomorphisms may be considered as analogues of the Atiyah-Hirzebruch isomorphism between complex $K-$theory and singular cohomology (cf. [BMS]).

For $i = 0, 1$, 3.16.2 is evident. For $i = 2$, it follows from 3.15.4 and isomorphism

$$K_3(k) \otimes \mathbf{Z}\left[\frac{1}{2}\right] \cong (K_3^M(k) \oplus K_3^{\mathrm{ind}}(k)) \otimes \mathbf{Z}\left[\frac{1}{2}\right]$$

proved by A. Suslin [S].

REFERENCES

[A] Aomoto, K., *Addition theorem of Abel type for hyper-logarithms*, Nagoya Math. J. **88** (1982), 55-71.

[B] Beilinson, A., *Notes on absolute Hodge cohomology*, Contemp. Math. **55** (1986), 35-68.

[Bl] Bloch, S., *Higher regulators, algebraic K-theory, and zeta functions of elliptic curves*, Lecture Notes, U.C. Irvine (1977).

[BMS] Beilinson, A., R. MacPherson, and V. Schechtman, *Notes on motivic cohomology*, Duke Math. J. **54**. No. 2, (1987), 679-710.

[C] Cartier, P., *Décomposition des polyèdres, le point sur le 3ème problème de Hilbert*, Seminaire Bourbaki **646** (1984-85).

[D] Deligne, P., *Théorie de Hodge* II, Publ. Math. IHES **40** (1970), 5-58.

[DS] Dupont, J. and C.-H. Sah, *Scissors congruences* II, J. Pure and Appl. Algebra **25** (1982), 159-195.

[Le] Lewin, L., *Dilogarithms and associated functions*, North Holland, 1981.

[Lo] Loday, J. L., *K-theorie algébrique et représentations de groupes*, Ann. Sci. ENS **9** (1976), 309-377.

[Ma] Manin, Yu. I., *Correspondences, motives and monoidal transformations*, (Russian), Mat. Sb. (N.S.) **119** (1968), 475-507.

[Mac] MacLane, S., *Homology*, Springer, 1963.

[Mi] Milnor, J., *Introduction to algebraic K-theory*, Ann. Math. Studies **72**. Princeton University Press, Princeton, 1971.

[Q1] Quillen, D., *Rational homotopy theory*, Ann. of Math. **90** (1969), 205-295.

[Q2] Quillen, D., *Higher K-theory* I, Lecture Notes in Math. **341** (1973), 85-147.

[R] Rogers, L. J., *On function sum theorems connected with the series* $\sum_1^\infty (x^n/n^2)$, Proc. London Math. Soc. **4** (1907), 169-189.

[S] Suslin, A., Talk at Berkeley ICM, 1986.

A. A. Beilinson
Department of Mathematics
M.I.T.
Cambridge MA 02133 USA

A. B. Goncharov,
Academy of Sciences of the USSR
Scientific Council on Cybernetics,
Moscow USSR

V. V. Schechtman
Institute of problems
of microelectronics technology
and superpure materials
Academy of Sciences of the USSR
Moscow USSR

A. N. Varchenko
Gubkin MING
Department of Mathematics,
Moscow USSR

Théorie de Dieudonné cristalline III : théorèmes d'équivalence et de pleine fidélité

PIERRE BERTHELOT et WILLIAM MESSING

A Alexandre Grothendieck,
pour son 60ème anniversaire

Introduction

Ce travail est le troisième d'une série consacrée à la théorie de Dieudonné cristalline. Le premier article [6] présentait les principaux résultats, sans en donner de démonstration. Dans le volume [7], écrit en collaboration avec Breen, les fondements de la théorie ont été exposés de manière systématique. Nous donnons ici la démonstration des résultats annoncés dans [6] et qui ne figurent pas dans [7], ainsi que certains compléments ; ces résultats regroupent essentiellement les énoncés actuellement connus sur la pleine fidélité ou la surjectivité essentielle du foncteur "cristal de Dieudonné" associé à un p-groupe fini localement libre, ou à un groupe p-divisible.

Essayer de donner un exposé historique détaillé de l'évolution et du développement des idées et techniques sous-jacentes à cette théorie irait malheureusement au-delà de notre compétence. Il est néanmoins clair que la vision initiale et les aspects les plus fondamentaux en sont dûs à Grothendieck. La naissance de la cohomologie cristalline et de la théorie de Dieudonné cristalline remonte à la lettre de Grothendieck à Tate (1966) qui commence ainsi :

"J'ai réfléchi aux groupes formels et à la cohomologie de de Rham, et suis arrivé à un projet de théorie ...". La théorie esquissée dans la lettre était une premiere tentative, et le "site cristallin" qui y était introduit fut rebaptisé peu de temps après "site infinitésimal" par Grothendieck, lorsqu'il s'aperçut de la nécessité d'y introduire des puissances divisées. Néanmoins, les idées de base y étaient présentes. En particulier, la lettre s'achevait en soulevant le problème de ce que Grothendieck allait bientôt appeler le "foncteur mystérieux", dont la construction devait être donnée par Fontaine onze ans plus tard [9].

La théorie de Dieudonné cristalline des groupes p-divisibles a été développée tout d'abord par Grothendieck (voir [12], [11]), puis par Messing [18] et Mazur-Messing [17]. Deux des problèmes fondamentaux soulevés, et non résolus, par Grothendieck, étaient les suivants :

(i) Développer une théorie de Dieudonné cristalline pour les p-groupes (commutatifs) finis localement libres ;

(ii) Etablir les propriétés de fidélité du foncteur de Dieudonné cristallin.

La réponse au premier de ces problèmes a été apportée par les auteurs en collaboration avec Breen, et est développée dans [7]. Dans cet article, nous étudions le second, en laissant néanmoins nombre de questions sans réponse.

Donnons maintenant un résumé rapide de ce qui suit.

La première partie est consacrée à donner une description terre-à-terre de la catégorie des cristaux lorsque l'on suppose que le schéma de base possède une p-base (en un sens plus restrictif que celui de [EGA], dans la mesure où l'existence d'une p-base au sens adopté ici entraîne que le schéma est réduit). Cette hypothèse de régularité très forte permet de donner une description des cristaux en termes de modules à connexion intégrable sur un relèvement, et en particulier de montrer la fidélité du foncteur image inverse pour la catégorie des cristaux dans certains cas importants. A part quelques applications aux cristaux de Dieudonné et à leur dualité, cette partie est indépendante de [7].

Dans la deuxième partie, nous abordons la théorie de Dieudonné proprement dite, en étudiant le cas des groupes étales ou de type multiplicatif sur un schéma de base quelconque S. Dans [7, ch. 5], nous avons montré que, pour tout p-groupe fini localement libre G sur S, de dual de Cartier G^*, la dualité classique entre les modules de Dieudonné, lorsque S est le spectre d'un corps parfait, peut être généralisée par un isomorphisme de dualité entre les complexes de Dieudonné cristallins $\Delta(G)^{\vee}[-1] \xrightarrow{\sim} \Delta(G^*)$. Nous explicitons tout d'abord ces complexes lorsque G est étale ou de type multiplicatif, et montrons que, dans ce cas, la dualité entre complexes de Dieudonné résulte de la dualité de Pontryagin. Le résultat principal de cette partie est que, pour tout schéma S sur lequel p est localement nilpotent, la catégorie des p-groupes finis étales (resp. de type multiplicatif) est équivalente, via le foncteur de Dieudonné $\mathbb{D}$, à la catégorie des cristaux en modules M sur S, qui sont de torsion, localement de présentation finie, et munis d'un isomorphisme $F : M^{\sigma} \xrightarrow{\sim} M$ (resp. $V : M \xrightarrow{\sim} M^{\sigma}$). Appliqué aux groupes p-divisibles, ce résultat étend au cas d'une base générale l'équivalence classique (sur une base lisse) entre $\mathbb{Z}_p$-faisceaux lisses et "F-cristaux unité".

La troisième partie est consacrée à la "théorie multiplicative". Connue de Grothendieck à la fin des années 60, et développée dans [18], la construction du cristal de Dieudonné associé à un groupe p-divisible par la méthode de l'exponentielle repose sur le fait que, sur le site cristallin nilpotent, l'extension universelle par un groupe vectoriel est "de nature cristalline". Nous donnons ici une autre démonstration de cette propriété, reposant sur une construction plus cohomologique, et nous comparons les deux constructions : c'est ce que nous appelerons "problème des deux fonctorialités". Plus précisément, considérons un épaississement $S_0 \hookrightarrow S$ muni de puissances divisées nilpotentes, et un morphisme $u_0 : G_0 \to H_0$ entre deux groupes p-divisibles sur S_0 ; chacune des deux constructions associe fonctoriellement à G_0 (resp. H_0) un cristal, de sorte que u_0 définit un homomorphisme entre les faisceaux sur S définis par ces cristaux. Si G (resp. H) est un relèvement sur S de G_0 (resp. H_0), ces faisceaux s'identifient tous deux à l'extension universelle de G (resp. H) par un groupe vectoriel, et le problème des deux fonctorialités est de montrer que les homomorphismes $E(G) \to E(H)$ définis par u_0 entre les extensions universelles sont les mêmes dans les deux constructions. L'importance de ce résultat vient de ce qu'il permet d'utiliser dans le cadre de la théorie de Dieudonné cristalline la théorie des déformations de [18].

Enfin, le chapitre 4 aborde le problème de la pleine fidélité du foncteur de Dieudonné cristallin, moyennant des hypothèses de régularité sur le schéma de base S. Nous traitons d'abord le cas où S est un schéma intègre, normal, et possédant localement une p-base. Il résulte de [4] que lorsque A est un anneau intègre, normal et parfait, le foncteur de Dieudonné est pleinement fidèle. Cela permet de se ramener au cas où la base est un corps, cas dans lequel on procède essentiellement par descente inséparable. Nous montrons ensuite comment étendre ce résultat à des bases plus générales en utilisant le théorème d'approximation d'Artin et des résultats liés. Le thème suivant consiste à montrer comment, dans le cas des groupes p-divisibles, les résultats précédents peuvent se prolonger par déformation infinitésimale à certaines intersections complètes, grâce à la théorie des déformations de [18]. En particulier, le foncteur de Dieudonné cristallin est pleinement fidèle pour les groupes p-divisibles sur un quotient artinien d'un anneau de valuation discrète d'égales caractéristiques p, ce qui répond affirmativement à une question de Raynaud. Ce chapitre s'achève par une application géométrique simple, montrant que sur une variété relevable, propre, lisse et simplement connexe, tout groupe p-divisible est "isoconstant", résultat dont la possibilité était déjà indiquée dans la lettre de Grothendieck mentionnée plus haut.

Signalons pour terminer que si, sur une base S vérifiant des hypothèses de régularité convenable,la théorie se comporte comme on pouvait l'espérer, il n'en est pas de même lorsque l'anneau des sections globales du faisceau structural de S possède une partie de p-torsion non triviale [8]. Il semble néanmoins naturel de demander si, sur une base quelconque, le foncteur défini sur la catégorie des groupes p-divisibles à isogénie près en prenant l'isocristal convergent associé au cristal de Dieudonné [5, (2.3)] ne serait pas pleinement fidèle.

Rappels et notations

D'une manière générale, nous reprendrons les notations et les conventions de [7] ; en particulier, tous les schémas en groupes considérés sont supposés commutatifs, et tous les schémas en groupes finis sont supposés de p-torsion, p étant un nombre premier fixé dans tout cet article. Nous renvoyons notamment à [7, 0.3] pour ce qui concerne les conventions de signes.

Dans tout l'article, $(\Sigma, \mathcal{I}, \gamma)$ désigne un $\mathbb{Z}_p$-schéma fixé, muni d'un PD-idéal quasi-cohérent dont les puissances divisées sont compatibles à celles de p; on posera $\Sigma_n = \Sigma \times \operatorname{Spec} \mathbb{Z}/p^n\mathbb{Z}$. En pratique, on prendra généralement $\Sigma = \operatorname{Spec} \mathbb{Z}_p$, et $\mathcal{I} = p.\mathcal{O}_\Sigma$, muni de ses puissance divisées canoniques. Nous désignerons par S un Σ-schéma de base sur lequel p est supposé localement nilpotent, et auquel les puissances divisées γ s'étendent (ce qui est automatique dans le cas précédent), et on notera $\mathrm{C{\scriptstyle RIS}}(S/\Sigma)$ le gros site cristallin de S relativement à $(\Sigma, \mathcal{I}, \gamma)$. En règle générale, la topologie utilisée sera la topologie fppf ; dans la deuxième partie, on peut indifféremment utiliser la topologie étale. Pour tout faisceau E sur $\mathrm{C{\scriptstyle RIS}}(S/\Sigma)$, et tout objet (U, T, δ) de $\mathrm{C{\scriptstyle RIS}}(S/\Sigma)$, nous noterons $E_{(U,T,\delta)}$, ou pour abréger $E_{(U,T)}$, voire E_T, le faisceau sur le petit site de T défini par E. Si S est de caractéristique p, et si M est un cristal sur $\mathrm{C{\scriptstyle RIS}}(S/\Sigma)$, nous désignerons par M^σ son image inverse par le Frobenius absolu de S.

Rappelons que, pour tout faisceau F sur S, on désigne par $\underline{F}$ le faisceau sur $\mathrm{C{\scriptstyle RIS}}(S/\Sigma)$ dont la valeur sur un épaississement (U, T, δ) est donnée par

$$\underline{F}(U, T, \delta) = F(U).$$

Si $\Sigma = \operatorname{Spec} \mathbb{Z}_p$, $\mathcal{I} = p\mathcal{O}_\Sigma$, si $\mathcal{O}_{S/\Sigma}$ désigne le faisceau structural de $\mathrm{C{\scriptstyle RIS}}(S/\Sigma)$, et si G est un schéma en groupes commutatifs, fini localement libre sur S, *le complexe de Dieudonné* de G est un complexe parfait de $\mathcal{O}_{S/\Sigma}$-modules, d'amplitude parfaite contenue dans $[0, 1]$, défini comme le tronqué au degré 1

$$\Delta(G) = t_{1]} \mathbb{R}\mathcal{H}om_{S/\Sigma}(\underline{G}, \mathcal{O}_{S/\Sigma}),$$

où $\mathcal{H}om_{S/\Sigma}$ désigne le faisceau des homomorphismes de groupes abéliens sur $\mathrm{C{\scriptstyle RIS}}(S/\Sigma)$. Le cristal de Dieudonné de G est le cristal en $\mathcal{O}_{S/\Sigma}$-modules localement de présentation finie, défini par

$$\mathbb{D}(G) = \mathcal{H}^1(\Delta(G)) = \mathcal{E}xt^1_{S/\Sigma}(\underline{G}, \mathcal{O}_{S/\Sigma}).$$

Lorsque G est un groupe p-divisible, le cristal de Dieudonné de G est défini de la même manière, et est alors localement libre de rang égal à la hauteur de G. Dans ces deux cas, il résulte de ce que $\mathbb{D}(G)$ est un cristal, et de sa définition comme foncteur dérivé du faisceau des homomorphismes sur le gros site cristallin, que, pour tout morphisme $u : (U', T', \delta') \to (U, T, \delta)$ de $\mathrm{C{\scriptstyle RIS}}(S/\Sigma)$, on a des isomorphismes canoniques

$$\mathbb{D}(G)_T \simeq \mathbb{D}(G_U)_T,$$
$$u^*(\mathbb{D}(G)_T) \xrightarrow{\sim} \mathbb{D}(G)_{T'} \simeq \mathbb{D}(G_{U'})_{T'}.$$

1. Cristaux sur un anneau possédant une p-base

Dans ce chapitre, de nature préliminaire, nous allons donner une description de la catégorie des cristaux sur un anneau possédant une p-base en termes de modules à connexion intégrable (cf. [7, 1.2.9]), plus agréable à manipuler que la description générale qui peut être donnée au moyen d'un plongement dans un schéma quasi-lisse [3, IV 1.6.3].

1.1. *Relèvements des anneaux ayant une p-base : existence.*

Nous utiliserons la définition suivante :

1.1.1. *Définition.* Soient A un anneau de caractéristique p, et $(x_i)_{i\in I}$ une famille d'éléments de A. On dit que (x_i) est une p-base de A si A, considéré comme A-module par l'intermédiaire du Frobenius $F_A : A \to A$, est un A-module libre admettant pour base les monômes $\underline{x}^{\underline{\alpha}}$, où $\underline{\alpha} = (\alpha_i)_{i\in I}$ est une famille d'entiers telle que pour tout i, on ait $0 \leq \alpha_i \leq p-1$, et $\alpha_i = 0$ sauf pour un nombre fini d'indices.

Par récurrence sur n, les monômes $\underline{x}^{\underline{\alpha}}$, où pour tout i on a $0 \leq \alpha_i \leq p^n - 1$, et $\alpha_i = 0$ sauf pour un nombre fini d'indices, forment alors une base de A considéré comme A-module par l'intermédiaire de F_A^n.

On observera que la définition adoptée ici est plus restrictive que celle de [EGA 0_{IV}, (21.1.9)], où l'on demande seulement que A soit libre de base les $\underline{x}^{\underline{\alpha}}$ sur le sous-anneau de A formé des puissances p-ièmes. Si A possède une p-base au sens adopté ici, F_A est injectif, et A est réduit.

Il résulte de [EGA 0_{IV}, (21.2.7)] qu'un anneau possédant une p-base est formellement lisse relativement à $\mathbf{F}_p$.

1.1.2. *Exemples.*

(i) Tout corps de caractéristique p possède une p-base [EGA 0_{IV} (21.4.2)].

(ii) Si A possède une p-base $(x_i)_{i\in I}$, et si A' est étale sur A, la famille $(x_i)_{i\in I}$ est aussi une p-base de A'.

(iii) Si A posséde une p-base $(x_i)_{i\in I}$, l'anneau $A' = A[T_j]_{j\in J}$ admet pour p-base la famille $(x_i, T_j)_{i\in I, j\in J}$. Tout schéma lisse sur un anneau ayant une p-base est localement spectre d'un anneau ayant une p-base.

(iv) Si A possède une p-base finie $(x_i)_{i\in I}$, et si J est un ensemble fini, l'anneau $A' = A[[Tj]]_{j\in J}$ admet pour p-base la famille $(x_i, T_j)_{i\in I, j\in J}$.

(v) Si A est un anneau local régulier essentiellement de type fini sur un corps de caractéristique p, alors A possède une p-base [16].

1.1.3. Si A est un anneau de caractéristique p, nous appelerons *relève-*

ment de A (resp. *relèvement modulo* p^n) la donnée d'une $\mathbf{Z}_p$-algèbre (resp. $\mathbf{Z}/p^n$-algèbre) plate B, complète pour la topologie p-adique, et d'un isomorphisme $B/pB \simeq A$. Nous allons montrer que tout anneau possédant une p-base possède un relèvement.

Soient donc A un anneau de caractéristique p, muni d'une p-base $(x_i)_{i\in I}$, et $W(A)$ (resp. $W_n(A)$) l'anneau des vecteurs de Witt à coefficients dans A (resp. de longueur n). Pour tout $x \in A$, on note $\tilde{x} = (x, 0, \ldots)$ le représentant de Teichmüller de x dans $W(A)$. Soit A_n le sous-anneau de $W_n(A)$ engendré par les éléments de la forme $p^j \tilde{a}^{p^{n-j}} \underline{\tilde{x}}^{\underline{\alpha}}$, avec $0 \leq j < n$, $\underline{\alpha} = (\alpha_i)_{i\in I}$, $0 \leq \alpha_i < p^{n-j}$, $\alpha_i = 0$ sauf pour un nombre fini d'indices, et $\underline{\tilde{x}}^{\underline{\alpha}} = \prod_i \tilde{x}_i^{\alpha_i}$.

1.1.4. Lemme. *Soient A un anneau de caractéristique p, muni d'une p-base $(x_i)_{i\in I}$, et k un entier.*

(i) *Soient $(a_{\underline{\alpha},j})$, $(b_{\underline{\alpha},j})$, $0 \leq j \leq k$, $0 \leq \alpha_i < p^{k-j}$, deux familles d'éléments de A. Il existe une famille $(c_{\underline{\alpha},j})$, $0 \leq j \leq k$, $0 \leq \alpha_i < p^{k-j}$, d'éléments de A, et un élément $y \in W(A)$, tels que*

$$\sum_{j=0}^{k} \left(\sum_{0\leq\alpha_i<p^{k-j}} p^j (\tilde{a}_{\underline{\alpha},j}^{p^{n-k}} - \tilde{b}_{\underline{\alpha},j}^{p^{n-k}}) \underline{\tilde{x}}^{\underline{\alpha}} \right) = \sum_{j=0}^{k} \left(\sum_{0\leq\alpha_i<p^{k-j}} p^j \tilde{c}_{\underline{\alpha},j}^{p^{k-j}} \underline{\tilde{x}}^{\underline{\alpha}} \right) + p^k V(y). \tag{1.1.4.1}$$

(ii) *Pour tout $a \in A$, il existe des éléments $b_{\underline{\beta},j} \in A$, $y \in W(A)$ tels que :*

$$\tilde{a}^{p^k} = \sum_{j=0}^{k} \left(\sum_{0\leq\beta_i<p^{k-j+1}} p^j (\tilde{b}_{\underline{\beta},j}^{p^{k-j+1}} \underline{\tilde{x}}^{\underline{\beta}} \right) + p^k V(y). \tag{1.1.4.2}$$

Montrons (i) par récurrence sur k. Pour $k = 0$, c'est clair. Supposons l'assertion prouvée pour $k-1$. Soit S le terme de gauche de (1.1.4.1) ; pour tout $\underline{\alpha}$, et tout $j < k$, posons $\underline{\alpha} = \underline{\beta}p + \underline{\rho}$, avec $0 \leq \rho_i < p$ pour tout i, et

$$S_\rho = \sum_{j=0}^{k-1} \left(\sum_{0\leq\beta_i<p^{k-j-1}} p^j (\tilde{a}_{\underline{\alpha},j}^{p^{k-j}} - \tilde{b}_{\underline{\alpha},j}^{p^{k-j}}) \underline{\tilde{x}}^{p\underline{\beta}} \right),$$

$$S'_\rho = \sum_{j=0}^{k-1} \left(\sum_{0\leq\beta_i<p^{k-j-1}} p^j (\tilde{a}_{\underline{\alpha},j}^{p^{k-j-1}} - \tilde{b}_{\underline{\alpha},j}^{p^{k-j-1}}) \underline{\tilde{x}}^{\underline{\beta}} \right).$$

L'hypothèse de récurrence entraîne qu'il existe des éléments $c_{\underline{\alpha},j} \in A$, $y_{\underline{\rho}} \in W(A)$ tels que

$$S'_{\rho} = \sum_{j=0}^{k-1} \left(\sum_{0 \leq \beta_i < p^{k-j-1}} p^j \tilde{c}_{\underline{\alpha},j}^{k-j-1} \underline{x}^{\underline{\beta}}) + p^{k-1} V(y_{\underline{\rho}}) \right).$$

Appliquant Frobenius à cette relation, on obtient la relation

$$S_{\rho} = \sum_{j=0}^{k-1} \left(\sum_{0 \leq \beta_i < p^{k-j-1}} p^j \tilde{c}_{\underline{\alpha},j}^{p^{k-j}} \underline{x}^{p\underline{\beta}} \right) + p^k y_{\underline{\rho}}.$$

Comme on a

$$S = \sum_{0 \leq \rho_i < p} S_{\underline{\rho}} \underline{\tilde{x}}^{\underline{\rho}} + p^k (\tilde{a}_k - \tilde{b}_k),$$

il suffit d'écrire

$$\tilde{a}_k - \tilde{b}_k + \sum_{0 \leq \rho_i < p} y_{\underline{\rho}} \underline{\tilde{x}}^{\underline{\rho}} = \tilde{c}_k + V(y)$$

pour en déduire la relation (1.1.4.1).

Pour prouver (ii), on procède encore par récurrence sur k. Pour $k = 0$, la relation se réduit à

$$\tilde{a} = \sum_{0 \leq \beta_i < p} \tilde{b}_{\underline{\beta}}^p \underline{\tilde{x}}^{\underline{\beta}} + V(y), \tag{1.1.4.3}$$

et résulte de ce que (x_i) est une p-base. Supposons la relation (1.1.4.2) vérifiée pour $k - 1$. En appliquant Frobenius, on obtient

$$\tilde{a}^{p^k} = \sum_{j=0}^{k-1} \left(\sum_{0 \leq \beta_i < p^{k-j}} p^j \tilde{b}_{\underline{\beta},j}^{p^{k-j+1}} \underline{\tilde{x}}^{p\underline{\beta}} \right) + p^k y.$$

On peut écrire $y = \tilde{a}' + V(z)$; en appliquant (1.1.4.3) à $\tilde{a}'$, on obtient la relation voulue.

1.1.5. Lemme. *Soit $k \geq n$ un entier. Avec les notations et les hypothèses de* (1.1.3), *tout élément y de A_n s'écrit de manière unique*

$$y = \sum_{j=0}^{n-1} \left(\sum_{0 \leq \alpha_i < p^{k-j}} p^j \tilde{a}_{\underline{\alpha},j}^{p^{k-j}} \underline{\tilde{x}}^{\underline{\alpha}} \right), \tag{1.1.5.1}$$

avec $a_{\underline{\alpha},j} \in A$.

Montrons par récurrence sur n que toute relation de la forme

$$(1.1.5.2)\qquad \sum_{j=0}^{n-1}\left(\sum_{0\leq\alpha_i<p^{k-j}} p^j\tilde{a}_{\underline{\alpha},j}^{p^{k-j}}\,\underline{\tilde{x}}^{\underline{\alpha}}\right)=\sum_{j=0}^{n-1}\left(\sum_{0\leq\alpha_i<p^{k-j}} p^j\tilde{b}_{\underline{\alpha},j}^{p^{k-j}}\,\underline{\tilde{x}}^{\underline{\alpha}}\right),$$

entraîne $a_{\underline{\alpha},j} = b_{\underline{\alpha},j}$ pour tout $\underline{\alpha}$ et tout j. Si $n = 1$, la relation (1.1.5.2) se réduit à

$$\sum_{0\leq\alpha_i<p^k} a_{\underline{\alpha}}^{p^k}\,\underline{x}^{\underline{\alpha}} = \sum_{0\leq\alpha_i<p^k} b_{\underline{\alpha}}^{p^k}\,\underline{x}^{\underline{\alpha}},$$

et entraîne donc que $a_{\underline{\alpha}} = b_{\underline{\alpha}}$ pour tout $\underline{\alpha}$. Si l'assertion pour $n-1$ est vraie, la relation (1.1.5.2) se réduit à

$$p^{n-1}\left(\sum_{0\leq\alpha_i<p^{k-n+1}} \tilde{a}_{\underline{\alpha},n-1}^{p^{k-n+1}}\,\underline{\tilde{x}}^{\underline{\alpha}}\right) = p^{n-1}\left(\sum_{0\leq\alpha_i<p^{k-n+1}} \tilde{b}_{\underline{\alpha},n-1}^{p^{k-n+1}}\,\underline{\tilde{x}}^{\underline{\alpha}}\right).$$

Comme on a

$$\left(\sum_{0\leq\alpha_i<p^{k-n+1}} \tilde{a}_{\underline{\alpha},n-1}^{p^{k-n+1}}\,\underline{\tilde{x}}^{\underline{\alpha}}\right) \equiv \left(\sum_{0\leq\alpha_i<p^{k-n+1}} a_{\underline{\alpha},n-1}^{p^{k-n+1}}\,\underline{x}^{\underline{\alpha}}\right)^{\sim} \mod VW_n(A),$$

et que, pour tout $c \in A$, on a $p^{n-1}\tilde{c} = (0, \ldots, c^{p^{n-1}})$, la relation précédente, jointe au fait que A est réduit, entraîne :

$$\sum_{0\leq\alpha_i<p^{k-n+1}} a_{\underline{\alpha},n-1}^{p^{k-n+1}}\,\underline{x}^{\underline{\alpha}} = \sum_{0\leq\alpha_i<p^{k-n+1}} b_{\underline{\alpha},n-1}^{p^{k-n+1}}\,\underline{x}^{\underline{\alpha}},$$

d'où l'unicité des $a_{\underline{\alpha},n-1}$.

Pour prouver que tout élément de A_n possède une telle écriture, considérons l'ensemble $A_{n,k}$ des éléments de $W_n(A)$ qui s'écrivent sous la forme (1.1.5.1). Il est clair que $A_{n,k} \subset A_n$ pour tout $k \geq n$, de sorte qu'il suffit de prouver que $A_{n,k}$ est un anneau, et qu'il contient les éléments de la forme $p^j\tilde{a}^{p^{n-j}}\underline{\tilde{x}}^{\underline{\alpha}}$, avec $0 \leq j < n$, $0 \leq \alpha_i < p^{n-j}$. Il résulte de (1.1.4) (i) que $A_{n,k}$ est un sous-groupe additif de A_n. Considérons d'autre part deux éléments de la forme

$$y = p^j\tilde{a}^{p^{k-j}}\underline{\tilde{x}}^{\underline{\alpha}},\ y' = p^{j'}\tilde{a}'^{p^{k-j'}}\underline{\tilde{x}}^{\underline{\alpha}'},$$

avec $j + j' \leq n$. Pour tout i, posons $\alpha_i + \alpha'_i = q_ip^{k-j-j'} + \alpha''_i$, avec $0 \leq \alpha''_i < p^{k-j-j'}$. On peut alors écrire

$$yy' = p^{j+j'}(\tilde{a}^{p^{j'}}\tilde{a}'^{p^j}\underline{\tilde{x}}^{\underline{q}})^{p^{k-j-j'}}\underline{\tilde{x}}^{\alpha''};$$

il en résulte que $A_{n,k}$ est bien un anneau.

Il est clair que $A_{n,n} = A_n$. Si $h \geq n$, on peut écrire grâce à (1.1.4) (ii)

$$p^j \tilde{a}^{p^{h-j}} \underline{\tilde{x}}^{\underline{\alpha}} = p^j \underline{\tilde{x}}^{\underline{\alpha}} \left(\sum_{j'=0}^{h-j} \left(\sum_{0 \leq \beta_i < p^{h-j-j'+1}} p^{j'} \tilde{b}_{\underline{\beta},j'}^{p^{h-j-j'+1}} \underline{\tilde{x}}^{\underline{\beta}} \right) + p^{h-j} V(y) \right)$$

$$= \sum_{j''=j}^{h} \left(\sum_{0 \leq \beta_i < p^{h-j''+1}} p^{j''} \tilde{b}_{\underline{\beta},j''-j}^{p^{h-j''+1}} \underline{\tilde{x}}^{\underline{\alpha}+\underline{\beta}} \right),$$

d'où l'on déduit que $p^j \tilde{a}^{p^{h-j}} \underline{\tilde{x}}^{\underline{\alpha}} \in A_{n,h+1}$, et, par récurrence, que $A_{n,k} = A_n$.

1.1.6. Lemme. *Avec les notations de* (1.1.2), *on a :*

(i) *Le noyau de l'homomorphisme canonique $A_n \to A$ est pA_n.*

(ii) *Le noyau de la multiplication par p^k sur A_n est $p^{n-k}A_n$.*

Ces deux assertions résultent de la relation générale

$$A_n \cap V^k W_n(A) = p^k A_n. \tag{1.1.6.1}$$

Pour la vérifier, observons qu'un élément de $A_n \cap V^k W_n(A)$ peut s'écrire, grâce à (1.1.5),

$$y = \sum_{j=0}^{n-1} \left(\sum_{0 \leq \alpha_i < p^{n+k-j}} p^j \tilde{a}_{\underline{\alpha},j}^{p^{n+k-j}} \underline{\tilde{x}}^{\underline{\alpha}} \right).$$

La nullité de l'image de y dans $W_k(A)$ entraîne, d'après l'unicité de l'écriture (1.1.5.1), que $a_{\underline{\alpha},j} = 0$ pour $j < k$, d'où :

$$y = p^k \sum_{j=k}^{n-1} \left(\sum_{0 \leq \alpha_i < p^{n+k-j}} p^{j-k} \tilde{a}_{\underline{\alpha},j}^{p^{n+k-j}} \underline{\tilde{x}}^{\underline{\alpha}} \right),$$

donc $y = p^k z$, avec $z \in A_n$.

1.1.7. Proposition. *Tout anneau de caractéristique p ayant une p-base possède un relèvement.*

Soit A un anneau de caractéristique p ayant une p-base ; choisissons une p-base $(x_i)_{i\in I}$, et soit $A_n \subset W_n(A)$ le sous-anneau défini en (1.1.3). Pour $k \geq n$, l'homomorphisme $W_k(A) \to W_n(A)$ envoie surjectivement A_k sur A_n d'après (1.1.5), et le noyau de $A_k \to A_n$ est $p^n A_k$ d'après (1.1.6.1). D'autre part, A_n est plat sur $\mathbb{Z}/p^n$ d'après (1.1.6) (ii). Par conséquent, l'anneau $A_\infty = \varprojlim_n A_n$ est un relèvement de A (cf. [7, 1.2.6]).

1.2. *Relèvement des anneaux ayant une p-base : propriétés.*

1.2.1. Nous allons maintenant donner quelques propriétés des relèvements d'un anneau ayant une p-base. On suppose donc donnés un anneau A de caractéristique p, une p-base $(x_i)_{i\in I}$ de A, et un relèvement A_n de A modulo p^n. Pour tout $m \leq n$, on notera A_m l'anneau $A_n/p^m A_n$.

Pour tout entier $m \geq 0$, on notera d'autre part $B_{n,m}$ le sous-anneau de A_n engendré par les éléments de la forme $p^{m-j}a^{p^j}$, pour $0 \leq j \leq m$, et $a \in A_n$; il peut aussi être engendré par les éléments de la forme $p^\alpha a^{p^\beta}$ avec $\alpha + \beta \geq m$, qui forment une famille multiplicative. Pour $m < 0$, on posera $B_{n,m} = B_{n,0}$. Les $B_{n,m}$ forment une famille décroissante, et on a également les inclusions

$$p^m A_n = p^m B_{n,0} \subset \ldots \subset pB_{n,m-1} \subset B_{n,m}.$$

1.2.2. Lemme. *Soient $x, y \in A_n$. Pour tout entier m, on a*

$$(x+y)^{p^m} \equiv x^{p^m} + y^{p^m} \bmod pB_{n,m-1}.$$

L'assertion étant claire pour $m = 1$, on procède par récurrence sur m. Supposons que

$$(x+y)^{p^{m-1}} = x^{p^{m-1}} + y^{p^{m-1}} + pz,$$

avec $z \in B_{n,m-2}$. On a alors

$$\begin{aligned}(x+y)^{p^m} =& x^{p^m} + y^{p^m} + \sum_{\alpha=1}^{p-1} \frac{p!}{\alpha!(p-\alpha)!}(x^{p^{m-1}})^\alpha (y^{p^{m-1}})^{p-\alpha} \\ &+ p^p z^p + \sum_{\substack{\alpha,\beta\gamma \\ \gamma\neq 0}} \frac{p!}{\alpha!\beta!\gamma!}(x^{p^{m-1}})^\alpha (y^{p^{m-1}})^\beta (pz)^\gamma.\end{aligned}$$

Dans cette expression, la première somme appartient à $pB_{n,m-1}$, et la deuxième, ainsi que le terme $p^p z^p$, à $p^2 B_{n,m-2} \subset pB_{n,m-1}$.

1.2.3. Proposition. *Avec les notations de* (1.2.1), *tout élément de* $B_{n,m}$ *s'écrit sous la forme*

$$x = a_0^{p^m} + pa_1^{p^{m-1}} + \ldots + p^m a_m,$$

et les éléments $p^j a_j^{p^{m-j}}$ *sont déterminés de manière unique si* $m \geq n$.

Montrons l'existence d'une telle expression par récurrence sur m, le cas $m = 0$ étant trivial. Comme les éléments de la forme $p^\alpha a^{p^\beta}$, avec $\alpha + \beta \geq m$, forment une famille multiplicative, $B_{n,m}$ est l'ensemble des sommes de la forme

$$x = \sum_k p^{\alpha_k} a_k^{p^{\beta_k}}$$

avec $\alpha_k + \beta_k \geq m$. On peut donc écrire

$$x \equiv a_1^{p^m} + \ldots + a_r^{p^m} \bmod pB_{n,m-1},$$

d'où, grâce au lemme précédent,

$$x \equiv (a_1 + \ldots + a_r)^{p^m} \bmod pB_{n,m-1}.$$

L'assertion en résulte en appliquant l'hypothèse de récurrence.

Pour prouver l'unicité de l'écriture précédente, on procède par récurrence sur n. Une égalité de la forme

$$\sum_{j \geq 0} p^j a_j^{p^{m-j}} = \sum_{j \geq 0} p^j a_j'^{p^{m-j}}$$

entraîne la congruence $a_0 \equiv a_0' \bmod p$, donc $a_0^{p^m} = a_0'^{p^m}$ pour $m \geq n$. On obtient donc

$$p\left(\sum_{j \geq 1} p^{j-1} a_j^{p^{m-j}}\right) = p\left(\sum_{j \geq 1} p^{j-1} a_j'^{p^{m-j}}\right),$$

d'où, par platitude de A_n,

$$\sum_{j \geq 1} p^{j-1} a_j^{p^{m-j}} = \sum_{j \geq 1} p^{j-1} a_j'^{p^{m-j}}$$

dans A_{n-1}. L'hypothèse de récurrence entraîne donc que dans A_n, on a la congruence $p^{j-1} a_j^{p^{m-j}} \equiv p^{j-1} a_j'^{p^{m-j}} \bmod p^{n-1} A_n$ pour tout $j \geq 1$, d'où $p^j a_j^{p^{m-j}} = p^j a_j'^{p^{m-j}}$ pour tout j.

Remarque. Soient S_j, P_j les polynômes universels à coefficients entiers définissant la somme et le produit dans l'anneau des vecteurs de Witt. Soient

$$x = a_0^{p^m} + pa_1^{p^{m-1}} + \ldots + p^m a_m, y = b_0^{p^m} + pb_1^{p^{m-1}} + \ldots + p^m b_m,$$

deux éléments de $B_{n,m}$. Par définition des S_k et des P_k, on a

$$(1.2.3.1) \qquad x + y = \sum_{j=0}^{m} p^j S_j(a_0, \ldots, a_j; b_0, \ldots, b_j)^{p^{m-j}},$$

$$(1.2.3.2) \qquad xy = \sum_{j=0}^{m} p^j P_j(a_0, \ldots, a_j; b_0, \ldots, b_j)^{p^{m-j}}.$$

1.2.4. Lemme. *Avec les notations de* (1.2.1), *on a quels que soient* $m, k \in \mathbf{N}$

$$(p^k A_n) \cap B_{n,m} = p^k B_{n,m-k}.$$

Pour $k \geq m$ et pour $k = 0$, l'assertion est triviale. Supposons la démontrée pour $k-1$. Soit $x \in p^k A_n \cap B_{n,m}$; on peut donc écrire $x = p^{k-1}x'$, où $x' \in B_{n,m-k+1}$. On a alors une égalité de la forme $p^{k-1}x' = p^k y$, avec $y \in A_n$. Par platitude, x' peut donc s'écrire $x' = py + p^{n-k+1}z$, avec $z \in A_n$. L'assertion étant triviale pour $k \geq n$, on peut supposer que $n - k + 1 \geq 1$, et écrire simplement $x' = py$. D'après (1.2.3), on peut d'autre part écrire

$$x' = a_0'^{p^{m-k+1}} + pa_1'^{p^{m-k}} + \ldots + p^{m-k+1}a'_{m-k+1} \equiv a_0'^{p^{m-k+1}} \bmod pB_{n,m-k}.$$

Par réduction modulo p, on voit que $a'_0 = 0$, de sorte que $a'_0 = pa''_0$. On obtient alors la relation $a_0'^{p^{m-k+1}} = p^{p^{m-k+1}} a_0''^{p^{m-k+1}}$, dont on déduit que $a_0'^{p^{m-k+1}} \in pB_{n,m-k+1}$. Par suite, $x' \in pB_{n,m-k}$, et $x \in p^k B_{n,m-k}$.

Supposons choisi pour tout $i \in I$ un relèvement $\tilde{x}_i$ de x_i dans A_n, et soit $(X_i)_{i \in I}$ une famille d'indéterminées ; notons

$$f : B_{n,m}[X_i]_{i \in I} \to A_n$$

l'homomorphisme envoyant X_i sur $\tilde{x}_i$.

1.2.5. Proposition. *Avec les notations précédentes, l'homomorphisme f induit un isomorphisme*

$$B_{n,m}[X_i]_{i\in I}/(X_i^{p^m} - \tilde{x}_i^{p^m}, pX_i^{p^{m-1}} - p\tilde{x}_i^{p^{m-1}}, \ldots, p^m X_i - p^m \tilde{x}_i)_{i\in I} \xrightarrow{\sim} A_n.$$

Pour $n = 1$, l'énoncé résulte de ce que $(x_i)_{i\in I}$ est une p-base. Supposons la proposition prouvée pour tout m, et tout $n' < n$. Le fait que f soit surjectif résulte immédiatement de ce que $(x_i)_{i\in I}$ est une p-base.

Soit $P(\underline{X}) \in B_{n,m}[X_i]_{i\in I}$ un polynôme d'image nulle dans A_n. L'hypothèse de récurrence permet d'écrire

$$P(\underline{X}) = \sum_{i\in I} \sum_{j=0}^{\mathrm{Inf}(m,n-2)} Q_{i,j}(\underline{X})(p^j X_i^{p^{m-j}} - p^j \tilde{x}_i^{p^{m-j}}) + P'(\underline{X}),$$

où $P'(\underline{X})$ est à coefficients dans

$$\mathrm{Ker}(B_{n,m} \to B_{n-1,m}) = B_{n,m} \cap p^{n-1}A_n = p^{n-1}B_{n,m-n+1}.$$

Soit donc $P' = p^{n-1}P''$, où P'' est à coefficients dans $B_{n,m-n+1}$. Comme $P(\underline{\tilde{x}}) = 0$, on obtient la relation $p^{n-1}P''(\underline{\tilde{x}}) = 0$, soit encore $P''(\underline{\tilde{x}}) \in pA_n$. Par réduction modulo p, le fait que $(x_i)_{i\in I}$ soit une p-base entraîne qu'il existe des polynômes $Q_i(\underline{X})$, $Q(\underline{X})$ tels que $P''(\underline{X}) = \sum_i Q_i(\underline{X})(X_i^{p^{m-n+1}} - \tilde{x}_i^{p^{m-n+1}}) + pQ(\underline{X})$, d'où

$$P'(\underline{X}) = \sum_i Q_i(\underline{X})(p^{n-1}X_i^{p^{m-n+1}} - p^{n-1}\tilde{x}_i^{p^{m-n+1}}),$$

ce qui achève la récurrence.

1.2.6. Proposition. *Soient A un anneau de caractéristique p ayant une p-base $(x_i)_{i\in I}$, A_n un relèvement de A modulo p^n, $\pi : A_n \twoheadrightarrow A$ l'homomorphisme canonique, $(\tilde{x}_i)_{i\in I}$ une famille d'éléments de A_n relevant $(x_i)_{i\in I}$, $\varphi : A \to A'$ un homomorphisme d'anneaux, $\pi' : C \twoheadrightarrow A'$ un homomorphisme surjectif de noyau J, où C est tel que $p^nC = 0$. On suppose qu'il existe un entier k tel que tout élément $t \in J$ vérifie la relation $t^{p^k} \in pC$. Alors, pour toute famille $(y_i)_{i\in I}$ d'éléments de C telle que, quel que soit i, $\pi'(y_i) = \varphi(x_i)$, il existe un unique homomorphisme $\psi : A_n \to C$ tel que $\psi(\tilde{x}_i) = y_i$.*

Observons d'abord que la restriction de ψ à $B_{n,n+k-1}$ est entièrement déterminée. Il suffit en effet que pour tout $x \in A_n$, et pour tout $j < n$, l'élément $\psi(p^j x^{p^{n+k-1-j}})$ soit entièrement déterminé. Or on doit avoir $\psi(p^j x^{p^{n+k-1-j}}) = p^j \psi(x)^{p^{n+k-1-j}}$; d'autre part, si $y, y+t$ sont deux relèvements de $\varphi(\pi(x))$, ils vérifient la congruence

$$y^{p^k} \equiv (y+t)^{p^k} \bmod pC,$$

grâce à la relation $t^{p^k} \in pC$; par suite, $p^j y^{p^{n+k-1-j}} \equiv p^j (y+t)^{p^{n+k-1-j}} \bmod p^n C$, et l'élément $p^j y^{p^{n+k-1-j}}$ de C est indépendant du choix du relèvement y.

La proposition (1.2.3) permet donc de définir $\psi : B_{n,n+k-1} \to C$ en écrivant un élément $x \in B_{n,n+k-1}$ sous la forme $x = \sum_j p^j a_j^{p^{n+k-1-j}}$, en choisissant des relèvements b_j des $\varphi(\pi(a_j))$ dans C, et en posant $\psi(x) = \sum_j p^j b_j^{p^{n+k-1-j}}$. Cet élément est bien défini d'après ce qui précède, et on vérifie facilement que ψ est un homomorphisme d'anneaux grâce aux relations (1.2.3.1) et (1.2.3.2).

La proposition (1.2.5) montre alors que prolonger φ équivaut à se donner une famille d'éléments $(y_i)_{i\in I}$ relevant $(\varphi(x_i))_{i\in I}$ et vérifiant les relations

$$p^j y_i^{p^{n+k-1-j}} = \psi(p^j \tilde{x}_i^{p^{n+k-1-j}})$$

pour tout $j < n$. Comme ces relations sont vérifiées d'après la définition de ψ sur $B_{n,n+k-1}$, l'existence et l'unicité de ψ en résultent.

1.2.7. Corollaire. *Soient A un anneau de caractéristique p ayant une p-base $(x_i)_{i\in I}$, A_n un relèvement de A modulo p^n, et $(\tilde{x}_i)_{i\in I}$ une famille de relèvements des x_i dans A_n.*

(i) *Si A'_n est un autre relèvement de A, A_n et A'_n sont isomorphes ; plus précisément, si $(\tilde{x}'_i)_{i\in I}$ est une famille de relèvements des x_i dans A'_n, il existe un unique isomorphisme $\iota : A_n \xrightarrow{\sim} A'_n$ tel que $\iota(\tilde{x}_i) = \tilde{x}'_i$ pour tout i.*

(ii) *La donnée d'un relèvement à A_n de l'endomorphisme de Frobenius de A équivaut à celle d'une famille $(y_i)_{i\in I}$ d'éléments de A_n tels que $y_i \equiv \tilde{x}_i^p \bmod pA_n$, et un tel relèvement est plat.*

En utilisant la platitude de A_n, on vérifie en effet facilement que les éléments $\underline{\tilde{x}}^{\underline{\alpha}}$, avec $0 \leq \alpha_i < p$, forment une base de A_n, considéré comme module sur lui-même grâce à un relèvement de Frobenius.

1.3. *Cristaux et modules à connexion.*

Nous allons utiliser l'existence de relèvements pour les anneaux ayant une p-base pour donner une description des cristaux sur un tel anneau en termes de modules à connexion.

1.3.1. Proposition. *Soient A un anneau de caractéristique p, $(x_i)_{i\in I}$ une p-base de A, A_n un relèvement de A modulo p^n, et $(\tilde{x}_i)_{i\in I}$ une famille d'éléments de A_n relevant les x_i. Alors, avec les notations de (1.2.1), l'homomorphisme naturel $\Omega^1_{A_n/\mathbf{Z}} \to \Omega^1_{A_n/B_{n,n}}$ est un isomorphisme, et les éléments $(d\tilde{x}_i)_{i\in I}$ forment une base de $\Omega^1_{A_n/\mathbf{Z}}$ en tant que A_n-module.*

Comme $d(p^j \tilde{x}^{p^{n-j}}) = 0$, la première assertion est claire. Pour prouver la seconde, il suffit d'après (1.2.5) de vérifier que

$$d(p^j X_i^{p^{n-j}} - p^j \tilde{x}_i^{p^{n-j}}) = 0,$$

ce qui est clair également.

1.3.2. Corollaire. *Sous les hypothèses précédentes, soit $D_{A_n/\mathbf{Z}}(1)$ l'enveloppe à puissances divisées de l'idéal diagonal dans $A_n \otimes_{\mathbf{Z}} A_n$. Alors :*

(i) *Pour chacune de ses deux structures naturelles de A_n-algèbre, $D_{A_n/\mathbf{Z}}(1)$ est une algèbre de polynômes à puissances divisées par rapport aux éléments $\xi_i = 1 \otimes \tilde{x}_i - \tilde{x}_i \otimes 1$.*

(ii) *Si M est un A_n-module, la donnée d'un isomorphisme*

$$\varepsilon : D_{A_n/\mathbf{Z}}(1) \otimes_{A_n} M \xrightarrow{\sim} M \otimes_{A_n} D_{A_n/\mathbf{Z}}(1)$$

se réduisant à l'identité par l'homomorphisme d'augmentation vers A_n, et vérifiant la condition de cocycle usuelle, est équivalente à celle d'une connexion intégrable et quasi-nilpotente

$$\nabla : M \to M \otimes_{A_n} \Omega^1_{A_n/\mathbf{Z}},$$

i.e. d'une famille de $\partial/\partial\tilde{x}_i$-dérivations $\nabla_i : M \to M$ telles que $[\nabla_i, \nabla_j] = 0$ pour tous i, j, que, pour tout $m \in M$, les $\nabla_i(m)$ soient nuls sauf un nombre fini, et que les ∇_i soient des endomorphismes quasi-nilpotents de M.

La première assertion se montre comme les assertions analogues de [3, II 4.2.7, 4.3.4] ; l'hypothèse de finitude faite sur la famille (x_i) dans [3] n'est pas nécessaire ici, l'expression d'un élément dans la p-base ne faisant intervenir qu'un nombre fini d'éléments x_i. De même, la seconde assertion se montre comme [3, II 4.3.10], l'hypothèse faite sur les $\nabla_i(m)$ permettant à nouveau de ne faire intervenir qu'un nombre fini d'éléments x_i.

Des résultats précédents, on déduit une description des cristaux sur A qui généralise la description classique pour les schémas lisses sur un corps parfait.

1.3.3. Proposition. *Soient A un anneau de caractéristique p, possédant une p-base, A_∞ un relèvement de A ; posons $S = \operatorname{Spec} A$, $\Sigma = \operatorname{Spec} \mathbf{Z}_p$. La catégorie des cristaux en modules quasi-cohérents sur $\mathrm{C\scriptstyle RIS}(S/\Sigma)$ est équivalente à la catégorie des A_∞-modules séparés et complets, munis d'une connexion intégrable et topologiquement quasi-nilpotente.*

Posons $S_n = \operatorname{Spec} A_n, \Sigma_n = \operatorname{Spec} \mathbf{Z}/p^n\mathbf{Z}$. Soit (U, T, δ) un objet de $\mathrm{C\scriptstyle RIS}(S/\Sigma)$ tel que $p^n\mathcal{O}_T = 0$. L'existence de la PD-structure δ entraîne que, pour toute section t de $\operatorname{Ker}(\mathcal{O}_T \to \mathcal{O}_U)$, on a $t^p = p!\delta_p(t) \equiv 0 \bmod p$. D'après (1.2.6), il existe donc, localement sur T, un morphisme $(U, T, \delta) \to (S, S_n)$ dans $\mathrm{C\scriptstyle RIS}(S/\Sigma)$. L'argument classique [3, IV 1.6.3] montre alors que la catégorie des cristaux en modules sur $\mathrm{C\scriptstyle RIS}(S/\Sigma_n)$ est équivalente à la catégorie des A_n-modules munis d'un isomorphisme ε comme en (1.3.2) (ii). Appliquant (1.3.2) et passant à la limite sur n, on obtient l'énoncé voulu.

1.3.4. Corollaire. *Avec les notations précédentes, soient $\varphi : A \hookrightarrow A'$ un homomorphisme injectif entre deux anneaux ayant une p-base, $S = \operatorname{Spec} A$, $S' = \operatorname{Spec} A'$. Alors le foncteur image inverse de la catégorie des cristaux en modules localement libres sur $\mathrm{C\scriptstyle RIS}(S/\Sigma)$ (resp. $\mathrm{C\scriptstyle RIS}(S/\Sigma_n)$) vers la catégorie analogue sur S' est fidèle.*

Soient A_∞, A'_∞ des relèvements de A et A' ; pour tout n, soient $A_n = A_\infty/p^n A_\infty$, $A'_n = A'_\infty/p^n A'_\infty$. D'après (1.2.6), il existe des homomorphismes $\varphi_n : A_n \to A'_n$ relevant φ, et compatibles entre eux. Utilisant la platitude de A'_n sur $\mathbf{Z}/p^n\mathbf{Z}$, on voit que l'injectivité de φ entraîne celle des φ_n pour tout n. Interprétant grâce à (1.3.3) les cristaux comme des A_∞-modules localement libres munis d'une connexion, i. e. comme des familles de A_n-modules à connexion M_n munis d'isomorphismes $M_n \otimes_{A_n} A_{n-1} \xrightarrow{\sim} M_{n-1}$, l'énoncé résulte de l'injectivité des φ_n.

1.3.5. *Exemples.*

(i) Soient A un anneau intègre ayant une p-base, $A' = K$ son corps des fractions. Le foncteur de restriction à la fibre générique est fidèle sur la catégorie des cristaux localement libres sur A.

(ii) Soient A un anneau ayant une p-base, et A' sa clôture parfaite, définie par

$$A' = \varinjlim(A \xrightarrow{F} A \xrightarrow{F} \cdots A \xrightarrow{F} \cdots).$$

Le foncteur de passage à la clôture parfaite est fidèle sur la catégorie des cristaux localement libres sur A.

1.4. *Relation entre complexe et cristal de Dieudonné.*

Nous allons ici voir comment la donnée du cristal de Dieudonné d'un groupe fini localement libre G permet, lorsque le schéma de base S vérifie certaines conditions, de reconstruire le complexe de Dieudonné. On note toujours $\Sigma = \operatorname{Spec} \mathbf{Z}_p$.

1.4.1. Proposition. *Soient S un schéma sur lequel p est nilpotent, et G un S-groupe fini localement libre. On suppose qu'il existe une PD-immersion fermée de S dans un schéma formel T sur $\mathbf{Z}_p$, tel que $\mathcal{O}_T$ soit sans p-torsion ; on note alors T_n la réduction de T modulo p^n, et on pose*

$$\mathbb{D}(G)_n = \mathbb{D}(G)_{(S,T_n)}, \mathbb{D}(G)_\infty = \varprojlim_n \mathbb{D}(G)_n.$$

Pour tout objet (U, Z, δ) de $\mathrm{CRIS}(S/\Sigma)$, et tout PD- morphisme $g : Z \to T$ prolongeant $U \to S$, il existe un isomorphisme canonique

$$\mathbf{L}g^*(\mathbb{D}(G)_\infty)[-1] \xrightarrow{\sim} \Delta(G)_{(U,Z,\delta)}. \tag{1.4.1.1}$$

Pour n assez grand, g se factorise à travers T_n ; comme l'homomorphisme canonique

$$\mathbf{L}g^*(\Delta(G)_{(S,T_n)}) \to \Delta(G)_{(U,Z,\delta)}$$

est un isomorphisme, il suffit de construire un isomorphisme

$$(\mathbb{D}(G)_\infty \overset{\mathbf{L}}{\otimes}_{\mathcal{O}_T} \mathcal{O}_{T_n})[-1] \to \Delta(G)_n \tag{1.4.1.2}$$

pour n assez grand.

On construit d'abord un morphisme

$$\mathbb{D}(G)_\infty[-1] \to \Delta(G)_n$$

induisant par composition avec le morphisme naturel $\Delta(G)_n \to \mathbb{D}(G)_n[-1]$ l'homomorphisme $\mathbb{D}(G)_\infty[-1] \to \mathbb{D}(G)_n[-1]$. Par définition, on peut représenter le complexe

$$\Delta(G) = t_{1]}\mathbf{R}\mathcal{H}om_{S/\Sigma}(\underline{G}, \mathcal{O}_{S/\Sigma})$$

par un complexe de longueur 1 de faisceaux sur $\text{Cris}(S/\Sigma)$:

$$\Delta(G) = [L^0 \to L^1],$$

donnant naissance pour tout n à une suite exacte de faisceaux sur T_n :

$$0 \to \mathcal{H}om_{S/\Sigma}(\underline{G}, \mathcal{O}_{S/\Sigma})_n \to L_n^0 \to L_n^1 \to \mathbb{D}(G)_n \to 0. \tag{1.4.1.3}$$

Dans le système projectif de suites exactes ainsi obtenues, les homomorphismes $\mathbb{D}(G)_m \to \mathbb{D}(G)_n$ sont des isomorphismes dès que p^n annule G. L'hypothèse faite sur T entraîne que le système des $\mathcal{H}om_{S/\Sigma}(\underline{G}, \mathcal{O}_{S/\Sigma})_n$ est essentiellement nul, d'après [7, 4.2.5]. Si on note N_m le noyau de $L_m^1 \to \mathbb{D}(G)_m$, le complexe $N_m \to L_m^1$ est une résolution de $\mathbb{D}(G)_m[-1]$, de sorte qu'on obtient pour m assez grand un morphisme dans la catégorie dérivée des $\mathcal{O}_{T_m}$-modules

$$\mathbb{D}(G)_m[-1] \to \Delta(G)_n,$$

donnant par composition avec $\Delta(G)_n \to \mathbb{D}(G)_n[-1]$ l'homomorphisme canonique $\mathbb{D}(G)_m[-1] \to \mathbb{D}(G)_n[-1]$. Par adjonction, on en tire le morphisme (1.4.1.2).

Pour vérifier que (1.4.1.2) est un isomorphisme, on peut localiser sur S et supposer que G est plongé dans un groupe p-divisible H sur S [7, 3.1.1] ; soit H' le groupe p-divisible quotient. Le complexe $\Delta(G)$ est alors représenté par le complexe de cristaux localement libres de rang fini $\mathbb{D}(H') \to \mathbb{D}(H)$. Par passage à la limite, la suite (1.4.1.3), dont tous les termes vérifient la condition de Mittag-Leffler, donne alors la suite exacte

$$0 \to \mathbb{D}(H')_\infty \to \mathbb{D}(H)_\infty \to \mathbb{D}(G)_\infty \to 0$$

qui constitue une résolution plate de $\mathbb{D}(G)_\infty$ sur $\mathcal{O}_T$. L'assertion en résulte immédiatement.

Remarque. Lorsque A est un anneau possédant une p-base, et $S = \operatorname{Spec} A$, on peut prendre pour T le spectre formel d'un relèvement A_∞ de A. Tout objet de $\text{Cris}(S/\Sigma)$ possède alors localement un morphisme dans T, ainsi qu'on l'a remarqué dans la démonstration de (1.3.3), et l'isomorphisme (1.4.1.1) fournit alors une description de $\Delta(G)$ en tant que "cristal en complexes de la catégorie dérivée", compte tenu des propriétés de platitude résultant de (1.3.2).

Cette propriété reste valable plus généralement si S est plat, d'intersection complète relative au-dessus d'un tel anneau, et si on prend pour T le voisinage à puissance divisées (complété) obtenu à partir d'un plongement de S dans un A_∞-schéma formel lisse, grâce au fait que S peut localement être relevé, et à la platitude de l'enveloppe à puissances divisées avec les hypothèses faites ici [7, 2.3.3].

Rappelons [7, 5.2.4] que, si G est un groupe fini localement libre sur une base S, de dual de Cartier G^*, il existe un isomorphisme canonique

$$\Phi_G : \Delta(G)^\vee[-1] \xrightarrow{\sim} \Delta(G^*).$$

De même, on dispose, pour un groupe p-divisible H, de l'isomorphisme

$$\Phi_H : \mathbb{D}(H)^\vee \xrightarrow{\sim} \mathbb{D}(H^*).$$

Lorsque S est le spectre d'un anneau A ayant une p-base, on peut reformuler cet isomorphisme à l'aide de la description donnée en (1.3.3) pour la catégorie des cristaux sur S en termes d'un relèvement A_∞ de A : par passage à la limite à partir des isomorphismes définis par Φ_H sur les A_n, on obtient un isomorphisme compatible aux connexions, ainsi qu'à F et V,

$$\Phi_H : D(H)_\infty^\vee \xrightarrow{\sim} D(H^*)_\infty.$$

Pour les groupes finis localement libres, on peut également donner un énoncé de dualité reliant les cristaux $\mathbb{D}(G)$ et $\mathbb{D}(G^*)$:

1.4.2. Proposition. *Soient A un anneau de caractéristique p, possédant une p-base, A_∞ un relèvement de A, $\Sigma = Spec\ \mathbb{Z}_p$, $S =$ Spec A, $T =$ Spf A_∞, et G un S-groupe fini localement libre, de dual de Cartier G^*. Avec les notations de* (1.4.1), *l'isomorphisme Φ_G définit un isomorphisme*

$$\mathcal{H}om_{\mathcal{O}_T}(\mathbb{D}(G)_\infty, \mathcal{O}_T \otimes \mathbb{Q}_p/\mathbb{Z}_p) \xrightarrow{\sim} \mathbb{D}(G^*)_\infty.$$

Soient $A_n = A_\infty/p^n A_\infty$, $T_n =$ Spec A_n, $\Delta(G)_n = \Delta(G)_{(S,T_n)}$. Comme $\mathbb{D}(G)_\infty$ est de torsion, on dispose d'un isomorphisme

$$\mathcal{H}om_{\mathcal{O}_T}(\mathbb{D}(G)_\infty, \mathcal{O}_T \otimes \mathbb{Q}_p/\mathbb{Z}_p) \xrightarrow{\sim} \mathcal{E}xt^1_{\mathcal{O}_T}(\mathbb{D}(G)_\infty, \mathcal{O}_T)$$

Par ailleurs, l'homomorphisme canonique

$$\mathcal{E}xt^1_{\mathcal{O}_T}(\mathbb{D}(G)_\infty, \mathcal{O}_T) \to \mathcal{E}xt^1_{\mathcal{O}_{T_n}}(\mathbb{D}(G)_\infty \overset{\mathbf{L}}{\otimes} \mathcal{O}_{T_n}, \mathcal{O}_{T_n})$$

est un isomorphisme dès que p^n annule G : cela résulte de ce que $\mathbb{D}(G)_\infty$ admet localement une résolution de longueur 1 par des A_∞-modules libres de rang fini, comme on le voit en écrivant localement G comme noyau d'un homomorphisme entre groupes p-divisibles. En composant avec l'isomorphisme déduit de (1.4.1.2), on obtient donc un isomorphisme

$$\mathcal{H}om_{\mathcal{O}_T}(\mathbb{D}(G)_\infty, \mathcal{O}_T \otimes \mathbb{Q}_p/\mathbb{Z}_p) \xrightarrow{\sim} \mathcal{E}xt^1_{\mathcal{O}_{T_n}}(\Delta_n[1], \mathcal{O}_{T_n}) \simeq \mathcal{E}xt^0_{\mathcal{O}_{T_n}}(\Delta(G)_n, \mathcal{O}_{T_n}).$$

L'isomorphisme Φ_G fournit d'autre part un isomorphisme

$$\mathcal{E}xt^0_{\mathcal{O}_{T_n}}(\Delta(G)_n, \mathcal{O}_{T_n}) \xrightarrow{\sim} \mathbb{D}(G^*)_n \ ;$$

comme $\mathbb{D}(G)_\infty \xrightarrow{\sim} \mathbb{D}(G)_n$ lorsque p^n annule G, on obtient l'isomorphisme voulu, manifestement indépendant de n. Il est facile de vérifier qu'il est compatible aux connexions, ainsi qu'à F et V.

Remarque. Comme en (1.4.1), l'énoncé s'étend aux schémas S qui sont plats d'intersection complète relative sur un anneau ayant une p-base, en prenant un plongement de S dans un schéma formel lisse Y sur A_∞, et en prenant pour T le voisinage à puissances divisées (complété) de S dans Y.

2. Le cristal de Dieudonné des groupes finis étales ou de type multiplicatif

Nous allons d'abord donner une description élémentaire du cristal de Dieudonné associé à un groupe fini étale ou de type multiplicatif. Il en résulte que, pour de tels groupes, et sur une base quelconque sur laquelle p est localement nilpotent, le cristal de Dieudonné fournit une équivalence avec une catégorie convenable de cristaux munis d'opérations F, V, généralisant la relation classique (cf [19, VI 3.1.2], [15, 5.5]) entre systèmes locaux p-adiques et F-cristaux unité sur les schémas lisses sur un corps.

2.1. *Complexe de Dieudonné d'un groupe fini étale.*

2.1.1. Soit G un faisceau abélien sur S. Nous noterons

$$G^\vee = \mathcal{H}om(G, \mathbb{Q}_p/\mathbb{Z}_p)$$

le dual de Pontryagin de G. La suite exacte de faisceaux abéliens sur $\mathrm{Cris}(S/\Sigma)$

$$0 \to \underline{\mathbb{Z}}_p \to \underline{\mathbb{Q}}_p \to \underline{\mathbb{Q}}_p/\underline{\mathbb{Z}}_p \to 0$$

fournit dans la catégorie dérivée un morphisme

$$\underline{\mathbb{Q}}_p/\underline{\mathbb{Z}}_p \to \underline{\mathbb{Z}}_p[1].$$

Par composition, on en tire un morphisme

$$\underline{\mathbb{Q}}_p/\underline{\mathbb{Z}}_p \to \mathcal{O}_{S/\Sigma}[1]. \tag{2.1.1.1}$$

On en déduit par fonctorialité un morphisme

$$\underline{G}^\vee \simeq \mathcal{H}om_{S/\Sigma}(\underline{G}, \underline{\mathbb{Q}}_p/\underline{\mathbb{Z}}_p) \to t_{0]}\mathbf{R}\mathcal{H}om_{S/\Sigma}(\underline{G}, \mathcal{O}_{S/\Sigma}[1]) \simeq \Delta(G)[1],$$

qui fournit par adjonction un morphisme $\mathcal{O}_{S/\Sigma}$-linéaire

$$u_G : \underline{G}^\vee \overset{\mathbf{L}}{\otimes} \mathcal{O}_{S/\Sigma} \to \Delta(G)[1]. \tag{2.1.1.2}$$

2.1.2. Proposition. *Pour tout groupe fini étale G, le morphisme u_G est un isomorphisme.*

L'assertion est locale pour la topologie étale sur $\text{Cris}(S/\Sigma)$. Si $S' \to S$ est un morphisme étale surjectif, et si (U, T, δ) est un objet de $\text{Cris}(S/\Sigma)$, il existe un unique morphisme étale $T' \to T$ relevant $S' \times_S U \to U$. Par platitude, δ s'étend à T, et (U', T', δ) est un objet de $\text{Cris}(S'/\Sigma)$, qui est un recouvrement de (U, T, δ). Si $G' = S' \times_S G$, l'image inverse sur T' de $u_{G(U,T,\delta)}$ s'identifie à $u_{G'(U',T',\delta)}$, car $\Delta(G)$ est un complexe parfait sur $\text{Cris}(S/\Sigma)$. On peut donc faire une localisation étale sur S.

On est ainsi ramené au cas où $G = \mathbb{Z}/p^n\mathbb{Z}$. Prenant la résolution de G définie par la multiplication par p^n sur $\mathbb{Z}$, il suffit par fonctorialité de vérifier que l'homomorphisme

$$\underline{\mathbb{Q}}_p/\underline{\mathbb{Z}}_p \overset{\mathbf{L}}{\otimes} \mathcal{O}_{S/\Sigma} \to \mathcal{O}_{S/\Sigma}[1]$$

défini par adjonction à partir de (2.1.1.1) est un isomorphisme. En remplaçant $\underline{\mathbb{Q}}_p/\underline{\mathbb{Z}}_p$ par la résolution plate $\underline{\mathbb{Z}}_p \to \underline{\mathbb{Q}}_p$, l'assertion résulte de ce que $\mathcal{O}_{S/\Sigma}$ est un faisceau de torsion.

En passant à la cohomologie, l'isomorphisme u_G fournit :

2.1.3. Corollaire. (i) *Pour tout groupe fini étale G, il existe un isomorphisme canonique*

$$u_G : \underline{G}^\vee \otimes \mathcal{O}_{S/\Sigma} \xrightarrow{\sim} \mathbb{D}(G), \tag{2.1.3.1}$$

tel que pour $G = \mathbb{Z}/p^n$, l'image de la section $1 \in \mathbb{Z}/p^n = \underline{G}^\vee$ soit la classe de l'extension déduite de l'extension

$$0 \to \underline{\mathbb{Z}}_p \xrightarrow{p^n} \underline{\mathbb{Z}}_p \to \underline{\mathbb{Z}}/p^n \to 0 \tag{2.1.3.2}$$

par l'homomorphisme naturel $\underline{\mathbb{Z}}_p \to \mathcal{O}_{S/\Sigma}$.

(ii) *Pour tout groupe p-divisible étale H, il existe un isomorphisme canonique*

$$u_H : \underline{H}^\vee \otimes \mathcal{O}_{S/\Sigma} \xrightarrow{\sim} \mathbb{D}(H).$$

La section $1 \in \mathbb{Z}/p^n = (\mathbb{Z}/p^n)^\vee$ correspond à l'inclusion naturelle i_n de $\mathbb{Z}/p^n$ dans $\mathbb{Q}_p/\mathbb{Z}_p$ (envoyant 1 sur p^{-n}). D'après la définition de u_G, il lui correspond la classe de l'extension déduite de

$$0 \to \underline{\mathbb{Z}}_p \to \underline{\mathbb{Q}}_p \to \underline{\mathbb{Q}}_p/\underline{\mathbb{Z}}_p \to 0$$

par image inverse par i_n, qui s'identifie à (2.1.3.2), puis par image directe par l'homomorphisme $\underline{\mathbb{Z}}_p \to \mathcal{O}_{S/\Sigma}$.

Pour vérifier l'assertion (ii), il suffit d'observer que, en notant $H(n)$ le noyau de la multiplication par p^n sur H, on a $\mathbb{D}(H) \simeq \varprojlim_n \mathbb{D}(H(n))$ d'après [7, 2.4.5 (ii)], et $\underline{H}^\vee \otimes \mathcal{O}_{S/\Sigma} \simeq \varprojlim_n (\underline{H}(n)^\vee \otimes \mathcal{O}_{S/\Sigma})$, car il suffit de le vérifier sur chaque épaississement (U, T, δ), et $\mathcal{O}_T$ est localement annulé par une puissance de p.

2.1.4. *Remarques.* (i) Comme, pour tout groupe fini étale G, on a

$$\mathbb{R}\mathcal{H}om_{S/\Sigma}(\underline{G}, \underline{\mathbb{Q}}_p) = 0, \ \mathcal{E}xt^i_{S/\Sigma}(\underline{G}, \underline{\mathbb{Q}}_p/\underline{\mathbb{Z}}_p) = 0 \text{ pour } i \neq 1,$$

l'homomorphisme de (2.1.1) induit un isomorphisme

$$\underline{G}^\vee \xrightarrow{\sim} \mathbb{R}\mathcal{H}om_{S/\Sigma}(\underline{G}, \underline{\mathbb{Z}}_p)[1];$$

compte tenu de ce que $\underline{G}^\vee \xrightarrow{\sim} \underline{G}^\vee \overset{\mathbb{L}}{\otimes} \underline{\mathbb{Z}}_p$, on peut ainsi identifier l'homomorphisme u_G à l'homomorphisme

$$\mathbb{R}\mathcal{H}om_{S/\Sigma}(\underline{G}, \underline{\mathbb{Z}}_p) \overset{\mathbb{L}}{\underset{\mathbb{Z}_p}{\otimes}} \mathcal{O}_{S/\Sigma}[1] \to \Delta(G)[1]$$

défini par fonctorialité, adjonction et décalage à partir de l'homomorphisme naturel $\underline{\mathbb{Z}}_p \to \mathcal{O}_{S/\Sigma}$.

(ii) Rappelons [7, 2.5.7] que le Frobenius de $\Delta(G)$ est défini en se ramenant au cas où S est de caractéristique p, et en prenant l'image inverse par le Frobenius absolu $\mathbf{f}_S$ (ce qui revient à localiser sur $\mathrm{C{\small RIS}}(S/\Sigma)$), puis par fonctorialité par rapport au Frobenius $F_G : G \to G^{(p/S)}$. On en déduit que, via l'isomorphisme u_G, le Frobenius de $\Delta(G)$correspond à $(F_G)^\vee \otimes 1$, soit aussi à $F_{G^\vee}^{-1} \otimes 1$.

2.2. *Complexe de Dieudonné d'un groupe fini de type multiplicatif.*

2.2.1. Rappelons [7, 5.2.1] que la suite exacte canonique de faisceaux sur $\text{Cris}(S/\Sigma)$

$$0 \to 1 + \mathcal{J}_{S/\Sigma} \to \mathcal{O}^*_{S/\Sigma} \to \underline{\mathbb{G}}_m \to 0 \tag{2.2.1.1}$$

induit, après composition avec l'homomorphisme logarithme

$$\log : 1 + \mathcal{J}_{S/\Sigma} \to \mathcal{O}_{S/\Sigma},$$

un morphisme canonique

$$\eta : \underline{\mathbb{G}}_m \to \mathcal{O}_{S/\Sigma}[1]. \tag{2.2.1.2}$$

Supposons maintenant que G soit un faisceau abélien sur S, et posons

$$G^D = \mathcal{H}om(G, \mathbb{G}_m);$$

lorsque G est un groupe fini localement libre, G^D est donc le dual de Cartier G^* de G. Par fonctorialité et adjonction, η définit un morphisme $\mathcal{O}_{S/\Sigma}$-linéaire

$$v_G : \underline{G}^D \overset{\mathbf{L}}{\otimes} \mathcal{O}_{S/\Sigma} \to \Delta(G)[1]. \tag{2.2.1.3}$$

2.2.2. Proposition. *Pour tout groupe fini localement libre de type multiplicatif G, le morphisme*

$$v_G : \underline{G}^* \overset{\mathbf{L}}{\otimes} \mathcal{O}_{S/\Sigma} \to \Delta(G)[1]$$

est un isomorphisme.

Par localisation étale, on peut supposer que $G = \boldsymbol{\mu}_{p^n}$, et même, grâce aux propriétés d'exactitude du complexe de Dieudonné [7, 3.1.6], que $G = \boldsymbol{\mu}_p$. En notant $\boldsymbol{\mu}_{p^\infty}$ le groupe p-divisible $\cup_n \boldsymbol{\mu}_{p^n}$, et en utilisant la résolution de $\boldsymbol{\mu}_p$ par la multiplication par p sur $\boldsymbol{\mu}_{p^\infty}$, on est ramené à vérifier que l'homomorphisme

$$\mathcal{H}om_{S/\Sigma}(\underline{\boldsymbol{\mu}}_{p^\infty}, \underline{\mathbb{G}}_m) \otimes \mathcal{O}_{S/\Sigma} \to \mathcal{E}xt^1_{S/\Sigma}(\underline{\boldsymbol{\mu}}_{p^\infty}, \mathcal{O}_{S/\Sigma})$$

est un isomorphisme. La source et le but étant localement libres de rang 1 sur $\mathcal{O}_{S/\Sigma}$, et le cristal de Dieudonné commutant au changement de base, le lemme de Nakayama ramène à faire cette vérification lorsque S est le spectre d'un corps k, et pour l'épaississement particulier $S \subset S$.

Il s'agit alors de montrer que, par composition avec le morphisme η, l'image de l'inclusion naturelle $i : \underline{\mu}_{p^\infty} \hookrightarrow \mathbb{G}_m$ est une base du k-espace vectoriel (de rang 1) $\mathcal{E}xt^1_{S/\Sigma}(\underline{\mu}_{p^\infty}, \mathcal{O}_{S/\Sigma})_S$, i.e. n'est pas nulle. La valeur sur S des faisceaux $\mathcal{E}xt^i_{S/\Sigma}$ ne dépendant que des épaississements situés au-dessus de S, il suffit pour faire le calcul de se placer sur le site $\mathrm{CRIS}(S/\Sigma_1)$, dont les objets sont de caractéristique p.

Pour la commodité des références ultérieures, nous allons décrire plus généralement l'image de i dans $\mathcal{E}xt^1_{S/\Sigma}(\underline{\mu}_{p^\infty}, \mathcal{O}_{S/\Sigma})$ au-dessus de $\sum_n$ pour tout n, S étant un schéma quelconque de caractéristique p. Considérons le diagramme commutatif de suites exactes

$$\begin{array}{lccccccccc}
(2.2.2.1) & 0 & \longrightarrow & \underline{\mu}_{p^n} & \longrightarrow & \underline{\mu}_{p^\infty} & \xrightarrow{p^n} & \underline{\mu}_{p^\infty} & \to & 0 \\
& & & \downarrow{\scriptstyle \varphi_n} & & \downarrow{\scriptstyle \varphi_n} & & \downarrow{\scriptstyle i} & & \\
(2.2.2.2) & 0 & \longrightarrow & 1+\mathcal{J}_{S/\Sigma_n} & \longrightarrow & \mathcal{O}^*_{S/\Sigma_n} & \longrightarrow & \mathbb{G}_m & \to & 0,
\end{array}$$

dans lequel les flèches verticales φ_n sont définies sur un épaississement (U, Z, δ) en relevant une section s de μ_{p^n} en une section $\tilde{s}$ de $\mathcal{O}^*_Z$, et en associant à s la section $\tilde{s}^{p^n}$; celle-ci ne dépend que de s à cause de la relation $t^p = p!\delta_p(t)$ vérifiée par toute section de $\mathcal{J}_{S/\Sigma_n}$, qui entraîne que $\tilde{s}^{p^n} \equiv (\tilde{s}+t)^{p^n} \bmod p^n$. Il en résulte que les extensions obtenues par image directe de (2.2.2.1) par φ_n et par image inverse de (2.2.2.2) par i sont égales. Si l'on note l_n l'homomorphisme composé

$$(2.2.2.3) \qquad l_n : \underline{\mu}_{p^n} \xrightarrow{\varphi_n} 1 + \mathcal{J}_{S/\Sigma} \xrightarrow{\log} \mathcal{O}_{S/\Sigma_n},$$

l'image de i dans $\mathcal{E}xt^1_{S/\Sigma}(\underline{\mu}_{p^\infty}, \mathcal{O}_{S/\Sigma})$ est donc, au-dessus de Σ_n, la classe de l'extension déduite de (2.2.2.1) par l'homomorphisme induit par l_n :

$$\mathcal{E}xt^1_{S/\Sigma_n}(\underline{\mu}_{p^\infty}, \underline{\mu}_{p^n}) \xrightarrow{l_n} \mathcal{E}xt^1_{S/\Sigma_n}(\underline{\mu}_{p^\infty}, \mathcal{O}_{S/\Sigma_n}).$$

Pour montrer que cette classe d'extension est non nulle, on peut supposer que $n = 1$. On observe d'abord que, puisque $\mathcal{H}om_{S/\Sigma}(\underline{\mu}_{p^\infty}, \mathcal{O}_{S/\Sigma}) = 0$, l'homomorphisme

$$\mathcal{H}om_{S/\Sigma}(\underline{\mu}_p, \mathcal{O}_{S/\Sigma}) \xrightarrow{\partial} \mathcal{E}xt^1_{S/\Sigma}(\underline{\mu}_{p^\infty}, \mathcal{O}_{S/\Sigma}),$$

est injectif, ce qui entraîne qu'il suffit de prouver que l_1 est non nul. Il suffit pour cela de construire un objet (U, Z, δ) de $\mathrm{CRIS}(S/\Sigma_1)$ tel que $l_1 : \mu_p(\mathcal{O}_U) \to \mathcal{O}_Z$ soit non nul. On prend alors $U = \mathrm{Spec}\ \mathcal{O}_S[t]/(t^p - 1)$, et $Z = \mathrm{Spec}\ \mathcal{D}$, où $\mathcal{D}$ est l'enveloppe à puissances divisées de l'idéal $(t^p - 1)$ dans $\mathcal{O}_S[t]$, et l'assertion résulte de ce que $\log(t^p) = \sum_i (-1)^{i-1}(i-1)!(t^p - 1)^{[i]}$ est non nul dans $\mathcal{D}$.

2.2.3. Corollaire. (i) *Pour tout groupe fini localement libre de type multiplicatif G, il existe un isomorphisme canonique*

$$v_G : \underline{G}^* \otimes \mathcal{O}_{S/\Sigma} \xrightarrow{\sim} \mathbb{D}(G), \tag{2.2.3.1}$$

tel que pour $G = \mu_{p^n}$, l'image de la section $1 \in \underline{\mathbb{Z}}/p^n = \underline{G}^*$ soit la classe de l'extension déduite de l'extension (2.2.2.1) par l'homomorphisme l_n défini en (2.2.2.3) et par l'inclusion $\mu_{p^n} \hookrightarrow \mu_{p^\infty}$.

(ii) *Pour tout groupe p-divisible de type multiplicatif H, il existe un isomorphisme canonique*

$$v_H : \underline{H}^D \otimes \mathcal{O}_{S/\Sigma} \xrightarrow{\sim} \mathbb{D}(H).$$

La première assertion résulte du calcul fait dans la démonstration précédente, via l'homomorphisme de fonctorialité

$$\mathcal{E}xt^1_{S/\Sigma}(\underline{\mu}_{p^\infty}, \mathcal{O}_{S/\Sigma}) \to \mathcal{E}xt^1_{S/\Sigma}(\underline{\mu}_{p^n}, \mathcal{O}_{S/\Sigma}).$$

L'énoncé pour les groupes p-divisibles résulte comme en (2.1.3) du cas des groupes finis.

Comme l'image directe sur $\mathrm{CRIS}(S/\Sigma)$ du faisceau $\mathcal{O}_{S/\Sigma_n}$ est le faisceau $\mathcal{O}_{S/\Sigma}/p^n\mathcal{O}_{S/\Sigma}$, nous noterons encore l_n l'homomorphisme

$$l_n : \underline{\mu}_{p^n} \xrightarrow{\varphi_n} 1 + (\mathcal{J}_{S/\Sigma}/p^n\mathcal{O}_{S/\Sigma}) \xrightarrow{\log} \mathcal{O}_{S/\Sigma}/p^n\mathcal{O}_{S/\Sigma}$$

image directe de (2.2.2.3), et qui peut être décrit de la même manière.

2.2.4. Corollaire. *Soient S un schéma de caractéristique p, $\zeta \in \Gamma(S, \mathcal{O}_S)$ une racine p^n-ième de l'unité, $u : \mathbb{Z}/p^n \to \mu_{p^n}$ l'homomorphisme envoyant 1 sur ζ. Avec les identifications*

$$u_{\mathbb{Z}/p^n} : \mathcal{O}_{S/\Sigma}/p^n\mathcal{O}_{S/\Sigma} \xrightarrow{\sim} \mathbb{D}(\mathbb{Z}/p^n),$$
$$v_{\mu_{p^n}} : \mathcal{O}_{S/\Sigma}/p^n\mathcal{O}_{S/\Sigma} \xrightarrow{\sim} \mathbb{D}(\mu_{p^n})$$

fournies par (2.1.3.1) *et* (2.2.3.1), *l'homomorphisme* $\mathbb{D}(u) : \mathbb{D}(\mu_{p^n}) \to \mathbb{D}(\mathbb{Z}/p^n)$ *défini par* u *est la multiplication par* $l_n(\zeta)$.

On peut se limiter à prouver ce corollaire sur le site $\mathrm{CRIS}(S/\Sigma_n)$. D'autre part, l'assertion est locale sur $\mathrm{CRIS}(S/\Sigma_n)$ pour la topologie fppf. Comme on peut extraire une racine p^n-ième de ζ en passant à un revêtement fppf de la forme Spec $\mathcal{O}_S[t]/(t^{p^n} - \zeta)$, et qu'un tel revêtement se relève localement sur T pour tout objet (U, T, δ) de $\mathrm{CRIS}(S/\Sigma_n)$, on peut supposer qu'il existe une racine p^{2n}-ième de l'unité ζ' telle que $\zeta'^{p^n} = \zeta$. On dispose alors d'un diagramme commutatif

$$\begin{array}{llcccccccc}
(2.2.4.1) & 0 \longrightarrow & \underline{\mathbb{Z}}/p^n & \xrightarrow{p^n} & \underline{\mathbb{Z}}/p^{2n} & \longrightarrow & \underline{\mathbb{Z}}/p^n & \longrightarrow 0 \\
& & \downarrow{\scriptstyle \underline{u}} & & \downarrow{\scriptstyle \underline{u}'} & & \downarrow{\scriptstyle \underline{u}} & \\
(2.2.4.2) & 0 \longrightarrow & \underline{\mu}_{p^n} & \longrightarrow & \underline{\mu}_{p^{2n}} & \xrightarrow{p^n} & \underline{\mu}_{p^n} & \longrightarrow 0
\end{array}$$

dans lequel $u'(1) = \zeta'$.

D'après (2.1.3), l'image de 1 dans $\mathbb{D}(\mathbb{Z}/p^n)$ est la classe de l'extension déduite de (2.2.4.1) par l'homomorphisme canonique $\mathbb{Z}/p^n \to \mathcal{O}_{S/\Sigma_n}$. D'après (2.2.3), celle de 1 dans $\mathbb{D}(\mu_{p^n})$ est la classe de l'extension déduite de (2.2.2.4) par l'homomorphisme $l_n : \underline{\mu}_{p^n} \to \mathcal{O}_{S/\Sigma_n}$. La commutativité du diagramme montre alors que, par composition avec u, cette extension donne celle qu'on déduit de (2.2.4.1) par l'homomorphisme $\mathbb{Z}/p^n \to \mathcal{O}_{S/\Sigma}$ défini par $l_n(\zeta)$, d'où le corollaire.

2.3. *Relation avec la dualité.*

2.3.1. Rappelons [7, 5.2.4] que, pour tout groupe fini localement libre G, de dual de Cartier G^*, il existe un isomorphisme de dualité

$$\Phi_G : \Delta(G)^\vee[-1] \xrightarrow{\sim} \Delta(G^*),$$

défini comme suit : le morphisme (2.2.1.2) $\eta : \underline{\mathbb{G}}_m \to \mathcal{O}_{S/\Sigma}[1]$ induit par fonctorialité un morphisme

$$\rho_G : \underline{G}^* \to t_{0]}\mathbf{R}\mathcal{H}om_{S/\Sigma}(\underline{G}, \mathcal{O}_{S/\Sigma}[1]) \simeq \Delta(G)[1],$$

fournissant par dualité le morphisme

$$\Phi_G : \Delta(G)^\vee[-1] \simeq \mathbf{R}\mathcal{H}om_{\mathcal{O}_{S/\Sigma}}(\Delta(G)[1], \mathcal{O}_{S/\Sigma}) \to \Delta(G^*).$$

Supposons maintenant que G soit un groupe étale. Les homomorphismes u_G et v_G définis plus haut fournissent également un isomorphisme de dualité

$$\Psi_G : \Delta(G)^\vee[-1] \xrightarrow{\sim} \Delta(G^*)$$

de la manière suivante. D'après la remarque (2.1.4) (i), le dual de u_G donne un isomorphisme

(2.3.1.1)

$$\begin{array}{ccc} \Delta(G)^\vee[-1] & \xrightarrow[\sim]{u_G^\vee} & \mathbf{R}\mathcal{H}om_{\mathcal{O}_{S/\Sigma}}(\mathbf{R}\mathcal{H}om(\underline{G},\underline{\mathbf{Z}}_p)[1] \overset{\mathbf{L}}{\underset{\mathbf{Z}_p}{\otimes}} \mathcal{O}_{S/\Sigma}, \mathcal{O}_{S/\Sigma}) \\ & & \downarrow \wr \\ & & \mathbf{R}\mathcal{H}om_{\mathcal{O}_{S/\Sigma}}(\mathbf{R}\mathcal{H}om(\underline{G},\underline{\mathbf{Z}}_p) \otimes \overset{\mathbf{L}}{\underset{\mathbf{Z}_p}{\otimes}} \mathcal{O}_{S/\Sigma}, \mathcal{O}_{S/\Sigma})[-1] \\ & & \downarrow \wr \\ & & \mathbf{R}\mathcal{H}om_{\mathbf{Z}_p}(\mathbf{R}\mathcal{H}om(\underline{G},\underline{\mathbf{Z}}_p), \mathcal{O}_{S/\Sigma})[-1]. \end{array}$$

Comme G est étale, le morphisme canonique

$$\underline{G} \overset{\mathbf{L}}{\otimes} \mathcal{O}_{S/\Sigma} \to \mathbf{R}\mathcal{H}om_{\mathbf{Z}_p}(\mathbf{R}\mathcal{H}om(\underline{G},\underline{\mathbf{Z}}_p), \mathcal{O}_{S/\Sigma}) \tag{2.3.1.2}$$

est un isomorphisme. Enfin, l'isomorphisme v_{G^*} fournit l'isomorphisme

$$\underline{G} \overset{\mathbf{L}}{\otimes} \mathcal{O}_{S/\Sigma}[-1] \xrightarrow{\sim} \underline{G}^{**} \overset{\mathbf{L}}{\otimes} \mathcal{O}_{S/\Sigma}[-1] \xrightarrow[\sim]{v_{G^*}[-1]} \Delta(G^*). \tag{2.3.1.3}$$

Par composition, on obtient Ψ_G.

2.3.2. Proposition. *Soit G un groupe fini étale. Les isomorphismes Φ_G et Ψ_G sont opposés.*

Notons

$$i_G : \underline{G} \to \mathbf{R}\mathcal{H}om_{\mathcal{O}_{S/\Sigma}}(t_{1]}\mathbf{R}\mathcal{H}om(\underline{G}, \mathcal{O}_{S/\Sigma}), \mathcal{O}_{S/\Sigma}) = \Delta(G)^\vee$$

le morphisme de bidualité, et

$$j_G : \underline{G} \to \underline{G} \overset{\mathbf{L}}{\otimes} \mathcal{O}_{S/\Sigma}$$

le morphisme d'extension des scalaires. L'argument de [7, 5.2.13] montre qu'il suffit de prouver que $\Phi_G \circ i_G[-1] = -\Psi_G \circ i_G[-1]$. Compte tenu de la remarque (2.1.4) (i), il est clair que le composé de $i_G[-1]$ et de (2.3.1.1) est le décalé du morphisme naturel

$$\underline{G} \to \mathbf{R}\mathcal{H}om_{\mathbf{Z}_p}(\mathbf{R}\mathcal{H}om(\underline{G}, \underline{\mathbf{Z}}_p), \mathcal{O}_{S/\Sigma}).$$

Son composé avec l'inverse décalé de (2.3.1.2) n'est donc autre que $j_G[-1]$. On est donc ramené à vérifier que le morphisme composé de ce dernier et de (2.3.1.3) est $-\Phi_G \circ i_G[-1]$.

La définition de v_G entraîne que le composé

$$\underline{G}[-1] \xrightarrow{j_G[-1]} \underline{G} \overset{\mathbf{L}}{\otimes} \mathcal{O}_{S/\Sigma}[-1] \xrightarrow{\sim} \underline{G}^{**} \overset{\mathbf{L}}{\otimes} \mathcal{O}_{S/\Sigma}[-1] \underset{\sim}{\xrightarrow{v_{G^*}[-1]}} \Delta(G^*)$$

n'est autre que le composé

$$\underline{G}[-1] \xrightarrow{\sim} \underline{G}^{**}[-1] \xrightarrow{\rho_{G^*}[-1]} \Delta(G^*),$$

c'est à dire le décalé du morphisme noté ρ'_G dans [7, 5.2.12]. D'autre part, $\Phi_G \circ i_G[-1]$ est par définition le composé

$$\begin{array}{ccccc} \underline{G}[-1] \xrightarrow{i_G[-1]} \mathbf{R}\mathcal{H}om_{\mathcal{O}_{S/\Sigma}}(\Delta(G), \mathcal{O}_{S/\Sigma})[-1] & \xrightarrow{\sim} & \mathbf{R}\mathcal{H}om_{\mathcal{O}_{S/\Sigma}}(\Delta(G)[1], \mathcal{O}_{S/\Sigma}) \\ & & \big\downarrow \rho_G^\vee \\ & & t_{1]}\mathbf{R}\mathcal{H}om_{S/\Sigma}(\underline{G}^*, \mathcal{O}_{S/\Sigma}). \end{array}$$

L'égalité à vérifier est donc, au décalage près, l'anticommutativité du diagramme [7, (5.2.12.1)] (dans lequel la flèche verticale de droite est en fait dirigée de bas en haut).

2.4. *Le théorème d'équivalence.*

Nous allons maintenant montrer que, sur une base S quelconque, la donnée d'un groupe fini étale, ou d'un groupe fini de type multiplicatif, équivaut à celle de son cristal de Dieudonné. On suppose ici que $\Sigma = \operatorname{Spec} \mathbf{Z}_p$, $\mathcal{J} = p.\mathcal{O}_\Sigma$, muni de ses puissances divisées canoniques. Soient $\mathbf{C}_S$ la catégorie des cristaux de présentation finie sur $\text{Cris}(S/\Sigma)$, munis d'homomorphismes

$$F : M^\sigma \to M, \ V : M \to M^\sigma$$

tels que $F \circ V = p$, $V \circ F = p$, et $\mathbf{C}_S^{\text{ét}}$ (resp. $\mathbf{C}_S^{\text{tm}}$) la sous-catégorie pleine de $\mathcal{C}_S$ formée des cristaux localement annulés par une puissance de p, et pour lesquels F (resp. V) est un isomorphisme. Nous utiliserons ici la topologie étale, plutôt que la topologie fppf : d'après [7, 2.3.11], cela ne change pas les cristaux de Dieudonné.

2.4.1. Théorème. *Pour tout schéma S localement annulé par une puissance de p, le foncteur $\mathbb{D}$ induit une équivalence de catégories entre la catégorie des groupes finis étales (resp. de type multiplicatif) sur S, et la catégorie $\mathcal{C}_S^{\text{ét}}$ (resp. $\mathcal{C}_S^{\text{tm}}$).*

Comme p est localement nilpotent sur S, les deux catégories de groupes considérées sont invariantes par réduction modulo p. On peut donc supposer que S est de caractéristique p, car le cristal de Dieudonné d'un groupe s'identifie à celui de sa réduction modulo p [7, 2.5.7]. Puisque les foncteurs $G \mapsto G^\vee$ et $G \mapsto G^*$ sont des équivalences de catégories, on est alors ramené à montrer que le foncteur

$$T(G) = \underline{G} \otimes \mathcal{O}_{S/\Sigma},$$

où $T(G)$ est muni du Frobenius défini par $F_G^{-1} : \underline{G}^\sigma \to \underline{G}$ et du Verschiebung défini par $pF_G : \underline{G} \to \underline{G}^\sigma$, est une équivalence entre la catégorie des groupes finis étales sur S et la catégorie $\mathcal{C}_S^{\text{ét}}$.

Nous allons d'abord donner quelques lemmes. Rappelons que, pour tout anneau A de caractéristique p, l'idéal $VW(A)$ de l'anneau $W(A)$ est muni de puissances divisées canoniques [13, 0 1.4].

2.4.2. Lemme. *Soient A un anneau de caractéristique p, B un anneau annulé par p^n, et muni d'un homomorphisme surjectif $B \twoheadrightarrow A$ dont le noyau J possède une structure de PD-idéal δ compatible aux puissances divisées de p. Il existe alors un PD-morphisme canonique $\theta_n : W_n(A) \to B$ s'insérant dans un carré commutatif*

$$\begin{array}{ccc} W_n(A) & \xrightarrow{\theta_n} & B \\ \downarrow & & \downarrow \\ A & \xrightarrow{F^n} & A, \end{array}$$

et tel que pour tout PD-morphisme $(B, J, \delta) \longrightarrow (B', J', \delta')$ le carré

$$\begin{array}{ccc} W_n(A) & \xrightarrow{\theta_n} & B \\ \downarrow & & \downarrow \\ W_n(A') & \xrightarrow{\theta_n} & B', \end{array}$$

avec $A' = B'/J'$, soit commutatif.

Soit $x=(x_0,x_1,\ldots,x_{n-1})\in Wn(A)$. On choisit des éléments $\tilde{x}_0,\tilde{x}_1,\ldots,$ $\tilde{x}_{n-1}$ de B relevant $x_0,x_1,\ldots,x_{n-1}$, et on pose

$$\theta_n(x) = \tilde{x}_0^{p^n} + p\tilde{x}_1^{p^{n-1}} + \cdots + p^{n-1}\tilde{x}_{n-1}^p.$$

L'application θ_n est indépendante du choix des relèvements : si y, $y' \in B$ sont tels que $y' = y+z$, avec $z \in J$, alors $y'^p \equiv y^p + z^p \bmod p$, et la relation $z^p = p!\delta_p(z)$ entraîne donc que $y'^p \equiv y^p$; l'indépendance de θ_n en résulte. Cette application peut encore être décrite comme provenant par passage au quotient de la composante fantôme d'indice n : $W_{n+1}(B) \to B$, ce qui montre que c'est un homomorphisme d'anneaux.

Si δ est compatible aux puissances divisées de p, θ_n est un *PD*-morphisme. En effet, si $x = Vy$, on a par définition

$$x^{[k]} = p^{k-1}V(y^k)/k! \;;$$

d'autre part, la définition de θ_n entraîne la relation

$$\theta_n(V(y)) = p\theta_{n-1}(y)$$

(où $\theta_{n-1}(y)$ désigne en fait un relèvement arbitraire de $\theta_{n-1}(y) \in B/p^{n-1}B$ dans B). On en déduit :

$$\theta_n(x^{[k]}) = p^{k-1}\theta_n(V(y^k))/k! = p^k\theta_{n-1}(y^k)/k!.$$

D'autre part, on a

$$\delta_k(\theta_n(x)) = \delta_k(\theta_n(V(y))) = \delta_k(p\theta_{n-1}(y)) = p^k\theta_{n-1}(y)^k/k!,$$

la dernière relation résultant de la compatibilité de δ aux puissances divisées de p.

Enfin, la fonctorialité de θ_n par rapport à (B,J) résulte immédiatement de sa définition.

Pour tout schéma S de caractéristique p, et tout entier n, nous noterons $W_n(S) = (|S|, W_n(\mathcal{O}_S))$ le spectre de l'anneau des vecteurs de Witt à cœfficients dans $\mathcal{O}_S$. On dispose ainsi d'une immersion fermée $S \hookrightarrow W_n(S)$, qui est un objet de $\mathrm{CRIS}(S/\Sigma)$ grâce aux puissances divisées canoniques de $W_n(S)$.

2.4.3. Proposition. *Soit S un schéma de caractéristique p. Le foncteur qui associe à un F-cristal M sur* $\mathrm{CRIS}(S/\Sigma)$ *le faisceau $M_{W_n(S)}$ défini* *par M sur $W_n(S)$ induit une équivalence de catégories entre la catégorie des F-cristaux quasi-cohérents annulés par p^n et tels que $F : M^\sigma \to M$ soit un isomorphisme, et la catégorie des $W_n(\mathcal{O}_S)$-modules quasi-cohérents E munis d'un isomorphisme $F : E^\sigma \xrightarrow{\sim} E$ (l'exposant σ désignant dans ce cas l'extension des scalaires par l'endomorphisme de Frobenius $F : W_n(\mathcal{O}_S) \to W_n(\mathcal{O}_S)$).*

Un cristal annulé par p^n est déterminé par sa restriction à $\mathrm{CRIS}(S/\Sigma_n)$, de sorte qu'on peut remplacer Σ par Σ_n dans l'énoncé.

Il résulte de (2.4.2) que, pour tout objet (U, T, δ) de $\mathrm{CRIS}(S/\Sigma)$, il existe un PD-morphisme canonique $\theta_n : Z \to W_n(U)$ relevant la puissance n-ième du Frobenius absolu de U. Si E est un $W_n(\mathcal{O}_S)$-module quasi-cohérent, la famille des $\theta_n^*(E)$ pour (U, T, δ) variable est alors un cristal E^θ sur $\mathrm{CRIS}(S/\Sigma_n)$, et, lorsque $E = M_{W_n(S)}$, pour un cristal M sur $\mathrm{CRIS}(S/\Sigma_n)$, on a, par définition de l'image inverse d'un cristal

$$M^{\sigma^n} \simeq (M_{W_n(S)})^\theta. \tag{2.4.3.1}$$

De même, on a un isomorphisme $M^{\sigma^{n+1}} \simeq (M^\sigma_{W_n(S)})^\theta$. Si M est muni d'une structure de F-cristal telle que $F : M^\sigma \to M$ soit un isomorphisme, on obtient alors par itération un isomorphisme $M \simeq (M_{W_n(S)})^\theta$ compatible à l'action de F. Il en résulte que le foncteur $E \mapsto E^\theta$ est quasi-inverse du foncteur $M \mapsto M_{W_n(S)}$.

Nous dirons qu'un F-cristal de $\mathrm{C}_S^{\text{ét}}$ est *trivial* (resp. *localement trivial*) s'il est isomorphe à une somme directe finie de cristaux de la forme $\mathcal{O}_{S/\Sigma}/p^n\mathcal{O}_{S/\Sigma}$, munis du Frobenius et du Verschiebung canoniques (i.e. $F = \mathrm{Id}, V = p \cdot \mathrm{Id}$) (resp. si un tel isomorphisme existe localement pour la topologie étale sur S). Nous noterons $\mathrm{C}_S'^{\text{ét}}$ la sous-catégorie pleine de $\mathrm{C}_S^{\text{ét}}$ formée des F-cristaux localement triviaux ; nous montrerons plus loin qu'on a en fait $\mathrm{C}_S'^{\text{ét}} = \mathrm{C}_S^{\text{ét}}$, de sorte que le théorème résulte de la proposition suivante :

2.4.4. Proposition. *Le foncteur T est une équivalence de catégories entre la catégorie des groupes finis étales sur S, et la catégorie $C_S'^{\text{ét}}$.*

Le fait que T soit à valeur dans $C_S'^{\text{ét}}$ résulte de ce que tout groupe fini étale est, localement pour la topologie étale, isomorphe à une somme directe de groupes $\mathbf{Z}/p^n\mathbf{Z}$.

En sens inverse, on peut associer à tout faisceau E sur $\mathrm{CRIS}(S/\Sigma)_{\text{ét}}$, muni d'un homomorphisme $F : E^\sigma \to E$, un faisceau $H(E)$ sur $S_{\text{ét}}$, défini de la manière suivante. Tout d'abord, on associe à E un faisceau $u_{S/\Sigma^\bullet}(E)$ sur $S_{\text{ét}}$ en posant, pour tout S-schéma U,

$$\Gamma(U, u_{S/\Sigma^\bullet}(E)) = \Gamma(U/\Sigma, E).$$

Pour vérifier que $u_{S/\Sigma^\bullet}(E)$ est bien un faisceau étale, on considère un morphisme étale surjectif $U' \to U$, le produit fibré $U'' = U' \times_U U'$, les deux projections p_1, p_2 de U'' sur U' ; soit $s \in \Gamma(U'/\Sigma, E)$ une section dont les deux images inverses sur U'' sont égales. Pour tout objet (V, Z, δ) de $\text{Cris}(U/\Sigma)$, l'immersion $V \hookrightarrow Z$ est une nilimmersion, de sorte que $U' \times_U V$ se relève un schéma Z' étale et surjectif sur Z ; soit $Z'' = Z' \times_Z Z'$. Par platitude, les puissances divisées de Z s'étendent à Z' et Z'', faisant respectivement des épaississements $U' \times_U V \hookrightarrow Z'$ et$U'' \times_U V \hookrightarrow Z''$ des objets de $\text{Cris}(U'/\Sigma)$ et $\text{Cris}(U''/\Sigma)$. La section $s \in \Gamma(U'/\Sigma, E)$ induit une section $s_{Z'}$ de $E_{Z'}$, dont les deux images inverses dans $E_{Z''}$ par les deux projections de Z'' sur Z' sont les sections induites sur $E_{Z''}$ par $p_{1\text{cris}}^*(s)$ et $p_{2\text{cris}}^*(s)$. Elles sont donc égales ; comme E est un faisceau étale sur $\text{Cris}(S/\Sigma)$, $s_{Z'}$ se redescend en une unique section s_Z de E_Z. Le même argument montre que, pour (V, Z, δ) variable, les sections s_Z se recollent, donc définissent une section $s \in \Gamma(U/\Sigma, E)$.

On considère alors l'endomorphisme

$$F : u_{S/\Sigma^\bullet}(E) \to u_{S/\Sigma^\bullet}(E)$$

qui, pour tout S-schéma U, est l'homomorphisme composé

$$\Gamma(U/\Sigma, E) \to \Gamma(U/\Sigma, E^\sigma) \xrightarrow{F} \Gamma(U/\Sigma, E),$$

(la première flèche étant l'homomorphisme qui, à une section, associe son image inverse par le Frobenius absolu de U). On pose alors

$$H(E) = u_{S/\Sigma^\bullet}(E)^{1-F} = \text{Ker}\ (1 - F : u_{S/\Sigma^\bullet}(E) \to u_{S/\Sigma^\bullet}(E)).$$

Supposons que M soit un F-cristal de la forme $T(G) = \underline{G} \otimes \mathcal{O}_{S/\Sigma}$, où G est un groupe fini étale sur S. L'homomorphisme $\Gamma(U/\Sigma, M) \to \Gamma(U/\Sigma, M^\sigma)$ est alors induit par l'homomorphisme de Frobenius $\Gamma(U, G) \to \Gamma(U, G^\sigma)$; il résulte donc de la définition de F qu'on dispose d'un homomorphisme canonique $G \to H(T(G))$. Réciproquement, l'extension des scalaires fournit un homomorphisme canonique $T(H(M)) \to M$. Pour montrer que T est une équivalence de catégories, il suffit de vérifier qu'on obtient deux isomorphismes inverses, et par localisation on peut se limiter au cas où $G = \mathbb{Z}/p^n\mathbb{Z}$. On est ainsi ramené à vérifier que, pour tout S, on a

$$\text{Ker}\,(1 - F : \Gamma(S/\Sigma, \mathcal{O}_{S/\Sigma}/p^n\mathcal{O}_{S/\Sigma}) \to \Gamma(S/\Sigma, \mathcal{O}_{S/\Sigma}/p^n\mathcal{O}_{S/\Sigma})) = \Gamma(S, \mathbb{Z}/p^n\mathbb{Z}).$$

On peut se limiter à considérer les épaississements annulés par p^n. Soient (U, Z, δ) un objet de $\text{Cris}(S/\Sigma_n)$ tel qu'il existe un endomorphisme σ de Z relevant l'endomorphisme de Frobenius de U, M un cristal sur $\text{Cris}(S/\Sigma_n)$, $s \in \Gamma(S/\Sigma, M)$ une section de M, $s' = F(s)$. Par définition de l'image inverse, la section s'_Z de M_Z induite par s' est alors l'image inverse de s_Z par σ. En appliquant cette remarque à l'endomorphisme de Frobenius de W_n, on voit que l'homomorphisme canonique

$$u^*_{S/\Sigma_n}(M) \to M_{W_n(S)} \tag{2.4.4.1}$$

qui associe à une section de M sa valeur sur $W_n(S)$ est compatible à l'action de F lorsque M est un F-cristal. Lorsque $M = \mathcal{O}_{S/\Sigma_n}$, on obtient en particulier un homomorphisme

$$u_{S/\Sigma_n *}(\mathcal{O}_{S/\Sigma_n}) \to W_n(\mathcal{O}_S) \tag{2.4.4.2}$$

compatible à l'action de F. Compte tenu de l'exactitude de la suite d'Artin-Schreier

$$0 \to \mathbb{Z}/p^n\mathbb{Z} \to W_n(\mathcal{O}_S) \xrightarrow{1-F} W_n(\mathcal{O}_S) \to 0,$$

l'image dans $W_n(\mathcal{O}_S)$ d'une section s invariante par F est dans $\mathbb{Z}/p^n\mathbb{Z}$. L'énoncé résulte alors de la proposition suivante :

2.4.5. Proposition. *Soient $\Sigma_n = \text{Spec}\,(\mathbb{Z}/p^n\mathbb{Z})$, S un schéma de caractéristique p, M un F-cristal sur $\text{Cris}(S/\Sigma_n)$. Alors l'homomorphisme*

$$u_{S/\Sigma_n *}(M)^{1-F} \to (M_{W_n(S)})^{1-F},$$

induit par (2.4.4.1) entre les invariants sous F, est un isomorphisme.

Si E est un $W_n(\mathcal{O}_S)$-module quasi-cohérent, et si E^θ est le cristal sur $\text{Cris}(S/\Sigma_n)$ défini en (2.4.3), toute section t de E au-dessus d'un S-schéma U définit une section $\theta^*_n(t)$ de E^θ sur $\text{Cris}(U/\Sigma_n)$; de plus, s'il existe un cristal tel que $E = M_{W_n(S)}$, et que $t = s_{W_n(U)}$, avec $s \in \Gamma(U/\Sigma, M)$, on a $\theta^*_n(s_{W_n(U)}) = F^{n\,*}_{\text{cris}}(s)$, compte tenu de l'isomorphisme (2.4.3.1). Si M est un F-cristal, et si on note $F_n : M^{\sigma^n} \to M$ l'homomorphisme obtenu par itération de F, on obtient ainsi un homomorphisme commutant à F

$$F_n \circ \theta^*_n : M_{W_n(S)} \to u_{S/\Sigma_n *}(M), \tag{2.4.5.1}$$

dont le composé avec (2.4.4.1) dans les deux sens est F^n. Sur le faisceau des sections invariantes par F, le composé dans les deux sens est donc l'identité, d'où l'énoncé.

La démonstration précédente fournit le corollaire suivant, qu'on rapprochera de l'énoncé analogue [13, II, théorème 5.2] pour les H^i d'un schéma propre et lisse sur un corps parfait, valable pour tout i :

2.4.6. Corollaire. *Soient S un schéma de carctéristique p, $\Sigma_n =$ Spec $\mathbf{Z}/p^n\mathbf{Z}$, $\Sigma =$ Spec $\mathbf{Z}_p$, F l'action de Frobenius par fonctorialité sur la cohomologie cristalline de S. On a alors les suites exactes*

$$0 \longrightarrow H^0(S, \mathbf{Z}/p^n\mathbf{Z}) \longrightarrow H^0(S/\Sigma_n, \mathcal{O}_{S/\Sigma_n}) \xrightarrow{1-F} H^0(S/\Sigma_n, \mathcal{O}_{S/\Sigma_n}),$$

$$0 \longrightarrow H^0(S, \mathbf{Z}_p) \longrightarrow H^0(S/\Sigma, \mathcal{O}_{S/\Sigma}) \xrightarrow{1-F} H^0(S/\Sigma, \mathcal{O}_{S/\Sigma}).$$

Il reste à montrer que les catégories $\mathcal{C}'^{\text{ét}}_S$ et $\mathcal{C}^{\text{ét}}_S$ sont égales. Donnons d'abord quelques résultats préliminaires.

2.4.7. Lemme. *Soient A un anneau local nœthérien de caractéristique p, d'idéal maximal $\mathfrak{m}$, et $\widehat{A}$ le séparé complété de A pour la topologie $\mathfrak{m}$-adique. Soient n un entier, $W_n(A)$ l'anneau des vecteurs de Witt de longueur n à coefficients dans A, $\mathfrak{m}'$ l'idéal maximal de $W_n(A)$; pour tout entier i, soit $\mathfrak{a}_i$ l'idéal de $W_n(A)$ formé par les vecteurs de Witt à coefficients dans $\mathfrak{m}^i$. Alors :*

(i) *La topologie définie sur $W_n(A)$ par les $\mathfrak{a}_i$ est égale à la topologie $\mathfrak{m}'$-adique.*

(ii) *Si A, considéré comme A-module par l'intermédiaire de l'endomorphisme de Frobenius, est un A-module de type fini, $W_n(\widehat{A})$ est fidèlement plat sur $W_n(A)$.*

L'idéal $\mathfrak{m}'$ est l'ensemble des vecteurs de la forme $x = (a_0, a_1, ..., a_{n-1})$, où $a_0 \in \mathfrak{m}$. Si $\tilde{a}_0$ est le représentant de Teichmüller de a_0, un tel vecteur s'écrit encore $x = \tilde{a}_0 + V(y)$, avec $y \in W_{n-1}(A)$. Si $x_1, \ldots, x_m \in \mathfrak{m}'$, et si $m \geq n$, on peut alors écrire

$$x_1 x_2 \ldots x_m = \tilde{x}_1 \tilde{x}_2 \ldots \tilde{x}_m + \sum_{i=1}^{n-1} \prod_{j_1 < \ldots < j_{m-i}} \tilde{x}_{j_1} \ldots \tilde{x}_{j_{m-i}} V^i(y_{j_1 \ldots j_{m-i}}),$$

de sorte que $\mathfrak{m}'^m \subset \mathfrak{a}_{m-n+1}$.

Soient $t_1, \ldots, t_r$ des générateurs de l'idéal $\mathfrak{m}$, et soit x un élément de l'idéal $V^i \mathfrak{a}_{mp^n r^{n-i}}$. On peut écrire x sous la forme $x = V^i(\tilde{a}) + V^{i+1}(y)$, avec $a \in \mathfrak{m}^{mp^n r^{n-i}}$ et $y \in \mathfrak{a}_{mp^n r^{n-i}}$. L'élément a peut s'écrire $a = \sum_j a_j t_j^{mp^n r^{n-i-1}}$, ce qui donne

$$\tilde{a} = \sum_j \tilde{a}_j \tilde{t}_j^{mp^n r^{n-i-1}} + V(z),$$

avec $z \in \mathfrak{a}_{mp^n r^{n-i-1}}$. On en déduit l'expression

$$x = \sum_j V^i(\tilde{a}_j \tilde{t}_j^{mp^n r^{n-i-1}}) + V^{i+1}(y+z) =$$

$$\sum_j \tilde{t}_j^{mp^{n-i} r^{n-i-1}} V^i(\tilde{a}_j) + V^{i+1}(y+z),$$

avec $y+z \in \mathfrak{a}_{mp^n r^{n-i-1}}$, de sorte que $x \in \mathfrak{m}'^{mp^{n-i} r^{n-i-1}} + V^{i+1}\mathfrak{a}_{mp^n r^{n-i-1}}$. Par récurrence sur i, il en résulte que $\mathfrak{a}_{mp^n r^n} \subset \mathfrak{m}'^m$, d'où l'assertion (i).

Sous l'hypothèse de (ii), on voit en utilisant la filtration de $W_n(A)$ par les $V^i W_n(A)$ que $W_n(A)$ est un anneau nœthérien. Comme c'est un anneau local, le séparé complété $W_n(A)^\wedge$ est fidèlement plat sur $W_n(A)$. L'assertion (ii) résulte alors de ce qu'on a d'après (i) :

$$W_n(A)^\wedge \simeq \varprojlim_m W_n(A)/\mathfrak{m}'^m \simeq \varprojlim_i W_n(A)/\mathfrak{a}_i \simeq \varprojlim_i W_n(A/\mathfrak{m}^i) \simeq W_n(\widehat{A}).$$

2.4.8. Lemme. *Soient S un schéma de caractéristique p, Y un $\mathbf{Z}/p^n\mathbf{Z}$-schéma tel qu'il existe un relèvement $\sigma : Y \to Y$ du Frobenius de $Y \times \operatorname{Spec} \mathbf{F}_p$, et $S \hookrightarrow Y$ une immersion fermée. On note D le voisinage à puissances divisées (compatibles à celles de p) de S dans Y, D_1 celui de S dans Y^2, et $p_1, p_2 : D_1 \to D$ les morphismes induits par les deux projections ; on note encore σ les endomorphismes de D et D_1 induits par σ. Soit M un $\mathcal{O}_D$-module, muni d'un isomorphisme $F : \sigma^* M \xrightarrow{\sim} M$. Alors :*

(i) *Il existe un unique isomorphisme $\varepsilon : p_2^* M \xrightarrow{\sim} p_1^* M$ compatible à F et induisant l'identité de M par restriction à D plongé diagonalement dans D_1.*

(ii) *Si M' est un second $\mathcal{O}_D$-module, également muni d'un isomorphisme $F' : \sigma^* M' \xrightarrow{\sim} M'$, qui définit $\varepsilon' : p_2^* M' \xrightarrow{\sim} p_1^* M'$ d'après (i), tout homomorphisme $u : M \to M'$ tel que $F' \circ \sigma^*(u) = u \circ F$ vérifie la relation $\varepsilon' \circ p_2^*(u) = p_1^*(u) \circ \varepsilon$.*

(iii) *On suppose que les morphismes σ, p_1, p_2 sont plats. Si $N \subset M$ est un sous-module stable par F, ε induit un isomorphisme $p_2^* N \xrightarrow{\sim} p_1^* N$.*

Nous noterons $\Delta : D \hookrightarrow D_1$ l'immersion induite par l'immersion diagonale. Il existe alors un morphisme $\Phi_n : D_1 \to D$ qui rende commutatif le diagramme

$$\begin{array}{ccc} D & \overset{\Delta}{\hookrightarrow} & D_1 \\ {\scriptstyle \sigma^n}\downarrow & \overset{\Phi_n}{\swarrow} & \downarrow{\scriptstyle \sigma^n \times \sigma^n} \\ D & \overset{\Delta}{\hookrightarrow} & D_1 . \end{array}$$

En effet, il résulte de [3, IV 1.3.5] que le noyau de l'homomorphisme $\mathcal{O}_{D_1} \twoheadrightarrow \mathcal{O}_D$ est l'idéal à puissances divisées engendré par les sections de la forme $1 \otimes y - y \otimes 1$, où y est une section de $\mathcal{O}_Y$. Pour toute section x de $\mathcal{O}_Y$, il existe une section y telle que $\sigma(x) = x^p + py$. Si K est l'idéal de la diagonale dans Y^2, on en tire

$$\sigma(1 \otimes x - x \otimes 1) = (1 \otimes x^p - x^p \otimes 1) + p(1 \otimes y - y \otimes 1) \equiv (1 \otimes x - x \otimes 1)^p \bmod pK$$

grâce aux égalités

$$\begin{aligned}(-1)^i(1 \otimes x)^i(x \otimes 1)^{p-i} &+ (-1)^{p-i}(1 \otimes x)^{p-i}(x \otimes 1)^i \\ &= (-1)^i x^i \otimes x^i(x^{p-2i} \otimes 1 - 1 \otimes x^{p-2i})\end{aligned}$$

pour p impair et $2i < p$, et

$$1 \otimes x^2 - x^2 \otimes 1 = (1 \otimes x - x \otimes 1)^2 + 2x(1 \otimes x - x \otimes 1)$$

pour $p = 2$. Si $\overline{K}$ est le PD-idéal de $\mathcal{O}_{D_1}$ engendré par K, on peut alors écrire pour tout $m \geq 1$

$$\begin{aligned}\sigma((1 \otimes x - x \otimes 1)^{[m]}) &= \sigma(1 \otimes x - x \otimes 1)^{[m]} \\ &\equiv (p!(1 \otimes x - x \otimes 1)^{[p]})^{[m]} \equiv 0 \bmod p\overline{K}.\end{aligned}$$

Par conséquent, on a $\sigma(\overline{K}) \subset p\overline{K}$, et $\sigma^n(\overline{K}) = 0$, d'où l'existence de Φ_n.

La compatibilité entre F et ε doit donner naissance au diagramme commutatif

$$\begin{array}{ccccc} \sigma^{n*}p_2^*M & \overset{\sim}{\longrightarrow} & p_2^*\sigma^{n*}M & \xrightarrow[\sim]{p_2^*(F^n)} & p_2^*M \\ \Big\downarrow {\scriptstyle \sigma^{n*}(\varepsilon)} & & & & \Big\downarrow {\scriptstyle \varepsilon} \\ \sigma^{n*}p_1^*M & \overset{\sim}{\longrightarrow} & p_1^*\sigma^{n*}M & \xrightarrow[\sim]{p_1^*(F^n)} & p_1^*M . \end{array}$$

D'après ce qui précède, on a pour $i = 1,2$:

$$\sigma^{n*}p_i^* M \simeq \Phi_n^* \Delta^* p_i^* M \simeq \Phi_n^* M \ ;$$

Comme $\Delta^*(\varepsilon)$ doit être l'identité de M, ε est caractérisé par la commutativité du carré

$$\begin{array}{ccc} \Phi_n^* M & \xrightarrow{\sim} & p_2^* M \\ \| & & \downarrow \varepsilon \\ \Phi_n^* M & \xrightarrow{\sim} & p_2^* M \end{array}$$

dans lequel les isomorphismes horizontaux sont définis par F^n. L'assertion (i) en résulte ; les assertions (ii) et (iii) se prouvent de la même manière.

2.4.9. Proposition. *Soient S un schéma de caractéristique p, et M un $\mathcal{O}_{S/\Sigma}$-module localement de présentation finie, et localement annulé par une puissance de p, muni d'un isomorphisme $F : M^\sigma \xrightarrow{\sim} M$. Alors M est localement trivial pour la topologie étale sur S.*

La démonstration s'effectue par dévissage sur le schéma S. On peut supposer que M est annulé par une puissance fixée p^n de p.

(i) *Cas où S est le spectre d'un corps séparablement clos k.* Nous allons montrer que M est alors trivial. Soient C un anneau de Cohen de k, et $\sigma : C \to C$ un relèvement du Frobenius de k. D'après (1.3.3), M peut s'interpréter comme étant un C-module de présentation finie (encore noté M), annulé par p^n, muni d'une connexion intégrable $\nabla : M \to M \otimes \Omega^1_C$, et d'un isomorphisme horizontal $F : M^\sigma \xrightarrow{\sim} M$; on notera encore $F : M \to M$ l'application définie par $F(m) = F(1 \otimes m)$. On raisonne par récurrence sur la longueur de M.

Supposons d'abord que M soit un k-espace vectoriel. D'après (2.4.8), il suffit de montrer que M possède une base de vecteurs fixes par F, et, par récurrence sur la dimension de M, il suffit de montrer qu'il existe un vecteur fixe x non nul. Si on fixe une base de M, et un élément $y \in M$, les équations que doivent satisfaire les coordonnées d'un vecteur x tel que $F(x) - x = y$ définissent un revêtement étale de degré $p^{\dim(M)}$ de Spec k, qui possède donc $p^{\dim(M)}$ points à valeurs dans k puisque k est séparablement clos ; en particulier, pour $y = 0$, il existe un tel $x \neq 0$, d'où l'assertion.

Dans le cas général, soit $M_1 = \mathrm{Ker}\,(p : M \to M)$; M_1 est stable par F et ∇, et M/M_1 est muni d'une structure analogue par passage au quotient. Supposons $M \neq M_1$. L'hypothèse de récurrence entraîne qu'il existe un élément $x \in M/M_1$ invariant par F et d'ordre maximal dans M/M_1. Soit $y \in M$ un élément d'ordre maximal relevant x. On peut modifier y par un élément de M_1 de manière à obtenir un élément invariant par F : en effet, $t = F(1 \otimes y) - y$ est un élément du k-espace vectoriel de dimension finie M_1 ; comme précédemment, l'application $F - \mathrm{Id}_{M_1}$ est surjective, donc il existe $z \in M_1$ tel que $F(1 \otimes z) - z = t$; comme x n'est pas nul, $y \notin M_1$, et $y - z$ est alors un élément d'ordre maximal, invariant par F, et relevant x.

Soit alors N le sous-module de M engendré par y. Comme $F(1 \otimes y) = y$, N est stable par F, donc l'assertion (iii) du lemme précédent entraîne (compte tenu de (1.2.7) et (1.3.2)) que N est un sous-cristal de M. D'autre part, puisque y est d'ordre maximal, N est facteur direct de M en tant que C-module ; fixons un scindage arbitraire $s : M/N \to M$. Montrons qu'on peut trouver $u : M/N \to N$ tel que $s + u$ soit un scindage commutant à F. Si $s^\sigma : (M/N)^\sigma \to M^\sigma$ est l'homomorphisme déduit de s par extension des scalaires, l'application $v = F_M \circ s^\sigma \circ F_{M/N}^{-1} - s$ est un élément du C-module de longueur finie $\mathrm{Hom}_C(M/N, N)$. Comme l'application qui à u associe $F_M \circ u^\sigma \circ F_{M/N}^{-1}$ est un endomorphisme σ-linéaire de $\mathrm{Hom}_C(M/N, N)$, l'application qui à u associe $F_M \circ u^\sigma \circ F_{M/N}^{-1} - u$ est surjective, de sorte qu'il existe u tel que $F_M \circ u^\sigma \circ F_{M/N}^{-1} - u = v$, et $s - u$ est alors un scindage commutant à F.

Il résulte alors de l'assertion (ii) du lemme précédent que ce scindage est un scindage en tant que cristal. Comme N est engendré par un élément invariant par F, N est isomorphe en tant que F-cristal à un cristal de la forme $\mathcal{O}_{S/\Sigma}/p^n\mathcal{O}_{S/\Sigma}$, muni de son Frobenius canonique. On conclut grâce à l'hypothèse de récurrence appliquée à M/N.

(ii) *Cas où S est le spectre d'un corps quelconque k.* Soit k_s une clôture séparable de k. Fixons un anneau de Cohen $C(k)$ de k, et un relèvement σ du Frobenius sur $C(k)$; posons $C_n(k) = C(k)/p^nC(k)$. Pour toute extension finie séparable $k' \subset k_s$ de k, il existe une unique $C_n(k)$-algèbre étale $C_n(k')$ relevant k', et un unique endomorphisme σ' de $C_n(k')$ relevant le Frobenius de k' et prolongeant σ. De plus, si $k'' \subset k_s$ est une extension finie de k', il existe un unique homomorphisme de $C_n(k)$-algèbres $C_n(k') \to C_n(k'')$ relevant l'homomorphisme $k' \to k''$. L'anneau $C(k') = \varprojlim_n C_n(k')$ est un anneau de Cohen de k', et si l'on pose $C_n(k_s) = \varinjlim_{k'} C_n(k')$, $C(k_s) = \varprojlim_n C_n(k_s)$, l'anneau $C(k_s)$ est un anneau de Cohen de k_s ; de même, les σ' définissent par passage à la limite un relèvement du Frobenius sur $C(k_s)$.

D'après le cas (i), l'image inverse de M sur $\mathrm{Spec}\, k_s$ est isomorphe à une somme directe de cristaux de la forme $\mathcal{O}_{S/\Sigma}/p^n\mathcal{O}_{S/\Sigma}$. Cet isomorphisme s'explicite comme un isomorphisme de $C_n(k_s)$-modules munis d'un isomorphisme $M^\sigma \xrightarrow{\sim} M$, pour n assez grand. Par suite, il se redescend sur $C_n(k')$ pour une extension finie convenable, ce qui revient à dire qu'après la localisation étale Spec $k' \to$ Spec k, M est trivial.

(iii) *Cas où S est le spectre d'un anneau artinien A.* Soient k le corps résiduel de A et i : Spec $k \hookrightarrow$ Spec $A = S$ l'immersion correspondante. Il existe un entier m tel que la puissance m-ième F^m du Frobenius de S se factorise par un homomorphisme $\Phi_m : S \to$ Spec k. On en déduit les isomorphismes

$$\Phi^*_{m\ \mathrm{cris}} \circ i^*_{\mathrm{cris}}(M) \simeq F^{m*}_{\mathrm{cris}}(M) = M^{\sigma^m} \xrightarrow{\sim} M.$$

D'après (ii), $i^*_{\mathrm{cris}}(M)$ est localement trivial ; l'énoncé pour M en résulte alors par image inverse par Φ_m.

(iv) *Cas où S est le spectre d'un anneau local nœthérien complet A, dont le corps résiduel possède une p-base finie.* Soient $\mathfrak{m}$ l'idéal maximal de A, et $k = A/\mathfrak{m}$; on notera encore i : Spec $k \hookrightarrow$ Spec A l'immersion correspondante. D'après (ii), il existe une extension finie séparable k' de k telle que M soit trivial sur k'. Soit S' le revêtement fini étale de S relevant Spec k' ; comme S' est encore le spectre d'un anneau local nœthérien complet, on est ramené à montrer que lorsque M est trivial sur k, M est trivial sur A.

Pour tout entier m, on posera $A_m = A/\mathfrak{m}^m$; on a alors

$$W_n(A) \simeq \varprojlim_m W_n(A_m).$$

Les hypothèses faites sur A entraînent que, pour tout i, A_m, considéré comme A_m-module par le i-ième itéré du Frobenius, est de type fini sur lui-même, donc artinien ; il en résulte que les $W_n(A_m)$ sont des anneaux artiniens. Considérons pour tout m le $W_n(A_m)$-module M_m défini par la valeur de M sur l'épaississement Spec $A_m \hookrightarrow$ Spec $W_n(A_m)$ de $\mathrm{C{\scriptstyle RIS}}(S/\Sigma)$; comme M est un cristal, on a $M_m = M_{W_n(A)} \otimes W_n(A_m)$. En prenant une présentation finie de M sur $W_n(A)$, en la tensorisant par $W_n(A_m)$, et en passant à la limite projective sur m (ce qui conserve l'exactitude puisque les modules considérés sont artiniens), on en déduit l'isomorphisme

$$M_{W_n(A)} \simeq \varprojlim_m M_m.$$

De même, on a pour tout i un isomorphisme

$$W_n(A)/p^i W_n(A) \simeq \varprojlim_m W_n(A_m)/p^i W_n(A_m).$$

Comme M est trivial sur k, on peut trouver un isomorphisme

$$\psi : \bigoplus_{j=1}^{r} W_n(k)/p^{n_j} W_n(k) \xrightarrow{\sim} M_1$$

compatible à l'action de F. Considérons alors pour tout m l'homomorphisme

$$\Phi_m : W_n(k) \to W_n(A_m)$$

induit par l'homomorphisme Φ_m factorisant F^m sur A_m comme en (iii), qui factorise σ^m sur $W_n(A_m)$. Pour m variable, on obtient un système transitif d'isomorphismes

$$\begin{array}{ccc} \bigoplus W_n(A_{m+1})/p^{n_j}W_n(A_{m+1}) & \xrightarrow[\Phi^*_{m+1}(\psi)]{\sim} & \Phi^*_{m+1}(M_1) \\ \downarrow & & \downarrow F \\ \bigoplus W_n(A_m)/p^{n_j}W_n(A_m) & \xrightarrow[\Phi^*_m(\psi)]{\sim} & \Phi^*_m(M_1), \end{array}$$

$$\begin{array}{ccccc} \Phi^*_{m+1}(M_1) & \xrightarrow{\sim} & \sigma^{m+1^*}(M_{m+1}) & \xrightarrow[F^{m+1}]{\sim} & M_{m+1} \\ \downarrow F & & \downarrow F & & \downarrow \\ \Phi^*_m(M_1) & \xrightarrow{\sim} & \sigma^{m*}(M_m) & \xrightarrow[F^m]{\sim} & M_m, \end{array}$$

dans lequel la commutativité du diagramme supérieur résulte de ce que ψ est compatible à l'action de F, avec l'action triviale sur le terme de gauche. Par passage à la limite, on obtient donc un isomorphisme

$$\Psi : \bigoplus_{j=1}^{r} W_n(A)/p^{n_j}W_n(A)) \xrightarrow{\sim} M_{W_n(A)}$$

compatible à F. Le lemme (2.4.3) entraîne alors que M est trivial.

(v) *Cas où S est le spectre d'un anneau local nœthérien hensélien A, dont le corps résiduel possède une p-base finie.* Comme les revêtements étales du corps résiduel k de A se relèvent en des revêtements étales de A, on peut encore supposer que M est trivial sur k. Soient $\widehat{A}$ le complété de A, $\widehat{S} = \operatorname{Spec} \widehat{A}$. Si on note $\widehat{M}$ l'image inverse de M sur $\widehat{S}$, il résulte des hypothèses faites sur A et du cas (iv) que $\widehat{M}$ est trivial. Il existe donc un isomorphisme

$$\bigoplus_{i=1}^{r} \mathbf{Z}/p^{n_i}\mathbf{Z} \xrightarrow{\sim} (\widehat{M}_{W_n(\widehat{S})})^{1-F}.$$

Montrons que l'homomorphisme canonique

$$\Gamma(S, (M_{W_n(S)})^{1-F}) \to \Gamma(\widehat{S}, (\widehat{M}_{W_n(\widehat{S})})^{1-F}) \tag{2.4.9.1}$$

est un isomorphisme. On peut d'abord observer qu'il en résultera que l'homomorphisme canonique

$$\Gamma(S, (M_{W_n(S)})^{1-F}) \otimes W_n(\mathcal{O}_S) \to M_{W_n(S)} \tag{2.4.9.2}$$

est un isomorphisme. En effet, il résulte de (2.4.7) que $W_n(\widehat{A})$ est fidèlement plat sur $W_n(A)$; on est ainsi ramené au même énoncé sur $W_n(\widehat{S})$, où $\widehat{M}$ est trivial.

Pour prouver que (2.4.9.1) est un isomorphisme, on procède par récurrence sur l'entier n. Si $n = 1$, cela revient à montrer que M_S est un $\mathcal{O}_S$-module libre, ayant une base invariante par F. Comme $\widehat{M}$ est trivial, $\widehat{M}_{\widehat{S}} = M_{\widehat{S}} = M_S \otimes_{\mathcal{O}_S} \mathcal{O}_{\widehat{S}}$ est libre, et il en est donc de même de M_S par fidèle platitude. Soit Φ la matrice de F dans une base $(e_i)_{i=1,\dots,m}$ de M_S. Construire une base de M invariante par F revient à trouver une matrice inversible Λ telle que $\Lambda^\sigma = \Phi^{-1}\Lambda$. Les équations correspondantes définissent un revêtement étale de S, qui possède un point à valeurs dans k correspondant à une matrice inversible, puisque M est trivial sur k ; comme A est hensélien, ce point se relève au-dessus de S, et le déterminant de la matrice correspondante est encore inversible.

Pour n quelconque, remarquons que, puisque $W_n(\widehat{A})$ est fidèlement plat sur $W_n(A)$, l'homomorphisme (2.4.9.1) est injectif. L'hypothèse de récurrence entraîne qu'on a des isomorphismes

$$\Gamma(S, (M_{W_{n-1}(S)})^{1-F}) \xrightarrow{\sim} \Gamma(\widehat{S}, (\widehat{M}_{W_{n-1}(\widehat{S})})^{1-F}) \xrightarrow{\sim} \bigoplus_{i=1}^{r} \mathbb{Z}/p^{n'_i}\mathbb{Z}),$$

où $n'_i = \inf(n_i, n-1)$. Soient $(x_i)_{i=1,\dots,r}$ des générateurs de $\Gamma(S, (M_{W_{n-1}(S)})^{1-F})$; d'après l'hypothèse de récurrence, les x_i sont aussi des générateurs de $M_{W_{n-1}(S)}$, de sorte qu'on peut les relever en une famille (y_i) de générateurs de $M_{W_n(S)}$. Il existe alors des éléments $\gamma_{ij} \in A$ tels que l'on ait $F(y_i) - y_i = \sum_j V^{n-1}(\gamma_{ij})y_j$ pour tout i. Pour trouver des générateurs de $M_{W_n(S)}$ invariants par F relevant les x_i, il suffit de trouver des éléments $\xi_{ij} \in A$ tels que, si l'on pose $z_i = \sum_j V^{n-1}(\xi_{ij})y_j$, on ait $F(y_i+z_i) = y_i+z_i$ pour tout i. En notant Γ, Ξ les matrices formées par les γ_{ij} et les ξ_{ij}, et en notant encore V l'action de V sur les matrices coefficient par coefficient, on voit qu'il suffit qu'on ait

$$(I + V^{n-1}(\Xi))^\sigma(I + V^{n-1}(\Gamma)) = I + V^{n-1}(\Xi),$$

soit encore $\Xi^\sigma + \Gamma = \Xi$. Le système d'équations correspondant définit un revêtement étale S' de S, au-dessus duquel les x_i se relèvent en une famille de générateurs de $M_{W_n(S')}$ invariants par F. En particulier, tout élément x de $\Gamma(S, (M_{W_{n-1}(S)})^{1-F})$ se relève dans $\Gamma(S', (M_{W_n(S')})^{1-F})$.

Soient alors $y \in \Gamma(\widehat{S}, (\widehat{M}_{W_n(\widehat{S})})^{1-F})$, et $\bar{y}$ l'image de y dans $\Gamma(\widehat{S}, (\widehat{M}_{W_{n-1}(\widehat{S})})^{1-F})$. D'après l'hypothèse de récurrence, il existe $\bar{x} \in \Gamma(S, (M_{W_{n-1}(S)})^{1-F})$ d'image $\bar{y}$. Soit $S' \to S$ le revêtement étale construit précédemment ; l'image $\bar{x}'$ de $\bar{x}$ se relève alors en $x' \in \Gamma(S', (M_{W_n(S')})^{1-F})$. Si $\widehat{S'} = S' \times_S \widehat{S}$, et si y' est l'image de y dans $\Gamma(\widehat{S'}, (\widehat{M}_{W_n(\widehat{S'})})^{1-F})$, on a donc $y' - x' \in (V^{n-1}W_n(\mathcal{O}_{\widehat{S'}}) \cdot \widehat{M}_{W_n(\widehat{S'})})^{1-F}$. Comme $\widehat{M}$ est trivial, on a

$$(V^{n-1}W_n(\mathcal{O}_{\widehat{S'}}) \cdot \widehat{M}_{W_n(\widehat{S'})})^{1-F} \simeq \bigoplus_{i=1}^{r} p^{n-1} \cdot \mathbf{Z}/p^{n_i}\mathbf{Z}),$$

ce qui permet d'écrire $y' - x' = p^{n-1}z'$, avec $z \in \Gamma(\widehat{S}, (\widehat{M}_{W_n(\widehat{S})})^{1-F})$. Appliquant le même argument à z', on voit qu'on peut choisir x' de manière à avoir $y' = x'$ dans $\Gamma(\widehat{S'}, (\widehat{M}_{W_n(\widehat{S'})})^{1-F})$. Remarquons que, puisque A est hensélien, S' est le spectre d'un anneau A', produit d'anneaux locaux henséliens finis et étales sur A, et qu'il en est de même de $S'' = S' \times_S S'$. Par suite, $\widehat{S''}$ est le spectre du complété de $A'' = A' \otimes_A A'$, ce qui entraîne par (2.4.7) que $W_n(\widehat{S''})$ est fidèlement plat sur $W_n(S'')$. Compte tenu de l'isomorphisme

$$M_{W_n(\widehat{S''})} \simeq M_{W_n(S'')} \otimes_{W_n(\mathcal{O}_{S''})} W_n(\mathcal{O}_{\widehat{S''}}),$$

le fait que les deux images inverses de y' dans $M_{W_n(\widehat{S''})}$ soient égales entraîne qu'il en est de même des deux images inverses de x' dans $M_{W_n(S'')}$. Comme $W_n(S')$ et $W_n(S'')$ sont les revêtements étales de $W_n(S)$ relevant S' et S'' [13, 0 1.5.8], l'élément x' se redescend dans $M_{W_n(S)}$ en un élément x d'image y, ce qui montre que (2.4.9.1) est un isomorphisme.

Compte tenu de (2.4.3), l'isomorphisme (2.4.9.2) entraîne alors que M est trivial.

(vi) *Cas général.* On peut supposer S affine, soit $S = \operatorname{Spec} A$. La proposition (2.4.3) entraîne que M est déterminé par le $W_n(A)$-module $M_{W_n(A)}$, muni de l'isomorphisme de Frobenius $F : M^\sigma_{W_n(A)} \xrightarrow{\sim} M_{W_n(A)}$, et qu'il suffit de montrer que, quitte à localiser sur S pour la topologie étale, $M_{W_n(A)}$ est isomorphe à une somme directe de modules de la forme $W_n(A)/p^{n_i}W_n(A)$, munis du Frobenius usuel. Comme la formation de $W_n(A)$ commute aux limites inductives filtrantes, et que M est de présentation finie, on peut redescendre M et supposer que A est une algèbre de type fini sur $\mathbf{F}_p$. De même, si l'assertion est prouvée pour les localisés de S, on peut trouver un voisinage de chaque point de S au-dessus duquel il existe un recouvrement étale sur lequel M est trivial. On peut donc supposer A local, nœthérien, à corps résiduel absolument de type fini.

Soit alors A^h l'hensélisé de A. D'après (v), il existe un revêtement étale de Spec A^h au-dessus duquel M est trivial. Ecrivant A^h comme limite inductive de localisés d'algèbres étales sur A, on peut redescendre ce revêtement et la trivialisation de M sur une A-algèbre dont le spectre fournit un recouvrement étale de Spec A, ce qui achève la démonstration.

Par passage à la limite, le théorème (2.4.1) entraîne le résultat analogue pour les groupes p-divisibles :

2.4.10. Corollaire. *Pour tout schéma S localement annulé par une puissance de p, le foncteur $\mathbb{D}$ induit une équivalence de catégories entre la catégorie des groupes p-divisibles étales (resp. de type multiplicatif) sur S, et la catégorie des F-cristaux unité sur S (resp. la catégorie des F-cristaux munis d'un isomorphisme $V : M \xrightarrow{\sim} M^{\sigma}$ tel que $F \circ V = p$, $V \circ F = p$).*

3. Théorie multiplicative : l'extension universelle en tant que cristal

Nous allons ici comparer le cristal de Dieudonné construit par voie cohomologique dans [7] avec la construction initiale de [18], basée sur la théorie de l'extension universelle d'un groupe p-divisible par un groupe vectoriel. Préalablement, il nous faudra montrer comment cette extension universelle peut elle-même être décrite à partir d'un faisceau sur le site cristallin.

Dans cette section, nous travaillerons systématiquement avec le site cristallin *nilpotent* $\mathrm{NC{\scriptstyle RIS}}(S/\Sigma)$; rappelons qu'il s'agit du sous-site plein de $\mathrm{C{\scriptstyle RIS}}(S/\Sigma)$ formé des objets (U,T,δ) tels que l'idéal $\mathcal{J}_T = \mathrm{Ker}(\mathcal{O}_T \to \mathcal{O}_U)$ soit PD-nilpotent, i.e. vérifie $\mathcal{J}_T^{[n]} = 0$ pour n assez grand. Le schéma de référence $(\Sigma, \mathcal{I}, \gamma)$ est quelconque ; en particulier, on peut prendre $\Sigma =$ Spec $\mathbf{Z}_p$, muni de l'idéal $\mathcal{I} = 0$.

3.1. *Interprétation cristalline de l'extension universelle.*

Si S est un schéma, et M un $\mathcal{O}_S$-module quasi-cohérent, nous noterons encore M le faisceau fppf associé à M, qui associe à un S-schéma U le groupe $\Gamma(U, M \otimes_{\mathcal{O}_S} \mathcal{O}_U)$. On supposera que p est localement nilpotent sur S. Rappelons (cf. [18], ou [17]), que, si G est un groupe p-divisible sur S, on a $\mathrm{Hom}_S(G, M) = 0$ pour tout $\mathcal{O}_S$-module quasi-cohérent M, et que le foncteur $M \mapsto \mathrm{Ext}^1_S(G, M)$ est représenté par le groupe vectoriel $V(G) = \omega_{G^*}$. Il existe donc une extension universelle

$$0 \to V(G) \to E(G) \to G \to 0.$$

La formation de $E(G)$ est fonctorielle en G, et compatible au changement de base.

Nous allons donner une description cristalline de $E(G^*)$, et nous aurons pour cela à travailler sur le site cristallin nilpotent. Observons d'abord que les faisceaux d'extensions que nous aurons à utiliser sont simplement les restrictions à $\mathrm{NCRIS}(S/\Sigma)$ des faisceaux usuels calculés sur $\mathrm{CRIS}(S/\Sigma)$:

3.1.1. Lemme. *Soient G un groupe fini localement libre, un groupe p-divisible, ou un schéma abélien sur S, i un entier ≤ 1, E un faisceau abélien sur $\mathrm{CRIS}(S/\Sigma)$, égal à $\mathcal{O}_{S/\Sigma}$, $\mathcal{J}_{S/\Sigma}$ ou à $\underline{\mathbb{G}}_a$. Alors les faisceaux $\mathcal{E}xt^i_{S/\Sigma}(\underline{G}, E)$ calculés sur $\mathrm{NCRIS}(S/\Sigma)$ sont les restrictions des faisceaux $\mathcal{E}xt^i_{S/\Sigma}(\underline{G}, E)$ calculés sur $\mathrm{CRIS}(S/\Sigma)$.*

Considérons d'abord le cas d'un schéma abélien, et supposons $i \leq 2$. Soient (U, T, δ) un objet de $\mathrm{NCRIS}(S/\Sigma)$, $\mathcal{J}_T$ l'idéal de U dans T, $\bar{\delta}$ la PD-structure de l'idéal $\mathcal{J}_T + \mathcal{I}\mathcal{O}_T$ prolongeant les PD-structures données ; $\bar{\delta}$ est encore PD-nilpotente. La suite spectrale [7, (2.2.2.1)], reliant via [7, (2.1.6.1)] les $\mathcal{E}xt^i_{S/\Sigma}(\underline{G}, E)_T$ aux images directes cristallines de la restriction de E au-dessus de puissances de G_U, ramène à montrer que la cohomologie des faisceaux $\mathcal{O}_{S/\Sigma}$, $\mathcal{J}_{S/\Sigma}$ et $\underline{\mathbb{G}}_a$ est la même sur $\mathrm{CRIS}(G_U/T)$ et sur $\mathrm{NCRIS}(G_U/T)$. En ce qui concerne $\mathcal{O}_{S/\Sigma}$, on observe que G_U peut localement être relevé en un schéma abélien G_T sur T, et que les cohomologies cristallines de G_U et G_T coïncident (y compris sur le site $\mathrm{NCRIS}(S/\Sigma)$, l'argument de [3, III 2.3.4] restant valable grâce à la PD-nilpotence de δ). On est donc ramené à la même assertion sur $\mathrm{CRIS}(G_T/T)$ et sur $\mathrm{NCRIS}(G_T/T)$, qui résulte de [3, V 2.4, p. 313]. Pour $\underline{\mathbb{G}}_a$, la démonstration de [7, 2.3.12] reste valable telle quelle pour $\mathrm{NCRIS}(S/\Sigma)$, ce qui implique le lemme. Enfin, le cas de $\mathcal{J}_{S/\Sigma}$ résulte formellement des précédents.

Si G est un groupe fini localement libre, il est possible, localement sur S, de le plonger dans un schéma abélien [7, 3.1.1], et l'énoncé pour $i \leq 1$ résulte donc de la validité de l'énoncé pour les schémas abéliens lorsque $i \leq 2$ grâce au lemme des cinq. Pour les groupes p-divisibles, l'énoncé se ramène au cas des groupes finis par passage à la limite.

3.1.2. Considérons sur $\mathrm{NCRIS}(S/\Sigma)$ l'extension

$$0 \to 1 + \mathcal{J}_{S/\Sigma} \to \mathcal{O}^*_{S/\Sigma} \to \underline{\mathbb{G}}_m \to 0. \tag{3.1.2.1}$$

Si G est un groupe p-divisible, et si, pour tout $n \geq 1$, on note $G(n)$ le noyau de la multiplication par p^n sur G, on en tire des suites exactes

(3.1.2.2)
$$0 \to \mathcal{H}om_{S/\Sigma}(\underline{G}(n), 1 + \mathcal{J}_{S/\Sigma}) \to \mathcal{H}om_{S/\Sigma}(\underline{G}(n), \mathcal{O}^*_{S/\Sigma}) \to \underline{G}^*(n).$$

Pour $m \geq n$, l'homomorphisme $p^{m-n} : \underline{G}(m) \to \underline{G}(n)$ induit un morphisme de suites exactes, et on obtient en passant à la limite une suite exacte

(3.1.2.3)
$$0 \to \varinjlim_n \mathcal{H}om_{S/\Sigma}(\underline{G}(n), 1 + \mathcal{J}_{S/\Sigma}) \to \varinjlim_n \mathcal{H}om_{S/\Sigma}(\underline{G}(n), \mathcal{O}^*_{S/\Sigma}) \to \underline{G}^*$$

3.1.3. Proposition. *L'application* $\varinjlim_n \mathcal{H}om_{S/\Sigma}(\underline{G}(n), \mathcal{O}^*_{S/\Sigma}) \to \underline{G}^*$ *est un épimorphisme.*

Sur NCRIS(S/Σ), l'exponentielle exp : $\mathcal{J}_{S/\Sigma} \to 1 + \mathcal{J}_{S/\Sigma}$ est définie et fournit un inverse au logarithme. On peut alors prolonger la suite exacte (3.1.2.2) sous la forme

$$\mathcal{H}om_{S/\Sigma}(\underline{G}(n), \mathcal{O}^*_{S/\Sigma}) \to \underline{G}^*(n) \to \mathcal{E}xt^1_{S/\Sigma}(\underline{G}(n), \mathcal{J}_{S/\Sigma}).$$

D'après (3.1.1), le faisceau $\mathcal{E}xt^1_{S/\Sigma}(\underline{G}(n), \mathcal{J}_{S/\Sigma})$ est la restriction à NCRIS (S/Σ) du faisceau analogue sur CRIS(S/Σ). Du diagramme commutatif

$$\begin{array}{ccccccccc}
0 & \longrightarrow & G(m) & \longrightarrow & G & \xrightarrow{p^m} & G & \longrightarrow & 0 \\
 & & \downarrow & & \downarrow{\scriptstyle p^{m-n}} & & \downarrow & & \\
0 & \longrightarrow & G(n) & \longrightarrow & G & \xrightarrow{p^n} & G & \longrightarrow & 0,
\end{array}$$

et de l'isomorphisme $\mathcal{E}xt^1_{S/\Sigma}(\underline{G}, \mathcal{J}_{S/\Sigma})_{(U,T)} \xrightarrow{\sim} \mathcal{E}xt^1_{S/\Sigma}(\underline{G}(n), \mathcal{J}_{S/\Sigma})_{(U,T)}$, valable dès que p^n annule $\mathcal{O}_T$ [7, 3.3.4 (ii)], on déduit que l'application

$$\underline{G}^*(n) \to \varinjlim_m \mathcal{E}xt^1_{S/\Sigma}(\underline{G}(m), \mathcal{J}_{S/\Sigma})$$

est nulle. La surjectivité annoncée en découle aussitôt.

3.1.4. Corollaire. *Soit* ω_G *le module des différentielles invariantes sur* G. *Il existe, sur le petit site fppf de* S, *une suite exacte courte*

(3.1.4.1) $$0 \to \omega_G \to \varinjlim_n \mathcal{H}om_{S/\Sigma}(\underline{G}(n), \mathcal{O}^*_{S/\Sigma})_S \to G^* \to 0.$$

D'après [7, 3.3.4], il existe un isomorphisme canonique

$$\omega_G \xrightarrow{\sim} \mathcal{E}xt^1_{S/\Sigma}(\underline{G}, \mathcal{J}_{S/\Sigma})_S. \tag{3.1.4.2}$$

Dès que $p^n\mathcal{O}_S = 0$, le cobord

$$\partial : \mathcal{H}om_{S/\Sigma}(\underline{G}(n), \mathcal{J}_{S/\Sigma})_S \to \mathcal{E}xt^1_{S/\Sigma}(\underline{G}, \mathcal{J}_{S/\Sigma})_S$$

est un isomorphisme. On définit alors un isomorphisme

$$\omega_G \xrightarrow{\sim} \varinjlim_n \mathcal{H}om_{S/\Sigma}(\underline{G}(n), \mathcal{J}_{S/\Sigma})_S \tag{3.1.4.3}$$

en composant (3.1.4.2) avec $-\partial^{-1}$ (le signe affecté à ∂^{-1} étant motivé par la démonstration de la proposition suivante). La suite (3.1.4.1) provient alors de (3.1.2.3) grâce à l'isomorphisme fourni par le logarithme.

3.1.5. Proposition. *L'extension* (3.1.4.1) *est canoniquement isomorphe à l'extension universelle $E(G^*)$ de G^* par un groupe vectoriel.*

On peut supposer que $p^n\mathcal{O}_S = 0$, ce qui entraîne notamment que $\omega_G \simeq \omega_{G(n)}$. Soit $\alpha : G^*(n) \to \omega_G$ l'homomorphisme universel (voir [18, IV (1.4)], ou [17, I (1.2)]). D'après la construction de l'extension universelle (cf. [18, IV (1.10)], ou [17]), il suffit, pour prouver la proposition, de construire un diagramme commutatif

$$\begin{array}{ccccccc}
0 \to G^*(n) & \longrightarrow & G^* & \xrightarrow{p^n} & G^* \to 0 \\
\Big\downarrow \alpha & & \Big\downarrow & & \Big\| \\
0 \to \omega_B \to & \varinjlim_m \mathcal{H}om_{S/\Sigma}(\underline{G}(m), \mathcal{O}^*_{S/\Sigma})_S & & \to & G^* \to 0.
\end{array}$$

Or on peut définir un homomorphisme

$$G^* \to \varinjlim_m \mathcal{H}om_{S/\Sigma}(\underline{G}(m), \mathcal{O}^*_{S/\Sigma})_S \tag{3.1.5.1}$$

comme suit : étant donnés un S-schéma plat U, et un homomorphisme $v : G_U(m) \to \mathbb{G}_{mU}$, on associe à v l'application composée

$$\underline{G}_U(m) \to \underline{\mathbb{G}}_{mU} \xrightarrow{\tau_n} \mathcal{O}^*_{U/U},$$

où, pour tout objet (V, T, δ) de NCRIS(S/Σ) tel que $p^n\mathcal{O}_T = 0$, τ_n associe à $\xi \in \mathcal{O}_V^*$ la section $\tilde{\xi}^{p^n}$ de $\mathcal{O}_T^*$, où $\tilde{\xi}$ est un relèvement quelconque de ξ. Il suffit donc de voir que (3.1.5.1) induit sur $G^*(n)$ l'homomorphisme α.

Rappelons que, pour tout S-groupe fini localement libre H, l'homomorphisme $\alpha : H^* \longrightarrow \omega_H$ est l'application qui associe à $h : H_U \longrightarrow \mathbb{G}_{mU}$ la forme différentielle $h^*(dt/t)$, où dt/t est la forme différentielle invariante canonique sur $\mathbb{G}_m$. Par fonctorialité, il suffit donc de vérifier que (3.1.5.1) induit α sur $G^*(n)$ lorsque $G(n) = \boldsymbol{\mu}_{p^n}$, c'est à dire que, dans l'identification

$$\begin{aligned}\mathcal{H}om(\underline{\boldsymbol{\mu}}_{p^n}, 1+\mathcal{J}_{S/\Sigma})_S &\xrightarrow[\log]{\sim} \mathcal{H}om(\underline{\boldsymbol{\mu}}_{p^n}, \mathcal{J}_{S/\Sigma})_S \\ &\xrightarrow[-\partial]{\sim} \mathcal{E}xt^1(\underline{\boldsymbol{\mu}}_{p^\infty}, \mathcal{J}_{S/\Sigma})_S \xleftarrow{\sim} \omega_{\mu_{p^\infty}},\end{aligned}$$

l'homomorphisme $\tau_n : \underline{\boldsymbol{\mu}}_{p^n} \to 1 + \mathcal{J}_{S/\Sigma}$ correspond à dt/t.

Comme $p^n\mathcal{O}_S = 0$, les homomorphismes canoniques

$$\mathcal{E}xt^1(\underline{\boldsymbol{\mu}}_{p^\infty}, \mathcal{J}_{S/\Sigma})_S \to \mathcal{E}xt^1(\underline{\boldsymbol{\mu}}_{p^n}, \mathcal{J}_{S/\Sigma})_S, \qquad \omega_{\boldsymbol{\mu}_{p^\infty}} \to \omega_{\boldsymbol{\mu}_{p^n}},$$

sont des isomorphismes, ce qui permet de remplacer $\boldsymbol{\mu}_{p^\infty}$ par $\boldsymbol{\mu}_{p^n}$ dans l'assertion précédente, l'homomorphisme ∂ étant alors le cobord provenant de la suite exacte

$$(3.1.5.2) \qquad 0 \to \boldsymbol{\mu}_{p^n} \to \boldsymbol{\mu}_{p^{2n}} \to \boldsymbol{\mu}_{p^n} \to 0.$$

D'après [7, 3.2.3], on dispose, pour tout groupe fini localement libre G, d'un isomorphisme

$$(3.1.5.3) \qquad \ell^G[-1] \xrightarrow{\sim} t_{1]}\mathbb{R}\mathcal{H}om_{S/\Sigma}(\underline{G}, \mathcal{J}_{S/\Sigma})_S,$$

où ℓ^G est le complexe de co-Lie de G, induisant en degré 1 l'isomorphisme

$$\omega_G \xrightarrow{\sim} \mathcal{E}xt^1(\underline{G}, \mathcal{J}_{S/\Sigma})_S$$

utilisé dans l'identification ci-dessus. Explicitons-en la construction dans le cas de $\boldsymbol{\mu}_{p^n}$. Soit D l'enveloppe à puissances divisées de l'idéal d'augmentation dans l'algèbre $R = \mathcal{O}_S[t, t^{-1}]$ de $\mathbb{G}_{mS}$. Si $\Gamma(\omega_{\mathbb{G}_m})$ désigne l'algèbre à puissances divisées du $\mathcal{O}_S$-module $\omega_{\mathbb{G}_m}$, il existe d'après [7, 3.2.2] un homomorphisme canonique $\varphi : \Gamma(\omega_{\mathbb{G}_m}) \to D$, associant à une forme différentielle $\omega \in \omega_{\mathbb{G}_m}$ l'unique élément f de D tel que $d(f) = 1 \otimes \omega \in D \otimes_R \Omega^1_{\mathbb{G}_m}$. On a donc en particulier

$$\varphi(dt/t) = \log(t) = \sum_{i \geq 1} (-1)^{i-1}(i-1)!(t-1)^{[i]}.$$

Considérons alors la résolution lisse

$$0 \to \boldsymbol{\mu}_{p^n} \to \mathbb{G}_m \xrightarrow{p^n} \mathbb{G}_m \to 0$$

de $\boldsymbol{\mu}_{p^n}$. Elle fournit d'une part un isomorphisme

$$\ell^{\mu_{p^n}}[-1] \xrightarrow{\sim} [\omega_{\mathbb{G}_m} \xrightarrow{-p^n} \omega_{\mathbb{G}_m}],$$

d'autre part une identification de $t_{1]}\mathbb{R}\mathcal{H}om_{S/\Sigma}(\underline{\boldsymbol{\mu}}_{p^n}, \mathcal{J}_{S/\Sigma})_S$ à un sous-complexe du complexe

$$[\bar{I} \xrightarrow{d} (D' \otimes_R \Omega^1_{\mathbb{G}_m}) \oplus D'_2],$$

où D' (resp. D'_2) est l'enveloppe à puissances divisées de l'idéal de $\boldsymbol{\mu}_{p^n}$ (resp. $\boldsymbol{\mu}_{p^n} \times \boldsymbol{\mu}_{p^n}$) dans l'algèbre de $\mathbb{G}_m$ (resp. $\mathbb{G}_m \times \mathbb{G}_m$), $\bar{I}$ son PD-idéal canonique, d la différentielle naturelle sur le premier facteur et l'homomorphisme $\mu^* - p_1^* - p_2^*$ sur le deuxième, μ, p_1 et p_2 étant les homomorphismes induits par la loi de groupe et les deux projections. Si $[-p^n]$ désigne l'homomorphisme d'algèbres à puissances divisées $D \to D'$ induit par la multiplication par $-p^n$ sur $\mathbb{G}_m$, l'isomorphisme (3.1.5.3) est alors défini par le carré commutatif

$$\begin{array}{ccc} \omega_{\mathbb{G}_m} & \xrightarrow{-p^n} & \omega_{\mathbb{G}_m} \\ {\scriptstyle [-p^n]\circ\varphi}\downarrow & & \downarrow \\ \bar{I} & \xrightarrow{d} & (D' \otimes_R \Omega^1_{\mathbb{G}_m}) \oplus D'_2, \end{array}$$

la flèche en degré 1 étant donnée par l'homomorphisme canonique (avec 0 sur la composante D'_2). En particulier, si $n_{\mu_{p^n}} = H^{-1}(\ell^{\mu_{p^n}})$, et si on note encore dt/t l'élément de $n_{\mu_{p^n}}$ correspondant à $dt/t \in \omega_{\mathbb{G}_m}$, l'image de dt/t dans $\bar{I}$ est $\log(t^{-p^n}) = -\log(t^{p^n})$.

La suite (3.1.5.2) fournit un morphisme de triangles distingués d'où l'on déduit le carré commutatif [7, 3.2.9]

$$\begin{array}{ccc} n_{\mu_{p^n}} & \xrightarrow[\sim]{-\partial} & \omega_{\mu_{p^n}} \\ \downarrow & & \downarrow \\ \mathcal{H}om(\underline{\boldsymbol{\mu}}_{p^n}, \mathcal{J}_{S/\Sigma})_S & \xrightarrow[\sim]{\partial} & \mathcal{E}xt^1(\underline{\boldsymbol{\mu}}_{p^n}, \mathcal{J}_{S/\Sigma})_S, \end{array}$$

dans lequel le cobord ∂ du haut provient de la suite exacte à six termes [18, II (3.3.5)]. On vérifie facilement que l'image par $-\partial$ de $dt/t \in n_{\mu_{p^n}}$ est

$dt/t \in \omega_{\mu_{p^n}}$, tandis que, d'après ce qui précède, la flèche verticale de gauche envoie dt/t sur l'homomorphisme $\underline{\mu}_{p^n} \to \mathcal{J}_{S/\Sigma}$ défini par $-\log(t^{p^n}) \in \Gamma(\mu_{p^n}/\Sigma, \mathcal{J}_{S/\Sigma})$. La flèche verticale de droite étant l'homomorphisme canonique, cela achève la démonstration, compte tenu du signe adopté dans la définition de (3.1.4.3).

3.1.6 Soient (U, T, δ) un objet de NCRIS(S/Σ), G un groupe p-divisible sur S, et G' un relèvement de G_U sur T. Soient $\mathcal{J}_T$ l'idéal de U dans T, $\overline{\delta}$ la PD-structure de l'idéal $\mathcal{J}_T + \mathcal{I}\mathcal{O}_T$ prolongeant les PD-structures données, et

$$i_{\text{cris}} : (U/(T,\overline{\delta}))_{\text{NCris}} \to (T/(T,\overline{\delta}))_{\text{NCris}} \tag{3.1.6.1}$$

le morphisme de topos canonique. Comme δ est PD-nilpotente, on a comme en [3, III 2.3.4] un isomorphisme $\mathcal{O}_{T/T} \xrightarrow{\sim} i_{\text{cris}*}(\mathcal{O}_{U/T})$, d'où l'on déduit l'isomorphisme d'adjonction

$$\text{Hom}_{U/T}(\underline{G_U}(n), \mathcal{O}^*_{U/T}) \xrightarrow{\sim} \text{Hom}_{T/T}(\underline{G'}(n), \mathcal{O}^*_{T/T}).$$

Passant aux faisceaux associés, et appliquant (3.1.5), on en déduit un isomorphisme de faisceaux sur le petit site fppf de T

$$\varinjlim_m \mathcal{H}om_{S/\Sigma}(\underline{G}(m), \mathcal{O}^*_{S/\Sigma})_{(U,T)} \xrightarrow{\sim} E(G'^*). \tag{3.1.6.2}$$

On obtient alors :

3.1.7. Théorème. *Soient $(S, \mathcal{I}, \gamma)$ vérifiant les hypothèses de* (3.1), *G un groupe p-divisible sur S. Pour tout objet (U, T, δ) de* NCRIS(S/Σ), *notons $\mathcal{E}(G)_T$ le faisceau fppf sur T défini par*

$$\mathcal{E}(G)_T(T') = \varinjlim_m \mathcal{H}om_{S/\Sigma}(\underline{G}(m), \mathcal{O}^*_{S/\Sigma})_{(U',T',\delta)},$$

avec $U' = U \times_T T'$.

(i) *Le foncteur $T \mapsto \mathcal{E}(G)_T$ définit un cristal en groupes ind-représentables $\mathcal{E}(G)$ sur* NCRIS(S/Σ). *La formation de ce cristal est fonctorielle en G, et commute aux changements de base $(S', \mathcal{I}', \gamma') \to (S, \mathcal{I}, \gamma)$.*

(ii) *Soient (U, T, δ) un objet de* NCRIS(S/Σ), *G', H' deux groupes p-divisibles sur T, et $v : G'_U \to H'_U$ un homomorphisme. Il existe un homomorphisme canonique $\tilde{v} : E(G') \to E(H')$ relevant $E_U(v)$. Si K' est un troisième groupe p-divisible, et $w : H'_U \to K'_U$ un homomorphisme, alors $(w \circ v)^\sim = \tilde{w} \circ \tilde{v}$. Si v est un isomorphisme, il en est de même pour $\tilde{v}$ (en particulier, pour $G' = H'$, $v = \text{Id}_{G'_U}$, on a $\tilde{v} = \text{Id}_{E(G')}$).*

(iii) *Si G' est un groupe p-divisible sur T relevant G_U, il existe un isomorphisme canonique*

$$(3.1.7.1) \qquad \mathcal{L}ie(E(G'^*)) \xrightarrow{\sim} \mathbb{D}(G_U)_{(U,T,\delta)}.$$

Plus généralement, l'"algèbre de Lie" du faisceau

$$\mathcal{E}(G) = \varinjlim_m \mathcal{H}om_{S/\Sigma}(\underline{G}(m), \mathcal{O}^*_{S/\Sigma})$$

est la restriction de $\mathbb{D}(G)$ à $N\mathrm{CRIS}(S/\Sigma)$.

(i) Pour montrer que $\mathcal{E}(G)_T$ est ind-représentable, il suffit de montrer que, pour tout n, le noyau $\mathcal{E}(G)_T(n)$ de la multiplication par p^n sur $\mathcal{E}(G)_T$ est représentable. Cette propriété étant locale sur T, on peut supposer [14, théorème 4.4] qu'il existe un groupe p-divisible G' sur T relevant G. Soit alors $\mathcal{E}'_T(n)$ l'image inverse de $G'^*(n)$ par l'homomorphisme $E(G'^*) \to G'^*$ et l'isomorphisme (3.1.6.2). Comme $\mathcal{E}'_T(n)$ est extension de $G'^*(n)$ par ω_G, il est représentable par un T-schéma en groupes plat de présentation finie ; $\mathcal{E}(G)_T(n)$, qui est le noyau de p^n sur $\mathcal{E}'_T(n)$, est donc lui aussi représentable. Soit maintenant $v : (U',T',\delta') \to (U,T,\delta)$ un morphisme de $N\mathrm{CRIS}(S/\Sigma)$. Il existe un morphisme naturel $v^{-1}\mathcal{E}(G)_T \to \mathcal{E}(G)_{T'}$, et, pour vérifier que c'est un isomorphisme, on peut encore supposer G relevé en G' sur T. Pour tout n, l'extension $\mathcal{E}'_T(n)$ définit par image inverse une extension de $G'^*_{T'}(n)$ par $\omega_{G'_{T'}}$, d'où une extension

$$0 \to \omega_{G'_{T'}} \to v^{-1}\mathcal{E}(G)_T \to G'^*_{T'} \to 0.$$

Cette extension s'envoie dans l'extension universelle de $G'^*_{T'}$ par un groupe vectoriel, par un homomorphisme qui est l'identité sur $\omega_{G'_{T'}}$ et $G'^*_{T'}$, et est compatible avec l'identification $\mathcal{E}(G)_{T'} \xrightarrow{\sim} E(G'^*_{T'}$, d'où l'assertion.

(ii) On passe aux groupes p-divisibles duaux. D'après (i), $\mathcal{E}(G)_T$ relève $\mathcal{E}(G)_U$. L'homomorphisme $v : G'_U \to H'_U$ induit un homomorphisme de faisceaux cristallins

$$\mathcal{E}(v) : \varinjlim_m \mathcal{H}om_{U/\Sigma}(\underline{H}'_U(m), \mathcal{O}^*_{U/\Sigma}) \to \varinjlim_m \mathcal{H}om_{U/\Sigma}(\underline{G}'_U(m), \mathcal{O}^*_{U/\Sigma});$$

en prenant sa restriction aux épaississements plats au-dessus de (U,T,δ), on obtient donc grâce à (3.1.5) un morphisme $E(H'^*) \to E(G'^*)$ relevant $E_U(v)$. Les autres assertions sont alors claires.

(iii) Pour tout faisceau de groupes F sur T, $\mathcal{L}ie\ (F)$ est défini par

$$\mathcal{L}ie(F)(T') = \mathrm{Ker}[F(T'[\varepsilon]) \to F(T')]$$

pour tout T' plat sur T, $T'[\varepsilon]$ désignant le schéma des nombres duaux sur T'. En particulier, on en déduit l'isomorphisme

$$\begin{aligned}\mathcal{L}ie(E(G'^*)) =& \mathcal{L}ie(\varinjlim_m \mathcal{H}om_{S/\Sigma}(\underline{G}(m), \mathcal{O}^*_{S/\Sigma})_{(U,T)})\\ &\simeq \varinjlim_m \mathcal{L}ie(\mathcal{H}om_{S/\Sigma}(\underline{G}(m), \mathcal{O}^*_{S/\Sigma})_{(U,T)}).\end{aligned}$$

Pour tout T', on a

$$\begin{aligned}&\Gamma((U',T',\delta), \mathcal{H}om_{S/\Sigma}(\underline{G}(m), \mathcal{O}^*_{S/\Sigma})) =\\ &\quad \operatorname{Ker}[H^0(G_{U'}(m)/T', \mathcal{O}^*_{S/\Sigma}) \to H^0(G_{U'}(m) \times_{U'} G_{U'}(m)/T', \mathcal{O}^*_{S/\Sigma})],\end{aligned}$$

la flèche étant $\mu^* - p_1^* - p_2^*$, où μ est la loi de groupe sur G, p_1 et p_2 les deux projections ; on dispose également d'une formule analogue pour $\mathcal{O}_{S/\Sigma}$. En calculant le H^0 cristallin comme le H^0_{DR} de l'enveloppe à puissances divisées de l'idéal de $G(m)$ dans un schéma lisse, on obtient aussitôt l'isomorphisme

$$\mathcal{L}ie(\mathcal{H}om_{S/\Sigma}(\underline{G}(m), \mathcal{O}^*_{S/\Sigma})_{(U,T)}) \simeq \mathcal{H}om_{S/\Sigma}(\underline{G}(m), \mathcal{O}_{S/\Sigma})_{(U,T)}. \tag{3.1.7.2}$$

Par suite, on a

$$\mathcal{L}ie(E(G'^*)) \simeq \varinjlim_m \mathcal{H}om_{S/\Sigma}(\underline{G}(m), \mathcal{O}_{S/\Sigma})_{(U,T)}.$$

Si $p^m \mathcal{O}_T = 0$, le cobord $\mathcal{H}om_{S/\Sigma}(\underline{G}(m), \mathcal{O}_{S/\Sigma})_{(U,T)} \to \mathbb{D}(G)_{(U,T)}$ est un isomorphisme, compatible avec les flèches de transition du système. On en déduit l'isomorphisme annoncé.

3.2. *Les deux fonctorialités.*

Rappelons que la thèse de Messing [18] fournit, pour tout groupe p-divisible G, un cristal sur $\mathrm{NCris}(S/\Sigma)$. La méthode de [18] consiste à construire un cristal en groupes ind-représentables $\mathbb{E}(G)$ sur $\mathrm{NCris}(S/\Sigma)$, puis à définir le cristal associé à G (également noté $\mathbb{D}(G)$ dans [18]) par

$$(U,T,\delta) \mapsto \mathcal{L}ie(\mathbb{E}(G))_{(U,T,\delta)}.$$

Cette construction permet d'obtenir une classification des déformations d'un groupe p-divisible le long d'un épaississement muni de puissances divisées nilpotentes au moyen d'un relèvement sur T de la filtration naturelle sur U

$$\omega_{G^*_U} \hookrightarrow \mathcal{L}ie(\mathbb{E}(G))_{(U,U)}.$$

Nous allons montrer dans cette section que le cristal ainsi obtenu s'identifie fonctoriellement à la restriction à NCRIS(S/Σ) de celui que nous utilisons ici (qui s'identifie lui-même avec celui de Mazur-Messing [17] d'après [7, 1.4]), ce qui permettra en particulier d'utiliser dans notre contexte les résultats de [18] sur la théorie des déformations de groupes p-divisibles.

3.2.1. Rappelons tout d'abord la construction de $\mathsf{E}(G)$ donnée dans [18]. Soient $(U,T,\delta) \in$ NCRIS(S/Σ), $i : U \hookrightarrow T$ l'immersion donnée, J son idéal, G_1, G_2 deux groupes p-divisibles sur T, $v : G_{1U} \to G_{2U}$ un homomorphisme. Il existe alors un unique homomorphisme $\widehat{v} : E(G_1) \to E(G_2)$ entre les extensions universelles satisfaisant les deux propriétés suivantes :

(i) $\widehat{v}$ relève v.

(ii) Si $w : \omega_{G_1^*} \to \omega_{G_2^*}$ est un homomorphisme (linéaire) quelconque relevant $\omega_{G_{1U}^*} \to \omega_{G_{2U}^*}$, alors la différence

$$d : \omega_{G_{1U}^*} \to \mathrm{Ker}[E(G_2) \to i_* E(G_{2U})],$$

mesurant le défaut de commutativité du diagramme

$$\begin{array}{ccc} \omega_{G_1^*} & \hookrightarrow & E(G_1) \\ {\scriptstyle w}\downarrow & & \downarrow{\scriptstyle \widehat{v}} \\ \omega_{G_2^*} & \hookrightarrow & E(G_2), \end{array}$$

est une *exponentielle*. Rappelons que cela signifie qu'il existe une application linéaire $\theta : \omega_{G_1^*} \to J \otimes \mathcal{L}ie(E(G_2))$, telle que $d = \exp \circ \theta$, en notant exp l'isomorphisme

$$\exp : J \otimes \mathcal{L}ie(E(G_2)) \xrightarrow{\sim} \mathrm{Ker}[E(G_2) \to i_* E(G_{2U})]$$

(cf. [18, III] pour plus de détails sur l'exponentielle).

D'autre part, nous avons vu en (3.1.7) (ii) qu'il existe un autre homomorphisme $\tilde{v} : E(G_1) \to E(G_2)$ relevant $E(v)$. Le *problème des deux fonctorialités* est de prouver que $\tilde{v} = \widehat{v}$. Comme $\widehat{v}$ est caractérisé axiomatiquement, il suffit de prouver que $\tilde{v}$ vérifie les axiomes.

Il sera commode de travailler avec les groupes duaux, et d'utiliser les notations de (3.1.7). Soient G', H' deux groupes p-divisibles sur T, et $u : H'_U \to G'_U$ un homomorphisme induisant une application

$$\tilde{u} : \varinjlim_m \mathcal{H}om_{U/\Sigma}(\underline{G'_U}(m), \mathcal{O}^*_{U/\Sigma})_{(U,T)} \to \varinjlim_m \mathcal{H}om_{U/\Sigma}(\underline{H'_U}(m), \mathcal{O}^*_{U/\Sigma})_{(U,T)}.$$

Soit $w : \omega_{G'} \to \omega_{H'}$ une application linéaire relevant l'application induite par u sur les espaces cotangents. Soient

$$j : \omega_{H'} \hookrightarrow \varinjlim_m \mathcal{H}om_{U/\Sigma}(\underline{H'_U}(m), \mathcal{O}^*_{U/\Sigma})_{(U,T)}$$

l'application définie par (3.1.4) et (3.1.6), et $d = \tilde{u}|\omega_{G'} - j \circ w$.

3.2.2. Théorème. *Avec les notations précédentes, l'application d est une exponentielle.*

Nous aurons besoin de quelques préliminaires, rappelant notamment dans ce contexte particulier différentes propriétés de l'exponentielle.

3.2.3. Lemme. *Si $p^n\mathcal{O}_T = 0$, le faisceau fppf $\mathcal{H}om_{T/\Sigma}(\underline{H}'(n), \mathcal{O}^*_{T/\Sigma})_{(T,T)}$ est représentable par un T-schéma en groupes affine.*

Considérons la suite exacte de faisceaux fppf sur T :

$$0 \to \mathcal{H}om(\underline{H}'(n), \mathcal{J})_T \to \mathcal{H}om(\underline{H}'(n), \mathcal{O}^*)_T \to H'(n)^* \xrightarrow{\partial} \mathcal{E}xt^1(\underline{H}'(n), \mathcal{J})_T.$$

Comme $\mathcal{E}xt^1(\underline{H}'(n), \mathcal{J})_T$ est isomorphe à $\omega_{H'(n)}$ [7, 3.3.4], le noyau de ∂ est un T-schéma en groupes affine. D'autre part, si $p^n\mathcal{O}_T = 0, \mathcal{H}om(\underline{H}'(n), \mathcal{J})_T$ est isomorphe à $\omega_{H'}$. Par descente fidèlement plate, il en résulte que $\mathcal{H}om(\underline{H}'(n), \mathcal{O}^*)_T$ est représentable par un schéma en groupes affine.

On remarquera que ce groupe n'est pas plat en général. Il sera donc plus commode de le considérer non pas comme un faisceau sur le petit site fppf de T, mais plutôt comme un faisceau sur le "grand" site $\mathrm{T}_{\mathrm{fppf},\delta}$ formé des T-schémas T' auxquels les puissances divisées δ s'étendent (cf. [7, 1.1.4]). Comme U appartient à ce site, ainsi que tous les U-schémas, la restriction à U de $\mathcal{H}om_{T/\Sigma}(\underline{H}'(n), \mathcal{O}^*_{T/\Sigma})_T$ (en tant que faisceau) correspond effectivement à l'image inverse par l'immersion $i : U \hookrightarrow T$.

Pour tout T' au-dessus de T auquel les puissances divisées δ s'étendent, on dispose de l'isomorphisme d'adjonction relatif à l'inclusion de $U' = U \times_T T'$ dans T' :

$$\mathrm{Hom}_{U'/T'}(\underline{H}'(n)_{U'}, \mathcal{O}^*_{U'/T'}) \xrightarrow{\sim} \mathrm{Hom}_{T'/T'}(\underline{H}'(n)_{T'}, \mathcal{O}^*_{T'/T'}).$$

Si T' est un U-schéma, on a $U' = T'$. En notant $i_{S/\Sigma}$, comme en [7, 1.1.4], le morphisme canonique $S_{\gamma,\mathrm{fppf}} \to (S/\Sigma)_{\mathrm{NCRIS}}$ on en déduit l'isomorphisme

$$i^* i^*_{T/T}\mathcal{H}om_{T/T}(\underline{H}'(n), \mathcal{O}^*_{T/T}) \xrightarrow{\sim} i^*_{U/U}\mathcal{H}om_{U/U}(\underline{H}'(n)_U, \mathcal{O}^*_{U/U}).$$

3.2.4. Lemme. *Soient X, X' deux T-schémas plats, d'intersection complète relative, et $X \hookrightarrow Y$, $X' \hookrightarrow Y'$ deux immersions fermées dans des T-schémas lisses. Il existe alors un isomorphisme canonique entre les voisinages à puissances divisées correspondants*

$$D_{X \times_T X'}(Y \times_T Y') \xrightarrow{\sim} D_X(Y) \times_T D_{X'}(Y').$$

D'après [7, 2.3.3], $D_X(Y)$ et $D_{X'}(Y')$ sont plats sur T. Si I, I' sont les idéaux définissant X, X' dans Y, Y', et si $\bar{I}, \bar{I}'$ sont les PD-idéaux qu'ils engendrent dans $\mathcal{O}_{D_X(Y)}$ et $\mathcal{O}_{D_{X'}(Y')}$, il suffit de montrer que les extensions de ces PD-idéaux à $\mathcal{O}_{D_X(Y)\times D_{X'}(Y')}$ ont des puissances divisées compatibles. Comme $X \times_T D_{X'}(Y')$ est plat sur T, l'intersection de ces deux idéaux est égale à leur produit, et la compatibilité de leurs puissances divisées est alors automatique.

3.2.5. Nous appliquerons le lemme précédent au schéma en groupes $H'(n)$, et à un plongement de $H'(n)$ dans un T-groupe lisse L : on peut prendre par exemple le plongement canonique dans le groupe lisse représentant les morphismes de schémas de $H'(n)^*$ dans $\mathbf{G}_m$ [18, II (3.2)]. Soient I l'idéal de $H'(n)$ dans L, D l'enveloppe à puissances divisées de I, $\bar{I}$ le PD-idéal engendré par I dans D. Le lemme précédent entraîne que Spec D est un T-schéma en groupes commutatif et plat, et on dispose d'une immersion fermée $H'(n) \hookrightarrow$ Spec D. Notons

$$\widehat{D} = \varprojlim_m D/\bar{I}^{[m]}$$

le PD-complété de D. Les points à valeurs dans T du groupe $\mathcal{H}om(\underline{H}'(n), \mathcal{O}^*_{T/T})$ sont donnés par

$$\{x \in \widehat{D}^* \mid x \in \operatorname{Ker}[d : \widehat{D} \to \widehat{D} \otimes \Omega^1_{L/T}], \text{ et } \mu^*(x) = x \otimes x\},$$

où μ^* est induit par la loi de groupe de L. La situation est compatible aux changements de base $T' \to T$ (pour un T-schéma T' auquel les puissances divisées s'étendent), et les points à valeurs dans T' de ce groupe peuvent être décrits comme les éléments de $\widehat{D}\widehat{\otimes}_{\mathcal{O}_T}\mathcal{O}_{T'}$ vérifiant l'analogue des conditions précédentes. Nous utiliserons cette description pour rendre explicite l'exponentielle du groupe $i^*_{T/T}\mathcal{H}om(\underline{H}'(n), \mathcal{O}^*_{T/T})$.

3.2.6. Lemme. *Si $p^n\mathcal{O}_T = 0$, il existe un isomorphisme canonique*

$$i^*_{T/T}(\mathcal{H}om_{T/T}(\underline{H}'(n), \mathcal{O}_{T/T})) \otimes J \xrightarrow{\sim} i^*_{T/T}(\mathcal{H}om_{T/T}(\underline{H}'(n), J \otimes \mathcal{O}_{T/T})).$$

Soit $0 \to H'(n) \to A \to B \to 0$ une résolution de $H'(n)$ par des schémas abéliens (il en existe localement sur T [7, 3.1.1]). Comme

$$\mathcal{E}xt^2_{S/\Sigma}(\underline{B}, \mathcal{O}_{T/T}) = 0,$$

et que $\mathcal{E}xt^1_{S/\Sigma}(\underline{H}'(n), \mathcal{O}_{T/T})$ est un $\mathcal{O}_{T/T}$-module localement libre, la suite exacte de cohomologie correspondante fournit une suite exacte

$$0 \to i^*_{T/T}\mathcal{H}om(\underline{H}'(n), \mathcal{O}_{T/T}) \otimes J \to i^*_{T/T}\mathcal{E}xt^1(\underline{B}, \mathcal{O}_{T/T}) \otimes J \\ \to i^*_{T/T}\mathcal{E}xt^1(\underline{A}, \mathcal{O}_{T/T}) \otimes J.$$

Comme on dispose également de la suite exacte

$$0 \to i^*_{T/T}\mathcal{H}om(\underline{H}'(n), J \otimes \mathcal{O}_{T/T}) \to i^*_{T/T}\mathcal{E}xt^1(\underline{B}, J \otimes \mathcal{O}_{T/T}) \\ \to i^*_{T/T}\mathcal{E}xt^1(\underline{A}, J \otimes \mathcal{O}_{T/T}),$$

et d'un morphisme de la première dans la seconde, on est ramené à prouver l'assertion correspondante pour les $\mathcal{E}xt^1$ des schémas abéliens. Or, pour tout schéma abélien $\pi : A \to T$, on a un isomorphisme

$$\mathcal{E}xt^1(\underline{A}, J \otimes \mathcal{O}_{T/T}) \xrightarrow{\sim} R^1\pi_{\mathrm{cris}*}(J \otimes \mathcal{O}_{T/T});$$

en effet, il suffit d'observer que la démonstration classique pour la cohomologie cohérente (cf. [20, VII th.7]) s'adapte immédiatement dans ce contexte, grâce à la relation $\pi_{\mathrm{cris}\,*}(J \otimes \mathcal{O}_{A/T}) \simeq J \otimes \mathcal{O}_{T/T}$, et à la formule de Künneth

$$R^1\pi_{A\times A\mathrm{cris}\,*}(J \otimes \mathcal{O}_{A\times A/T}) \\ \xrightarrow{\sim} R^1\pi_{\mathrm{cris}*}(J \otimes \mathcal{O}_{A/T}) \oplus R^1\pi_{\mathrm{cris}\,*}(J \otimes \mathcal{O}_{A/T}),$$

résultant de ce que la cohomologie de A est localement libre. Le lemme découle alors de l'isomorphisme

$$R^1\pi_{\mathrm{cris}\,*}(\mathcal{O}_{A/T}) \otimes J \xrightarrow{\sim} R^1\pi_{\mathrm{cris}\,*}(J \otimes \mathcal{O}_{A/T}).$$

3.2.7. Nous noterons e le morphisme de topos $(U/U)_{\mathrm{Ncris}} \longrightarrow (T/T)_{\mathrm{Ncris}}$. On a donc

$$e_*(\mathcal{O}_{U/U})_{(U',T',\delta')} = \mathcal{O}_{T'\times_T U},$$

ce qui, grâce à l'exponentielle, fournit une suite exacte courte sur NCRIS(S/Σ)

$$0 \to J\mathcal{O}_{T/T} \to \mathcal{O}^*_{T/T} \to e_*(\mathcal{O}^*_{U/U}) \to 0.$$

En lui appliquant le foncteur $\mathcal{H}om(\underline{H}'(n), -)$, on obtient la suite exacte

$$0 \to \mathcal{H}om(\underline{H}'(n), J\mathcal{O}_{T/T}) \to \mathcal{H}om(\underline{H}'(n), \mathcal{O}^*_{T/T}) \to e_*\mathcal{H}om(\underline{H}'(n), \mathcal{O}^*_{U/U}).$$

On vérifie immédiatement que $i^*_{T/T} \circ e_* = i_* \circ i^*_{U/U}$, et on en déduit donc la suite exacte de faisceaux fppf sur T

$$0 \to i^*_{T/T}\mathcal{H}om(\underline{H}'(n), J\mathcal{O}_{T/T}) \to i^*_{T/T}\mathcal{H}om(\underline{H}'(n), \mathcal{O}^*_{T/T}) \to i_* i^*_{U/U}\mathcal{H}om(\underline{H}'(n), \mathcal{O}^*_{U/U}).$$

Compte tenu de (3.1.7.2), le lemme (3.2.6) fournit alors les isomorphismes

$$\mathcal{L}ie\ (i^*_{T/T}\mathcal{H}om(\underline{H}'(n), \mathcal{O}^*_{T/T})) \otimes J \simeq i^*_{T/T}\mathcal{H}om(\underline{H}'(n), \mathcal{O}_{T/T}) \otimes J \simeq i^*_{T/T}\mathcal{H}om(\underline{H}'(n), J \otimes \mathcal{O}_{T/T}),$$

d'où un homomorphisme naturel

$$i^*_{T/T}\mathcal{H}om(\underline{H}'(n), J \otimes \mathcal{O}_{T/T}) \to \mathrm{Ker}\ [i^*_{T/T}\mathcal{H}om(\underline{H}'(n), \mathcal{O}^*_{T/T}) \to i_* i^*_{U/U}\mathcal{H}om(\underline{H}'(n), \mathcal{O}^*_{U/U})].$$

Nous avons considéré ici les schémas en groupes obtenus en appliquant $i^*_{T/T}$ (resp. $i^*_{U/U}$) aux faisceaux cristallins appropriés. Cette méthode est rendue nécessaire par la non-platitude de ces schémas en groupes. Mais, si on s'intéresse par exemple seulement aux homomorphismes de $\mathbb{G}_a$ dans $i^*_{T/T}\mathcal{H}om(\underline{H}'(n), \mathcal{O}^*_{T/T})$, il est possible de se restreindre au petit site fppf de T. Supposons donc que T' soit un T-schéma plat. On a alors

$$\Gamma((T', T'), \mathcal{H}om(\underline{H}'(n), J\mathcal{O}_{T/T})) = \mathrm{Hom}(\underline{H}'_{T'}(n), J\mathcal{O}_{T'/T'}).$$

Avec les notations de (3.2.5), on dispose du plongement $H'_{T'}(n) \hookrightarrow L_{T'}$, et de l'enveloppe à puissances divisées correspondante $D_{T'}$. Se donner une section d'un faisceau F sur NCRIS$(H'_{T'}(n)/T')$ équivaut à se donner une famille compatible d'éléments "horizontaux" des $F(\mathrm{Spec}\ (D_{T'}/\bar{I}^{[m]}_{T'}))$ pour $m \geq 1$. Pour $F = J\mathcal{O}_{T'/T'}$, la platitude de T' sur T, et de $D_{T'}/\bar{I}^{[m]}_{T'}$ sur T' montre que se donner une telle section revient à se donner une famille compatible d'éléments horizontaux des $J \otimes D_{T'}/\bar{I}^{[m]}_{T'}$, pour $m \geq 1$. Comme un élément de Hom $(H'_{T'}(n), J\mathcal{O}_{T'/T'})$ est une section particulière, on voit que, pour les T-schémas plats T', on peut remplacer $J\mathcal{O}_{T'/T'}$ par $J \otimes \mathcal{O}_{T'/T'}$. On obtient donc :

3.2.8. Lemme. *La restriction de $i^*_{T/T}\mathcal{H}om(\underline{H}'(n), J\mathcal{O}_{T/T})$ au petit site fppf de T (i.e. $\mathcal{H}om(\underline{H}'(n), J\mathcal{O}_{T/T})_T$) est canoniquement isomorphe à $\mathcal{H}om(\underline{H}'(n), J \otimes \mathcal{O}_{T/T})_T$.*

3.2.9. Proposition. *Soit H le schéma en groupes affine $i^*_{T/T}\mathcal{H}om(\underline{H}'(n), \mathcal{O}^*_{T/T})$. L'application exponentielle* [18, III (2.3.3.1)]

$$\mathrm{Hom}_{\mathcal{O}_T}(\mathbb{G}_a, J \otimes \mathcal{L}ie(H)) \to \mathrm{Ker}\,[\mathrm{Hom}\,(\mathbb{G}_a, H) \to \mathrm{Hom}\,(\mathbb{G}_{aU}, H_U)]$$

est induite par l'application exponentielle ordinaire.

L'application exponentielle est fonctorielle. Comme nous l'avons vu en (3.2.5), le groupe H est un sous-groupe du T-groupe affine et plat Spec $(\widehat{D}^\vee)$, où on note $\widehat{D}^\vee = \varinjlim_m (D/\bar{I}^{[m]})^\vee$. Il suffit donc de vérifier que, pour le groupe plus grand $K = \mathcal{H}om(\mathrm{Spf}(\widehat{D}), \mathbb{G}_m)$, l'exponentielle est induite par l'exponentielle ordinaire $\exp : J \otimes \mathbb{G}_a \to \mathbb{G}_m$.

Notons $B = \widehat{D}^\vee$, et rappelons d'abord la construction qui associe à toute application linéaire $\theta : \mathbb{G}_a \to J \otimes \mathcal{L}ie(K)$, avec $\theta(1) = \Sigma_\alpha j\alpha \otimes \partial_\alpha$, l'exponentielle $\exp(\theta)$. En supposant T affine, soit $T = \mathrm{Spec}\, R$, il faut associer à θ un homomorphisme de bigèbres $B \to R[X]$. Les ∂_α appartiennent à $\mathcal{L}ie(K)$, et sont des éléments primitifs de $\widehat{D}$. Chacun des quotients $D/\bar{I}^{[m]}$ est plat sur R, et par suite l'image de $\Sigma_\alpha j_\alpha \partial_\alpha$ dans un tel quotient possède des puissances divisées (nilpotentes). Par passage à la limite sur m, on peut donc définir pour tout r l'élément $(\Sigma_\alpha j_\alpha \partial_\alpha)^{[r]}$ de $\widehat{D}$. En développant $(\Sigma_\alpha j_\alpha \partial_\alpha)^{[r]}$ comme une somme de produits de termes de la forme $j_\alpha^{[k]} \partial_\alpha^k$, on obtient une expression dans laquelle chacun des termes ∂_α^k peut être considéré comme la k-ième itérée d'une dérivation invariante par translation $\tilde{\partial}_\alpha$ sur B (grâce au dictionnaire classique associant à $\partial_\alpha \in \mathcal{L}ie(K)$ la dérivation $\tilde{\partial}_\alpha : B \to B \otimes B \xrightarrow{\partial_\alpha \otimes 1} B$). On obtient ainsi un opérateur différentiel bien défini $(\Sigma_\alpha j_\alpha \partial_\alpha)^{[r]} : B \to B$. On définit alors $\exp(\theta)$ comme l'homomorphisme de bigèbres $B \to R[X]$ donné par

$$b \mapsto \sum_{r \geq 0} \eta\left((\Sigma_\alpha j_\alpha \partial_\alpha)^{[r]}(b)\right) X^r,$$

en notant $\eta : B \to R$ l'homomorphisme d'augmentation.

Soient maintenant R' une R-algèbre plate, et $x \in \mathbb{G}_a(R')$. En effectuant un changement de base de T à $T' = \mathrm{Spec}\, R'$, on se ramène au cas où $R = R'$. On veut calculer $\exp(\theta)(x) \in K(R) = \mathrm{Hom}\,(\mathrm{Spf}\,(\widehat{D}), \mathbb{G}_m)$. Comme homomorphisme de B dans R, $\exp(\theta)(x)$ est donné par

$$b \mapsto \sum_{r \geq 0} \Gamma((\Sigma_\alpha j_\alpha \partial_\alpha)^{[r]}(b)) x^r = \sum_{r \geq 0} \eta((x \cdot \Sigma_\alpha j_\alpha \partial_\alpha)^{[r]}(b)).$$

Or l'élément $\eta(\tilde{\partial}_\alpha^k(b)) \in R$ est simplement égal à $b(\partial_\alpha^k)$. Cette somme est donc égale à $b(\Sigma_r (x \cdot \Sigma_\alpha j_\alpha \partial_\alpha)^{[r]})$. Par suite, l'homomorphisme considéré de B dans R est égal à l'élément $\Sigma_r(x \cdot \Sigma_\alpha j_\alpha \partial_\alpha)^{[r]}$ de $B^\vee = \widehat{D}$. Pour le décrire en tant qu'homomorphisme de groupes formels $\mathrm{Spf}(\widehat{D}) \to \mathbb{G}_m$, il faut prendre sa valeur en un point à valeurs dans R' de $\mathrm{Spf}(\widehat{D})$, pour toute R-algèbre discrète R' à laquelle δ s'étend. En effet, comme chaque quotient $D/\bar{I}^{[m]}$ est plat sur R, l'application de restriction donne une bijection entre $\mathrm{Hom}(\mathrm{Spf}(\widehat{D}), \mathbb{G}_m)$ et l'ensemble des homomorphismes entre ces groupes sur le petit site fppf. Si $\tau : \widehat{D} \to R'$ est le point considéré, on obtient donc

$$\begin{aligned}\exp(\theta)(x)(\tau) = \tau(\Sigma_r(x \cdot \Sigma_\alpha j_\alpha \partial_\alpha)^{[r]}) &= \Sigma_r(\tau(x \cdot \Sigma_\alpha j_\alpha \partial_\alpha))^{[r]} \\ &= \exp(\tau(x \cdot \Sigma_\alpha j_\alpha \partial_\alpha)),\end{aligned}$$

ce qui achève la démonstration.

3.2.10. Revenons à la démonstration du théorème (3.2.2). Supposons T affine, et posons $T = \mathrm{Spec} R$, $G'(n) = \mathrm{Spec} A$, $H'(n) = \mathrm{Spec} B$. Soit A' (resp. B') l'algèbre symétrique du R-module A (resp. B) ; A' et B' sont des R-algèbres lisses, et il existe des surjections canoniques $A' \twoheadrightarrow A$, $B' \twoheadrightarrow B$. L'homomorphisme donné $u : H'_U \to G'_U$ induit un morphisme de bigèbres $A/JA \to B/JB$. Soient $u' : A \to B$ un relèvement linéaire quelconque de ce morphisme, et $v : A' \to B'$ l'application induite entre les algèbres symétriques. Notons D_A, D_B les enveloppes à puissances divisées de $\mathrm{Ker}(A' \twoheadrightarrow A)$ et $\mathrm{Ker}(B' \twoheadrightarrow B)$, et $\bar{J}_A$, $\bar{J}_B$ les idéaux à puissances divisées correspondants. Posons $KA = \bar{J}_A + J \cdot D_A$, $K_B = \bar{J}_B + J \cdot D_B$. La nilpotence des puissances divisées sur J entraîne que

$$\varprojlim_m D_A/\bar{J}_A^{[m]} \simeq \varprojlim_m D_A/K_A^{[m]} = \widehat{D}_A,$$

et de même pour D_B. L'application $v : A' \to B'$ s'étend en un PD-morphisme envoyant K_A dans K_B, et induit donc une application $\widehat{D}_A \to \widehat{D}_B$.

Le groupe $\mathrm{Hom}(\underline{G}_U(n), \mathcal{O}^*_{U/T})$ est égal au H^0 du complexe simple associé au bicomplexe suivant :

$$\begin{array}{ccccc}
\widehat{D}_A & \xrightarrow{\mathrm{dlog}} & \widehat{D}_A \otimes_{A'} \Omega^1_{A'/R} & \longrightarrow & \cdots \\
\downarrow & & \downarrow & & \\
(D_A \otimes D_A)^\wedge & \longrightarrow & (D_A \otimes D_A)^\wedge \otimes_{A' \otimes A'} \Omega^1_{A' \otimes A'/R} & \longrightarrow & \cdots \\
\downarrow & & \downarrow & & \\
\vdots & & \vdots & &
\end{array}$$

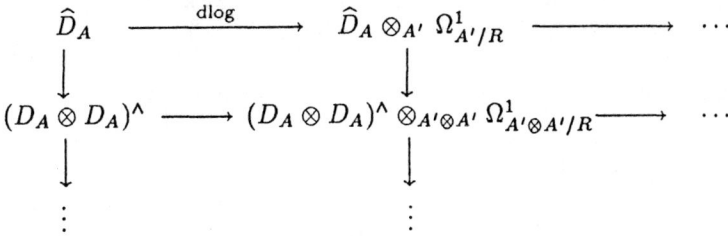

De même, $\mathrm{Hom}(\underline{G}'(n), \mathcal{J}_{T/T})$ est le H^0 du complexe simple associé au bicomplexe

$$\begin{array}{ccc} \widehat{J}_A & \xrightarrow{d} & \widehat{D}_A \otimes_{A'} \Omega^1_A \longrightarrow \cdots \\ \downarrow & & \downarrow \\ (\bar{J}_A \otimes D_A + D_A \otimes \bar{J}_A)^\wedge & \longrightarrow & \cdots \end{array}$$

Avec des changements de notations évidents, on dispose de complexes analogues pour calculer les points à valeurs dans T' de $\mathcal{H}om(\underline{G}(n)_U, \mathcal{O}^*_{U/T})_T$ (resp. ...).

Il faut montrer que dans le diagramme

$$\begin{array}{ccc} \mathcal{H}om(\underline{G}'(n), \mathcal{J}_{T/T})_T & \hookrightarrow & \mathcal{H}om(\underline{G}'(n)_U, \mathcal{O}^*_{U/T})_T \\ {\scriptstyle w}\downarrow & & \downarrow{\scriptstyle v} \\ \mathcal{H}om(\underline{H}'(n), \mathcal{J}_{T/T})_T & \hookrightarrow & \mathcal{H}om(\underline{H}'(n)_U, \mathcal{O}^*_{U/T})_T \end{array} \tag{3.2.10.1}$$

la différence entre les deux applications composées est une exponentielle. Comme cette différence se réduit à 0 modulo J, lorsqu'elle est considérée comme application $\omega_{G'} \to E(H'^*)$, elle se factorise en un homomorphisme

$$\omega_{G'} \xrightarrow{\theta} \mathcal{L}ie(E(H'^*)) \otimes J \xhookrightarrow{\exp} E(H'^*),$$

et il faut montrer que θ est une application linéaire. D'après (3.2.6), le but de θ est égal à

$$\mathcal{H}om(\underline{H}'(n), \mathcal{O}_{T/T})_T \otimes J = \mathcal{H}om(\underline{H}'(n), J \otimes \mathcal{O}_{T/T})_T.$$

Comme $J \otimes \mathcal{O}_{T/T}$ est un cristal, on peut par adjonction écrire

$$\mathcal{H}om(\underline{H}'(n), J \otimes \mathcal{O}_{T/T})_T \simeq \mathcal{H}om(\underline{H}'_U(n), J \otimes \mathcal{O}_{U/T})_{(U,T)}.$$

Sur les objets (U', T', δ') de NCRIS(U/T) tels que T' soit plat sur T, l'homomorphisme $J \otimes \mathcal{O}_{U/T} \to \mathcal{J}_{U/T}$ est injectif, de sorte que $\mathcal{H}om(\underline{H}'_U(n), J \otimes \mathcal{O}_{U/T})_{(U,T)}$ est inclus dans $\mathcal{H}om(\underline{H}'_U(n), \mathcal{J}_{U/T})_{(U,T)}$. Pour vérifier la linéarité de θ, on peut donc considérer que son but est $\mathcal{H}om(\underline{H}'_U(n), \mathcal{J}_{U/T})_{(U,T)}$.

Soit alors $x \in \mathrm{Hom}(\underline{G}'(n), \mathcal{J}_{T/T})$. Son image par l'inclusion

$$\mathrm{Hom}(\underline{G}'(n), \mathcal{J}_{T/T}) \hookrightarrow \mathrm{Hom}(\underline{G}'(n), \mathcal{O}^*_{T/T}) = \mathrm{Hom}(\underline{G}'_U(n), \mathcal{O}^*_{U/T})$$

est donnée par $x \in \widehat{\bar{J}}_A \mapsto \exp(x) \in \widehat{D}^*_A$. En appliquant v, on obtient pour image de x par la droite du carré (3.2.10.1) l'élément $v(\exp(x)) \in \widehat{D}^*_B$. Par la gauche du carré, x est envoyé sur $\exp(w(x))$. Observons que ces deux éléments appartiennent à $1+\widehat{K}_B$, donc sont éléments de $\mathrm{Hom}(\underline{H}'_U(n), 1+\mathcal{J}_{U/T})$. En leur appliquant le logarithme, on obtient donc des éléments de $\mathrm{Hom}(\underline{H}'_U(n), \mathcal{J}_{U/T})$. Si on compose $\theta : \omega_{G'} \to \mathcal{H}om(\underline{H}'(n), \mathcal{J}_{T/T})_T \otimes J$ avec l'inclusion

$$\mathcal{H}om(\underline{H}'(n), \mathcal{J}_{T/T})_T \otimes J \hookrightarrow \mathcal{H}om(\underline{H}'_U(n), \mathcal{J}_{U/T})_{(U,T)},$$

on obtient d'après (3.2.9) la relation

$$\theta(x) = \log(v(\exp(x))) - \log(\exp(w(x))).$$

La commutativité du diagramme

$$\begin{array}{ccc} \widehat{K}_A & \xrightarrow{\exp} & 1+\widehat{K}_A \\ {\scriptstyle v}\downarrow & & \downarrow{\scriptstyle v} \\ \widehat{K}_B & \xrightarrow{\exp} & 1+\widehat{K}_B \end{array}$$

entraîne alors immédiatement que θ est une application linéaire, achevant ainsi la démonstration de (3.2.2). On en déduit aussitôt le corollaire :

3.2.11. Corollaire. *En restriction à* $\mathrm{NC{\scriptstyle RIS}}(S/\Sigma)$, *les deux foncteurs contravariants de la catégorie des groupes p-divisibles sur S dans la catégorie des cristaux en modules localement libres donnés respectivement par* $G \mapsto \mathbb{D}(G)$ *(avec la notation de* [7]*) et* $G \mapsto \mathcal{L}ie(E(G^*))$ *(avec la notation de* [18]*) sont canoniquement isomorphes. De plus, pour tout épaississement* (U, T, δ) *de* $\mathrm{NC{\scriptstyle RIS}}(S/\Sigma)$ *et tout relèvement de* G_U *en un groupe p-divisible* G' *sur* T, *cet isomorphisme est compatible avec les filtrations de Hodge définies par* G'.

4. Propriétés de pleine fidélité du foncteur de Dieudonné

Nous donnerons ici divers cas dans lesquels il est possible de démontrer que le foncteur de Dieudonné cristallin $\mathbb{D}$ est pleinement fidèle sur la catégorie des p-groupes finis, ou des groupes p-divisibles.

Dans ce chapitre, Σ désigne $\mathrm{Spec}\,\mathbb{Z}_p$, muni du PD-idéal canonique $p\mathbb{Z}_p$, et $\Sigma_m = \mathrm{Spec}\,\mathbb{Z}/p^m\mathbb{Z}$.

4.1. ***Cas d'un schéma possédant localement une p-base.***

Le principal résultat que nous montrerons ici est le théorème de pleine fidélité suivant :

4.1.1. Théorème. *Soient S un schéma de caractéristique p, normal, localement irréductible, et possédant localement une p-base. Le foncteur de Dieudonné cristallin $\mathbb{D}$ est un foncteur pleinement fidèle de la catégorie des p-groupes finis localement libres sur S (resp. des groupes p-divisibles sur S) dans celle des cristaux de Dieudonné sur S.*

Comme, pour tout groupe p-divisible H, on a $\mathbb{D}(H(n)) = \mathbb{D}(H)/p^n\mathbb{D}(H)$, il suffit de prouver l'assertion relative aux groupes finis localement libres. La fidélité de $\mathbb{D}$ est immédiate, car S est localement spectre d'un anneau intègre, de sorte qu'il suffit, pour verifier la nullité d'un morphisme entre groupes finis, de le vérifier après passage à la clôture parfaite du corps des fractions du point générique, et cela résulte alors de l'identification entre le cristal de Dieudonné et le module de Dieudonné classique [7, 4.2.14].

4.1.2. *Réduction au cas d'un corps de base.* On peut supposer S de la forme $S = \operatorname{Spec} A$, où A est un anneau intègre normal admettant une p-base. Pour tout morphisme $\operatorname{Spec} B \to \operatorname{Spec} A$, nous noterons par l'indice B les objets déduits d'objets sur $\operatorname{Spec} A$ par changement de base. Soient G, H deux A-groupes finis localement libres, et $v : \mathbb{D}(H) \to \mathbb{D}(G)$ un morphisme entre leurs cristaux de Dieudonné. Quitte à localiser davantage, on peut supposer que les algèbres de G et H sont libres sur A. Soient k le corps des fractions de A, et K la clôture parfaite de k. Si A' est un anneau de valuation de k contenant A, sa clôture intégrale A'' dans K est alors un anneau de valuation parfait. Comme le foncteur de Dieudonné est pleinement fidèle lorsque la base est un anneau de valuation parfait [4, 3.4.1 et 4.3.4], il existe un homomorphisme $u_1 : G_{A''} \to H_{A''}$ tel que $\mathbb{D}(u_1) = v_{A''}$. Si l'on suppose le théorème démontré lorsque $S = \operatorname{Spec} k$, il existe aussi un homomorphisme $u_2 : G_k \to H_k$ tel que $\mathbb{D}(u_2) = v_k$. Sur K, on peut comparer u_1 et u_2 : en appliquant $\mathbb{D}$, on obtient $\mathbb{D}(u_{1K}) = (v_{A''})_K = (v_k)_K = \mathbb{D}(u_{2K})$. Puisque K est parfait, et que le foncteur $\mathbb{D}$ est isomorphe sur un corps parfait au foncteur de Dieudonné usuel [7, 4.2.14], on a donc $u_{1K} = u_{2K}$. En utilisant des bases sur A des algèbres de G et H, l'homomorphisme u_1 (resp. u_2) est défini par une matrice à coefficients dans A'' (resp. k), et il en résulte donc que cette matrice est à coefficients dans $A'' \cap k = A'$. Mais A est normal, et est donc l'intersection des anneaux de valuation de k le contenant, si bien que la matrice obtenue est en fait à coefficients dans A. Elle définit alors un homomorphisme de bigèbres, car il suffit de le vérifier en passant au corps des fractions k, puisque A est intègre. On obtient ainsi un homomorphisme $u : G \to H$, et il reste à s'assurer que $\mathbb{D}(u) = v$; comme $\mathbb{D}(u_k) = v_k$ par construction, cela résulte de ce que, A ayant une p-base, le passage au corps des fractions est fidèle sur la catégorie des cristaux, d'après (1.3.5) (i).

4.1.3. *Cas d'un corps de base.* Nous supposons donc dans la suite de la démonstration que $S = \operatorname{Spec} k$, où k est un corps quelconque de caractéristique p. Nous noterons encore K la clôture parfaite de k. Utilisant la théorie de Dieudonné classique sur K, et le théorème de comparaison [7, 4.2.14], on peut trouver un homomorphisme $u'_K : G_K \to H_K$ tel que $\mathbb{D}(u'_K) = v_K$. Il existe une extension finie de k de la forme $k' = k(x_1^{p^{-N}}, \ldots, x_r^{p^{-N}})$, où $x_1, \ldots, x_r$ sont des éléments p-indépendants de k, telle que u'_K provienne de $u' : G_{k'} \to H_{k'}$ par changement de base. De plus, $\mathbb{D}(u') = v_{k'}$, car l'extension des scalaires de k' à K est un foncteur fidèle sur la catégorie des cristaux d'après (1.3.5) (ii). Le problème est alors de redescendre u' sur k : s'il existe $u : G \to H$ tel que $u_{k'} = u'$, alors $\mathbb{D}(u)_{k'} = \mathbb{D}(u_{k'}) = v_{k'}$, et par suite $\mathbb{D}(u) = v$, en utilisant encore la fidélité de l'extension du corps de base sur la catégorie des cristaux.

Soit $k'' = k' \otimes_k k'$; alors k'' est une intersection complète de la forme

$$k'' = k'[Y_1, \ldots, Y_r]/(Y_1^{p^N} - x_1, \ldots, Y_r^{p^N} - x_r).$$

Si $S' = \operatorname{Spec} k'$, $S'' = \operatorname{Spec} k''$, et si p_1, $p_2 : S'' \to S'$ sont les deux projections, il suffit de montrer que l'homomorphisme $w = p_1^*(u') - p_2^*(u')$ est nul. Par hypothèse, on a :

$$\mathbb{D}(w) = \mathbb{D}(p_1^*(u') - p_2^*(u')) = p_1^*(\mathbb{D}(u')) - p_2^*(\mathbb{D}(u')) = p_1^*(v_{k'}) - p_2^*(v_{k'}) = 0.$$

Rappelons [7, 4.3.4] que, pour tous schémas S, T de caractéristique p, et toute immersion fermée $S \hookrightarrow T$ définie par un idéal dont les éléments sont de puissance p-ième nulle, il existe un morphisme canonique $\Phi : T \to S$ factorisant les Frobenius de S et T. Prenant un épaississement (U, T, δ) variable dans $\mathrm{CRIS}(S/\mathbb{F}_p)$, on obtient ainsi un foncteur Φ^* associant à tout $\mathcal{O}_S$-module quasi-cohérent un cristal sur $\mathrm{CRIS}(S/\mathbb{F}_p)$.

Nous utiliserons le lemme suivant :

4.1.4. Lemme. *Soient S' un schéma de caractéristique p, ayant localement une p-base, et $S'' \to S'$ un morphisme plat localement d'intersection complète. Alors le foncteur Φ^* est un foncteur fidèle et exact de la catégorie des $\mathcal{O}_{S''}$-modules quasi-cohérents dans celle des cristaux sur $\mathrm{CRIS}(S''/\mathbb{F}_p)$.*

L'assertion étant locale, on peut supposer que S' et S'' sont affines, que $S' = \operatorname{Spec} A$, où A possède une p-base, et qu'on dispose d'une immersion fermée de S'' dans un S'-schéma lisse L, définie par une suite régulière $(f_1, \ldots, f_r)$. Soit $D_{S''}(L)$ le voisinage à puissances divisées de S'' dans L. La lissité de L entraîne que, pour tout objet (U, T, δ) de $\mathrm{CRIS}(S''/\mathbf{F}_p)$ avec T affine, il existe un PD-morphisme $T \to D_{S''}(L)$ prolongeant $U \to S''$. Il suffit donc de prouver que $\Phi : \mathcal{O}_{S''} \to \mathcal{O}_{D_{S''}(L)}$ est fidèlement plat. Mais, si $L = \operatorname{Spec} B$, il résulte de [7, preuve de 2.3.3] et [3, VI 3.2.5] que $\mathcal{O}_{D_{S''}(L)}$ est libre de base $\{f_1^{[pq_1]} \ldots f_r^{[pq_r]}\}_{q \in \mathbf{N}^r}$ sur $B/(f_1^p, \ldots, f_r^p)$. Comme B est lisse sur A, son endomorphisme de Frobenius absolu est fidèlement plat, et il en est donc de même de l'homomorphisme $\Phi : B/(f_1, \ldots, f_r) \to B/(f_1^p, \ldots, f_r^p)$.

4.1.5. *Fin de la démonstration de* (4.1.1). Nous appliquerons le lemme précédent de la manière suivante. Soit ${}_FH$ le noyau de F sur H. Comme H est défini sur k, ${}_FH$ est plat, et l'homomorphisme $\mathbb{D}(H_{k''}) \to \mathbb{D}({}_FH_{k''})$ est surjectif. Par suite, l'application $\tilde{w} : {}_FG_{k''} \to {}_FH_{k''}$ induite par w est telle que $\mathbb{D}(\tilde{w}) = 0$. Mais, d'après [7, 4.3.11], le cristal de Dieudonné d'un groupe E annulé par F s'identifie (en restriction à $\mathrm{CRIS}(S''/\mathbf{F}_p)$) au cristal $\Phi^*(\omega_E)$. Par suite, l'application $\Phi^*(\omega_{H_{k''}}) \to \Phi^*(\omega_{G_{k''}})$ induite par w est nulle, ce qui entraîne d'après le lemme que $\omega_{H_{k''}} \to \omega_{G_{k''}}$ l'est également. Il en résulte que $\tilde{w} : {}_FG_{k''} \to {}_FH_{k''}$ est nulle [SGA3, VIIA].

Avec les notations de (4.1.4), soient A'_∞ un relèvement de A (il en existe d'après (1.1.7)), et B_∞ une A'_∞-algèbre formellement lisse relevant B. D'après [7, 2.3.3], la séparée complétée p-adique de l'enveloppe à puissances divisées de l'idéal de S'' dans B_∞ est sans p-torsion ; son spectre formel T définit alors un épaississement de S'' tel que tout objet de $\mathrm{CRIS}(S''/\Sigma)$ possède localement un PD-morphisme dans (S'', T). Il résulte alors de [7, 4.2.7] que $\mathbb{D}$ est un foncteur exact de la catégorie des S''-groupes finis localement libres dans celle des cristaux sur S''/Σ. Prenant alors S' et S'' comme en (4.1.3), on en conclut que l'application $F(G_{k''}) \to H_{k''}$ factorisant w induit encore 0 sur les cristaux de Dieudonné. On voit ainsi par récurrence que w s'annule sur la composante connexe de $G_{k''}$, et se factorise donc par son quotient étale $G_{k''}^{\text{ét}}$. On peut donc supposer que G est étale.

D'après la remarque de (1.4.2), la nullité de $\mathbb{D}(w)$ entraîne celle de $\mathbb{D}(w^*)$, où w^* est l'application induite par w entre les duaux de Cartier. Par suite, w^* se factorise également par le quotient étale de $H_{k''}^*$, ce qui permet de supposer que H est de type multiplicatif. Quitte à tensoriser toute la situation par une clôture séparable de k, on peut supposer que k est séparablement clos, de sorte que les groupes G et H^* sont alors constants sur k. Supposons que $w \neq 0$, et soit n le plus petit entier tel que la restriction de w au noyau $G(n)_{k''}$ de p^n sur $G_{k''}$ soit non nulle. En remplaçant G par $G(n)/G(n-1)$, on est ramené au cas où G est annulé par p, donc au cas où $G = \mathbb{Z}/p$. Comme $H(1)$ est somme directe de copies de μ_p, on peut supposer que $H = \mu_p$. Si w est défini par $\zeta = w(1) \in \mu_p(k'')$, avec $\zeta \neq 1$, $\mathbb{D}(w)$ s'identifie à la multiplication par $l_1(\zeta)$ sur $\mathcal{O}_{S''/\Sigma_1}$ d'après (2.2.4). L'hypothèse entraîne donc que $l_1(\zeta) = 0$ dans $\Gamma(S''/\Sigma_1, \mathcal{O}_{S''/\Sigma_1})$.

Soient donc $\zeta \in \mu_p(B/(f_1, \ldots, f_r))$, $\zeta \neq 1$, et $\widehat{\zeta} \in B$ un relèvement de ζ. Montrons que la nullité de $l_1(\zeta) = \log(\widehat{\zeta}^p)$ dans l'enveloppe à puissances divisées $D_B(f_1, \ldots, f_r)$, fournit une contradiction, entraînant que $w = 0$. Comme cette enveloppe est un $B/(f_1^p, \ldots, f_r^p)$-module libre ayant pour base les éléments $f_1^{[pq_1]} \ldots f_r^{[pq_r]}$, il est facile de voir que la PD-complétion de cet anneau est, en tant que $B/(f_1^p, \ldots, f_r^p)$-module, le produit d'une famille dénombrable, indexée par $\mathbb{N}^r$, de copies de $B/(f_1^p, \ldots, f_r^p)$. En particulier, $D_B(f_1, \ldots, f_r)$ s'injecte dans cette complétion ; comme l'exponentielle est définie sur la PD-complétion, cela montre que la nullité de $\log(\widehat{\zeta}^p)$ entraîne celle de $\widehat{\zeta}^p - 1$ dans $B/(f_1^p, \ldots, f_r^p)$. Comme Φ est injectif, on en déduit la nullité de $\zeta - 1$ dans $B/(f_1, \ldots, f_r)$, d'où la contradiction.

4.2. *Cas d'un anneau local régulier excellent.*

Soit A un anneau local régulier excellent de caractéristique p. On conjecture [1] que son hensélisé vérifie la conclusion du théorème d'approximation d'Artin, et c'est un résultat connu dans certains cas de petite dimension, ainsi que dans la situation, étudiée par Artin, d'une algèbre essentiellement de type fini sur un anneau de Dedekind excellent. Nous allons voir que les conclusions de (4.1.1) restent valables sur un anneau de base vérifiant cette conjecture.

4.2.1. Théorème. *Soit A un anneau local régulier de caractéristique p, excellent, et dont l'hensélisé possède la propriété d'approximation d'Artin* [1, (1.7)]. *Alors le foncteur de Dieudonné cristallin $\mathbb{D}$ est un foncteur pleinement fidèle de la catégorie des p-groupes finis et plats sur A (resp. des groupes p-divisibles sur A) vers celle des cristaux de Dieudonné sur A.*

La fidélité du foncteur $\mathbb{D}$ est encore immédiate.

4.2.2. *Réduction au cas d'un anneau local régulier hensélien.* Soient G, H deux groupes finis plats sur A, et $v : \mathbb{D}(H) \to \mathbb{D}(G)$ un homomorphisme entre leurs cristaux de Dieudonné. Soient $\tilde{A}$ l'hensélisé de A, $\tilde{G}$, $\tilde{H}$, $\tilde{v}$ les images inverses de G, H, v sur $\tilde{A}$. Supposant l'énoncé prouvé dans le cas hensélien, il existe sur $\tilde{A}$ un homomorphisme $\tilde{u} : \tilde{G} \to \tilde{H}$ tel que $\mathbb{D}(\tilde{u}) = \tilde{v}$. En écrivant $\tilde{A} = \varinjlim_\lambda A_\lambda$, où les A_λ sont des A-algèbres locales essentiellement étales, on peut redescendre $\tilde{u}$ en $u_\lambda : G_\lambda \to H_\lambda$ pour λ assez grand. Si p^m annule G et H, leurs cristaux de Dieudonné relatifs à Σ sont déterminés par leur restriction à $\mathrm{C}\small{RIS}(S/\Sigma_m)$. Pour tout λ, soient B_λ l'anneau des polynômes à coefficients dans $\mathbf{Z}/p^m$ en une famille d'indéterminées indexée par A_λ, et D_λ l'enveloppe à puissances divisées de $\mathrm{Ker}(B_\lambda \twoheadrightarrow A_\lambda)$. Posons $\tilde{B} = \varinjlim_\lambda B_\lambda, \tilde{D} = \varinjlim_\lambda D_\lambda$; la formation des enveloppes à puissances divisées commutant aux limites inductives, $\tilde{D}$ est l'enveloppe à puissances divisées de $\mathrm{Ker}(\tilde{B} \twoheadrightarrow \tilde{A})$. Comme $\tilde{B}$ (resp. B_λ) est une algèbre de polynômes, la catégorie des cristaux de Dieudonné sur $\tilde{A}$ (resp. A_λ) relativement à Σ_m est équivalente à celle des $\tilde{D}$-modules (resp. D_λ) de présentation finie munis de certaines structures supplémentaires, de sorte qu'il existe λ tel que $\mathbb{D}(u_\lambda) = v_\lambda$.

Soient alors $C = A_\lambda \otimes_A A_\lambda$, et $w = \mathrm{Id} \otimes u_\lambda - u_\lambda \otimes \mathrm{Id}$, de sorte que $\mathbb{D}(w) = 0$. Comme C est un produit fini d'anneaux intègres, il en résulte que $w = 0$. On peut donc redescendre u_λ en $u : G \to H$. Pour vérifier que $\mathbb{D}(u) = v$, on peut observer que, pour tout PD-épaississement T de $\mathrm{Spec}\, A$, il existe un unique T-schéma T_λ relevant $\mathrm{Spec}\, A_\lambda$ sur T. Comme $(\mathbb{D}(u) - v)_{T_\lambda} = 0$, on en déduit par descente fidèlement plate que $\mathbb{D}(u)_T = v_T$, de sorte que $\mathbb{D}(u) = v$.

4.2.3. *Réduction au cas d'un anneau local régulier complet.* La démonstration est analogue à celle qui précède. Soit A un anneau local, régulier, excellent et hensélien, vérifiant la propriété d'approximation, et soit A' son complété ; si G, H, v sont donnés comme précédemment sur A, notons par $'$ l'image inverse sur A'. Si on suppose $\mathbb{D}$ pleinement fidèle sur A', il existe un morphisme $u' : G' \to H'$ tel que $\mathbb{D}(u') = v'$. Posons $A'' = A' \otimes_A A'$, et $w = u' \otimes \mathrm{Id} - \mathrm{Id} \otimes u'$; on a donc $\mathbb{D}(w) = 0$. Comme A est excellent, ses fibres formelles sont géométriquement régulières et en particulier normales. Il résulte alors de [EGA, IV (6.14.1)] que A'' est normal, i.e. que pour tout idéal premier $\mathfrak{p}$ de A'', $A''_\mathfrak{p}$ est normal. En particulier, le fait que $A''_\mathfrak{p}$ soit intègre implique que $w_\mathfrak{p} = 0$. Par suite, $w = 0$, et u' se redescend en $u : G \to H$, et il reste à voir que $\mathbb{D}(u) = v$. Ecrivons A' comme limite inductive de ses sous-A-algèbres de présentation finie, et introduisons comme en (4.2.2) les systèmes inductifs B_λ, D_λ, et $D' = \varinjlim_\lambda D_\lambda$. Utilisant le fait que les cristaux considérés sont définis par des D_λ-modules de présentation finie, on voit qu'on a encore $\mathbb{D}(u_\lambda) = v_\lambda$ pour λ assez grand. L'hypothèse que A vérifie la propriété d'approximation entraîne l'existence d'un homomorphisme de A-algèbres $A_\lambda \to A$, de sorte que $\mathbb{D}(u) = v$.

4.2.4. *Cas d'un anneau local régulier complet.* On peut écrire A sous la forme $k[[t_1, \ldots, t_n]]$, k étant un corps. Si $[k : k^p]$ est fini, A possède une p-base, et on est sous les hypothèses de (4.1.1). Dans le cas général, on peut d'après [2] écrire $A = \varinjlim_\lambda A_\lambda$, les A_λ étant des k-algèbres lisses. Soient $G, H, v : \mathbb{D}(H) \to \mathbb{D}(G)$ donnés comme précédemment sur A, n un entier tel que $p^n G = p^n H = 0$. Introduisons encore sur $\mathbb{Z}/p^n$ les algèbres de polynômes B_λ de quotient A_λ, $B = \varinjlim_\lambda B_\lambda$, l'enveloppe à puissances divisées D_λ de l'idéal de A_λ dans B_λ, et $D = \varinjlim_\lambda D_\lambda$. Pour λ assez grand, G et H se redescendent en des groupes finis localement libres G_λ et H_λ sur A_λ. Si $D_\lambda(1)$ (resp. $D(1)$) désigne l'enveloppe à puissances divisées de l'idéal de A_λ dans $B_\lambda \otimes B_\lambda$ (resp. de A dans $B \otimes B$), les cristaux de Dieudonné correspondants peuvent être vus comme des D_λ-modules (resp. D-modules) M munis d'un isomorphisme $D_\lambda(1) \otimes M \xrightarrow{\sim} M \otimes D_\lambda(1)$ (resp. $D(1) \otimes M \xrightarrow{\sim} M \otimes D(1)$) (vérifiant la condition de cocycle, etc). Ces données étant de présentation finie, on peut redescendre v en v_λ : $\mathbb{D}(H_\lambda) \to \mathbb{D}(G_\lambda)$ pour λ assez grand, en un morphisme qu'on peut supposer compatible à F et V. Appliquant alors (4.1.1) sur $\operatorname{Spec} A_\lambda$, on obtient $u_\lambda : G_\lambda \to H_\lambda$ tel que $\mathbb{D}(u_\lambda) = v_\lambda$, qui fournit par changement de base l'homomorphisme $u : G \to H$ cherché.

4.2.5. Comme on l'a observé en (1.1.2) (v), un anneau local régulier essentiellement de type fini sur un corps possède une p-base. Par un argument de passage à la limite analogue à celui de (4.2.4), on déduit de (4.1.1) la proposition suivante :

4.2.6. Proposition. *Soit S un schéma régulier localement essentiellement de type fini sur un corps de caractéristique p. Alors le foncteur de Dieudonné cristallin est pleinement fidèle de la catégorie des p-groupes finis et plats sur S (resp. des groupes p-divisibles sur S) vers celle des cristaux de Dieudonné sur S.*

4.3. **Extension de la pleine fidélité par déformation.**

Soit S un schéma plat d'intersection complète relative sur un schéma possédant localement une p-base. Nous allons voir ici comment, grâce aux résultats du Section 3, la théorie des déformations permet de ramener la pleine fidélité du foncteur de Dieudonné pour la catégorie des groupes p-divisibles sur S à la même propriété sur un sous-schéma S_0 de S défini par un idéal muni d'une PD-structure nilpotente.

4.3.1. Théorème. *Soient Z un schéma de caractéristique p possédant localement une p-base, $S_0 \hookrightarrow S$ une Z-immersion définie par un idéal de S possédant une structure d'idéal à puissances divisées nilpotente. On suppose que S est plat d'intersection complète relative sur Z. Alors, si le* *foncteur de Dieudonné cristallin pour la catégorie des groupes p-divisibles sur S_0 est pleinement fidèle, il en est de même sur S.*

Soient G, H deux groupes p-divisibles sur S, et $v : \mathbb{D}(H) \to \mathbb{D}(G)$ un morphisme entre leurs cristaux de Dieudonné ; notons par l'indice 0 les réductions sur S_0. Par hypothèse, il existe un morphisme $u_0 : G_0 \to H_0$ tel que $\mathbb{D}(u_0) = v_0$. Comme le foncteur $\mathbb{D}$ est fidèle par hypothèse sur S_0, et que la réduction modulo un idéal nilpotent est également fidèle pour les groupes p-divisibles, la construction d'un morphisme $u : G \to H$ tel que $\mathbb{D}(u) = v$ est un problème local sur S. On peut donc supposer que Z et S sont affines, et poser $Z = \operatorname{Spec} R$, $S = \operatorname{Spec} A$, l'anneau R possédant une p-base. Considérons l'immersion $S_0 \hookrightarrow S$ comme un objet de $\mathrm{NC{\small RIS}}(S/\Sigma)$ grâce à la PD-structure nilpotente dont on a supposé l'existence, et montrons que $\mathbb{D}(u_0)_S$ envoie $\omega_H \subset \mathbb{D}(H)_S \simeq \mathbb{D}(H_0)_S$ dans $\omega_G \subset \mathbb{D}(G)_S \simeq \mathbb{D}(G_0)_S$. D'après [7, 4.3.10], il existe sur $\mathrm{C{\small RIS}}(S/\Sigma_1)$ un diagramme commutatif, dans lequel la ligne supérieure est exacte :

$$\begin{array}{ccccccc}
\mathbb{D}(G)^\sigma & \xrightarrow{F} & \mathbb{D}(G) & \longrightarrow & \Phi^*(\omega_G) & \longrightarrow & 0 \\
 & & \big\downarrow{\scriptstyle V} & & \big\downarrow\cap & & \\
 & & \mathbb{D}(G)^\sigma & \xleftarrow{\sim} & \Phi^*(\mathbb{D}(G)_S) & &
\end{array}$$

Ce diagramme montre que l'image de V est $\Phi^*(\omega_G)$. En utilisant la compatibilité de v avec V, et le diagramme analogue pour H, il en résulte que $\Phi^*(v_S)$ envoie $\Phi^*(\omega_H)$ dans $\Phi^*(\omega_G)$. Choisissons une présentation $A \simeq B/I$ de A comme quotient d'une R-algèbre lisse B par l'idéal engendré par une suite régulière $(f_1, \ldots, f_r)$. D'après [7, 2.3.3] et [3, VI 3.2.5], l'enveloppe à puissances divisées de l'idéal de A dans B est une $B/(f_1^p, \ldots, f_r^p)$-algèbre, libre comme $B/(f_1^p, \ldots, f_r^p)$-module, de base les $f_1^{[pq_1]} \ldots f_r^{[pq_r]}$. Comme Z possède une p-base, le Frobenius absolu de Z est fidèlement plat, et il en est de même pour B. Il s'ensuit que l'homomorphisme $\Phi : A \to B/(f_1^p, \ldots, f_r^p)$ est également fidèlement plat, ainsi que $\Phi : A \to D_B(I)$. Donc v_S envoie ω_H dans ω_G, d'où l'assertion, puisque par construction $\mathbb{D}(u_0)_S = v_{0S} = v_S$.

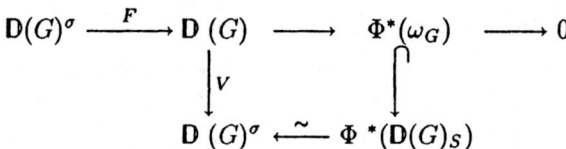

Il résulte alors de (3.2.11) et [18, V 1.6] qu'il existe un homomorphisme $u : G \to H$ tel que $u_{S_0} = u_0$ et $\mathbb{D}(u)_S = v_S$, et il reste à voir que $\mathbb{D}(u) = v$. La différence $w = \mathbb{D}(u) - v$ est un morphisme de F-cristaux. Le Frobenius absolu de S se factorise par S_0, de sorte que $w^\sigma = 0$. Comme w est compatible à F, on en déduit que $w \circ F_{\mathbb{D}(H)} = 0$, et a fortiori que $pw = 0$. Choisissons d'après (1.1.7) un relèvement R_∞ de R, et un relèvement B_∞ de B en une R_∞-algèbre p-adiquement complète et formellement lisse. D'après [7, 2.3.3 et 1.2.6], l'enveloppe à puissances divisées complétée D_∞ de l'idéal de A dans B_∞ est sans p-torsion. Comme w est déterminé par sa valeur sur D_∞, et que $\mathbb{D}(G)$ est localement libre, il en résulte que $w = 0$, ce qui achève la démonstration.

4.3.2. *Remarques.* (i) Supposons que S soit un Z-schéma plat d'intersection complète relative, possédant une suite croissante de sous-schémas fermés

$$S_0 \subset S_1 \subset \ldots \subset S_m = S,$$

telle que, pour $i > 0$, chaque S_i soit plat d'intersection complète relative sur Z, que, pour tout $i \geq 0$, l'idéal définissant S_i dans S_{i+1} possède une structure d'idéal à puissances divisées nilpotente, et que le foncteur $\mathbb{D}$ soit pleinement fidèle pour les groupes p-divisibles sur S_0. Alors $\mathbb{D}$ est pleinement fidèle pour les groupes p-divisibles sur S.

On notera que l'hypothèse sur l'existence de puissances divisées est satisfaite lorsque les idéaux définissant les immersions $S_i \subset S_{i+1}$ sont de carré nul, un tel idéal possédant toujours une PD-structure triviale. Par exemple, si k est un corps, on obtient ainsi la pleine fidélité de $\mathbb{D}$ pour les groupes p-divisibles sur le spectre d'un anneau local artinien de la forme

$$A = k[t_1, \ldots, t_n]/(t_1^{r_1}, \ldots, t_n^{r_n}).$$

(ii) Par contre, Gabber a montré qu'il n'est pas toujours possible de filtrer un anneau artinien intersection complète de manière à avoir des quotients intersections complètes et des idéaux intermédiaires de carré nul. Un cas particulier de son contre-exemple est (en caractéristique $\neq 2$)

$$A = k[x, y, z]/(x^2 + yz, y^2 + zx, z^2 + xy).$$

(iii) Nous ne connaissons pas d'exemple d'anneau artinien intersection complète pour lequel l'hypothèse de (i) n'est pas vérifiée.

4.4. ***Questions et contre-exemples.***

De nombreux problèmes restent ouverts concernant les propriétés de fidélité du foncteur de Dieudonné cristallin $\mathbb{D}$. Nous en indiquons quelques-uns ici, ainsi que certains contre-exemples sur des bases ne vérifiant pas d'hypothèses de régularité du type de celles que nous avons considérées plus haut.

4.4.1. Soient Z un schéma de caractéristique p possédant localement une p-base, et S un Z-schéma qui soit plat, intersection complète relative sur Z. Est-ce que le foncteur $\mathbb{D}$ est fidèle sur la catégorie des p-groupes finis localement libres sur S ?

L'exemple suivant, dû à Koblitz et Ogus, montre qu'une hypothèse de régularité est nécessaire pour pouvoir espérer la fidélité de $\mathbb{D}$ pour les groupes finis. Soient en effet $A = k[x_1, \ldots, x_6]/(x_1^p, \ldots, x_6^p, x_1x_2 + x_3x_4 + x_5x_6)$, $Z = \operatorname{Spec} k$, $S = \operatorname{Spec} A$. L'élément

$$z = \sum_{\substack{0 \le i_1, i_2, i_3 < p \\ i_1+i_2+i_3=p}} (x_1x_4)^{i_1}(x_2x_5)^{i_2}(x_3x_6)^{i_3}$$

est non nul, mais un calcul direct montre que son image par l'application

$$\Phi^* : \Gamma(S, \mathcal{O}_S) \to \Gamma(S/\mathbf{F}_p, \mathcal{O}_{S/\mathbf{F}_p})$$

est nulle. L'élément z définit un endomorphisme u non nul du groupe α_p sur S, mais $\mathbb{D}(u) = 0$, car d'après [7, 4.3.6] il existe sur $\mathrm{CRIS}(S/\mathbf{F}_p)$ un isomorphisme fonctoriel

$$(4.4.1.1) \qquad \Phi^*(\mathcal{L}ie(\alpha_p)) \xrightarrow{\sim} \mathbb{D}(\alpha_p).$$

Le foncteur $\mathbb{D}$ n'est donc pas fidèle sur la catégorie des groupes finis et plats sur S.

On observera que le même exemple montre que, si M est un cristal localement libre de rang fini sur S, la donnée d'un facteur direct $\omega \subset M_S$ n'est pas déterminée par celle de $\Phi^*(\omega) \subset M^\sigma$, contrairement à ce qui est énoncé en [12, V Section 3].

Par ailleurs, même si l'on suppose A artinien, intersection complète relative sur k, il n'est pas vrai en général que $\mathbb{D}$ soit pleinement fidèle sur la catégorie des p-groupes finis et plats sur A. Il suffit pour s'en convaincre de regarder le cas où $A = k[t]/(t^p)$, et de prendre le groupe α_p. Puisque F et V sont nuls, un endomorphisme de $\mathbb{D}(\alpha_p)$ est simplement une section de $\mathcal{O}_{S/\Sigma_1}$, c'est à dire un élément $P(t)$ de l'enveloppe à puissances divisées $D_{k[t]}(t^p)$ tel que $d(P(t)) = 0$. Mais il résulte de (4.4.1.1) qu'un endomorphisme provenant d'un endomorphisme de α_p est de la forme $\Phi^*(u)$, donc est une puissance p-ième d'un élément de $k[t]$. La section $P(t) = (t^p)^{[n]}$, où n est un entier quelconque $\geq p$ fournit alors un contre-exemple.

4.4.2. Supposons encore que S soit plat d'intersection complète relative sur un schéma Z possédant localement une p-base. Le foncteur de Dieudonné est-il pleinement fidèle sur la catégorie des groupes p-divisibles sur S ?

Ce résultat peut être démontré dans certains cas en utilisant les méthodes de (4.3). On observera que, lorsqu'il est vrai, la réponse à la question (4.4.1) est également affirmative : soit en effet $u : G \to H$ un morphisme tel que $\mathbb{D}(u) = 0$. Comme la nullité de u est une propriété locale sur S, on peut supposer que Z et S sont affines, que G et H sont plongés dans des groupes p-divisibles G' et H', et, quitte à remplacer G' par $G' \times H'$, que u se prolonge en $u' : G' \to H'$. Soient G'', H'' les groupes p-divisibles quotients. Soit D_∞ le complété p-adique de l'enveloppe à puissances divisées de l'idéal de S dans un schéma formel lisse sur un relèvement de Z. Les cristaux sur S s'interprètent comme des modules à connexion sur D_∞, et on dispose d'un diagramme commutatif à lignes exactes

$$\begin{array}{ccccccccc} 0 & \longrightarrow & \mathbb{D}(H'')_{D_\infty} & \longrightarrow & \mathbb{D}(H')_{D_\infty} & \longrightarrow & \mathbb{D}(H)_{D_\infty} & \longrightarrow & 0 \\ & & \downarrow & & \downarrow{\scriptstyle \mathbb{D}(u')} & & \downarrow{\scriptstyle \mathbb{D}(u)} & & \\ 0 & \longrightarrow & \mathbb{D}(G'')_{D_\infty} & \longrightarrow & \mathbb{D}(G')_{D_\infty} & \longrightarrow & \mathbb{D}(G)_{D_\infty} & \longrightarrow & 0 \end{array}$$

dont l'exactitude à gauche vient de ce que, compte tenu des hypothèses faites sur Z et S, D_∞ est sans p-torsion [7, 4.2.5]. La nullité de $\mathbb{D}(u)$ entraîne alors que $\mathbb{D}(u')$ se factorise en un morphisme de cristaux de Dieudonné

$$v : \mathbb{D}(H') \to \mathbb{D}(G'').$$

Si on dispose de la pleine fidélité, on en déduit que u' se factorise par G'', d'où la nullité de u.

Rappelons enfin que, sans hypothèse sur S, le foncteur $\mathbb{D}$ n'est pas pleinement fidèle en général pour les groupes p-divisibles. On obtient déjà un contre-exemple en prenant $S = \operatorname{Spec} k[x,y]/(x^2, xy, y^2)$, schéma dont le H^0 cristallin possède des éléments de torsion [8, A.4].

4.4.3. Toujours sous les mêmes hypothèses, on peut demander si $\mathbb{D}$ est pleinement fidèle pour les groupes de Barsotti-Tate tronqués [18, I 1.2], dont les propriétés sont beaucoup plus fortes que celles des groupes finis localement libres quelconques. Cette question est liée à celle de la caractérisation de l'image essentielle du foncteur $\mathbb{D}$, pour laquelle la réponse n'est connue dans aucun cas, à l'exception de celui des corps parfaits. En effet, une réponse positive à la question concernant les groupes de Barsotti-Tate tronqués permettrait de montrer que lorsque la base est le spectre d'un corps k (non nécessairement parfait), $\mathbb{D}$ est un foncteur essentiellement surjectif de la catégorie des groupes p-divisibles dans celle des cristaux de Dieudonné libres sur $\mathcal{O}_{S/\Sigma}$. Dans ce cas, la technique de descente de la clôture parfaite K à k employée en (4.1.3) ramène à montrer que, si G est un groupe p-divisible sur K, une donnée de descente de K à k sur $\mathbb{D}(G)$ est induite par une donnée de descente sur G. En écrivant K comme limite inductive d'extensions finies radicielles de k, il est possible, pour n fixé, de redescendre $G(n)$ sur une extension finie k' de k, ce qui ramène le problème à un problème de pleine fidélité pour les images inverses de $G(n)$

sur l'anneau artinien $k' \otimes_k k'$, qui est intersection complète relative sur k (et admet même une filtration comme en (4.3.2) (i)).

4.5. *Groupes p-divisibles sur un schéma simplement connexe.*

Nous appliquerons ici le théorème de pleine fidélité (4.1.1) pour montrer que, sur une certaine classe de variétés simplement connexes, tout groupe p-divisible est isogène à un groupe p-divisible constant.

4.5.1. Soient k un corps parfait de caractéristique p, $W = W(k)$, $K = \mathrm{Frac}(W)$, $\overline{K}$ une clôture algébrique de K, et X un k-schéma propre, lisse et géométriquement connexe. On suppose qu'il existe une extension finie K' de K, de corps résiduel k', et un schéma propre et lisse $\mathfrak{X}$ sur l'anneau V' des entiers de K' dont la réduction X' soit isomorphe à $X_{k'}$. On suppose enfin que $\pi_1(\mathfrak{X}_{\overline{K}}) = 0$.

4.5.2. Théorème. *Sous les hypothèses de (4.5.1), tout groupe p-divisible G sur X est isogène à un groupe de la forme H_X, où H est un groupe p-divisible sur k.*

Le foncteur de Dieudonné est essentiellement surjectif sur k, et pleinement fidèle sur X, d'après (4.1.1). Il suffit donc de montrer l'énoncé analogue pour les cristaux de Dieudonné, ce qui sera fait en (4.5.5). Montrons d'abord la proposition suivante :

4.5.3. Proposition. *Soient K un corps de caractéristique* 0, $\overline{K}$ *une clôture algébrique de K, X un K-schéma propre, lisse et géométriquement connexe, tel que $\pi_1(X_{\overline{K}}) = 0$. Alors le foncteur qui à un K-espace vectoriel de dimension finie E associe le $\mathcal{O}_X$-module $\mathcal{E} = E \otimes_K \mathcal{O}_X$, muni de la connexion intégrable triviale, est une équivalence de la catégorie des K-espaces vectoriels dans celle des $\mathcal{O}_X$-modules cohérents à connexion intégrable, et admet pour foncteur quasi-inverse le foncteur qui associe à $\mathcal{E}$ le K-espace vectoriel $H^0(X, \mathcal{E} \otimes_{\mathcal{O}_X} \Omega^{\cdot}_X)$.*

Comme $\Gamma(X, \mathcal{O}_X) = K$, le foncteur qui à un K-espace vectoriel E associe le fibré $E \otimes_K \mathcal{O}_X$ muni de la connexion triviale est pleinement fidèle. Il suffit donc de montrer qu'il est essentiellement surjectif. Soit $\mathcal{E}$ un fibré à connexion intégrable sur X ; le schéma X, muni du fibré à connexion $\mathcal{E}$, peut être redescendu sur un sous-corps L de K possédant un plongement dans $\mathbf{C}$. Puisque le groupe fondamental est invariant par extension algébriquement close du corps de base, on a encore $\pi_1(X_{\overline{L}}) = 0$; on peut donc supposer qu'il existe un plongement de K dans $\mathbf{C}$.

Soit $E = H^0(X, \mathcal{E} \otimes_{\mathcal{O}_X} \Omega^{\cdot}_X)$; il existe un homomorphisme horizontal canonique $E \otimes_K \mathcal{O}_X \to \mathcal{E}$, et il suffit de montrer que c'est un isomorphisme. Par descente de $\mathbf{C}$ à K, il suffit qu'il en soit ainsi sur $\mathbf{C}$. D'après le théorème de Grothendieck [10, th. 4.2], et GAGA, la nullité de $\pi_1(X_{\mathbf{C}})$ entraîne que $\mathcal{E} \otimes \mathbf{C}$ est un fibré trivial, muni d'une connexion triviale. On est alors ramené au cas où $\mathcal{E} = \mathcal{O}_X$, qui résulte à nouveau de ce que $\Gamma(X, \mathcal{O}_X) = K$.

4.5.4. Proposition. *Sous les hypothèses de* (4.5.1), *le foncteur qui associe à un K-espace vectoriel de dimension finie E l'isocristal convergent $E \otimes_K \mathcal{O}_{X/K}$* [5, (2.3)] *est une équivalence de catégories, et admet pour foncteur quasi-inverse celui qui associe à un isocristal convergent $\mathcal{E}$ l'espace $\Gamma(X/K, \mathcal{E})$ des sections globales de $\mathcal{E}$.*

Comme X est géométriquement connexe, l'espace $\Gamma(X/K, \mathcal{O}_{X/K})$ des sections globales de l'isocristal "constant" $\mathcal{O}_{X/K}$ est réduit à K. Il en résulte que le foncteur considéré est pleinement fidèle. Considérons alors un isocristal convergent $\mathcal{E}$ sur X/K, et soit $E = \Gamma(X/K, \mathcal{E})$ l'espace des sections globales de $\mathcal{E}$. Il existe un morphisme canonique $\varphi : E \otimes_K \mathcal{O}_{X/K} \longrightarrow \mathcal{E}$, et il suffit de montrer que E est de dimension finie, et φ un isomorphisme.

Soit $\mathcal{E}'$ l'image inverse sur X'/K' de $\mathcal{E}$. Comme K' est une extension finie de K, l'homomorphisme canonique $\Gamma(X/K, \mathcal{E}) \otimes_K K' \to \Gamma(X'/K', \mathcal{E}')$ est un isomorphisme [5, (3.2)]. Comme X' possède un relèvement propre et lisse $\mathfrak{X}$ sur V', l'isocristal $\mathcal{E}'$ peut être réalisé comme un fibré analytique $\mathcal{E}'^{an}$ muni d'une connexion intégrable (vérifiant certaines conditions) sur l'espace analytique rigide $\mathfrak{X}_{K'}^{an}$ associé à la fibre générique $\mathfrak{X}_{K'}$ de $\mathfrak{X}$. Par GAGA (rigide), on peut redescendre ce fibré en un fibré algébrique à connexion intégrable sur $\mathfrak{X}_{K'}$, qu'on notera $\tilde{\mathcal{E}}'$. Par définition, l'espace des sections globales du cristal convergent $\mathcal{E}'$ est alors $H^0(\mathfrak{X}_{K'}^{an}, \mathcal{E}'^{an} \otimes \Omega^{\cdot}_{\mathfrak{X}_{K'}^{an}})$, isomorphe à $H^0(\mathfrak{X}_{K'}, \tilde{\mathcal{E}}' \otimes \Omega^{\cdot}_{\mathfrak{X}_{K'}})$ par GAGA, et donc de dimension finie. Il en est donc de même de $\Gamma(X/K, \mathcal{E})$.

La catégorie des isocristaux convergents est un champ sur X, de sorte qu'il suffit, pour montrer que φ est un isomorphisme, de le montrer localement. Localement, on peut réaliser les isocristaux convergents comme des fibrés à connexion intégrable sur la fibre générique d'un relèvement formel de X ; comme l'extension K' de K est finie, on voit donc qu'un homomorphisme entre isocristaux convergents sur X/K est un isomorphisme si et seulement si son image inverse sur X'/K' est un isomorphisme. Compte tenu de l'isomorphisme de changement de corps de base, il suffit alors de montrer que l'analogue de φ pour $\mathcal{E}'$ est un isomorphisme. En utilisant la description de $\mathcal{E}'$ et de l'espace de ses sections globales donnée plus haut, cette assertion résulte de (4.5.3) appliqué au fibré à connexion intégrable algébrique $\tilde{\mathcal{E}}'$.

4.5.5. Proposition. *Sous les hypothèses de* (4.5.1), *soient* $\mathcal{E}$ *un cristal de Dieudonné sur* X, *et* $M = \Gamma(X/W, \mathcal{E})$. *Alors* M *est un module de Dieudonné (sans torsion), et* $\mathcal{E}$ *est isogène au cristal de Dieudonné "constant"* $M \otimes_W \mathcal{O}_{X/\Sigma}$.

Comme X est propre et lisse, M est de type fini sur W et sans torsion. L'homomorphisme canonique $M^\sigma \to \Gamma(X/W, \mathcal{E}^\sigma)$ est un isomorphisme, de sorte que les applications $F : \mathcal{E}^\sigma \to \mathcal{E}$ et $V : \mathcal{E} \to \mathcal{E}^\sigma$ munissent M d'une structure de module de Dieudonné. Il suffit alors de montrer que l'homomorphisme canonique $\varphi : M \otimes_W \mathcal{O}_{X/\Sigma} \to \mathcal{E}$ est une isogénie.

Soit $\varphi_K : M \otimes_W \mathcal{O}_{X/K} \simeq (M \otimes_W K) \otimes_K \mathcal{O}_{X/K} \to \mathcal{E}_K$ le morphisme induit par φ entre les isocristaux convergents associés [5, (2.3)]. On dispose d'un isomorphisme canonique $M \otimes_W K \xrightarrow{\sim} \Gamma(X/K, \mathcal{E}_K)$ [5, (3.3)], qui identifie φ_K au morphisme canonique $\Gamma(X/K, \mathcal{E}_K) \otimes_K \mathcal{O}_{X/K} \to \mathcal{E}_K$. Ce dernier est un isomorphisme d'après (4.5.4), ce qui achève la démonstration.

Remarque. On observera que la démonstration de (4.5.2) fournit en fait un groupe p-divisible $H(G)$ sur k fonctoriellement associé à un groupe p-divisible G sur X, tel que G soit isogène à $H(G)_X$: il suffit de prendre le groupe associé au module de Dieudonné $\Gamma(X/W, \mathbb{D}(G))$ par la théorie de Dieudonné classique sur le corps k.

BIBLIOGRAPHIE

[EGA] A. Grothendieck, *Eléments de Géométrie Algébrique*, rédigé avec la collaboration de J. Dieudonné, Publ. Math. I.H.E.S. ; ch. 0_{IV} : n° **20** (1964) ; ch. IV, §6 : n° **24** (1965).

[SGA3] *Schémas en groupes I*, Séminaire de Géométrie Algébrique du Bois Marie, 1962-64, dir. par M. Demazure et A. Grothendieck, Lecture Notes in Math. **151**, Springer Verlag (1970).

[1] M. Artin, *Algebraic approximation of structures over complete local rings*, Publ. Math. I.H.E.S. **36**, (1969), 23-58.

[2] M. Artin, C. Rotthaus, *A structure theorem for power series rings*, in *Algebraic Geometry and Commutative Algebra I*, Kinokuniya Co. (1988), 35-44.

[3] P. Berthelot, *Cohomologie cristalline des schémas de caractéristique* $p > 0$, Lecture Notes in Math. **407**, Springer Verlag (1974).

[4] P. Berthelot, *Théorie de Dieudonné sur un anneau de valuation parfait*, Ann. Scient. Éc. Norm. Sup. **13**, (1980), 225-268.

[5] P. Berthelot, *Cohomologie rigide et cohomologie rigide à support propre*, à paraître.

[6] P. Berthelot, W. Messing, *Théorie de Dieudonné cristalline I*, in *Journées de Géométrie Algébrique de Rennes I*, Astérisque **63**, (1979), 17-37.

[7] P. Berthelot, L. Breen, W. Messing, *Théorie de Dieudonné cristalline II*, Lecture Notes in Math. **930**, Springer Verlag (1982).

[8] P. Berthelot, A. Ogus, *F-isocrystals and De Rham Cohomology I*, Invent. Math. **72**, (1983), 159-199.

[9] J.-M. Fontaine, *Modules galoisiens, modules filtrés et anneaux de Barsotti-Tate*, in *Journées de Géométrie Algébrique de Rennes III*, Astérisque **65**, (1979), 3-80.

[10] A. Grothendieck, *Représentations linéaires et compactification profinie des groupes discrets*, Manuscr. Math. **2**, (1970), 375-396.

[11] A. Grothendieck, *Groupes de Barsotti-Tate et cristaux*, Actes, Congrès intern. Math. 1970, **1**, Gauthier-Villars (1970), 431-436.

[12] A. Grothendieck, *Groupes de Barsotti-Tate et cristaux de Dieudonné*, Sém. Math. Sup. **45**, Presses de l'Université de Montréal (1974).

[13] L. Illusie, *Complexe de de Rham-Witt et cohomologie cristalline*, Ann. Scient. Éc. Norm. Sup. **12**, (1979), 501-661.

[14] L. Illusie, *Déformations de groupes de Barsotti-Tate, d'après A. Grothendieck*, in *Séminaire sur les pinceaux arithmétiques : la conjecture de Mordell*, Astérisque **127**, (1985), 151-198.

[15] N. Katz, *Travaux de Dwork*, in *Séminaire Bourbaki* 1971/72, exp. 409, Lecture Notes in Math. **317**, (1973), 167-200.

[16] T. Kimura, H. Niitsuma, *Regular local rings of characteristic p and p-basis*, J. Math. Soc. Japan **32**, (1980), 363-371.

[17] B. Mazur, W. Messing, *Universal extensions and one dimensional crystalline cohomology*, Lecture Notes in Math. **370**, Springer Verlag (1974).

[18] W. Messing, *The crystals associated to Barsotti-Tate groups : with applications to abelian schemes*, Lecture Notes in Math. **264**, Springer Verlag (1972).

[19] N. Saavedra Rivano, *Catégories tannakiennes*, Lecture Notes in Math. **265**, Springer Verlag (1972).

[20] J.-P. Serre, *Groupes algébriques et corps de classes*, Hermann (1959).

Pierre Berthelot
IRMAR
Université de Rennes
Campus de Beaulieu
35042 Rennes cedex
France

William Messing
School of Mathematics
University of Minnesota
Minneapolis, MN 55455

Complex Immersions and Arakelov Geometry

JEAN-MICHEL BISMUT, HENRI GILLET
and CHRISTOPHE SOULÉ

Contents

Abstract

In this paper we establish an arithmetic Riemann–Roch–Grothendieck Theorem for immersions. Our final formula involves the Bott–Chern currents attached to certain holomorphic complexes of Hermitian vector bundles, which were previously introduced by the authors. The functorial properties of such currents are studied. Explicit formulas are given for Koszul complexes.

In [BGS3] and in [GS3], the direct image by a submersion of an Hermitian vector bundle on an arithmetic variety was defined. One may wonder [Ma], [GS3] whether a Riemann–Roch–Grothendieck formula holds for such direct images, which would involve the characteristic classes defined in [GS2], with values in the arithmetic Chow groups of [GS1]. Such a formula would be stronger than the corresponding Riemann–Roch–Grothendieck theorem with values in Chow groups [SGA6] and the Riemann–Roch–Grothendieck theorem at the level of differential forms obtained in [BGS2,3] for submersions, using Quillen metrics [Q2].

When proving his Riemann–Roch theorem with values in Chow groups, Grothendieck used the factorization of any projective map between smooth manifolds as a regular closed immersion followed by the projection attached to a projective bundle [SGA6]. The main step of the proof is then to study the direct image by a regular immersion by blowing up the subvariety. It seems that the same reduction to the immersion case will be necessary for proving the arithmetic Riemann–Roch–Grothendieck theorem conjecture of [Ma] and [GS3].

The purpose of this paper is to prove an arithmetic Riemann–Roch–Grothendieck theorem for immersions, in which our datas are:

- an immersion $i: Y \to X$ of arithmetic varieties;
- a vector bundle η on Y, equipped with an Hermitian metric g^η;
- a resolution of the sheaf $i_*\eta$ by a complex of vector bundles (ξ, v) on X, equipped with an Hermitian metric h^ξ.

Our main result in Theorem 4.13 calculates the arithmetic Chern character of ξ [GS2] in terms of the Chern character of η, of the Todd genus of the normal bundle N of Y in X which is equipped with a metric g^N, and of a secondary invariant $T(h^\xi)$ attached to the Hermitian chain complex (ξ, v) introduced in Bismut–Gillet–Soulé [BGS4] under the name of a Bott–Chern singular current. In fact the current $T(h^\xi)$ solves the equations of currents on X

$$\frac{\bar{\partial}\partial}{2i\pi} T(h^\xi) = Td^{-1}(N, g^N) ch(\eta, g^\eta) \delta_Y - ch(\xi, h^\xi) \tag{0.1}$$

where in (0.1), the various characteristic classes are calculated by Chern–Weil theory using the holomorphic Hermitian connections associated with the corresponding metrics. The construction of the current $T(h^\xi)$ given in [BGS4] uses in an essential way Quillen's superconnections [Q1], and a result proved in Bismut [B2, Theorem 3.2] on the large parameter behaviour of Quillen's superconnection Chern character forms. When $Y = \emptyset$, i.e. when the complex (ξ, v) is acyclic, the currents $T(h^\xi)$ have already been introduced in Bismut–Gillet–Soulé [BGS1] and were shown to coincide with objects considered earlier by Bott–Chern [BoC] and Donaldson [D].

An important intermediary result, proved in Theorem 2.7, concerns the behaviour of the currents $T(h^\xi)$ under composition of immersions. In fact any result of the type of the Riemann–Roch–Grothendieck theorem on direct images implies that the considered objects behave functorially with respect to the composition of maps. Here the functorial behaviour of the currents $T(h^\xi)$ under composition of immersions is one of the key tools by which we prove our Riemann–Roch–Grothendieck theorem for immersions.

A second key instrument is the explicit computation of the currents $T(h^\xi)$ associated with Koszul complexes. This computation makes use of the formalism of Mathai–Quillen [MQ] in a complex setting. We also introduce Euler–Green currents in the sense of [GS1], and we relate them to the currents $T(h^\xi)$ of Koszul complexes. Such Euler–Green currents are obtained by a double transgression formula from the Mathai–Quillen–Thom forms [MQ].

Let us here point out that Bott–Chern singular currents can be constructed in an enormous variety of ways, and that much of our work consists in showing that two Bott–Chern currents which solve equation (0.1) differ by ∂ or $\bar{\partial}$ coboundaries, i.e. they represent the same class in Bott–Chern theory.

A third instrument is the deformation to the normal cone described in Baum–Fulton–MacPherson [BaFM], by which we show that our currents $T(h^\xi)$ are related to the arithmetic characteristic classes of Gillet–Soulé [GS1].

A common technical feature of our proofs is that we use the properties of the wave front set of the considered currents. In fact it was shown in [BGS4], by using microlocal estimates of [B2], that the wave front set of the current $T(h^\xi)$ is included in the conormal bundle to Y in X. By using standard arguments in [H, Chapter 8], we can in particular multiply two currents $T(h^\xi)$ associated with transversal submanifolds in X.

Our paper is organized as follows. In Section 1, we recall the main results of [B2] and [BGS4] concerning the Quillen's Chern character currents and the Bott–Chern singular currents. In Section 2, we study various functorial properties of our Bott–Chern singular currents, including a transitivity property under composition of immersions.

In Section 3, we calculate the Bott–Chern singular current associated with a Koszul complex, and we compare this current with the Euler–Green current, which is one component of the arithmetic Euler class defined in [GS2]. Notice that this is one of the very few cases where an explicit formula can be given for a Green current. The transitivity property established in Section 2 is used here to check the multiplicativity of the Euler–Green current of an orthogonal direct sum. Finally, in Section 4, we study the deformation to the normal cone (a variant of Grothendieck's blowing-up introduced in [BaFM]) to prove our Riemann–Roch–Grothendieck theorem for immersions. Notice however that our final formulas given in Theorem 4.13 does not lie in the arithmetic Chow group of the ambiant variety, but rather is integrated down to a base on which both the variety and the subvariety project smoothly (we do not define direct images for immersions in arithmetic Chow groups and arithmetic Grothendieck groups).

The results obtained in this paper were announced in [BGS5].

The authors are indebted to G. Lebeau for helpful discussions.

1. A Bott–Chern Singular Current

Let $i: M' \to M$ be an immersion of complex manifolds. Let η be a holomorphic vector bundle on M', let (ξ, v) be a holomorphic chain complex of vector bundles on M which is such that there is an exact sequence of sheaves

$$\mathcal{O}_M(\xi) \to i_* \mathcal{O}_{M'}(\eta) \to 0.$$

We assume that the vector bundles in ξ are equipped with Hermitian metrics.

In this section, we recall the main results of Bismut [B2] and Bismut–Gillet–Soulé [BGS4] which concern:

- The asymptotic behavior of the Quillen superconnection forms naturally associated with the Hermitian complex (ξ, v) [B2].
- The corresponding construction of singular Bott–Chern currents [BGS4].

This section is organized as follows. In a), we give our main assumptions and notations. In b), we introduce assumption (A) for the metrics on the complex (ξ, v). In c), we briefly recall elementary properties of wave front sets. In d), we review Quillen's superconnections [Q1]. In e), we recall the double transgression formulas of [BGS1]. In f), we state the results of convergence of superconnection currents which were proved in [B2]. In g), we recall our construction of Bott–Chern singular currents [BGS4].

(a) *A holomorphic chain complex.* Let M be a compact connected complex manifold of complex dimension ℓ. Let $M' = \bigcup_1^n M'_j$ be a finite union of compact connected complex submanifolds of M, such that, for $j \neq j'$, $M'_j \cap M'_{j'} = \emptyset$. Let i be the embedding $M' \to M$. Let N be the normal vector bundle to M' in M, and let N^* be its dual.

Let

$$(\xi, v): 0 \to \xi_m \to_v \xi_{m-1} \cdots \to_v \xi_0 \to 0 \tag{1.1}$$

be a holomorphic chain complex of vector bundles on M.

Let η be a holomorphic vector bundle on M'. We assume there is a holomorphic restriction map $r: \xi_{0|M'} \to \eta$ which is such that we have an exact sequence of sheaves

$$0 \to \mathcal{O}_M(\xi_m) \to_v \mathcal{O}_M(\xi_{m-1}) \to \cdots \to_v \mathcal{O}_M(\xi_0) \to_r i_* \mathcal{O}_{M'}(\eta) \to 0. \tag{1.2}$$

In particular the complex (ξ, v) is acyclic on $M \setminus M'$.

For $x \in M'$, $0 \leq k \leq m$, let $F_{k,x}$ be the kth homology group of the complex $(\xi, v)_x$. Set $F_x = \bigoplus_0^m F_{k,x}$.

The following results are consequences of the local uniqueness of resolutions (see Serre [S, IV Appendix 1] and Eilenberg [E, Theorem 8]) and are proved in [B2, Section 1]:

- For $k = 0, \ldots, m$, $x \in M'$, the dimension of $F_{k,x}$ is constant on each M'_j, so that F_k is a holomorphic vector bundle on M'.
- For $x \in M'$, $U \in T_x M$, let $\partial_U v(x)$ be the derivative of the chain map v calculated in any given local holomorphic trivialization of (ξ, v) near x. Then $\partial_U v(x)$ acts on F_x. When acting on F_x, $\partial_U v(x)$ only depends on the image y of U in N_x. So we now write $\partial_y v(x)$ instead of $\partial_U v(x)$.
- For any $x \in M'$, $y \in N$, $(\partial_y v)^2(x) = 0$. If $y \in N$, let i_y be the interior multiplication operator by y acting on the exterior algebra $\Lambda(N^*)$. Let i_y act like $i_y \otimes 1$ on $\Lambda N^* \otimes \eta$. Then the graded holomorphic complex $(F, \partial_y v)$ on the total space of the vector bundle N is canonically isomorphic to the Koszul complex $(\Lambda N^* \otimes \eta, i_y)$.

(b) *Assumption (A) on the Hermitian metrics of a chain complex.* We now assume that $\xi_0, \ldots, \xi_m$ are equipped with smooth Hermitian metrics $h^{\xi_0}, \ldots, h^{\xi_m}$. We equip $\xi = \bigoplus_{k=0}^m \xi_k$ with the metric h^ξ which is the orthogonal sum of the metrics $h^{\xi_0}, \ldots, h^{\xi_m}$. Let v^* be the adjoint of v with respect to the metric h^ξ. Using finite dimensional Hodge theory, we get the identification of smooth vector bundles on M' for $0 \leq k \leq m$

$$F_k \cong \{f \in \xi_k; vf = 0, v^* f = 0\}. \tag{1.3}$$

As a smooth subvector bundle of ξ_k, the right hand side of (1.3) inherits a Hermitian metric from the metric h^{ξ_k}. Using the identification (1.3), we find that for every $k = 0, \ldots, m$, F_k is a holomorphic Hermitian vector bundle on M'. Let h^{F_k} denote the Hermitian metric on F_k. We equip $F = \bigoplus_0^m F_k$ with the metric h^F which is the orthogonal sum of the metrics $h^{F_0}, \ldots, h^{F_m}$.

Let g^N, g^η be Hermitian metrics on the vector bundles N, η. We equip the vector bundle $\Lambda N^* \otimes \eta$ with the tensor product of the metric induced by g^N on $\Lambda(N^*)$ and of the metric g^η.

Definition 1.1. Given metrics g^N, g^η on N, η, we will say that the metrics $h^{\xi_0}, \ldots, h^{\xi_m}$ on $\xi_0, \ldots, \xi_m$ verify assumption (A) with respect to g^N, g^η if the canonical identification of holomorphic chain complexes on the total space of N

$$(F, \partial_y v) \cong (\Lambda N^* \otimes \eta, i_y) \tag{1.4}$$

also identifies the metrics.

Proposition 1.2. *Given metrics g^N, g^η, on N, η, there exist metrics $h^{\xi_0}, \ldots, h^{\xi_m}$ on $\xi_0, \ldots, \xi_m$ which verify assumption* (A) *with respect to g^N, g^η.*

Proof. This result is proved in [B2, Proposition 1.6]. □.

(c) *Wave front sets.* If γ is a current on M, we note $WF(\gamma)$ the wave front set of γ. For the definition and properties of wave front sets, we refer to Hörmander [H, Chapter VIII]. Let us just recall that $WF(\gamma)$ is a closed conic subset of $T^*_R M \setminus \{0\}$. Also if p is the projection $T^*_R M \to M$, $p(WF(\gamma))$ is exactly the singular support of γ, whose complement in M is the set of points x such that γ is C^∞ on a neighborhood of x.

Let $\mathcal{D}'_{N^*_R}$ be the set of currents γ on M which are such that $WF(\gamma) \subset N^*_R$. In particular currents in $\mathcal{D}'_{N^*_R}$ are smooth on $M \setminus M'$. By [H, Definition 8.2.2], $\mathcal{D}'_{N^*_R}$ has a natural topology which we now describe.

Let U be a small open set in M, which we identify with an open ball in $R^{2\ell}$. Over U, we identify $T^*_R M$ with $U \times R^{2\ell}$. Let Γ be a closed conic set in $R^{2\ell}$ such that if $x \in U$, $\Gamma \cap N^*_{R,x} = \emptyset$. Let φ be a smooth current on $R^{2\ell}$ with compact support included in U and let m be an integer. If γ is a current, let $\widehat{\varphi\gamma}(\xi)$ be the Fourier transform of $\varphi\gamma$ (which is here considered as a current on $R^{2\ell}$). If $\gamma \in \mathcal{D}'_{N^*_R}$, set

$$p_{U,\Gamma,\varphi,m}(\gamma) = \sup_{\xi \in \Gamma} |\xi|^m |\widehat{\varphi\gamma}(\xi)|. \tag{1.5}$$

If γ_n is a sequence of currents on $\mathcal{D}'_{N^*_R}$, we say that γ_n converges to $\gamma \in \mathcal{D}'_{N^*_R}$ if

- $\gamma_n \to \gamma$ in the sense of distributions.
- If U, Γ, φ, m are taken as before

$$p_{U,\Gamma,\varphi,m}(\gamma_n - \gamma) \to 0. \tag{1.6}$$

Definition 1.3. $P^M_{M'}$ denotes the vector space of currents ω on M which have the following two properties:

- ω is a sum of currents of type (p,p).
- The wave front set of ω is included in N^*_R.

$P^{M,0}_{M'}$ is the vector space of current $\omega \in P^M_{M'}$ which are such that there exist currents $\alpha, \beta \in \mathcal{D}'_{N^*_R}$ for which $\omega = \partial\alpha + \bar{\partial}\beta$.

We equip $P^M_{M'}$ with the topology induced by $\mathcal{D}'_{N^*_R}(M)$.

If $M' = \emptyset$, we will write P^M, $P^{M,0}$ instead of $P^M_{M'}$, $P^{M,0}_{M'}$.

(d) *Quillen's superconnections.* We now assume that $\xi_0, \ldots, \xi_m$ are equipped with Hermitian metrics $h^{\xi_0}, \ldots, h^{\xi_m}$. We otherwise use the notations of Section 1b).

Set

$$\xi_+ = \bigoplus_{k \text{ even}} \xi_k \qquad \xi_- = \bigoplus_{k \text{ odd}} \xi_k. \tag{1.7}$$

Then $\xi = \xi_+ \oplus \xi_-$ is a Z_2-graded Hermitian vector bundle. $\operatorname{End}\xi$ is naturally Z_2-graded, the even (resp. odd) elements in $\operatorname{End}\xi$ commuting (resp. anticommuting) with the operator $\tau = \pm 1$ on $\xi_\pm$ which defines the Z_2-grading.

For $0 \leq k \leq m$, let ∇^{ξ_k} be the holomorphic Hermitian connection on ξ_k. Then $\nabla^{\xi} = \bigoplus_0^m \nabla^{\xi_k}$ is the holomorphic Hermitian connection on the vector bundle ξ.

We now briefly recall the definition of a superconnection in the sense of Quillen [Q1]. The algebra $\Lambda(T^*_R M) \hat{\otimes} \operatorname{End}\xi$ is naturally Z_2-graded. Let S be a smooth section of $(\Lambda(T^*_R M) \hat{\otimes} \operatorname{End}\xi)^{\text{odd}}$. Then by definition $\nabla^\xi + S$ is a superconnection on the Z_2-graded vector bundle ξ.

In the sequel ∇^ξ will be considered as a first order differential operator acting on the set of smooth sections of $\Lambda(T^*_R M) \hat{\otimes} \operatorname{End}\xi$. The curvature $(\nabla^\xi + S)^2$ of the superconnection $\nabla^\xi + S$ is then a smooth section of $(\Lambda(T^*_R M) \hat{\otimes} \operatorname{End}\xi)^{\text{even}}$.

If $A \in \operatorname{End} \xi$, we define its supertrace $Tr_s[A] \in \mathbb{C}$ by

$$Tr_s[A] = Tr[\tau A]. \tag{1.8}$$

We extend Tr_s as a linear map from $\Lambda(T_R^* M) \hat{\otimes} \operatorname{End} \xi$ into $\Lambda(T_R^* M)$, with the convention that if $\omega \in \Lambda(T_R^* M)$, $A \in \operatorname{End} \xi$

$$Tr_s[\omega A] = \omega Tr_s[A].$$

If $B, B' \in \Lambda(T_R^* M) \hat{\otimes} \operatorname{End} \xi$, let $[B, B']$ be the supercommutator

$$[B, B'] = BB' - (-1)^{\deg B \deg B'} B'B.$$

Then by [Q1], Tr_s vanishes on supercommutators. Let φ be the homomorphism of $\Lambda^{\text{even}}(T_R^* M)$ into itself which to $\omega \in \Lambda^{2p}(T_R^* M)$ associates $(2\pi i)^{-p}\omega$.

Let S be an odd smooth section of $\Lambda(T_R^* M) \hat{\otimes} \operatorname{End} \xi$. The basic result of Quillen [Q1] asserts that the form $\varphi(Tr_s[\exp(-(\nabla^\xi + S)^2)])$ is closed and represents in cohomology the Chern character of $\xi_0 - \xi_1 + \cdots + (-1)^m \xi_m$.

(e) *Double transgression formulas.* We make the same assumptions as in Section 1b). Set

$$V = v + v^*. \tag{1.9}$$

Then V is a smooth section of $\operatorname{End}^{\text{odd}} \xi$. For $u \geq 0$, let A_u be the superconnection on ξ

$$A_u = \nabla^\xi + \sqrt{u}\, V. \tag{1.10}$$

Then the curvature A_u^2 of A_u is a smooth section of $\left(\Lambda(T_R^* M) \hat{\otimes} \operatorname{End} \xi\right)^{\text{even}}$.

Let N_H be the number operator of the complex (ξ, v). Namely N_H acts on ξ_k $(0 \leq k \leq m)$ by multiplication by k. We now recall a result of [BGS1].

Theorem 1.4. *The forms $Tr_s[\exp(-A_u^2)]$ and $Tr_s[N_H \exp(-A_u^2)]$ lie in P^M, and depend smoothly on $u \geq 0$. Moreover, for $u > 0$, the following identities hold*

$$\begin{aligned} \frac{\partial}{\partial u} Tr_s[\exp(A_u^2)] &= -d\, Tr_s\left[\frac{V}{2\sqrt{u}} \exp(-A_u^2)\right] \\ Tr_s\left[\frac{V}{\sqrt{u}} \exp(-A_u^2)\right] &= (\bar{\partial} - \partial) Tr_s\left[\frac{N_H}{u} \exp(-A_u^2)\right] \end{aligned} \tag{1.11}$$

In particular

$$\frac{\partial}{\partial u} Tr_s\left[\exp(-A_u^2)\right] = \frac{1}{u}\bar{\partial}\partial\, Tr_s\left[N_H \exp(-A_u^2)\right]. \tag{1.12}$$

Proof. These formulas are proved in [BGS1, Theorem 1.15], or in [B2, Theorem 2.4]. Observe that signs have been changed with respect to [BGS1], since here v decreases the grading in ξ by 1, while in [BGS1], v increases the grading in ξ by 1. □

(f) *Convergence of superconnection currents.* Set

$$F_+ = \bigoplus_{k \text{ even}} F_k \qquad F_- = \bigoplus_{k \text{ odd}} F_k.$$

$F = F_+ \oplus F_-$ is a Hermitian Z_2-graded vector bundle. If $y \in N$, let $\bar{y}$ be the conjugate element of y in $\bar{N}$. Then $y \in N$ represents $Y = y + \bar{y} \in N_R$. In particular if N is equipped with a metric g^N, $|Y|^2 = 2|y|^2$.

The superconnection formalism of Quillen can also be applied to the Z_2-graded vector bundle $F = F_+ \oplus F_-$. Let $(\partial_y v)^*$ be the adjoint of $\partial_y v$ with respect to the metric h^F on F. Then $(\partial_y v)^*$ is a antiholomorphic function of y. Set

$$\partial_Y V = \partial_y v + (\partial_y v)^*. \tag{1.13}$$

$\partial_Y V$ is an odd section of End F. If we use the canonical identification (1.4), then

$$\partial_Y V = i_y + i_y^*. \tag{1.14}$$

For $0 \le k \le m$, let ∇^{F_k} be the holomorphic Hermitian connection on the vector bundle F_k. Then $\nabla^F = \bigoplus_{k=0}^{m} \nabla^{F_k}$ is the holomorphic Hermitian connection on F. Let B be the superconnection on F

$$B = \nabla^F + \partial_Y V. \tag{1.15}$$

Then B^2 is the curvature of the superconnection B. B^2 is a smooth section of $\left(\Lambda(T_R^* N) \hat{\otimes} \text{End}\, F\right)^{\text{even}}$.

N_H acts naturally on F, i.e. if $0 \le k \le m$, $f \in F_k$, then

$$N_H f = kf. \tag{1.16}$$

By [Q1], for any $u \geq 0$, the form $Tr_s[\exp(-A_u^2)]$ is closed and

$$\varphi(Tr_s[\exp(-A_u^2)])$$

represents in cohomology the Chern character of $\xi_0 - \xi_1 + \cdots + (-1)^m \xi_m$.

$\delta_{M'}$ denotes the current of integration on the oriented manifold M'.

Let $C^1(M)$ be the set of continuous differential forms on M which have continuous first derivatives. Les $\| \; \|_{C^1(M)}$ be a norm on $C^1(M)$ such that $\|\mu^n\|_{C^1(M)} \to 0$ if and only if μ^n tends to 0 uniformly on M together with its first derivatives.

We now recall the result of Bismut announced in [B1] and proved in [B2, Theorem 3.2]

Theorem 1.5. *As $u \to \infty$, we have the following convergence of currents on M*

$$(1.17)\quad \begin{aligned} Tr_s\big[\exp(-A_u^2)\big] &\to \Big[\int_N Tr_s\big[\exp(-B^2)\big]\Big]\delta_{M'} \text{ in } P_{M'}^M \\ Tr_s\big[N_H\exp(-A_u^2)\big] &\to \Big[\int_N Tr_s\big[N_H\exp(-B^2)\big]\Big]\delta_{M'} \text{ in } P_{M'}^M. \end{aligned}$$

There exists $C > 0$ such that if μ is a smooth differential form on M, then for $u \geq 1$

$$(1.18)\quad \begin{aligned} &\Big|\int_M \mu\Big\{Tr_s\big[\exp(-A_u^2)\big] - \Big[\int_N Tr_s\big[\exp(-B^2)\big]\Big]\delta_{M'}\Big\}\Big| \\ &\qquad \leq \frac{C}{\sqrt{u}}\|\mu\|_{C^1(M)}. \\ &\Big|\int_M \mu\Big\{Tr_s\big[N_H\exp(-A_u^2)\big] - \Big[\int_N Tr_s\big[N_H\exp(-B^2)\big]\Big]\delta_{M'}\Big\}\Big| \\ &\qquad \leq \frac{C}{\sqrt{u}}\|\mu\|_{C^1(M)}. \end{aligned}$$

If U, Γ, φ, m are taken as in Section 1c), there exists $C > 0$ such that for $u \geq 1$

(1.19)

$$\begin{aligned} p_{U,\Gamma,\varphi,m}\Big(Tr_s\big[\exp(-A_u^2)\big] - \Big[\int_N Tr_s\big[\exp(-B^2)\big]\Big]\delta_{M'}\Big) &\leq \frac{C}{\sqrt{u}} \\ p_{U,\Gamma,\varphi,m}\Big(Tr_s\big[N_H\exp(-A_u^2)\big] - \Big[\int_N Tr_s\big[N_H\exp(-B^2)\big]\Big]\delta_{M'}\Big) &\leq \frac{C}{\sqrt{u}}. \end{aligned}$$

Proof. (1.17), (1.18), (1.19) are proved in [B2, Theorems 3.2 and 4.3]. □

Recall that the Todd polynomial is an ad-invariant polynomial on matrices which is such that if C is a diagonal matrix with diagonal entries $x_1, \ldots, x_p$, then

$$Td(C) = \prod_1^p \frac{x_i}{1-e^{-x_i}}. \tag{1.20}$$

Let $(Td^{-1})'$ be the ad-invariant polynomial which is such that if C is taken as before then

$$(Td^{-1})'(C) = \frac{\partial}{\partial b}\left\{\frac{\prod_1^p(1-e^{-(x_i+b)})}{x_i+b}\right\}_{b=0}. \tag{1.21}$$

Let E be a holomorphic vector bundle of dimension k on M. Let Q be an ad-invariant polynomial on (k, k) matrices. If h^E is a Hermitian metric on E, and if Ω is the curvature of the corresponding holomorphic connection, we use the notation

$$Q(h^E) = Q\left(-\frac{\Omega}{2i\pi}\right). \tag{1.22}$$

Recall that ch is the polynomial $A \to Tr[\exp(A)]$.

Our definition of φ extends to an arbitrary manifold and in particular to the total space of N.

Another result of [B2] is as follows.

Theorem 1.6. *The form on M $Tr_s[N_H \exp(-A_0^2)]$ is closed. The form on M' $\int_N Tr_s[N_H \exp(-B^2)]$ is closed.*

If the metrics $h^{\xi_0}, \ldots, h^{\xi_m}$ verify assumption (A) with respect to metrics g^N, g^η on N, η, then

$$\begin{aligned} \int_N \varphi(Tr_s[\exp(-B^2)]) &= Td^{-1}(g^N)ch(g^\eta) \\ \int_N \varphi(Tr_s[N_H \exp(-B^2)]) &= -(Td^{-1})'(g^N)ch(g^\eta). \end{aligned} \tag{1.23}$$

Proof. Clearly

$$Tr_s[N_H \exp(-A_0^2)] = \sum_0^m (-1)^k k\, Tr_s[\exp(-(\nabla^{\xi_k})^2)] \tag{1.24}$$

and so the form (1.24) is closed on M. The form $\int_N Tr_s[N_H \exp(-B^2)]$ is closed by [B2, Theorem 4.3]. The first line of (1.23) is a result of Mathai–Quillen [MQ, Theorem 4.5] of which a related proof is given in [B2, Theorem 3.2]. The second line of (1.23) is proved in [B2, Theorem 4.3]. □

(g) *A singular Bott–Chern current.* We make the same assumptions as in Section 1f). We now recall our definition [BGS4] of a singular Bott–Chern current associated with the Hermitian chain complex (ξ, v).

Definition 1.7. For $0 < \mathrm{Re}(s) < \frac{1}{2}$, $1 \leq A \leq +\infty$, let $\zeta_\xi^A(s)$ be the even current on M defined by the formula

$$\zeta_\xi^A(s) = \frac{1}{\Gamma(s)} \int_0^A u^{s-1} \Big\{ Tr_s\big[N_H \exp(-A_u^2)\big] - \Big[\int_N Tr_s\big[N_H \exp(-B^2)\big]\Big] \delta_{M'} \Big\} du. \tag{1.25}$$

By Theorem 1.5, it is clear that the current $\zeta_\xi^A(s)$ is well-defined. Also one verifies easily in [BGS4] that $\zeta_\xi^A(s)$ extends into a current depending holomorphically on s near 0.

In particular by [BGS4, Section 2a)], the current $\zeta_\xi^{A'}(0) = \dfrac{\partial \zeta_\xi^A(0)}{\partial s}$ is given by

(1.26)

$$\begin{aligned} \zeta_\xi^{A'}(0) = & \int_0^1 Tr_s\big[N_H(\exp(-A_u^2) - \exp(-A_0^2))\big] \frac{du}{u} \\ & + \int_1^A \Big\{ Tr_s\big[N_H \exp(-A_u^2)\big] \\ & - \Big[\int_N Tr_s\big[N_H \exp(-B^2)\big]\Big] \delta_{M'} \Big\} \frac{du}{u} \\ & - \Gamma'(1) \Big\{ Tr_s\big[N_H \exp(-A_0^2)\big] - \Big[\int_N Tr_s\big[N_H \exp(-B^2)\big]\Big] \delta_{M'} \Big\}. \end{aligned}$$

Of course in the case where (ξ, v) is acyclic, i.e. if M' is empty, for $A = +\infty$, the current $\zeta_\xi^{A'}(0)$ coincides with the smooth current defined in our earlier work [BGS1, Section 1c)].

Remember that the metric h^ξ on ξ is the orthogonal sum of the metrics $h^{\xi_0}, \dots, h^{\xi_m}$ on $\xi_0, \dots, \xi_m$.

Note that the map φ extends to even currents in the obvious way.

Definition 1.8. For $1 \leq A \leq +\infty$, let $T^A(h^\xi)$ be the current

$$T^A(h^\xi) = \varphi(\zeta_\xi^{A'}(0)). \tag{1.27}$$

Set

$$T(h^\xi) = T^\infty(h^\xi). \tag{1.28}$$

We define $ch(h^\xi)$ by the formula

$$ch(h^\xi) = \sum_0^m (-1)^k ch(h^{\xi_k}). \tag{1.29}$$

Theorem 1.9. *For $1 \le A \le +\infty$, the current $T^A(h^\xi)$ lies in $P_{M'}^M$. In particular $T(h^\xi) \in P_{M'}^M$, so that $WF(T(h^\xi)) \subset N_R^*$. As $A \to +\infty$*

$$T^A(h^\xi) \to T(h^\xi) \quad \text{in } P_{M'}^M. \tag{1.30}$$

Also the current $T(h^\xi)$ verifies the equation of currents

$$\frac{\bar\partial\partial}{2i\pi} T(h^\xi) = \left[\int_N \varphi(Tr_s[\exp(-B^2)])\right] \delta_{M'} - ch(h^\xi). \tag{1.31}$$

In particular, if the metrics $h^{\xi_0}, \ldots, h^{\xi_m}$ verify assumption (A) *with respect to the metrics g^N, g^η on N, η, then*

$$\frac{\bar\partial\partial}{2i\pi} T(h^\xi) = Td^{-1}(g^N) ch(g^\eta) \delta_{M'} - ch(h^\xi). \tag{1.32}$$

Proof. Theorem 1.9 is proved in [BGS4, Theorem 2.5].

Let $\tilde M$ be a compact complex manifold, let $f: \tilde M \to M$ be a holomorphic map. We assume that f is transversal to M', i.e. if $\tilde M' = f^{-1}(M')$, if $x \in \tilde M'$, then $Im[df(x)] + T_{f(x)}M' = T_x M$. Then $\tilde M'$ is a finite union of complex submanifolds of M.

Let $\tilde i$ be the embedding $\tilde M' \to \tilde M$. We still denote by f the restriction of f to $\tilde M'$. Using the local uniqueness of resolutions, we find that we have the exact sequence of sheaves

$$0 \to \mathcal{O}_{\tilde M}(f^*\xi_m) \to_{f^*v} \cdots \to_{f^*v} \mathcal{O}_{\tilde M}(f^*\xi_0) \to_{f^*r} \tilde i_* \mathcal{O}_{\tilde M'} f^*\eta \to 0. \tag{1.33}$$

Let $T(f^*h^\xi)$ be the current on $\tilde M$ constructed as before, which is associated with the Hermitian chain complex $(f^*(\xi, v), f^*h^\xi)$.

Since $WF(T(h^\xi)) \subset N_R^*$, and since f is transversal to M', by [H, Theorem 8.2.4], the pulled-back current $f^*T(h^\xi)$ on $\tilde M$ is well defined.

Theorem 1.10. *The following identity holds*

$$T(f^* h^\xi) = f^* T(h^\xi). \tag{1.34}$$

Proof. This result is proved in [BGS4, Theorem 2.7]. □

2. Metric and geometric properties of Bott–Chern currents

The purpose of this section is to establish the behaviour of the singular Bott–Chern currents $T(h^\xi)$ considered in [BGS4] and in Section 1g) under natural modifications of our datas, which are here an immersion of complex manifolds, a chain complex of holomorphic vector bundles, and metrics on these vector bundles.

In particular, we establish in this section a key transitivity property, which describes the behaviour of the singular Bott–Chern currents under composition of immersions.

Note that in the whole section, the microlocal properties of the considered currents play a key role in the formulation of the results and in the proofs themselves.

This section is organized as follows. In a) we study the dependence of the current $T(h^\xi)$ on the metrics h^ξ. In b), we establish a transivity formula for Bott–Chern currents associated with a commutative diagram of immersions. In c), we assume that our vector bundles are themselves replaced by acyclic complexes, and we study the corresponding Bott–Chern currents.

In this section, we make the same assumptions as in Section 1, and we use the same notations.

(a) *The current $T(h^\xi)$ as a function of h^ξ.* Let E be a holomorphic vector bundle on M of dimension k. Let Q be an ad-invariant polynomial on (k,k) matrices. We now use the same notations as in (1.22). Let h_0^E, h_1^E be two Hermitian metrics on E. By [BGS1, Theorem 1.29], there is a well-defined class of smooth forms $\tilde{Q}(h_0^E, h_1^E) \in P^M/P^{M,0}$ such that

$$\frac{\bar\partial\partial}{2i\pi}\tilde{Q}(h_0^E, h_1^E) = Q(h_1^E) - Q(h_0^E). \tag{2.1}$$

$\tilde{Q}(h_0^E, h_1^E)$ is the axiomatically defined Bott–Chern class of [BGS1, Section 1f)] associated with the exact sequence $0 \to (E, h_0^E) \to (E, h_1^E) \to 0$.

We now make the same assumptions as in Section 1a). Let $\mathcal{M}^\xi$ (resp. $\mathcal{M}^F$) be the set of Hermitian metrics $h^\xi = (h^{\xi_0}, \dots, h^{\xi_m})$ (resp. $h^F = (h^{F_0}, \dots, h^{F_m})$) on the vector bundles $\xi_0, \dots, \xi_m$ (resp. $F_0, \dots, F_m$).

If $h_0^\xi = (h_0^{\xi_0}, \dots, h_0^{\xi_m})$, $h_1^\xi = (h_1^{\xi_0}, \dots, h_1^{\xi_m})$ are elements of $\mathcal{M}^\xi$, set

$$\widetilde{ch}(h_0^\xi, h_1^\xi) = \sum_0^m (-1)^k \widetilde{ch}(h_0^{\xi_k}, h_1^{\xi_k}). \tag{2.2}$$

Then by (2.1)

$$\frac{\bar{\partial}\partial}{2i\pi} \widetilde{ch}(h_0^\xi, h_1^\xi) = ch(h_1^\xi) - ch(h_0^\xi). \tag{2.3}$$

Take now $h^F = (h^{F_0}, \dots, h^{F_m}) \in \mathcal{M}^F$. We equip F with the orthogonal sum of the metrics $(h^{F_k})_{0 \le k \le m}$.

Let g^N, g^η be Hermitian metrics on the vector bundles N, η. We equip $\Lambda N^* \otimes \eta$ with the tensor product of the metric induced on ΛN^* by g^N and of the metric g^η.

Definition 2.1. We will say that the metrics $h^F = (h^{F_0}, \dots, h^{F_m})$ verify assumption (A) with respect to the metrics g^N, g^η on N, η if the identification of Z-graded complexes $(F, \partial_y v) \cong (\Lambda N^* \otimes \eta, i_y)$ also identifies the metrics.

Let ∇^F be the holomorphic Hermitian connection on F. For $y \in V$, let $(\partial_y v)^*$ be the adjoint of $\partial_y v$. Set as in (1.13)

$$\partial_Y V = \partial_y v + (\partial_y v)^*. \tag{2.4}$$

Let B be the superconnection

$$B = \nabla^F + \partial_Y V. \tag{2.5}$$

Definition 2.2. Let $\theta(h^F)$ be the smooth form on M'

$$\theta(h^F) = \int_N \varphi\big(Tr_s\big[\exp(-B^2)\big]\big).$$

The form $\theta(h^F)$ is closed. By [BGS1, Proposition 1.8], $\theta(h^F) \in P^{M'}$. Also if the metrics $h^F = (h^{F_0}, \dots, h^{F_m})$ verify assumption (A) with respect to the metrics g^N, g^η, then by Theorem 1.6

$$\theta(h^F) = Td^{-1}(g^N) ch(g^\eta). \tag{2.6}$$

If $\ell \in R \to h_\ell^F = (h_\ell^{F_0}, \ldots, h_\ell^{F_m}) \in \mathcal{M}^F$ is a smooth map from R into $\mathcal{M}^F$, let $\ell \in R \to B_\ell$ be the corresponding family of superconnections on the graded vector bundle F. For any $\ell \in R$

$$(2.7)\qquad (h_\ell^F)^{-1}\frac{\partial h_\ell^F}{\partial \ell} = \left((h_\ell^{F_0})^{-1}\frac{\partial h_\ell^{F_0}}{\partial \ell}, \ldots, (h_\ell^{F_m})^{-1}\frac{\partial h_\ell^{F_m}}{\partial \ell}\right)$$

is a smooth section of End F.

Definition 2.3. Let χ be the smooth form

$$(2.8)\qquad \chi = -\int_0^1 d\ell \int_N \varphi\left(Tr_s\left[(h_\ell^F)^{-1}\frac{\partial h^F}{\partial \ell}\exp(-B_\ell^2)\right]\right) d\ell.$$

Theorem 2.4. *The form χ lies in $P^{M'}$ and its class in $P^{M'}/P^{M',0}$ only depends on h_0^F, h_1^F. Moreover*

$$(2.9)\qquad \frac{\bar{\partial}\partial}{2i\pi}\chi = \theta(h_1^F) - \theta(h_0^F).$$

If the metrics h_0^F and h_1^F verify assumption (A) *with respect to metrics (g_0^N, g_0^η) and (g_1^N, g_1^η) on N, η, then*

$$(2.10)\quad \chi = \widetilde{Td^{-1}}(g_0^N, g_1^N)ch(g_0^\eta) + Td^{-1}(g_1^N)\widetilde{ch}(g_0^\eta, g_1^\eta) \ \ in\ P^{M'}/P^{M',0}.$$

Proof. The proof of (2.9) follows from the extension to superconnections of a formula of Bott and Chern [BoC, 3.28]. This extension was proved in Bismut [B2, Theorem 2.1]. Using this formula, the proof of (2.9) is strictly identical to the proof of [BGS1, Theorem 1.27]. The fact that the class of χ in $P^{M'}/P^{M',0}$ only depends on h_0^F and h_1^F follows from the analogue of the formulas in [BGS1, Theorem 1.25] for superconnections which was proved in [B2, Theorem 2.2]. Of course the formulas in [B2] must be integrated along the fibre N. We can then proceed as in [BGS1, Section 1e)].

If h_0^F, h_1^F verify assumption (A) with respect to (g_0^N, g_0^η) and (g_1^N, g_1^η), we can find a smooth family of metrics $\ell \to (g_\ell^N, g_\ell^\eta)$ on N, η which interpolates between (g_0^N, g_0^η), and (g_1^N, g_1^η). If h_ℓ^F is the metric on F associated with the metric (g_ℓ^N, g_ℓ^η), the family $\ell \to h_\ell^F$ interpolates between h_0^F and h_1^F.

The operator $(g_\ell^N)^{-1}\frac{\partial g_\ell^N}{\partial \ell}$ acts naturally on the exterior algebra ΛN^*. One verifies easily that since $F = \Lambda N^* \otimes \eta$,

$$(h_\ell^F)^{-1}\frac{\partial h_\ell^F}{\partial \ell} = (g_\ell^N)^{-1}\frac{\partial g_\ell^N}{\partial \ell} \otimes 1 + 1 \otimes (g_\ell^\eta)^{-1}\frac{\partial g_\ell^\eta}{\partial \ell}. \tag{2.11}$$

Let $(\nabla_\ell^N)^2$, $(\nabla_\ell^\eta)^2$ be the curvatures of the holomorphic Hermitian connections on the holomorphic Hermitian vector bundles (N, g_ℓ^N), (η, g_ℓ^η). By proceeding as in Mathai–Quillen [MQ, Theorem 4.5], Bismut [B2, Theorem 3.2], we find that

$$\begin{aligned}
-\int_N Tr_s\left[(h_\ell^F)\frac{\partial h_\ell^F}{\partial \ell}\exp(-B_\ell^2)\right] &= (2i\pi)^{\dim N}\frac{\partial}{\partial b}\Big\{Td^{-1}\Big(-(\nabla_\ell^N)^2 \\
&- b(g_\ell^N)^{-1}\frac{\partial g_\ell^N}{\partial \ell}\Big)Tr\left[\exp(-(\nabla_\ell^\eta)^2)\right] \\
&+ Td^{-1}(-(\nabla_\ell^N)^2)Tr\left[\exp\Big(-(\nabla_\ell^\eta)^2 - b(g_\ell^\eta)^{-1}\frac{\partial g_\ell^\eta}{\partial \ell}\Big)\right]\Big\}_{b=0}.
\end{aligned} \tag{2.12}$$

Using (2.8), (2.12) and also [BGS1, Remark 1.28, Corollary 1.30 and Remark 1.31] we find that

$$\chi = \widetilde{Td^{-1} \otimes ch}((g_0^N, g_0^\eta), (g_1^N, g_1^\eta)) \quad \text{in } P^{M'}/P^{M',0}. \tag{2.13}$$

Our theorem is proved. □

We now will write $\chi(h_0^F, h_1^F)$ instead of χ the class of forms in $P^{M'}/P^{M',0}$ defined in (2.8).

Let $h_0^\xi = (h_0^{\xi_0}, \dots, h_0^{\xi_m})$, $h_1^\xi = (h_1^{\xi_0}, \dots, h_1^{\xi_m})$ be two elements of $\mathcal{M}^\xi$. Let $h_0^F = (h_0^{F_0}, \dots, h_0^{F_m})$, $h_1^F = (h_1^{F_0}, \dots, h_1^{F_m})$ be the elements of $\mathcal{M}^F$ respectively associated with h_0^ξ, h_1^ξ by the construction of Section 1b).

Theorem 2.5. *The following identity holds*

$$T(h_1^\xi) - T(h_0^\xi) = \chi(h_0^F, h_1^F)\delta_{M'} - \widetilde{ch}(h_0^\xi, h_1^\xi) \quad \textit{in } P_{M'}^M / P_{M'}^{M,0}. \tag{2.14}$$

Proof. Let $\mathbf{P}^1$ be the complex projective plane equipped with the meromorphic coordinate z and the two distinguished points $\{0\}$ and $\{\infty\}$.

The complex (ξ, v) lifts naturally to $M \times \mathbf{P}^1$. On $M \times \mathbf{P}^1$, we equip $(\xi^0, \dots, \xi^m)$ with Hermitian metrics $h^\xi = (h^{\xi_0}, \dots, h^{\xi_m})$ which restrict to the metrics $(h_0^{\xi_0}, \dots, h_0^{\xi_m})$ and $(h_1^{\xi_0}, \dots, h_1^{\xi_m})$ on $M \times \{0\}$ and $M \times \{\infty\}$ respectively.

The vector bundle N_R^* on M' lifts naturally to $M' \times \mathbf{P}^1$. By Theorem 1.9, the wave front set $WF(T(h^\xi))$, of the current $T(h^\xi)$ on $M \times \mathbf{P}^1$ is included in N_R^*. Also the wave front set $WF(\mathrm{Log}|z|^2)$ of the distribution $\mathrm{Log}|z|^2$ on $\mathbf{P}^1$ is equal to $(T_R^*\mathbf{P}^1 \setminus \{0\})_0 \cup (T_R^*\mathbf{P}^1 \setminus \{0\})_\infty$. The current $\mathrm{Log}|z|^2$ lifts to $M \times \mathbf{P}^1$, and its wave front set $WF(\mathrm{Log}|z|^2)$ also lifts in the obvious way.

Since $WF(T(h^\xi)) \cap (-WF(\mathrm{Log}|z|^2)) = \emptyset$, by [H, Theorem 8.2.10], the product of currents $\mathrm{Log}|z|^2 T(h^\xi)$ is well defined. Also the usual rules of differential calculus can still be used. In particular

$$(2.15)\quad \begin{aligned}\frac{\bar\partial\partial}{2i\pi}(\mathrm{Log}|z|^2)T(h^\xi) - \mathrm{Log}|z|^2\frac{\bar\partial\partial}{2i\pi}T(h^\xi) &= \frac{\bar\partial}{2i\pi}(\partial(\mathrm{Log}|z|^2)T(h^\xi)) \\ &+ \frac{\partial}{2i\pi}(\mathrm{Log}|z|^2\bar\partial T(h^\xi)).\end{aligned}$$

It is well-known that

$$(2.16)\quad \frac{1}{2i\pi}\bar\partial\partial\,\mathrm{Log}|z|^2 = \delta_0 - \delta_\infty.$$

Let $h^F = (h^{F_0}, \dots, h^{F_m})$ be the metrics on the vector bundles $F_0, \dots, F_m$ on $M' \times \mathbf{P}^1$ induced by the metrics $h^\xi = (h^{\xi_0}, \dots, h_0^{\xi_m})$. Using Theorem 1.9, and (2.15), (2.16) we find that

$$(2.17)\quad \begin{aligned}&T(h^\xi)\delta_{M\times\{0\}} - T(h^\xi)\delta_{M\times\{\infty\}} - \mathrm{Log}|z|^2(\theta(h^F)\delta_{M'\times\mathbf{P}^1} - ch(h^\xi)) \\ &= \frac{\bar\partial}{2i\pi}(\partial(\mathrm{Log}|z|^2)T(h^\xi)) + \frac{\partial}{2i\pi}(\mathrm{Log}|z|^2\bar\partial T(h^\xi)).\end{aligned}$$

We now integrate (2.17) along the fiber of $M \times \mathbf{P}^1 \to M$. By [H, Theorem 8.2.10], the wave front sets of the currents $\partial(\mathrm{Log}|z|^2)T(h^\xi)$ and $\mathrm{Log}|z|^2\bar\partial T(h^\xi)$ are included in the sum of the duals of the real normal bundles to $M', M \times \{0\}, M \times \{\infty\}$ in $M \times \mathbf{P}^1$. By [H, Theorem 8.2.13], the wave front set of their integral along the fibers is included in N_R^*. Also ∂ and $\bar\partial$ commute with integration along the fiber.

From (2.17), we deduce in particular that

$$(2.18)\quad T(h_0^\xi) - T(h_1^\xi) - \left[\int_{\mathbf{P}^1}\mathrm{Log}|z|^2\theta(h^F)\right]\delta_{M'} + \int_{\mathbf{P}^1}\mathrm{Log}|z|^2 ch(h^\xi) \in P_{M'}^{M,0}.$$

By [BGS1, Theorem 1.29], we know that

$$(2.19)\quad \int_{\mathbf{P}^1}\mathrm{Log}|z|^2 ch(h^\xi) = -\widetilde{ch}(h_0^\xi, h_1^\xi) \quad \text{in } P^M/P^{M,0}.$$

The metric h_0^F lifts into a Hermitian metric on the vector bundle F on $M' \times \mathbf{P}^1$. Clearly

$$\begin{aligned}(2.20)\qquad \frac{\bar\partial\partial}{2i\pi}(\mathrm{Log}|z|^2)\chi(h_0^F,h^F) \quad &- \quad \mathrm{Log}|z|^2\frac{\bar\partial\partial}{2i\pi}\chi(h_0^F,h^F)\\ &= \frac{\bar\partial}{2i\pi}(\partial\,\mathrm{Log}|z|^2\chi(h_0^F,h^F))\\ &+ \frac{\partial}{2i\pi}(\mathrm{Log}|z|^2\bar\partial\chi(h_0^F,h^F)).\end{aligned}$$

By Theorem 2.4, we know that

$$(2.21)\qquad \frac{\bar\partial\partial}{2i\pi}\chi(h_0^F,h^F)=\theta(h^F)-\theta(h_0^F).$$

Observe that since h^F coincides with h_0^F on $M\times\{0\}$, the restriction of $\chi(h_0^F,h^F)$ to $M'\times\{0\}$ vanishes in $P^{M'}/P^{M',0}$. Using (2.16), (2.21) and integrating (2.20) along the fiber of $M'\times\mathbf{P}^1\to M'$, we get

$$(2.22)\qquad \chi(h_0^F,h_1^F)+\int_{\mathbf{P}^1}\mathrm{Log}|z|^2\theta(h^F)\in P^{M',0}.$$

From (2.18)–(2.22), we deduce (2.14). □

(b) *A transitivity property of Bott–Chern singular currents.* We make the same assumptions as in Section 1a).

Let $\tilde M' = \bigcup_{j'=1}^{n'}\tilde M'_{j'}$, be another finite union of compact connected complex submanifolds of M such that if $j\neq j'$, $\tilde M'_j\cap\tilde M'_{j'}=\emptyset$. Let $\tilde i$ be the embedding $\tilde M'\to M$.

We otherwise assume that to the pair M, $\tilde M'$, we have associated the analogue of the objects which were associated with the pair M, M'. These objects will be denoted with a $\sim$. In particular:

- $(\tilde\xi,\tilde v)$ is a complex of holomorphic vector bundles on M

$$(2.23)\qquad (\tilde\xi,\tilde v):0\to\tilde\xi_{\tilde m}\to_{\tilde v}\cdots\to_{\tilde v}\tilde\xi_0\to 0.$$

- $\tilde\eta$ is a holomorphic vector bundle on $\tilde M'$.
- There exists a holomorphic restriction map $\tilde r:\tilde\xi_{0|\tilde M'}\to\tilde\eta$ such that we have the exact sequence of sheaves on M

$$(2.24)\qquad 0\to\mathcal{O}_M(\tilde\xi_{\tilde m})\to_{\tilde v}\cdots\to\mathcal{O}_M(\tilde\xi_0)\to_{\tilde r}\tilde i_*\,\mathcal{O}_{\tilde M'}(\tilde\eta)\to 0.$$

$\tilde{N}$ denotes the normal bundle to $\tilde{M}'$ in M.

We now make the fundamental assumption that M' and $\tilde{M}'$ intersect transversally, i.e. if $x \in M' \cap \tilde{M}'$, then $T_xM' + T_x\tilde{M}' = T_xM$.

Let M'' be the complex submanifold of M $M'' = M' \cap \tilde{M}'$. Then if $x \in M''$, $T_xM'' = T_xM' \cap T_x\tilde{M}'$. Let i'' be the embedding $M'' \to M$. Let $j, \tilde{j}$ be the embeddings $j: M'' \to M'$, $\tilde{j}: M'' \to \tilde{M}'$.

We will denote with a $''$ the objects naturally associated with the pair M, M''. In particular, N'' is the normal bundle to M'' in M. Observe that since M' and $\tilde{M}'$ are transversal, the vector bundle N'' splits holomorphically into

$$N'' = N_{|M''} \oplus \tilde{N}_{|M''}. \tag{2.25}$$

Also $N_{|M''}$ (resp. $\tilde{N}_{|M''}$) is exactly the normal bundle to M'' in $\tilde{M}'$ (resp. in M').

Let (ξ'', v'') be the double complex

$$(\xi'', v'') = (\xi \mathbin{\hat{\otimes}} \tilde{\xi}, v \mathbin{\hat{\otimes}} 1 + 1 \mathbin{\hat{\otimes}} \tilde{v}).$$

Then

• ξ'' is the graded tensor product of the graded vector bundles ξ and $\tilde{\xi}$, and so for $0 \leq k \leq m + \tilde{m}$

$$\xi''_k = \bigoplus_{j+j'=k} (\xi_j \mathbin{\hat{\otimes}} \tilde{\xi}_{j'}) \tag{2.26}$$

• $v \mathbin{\hat{\otimes}} 1$ acts on $\xi \mathbin{\hat{\otimes}} \tilde{\xi}$ like $v \otimes 1$, and moreover, if $f \in \xi_j$, $\tilde{f} \in \tilde{\xi}_{j'}$, then

$$(1 \mathbin{\hat{\otimes}} \tilde{v})(f \mathbin{\hat{\otimes}} f) = (-1)^{\deg f} f \mathbin{\hat{\otimes}} \tilde{v}\tilde{f}. \tag{2.27}$$

One easily verifies that $v''^2 = 0$. To simplify our notations, we will consider v and $\tilde{v}$ as odd anticommuting elements in $\operatorname{End} \xi''$, acting on ξ'' like $v \mathbin{\hat{\otimes}} 1$ and $1 \mathbin{\hat{\otimes}} \tilde{v}$ so that $v'' = v + \tilde{v}$.

Let $\tilde{\eta}$ be the holomorphic vector bundle on M'' $\eta'' = \eta_{|M''} \otimes \tilde{\eta}_{|M''}$. In the sequel, it will be convenient to consider η, $\tilde{\eta}$ as even vector bundles, so that $\eta'' = \eta_{|M''} \mathbin{\hat{\otimes}} \tilde{\eta}_{|M''}$.

Let r'' be the restriction map

$$r'': \xi''_{0|M''} = (\xi_0 \mathbin{\hat{\otimes}} \tilde{\xi}_0)\big|_{M''} \xrightarrow[r \hat{\otimes} r']{} \eta''. \tag{2.28}$$

Using again the local uniqueness of resolutions (see Serre [S, Chapter IV, Appendix 1], Eilenberg [E, Theorem 8]) one verifies easily that we have the exact sequence of sheaves

$$
\begin{aligned}
&0 \to \mathcal{O}_M(\xi'', v'') \to_{r''} i''_* \mathcal{O}_{M''}(\eta'') \to 0 \\
(2.29) \qquad &0 \to \mathcal{O}_{\tilde{M}'}((\xi, v)_{|\tilde{M}'}) \to_{r} \tilde{j}_* \mathcal{O}_{M''}(\eta_{|M''}) \to 0 \\
&0 \to \mathcal{O}_{M'}((\tilde{\xi}, \tilde{v})_{|M'}) \to_{r'} j_* \mathcal{O}_{M''}(\eta'_{|M''}) \to 0.
\end{aligned}
$$

By [B2, Section 1b)], we know that if F (resp. $\tilde{F}$, F'') is the direct sum of the homology groups of the complex $(\xi, v)_{|M'}$ (resp. $(\tilde{\xi}, \tilde{v})_{|\tilde{M}'}$, $(\xi'', v'')_{|M''}$), then F (resp. $\tilde{F}$, F'') is a Z-graded holomorphic vector bundle. Moreover we have the identification of holomorphic vector bundles

$$(2.30) \qquad F'' = F_{|M''} \hat{\otimes} \tilde{F}_{|M''}.$$

In the sequel, we always equip the direct sum of Hermitian vector bundles with the orthogonal sum of the considered metrics, and tensor products of vector bundles with the tensor product of the Hermitian metrics.

Let now $h^{\xi} = (h^{\xi_0}, \dots, h^{\xi_m}) \in \mathcal{M}^{\xi}$, $h^{\tilde{\xi}} = (h^{\tilde{\xi}_0}, \dots, h^{\tilde{\xi}_{\tilde{m}}}) \in \mathcal{M}^{\tilde{\xi}}$. Let $h^{\xi''} = (h^{\xi''_0}, \dots, h^{\xi''_{m+\tilde{m}}}) \in \mathcal{M}^{\xi''}$ be the corresponding family of Hermitian metrics on $\xi''_0, \dots, \xi''_{m+\tilde{m}}$. Let h^F (resp. $h^{\tilde{F}}$, $h^{F''}$) be the family of metrics on F (resp. $\tilde{F}$, F'') induced by h^{ξ} (resp. $h^{\tilde{\xi}}$, $h^{\xi''}$). Then (2.30) is an identification of holomorphic Hermitian vector bundles.

Note that if the metrics h^{ξ} and $h^{\tilde{\xi}}$ verify assumption (A) with respect to metrics (g^N, g^{η}) and $(g^{\tilde{N}}, g^{\tilde{\eta}})$ on N, η and $\tilde{N}, \tilde{\eta}$, then $h^{\xi''}$ verifies assumption (A) with respect to the metrics $(g^N \oplus g^{\tilde{N}}, g^{\eta} \otimes g^{\tilde{\eta}})$ on (N'', η'').

We now construct the currents $T(h^{\xi})$, $T(h^{\tilde{\xi}})$, $T(h^{\xi''})$ associated with the holomorphic Hermitian chain complexes $(\xi, v), (\tilde{\xi}, \tilde{v}), (\xi'', v'')$. By [BGS4, Theorem 2.5] or by Theorem 1.9, we know that

$$(2.31) \qquad WF(T(h^{\xi})) \subset N^*_R; WF(T(h^{\tilde{\xi}})) \subset \tilde{N}^*_R; WF(T(h^{\xi''})) \subset N''^*_R.$$

Recall that M' and $\tilde{M}'$ are transversal. Using (2.31) and [H, Theorem 8.2.4], we know that the pulled back currents $i^* T(h^{\tilde{\xi}})$ and $\tilde{i}^* T(h^{\xi})$ on M' and $\tilde{M}'$ are well-defined.

By (2.29), the complex $(\xi, v)_{|\tilde{M}}$ provides a resolution on $\tilde{M}'$ of the sheaf of holomorphic sections of $\eta_{|M''}$. Therefore, if we denote by $\tilde{i}^* h^{\xi}$ the family of Hermitian metrics induced by h^{ξ} on $\xi_{|\tilde{M}'}$, we can define the current $T(\tilde{i}^* h^{\xi})$ on $\tilde{M}'$. Similarly, we also consider the current $T(i^* h^{\tilde{\xi}})$ on M'.

By [BGS4, Theorem 2.7] or by Theorem 1.10, we know that

$$
\begin{aligned}
T(\tilde{i}^* h^{\xi}) &= \tilde{i}^* T(h^{\xi}) \\
T(i^* h^{\tilde{\xi}}) &= i^* T(h^{\tilde{\xi}}).
\end{aligned}
\tag{2.32}
$$

The push-forward of a current by a smooth map is always well defined. Then using the fact that M' and $\tilde{M}'$ intersect transversally, (2.31) and [H, Theorem 8.2.10], we find that the product currents $T(h^{\xi})\delta_{M'}$, and $T(h^{\tilde{\xi}})\delta_{\tilde{M}'}$ are well defined. One easily verifies that

$$
\begin{aligned}
T(h^{\tilde{\xi}})\delta_{M'} &= i_*\left(i^* T(h^{\tilde{\xi}})\right) \\
T(h^{\xi})\delta_{\tilde{M}'} &= \tilde{i}_*\left(\tilde{i}^* T(h^{\xi})\right).
\end{aligned}
\tag{2.33}
$$

Note the commutative diagram of immersions

$$
\begin{array}{ccc}
M'' & \xrightarrow{\ j\ } & M' \\
\downarrow \tilde{j} & \searrow^{i''} & \downarrow i \\
\tilde{M}' & \xrightarrow[\tilde{i}]{} & M
\end{array}
\tag{2.34}
$$

We now will prove that to (2.34), we can associate a corresponding transitivity result for singular Bott–Chern currents.

If $x \in M' \cup \tilde{M}'$, set

$$
\begin{aligned}
(N_R^* + \tilde{N}_R^*)_x &= N_R^* \quad \text{if } x \in M' \setminus M'' \\
&= \tilde{N}_R^* \quad \text{if } x \in \tilde{M}' \setminus M'' \\
&= \tilde{N}''^*_R \quad \text{if } x \in M''.
\end{aligned}
\tag{2.35}
$$

Definition 2.6. $\mathcal{D}'_{N_R^* + \tilde{N}_R^*}(M)$ denotes the set of currents on M whose wave front set is included in $N_R^* + \tilde{N}_R^*$. $P^M_{M' \cup \tilde{M}'}$ is the set of currents ω on M which have the following two properties.

- ω is a sum of currents of type (p, p).
- The wave front set of ω is included in $N_R^* + \tilde{N}_R^*$.

$P^{M,0}_{M' \cup \tilde{M}'}$ is the set of currents $\omega \in P^M_{M' \cup \tilde{M}'}$, which are such that there exist currents $\alpha, \beta \in \mathcal{D}'_{N_R^* + \tilde{N}_R^*}(M)$ for which $\omega = \partial\alpha + \bar{\partial}\beta$.

Theorem 2.7. *The following identities hold*

$$\text{(2.36)}\qquad \begin{aligned} T(h^{\xi''}) &= ch(h^{\tilde{\xi}})T(h^{\xi}) + i_*\big(\theta(h^F)T(i^*h^{\tilde{\xi}})\big) \ \text{ in } P^{M}_{M'\cup\tilde{M}'}/P^{M,0}_{M'\cup\tilde{M}'} \\ T(h^{\xi''}) &= ch(h^{\xi})T(h^{\tilde{\xi}}) + \tilde{i}_*\big(\theta(h^{\tilde{F}})T(\tilde{i}^*h^{\xi})\big) \ \text{ in } P^{M}_{M'\cup\tilde{M}'}/P^{M,0}_{M'\cup\tilde{M}'}. \end{aligned}$$

Proof. We first show that the equalities in (2.36) make sense. By [BGS4, Theorem 2.5] or by Theorem 1.8, we know that the wave front set of the current $i^*T(h^{\tilde{\xi}}) = T(i^*h^{\tilde{\xi}})$ is included in the real conormal bundle $\tilde{N}^*_R$ to M'' in M'. The current $\theta(h^F)$ on M' is smooth. By [H, Theorem 8.2.13], we find that the wave front set of the current $i_*\big(\theta(h^F)i^*T(h^{\tilde{\xi}})\big)$ is included in $N^*_R + \tilde{N}^*_R$. So the wave front set of the current $ch(h^{\tilde{\xi}})T(h^{\xi}) + i_*\big(\theta(h^F)i^*T(h^{\tilde{\xi}})\big)$ is included in $N^*_R + \tilde{N}^*_R$. Therefore this current lies in $P^M_{M'\cup\tilde{M}'}$. The first line in (2.36) is then an identity of currents in $P^M_{M'\cup\tilde{M}'}/P^{M,0}_{M'\cup\tilde{M}'}$. The same is true for the second line in (2.36).

We now prove (2.36). To the Hermitian chain complexes (ξ, v), $(\tilde{\xi}, \tilde{v})$, (ξ'', v''), we associate linear maps V, $\tilde{V}$, V'' as in (1.9). For $u \geq 0$, let A_u, $\tilde{A}_u$, A''_u be the superconnections on the vector bundles ξ, $\tilde{\xi}$, ξ''

$$\text{(2.37)}\qquad \begin{aligned} A_u &= \nabla^{\xi} + \sqrt{u}\,V \\ \tilde{A}_u &= \nabla^{\tilde{\xi}} + \sqrt{u}\,\tilde{V} \\ A''_u &= \nabla^{\xi''} + \sqrt{u}\,V''. \end{aligned}$$

Let N_H, $\tilde{N}_H$, N''_H be the operators defining the Z-grading of the complexes (ξ, v), $(\tilde{\xi}, \tilde{v})$, (ξ'', v''). Clearly

$$\text{(2.38)}\qquad N''_H = N_H \,\hat{\otimes}\, 1 + 1 \,\hat{\otimes}\, \tilde{N}_H.$$

For $0 \leq u < +\infty$, set

$$\text{(2.39)}\qquad \begin{aligned} \alpha_u &= Tr_s\big[\exp(-A_u^2)\big] \quad ; \quad \beta_u = Tr_s\big[N_H \exp(-A_u^2)\big] \\ \tilde{\alpha}_u &= Tr_s\big[\exp(-\tilde{A}_u^2)\big] \quad ; \quad \tilde{\beta}_u = Tr_s\big[\tilde{N}_H \exp(-\tilde{A}_u^2)\big] \\ \alpha''_u &= Tr_s\big[\exp(-A_u''^2)\big] \quad ; \quad \beta''_u = Tr_s\big[N''_H \exp(-A_u''^2)\big]. \end{aligned}$$

By the construction of the metrics $h^{\xi''} \in \mathcal{M}^{\xi''}$ and by (2.38), one finds easily that

$$\text{(2.40)}\qquad \begin{aligned} \alpha''_u &= \alpha_u \tilde{\alpha}_u \\ \beta''_u &= \alpha_u \tilde{\beta}_u + \tilde{\alpha}_u \beta_u. \end{aligned}$$

By [B2, Theorems 3.2 and 4.3], or by Theorem 1.5, we know that as $u \to +\infty$, the currents α_u, β_u, $\tilde{\alpha}_u$, $\tilde{\beta}_u$, α''_u, β''_u have limits α_∞, β_∞, $\tilde{\alpha}_\infty$, $\tilde{\beta}_\infty$, α''_∞, β''_∞ which are explicitly known. By [B2, Theorem 4.3], or by Theorem 1.6, we know that the currents β_0, β_∞, $\tilde{\beta}_0$, $\tilde{\beta}_\infty$, β''_0, β''_∞, are closed. Set

$$
\begin{aligned}
&\bullet \text{ for } u \leq 1 \quad \tilde{\eta}_u = \int_0^u (\tilde{\beta}_v - \tilde{\beta}_0)\frac{dv}{v} \\
&\bullet \text{ for } 1 \leq u \leq +\infty \quad \tilde{\eta}_u = \int_0^1 (\tilde{\beta}_v - \tilde{\beta}_0)\frac{dv}{v} + \int_1^u (\tilde{\beta}_v - \tilde{\beta}_\infty)\frac{dv}{v}.
\end{aligned} \tag{2.41}
$$

By Theorem 1.9, for any $u \in [0, +\infty]$, $\tilde{\eta}_u \in \mathcal{D}'_{\tilde{N}^*_R}(M)$ and the map $u \in [0, +\infty] \to \tilde{\eta}_u \in \mathcal{D}'_{\tilde{N}^*_R}(M)$ is continuous.

By [BGS1, Theorem 1.15], [B2, Theorem 2.4] or by Theorem 1.4, we know that

$$
\frac{\partial}{\partial u}\tilde{\alpha}_u = \frac{1}{u}\bar{\partial}\partial\,\tilde{\beta}_u. \tag{2.42}
$$

Since $\tilde{\beta}_0$ and $\tilde{\beta}_\infty$ are closed currents, we deduce from (2.42) that for $0 \leq u < +\infty$

$$
\tilde{\alpha}_u = \tilde{\alpha}_0 + \bar{\partial}\partial\,\tilde{\eta}_u. \tag{2.43}
$$

Also by (2.40) we have

$$
\beta''_u - \beta''_0 = \tilde{\alpha}_u\beta_u - \tilde{\alpha}_0\beta_0 + \alpha_u\tilde{\beta}_u - \alpha_0\tilde{\beta}_0. \tag{2.44}
$$

Using (2.43), (2.44), we find that

$$
\begin{aligned}
\beta''_u - \beta''_0 &= \tilde{\alpha}_0(\beta_u - \beta_0) + \alpha_u\tilde{\beta}_u - \alpha_0\tilde{\beta}_0 \\
&\quad + \left(u\frac{\partial}{\partial u}\alpha_u\right)\tilde{\eta}_u - \bar{\partial}(\partial(\beta_u)\tilde{\eta}_u) - \partial(\beta_u\bar{\partial}\tilde{\eta}_u).
\end{aligned} \tag{2.45}
$$

Therefore from (2.45), we deduce that

$$
\begin{aligned}
\int_0^1 (\beta''_u - \beta''_0)\frac{du}{u} &= \tilde{\alpha}_0\int_0^1 (\beta_u - \beta_0)\frac{du}{u} + \alpha_1\tilde{\eta}_1 \\
&\quad + \tilde{\beta}_0\int_0^1 (\alpha_u - \alpha_0)\frac{du}{u} \\
&\quad - \bar{\partial}\int_0^1 \partial(\beta_u)\tilde{\eta}_u\frac{du}{u} \\
&\quad - \partial\int_0^1 \beta_u\bar{\partial}\,\tilde{\eta}_u\frac{du}{u}.
\end{aligned} \tag{2.46}
$$

Clearly, the integrals in (2.46) define smooth currents on M.

We can define currents η_u associated to the family of currents β_u by a formula similar to (2.41). In particular the analogue of (2.43) is

$$\alpha_u = \alpha_0 + \bar{\partial}\partial\,\eta_u. \tag{2.47}$$

Since $\tilde{\beta}_0$ is closed, we deduce from (2.46), (2.47), that

(2.48)
$$\int_0^1 (\beta_u'' - \beta_0'')\frac{du}{u} = \tilde{\alpha}_0 \int_0^1 (\beta_u - \beta_0)\frac{du}{u} + \alpha_1\tilde{\eta}_1 + \bar{\partial}\partial\left(\tilde{\beta}_0 \int_0^1 \eta_u \frac{du}{u}\right)$$
$$- \bar{\partial}\int_0^1 \partial(\beta_u)\tilde{\eta}_u \frac{du}{u} - \partial\int_0^1 \beta_u \bar{\partial}\,\tilde{\eta}_u \frac{du}{u}.$$

Remember that by [H, Theorem 8.2.10], if $\omega \in \mathcal{D}'_{N^*_R}(M)$, $\tilde{\omega} \in \mathcal{D}'_{\tilde{N}^*_R}(M)$, since $N^*_{R|M''} \cap \tilde{N}^*_{R|M''} = \{0\}$, we can form the product $\omega\tilde{\omega} \in \mathcal{D}'_{N^*_R+\tilde{N}^*_R}(M)$, and the map $(\omega,\tilde{\omega}) \in \mathcal{D}'_{N^*_R}(M) \times \mathcal{D}'_{\tilde{N}^*_R}(M) \to \omega\tilde{\omega} \in \mathcal{D}'_{N^*_R+\tilde{N}^*_R}(M)$ is continuous. Also the usual rules of differential calculus can be used on such products.

In view of (2.30), it is clear that

$$\begin{aligned} \alpha''_\infty &= \alpha_\infty\tilde{\alpha}_\infty \\ \beta''_\infty &= \alpha_\infty\tilde{\beta}_\infty + \tilde{\alpha}_\infty\beta_\infty. \end{aligned} \tag{2.49}$$

Remember that by Theorem 1.5, as $u \to +\infty$, α_u, β_u converge to α_∞, β_∞ in $\mathcal{D}'_{N^*_R}(M)$, $\tilde{\alpha}_u$, $\tilde{\beta}_u$ converge to $\tilde{\alpha}_\infty$, $\tilde{\beta}_\infty$ in $\mathcal{D}'_{\tilde{N}^*_R}(M)$. So (2.49) can be considered as a consequence of (2.40) and of the previous considerations on the products of currents.

By proceeding formally as in (2.44) and using (2.43) and (2.47), we find that

$$\begin{aligned} \beta''_u - \beta''_\infty &= \tilde{\alpha}_0(\beta_u - \beta_\infty) + \alpha_u\tilde{\beta}_u - \alpha_\infty\tilde{\beta}_\infty \\ &+ u\left(\frac{\partial}{\partial u}\alpha_u\right)\tilde{\eta}_u + (\tilde{\alpha}_0 - \tilde{\alpha}_\infty)\beta_\infty \\ &- \bar{\partial}((\partial\beta_u)\tilde{\eta}_u) - \partial(\beta_u\bar{\partial}\,\tilde{\eta}_u). \end{aligned} \tag{2.50}$$

Remember that by Theorem 1.6, the current β_∞ is closed, so that using (2.43)

$$\begin{aligned} (\tilde{\alpha}_0 - \tilde{\alpha}_\infty)\beta_\infty &= -(\bar{\partial}\partial\,\tilde{\eta}_\infty)\beta_\infty = \partial((\bar{\partial}\,\tilde{\eta}_\infty)\beta_\infty) \\ \bar{\partial}((\partial\beta_u)\tilde{\eta}_u) &= \bar{\partial}(\partial(\beta_u - \beta_\infty)\tilde{\eta}_u). \end{aligned} \tag{2.51}$$

From (2.50), (2.51), we deduce that for any $A \in [1, +\infty[$

$$\begin{aligned}\int_1^A (\beta''_u - \beta''_\infty)\frac{du}{u} &= \tilde{\alpha}_0 \int_1^A (\beta_u - \beta_\infty)\frac{du}{u} + \tilde{\alpha}_A \tilde{\eta}_A - \tilde{\alpha}_1 \tilde{\eta}_1 \\ &\quad + \int_1^A (\alpha_u - \alpha_\infty)\tilde{\beta}_\infty \frac{du}{u} \\ &\quad - \partial \int_1^A (\beta_u \bar{\partial}\tilde{\eta}_u - \beta_\infty \bar{\partial}\tilde{\eta}_\infty)\frac{du}{u} \\ &\quad - \bar{\partial} \int_1^A \partial(\beta_u - \beta_\infty)\tilde{\eta}_u \frac{du}{u}.\end{aligned} \tag{2.52}$$

Using (2.47) and the fact that the current $\tilde{\beta}_\infty$ is closed, we find that if $A \in [1, +\infty[$

$$\int_1^A (\alpha_u - \alpha_\infty)\tilde{\beta}_\infty \frac{du}{u} = \bar{\partial}\partial \int_1^A (\eta_u - \eta_\infty)\tilde{\beta}_\infty \frac{du}{u}. \tag{2.53}$$

We make $A \to +\infty$ in (2.52). By Theorem 1.9, we have

$$\begin{aligned}\int_1^A (\beta''_u - \beta''_\infty)\frac{du}{u} &\to \int_1^{+\infty} (\beta''_u - \beta''_\infty)\frac{du}{u} \text{ in } \mathcal{D}'_{N''^{\bullet}_R}(M). \\ \tilde{\alpha}_0 \int_1^A (\beta_u - \beta_\infty)\frac{du}{u} &\to \tilde{\alpha}_0 \int_1^{+\infty} (\beta_u - \beta_\infty)\frac{du}{u} \text{ in } \mathcal{D}'_{N^{\bullet}_R}(M).\end{aligned} \tag{2.54}$$

Also, again using Theorem 1.9 and the properties of the product of currents explained after equation (2.48), we find that as $A \to +\infty$

$$\begin{aligned}\alpha_A \tilde{\eta}_A &\to \alpha_\infty \tilde{\eta}_\infty \text{ in } \mathcal{D}'_{N^{\bullet}_R + \tilde{N}^{\bullet}_R}(M). \\ \int_1^A (\alpha_u - \alpha_\infty)\tilde{\beta}_\infty \frac{du}{u} &\to \int_1^{+\infty} (\alpha_u - \alpha_\infty)\tilde{\beta}_\infty \frac{du}{u} \text{ in } \mathcal{D}'_{N^{\bullet}_R + \tilde{N}^{\bullet}_R}(M).\end{aligned} \tag{2.55}$$

In the sequel, the constant C may vary from line to line.

Using (1.18) and (2.41), we find that if μ is a smooth differential form on M, then if $u \geq 1$

$$\left| \int_M \mu(\eta_u - \eta_\infty) \right| \leq C \|\mu\|_{C^1(M)} \int_u^{+\infty} \frac{dv}{v^{3/2}}$$

or equivalently that if $u \geq 1$

$$\left| \int_M \mu(\eta_u - \eta_\infty) \right| \leq C \frac{\|\mu\|_{C^1(M)}}{\sqrt{u}}. \tag{2.56}$$

From (2.56), we get

$$\int_1^A (\eta_u - \eta_\infty)\frac{du}{u} \to \int_1^{+\infty} (\eta_u - \eta_\infty)\frac{du}{u} \text{ in } \mathcal{D}'(M). \tag{2.57}$$

Also if U, Γ, φ, m are taken as in Section 1c) with respect to the submanifold M', using (1.19), we find that

$$p_{U,\Gamma,\varphi,m}(\eta_u - \eta_\infty) \leq C\int_u^{+\infty} \frac{dv}{v^{3/2}}$$

or equivalently

$$p_{U,\Gamma,\varphi,m}(\eta_u - \eta_\infty) \leq \frac{C}{\sqrt{u}}. \tag{2.58}$$

From (2.57), (2.58), we deduce that

$$\int_1^A (\eta_u - \eta_\infty)\frac{du}{u} \to \int_1^{+\infty} (\eta_u - \eta_\infty)\frac{du}{u} \text{ in } \mathcal{D}'_{N^\bullet_R}(M). \tag{2.59}$$

Since M' and $\tilde{M}'$ are transversal, we deduce from (2.59) that

$$\int_1^A (\eta_u - \eta_\infty)\tilde{\beta}_\infty\frac{du}{u} \to \int_1^{+\infty} (\eta_u - \eta_\infty)\tilde{\beta}_\infty\frac{du}{u} \text{ in } \mathcal{D}'_{N^\bullet_R+\tilde{N}^\bullet_R}(M). \tag{2.60}$$

From (2.60), we get

$$\bar{\partial}\partial\int_1^A (\eta_u - \eta_\infty)\tilde{\beta}_\infty\frac{du}{u} \to \bar{\partial}\partial\int_1^{+\infty} (\eta_u - \eta_\infty)\tilde{\beta}_\infty\frac{du}{u} \text{ in } \mathcal{D}'_{N^\bullet_R+\tilde{N}^\bullet_R}(M). \tag{2.61}$$

Clearly, since β_∞ is closed

$$\beta_u\bar{\partial}\,\tilde{\eta}_u - \beta_\infty\bar{\partial}\,\tilde{\eta}_\infty = (\beta_u - \beta_\infty)\bar{\partial}\tilde{\eta}_u + \bar{\partial}\big(\beta_\infty(\tilde{\eta}_u - \tilde{\eta}_\infty)\big). \tag{2.62}$$

The same arguments as in (2.60) show that

$$\int_1^A \beta_\infty(\tilde{\eta}_u - \tilde{\eta}_\infty)\frac{du}{u} \to \int_1^{+\infty} \beta_\infty(\tilde{\eta}_u - \tilde{\eta}_\infty)\frac{du}{u} \text{ in } \mathcal{D}'_{N^\bullet_R+\tilde{N}^\bullet_R}(M). \tag{2.63}$$

and so

$$\bar{\partial}\partial\int_1^A \beta_\infty(\tilde{\eta}_u - \tilde{\eta}_\infty)\frac{du}{u} \to \bar{\partial}\partial\int_1^{+\infty} \beta_\infty(\tilde{\eta}_u - \tilde{\eta}_\infty)\frac{du}{u} \text{ in } \mathcal{D}'_{N^\bullet_R+\tilde{N}^\bullet_R}(M). \tag{2.64}$$

We now study the behaviour of $A \to +\infty$ of the current

$$\int_1^A (\beta_u - \beta_\infty)\bar{\partial}\,\tilde{\eta}_u \frac{du}{u}. \tag{2.65}$$

Here the situation is slightly more involved because the current $\bar{\partial}\,\tilde{\eta}_u$ depends on u. As $u \to \infty$, the current $\tilde{\eta}_u$ converges as a smooth current on $M \setminus \tilde{M}'$. Using the estimates in Theorem 1.5, it is clear that difficulties in the convergence of the currents (2.65) may occur only near M''.

Take $x_0 \in M''$. Let U be a small open neighborhood of x_0, let Γ, φ, m as in Section 1c) with respect to the submanifold M'. Similarly, given U, and replacing M' by $\tilde{M}'$, we choose $\tilde{\Gamma}$, $\tilde{\varphi}$, $\tilde{m}$ as in Section 1c).

Then by Theorem 1.5, we know that if $u \geq 1$

$$\begin{aligned} \left|\widehat{\varphi(\beta_u - \beta_\infty)}(\xi)\right| &\leq \frac{C}{\sqrt{u}}(1+|\xi|) \\ p_{U,\Gamma,\varphi,m}(\beta_u - \beta_\infty) &\leq \frac{C}{\sqrt{u}} \\ \left|\widehat{\tilde{\varphi}\,\bar{\partial}\,\tilde{\eta}_u}(\xi)\right| &\leq C(1+|\xi|^2) \\ p_{U,\tilde{\Gamma},\tilde{\varphi},\tilde{m}}(\tilde{\eta}_u) &\leq C. \end{aligned} \tag{2.66}$$

By proceeding as in Hörmander [H, Theorem 8.2.4] and using the transversality assumption $N^*_{R|M'} \cap \tilde{N}^*_{R|M'} = \{0\}$, we can easily prove the following estimates.

- There exists $k \in N$ such that for $u \geq 1$

$$\left|\widehat{\varphi(\beta_u - \beta_\infty)\tilde{\varphi}\,\bar{\partial}\,\tilde{\eta}_u}(\xi)\right| \leq \frac{C}{\sqrt{u}}(1+|\xi|)^k. \tag{2.67}$$

From (2.67), we immediately deduce that as $A \to +\infty$, the currents (2.65) converge, i.e.

$$\int_1^A (\beta_u - \beta_\infty)\bar{\partial}\,\tilde{\eta}_u \frac{du}{u} \to \int_1^{+\infty} (\beta_u - \beta_\infty)\bar{\partial}\,\tilde{\eta}_u \frac{du}{u} \quad \text{in } \mathcal{D}'(M). \tag{2.68}$$

- If θ is a smooth current with compact support in U, if Δ is a closed cone in $R^{2\ell}$ such that $\Delta \cap \left(N^*_R + \tilde{N}^*_R\right) = \{0\}$ on $(M' \cup \tilde{M}') \cap U$, if m'' is an integer, then for $u \geq 1$

$$p_{U,\Delta,\theta,m''}\left((\beta_u - \beta_\infty)\bar{\partial}\,\tilde{\eta}_u\right) \leq \frac{C}{\sqrt{u}}. \tag{2.69}$$

To prove (2.69), we break a convolution integral in the ξ variable into three pieces, which are separately estimated using (2.66). Here again it is crucial that $N^*_{R|M''} \cap \tilde{N}^*_{R|M''} = \{0\}$.

From (2.68), (2.69), it is now clear that

$$\int_1^A (\beta_u - \beta_\infty)\bar{\partial}\,\tilde{\eta}_u \frac{du}{u} \to \int_1^{+\infty} (\beta_u - \beta_\infty)\bar{\partial}\,\tilde{\eta}_u \frac{du}{u} \text{ in } \mathcal{D}'_{N^*_R + \tilde{N}^*_R}(M) \tag{2.70}$$

and so

$$\partial \int_1^A (\beta_u - \beta_\infty)\bar{\partial}\,\tilde{\eta}_u \frac{du}{u} \to \partial \int_1^{+\infty} (\beta_u - \beta_\infty)\bar{\partial}\,\tilde{\eta}_u \frac{du}{u} \text{ in } \mathcal{D}'_{N^*_R + \tilde{N}^*_R}(M). \tag{2.71}$$

We can prove in the same way that as $A \to +\infty$

$$\begin{gathered}\int_1^A \partial(\beta_u - \beta_\infty)\tilde{\eta}_u \frac{du}{u} \to \int_1^{+\infty} \partial(\beta_u - \beta_\infty)\tilde{\eta}_u \frac{du}{u} \text{ in } \mathcal{D}'_{N^*_R + \tilde{N}^*_R}(M). \\ \bar{\partial} \int_1^A \partial(\beta_u - \beta_\infty)\tilde{\eta}_u \frac{du}{u} \to \bar{\partial} \int_1^{+\infty} \partial(\beta_u - \beta_\infty)\tilde{\eta}_u \frac{du}{u} \text{ in } \mathcal{D}'_{N^*_R + \tilde{N}^*_R}(M).\end{gathered} \tag{2.72}$$

From (2.53), (2.54), (2.55), (2.61), (2.71), (2.72) we deduce that we can take the obvious limits in (2.52).

Now note that the currents which appear in (2.46) after the operators ∂, $\bar{\partial}$ or $\bar{\partial}\partial$ are smooth. Similarly as we saw in (2.60), (2.63), (2.70), (2.72), we find in particular that the wave front sets of the currents which appear after operators ∂, $\bar{\partial}$ or $\bar{\partial}\partial$ in (2.52) for $A = +\infty$ are included in $N^*_R + \tilde{N}^*_R$. From (2.46), (2.52), we get

$$\begin{aligned}\int_0^1 (\beta''_u - \beta''_0)\frac{du}{u} &+ \int_1^{+\infty} (\beta''_u - \beta''_\infty)\frac{du}{u} \\ &= \tilde{\alpha}_0 \left\{ \int_0^1 (\beta_u - \beta_0)\frac{du}{u} + \int_1^{+\infty} (\beta_u - \beta_\infty)\frac{du}{u} \right\} \\ &+ \alpha_\infty \left\{ \int_0^1 (\tilde{\beta}_u - \tilde{\beta}_0)\frac{du}{u} + \int_1^{+\infty} (\tilde{\beta}_u - \tilde{\beta}_\infty)\frac{du}{u} \right\} \\ &\text{in } P^M_{M'\cup\tilde{M}'}/P^{M,0}_{M'\cup\tilde{M}'}.\end{aligned} \tag{2.73}$$

Also using (2.40), (2.49), we know that

$$\beta''_\infty - \beta''_0 = \tilde{\alpha}_0(\beta_\infty - \beta_0) + \alpha_\infty(\tilde{\beta}_\infty - \tilde{\beta}_0) + \beta_\infty(\tilde{\alpha}_\infty - \tilde{\alpha}_0) + \tilde{\beta}_0(\alpha_\infty - \alpha_0). \tag{2.74}$$

Remember that the currents $\tilde{\beta}_0$ and β_∞ are closed. Therefore using (2.42), (2.47), we find that

$$\beta_\infty(\tilde{\alpha}_\infty - \tilde{\alpha}_0) = \bar{\partial}\partial(\beta_\infty\,\tilde{\eta}_\infty). \tag{2.75}$$

The currents $\beta_\infty\,\tilde{\eta}_\infty$ and $\tilde{\beta}_0\,\eta_\infty$ lie in $\mathcal{D}'_{N^\bullet_R+\tilde{N}^\bullet_R}(M)$. From (2.74), (2.75), we find that

$$\beta''_\infty - \beta''_0 = \tilde{\alpha}_0(\beta_\infty - \beta_0) + \alpha_\infty(\tilde{\beta}_\infty - \tilde{\beta}_0) \ \text{ in } P^M_{M'\cup\tilde{M}'}/P^{M,0}_{M'\cup\tilde{M}'}. \tag{2.76}$$

Using now formula (1.26), together with (2.73), (2.76), we deduce the first line of (2.36). By interchanging the roles of M' and $\tilde{M}'$, we also obtain the second line of (2.36). □

Remark 2.8. It is much easier to prove directly that the right hand sides of both lines of (2.36) coincide in $P^M_{M'\cup\tilde{M}'}/P^{M,0}_{M'\cup\tilde{M}'}$. In fact

$$\begin{aligned}&\bar{\partial}\partial(T(h^\xi))T(h^{\tilde{\xi}}) - T(h^\xi)\bar{\partial}\partial T(h^{\tilde{\xi}}) = \\ &\bar{\partial}(\partial T(h^\xi))T(h^{\tilde{\xi}}) + \partial(T(h^\xi)\bar{\partial}T(h^{\tilde{\xi}})).\end{aligned} \tag{2.77}$$

Using now equation (1.31) for $\frac{\bar{\partial}\partial}{2i\pi}T(h^\xi)$ and its analogue for $\frac{\bar{\partial}\partial}{2i\pi}T(h^{\tilde{\xi}})$, we get from (2.77) that

$$\begin{aligned}T(h^{\tilde{\xi}})\theta(h^F)\delta_{M'} - ch(h^\xi)T(h^{\tilde{\xi}}) - T(h^\xi)\theta(h^{\tilde{F}})\delta_{\tilde{M}'}&\\ + ch(h^{\tilde{\xi}})T(h^\xi) \in P^{M,0}_{M'\cup\tilde{M}'},&\end{aligned} \tag{2.78}$$

which is equivalent to the equality of the right-hand sides of (2.36). Note that in (2.78), only equation (1.31) and the wave front set properties of the currents $T(h^\xi)$ and $T(h^{\tilde{\xi}})$ have been used. The identities (2.36) are much deeper, since they involve the explicit form of the currents $T(h^\xi)$, $T(h^{\tilde{\xi}})$, $T(h^{\xi''})$.

Theorem 2.7 plays a fundamental role in the sequel, in particular in our construction of singular currents associated with Koszul complexes. Also as pointed out in the introduction, it should be deeply related to a version with metrics of the Theorem of Grothendieck–Riemann–Roch.

(c) *Bott–Chern currents and double complexes.* Let now $(\xi^j, v)_{1\le j\le p}$ be p holomorphic chain complexes of vector bundles $\xi^j_0,\dots,\xi^j_m$ on M, which provide resolutions of the sheafs $\mathcal{O}_{M'}(\eta^0),\dots,\mathcal{O}_{M'}(\eta^p)$, where $\eta^0,\dots,\eta^p$

are holomorphic vector bundles on M'. Let r denote the restriction map $\xi^j_{0|M'} \to \eta^j$ $(1 \le j \le p)$.

We also assume that there are holomorphic chain maps $\tilde{v}: \xi^j \to \xi^{j-1}$, $\eta^j \to \eta^{j-1}$ $(1 \le j \le p)$, which are such that the following diagram commutes

$$
\begin{array}{ccccccccccc}
 & & 0 & & & & 0 & & 0 & & \\
 & & \uparrow & & & & \uparrow & & \uparrow & & \\
0 & \to & \xi^0_m & \to_v & \cdots & \to_v & \xi^0_0 & \to_r & \eta^0 & \to & 0 \\
 & & \uparrow \tilde{v} & & & & \uparrow \tilde{v} & & \uparrow \tilde{v} & & \\
0 & \to & \xi^1_m & \to_v & \cdots & \to_v & \xi^1_0 & \to_r & \eta^1 & \to & 0 \\
 & & \uparrow & & & & & & \uparrow & & \\
 & & \vdots & & & & & & \vdots & & \\
 & & \uparrow \tilde{v} & & & & & & \uparrow \tilde{v} & & \\
0 & \to & \xi^p_m & \to_v & \cdots & \to_v & \xi^p_0 & \to_r & \eta^p & \to & 0 \\
 & & \uparrow & & & & \uparrow & & \uparrow & & \\
 & & 0 & & & & 0 & & 0 & &
\end{array}
\tag{2.79}
$$

Let $(\xi_m, \tilde{v}), \ldots, (\xi_0, \tilde{v})$, $(\eta, \tilde{v})$ denote the vertical holomorphic complexes of vector bundles in (2.79).

We now make the fundamental assumption that the complexes $(\xi_m, \tilde{v}), \ldots, (\xi_0, \tilde{v})$, $(\eta, \tilde{v})$ are acyclic.

We equip the $(\xi^j_i)_{\substack{1\le i\le m\\ 1\le j\le p}}$, and the $(\eta^j)_{1\le j\le p}$ with Hermitian metrics $(h^{\xi^j_i})_{\substack{1\le i\le m\\ 1\le j\le p}}$, and g^{η_j} $(1 \le j \le p)$, and we equip the normal bundle N with a Hermitian metric g^N. Let h^{ξ_i}, h^{ξ^j} be the induced Hermitian metric on ξ_i, ξ^j, and let g^η be the induced Hermitian metric on η.

We also make the assumption that for every $j (1 \le j \le p)$, the metrics h^{ξ^j} verify assumption (A) with respect to the metrics g^N, g^{η^j}.

For any $j (1 \le j \le p)$, we can define the currents $T(h^{\xi^j})$, whose wave front sets are included in N^*_R. Similarly, since the complexes $(\xi_i, \tilde{v})_{0\le i\le m}$ and $(\eta, \tilde{v})$ are acyclic, we can construct the associated smooth currents $T(h^{\xi_i})$ and $T(g^\eta)$ on M and M' respectively. Note that such smooth currents were already constructed in [BGS1, Section 1c)].

Theorem 2.9. *The following equality holds*

$$
\sum_{j=0}^{p} (-1)^j T(h^{\xi^j}) = \sum_{i=0}^{m} (-1)^i T(h^{\xi_i}) - i_*\left(Td^{-1}(g^N) T(g^\eta)\right) \tag{2.80}
$$

$$
\text{in } P^M_{M'} / P^{M,0}_{M'}.
$$

Proof. Let $\mathbf{P}^1$ be the one dimensional projective plane with distinguished points 0 and ∞, and let z be the canonical meromorphic coordinate on $\mathbf{P}^1$. By [BGS1, Section 1f)] or by the Grassman graph construction of [BaFM, Chapter II] which will be explained in more detail in Section 4, one finds easily that there exist a double complex on $M \times \mathbf{P}^1$ which will be noted as in (2.79), having the following three properties.

• The restriction of the new complex to $M \times \{0\}$ coincides with the complex (2.79).

• The rows in the double complex provide resolution of sheaves of holomorphic sections of vector bundles on $M' \times \mathbf{P}^1$, and the columns are acyclic.

• On $M' \times \{\infty\}$, the new double complex splits vertically. In particular if $p = 2$, on $M' \times \{\infty\}$, there is an identification of holomorphic chain complexes

$$(\xi^1, \eta^1) = (\xi^0, \eta^0) \oplus (\xi^2, \eta^2)$$

and $\tilde{v}$ acts as the obvious map. If $p \geq 2$, the complex breaks vertically into short exact sequences of complexes which are split vertically.

Let $(\bar{h}^{\xi_i^j})_{\substack{0 \leq i \leq m \\ 0 \leq j \leq p}}$ be Hermitian metrics on the $(\xi_i^j)_{\substack{0 \leq i \leq m \\ 0 \leq j \leq p}}$ on $M \times \mathbf{P}^1$ which have the following two properties:

• They restrict to the given metrics $(h^{\xi_i^j})$ on $M \times \{0\}$.

• On $M' \times \{\infty\}$, the Hermitian chain complexes $(\xi_i, \tilde{v})_{0 \leq i \leq m}$ are split in the sense of [BGS1, Section 1f)]. If $p = 2$, this means that $\xi_i^1 = \xi_i^0 \oplus \xi_i^2$ $(0 \leq i \leq m)$, and that the previous splittings are orthogonal.

For $0 \leq j \leq p$, let $T(\bar{h}^{\xi^j})$ be the current on $M \times \mathbf{P}^1$ associated with the Hermitian chain complex ξ^j equipped with the metrics $(\bar{h}^{\xi^j})$.

The wave front set of the current $\mathrm{Log}|z|^2$ was calculated in the proof of Theorem 2.5. By Theorem 1.9, the wave front set of the current $T(\bar{h}^{\xi^j})$ is included in the normal bundle to $M' \times \mathbf{P}^1$ in $M \times \mathbf{P}^1$, which is the natural lift of N to $M' \times \mathbf{P}^1$. As in the proof of Theorem 2.5, we deduce that the product of currents $\mathrm{Log}|z|^2 T(\bar{h}^{\xi^j})$ is well defined. Then we have the analogue of (2.15) i.e.

$$\begin{aligned} &\frac{\bar{\partial}\partial}{2i\pi}(\mathrm{Log}|z|^2)T(\bar{h}^{\xi^j}) - \mathrm{Log}|z|^2 \frac{\bar{\partial}\partial}{2i\pi} T(\bar{h}^{\xi^j}) \\ &\qquad = \frac{\bar{\partial}}{2i\pi}(\partial(\mathrm{Log}|z|^2)T(\bar{h}^{\xi^j})) \\ &\qquad + \frac{\partial}{2i\pi}(\mathrm{Log}|z|^2 \bar{\partial} T(\bar{h}^{\xi^j})). \end{aligned} \tag{2.81}$$

For $0 \leq j \leq p$, let $\bar{h}^{F^j}$ be the metric induced on the direct sum F^j of the homology groups of $(\xi^j, v)_{|M'}$ by the metrics $\bar{h}^{\xi^j}$

Using Theorem 1.9 and (2.81), we get

$$\begin{aligned}
&\mathrm{Log}|z|^2 \sum_{j=0}^{p}(-1)^j \theta(\bar{h}^{F^j})\delta_{M'\times \mathbf{P}_1} - \mathrm{Log}|z|^2 \sum_{j=0}^{p}(-1)^j ch(\bar{h}^{\xi^j}) \\
&\quad + \sum_{j=0}^{p}(-1)^j T(\bar{h}^{\xi^j})\delta_{M\times\{\infty\}} - \sum_{j=0}^{p}(-1)^j T(\bar{h}^{\xi^j})\delta_{M\times\{0\}} \\
&= -\frac{\bar{\partial}}{2i\pi}\left(\partial(\mathrm{Log}|z|^2)\left(\sum_{j=0}^{p}(-1)^j T(\bar{h}^{\xi^j})\right)\right) \\
&\quad - \frac{\partial}{2i\pi}\left((\mathrm{Log}|z|^2)\bar{\partial}\left(\sum_{j=0}^{p}(-1)^j T(\bar{h}^{\xi^j})\right)\right)
\end{aligned} \tag{2.82}$$

We claim that

$$\sum_{j=0}^{P}(-1)^j T(\bar{h}^{\xi^j})\delta_{M\times\{\infty\}} = 0. \tag{2.83}$$

In fact by the analogue of (2.32), the current $T(\bar{h}^{\xi^j})\delta_{M\times\{\infty\}}$ is the image of the restriction to $M \times \{\infty\}$ of $T(\bar{h}^{\xi^j})$ by the embedding $M \times \{\infty\} \to M \times \mathbf{P}^1$. By [BGS4, Theorem 2.7] or by Theorem 1.10, the restriction of $T(\bar{h}^{\xi^j})$ to $M \times \{\infty\}$ is the Bott–Chern current associated with the chain complex $(\xi^j, v)_{|M\times\{\infty\}}$. Since on $M \times \{\infty\}$, for $1 \leq i \leq m$, the Hermitian chain complex $(\xi_i, \tilde{v})$ splits, it is elementary to verify that for any $u \geq 0$, if $A_u^{\xi^j}$ is the superconnection (1.10) associated with the Hermitian chain complex (ξ^j, v) and if $N_H^{\xi^j}$ is the corresponding number operator, then the restriction to $M \times \{\infty\}$ of the form

$$\sum_{j=0}^{p}(-1)^j Tr_s[N_H^{\xi^j}\exp(-A_u^{\xi^j})^2]$$

vanishes identically. (2.83) is now obvious using formula (1.26).

We now integrate (2.82) along the fiber of the projection map $M \times \mathbf{P}^1 \to M$. Taking (2.83) into account, we get

$$
\begin{aligned}
\sum_{j=0}^{p}(-1)^j T(h^{\xi^j}) = &- \int_{\mathbf{P}^1} \mathrm{Log}|z|^2 \sum_{j=0}^{p}(-1)^j ch(\bar{h}^{\xi^j}) \\
&+ \int_{\mathbf{P}^1} \mathrm{Log}|z|^2 \sum_{j=0}^{p}(-1)^j \theta(\bar{h}^{F^j})\delta_{M'\times\mathbf{P}^1} \\
&+ \frac{\bar{\partial}}{2i\pi}\int_{\mathbf{P}^1} \partial(\mathrm{Log}|z|^2)\sum_{j=0}^{p}(-1)^j T(\bar{h}^{\xi^j}) \\
&+ \frac{\partial}{2i\pi}\int_{\mathbf{P}^1} \mathrm{Log}|z|^2\bar{\partial}\left(\sum_{j=0}^{p}(-1)^j T(\bar{h}^{\xi^j})\right).
\end{aligned}
\tag{2.84}
$$

By [BGS1, proof of Theorem 1.29], we know that

$$
\sum_{i=0}^{m}(-1)^i T(h^{\xi_i}) = - \int_{\mathbf{P}^1} \mathrm{Log}|z|^2 \sum_{i=0}^{m}(-1)^i ch(\bar{h}^{\xi_i}) \quad \text{in } P^M/P^{M,0}.
\tag{2.85}
$$

Note the trivial relation

$$
\sum_{j=0}^{p}(-1)^j ch(\bar{h}^{\xi^j}) = \sum_{i=0}^{m}(-1)^i ch(\bar{h}^{\xi_i}).
\tag{2.86}
$$

Let $(\tilde{g}^{\eta^j})_{0\le j\le p}$ be Hermitian metrics on the $(\eta^j)_{0\le j\le p}$ on $M'\times\mathbf{P}^1$ which have the following properties:

- They coincide with the metrics $(g^{\eta^j})_{0\le j\le p}$ on $M'\times\{0\}$.
- On $M'\times\{\infty\}$, the Hermitian chain complex $(\eta, \tilde{v})$ splits.

For $0\le j\le p$, let $(\tilde{h}^{F^j})_{0\le j\le p}$ be the metric on $F^j = \Lambda N^*\otimes\eta^j$ associated with the given metric g^N on N and with the metric $\tilde{g}^{\eta^j}$. Note that

$$
\tilde{h}^{F^j} = h^{F^j} \quad \text{on} \quad M'\times\{0\}.
\tag{2.87}
$$

In Theorem 2.4, a class of forms $\chi(\tilde{h}^{F^j}, \bar{h}^{F^j}) \in P^{M'\times\mathbf{P}^1}/P^{M'\times\mathbf{P}^1,0}$ was constructed such that

$$
\frac{\bar{\partial}\partial}{2i\pi}\chi(\tilde{h}^{F^j}, \bar{h}^{F^j}) = \theta(\bar{h}^{F^j}) - \theta(\tilde{h}^{F^j}).
\tag{2.88}
$$

Using (2.87), we find from Definition 2.3 that $\chi(\tilde{h}^{F^j}, \bar{h}^{F^j})$ restricts to the zero class of forms on $M' \times \{0\}$. For $0 \leq j \leq p$, let $\chi_\infty(\tilde{h}^{F^j}, \bar{h}^{F^j})$ be the restriction of $\chi(\tilde{h}^{F^j}, \bar{h}^{F^j})$ to $M' \times \{\infty\}$ which we identify with M', so that $\chi_\infty(\tilde{h}^{F^j}, \bar{h}^{F^j})$ is now a form on M'. By proceeding as in (2.15)–(2.19), we get

$$\begin{aligned}\int_{\mathbb{P}^1} \mathrm{Log}|z|^2 \Big(\sum_{j=0}^{p}(-1)^j \theta(\bar{h}^{F^j})\Big) = \int_{\mathbb{P}^1} \mathrm{Log}|z|^2 \Big(\sum_{j=0}^{p}(-1)^j \theta(\tilde{h}^{F^j})\Big) \\ - \sum_{j=0}^{p}(-1)^j \chi_\infty(\tilde{h}^{F^j}, \bar{h}^{F^j}) \text{ in } P^{M'}/P^{M',0}.\end{aligned} \tag{2.89}$$

By (2.6), we know that

$$\theta(\tilde{h}^{F^j}) = Td^{-1}(g^N) ch(\tilde{g}^{\eta^j}). \tag{2.90}$$

Using Theorem 1.9, (2.16), (2.90), we find that

$$\int_{\mathbb{P}^1} \mathrm{Log}|z|^2 \left(\sum_{j=0}^{p}(-1)^j \theta(\tilde{h}^{F^j})\right) = -Td^{-1}(g^N) T(g^\eta) \text{ in } P^{M'}/P^{M',0}. \tag{2.91}$$

Using (2.84), (2.85), (2.86), (2.89), (2.91) and proceeding as in the proof of Theorem 2.5, we get

$$\begin{aligned}\sum_{j=0}^{p}(-1)^j T(h^{\xi^j}) = \sum_{i=0}^{m}(-1)^i T(h^{\xi_i}) - i_* Td^{-1}(g^N) T(g^\eta) \\ - \sum_{j=0}^{p}(-1)^j i_* \chi_\infty(\tilde{h}^{F^j}, \bar{h}^{F^j}) \text{ in } P^M_{M'}/\tilde{P}^{M,0}_{M'}.\end{aligned} \tag{2.92}$$

To prove Theorem 2.9, we only need to show that

$$\sum_{j=0}^{p}(-1)^j \chi_\infty(\tilde{h}^{F^j}, \bar{h}^{F^j}) = 0 \quad \text{in } P^{M'}/P^{M',0}. \tag{2.93}$$

We now use the notations of Section 1a). Note that since $v\tilde{v} + \tilde{v}v = 0$, if $y \in TM$, then

$$\partial_y v\tilde{v} + \tilde{v}\partial_y v + v\partial_y \tilde{v} + \partial_y \tilde{v} v = 0. \tag{2.94}$$

Clearly $\tilde{v}$ acts on the F^j's and maps F^j into F^{j-1} $(0 \le j \le p)$. Using (2.94), we find that if $y \in N$, we have the identity of operators acting on the F^j's

$$\partial_y v\tilde{v} + \tilde{v}\partial_y v = 0. \tag{2.95}$$

Also remember that for any j $(0 \le j \le p)$, we have a canonical isomorphism of Z-graded complexes

$$(F^j, \partial_y v) \cong (\Lambda N^* \otimes \eta^j, i_y). \tag{2.96}$$

In particular r provides the canonical isomorphism $F_0^j \cong \eta^j$. Tautologically, the action of $\tilde{v}$ on the F_0^j's coincides with the action of $\tilde{v}$ on the η^j's, i.e.

$$(F_0, \tilde{v}) \cong (\eta, \tilde{v}). \tag{2.97}$$

Using the canonical isomorphism (2.96), (2.95) can be rewritten in the form

$$i_y \tilde{v} + \tilde{v} i_y = 0. \tag{2.98}$$

From (2.96) and (2.98), we deduce easily that $\tilde{v}$ acts on $\Lambda N^* \hat{\otimes} \eta$ like $1 \hat{\otimes} \tilde{v}$.

Let $\bar{h}^F, \tilde{h}^F$ be the metrics on $F = \bigoplus_{j=0}^{p} F^j$ which are the orthogonal sums of the metrics $\bar{h}^{F^j}$, $\tilde{h}^{F^j}$ respectively. Remember that on $M' \times \{\infty\}$, the holomorphic Hermitian complex (2.79) splits vertically. Therefore on $M' \times \{\infty\}$, when equipped with the metric $\bar{h}^F$, the complex $(\Lambda N^* \hat{\otimes} \eta, i_y + \tilde{v})$ also splits vertically. On the other hand, since on $M' \times \{\infty\}$ the Hermitian complex $(\tilde{\eta}, \tilde{v})$ is split, on $M' \times \{\infty\}$ the complex $(\Lambda N^*, i_y + \tilde{v})$ equipped with the metric $\tilde{h}^F$ is also split. For $0 \le \ell \le 1$, set

$$\tilde{h}_\ell^F = (1-\ell)\bar{h}^F + \ell\tilde{h}^F. \tag{2.99}$$

Then the complex $(\Lambda N^* \otimes \eta, i_y + \tilde{v})$ still splits vertically as a Hermitian complex when equipped with the metric $\tilde{h}_\ell^F$. Using the notations in Section 2a), we thus find that for $\ell \in [0,1]$

$$\sum_{j=0}^{p} (-1)^j Tr_s \left[(\tilde{h}_\ell^{F^j})^{-1} \frac{\partial \tilde{h}_\ell^{F^j}}{\partial \ell} \exp(-(B_\ell^j)^2) \right] = 0. \tag{2.100}$$

and so from (2.8) and (2.100) we get (2.93).

Our Theorem is proved. □

3. Bott–Chern currents, Euler–Green currents and Koszul complexes

Let M be a complex manifold, let E be a holomorphic Hermitian vector bundle on M. If M^E is the total space of M, if i is the embedding $M \to M^E$, the Koszul complex on $M^E(\Lambda E^*, i_y)$ provides a resolution of the sheaf $i_* \mathcal{O}_M$.

If g^E is the metric of E, and if h^E is the metric induced by g^E on ΛE^*, we first calculate the Bott–Chern current $\tilde{T}(g^E) = T(h^E)$ on M^E. If $e(g^E)$ is the Chern–Weil representative of the highest Chern class of E, then

$$\frac{\bar{\partial}\partial}{2i\pi}\tilde{T}(g^E) = Td^{-1}(g^E)(\delta_M - e(g^E)). \tag{3.1}$$

A simpler equation of currents on M^E is given by

$$\frac{\bar{\partial}\partial}{2i\pi}\tilde{e}(g^E) = \delta_M - e(g^E), \tag{3.2}$$

so that $-\tilde{e}(g^E)$ is a Green current in the sense of [GS1].

By extending the formalism of Mathai-Quillen [MQ] in a complex geometric framework, we exhibit an explicit locally integrable current $\tilde{e}(g^E)$ on M^E which solves equations (3.2). If $\dim E = 1$, $\tilde{e}(g^E)$ coincides with the obvious solution $\operatorname{Log}|y|^2$.

This section is organized as follows. In a), we recall results of [BGS4] which express the currents $T(h^\xi)$ as a principal part of a nonintegrable current. In b), we calculate the Bott–Chern current of a Koszul complex. In c), we briefly develop a differential geometric setting for the de Rham theory of M^E. In d), we give a double transgression formula for the canonical Thom form of (E, g^E) constructed in the Mathai–Quillen [MQ]. In e), we study the asymptotic behaviour of Mathai–Quillen currents, which here depend on a parameter $u \geq 0$. In f), we explicitly construct a Green current $-\tilde{e}(g^E)$. In g) we compare $\tilde{T}(g^E)$ with $\tilde{e}(g^E)$. In h), we describe $\tilde{e}(g^E)$ as a function of g^E. Finally in i), we study the behaviour of $\tilde{e}(g^E)$ in direct sums.

This section relies on the formalism of Mathai–Quillen [MQ], to which the reader is referred when necessary.

(a) *The current $T(h^\xi)$ as a finite part.* Our assumptions are the same as in Section 1g).

Let $\omega(h^\xi)$ be the restriction of the current $T(h^\xi)$ to the open set $M \backslash M'$. Then by [BGS4, Theorem 2.5], we know that $\omega(h^\xi)$ is a smooth current on $M \setminus M'$. By [BGS4, Theorem 3.3], we know that, in general, $\omega(h^\xi)$ is nonintegrable on $M \setminus M'$. We now briefly recall a result of [BGS4].

Let g^M be a Hermitian metric on TM. We identify N with the vector bundle orthogonal to TM' in TM. Then g^M induces on N a metric g^N. Let g^η be a Hermitian metric on η.

For $\varepsilon > 0$, let M_ε be the set of points of M whose Riemannian distance to M' is larger than ε.

Theorem 3.1. *Let μ be a smooth differential form on M. Then*

(3.3)
$$\begin{aligned}\int_M \mu T(h^\xi) = \lim_{\varepsilon\to 0}\Big\{\int_{M_\varepsilon} \mu\omega(h^\xi) \\ + 2\operatorname{Log}\varepsilon \int_{M'} i^*\mu \int_N \varphi(Tr_s[N_H \exp(-B^2)])\Big\} \\ - \int_{M'} i^*\mu \int_N (2\operatorname{Log}(|Y|_{g^N}) - \Gamma'(1))\varphi(Tr_s[N_H\exp(-B^2)])\end{aligned}$$

If the metrics $h^\xi = (h^{\xi_0}, \dots, h^{\xi_m})$ verify assumption (A) with respect to the metrics g^N, g^η, then

$$\int_N \varphi(Tr_s[N_H \exp(-B^2)]) = -(Td^{-1})'(g^N)ch(g^\eta)$$

$$\begin{aligned}\int_N (2\operatorname{Log}(|Y|_{g^N}) - \Gamma'(1))\varphi(Tr_s[N_H\exp(-B^2)]) \\ = -\left(\sum_{k=1}^{\dim N-1}\frac{1}{k} + \operatorname{Log}2\right)(Td^{-1})'(g^N)ch(g^\eta).\end{aligned} \tag{3.4}$$

Proof. This result was proved in [BGS4, Theorem 3.4]. □

(b) *The singular Bott–Chern current of a Koszul complex.* Let now M be a complex compact connected manifold of dimension ℓ. Let E be a holomorphic vector bundle on M of dimension k. Let M^E be the total space of E. Let p be the projection $M^E \to M$. We embed M in M^E as the zero section of M^E. Let i be this embedding.

Let E^* be the dual of E, and let $\Lambda E^* = \bigoplus_{j=0}^{k} \Lambda^j E^*$ be the exterior algebra of E^*. ΛE^* is a Z-graded vector bundle on M, which lifts to M^E.

If $y \in E$, the interior multiplication operator i_y acts naturally on $(\Lambda E^*)_{py}$. Then the Koszul complex $(\Lambda E^*, i_y)$ provides a resolution of the sheaf $i_*\mathcal{O}_M$, i.e. we have the exact sequence of sheaves

$$0 \to \mathcal{O}_{M^E}(\Lambda^k E^*) \to_{i_y} \cdots \to_{i_y} \mathcal{O}_{M^E} \to_r i_*\mathcal{O}_M \to 0.$$

Assume now that g^E is a Hermitian metric on E. Then g^E induces metrics $h^E = \left(h^{\Lambda^0 E^*}, \dots, h^{\Lambda^k E^*}\right)$ on $\Lambda^0 E^*, \dots, \Lambda^k E^*$. For $y \in E$, the adjoint i_y^* of i_y is given by $i_y^* = \bar{y}\wedge$.

The normal bundle N to M in M^E is exactly E. Note that since E is equipped with the metric g^E, assumption (A) of Section 1b) is verified.

The manifold M^E is noncompact. Still currents on M^E are paired with smooth compactly supported forms on M^E. It is then quite easy to verify that all the results of [B2], [BGS4] and of Sections 1 and 2 are still valid on M^E.

In particular, let $T(h^E)$ be the current on M^E associated with the holomorphic Hermitian chain complex $(\Lambda E^*, i_y)$. To note that $T(h^E)$ depends only on g^E, we will use the notation $\tilde{T}(g^E) = T(h^E)$.

Let e be the ad invariant polynomial on (k,k) matrices $e\colon A \to \det A$. Then using the notations of Section 1f), $e(g^E)$, $Td(g^E)$ are smooth forms on M. When lifting such forms to M^E, we omit the notation p^*.

Theorem 3.2. *The current $\tilde{T}(g^E)$ lies in $P_M^{M^E}$. Also*

$$\frac{\bar{\partial}\partial}{2i\pi}\tilde{T}(g^E) = Td^{-1}(g^E)(\delta_M - e(g^E)). \tag{3.5}$$

Proof. Theorem 3.2 is a consequence of Theorem 1.9 and of the classical relation

$$ch(h^E) = e(g^E)Td^{-1}(g^E).$$ □

Let now $\omega(g^E)$ be the restriction of $\tilde{T}(g^E)$ to M^E/M. Then $\omega(g^E)$ is smooth on M^E/M. Also by Theorem 3.1, $\omega(g^E)$ entirely determines $\tilde{T}(g^E)$.

Let ∇^E be the holomorphic Hermitian connection on (E, g^E), and let $\Omega^E = (\nabla^E)^2$ be its curvature.

Before we proceed, we will explain the notations and conventions of Mathai–Quillen [MQ], which will be used in the sequel. First observe that the connection ∇^E defines a horizontal subspace $T^H M^E$ in TM^E so that

$$TM^E = T^H M^E \oplus E. \tag{3.6}$$

If $Y \in T_R E$, let Y^V be its component in E_R with respect to the splitting (3.6). If A is an antisymmetric tensor in $\mathrm{End}\, E_R$, we identify A with the two form on $T_R M^E$

$$Y, Z \in T_R M^E \to \langle Y^V, AZ^V \rangle. \tag{3.7}$$

The two form (3.7) will still be noted A, and its exterior powers $A^2, \ldots, A^k$. Note that we omit the wedge product sign.

We now follow Mathai–Quillen [MQ]. Namely assume temporarily that A is invertible. Then we can define the forms A^{-1}, $(A^{-1})^2, \ldots, (A^{-1})^k$ which are the powers in $\Lambda(E_R^*)$ of the two form A^{-1}. If $Pf(A)$ is the Pfaffian of A, the forms $Pf(A)A^{-1}, \ldots, Pf(A)(A^{-1})^k$ are in fact rational functions of A, which can be extended by continuity to an arbitrary A. We will still note them this way, even if A is noninvertible.

If $A \in \mathrm{End}\, E$, A induces the $(1,1)$ form on M^E, $(Y, Z) \in T^{(0,1)} M^E \times T^{(1,0)} M^E \to \langle Y^V, AZ^V \rangle$ which we still note A. Therefore, we can extend to any such A the previous considerations. In particular, if I_E is the identity map of E, we identify I_E with the $(1,1)$ form on M^E defined by $(Y, Z) \in T^{(0,1)} M^E \times T^{(1,0)} M^E \to \langle Y^V, Z^V \rangle$.

Let J_E be the complex structure of E_R. Then, with the previous conventions $J_E = \sqrt{-1} I_E$.

If like in usual Chern–Weil's theory, we replace formally A by the $(1,1)$ form Ω^E on M which takes values in $\mathrm{End}\, E$, we get a form on M with values in forms on M^E. By antisymmetrization in all the indices, we finally get a differential form on the manifold M^E.

Observe that the formal forms $(\Omega^E)^{-1}, \ldots, ((\Omega^E)^{-1})^k$ should be considered as forms of total degree 0.

Similarly if $A \in \mathrm{End}\, E_R$, and $Y \in E_R$, we identify AY with the one form $Z \in T_R M^E \to \langle AY, Z^V \rangle$. Then the form $Pf(\Omega^E)(\Omega^E)^{-1} Y$ is well defined. Note that the formal degree of the form $(\Omega^E)^{-1} Y$ is -1.

In the previous constructions, for any $b \in \mathbf{C}$, we can replace Ω^E by $\Omega^E + bJ_E$, and we still get a meaningful form on M^E.

For instance if $\det(B)$ is the determinant of $B \in \mathrm{End}\, E$, then

$$\det\left(\frac{\Omega^E}{2i\pi} + bI_E\right)\left(\frac{\Omega^E}{2\pi} + bJ_E\right)^{-j} \qquad (0 \le j \le k)$$

is a well defined series of forms on M^E.

If $y \in E$, y represents $Y = y + \bar{y} \in E_R$, so that $|Y|^2 = 2|y|^2$.

Let φ be the homomorphism from $\Lambda^{\mathrm{even}}(T_R^* M_E)$ into itself which to $\alpha \in \Lambda^{2j}(T_R^* M_E)$ associates $(2\pi i)^{-j}\alpha$.

Theorem 3.3. *On $M^E \setminus M$, we have the following identity of differential forms*

$$\omega(g^E) = -\frac{\partial}{\partial b}\Big[\det\Big(I_E - \exp\Big(\frac{\Omega^E}{2i\pi} + bI_E\Big)\Big) \cdot \operatorname{Log}\Big(\frac{|Y|^2}{2} + \Big(2\pi\Big(\frac{\Omega^E}{2\pi} + bJ_E\Big)\Big)^{-1}\Big)\Big]_{b=0}. \tag{3.8}$$

Equivalently

$$\omega(g^E) = -\frac{\partial}{\partial b}\Big[\det\Big(I_E - \exp\Big(\frac{\Omega^E}{2i\pi} + bI_E\Big)\Big) \Big\{\operatorname{Log}\Big(\frac{|Y|^2}{2}\Big) + \sum_{j=1}^{\dim E} \frac{2^j}{j|Y|^{2j}} \Big(\Big(-2\pi\Big(\frac{\Omega^E}{2\pi} + bJ_E\Big)\Big)^{-1}\Big)^j\Big\}\Big]_{b=0}. \tag{3.9}$$

Let $\gamma(g^E)$ be the current on M^E

$$\gamma(g^E) = \frac{1}{2}\Big(\frac{\bar\partial - \partial}{2i\pi}\Big)\tilde{T}(g^E). \tag{3.10}$$

Then $\gamma(g^E)$ is a locally integrable current on M^E given by the formula

$$\gamma(g^E) = \frac{1}{2}\det\Big(I_E - \exp\Big(\frac{\Omega^E}{2i\pi}\Big)\Big)(-\Omega^E)^{-1}Y\Big(\frac{|Y|^2}{2} + (\Omega^E)^{-1}\Big)^{-1} \tag{3.11}$$

or equivalently

$$\gamma(g^E) = \frac{1}{2}\det\Big(I_E - \exp\Big(\frac{\Omega^E}{2i\pi}\Big)\Big)(-\Omega^E)^{-1}Y\sum_{j=1}^{\dim E}\frac{2^j}{|Y|^{2j}}((-\Omega^E)^{-1})^{j-1}. \tag{3.12}$$

Also

$$d\gamma(g^E) = Td^{-1}(g^E)\big(e(g^E) - \delta_M\big). \tag{3.13}$$

Proof. Let N_H be the operator acting on $\Lambda^j(E^*)$ $(0 \le j \le k)$ by multiplication by j. We still denote by ∇^E the holomorphic Hermitian connection on ΛE^*. Set $V = i_y + \bar{y}\wedge$. For $u > 0$, set $A_u = \nabla^E + \sqrt{u}V$. By [B2, equations (3.142), (4.21) and (4.22)], we know that

$$Tr_s\left[N_H \exp(-A_u^2)\right] = \frac{\partial}{\partial b}\Big[\det\left(I - \exp(\Omega^E + bI_E)\right) \cdot \exp\left\{-u\left(\frac{|Y|^2}{2} + (\Omega^E + bI_E)^{-1}\right)\right\}\Big]_{b=0}. \tag{3.14}$$

From (3.14), we deduce that on $M^E \setminus M$, if $s \in \mathbf{C}$, $\mathrm{Re}(s) > 0$

$$\frac{1}{\Gamma(s)} \int_0^{+\infty} u^{s-1} Tr_s\left[N_H \exp(-A_u^2)\right] du = \frac{\partial}{\partial b}\left[\det\left(I - \exp(\Omega^E + bI_E)\right)\left\{\frac{|Y|^2}{2} + (\Omega^E + bI_E)^{-1}\right\}^{-s}\right]_{b=0}. \tag{3.15}$$

and so

$$\left[\frac{1}{\Gamma(s)} \int_0^{+\infty} u^{s-1} Tr_s\left[N_H \exp(-A_u^2)\right] du\right]'(0) = -\frac{\partial}{\partial b}\left[\det\left(I - \exp(\Omega^E + bI_E)\right) \mathrm{Log}\left(\frac{|Y|^2}{2} + (\Omega^E + bI_E)^{-1}\right)\right]_{b=0}. \tag{3.16}$$

Using (3.16) we obtain (3.8). Note here that as a two form, Ω^E gets rescaled by a factor $\frac{1}{2i\pi}$, and that as a two form of partial vertical degree 2, $(\Omega^E + bI_E)^{-1}$ is also rescaled by the same factor. By expanding the Log in (3.8), we obtain (3.9). Note that the expansion terminates at $k = \dim E$, since this corresponds to the highest possible power of a two form on E_R.

By [BGS4, Theorem 3.3], we know that the current $\gamma(g^E)$ is locally integrable. We temporarily fix an arbitrary square root of $i = \sqrt{-1}$. We extend the map φ to the whole $\Lambda(T_R^* M^E)$ so that if $\alpha \in \Lambda^k(T_R^* M^E)$, then $\varphi(\alpha) = (2\pi i)^{-k/2}\alpha$. Set

$$\eta = \int_0^{+\infty} Tr_s\left[\sqrt{u}\, V \exp(-A_u^2)\right] \frac{du}{2u}. \tag{3.17}$$

Then by [BGS4, Theorem 2.5], η defines a locally integrable current M^E, and moreover

$$\frac{1}{2}(\bar{\partial} - \partial)\tilde{T}(g^E) = (2i\pi)^{1/2}\varphi(\eta). \tag{3.18}$$

By [B2, equations (4.9) and (4.11)], we know that

$$Tr_s\left[\sqrt{u}\, V \exp(-A_u^2)\right] = -i_Y Tr_s\left[\exp(-A_u^2)\right].$$

Using [MQ, Theorem 4.5], or [B2, equations (3.138), (3.139)] we find that

$$i_Y Tr_s\left[\exp(-A_u^2)\right] = i_Y \det\left(I - \exp(\Omega^E)\right) \exp\left\{-u\left(\frac{|Y|^2}{2} + (\Omega^E)^{-1}\right)\right\}. \tag{3.19}$$

Clearly, $i_Y(\Omega^E)^{-1} = -(\Omega^E)^{-1}Y$, and so from (3.19), we find that

$$i_Y Tr_s\left[\exp(-A_u^2)\right] = u \det\left(I - \exp(\Omega^E)\right)(\Omega^E)^{-1}Y \cdot \exp\left\{-u\left(\frac{|Y|^2}{2} + (\Omega^E)^{-1}\right)\right\}. \tag{3.20}$$

Using (3.17) and (3.20) we find that

$$\eta = -\frac{1}{2}\det\left(I - \exp(\Omega^E)\right)(\Omega^E)^{-1}Y\left(\frac{|Y|^2}{2} + (\Omega^E)^{-1}\right)^{-1}.$$

Since the form $(\Omega^E)^{-1}Y$ is of formal degree -1, we then get

$$\varphi(\eta) = -\frac{1}{2}\det\left(I - \exp\left(\frac{\Omega^E}{2i\pi}\right)\right)(2i\pi)^{1/2}(\Omega^E)^{-1}Y \cdot \left(\frac{|Y|^2}{2} + (\Omega^E)^{-1}\right)^{-1}. \tag{3.21}$$

From (3.18), (3.21) we obtain (3.11). By expanding (3.11), we get (3.12). (3.13) now follows from (3.5) and (3.10).

Remark 3.4. By formula (3.9), it is clear that in general the current $\omega(g^E)$ is not locally integrable, since its singularity near M is of the form $|Y|^{-2\dim E}$. The singularity of $\gamma(g^E)$ near M^E is controlled by $|Y|^{-2\dim E+1}$ which is integrable near 0.

Remark 3.5. Let s be a holomorphic section of E on M which is transversal to M in M^E. Namely we assume that if $x \in M$ is such that $s(x) = 0$, if $ds(x)$ is the differential of s at x, then $\operatorname{Im}[ds(x)] = E$. Set

$$M' = \{x \in M; s(x) = 0\}. \tag{3.22}$$

Then on M', ds identifies $E_{|M'}$ with the normal bundle N to M'. Let i be the embedding $M' \to M$.

The chain complex $(\Lambda E^*, i_s) = s^*(\Lambda E^*, i_y)$ provides a resolution of the sheaf $i_* \mathcal{O}_{M'}$, i.e. we have the exact sequence of sheaves

$$\mathcal{O}_M(\Lambda E^*, i_s) \to_r i_* \mathcal{O}_{M'} \to 0.$$

Let h^E be the metric induced by g^E on ΛE^*. Let $\tilde{T}(g^E, s) = T(h^E, s)$ be the singular Bott–Chern current on M associated with the complex $(\Lambda E^*, i_s)$. Remember that $WF(\tilde{T}(g^E)) \subset E_R$. By [H, Theorem 8.2.4], the pulled-back current $s^* \tilde{T}(g^E)$ on M is well-defined. It follows from [BGS4, Theorem 2.7] or from Theorem 2.10 that

$$\tilde{T}(g^E, s) = s^* \tilde{T}(g^E). \tag{3.23}$$

(c) *Equivariant cohomology and differential forms.* We make the same assumptions as in Section 3b), and we use the same notations.

Clearly if p is the projection $M^E \to M$, then $T^H M^E \cong p^* TM$. In the sequel, we will generally omit the notation p^*. From (3.6), we deduce an isomorphism of smooth vector bundles

$$\Lambda(T_R^* M^E) \cong \Lambda(T_R^* M) \hat{\otimes} \Lambda(E_R^*). \tag{3.24}$$

If β is a smooth section of $\Lambda^j(E_R^*)$ on M^E, $\nabla^E \beta$ is a smooth section of $\Lambda^1(T_R^* M^E) \hat{\otimes} \Lambda^j(E_R^*)$. We denote by ${}^a\nabla^E \beta$ the corresponding $j+1$ form on M^E which is obtained by using (3.24) and antisymmetrization in the indices of $\nabla^E \beta$.

If α and β are smooth sections on M^E of $\Lambda(T_R^* M)$ and $\Lambda(E_R^*)$ respectively, set

$${}^a\nabla^E(\alpha\beta) = (d\alpha)\beta + (-1)^{\deg \alpha} \alpha({}^a\nabla^E \beta). \tag{3.25}$$

From (3.24), we deduce that ${}^a\nabla^E$ acts on the smooth sections on M^E of $\Lambda(T_R^* M^E)$.

$\Omega^E Y$ is a two form on M with values in vectors in $E_R \subset T_R M_E$. Therefore the operator $i_{\Omega^E Y}$ acts on $\Lambda(T_R^* M^E)$ and increases the total degree in $\Lambda(T_R^* M^E)$ by 1.

In the sequel d denotes exterior differentiation acting on smooth sections of $\Lambda(T_R^* M^E)$ on M^E.

Proposition 3.6. *The following identity holds*

$$d = {}^a\nabla^E + i_{\Omega^E Y}. \tag{3.26}$$

Proof. Let ∇^M be any torsion free connection on $T_R M$. ∇^M lifts into a connection on $T^H M^E$, which we still note ∇^M. Using the splitting (3.6) of TM^E, the connection $\nabla = \nabla^M \oplus \nabla^E$ defines a connection on $T_R M^E$. Let T be the torsion of ∇. One verifies easily that if $Y \in M^E$, $U, U' \in (T_R^* M^E)_Y$, then

$$T_Y(U, U') = \Omega^E(p_* U, p_* U')Y. \tag{3.27}$$

The connection ∇ maps smooth sections of $\Lambda^j(T_R^* M^E)$ into smooth sections of $\Lambda^1(T_R^* M^E) \hat{\otimes} \Lambda^j(T_R^* M^E)$. By antisymmetrization in all the indices, we obtain an operator ${}^a\nabla$ which map smooth sections of $\Lambda^j(T_R^* M^E)$ into smooth sections of $\Lambda^{j+1}(T_R^* M^E)$. Classically

$$d = {}^a\nabla + i_T. \tag{3.28}$$

Using (3.27), we get

$$d = {}^a\nabla + i_{\Omega^E Y}. \tag{3.29}$$

We now take α, β as in (3.25). Since $\Omega^E Y$ takes its values in E_R, $i_{\Omega^E Y}\alpha = 0$ and so

$${}^a\nabla\alpha = d\alpha. \tag{3.30}$$

Similarly using (3.29), we get

$$({}^a\nabla + i_{\Omega^E Y})\beta = d\beta. \tag{3.31}$$

From (3.30), (3.31), we obtain (3.26). □

Remark 3.7. Equation (3.26) is the differential geometric counterpart to the algebraic considerations in [MQ, Section 5].

(d) *Double transgression formulas for the Mathai–Quillen Thom form.* We now recall the basic result of Mathai–Quillen [MQ, Theorem 4.10] in the context of complex geometry.

Theorem 3.8. *For any $u > 0$, let a_u be the form on M^E*

$$a_u = \det\left(\frac{-\Omega^E}{2i\pi}\right) \exp\left\{-u\left(\frac{|Y|^2}{2} + (\Omega^E)^{-1}\right)\right\}. \tag{3.32}$$

Then for any $u > 0$, the form a_u is closed and lies in $\mathbf{P}^{M^E}$. It is integrable on M^E and represents the Thom class of E_R.

Proof. As a two form on M with values in End E, Ω^E is of type $(1,1)$. Also since Ω^E is a complex endomorphism of E_R, for any $U, V \in T_R M$, $\Omega^E(U,V)$ induces a two form on E_R of type $(1,1)$. It is now clear that a_u is a form of type (k,k).

By Bianchi's identities, ${}^a\nabla^E \Omega^E = 0$. We now use the rotations $(\Omega^E)^{-1}$ with the conventions described before. We find that ${}^a\nabla(\Omega^E)^{-1} = 0$. $(\Omega^E)^{-1}Y$ is a form of degree -1 taking values in E. Using Proposition 3.6, we find that

$$d(\Omega^E)^{-1}Y = -2(\Omega^E)^{-1} - |Y|^2 \tag{3.33}$$

and so

$$d\left(\frac{|Y|^2}{2} + (\Omega^E)^{-1}\right) = 0. \tag{3.34}$$

We thus find that $da_u = 0$. By proceeding as in [MQ, Theorem 4.10], we find that the integral along the fiber $\int_{E_R} a_u$ is equal to 1. Therefore a_u represents the Thom class of E_R. □

In [MQ, Theorem 7.6], Mathai–Quillen gave a formula for the Euler form of a real oriented Euclidean vector bundle, by transgressing their formula (3.32) for a_u. We will here reprove this result, and besides obtain a double transgression formula.

Definition 3.9. For $u \geq 0$, let b_u and c_u be the forms on M^E

$$\begin{aligned} b_u &= \frac{1}{2}\det\left(\frac{-\Omega^E}{2i\pi}\right)(-\Omega^E)^{-1}Y \exp\left\{-u\left(\frac{|Y|^2}{2} + (\Omega^E)^{-1}\right)\right\}. \\ c_u &= \frac{\partial}{\partial b}\left[\det\left(-\left(\frac{\Omega^E}{2i\pi} + bI_E\right)\right)\right. \\ &\quad \left.\cdot \exp\left\{-u\left(\frac{|Y|^2}{2} + \left(2\pi\left(\frac{\Omega^E}{2\pi} + bJ_E\right)\right)^{-1}\right)\right\}\right]_{b=0}. \end{aligned} \tag{3.35}$$

Theorem 3.10. *The form b_u is of total degree* $2\dim E - 1$, *and the form c_u is of complex type* $(\dim E - 1, \dim E - 1)$. *For any $u > 0$*

$$\begin{aligned} \frac{\partial}{\partial u} a_u &= -db_u \\ b_u &= \frac{1}{2u}\left(\frac{\bar\partial - \partial}{2i\pi}\right) c_u. \end{aligned} \tag{3.36}$$

In particular for $u > 0$

$$\frac{\partial}{\partial u} a_u = \frac{\bar{\partial}\partial}{2i\pi}\left(\frac{c_u}{u}\right). \tag{3.37}$$

Proof. Equation (3.36) is proved in Mathai–Quillen [MQ, Section 7]. In fact using (3.33), we have

$$a_u = \det\left(-\frac{\Omega^E}{2i\pi}\right)\exp\left(\frac{ud}{2}(\Omega^E)^{-1}Y\right) \tag{3.38}$$

and so

$$\begin{aligned}\frac{\partial}{\partial u} a_u &= \frac{1}{2}\det\left(-\frac{\Omega^E}{2i\pi}\right) d(\Omega^E)^{-1}Y \exp\left\{-u\left(\frac{|Y|^2}{2} + (\Omega^E)^{-1}\right)\right\} \\ &= d\left(\frac{1}{2}(\Omega^E)^{-1}Y a_u\right).\end{aligned} \tag{3.39}$$

Let θ be the Kähler form of E_R. If $X, Y \in E_R$, then $\theta(X,Y) = \langle X, J_E Y\rangle$. If $X \in E_R$, the element of E_R^* which corresponds to X by the metric is given by

$$\sqrt{-1}(-i_{X^{(0,1)}} + i_{X^{(1,0)}})\theta.$$

Let A be an invertible skew-adjoint matrix in $\operatorname{End} E$. Let θ^A be the $(1,1)$ form on E

$$\begin{aligned}(U,V) \to \theta^A(U,V) = \theta(A^{-1}U, A^{-1}V) &= -\theta(A^{-2}U, V) \\ &= -\theta(U, A^{-2}V).\end{aligned} \tag{3.40}$$

From (3.40), we deduce that if $Y \in E_R$, if $A^{-1}Y \in E_R$ is identified with the corresponding element in E_R^*, then

$$A^{-1}Y = \sqrt{-1}(i_{AY^{(0,1)}} - i_{AY^{(1,0)}})\theta^A. \tag{3.41}$$

If $d^E = \partial^E + \bar{\partial}^E$ is the exterior differential on one fiber E, set

$$\partial_A^E = \partial^E + i_{AY^{(0,1)}};\ \bar{\partial}_A^E = \bar{\partial}^E + i_{AY^{(1,0)}}. \tag{3.42}$$

Clearly

$$(\partial_A^E)^2 = 0;\ (\bar{\partial}_A^E)^2 = 0. \tag{3.43}$$

Also the Lie derivative with respect to AY is given by

$$L_{AY} = (d^E + i_{AY})^2. \tag{3.44}$$

Using (3.43), (3.44), we get

$$L_{AY} = \bar{\partial}^E_A \partial^E_A + \partial^E_A \bar{\partial}^E_A. \tag{3.45}$$

Clearly $L_{AY}\theta^A = 0$, and so

$$\bar{\partial}^E_A \partial^E_A \theta^A = -\partial^E_A \bar{\partial}^E_A \theta^A. \tag{3.46}$$

Since the form θ^A is ∂^E and $\bar{\partial}^E$ closed, we obtain from (3.41) that

$$\sqrt{-1}(\partial^E_A - \bar{\partial}^E_A)\theta^A = A^{-1}Y. \tag{3.47}$$

Also

$$\frac{|Y|^2}{2} + A^{-1} = -(\partial^E_A + \bar{\partial}^E_A)\frac{A^{-1}Y}{2}. \tag{3.48}$$

From (3.46), (3.48), we get

$$\frac{|Y|^2}{2} + A^{-1} = -\sqrt{-1}\,\bar{\partial}^E_A \partial^E_A \theta^A. \tag{3.49}$$

In particular

$$\partial^E_A\left(\frac{|Y|^2}{2} + A^{-1}\right) = \bar{\partial}^E_A\left(\frac{|Y|^2}{2} + A^{-1}\right) = 0. \tag{3.50}$$

We deduce from (3.47), (3.50) that

$$\begin{aligned} &i \det\left(\frac{-A}{2i\pi}\right)(-A)^{-1}Y \exp\left\{-u\left(\frac{|Y|^2}{2} + A^{-1}\right)\right\} \\ &= (\partial^E_A - \bar{\partial}^E_A)\left[\det\left(\frac{-A}{2i\pi}\right)\theta^A \exp\left\{-u\left(\frac{|Y|^2}{2} + A^{-1}\right)\right\}\right]. \end{aligned} \tag{3.51}$$

In general, the form $\det\left(\frac{-A}{2i\pi}\right)\theta^A \exp\left\{-u\left(\frac{|Y|^2}{2} + A^{-1}\right)\right\}$ cannot be extended by continuity into a well-defined form when A is noninvertible. On the contrary the form

(3.52)

$$\begin{aligned} &\frac{\partial}{\partial b}\left[\det\left(-\left(\frac{A}{2i\pi} + bI_E\right)\right)\exp\left\{-u\left(\frac{|Y|^2}{2} + \left(2\pi\left(\frac{A}{2\pi} + bJ_E\right)\right)^{-1}\right)\right\}\right]_{b=0} \\ &= \frac{\partial}{\partial b}\left[\det\left(-\left(\frac{A}{2i\pi} + bI_E\right)\right)\right]_{b=0} \exp\left\{-u\left(\frac{|Y|^2}{2} + A^{-1}\right)\right\} \\ &- u \det\left(\frac{-A}{2i\pi}\right)2\pi\theta^A \exp\left\{-u\left(\frac{|Y|^2}{2} + A^{-1}\right)\right\} \end{aligned}$$

extends by continuity to arbitrary A. Also it is clear that the first form in the right-hand side of (3.52) is ∂_A^E and $\bar\partial_A^E$ closed. From (3.51), (3.52), we find that

$$\begin{aligned}
&\det\left(\frac{-A}{2i\pi}\right)(-A)^{-1}Y\exp\left\{-u\left(\frac{|Y|^2}{2}+A^{-1}\right)\right\}\\
&= \frac{1}{u}\left(\frac{\bar\partial_A^E-\partial_A^E}{2i\pi}\right)\frac{\partial}{\partial b}\left[\det\left(-\left(\frac{A}{2i\pi}+bI_E\right)\right)\exp\left\{-u\left(\frac{|Y|^2}{2}\right.\right.\right.\\
&\left.\left.\left.+\left(2\pi\left(\frac{A}{2\pi}+bJ_E\right)^{-1}\right)^{-1}\right)\right\}\right]_{b=0}
\end{aligned} \tag{3.53}$$

and both sides of (3.53) extend to arbitrary A. Using Proposition 3.6 and (3.53), we obtain (3.36). (3.37) follows from (3.36).

Since $(-\Omega^E)^{-1}Y$ is of degree -1, the form b_u is of degree $2\dim E-1$. If we give the degree two to the variable b, the form

$$\begin{aligned}
d_u(b) = \det\left(-\left(\frac{\Omega^E}{2i\pi}+bI_E\right)\right)\exp\left(-u\left(\left(\frac{|Y|^2}{2}\right.\right.\right.\\
\left.\left.\left.+\ 2\pi\left(\frac{\Omega^E}{2\pi}+bJ_E\right)^{-1}\right)\right)\right)
\end{aligned} \tag{3.54}$$

has total degree $2\dim E$. When differentiating this form with respect to b at $b=0$, we get a form of total degree $2\dim E-2$. Since $d_u(b)$ is a sum of forms of type (p,p), c_u is of type $(\dim E-1, \dim E-1)$. Our theorem is proved. □

Remark 3.11. (3.36), (3.37) can also be derived from Theorem 1.4 and from equalities (3.14), (3.17), (3.20). Also it is no accident that the Kähler form θ appears in (3.40), (3.51). For the role of Kähler forms in double transgression formulas, we refer to [BGS2] and to [B3]. In particular the computations in (3.42)–(3.51) are related to complex equivariant cohomology and appear in another form in [B3], in relation with Quillen metrics and analytic torsion.

(e) *Convergence of Mathai–Quillen currents.* We make the same assumptions as in Section 3d), and we also use the same notations.

Theorem 3.12. *Take $n \in \mathbb{N}$. Then there exists a constant $C>0$ such that if μ is a smooth differential form on M^E with compact support* *included in* $B_n = \{Y \in M^E; |Y| \leq n\}$, *for* $u \geq 1$

$$\left|\int_{M^E} \mu(a_u - \delta_M)\right| \leq \frac{C}{\sqrt{u}} \|\mu\|_{C^1(M^E)}$$
$$(3.55) \qquad \left|u \int_{M^E} \mu b_u\right| \leq \frac{C}{\sqrt{u}} \|\mu\|_{C^1(M^E)}$$
$$\left|\int_{M^E} \mu c_u\right| \leq \frac{C}{\sqrt{u}} \|\mu\|_{C^1(M^E)}.$$

Moreover, if U, Γ, φ, m *are taken as in Section* 1c) *with respect to the embedding* $i: M \to M^E$, *there exists* $C' > 0$ *such that for* $u \geq 1$

$$p_{U,\Gamma,\varphi,m}(a_u - \delta_M) \leq \frac{C}{\sqrt{u}}$$
$$(3.56) \qquad p_{U,\Gamma,\varphi,m}(u b_u) \leq \frac{C}{\sqrt{u}}$$
$$p_{U,\Gamma,\varphi,m}(c_u) \leq \frac{C}{\sqrt{u}}.$$

Proof. Let τ_u be the map $Y \to \sqrt{u} Y$. Then $a_u = \tau_u^* a_1$, $u b_u = \tau_u^* b_1$, $c_u = \tau_u^* c_1$. If $\sigma_u = \tau_u^{-1}$ then

$$\int_{M^E} \mu\, a_u = \int_{M^E} (\sigma_u^* \mu) a_1$$
$$\int_{M^E} \mu\, u b_u = \int_{M^E} (\sigma_u^* \mu) b_1$$
$$\int_{M^E} \mu\, c_u = \int_{M^E} (\sigma_u^* \mu) c_1.$$

We thus deduce easily that as $u \to \infty$

$$\int_{M^E} \mu\, a_u \to \int_M (i^* \mu) \int_E a_1$$
$$(3.57) \qquad \int_{M^E} \mu\, u b_u \to \int_M (i^* \mu) \int_E b_1$$
$$\int_{M^E} \mu\, c_u \to \int_M (i^* \mu) \int_E c_1.$$

Moreover, it is clear from (3.32) that $\int_E a_1 = 1$. Also since $i_Y (\Omega^E)^{-1} =$

$-(\Omega^E)^{-1}Y$, we find that $b_1 = -\frac{1}{2} i_Y a_1$ and so $\int_E b_1 = 0$. Finally

$$\int_E c_1 = (2\pi)^{-\dim E} \frac{\partial}{\partial b} \Big[\det\Big(-\Big(\frac{\Omega^E}{2i\pi} + bI_E \Big)\Big) \cdot \det\Big(-\Big(\frac{\Omega^E}{2i\pi} + bI_E \Big)\Big)^{-1} \Big]_{b=0} = 0. \tag{3.58}$$

The estimates in (3.55) obviously follow from (3.57). Finally the estimates in (3.56) follow easily from the methods of [B2, proof of Theorem 3.3]. □

(f) *An Euler–Green current.* Note that the form c_0 is clearly closed.

Definition 3.13. For $s \in \mathbb{C}$, $0 < \mathrm{Re}(s) < \frac{1}{2}$, let $\rho(s)$ be the current on M^E

$$\rho(s) = \frac{1}{\Gamma(s)} \int_0^{+\infty} u^{s-1} c_u du. \tag{3.59}$$

By (3.55), one verifies that the current $\rho(s)$ is well defined. Also $\rho(s)$ extends into a meromorphic function of $s \in \mathbb{C}$ which is holomorphic at $s = 0$.

Let $\tilde{e}(g^E)$ be the current

$$\tilde{e}(g^E) = \rho'(0). \tag{3.60}$$

Equivalently

$$\tilde{e}(g^E) = \int_0^1 (c_u - c_0) \frac{du}{u} + \int_1^{+\infty} c_u \frac{du}{u} - \Gamma'(1) c_0. \tag{3.61}$$

Finally, let $\psi(g^E)$ be the current

$$\psi(g^E) = \int_0^{+\infty} b_u du. \tag{3.62}$$

By (3.55), the current $\psi(g^E)$ is also well-defined.

Remember that E is the normal bundle to M in M^E.

Theorem 3.14. *The total degree of the current* $\psi(g^E)$ *is* $2\dim E - 1$. *The current* $\tilde{e}(g^E)$ *is of complex type* $(\dim E - 1, \dim E - 1)$. *The wave* *front sets of the currents $\tilde{e}(g^E)$ and $\psi(g^E)$ are included in $E^*_{\mathbf{R}}$. In particular $\tilde{e}(g^E) \in P_M^{M^E}$. The following equations of currents hold on M^E*

$$\begin{aligned} \psi(g^E) &= \frac{1}{2}\left(\frac{\bar{\partial}-\partial}{2i\pi}\right)\tilde{e}(g^E) \\ d\psi(g^E) &= e(g^E) - \delta_M. \end{aligned} \tag{3.63}$$

In particular

$$\frac{\bar{\partial}\partial}{2i\pi}\tilde{e}(g^E) = \delta_M - e(g^E). \tag{3.64}$$

Proof. The degree of the currents $\psi(g^E)$ and $\tilde{e}(g^E)$ can be calculated by using Theorem 3.10. The wave front set properties of our currents follow easily from Theorem 3.12. The first line of (3.63) follows from (3.36), (3.61), (3.62) and from the fact that the form c_0 is closed. The second line of (3.63) also follows from the same equations. □

Theorem 3.15. *The currents $\tilde{e}(g^E)$ and $\psi(g^E)$ are locally integrable. Also we have the formulas*

$$\begin{aligned} \tilde{e}(g^E) &= -\frac{\partial}{\partial b}\left[\det\left(-\left(\frac{\Omega^E}{2i\pi}+bI_E\right)\right)\operatorname{Log}\left(\frac{|Y|^2}{2}\right.\right. \\ &\qquad \left.\left. + \left(2\pi\left(\frac{\Omega^E}{2\pi}+bJ_E\right)\right)^{-1}\right)\right]_{b=0} \\ \psi(g^E) &= \frac{1}{2}\det\left(-\frac{\Omega^E}{2i\pi}\right)(-\Omega^E)^{-1}Y\left(\frac{|Y|^2}{2}+(\Omega^E)^{-1}\right)^{-1}. \end{aligned} \tag{3.65}$$

Equivalently

(3.66)

$$\begin{aligned} \tilde{e}(g^E) &= -\frac{\partial}{\partial b}\left[\det\left(-\left(\frac{\Omega^E}{2i\pi}+bI_E\right)\right)\right. \\ &\qquad \left.\cdot\left\{\operatorname{Log}\left(\frac{|Y|^2}{2}\right)+\sum_{j=1}^{\dim E-1}\frac{2^j}{j|Y|^{2j}}\left(\left(-2\pi\left(\frac{\Omega^E}{2\pi}+bJ_E\right)\right)^{-1}\right)^j\right\}\right]_{b=0} \\ \psi(g^E) &= \frac{1}{2}\det\left(-\frac{\Omega^E}{2i\pi}\right)(-\Omega^E)^{-1}Y\sum_{j=1}^{\dim E}\frac{2^j}{|Y|^{2j}}((-\Omega^E)^{-1})^{j-1}. \end{aligned}$$

In particular, if $\dim E = 1$, *then*

$$\tilde{e}(g^E) = \operatorname{Log}(|y|^2). \tag{3.67}$$

Proof. Using formula (3.61), it is clear that in order to prove that the current $\tilde{e}(g^E)$ is locally integrable, we only need to show that the current $\int_1^{+\infty} c_u \frac{du}{u}$ is locally integrable. From (3.35), we get

$$c_u = \exp\left(-\frac{u|Y|^2}{2}\right)\frac{\partial}{\partial b}\Bigg[\det\Bigg(-\Bigg(\frac{\Omega^E}{2i\pi} + bI_E\Bigg)\Bigg)\frac{\sum_{j=0}^{\dim E}\left((-2\pi((\Omega^E/2\pi)+bJ_E))^{-1}\right)^j u^j}{j!}\Bigg]_{b=0}. \tag{3.68}$$

Let us now again make the key observation (which was already made in another form in (3.58)) that in (3.68), the last index in the sum is in fact $\dim E - 1$ and not $\dim E$. Also for $k \geq 0$

$$\begin{aligned}\int_1^{+\infty}\exp\left(-\frac{u|Y|^2}{2}\right)u^{k-1}du &\leq C\left(1+\mathrm{Log}\frac{1}{|Y|}\right) \quad \text{if } k=0\\ &\leq |Y|^{-2k} \quad \text{if } k>0.\end{aligned} \tag{3.69}$$

Now the function $\mathrm{Log}|Y|$ is locally integrable on M^E. Similarly for $1 \leq k \leq \dim E - 1$, the function $|Y|^{-2k}$ is locally integrable. From (3.69), we deduce that the current $\tilde{e}(g^E)$ is locally integrable on M^E. The explicit calculation of $\tilde{e}(g^E)$ can then be done as in (3.16). The properties of the current $\psi(g^E)$ can be proved in the same way.

Remember that $|Y|^2 = 2|y|^2$. If $\dim E = 1$, again using (3.58), we find that

$$\tilde{e}(g^E) = \left[\frac{\partial}{\partial b}\left(\frac{\Omega^E}{2i\pi}+b\right)\right]_{b=0}\mathrm{Log}|y|^2 = \mathrm{Log}|y|^2.$$

Our theorem is proved. □

Remark 3.16. We make the same assumptions as in Remark 3.5. Then for essentially the same reasons as in Remark 3.5, the current $\tilde{e}(g^E)$ can be pulled back by the section s. $s^*\tilde{e}(g^E)$ is now a current on M which lies in $P_{M'}^M$ and also

$$\frac{\bar{\partial}\partial}{2i\pi}s^*\tilde{e}(g^E) = \delta_{M'} - e(g^E).$$

(g) *Comparison of the currents $\tilde{T}(g^E)$ and $\tilde{e}(g^E)$.*

Theorem 3.17. *The following equation holds*

$$\tilde{T}(g^E) - Td^{-1}(g^E)\tilde{e}(g^E) \in P_M^{M^E,0}. \tag{3.70}$$

Proof. We use the notations of the proof of Theorem 3.3. By (3.14), we know that

$$\begin{aligned}\varphi(Tr_s[N_H \exp(-A_u^2)]) = \frac{\partial}{\partial b}\Bigg[Td^{-1}\Big(-\Big(\frac{\Omega^E}{2i\pi}+bI_E\Big)\Big)\\ \cdot \det\Big(-\Big(\frac{\Omega^E}{2i\pi}+bI_E\Big)\Big)\exp\Big\{-u\Big(\frac{|Y|^2}{2}\\ + \Big(2\pi\Big(\frac{\Omega^E}{2\pi}+bJ_E\Big)^{-1}\Big)\Big)\Big\}\Bigg]_{b=0}.\end{aligned} \tag{3.71}$$

Therefore

$$\varphi(Tr_s[N_H \exp(-A_u^2)]) - Td^{-1}(g^E)c_u = -(Td^{-1})'(g^E)a_u. \tag{3.72}$$

For $s \in \mathbf{C}$, $0 < \mathrm{Re}(s) < \frac{1}{2}$, let $\delta(s)$ be the current

$$\begin{aligned}\delta(s) = \frac{1}{\Gamma(s)}\int_0^{+\infty} u^{s-1}\{\varphi(Tr_s[N_H \exp(-A_u^2)])\\ + (Td^{-1})'(g^E)\delta_M - Td^{-1}(g^E)c_u\}du.\end{aligned} \tag{3.73}$$

By Section 1g) and by Theorem 3.12, we know that $\delta(s)$ is a well-defined current on M^E, which extends into a current which is a meromorphic function of $s \in \mathbf{C}$. This function is holomorphic near $s = 0$. In particular

$$\delta'(0) = \tilde{T}(g^E) - Td^{-1}(g^E)\tilde{e}(g^E). \tag{3.74}$$

On the other hand, by (3.72), we find that if $s \in \mathbf{C}$, $0 < \mathrm{Re}(s) < \frac{1}{2}$, then

$$\delta(s) = -(Td^{-1})'(g^E)\frac{1}{\Gamma(s)}\int_0^{+\infty} u^{s-1}(a_u - \delta_M)du. \tag{3.75}$$

By Theorem 3.10, we know that

$$a_u - \delta_M = -\frac{1}{2i\pi}\bar{\partial}\partial\int_u^{+\infty} c_v\frac{dv}{v}. \tag{3.76}$$

Using (3.75), (3.76), we find that if $0 < \mathrm{Re}(s) < \frac{1}{2}$, then

$$\delta(s) = (Td^{-1})'(g^E)\frac{\bar{\partial}\partial}{2i\pi}\left[\frac{1}{\Gamma(s+1)}\int_0^{+\infty} u^{s-1}c_u du\right]. \tag{3.77}$$

Clearly

$$\frac{1}{\Gamma(s+1)}\int_0^{+\infty} u^{s-1} c_u du = \frac{\rho(s)}{s}. \tag{3.78}$$

On the other hand, $\rho(0) = c_0$ and c_0 is a closed form. So from (3.77), (3.78), we deduce that

$$\delta(s) = (Td^{-1})'(g^E)\frac{\bar{\partial}\partial}{2i\pi}\left(\frac{\rho(s)-\rho(0)}{s}\right). \tag{3.79}$$

From (3.79), we get

$$\delta'(0) = \frac{\bar{\partial}\partial}{2i\pi}\left\{(Td^{-1})'(g^E)\frac{\rho''(0)}{2}\right\}. \tag{3.80}$$

Using (3.74), (3.80), we find that

$$\tilde{T}(g^E) - Td^{-1}(g^E)\tilde{e}(g^E) = \frac{\bar{\partial}\partial}{2i\pi}\left\{(Td^{-1})'(g^E)\frac{\rho''(0)}{2}\right\}. \tag{3.81}$$

Also, it is easy to verify that the wave front set of the current $\rho''(0)$ is included in E. From (3.81), we deduce (3.70). □

Remark 3.18. It should be pointed out that although the current $\tilde{T}(g^E)$ is in general not locally integrable, the current $Td^{-1}(g^E)\tilde{e}(g^E)$ is locally integrable.

(h) *The current $\tilde{e}(g^E)$ as a function of g^E.* Let now g^E, g'^E be two Hermitian metrics on E. In [BGS1, Theorem 1.29], we have defined a unique class $\tilde{e}(g^E, g'^E) \in P^M/P^{M,0}$ which is such that

$$\frac{\bar{\partial}\partial}{2i\pi}\tilde{e}(g^E, g'^E) = e(g'^E) - e(g^E). \tag{3.82}$$

$\tilde{e}(g^E, g'^E)$ lifts naturally to M^E.

Theorem 3.19. *The following identity holds*

$$\tilde{e}(g'^E) - \tilde{e}(g^E) = -\tilde{e}(g^E, g'^E) \quad \text{in } P_M^{M^E}/P_M^{M^{E,0}} \tag{3.83}$$

Proof. The proof of Theorem 3.18 is closely related to the proof of Theorem 2.5. In fact we lift E to $M \times \mathbb{P}^1$, and we consider a metric g on

E which restricts to g^E, g'^E on $M \times \{0\}$, $M \times \{\infty\}$ respectively. We then use equation (3.64) for $\tilde{e}(g^E)$ on $M \times \mathbf{P}^1$ in the obvious analogue of (2.15). To obtain (3.83), the key fact is that $\int_{\mathbf{P}^1} \mathrm{Log}|z|^2 \delta_M = 0$. Details are left to the reader. □

(i) *A transitivity property of the currents* $\tilde{e}(g^E)$. Let now E and E' be two holomorphic vector bundles on M. E and E' are vector subbundles of $E \oplus E'$. Then the manifolds M^E and $M^{E'}$ are vector submanifolds of $M^{E\oplus E'}$ which intersect transversally, and $M^E \cap M^{E'} = M$.

The vector bundles E and E' lift naturally to $M^{E\oplus E'}$. If $z = (y, y') \in E \oplus E'$, set

$$\sigma(z) = y; \qquad \sigma'(z) = y'$$

Then σ and σ' are holomorphic sections of E and E', which exactly vanish on $M^{E'}$ and M^E respectively. On $M^{E\oplus E'}$, the sections σ and σ' clearly possess the transversality properties which were described in Remark 3.5.

Let g^E and $g^{E'}$ be Hermitian matrics on E and E'. We equip $E \oplus E'$ with the metric $g^{E\oplus E'}$, which is the orthogonal sum of the metrics g^E and $g^{E'}$. By Remark 3.16, the currents $\tilde{\sigma}^* \tilde{e}(g^E)$ and $\tilde{\sigma}'^* \tilde{e}(g^{E'})$ are well-defined currents on $M^{E\oplus E'}$. Also the forms $e(g^E)$ and $e(g^{E'})$ lift naturally to $M^{E\oplus E'}$.

Theorem 3.20. *The following identities of currents hold*

$$\begin{aligned}
\tilde{e}(g^{E\oplus E'}) &= e(g^{E'})\sigma^* \tilde{e}(g^E) + \tilde{e}(g^{E'})\delta_{M^{E'}} \\
&\qquad \text{in } P^{M^{E\oplus E'}}_{M^E \cup M^{E'}} / P^{M^{E\oplus E'},0}_{M^E \cup M^{E'}}. \\
\tilde{e}(g^{E\oplus E'}) &= e(g^{E})\sigma'^* \tilde{e}(g^{E'}) + \tilde{e}(g^{E})\delta_{M^{E}} \\
&\qquad \text{in } P^{M^{E\oplus E'}}_{M^E \cup M^{E'}} / P^{M^{E\oplus E'},0}_{M^E \cup M^{E'}}.
\end{aligned} \tag{3.84}$$

Proof. Our Theorem can be given a direct proof by using formula (3.61) and by proceeding as in the proof of Theorem 2.7. We will instead use our previous results. Namely by Theorem 3.17, we know that

$$\tilde{e}(g^{E\oplus E'}) = Td(g^{E\oplus E'})\tilde{T}(g^{E\oplus E'}) \text{ in } P^{M^{E\oplus E'}}_{M^E \cup M^{E'}} / P^{M^{E\oplus E'},0}_{M^E \cup M^{E'}}. \tag{3.85}$$

and that corresponding equalities hold for the currents $\tilde{e}(g^E)$ and $\tilde{e}(g^{E'})$.

Also we have the obvious

$$Td(g^{E\oplus E'}) = Td(g^E)Td(g^{E'}). \tag{3.86}$$

Finally, by Theorem 2.7, we know that

(3.87)
$$\tilde{T}(g^{E\oplus E'}) = \det\left(I - \exp\left(\frac{\Omega^{E'}}{2i\pi}\right)\right)\sigma^*\tilde{T}(g^E) + (Td^{-1})(g^E)\tilde{T}(g^{E'})\delta_{M^{E'}}$$
$$\text{in } P^{M^{E\oplus E'}}_{M^E\cup M^{E'}}/P^{M^{E\oplus E'},0}_{M^E\cup M^{E'}}.$$

Using (3.85)–(3.87), the first line of (3.84) follows. The second line is obvious by interchanging the roles of E and E'. □

Remark 3.21. Let s, s' be holomorphic sections of E, E' on M. Set

$$M' = \{x \in M; s(x) = 0\}$$
$$\tilde{M}' = \{x \in M; s'(x) = 0\}.$$

Assume that s, s' both verify the assumptions of Remark 3.5 and that M' and $\tilde{M}'$ intersect transversally. Then the section $s'' = (s, s')$ of $E \oplus E'$ also verifies the assumptions of Remark 3.5. Set $M'' = M' \cap \tilde{M}'$. From (3.84), we deduce

(3.88)
$$s''^*\tilde{e}(g^{E\oplus E'}) = e(g^{E'})s^*(\tilde{e}(g^E)) + s'^*(\tilde{e}(g^{E'}))\delta_{M'} \text{ in } P^M_{M'\cup\tilde{M}'}/P^{M,0}_{M'\cup\tilde{M}'}.$$

(j) *The arithmetic Euler class.* We first recall some definitions from [GS1] and [GS2]. Let (A, Σ, F_∞) be an *arithmetic ring* i.e. A is an excellent regular noetherian integral domain, Σ a finite nonempty set of monomorphisms $\sigma: A \to \mathbb{C}$, and F_∞ a conjugate linear involution of the $\mathbb{C}$-algebra $\mathbb{C}^\Sigma$ which fixes A (embedded diagonally into $\mathbb{C}^\Sigma$).

Let X be an *arithmetic variety* over A, i.e. a regular quasi-projective flat scheme over $\operatorname{Spec} A$. Denote by $X_\sigma = X \otimes_\sigma \mathbb{C}$ the complex variety obtained by extending scalars from A to $\mathbb{C}$ by $\sigma \in \Sigma$, and by X_∞ the analytic space $X_\infty = \coprod_{\sigma\in\Sigma} X_\sigma(\mathbb{C})$.

For any integer $p \geq 0$, let $A^{pp}(X_{\mathbb{R}})$ (resp. $\mathcal{D}^{pp}(X_{\mathbb{R}})$) be the real vector space of real forms (resp. currents) α of type (p,p) on X_∞ such that $F^*_\infty(\alpha) = (-1)^p\alpha$. We also introduce the quotient spaces

$$(3.89)\qquad \begin{aligned}\tilde{A}^{pp}(X_{\mathbb{R}}) &= A^{pp}(X_{\mathbb{R}})/(\operatorname{Im}\partial + \operatorname{Im}\bar{\partial})\\ \tilde{\mathcal{D}}^{pp}(X_{\mathbb{R}}) &= \mathcal{D}^{pp}(X_{\mathbb{R}})/(\operatorname{Im}\partial + \operatorname{Im}\bar{\partial}).\end{aligned}$$

Let now X^p be the set of points of codimension p of the scheme X and $Z^p(X)$ the group of codimension p cycles on X, i.e. the free abelian group generated by X^p. Given $Z \in Z^p(X)$, let Z_∞ be the analytic cycle on X_∞

attached to Z, and $\delta_Z \in \mathcal{D}^{pp}(X_{\mathbb{R}})$ the current given by integration on Z_∞: if $Z = \sum_\alpha n_\alpha Z_\alpha$, with $Z_\alpha \subset X$ and $n_\alpha \in \mathbb{Z}$, for any smooth form ω on X_∞,

$$\delta_Z(\omega) = \sum_\alpha n_\alpha \int_{Z_{\alpha,\infty}} \omega. \tag{3.90}$$

A *Green current* for $Z \in Z^p(X)$ is an element $g \in \tilde{\mathcal{D}}^{p-1,p-1}(X_{\mathbb{R}})$ such that

$$-dd^c g = \delta_Z - \omega,$$

with ω a smooth form (here, as usual, $d = \partial + \bar{\partial}$ and $d^c = (\partial - \bar{\partial})/(4i\pi)$). The pair (Z, g) is called an *arithmetic cycle.* The *arithmetic Chow group* $\widehat{CH}^p(X)$ is the group generated by codimension p arithmetic cycles with the relation $(\operatorname{div}(f), -\operatorname{Log}|f|^2) = 0$, for every nonzero rational function f on a closed irreducible subscheme Y of codimension $p-1$ on X, where $\log|f|^2$ is the current defined by the formula

$$\operatorname{Log}|f|^2(\omega) = \int_{Y_\infty^{n-s}} \operatorname{Log}|f|^2 \omega, \tag{3.91}$$

Y_∞^{n-s} denoting the smooth locus of Y_∞. We refer to [GS1] for the study of these groups. Let

$$\widehat{CH}(X) = \bigoplus_{p \geq 0} \widehat{CH}^p(X).$$

If F is the fraction field of A, $X_F = X \bigotimes_A F$ the generic fiber of X and $CH_f^p(X)$ the Chow group of X with support in special fibers, we may view $\widehat{CH}^p(X)$ as a subquotient of

$$CH_f^p(X) \oplus Z^p(X_F) \oplus \tilde{\mathcal{D}}^{p-1,p-1}(X_{\mathbb{R}})$$

(see [GS1]).

Now let (E, h) be an *Hermitian vector bundle* on X, i.e. E is an algebraic vector bundle on X and h is an Hermitian metric on the holomorphic vector bundle E_∞ on X_∞ attached to E; furthermore h is invariant by F_∞. Let r be the rank of E and $\Phi(T_1, \ldots, T_r)$ a symmetric polynomial in r variables with coefficients in a ring B. In [GS2] a characteristic class

$$\hat{\Phi}(E, h) \in \widehat{CH}(X)_B = \widehat{CH}(X) \bigotimes_{\mathbb{Z}} B$$

is defined. In particular we get Chern classes $\hat{c}_p(E,h) \in \widehat{CH}^p(X)$ where $c_p(T_1,\ldots,T_r)$ is the p-th elementary symmetric polynomial.

Even if E is generated by its global sections, we do not know in general an explicit arithmetic cycle representing $\hat{\Phi}(E,h)$ (see [GS2], §5). However, we shall give below a description of the *Euler class* (i.e. the top Chern class) $\hat{c}_r(E,h) \in \widehat{CH}^r(X)$.

For this, let us assume we are given an algebraic section s of E over X which satisfies over X_∞ the conditions of Remark 3.5. Call $Z_F \in Z^r(X_F)$ the zero set of s on the generic fiber. Using Remark 3.16 we see that $g = -s_\infty^*(\tilde{e}(h))$ is a Green current for Z_F. On the other hand, viewing s as a morphism from X to the total space X^E of E, we may consider $e_f = s^*([X]) \in CH_f^r(X)$, where $[X] \in CH_f^r(X^E)$ is the part of the zero section supported on special fibers. The triple (e_f, Z_F, g) defines a class $\hat{e}(E,h)$ in $\widehat{CH}^r(X)$.

Theorem 3.22. *Under the above hypotheses,*

$$\hat{e}(E,h) = \hat{c}_r(E,h) \quad \text{in } \widehat{CH}^r(X). \tag{3.92}$$

Proof. Let h_0, h_1 be two Hermitian metrics on E. We will first prove that

$$\hat{e}(E,h_0) - \hat{e}(E,h_1) = \hat{c}_r(E,h_0) - \hat{c}_r(E,h_1). \tag{3.93}$$

Let $\tilde{e}(h_0,h_1) \in \tilde{A}^{r-1,r-1}(X_{\mathbb{R}})$ be the Bott–Chern class of forms constructed in [BGS1] associated with the polynomial $e(A) = \det(A)$ and the metrics h_0 and h_1. In particular, we know that

$$\frac{\bar{\partial}\partial}{2i\pi}\tilde{e}(h_0,h_1) = e(h_1) - e(h_0).$$

Let X^E be the total space of E. Let p be the projection $X^E \to X$. By Theorem 3.19, we know that

$$\tilde{e}(h_1) - \tilde{e}(h_0) = -p^*\tilde{e}(h_0,h_1). \tag{3.94}$$

By using Remark 3.5, we get from (3.94)

$$s^*\tilde{e}(h_1) - s^*\tilde{e}(h_0) = -\tilde{e}(h_0,h_1).$$

By using [GS2, Theorem 4.8], we obtain (3.93).

We now prove that $\hat{e}(E,h)$ does not depend on the choice of s. On X^E, the element $\hat{e}(p^*E,\ p^*h)$ of $\widehat{CH}(X^E)$ is the class of the arithmetic cycle $(s_0(X), -\tilde{e}(h))$. Using (3.93) we shall prove in 4e) below that

$$\hat{e}(E,h) = s_0^*\big(\hat{e}(p^*E, p^*h)\big) \tag{3.95}$$

where s_0^* on $\widehat{CH}$ is defined as in [GS1]. Therefore we conclude that

$$\hat{e}(E,h) = s^*\big(\hat{e}(p^*E,p^*h)\big) = s_0^*\big(\hat{e}(p^*E,p^*h)\big)$$

does not depend on the choice of s. The definition of the class $\hat{e}(E,h) = s_0^*\big(\hat{e}(p^*E,p^*h)\big)$ makes sense for any Hermitian vector bundle (E,h) on X (without assuming that E has a good global section). We shall see that this class $\hat{e}(E,h)$ has the same properties as $\hat{c}_r(E,h)$ ([GS2, par. 4]).

Property 1: the class $\hat{e}(E,h)$ is functorial. That is, given a morphism $f\colon Y \to X$ of arithmetic varieties,

$$f^*\hat{e}(E,h) = \hat{e}(f^*E, f^*h).$$

To see that, we first consider the commutative diagram

$$\begin{array}{ccc} Y^{f^*(E)} & \xrightarrow{\tilde{f}} & X^E \\ \downarrow & & \downarrow \\ Y & \xrightarrow{f} & X \end{array}$$

so that we may replace f by $\tilde{f}$. The functoriality follows from the formula for $\tilde{e}$ in Theorem 3.15.

Property 2: $\hat{e}(E,h)$ is multiplicative under direct sum. Let (E,h) and (E',h') be two hermitian vector bundles on X, and $(E \oplus E', h \oplus h')$ their orthogonal direct sum. Then $\hat{e}(E\oplus E',\ h\oplus h') = \hat{e}(E,h)\hat{e}(E',h')$ in $\widehat{CH}(X)$. By the definition of the product in $\widehat{CH}$ (see [GS1]) this follows from Remark 3.21.

Property 3: The image of $\hat{e}(E,h)$ into the (classical) Chow group $CH^r(X)$ is the Euler class of E. This is clear since it is given as the self-intersection of X in X^E.

Property 4: Let $\omega\colon \widehat{CH}^r \to A^{rr}(X_{\mathbf{R}})$ be the map sending the class of (Z,g) to the smooth form $\delta_Z + dd^c g$. Then $\omega(\hat{e}(E,h)) = e(E,h)$ is the Euler form of (E,h). Indeed, this follows from (3.64).

Property 5: When (L,h) is an Hermitian line bundle, $\hat{e}(L,h) = s_0^*(\mathrm{div}(y), -\mathrm{Log}\, |y|^2)$, where y is the tautological section of p^*L on X^L. This follows from (3.67) and the definition of $\hat{e}$ given above.

From [GS2] we know that $\hat{c}_r(E,h)$ enjoys also the properties 1, 2, 3, 4, 5. From this we shall conclude that $\hat{e}(E,h) = \hat{c}_r(E,h)$. Since X has ample line bundles, any bundle on X is the pull-back of a bundle on the product of a Grassmannian by a projective space over A, hence, using Property 1, we may assume that X is a complete variety. Let then

$$H^{r-1,r-1}(X_{\mathbf{R}}) = \mathrm{Ker}\big(dd^c\colon \tilde{A}^{r-1,r-1}(X_{\mathbf{R}}) \to A^{rr}(X_{\mathbf{R}})\big).$$

Since X_∞ is Kähler, this group is the subspace of the real cohomology of X_∞ of degree $(r-1, r-1)$ where F_∞^* is $(-1)^{r-1}$. We define a map

$$a\colon H^{r-1,r-1}(X_{\mathbf{R}}) \to \widehat{CH}^r(X)$$

by sending the element η to the class of $(0,\eta)$. From Properties 3 and 4 and the exact sequence of [GS1, Theorem 3.3.5] we conclude that

$$\delta(E) = \hat{e}(E,h) - \hat{c}_r(E,h)$$

lies in the image of a.

Let

$$0 \to S \to E \to Q \to 0$$

be an exact sequence of vector bundles on X with metrics h', h, h'' respectively. We shall prove that

$$\delta(E) = \delta(S) + \delta(Q). \tag{3.96}$$

From Property 2 we know that (3.96) holds when (E,h) is the orthogonal direct sum of (S,h') with (Q,h''). In general, on $X \times \mathbf{P}^1$ there is an Hermitian vector bundle $(\tilde{E},\tilde{h})$ whose restriction to $X \times \{0\}$ (resp. $X \times \{\infty\}$) is (E,h) (resp. $(S,h') \oplus (Q,h'')$) (see [BGS1, Theorem 1.29]). Since the restriction maps i_0^* and i_∞^* from $H^{r-1,r-1}((X \times \mathbf{P}^1)_\infty)$ to $H^{r-1,r-1}(X_\infty)$ coincide, we get

$$\delta(E) = \delta((S,h') \oplus (Q,h'')) = \delta(S) + \delta(Q).$$

In particular $\delta(E)$ does not depend on h.

(6) From Property 5 we know that $\delta(E) = 0$ when E has rank one. By induction on the rank of E we shall conclude that $\delta(E) = 0$. Indeed let $\mathbf{P}(E)$

be the projective space of E and $p: \mathbf{P}(E) \to X$ the projection. Since p^* is injective on $\widehat{CH}(X)$ ([GS2]3) we just need to check that $p^*(\delta(E)) = \delta(p^*E)$ vanishes. Using the standard exact sequence

$$0 \to S \to p^*E \to \mathcal{O}(1) \to 0$$

we get $\delta(p^*E) = \delta(\mathcal{O}(1)) + \delta(S) = 0$ by induction hypothesis. This concludes the proof of Theorem 3.22. □

4. Deformation to the normal cone

In this paragraph we study the deformation W to the normal cone of a closed immersion $i: Y \to X$ of complex manifolds. This construction is due to MacPherson [BaFM]. The variety W maps onto $\mathbf{P}^1$ and all its fibers W_t, $t \in \mathbf{P}^1$, except the fiber at infinity, are isomorphic to X. We prove in Theorem 4.8 that, given a bundle η on Y and a resolution ξ of $i_*\eta$ on X, there exists a complex $\tilde{\xi}$ of bundles on W such that $\tilde{\xi}$ restricted to $W_0 = X$ is ξ and $\tilde{\xi}$ restricted to $\mathbf{P}(N \oplus \mathbf{C}) \subset W_\infty$ (where N is the normal bundle to Y in X) maps onto the Koszul resolution of the direct image of η, with acyclic kernel. We then show in Theorem 4.11 how $T(h^\xi)$, for appropriate choices of metrics, relates to the singular Bott-Chern current $T(h^{\tilde{\xi}})$ and an Euler-Green current $\tilde{e}(g^{H^*})$. In e) we use deformation to the normal cone to end the proof of (3.95), which had been delayed in the previous paragraph. Finally, with Theorem 4.13, we get a version of an arithmetic Riemann-Roch-Grothendieck Theorem for immersions. The proof relies upon essentially all our previous results.

(a) *Projective bundles and Koszul complexes.* This section and the next one is a brief review of material from [SGA6], [Ha], and [BaFM]. If $Y \subset X$ is a closed imbedding of complex manifolds we write $\mathcal{I}_{Y/X}$ for the sheaf of ideals of Y in X; if Y is a divisor we also write $\mathcal{O}_X(-Y)$. Note that $\mathcal{O}_X(nY) = \mathcal{O}_X(-Y)^{\otimes -n}$, that $\mathcal{O}_X$ is a subsheaf of $\mathcal{O}_X(Y)$, and that $\mathcal{O}_X \subset \mathcal{O}_X(Y)$ canonically.

The conormal sheaf $\mathcal{N}^*_{Y/X} \cong \mathcal{I}_{Y/X}/\mathcal{I}^2_{Y/X}$ is the dual of the sheaf of sections of the normal bundle $N_{Y/X}$. (Here we follow [Ha] rather than [SGA6].) Note also $\mathcal{N}^*_{Y/X} \cong i^*\mathcal{I}_{Y/X}$ where $i: Y \to X$ is the inclusion. More generally, if Y is a submanifold of X, there is an isomorphism

$$Tor_i^{\mathcal{O}_X}(\mathcal{O}_Y, \mathcal{O}_Y) = \Lambda^i \mathcal{N}_{Y/X}.$$

of sheaves on Y [SGA6].

If $p: E \to X$ is a vector bundle with sheaf of holomorphic sections $\mathcal{E}$, there is a canonical homomorphism $\iota: p^*\mathcal{E}^* \to \mathcal{O}_E$ dual to the tautological section of p^*E. This homomorphism vanishes along the zero section $s: X \to E$, and the associated Koszul complex $K(\iota) = (\Lambda^\bullet p^*\mathcal{E}^*)$ is a resolution of $s_*\mathcal{O}_X$. If $\varphi: \mathcal{V} \to \mathcal{O}_X$ is a homomorphism with $\mathcal{V}$ locally free, the differential in the associated Koszul complex $K(\varphi)$ shall be

$$d_i^\varphi: \Lambda^i\mathcal{V} \to \Lambda^{i-1}\mathcal{V}$$

$$e_1 \wedge \cdots \wedge e_i \to \sum_{k=1}^{i} (-1)^{k-1}\varphi(e_k) e_1 \wedge \cdots \wedge \hat{e}_k \wedge \cdots \wedge e_i$$

so that the chain map d_i^φ is interior multiplication by φ.

Let $P = \mathbf{P}(E \oplus \mathbf{C})$. This parametrizes lines in $E \oplus \mathbf{C}$ or, equivalently, rank one quotients of $\mathcal{E}^* \oplus \mathcal{O}_X$. On P we have the universal exact sequence:

$$0 \to \mathcal{H} \to \hat{p}^*\mathcal{E}^* \oplus \mathcal{O}_P \to \mathcal{O}_P(1) \to 0$$

($\hat{p}: \mathbf{P}(E \oplus \mathbf{C}) \to X$ is the projection). $\mathbf{P}(E)$ embeds as a divisor in $\mathbf{P}(E \oplus \mathbf{C})$, given by the vanishing of the map $\mathcal{O}_P \to \mathcal{O}_P(1)$ induced by the inclusion $\mathcal{O}_P \subset \hat{p}^*\mathcal{E}^* \oplus \mathcal{O}_P$.

Equivalently, $\mathbf{P}(E)$ is the locus on which $\mathcal{O}_P \subset \mathcal{H}$. The complement of $\mathbf{P}(E)$ is canonically isomorphic to E, since a line $L \subset E \oplus \mathbf{C}$ which maps surjectively to $\mathbf{C}$ is equivalent to a homomorphism $\mathbf{C} \to E$. Specifically a homomorphism $\ell: \mathbf{C} \to E$ determines an injective homomorphism $\hat{e}: \mathbf{C} \to \mathbf{E} \oplus \mathbf{C}$, by $\hat{\ell}(x) = (\ell(x), x)$ i.e. we embed $E \subset \mathbf{P}(E \oplus \mathbf{C})$ by $\underline{x} \to (\underline{x}, 1)$.

The map $\theta: \mathcal{H} \to \hat{p}^*\mathcal{E}^*$ induced by the projection from $\hat{p}^*\mathcal{E}^* \oplus \mathcal{O}_P$ is an isomorphism on E, and if, on E, we compose θ^{-1} with the homomorphism induced by *minus* the second projection, we obtain ι. Let $\varphi: \mathcal{H} \to \mathcal{O}_P$ be the homomorphism induced by minus the projection $\hat{p}^*\mathcal{E}^* \oplus \mathcal{O}_P \to \mathcal{O}_P$. Since it is surjective on $\mathbf{P}(E)$ and equal to ι on E, the associated Koszul complex $K(\varphi)$ is a resolution of $s_*\mathcal{O}_X$, where $s: X \to E \subset \mathbf{P}(E \oplus \mathbf{C})$ is the zero section.

(b) *Deformation to the normal bundle.* Let $i: Y \to X$ be a closed embedding of complex manifolds. We suppose that Y is of pure codimension n in X.

Let $W = W_{Y/X}$ be the blow up of $X \times \mathbf{P}^1$ along $Y \times \{\infty\}$. Since Y and $X \times \mathbf{P}^1$ are manifolds, W is itself a manifold. The map $\pi: W \to X \times \mathbf{P}^1$ is an isomorphism away from $Y \times \{\infty\}$; we shall write P for the exceptional divisor in W of the blow up. Recall that

$$P \cong \mathbf{P}(N_{Y \times \{\infty\}/X \times \mathbf{P}^1}).$$

Since $N_{Y\times\{\infty\}/X\times\mathbf{P}^1} \cong p_Y^* N_{Y/X} \oplus p_\infty^* N_{\infty/\mathbf{P}^1}$, where $p_Y\colon Y\times\{\infty\}\to Y$ and $p_\infty\colon Y\times\{\infty\}\to\{\infty\}$ are the projections, we may also write

$$P \cong \mathbf{P}\big(N_{Y/X}\otimes N_{\infty/\mathbf{P}^1}^{-1}\oplus \mathbf{C}\big).$$

Note that the bundle $N_{\infty/\mathbf{P}^1}$, while trivial, is not canonically trivial. Hence P is the projective completion of $N_{Y/X}\otimes N_{\infty/\mathbf{P}^1}^{-1}$ with divisor at ∞, $\mathbf{P}\big(N_{Y/X}\otimes N_{\infty/\mathbf{P}^1}^{-1}\big)\cong\mathbf{P}(N_{Y/X})$.

The map $q_W : W\to\mathbf{P}^1$, obtained by composing π with the projection $q\colon X\times\mathbf{P}^1\to\mathbf{P}^1$, is flat, and for $z\in\mathbf{P}^1$:

$$q^{-1}(z)=\begin{cases} X & z\neq\infty\\ P\cup\tilde{X} & z=\infty\,.\end{cases}$$

Here $\tilde{X}$ is the blow up of X along Y, and $P\cap\tilde{X}$ is the divisor at ∞ on P, identified with the exceptional divisor on $\tilde{X}$.

We shall use the following notations:

$$\begin{array}{ccc} P & \xrightarrow{f} & W\\ {\scriptstyle\pi_P}\downarrow & & \downarrow{\scriptstyle\pi}\\ Y\times\{\infty\} & \xrightarrow{i_\infty} & X\times\mathbf{P}^1 \end{array}$$

$$\begin{array}{ll} i\colon Y\to X & \\ W=\pi^{-1}(\infty)=P\cup\tilde{X}. & \\ q\colon X\times\mathbf{P}^1\to\mathbf{P}^1 & \text{the projection,}\\ q_W=q\cdot\pi, & \\ p\colon X\times\mathbf{P}^1\to X & \text{the projection,}\\ p_W=p\cdot\pi, & \\ q_Y\colon Y\times\mathbf{P}^1\to\mathbf{P}^1 & \text{the natural projections}\\ p_Y\colon Y\times\mathbf{P}^1\to Y & \text{the natural projections} \end{array}$$

Note that the proper transform of $Y\times\mathbf{P}^1\subset X\times\mathbf{P}^1$ is isomorphic to $Y\times\mathbf{P}^1$. We write

$$j\colon Y\times\mathbf{P}^1\to W$$

for the induced map.

(c) *Deformation of resolutions.* Now suppose that η is a locally free sheaf on Y and that $\xi\to i_*\eta$ is a resolution of η by a bounded complex of locally free sheaves on X. We shall deform ξ through a complex $\tilde{\xi}$ on W to a Koszul type resolution of $s_*\eta$ on P, where $s: Y \to N_{Y/X} \otimes N^{-1}_{\infty/\mathbf{P}^1} \subset P$ is the zero section.

First observe that there is an exact sequence of sheaves on $\mathbf{P}^1$:

$$0 \to \mathcal{O}_{\mathbf{P}^1}(-\infty) \to \mathcal{O}_{\mathbf{P}^1} \to \mathcal{O}_\infty \to 0. \tag{4.3}$$

Hence if we write K^∞ for the complex $\mathcal{O}_{\mathbf{P}^1}(-\infty) \to \mathcal{O}_{\mathbf{P}^1}$, with $\mathcal{O}_{\mathbf{P}^1}$ in degree zero, $p^*\xi \otimes q^* K^\infty$ is a resolution of $i_{\infty *}\eta$. Let us write $\mathcal{G} = \pi^*(p^*\xi \otimes q^* K^\infty)$ which is a bounded complex of locally free sheaves on W.

Lemma 4.1. *Let $\mathcal{H}$ be the locally free sheaf on P which is the kernel of the map*

$$\pi_P^*(\mathcal{N}^*_{Y/X} \oplus \mathcal{N}^*_{\infty/\mathbf{P}^1}) \to \mathcal{O}_P(1).$$

Then

$$\mathcal{H}_p(\mathcal{G}) \cong f_*(\Lambda^p \mathcal{H} \otimes_{\mathcal{O}_P} \pi_P^*\eta).$$

Proof. This follows immediately from [SGA6, exp. VII, lemma 3.2]. □

The sheaves $f_*(\Lambda^p \mathcal{H} \otimes_{\mathcal{O}_P} \pi_p^*\eta)$ are locally direct sums of copies of $f_*\mathcal{O}_P$ hence, since P is a divisor in W, they are locally of projective dimension one. Therefore if

$$a: \mathcal{E} \to f_*(\Lambda^p \mathcal{H} \otimes_{\mathcal{O}_P} \pi_P^*\eta)$$

is an epimorphism, with $\mathcal{E}$ a locally free $\mathcal{O}_W$- module the kernel of α is again locally free. It follows, by induction on $i \geq 0$, that $\mathrm{Ker}(d_i^{\mathcal{G}})$ is a locally free sheaf on W.

Definition 4.2. For $i \geq 0$, $\tilde{\xi}_i$ is the locally free $\mathcal{O}_W$ module $\ker(d_i^{\mathcal{G}}) \otimes_{\mathcal{O}_W} \mathcal{O}_W(i\infty)$. Here $\mathcal{O}_W(\infty) = q_W^* \mathcal{O}_{\mathbf{P}^1}(\infty)$ and $\mathcal{O}_W(i\infty) = \mathcal{O}_W(\infty)^{\otimes i}$ contains $\mathcal{O}_W$ naturally as a subsheaf.

Observe that $\mathcal{G}_i \cong p_W^*\xi_i \oplus p_W^*\xi_{i-1}(-\infty)$, and that $d_i^{\mathcal{G}}(x, y) = (d_i^\xi(x) + (-1)^i y,\ d_{i-1}^\xi(y))$, where we identify $y \in p_W^*\xi_{i-1}(-\infty)$ with its image in $p_W^*\xi_i$ under the natural inclusion. The projection $\mathcal{G}_i \to \xi_i$ identifies $\mathrm{Ker}(d_i^{\mathcal{G}})$ with the subsheaf of $p_W^*\xi_i$ consisting of those sections x such that dx lies in $p_W^*\xi_{i-1}(-\infty)$. Hence $\tilde{\xi}_i$ is isomorphic to the fibre product of the diagram:

$$\begin{array}{ccc}
\tilde{\xi}_i & \longrightarrow & p_W^*\xi_{i-1}((i-1)\infty) \\
\Big| & & \Big| \\
p_W^*\xi_i(i\infty) & \xrightarrow{d_i^\xi} & p_W^*\xi_{i-1}(i\infty)
\end{array}$$

Remark. (i) $\tilde{\xi}_i$ contains $p_W^*\xi_i$ as a subsheaf, the inclusion being induced by the "graph" map

$$p_W^*\xi_i \to p_W^*\xi_i(i\infty) \oplus p_W^*\xi_{i-1}((i-1)\infty)$$
$$x \to (x, (-1)^i d_i^\xi(x)).$$

Clearly $\tilde{\xi}_i|_{W-W_\infty} = p_W^*\xi_i|_{W-W_\infty}$.

(ii) There is an epimorphism of locally free $\mathcal{O}_P$ modules $f^*\tilde{\xi}_i \to \Lambda^i(\mathcal{H}\otimes \pi_P^*\eta \otimes f^*\mathcal{O}_W(\infty))$.

(iii) There is an exact sequence

$$(4.4) \qquad 0 \to \mathrm{Ker}(p_W^* d_i^\xi)(i\infty) \to \tilde{\xi}_i \to \mathrm{Im}(p_W^* d_i^\xi)((i-1)\infty) \to 0.$$

Lemma 4.3. *The differential $p_W^* d_i : p_W^*\xi_i \to p_W^*\xi_{i-1}$ extends to a homomorphism $d_i^{\tilde{\xi}} : \tilde{\xi}_i \to \tilde{\xi}_{i-1}$ which makes $(\tilde{\xi}_\bullet, d_\bullet^{\tilde{\xi}})$ a complex of $\mathcal{O}_W$ modules.*

Proof. $p_W^* d_i$ extends trivially to the homomorphism

$$p_W^* d_i \otimes \mathrm{Id}_{\mathcal{O}_W(i\infty)} : p_W^*\xi_i(i\infty) \to p_W^*\xi_{i-1}(i\infty).$$

By definition this homomorphism, restricted to $\tilde{\xi}_i$, has image lying in $p_W^*\xi_{i-1}((i-1)\infty)$. Since $d_i^{\xi 2} = 0$, the image is contained in $\tilde{\xi}_{i-1} \subset p_W^*((i-1)\infty)$. Since $d_i^{\tilde{\xi}}|_{W-W_\infty} = p_W^*(d_i^\xi)$, $d_i^{\tilde{\xi}2}$ vanishes on $W - W_\infty$ and hence on W. □

Remark 4.4. The construction of the complex $\tilde{\xi}_\bullet = (\tilde{\xi}_\bullet, d_\bullet^{\tilde{\xi}})$ is local on X and is compatible with finite sums since both twisting by $\mathcal{O}_W(\infty)$ and fibre products are local operations compatible with finite sums.

Lemma 4.5. *If $\xi_\bullet$ is acyclic (i.e. $\eta \cong 0$ or Y is empty), then $\tilde{\xi}_\bullet$ is the pullback, via π, of a complex on $X \times \mathbb{P}^1$ which becomes split acyclic when restricted to $X \times \{\infty\}$.*

Proof. If $\xi_\bullet$ is acyclic, then

$$0 \to \mathrm{Im}(d_{i+1}) \to \xi_i \to \mathrm{Im}(d_i) \to 0$$

is an exact sequence of locally free sheaves. Hence the fibre product of the diagram

$$\begin{array}{ccc} & & q^*\mathrm{Im}(d_i)(-\infty) \\ & & \downarrow \\ q^*(\xi_i) & \xrightarrow{d_i} & q^*\mathrm{Im}(d_i) \end{array}$$

is already locally free on $X \times \mathbb{P}^1$, and is preserved under π_W^*. As discussed in [BGS1, Theorem 1.29] this pullback has the property that over $X \times \{\infty\}$ it splits as the direct sum $q^*\mathrm{Im}(d_{i+1}) \oplus q^*\mathrm{Im}(d_i)(-\infty)$. □

Corollary 4.6. *If $U = X - Y$, the restriction of $\tilde{\xi}$ to $U \times \mathbb{P}^1 \subset W$ restricts to a split acyclic complex on $U \times \{\infty\} \subset \tilde{X} \subset W_\infty$.*

Since $j_*\mathcal{O}_{Y\times\mathbb{P}^1}$ and $\mathcal{O}_{W_\infty}$ are Tor independent, the formation of $\tilde{\xi}$ commutes with restriction to $Y \times \mathbb{P}^1 \subset W$. The restriction of $\xi_\bullet$ to Y has locally free homology:

$$\begin{aligned}\mathcal{H}_p(i^*\xi_\bullet) &= \mathcal{T}or_p^{\mathcal{O}_X}(\mathcal{O}_Y, \eta)\\ &\cong \mathcal{T}or_p^{\mathcal{O}_X}(\mathcal{O}_Y, \mathcal{O}_Y) \otimes_{\mathcal{O}_Y} \eta\\ &\cong \Lambda^p \mathcal{N}^*_{Y/X} \otimes_{\mathcal{O}_Y} \eta.\end{aligned}$$

Hence $i^*\xi_\bullet$ breaks up into short exact sequences of locally free sheaves:

$$\begin{gathered}0 \to \mathcal{Z}_i \to i^*\xi_i \to \mathcal{B}_i \to 0\\ 0 \to \mathcal{B}_{i+1} \to \mathcal{Z}_i \to \mathcal{H}_i(i^*\xi_\bullet) \to 0.\end{gathered}$$

Therefore $\tilde{\xi}_i\big|_{Y\times\mathbb{P}^1}$ is obtained by pulling back the extension (or rather its inverse image under p_Y^*) by the inclusion $\mathcal{B}_i(-\infty) \to \mathcal{B}_i$ and twisting by $\mathcal{O}(i\infty)$. On restricting to $Y \times \{\infty\} \subset Y \times \mathbb{P}^1$, we find that, since $\mathcal{O}_{\mathbb{P}^1}(\infty)\big|_\infty \cong \mathcal{N}_{\infty/\mathbb{P}^1}$,

$$\tilde{\xi}_i\big|_{Y\times\mathbb{P}^1} \cong \left(\mathcal{Z}_i \otimes \mathcal{N}^i_{\infty/\mathbb{P}^1}\right) \oplus \left(\mathcal{B}_i \otimes \mathcal{N}^{i-1}_{\infty/\mathbb{P}^1}\right). \tag{4.5}$$

The differential $d_i^{\tilde{\xi}}$ restricts to the map $d_i\colon (x,y) \to (\gamma_i(y), 0)$, where

$$\gamma_i\colon \mathcal{B}_i \otimes \mathcal{N}^{i-1}_{\infty/\mathbb{P}^1} \to \mathcal{Z}^{i-1}_{i-1} \otimes \mathcal{N}^{i-1}_{\infty/\mathbb{P}^1}$$

is the natural inclusion. If we set $\mathcal{L}$ equal to the split acyclic complex with:

$$\mathcal{L}_i = \left(\mathcal{B}_{i+1} \otimes \mathcal{N}^i_{\infty/\mathbb{P}^1}\right) \oplus \left(\mathcal{B}_i \otimes \mathcal{N}^{i-1}_{\infty/\mathbb{P}^1}\right) \tag{4.6}$$

with differential $d_i(x,y) = (y, 0)$, then there is an obvious inclusion $\mathcal{L}_\bullet \to \tilde{\xi}\big|_{Y\times\{\infty\}}$ with cokernel the direct sum $\bigoplus_{i\geq 0} H_i(i^*\xi_\bullet) \otimes \mathcal{N}^i_{\infty/\mathbb{P}^1}$ viewed as a complex with zero differentials. Hence we have an exact sequence:

$$0 \to \mathcal{L}_\bullet \to \tilde{\xi}\big|_{Y\times\{\infty\}} \to \bigoplus_{i\geq 0} \Lambda^i\left(\mathcal{N}^*_{Y/X} \otimes \mathcal{N}_{\infty/\mathbb{P}^1}\right) \otimes_{\mathcal{O}_Y} \eta \to 0. \tag{4.7}$$

Suppose that $i: Y \to X$ is defined by equations $x_1 = \ldots = x_n = 0$ with the x_i part of a system of coordinates on X. Let $\xi_\bullet = K_\bullet(\underline{x})$ be the Koszul complex associated to the map $\mathcal{O}_X^n \to \mathcal{O}_X$ sending $\underline{a}$ to $\underline{a} \cdot \underline{x}$, where $\underline{x} = (x_1, \ldots, x_n)$. Then $K_\bullet(\underline{x})$ is a resolution of $i_* \mathcal{O}_Y$. On W we have an epimorphism

$$\mathcal{O}_W^n \oplus \mathcal{O}_W(-\infty) \to \mathcal{O}_W(-P)$$
$$(\underline{a}, b) \to \underline{a} \cdot \underline{x} + b.$$

Since $\mathcal{O}_W(-P)$ is invertible, the kernel $\mathcal{I}$ of this map is a locally free sheaf.

Lemma 4.7. *There is a canonical isomorphism of complexes* $\tilde{\xi}_\bullet \to K_\bullet(\varphi)$ *where* $\varphi: p_w^* \mathcal{I}(\infty) \to \mathcal{O}_W$ *is the map induced by minus the projection map* $\mathcal{O}_W^n \oplus \mathcal{O}_W(-\infty) \to \mathcal{O}_W(-\infty)$. *Furthermore*

(i) $\tilde{\xi}_\bullet\big|_{\tilde{X}}$ *is split acyclic.*

(ii) $\tilde{\xi}_\bullet\big|_{N_{Y/X} \otimes N^{-1}_{\infty/\mathbf{P} \subset P}}$ *is the tautological Koszul complex.*

(iii) $\tilde{\xi}_\bullet$ *is a resolution of* $j_* \mathcal{O}_{Y \times \mathbf{P}^1}$.

Proof. The complex $p_W^* \xi_\bullet \otimes K_\bullet^\infty$ is the Koszul complex associated to the map of sheaves $\mathcal{O}_W^n \oplus \mathcal{O}_W(-\infty) \to \mathcal{O}_W$ induced by the inclusion $\mathcal{O}_W(-P) \subset \mathcal{O}_W$. Hence the Koszul differential

$$\Lambda^i(\mathcal{O}_W^n \oplus \mathcal{O}_W(-\infty)) \to \Lambda^{i-1}(\mathcal{O}_W^n \oplus \mathcal{O}_W(-\infty))$$

is the composition of the canonical map

$$\Lambda^i(\mathcal{O}_W^n \oplus \mathcal{O}_W(-\infty)) \to \Lambda^{i-1}(\mathcal{O}_W^n \oplus \mathcal{O}_W(-P))$$

which has kernel $\Lambda^i \mathcal{I}$, the injective multiplication map

$$\Lambda^{i-1}(\mathcal{I}) \otimes \mathcal{O}_W(-P) \to \Lambda^{i-1}(\mathcal{I})$$

and the natural inclusion

$$\Lambda^{i-1}(\mathcal{I}) \to \Lambda^{i-1}(\mathcal{O}_W^n \oplus \mathcal{O}_W(-\infty)).$$

Hence the inclusion $\Lambda^i(\mathcal{I}) \to \Lambda^i(\mathcal{O}_W^n \oplus \mathcal{O}_W(-\infty))$ identifies $\Lambda^i(\mathcal{I})$ with $\mathrm{Ker}(d_i^{\mathcal{G}})$ and therefore, equivalently, identifies $\Lambda^i(\mathcal{I}(\infty))$ with $\tilde{\xi}_i$.

Minus the projection $\mathcal{O}_W^n \oplus \mathcal{O}_W(-\infty) \to \mathcal{O}_W(-\infty)$ induces a surjective map $\mathcal{O}_W(-\infty) \oplus \mathcal{O}_W \to \mathcal{O}_W$ and hence a map $\mathcal{I}(\infty) \to \mathcal{O}_W$. There is an induced map between Koszul complexes $\Lambda^\bullet(\mathcal{I}(\infty)) \to \Lambda^\bullet(\mathcal{O}_W^n(\infty) \oplus \mathcal{O}_W)$ which gives a commutative diagram

$$\begin{array}{ccccc}
\Lambda^i(\mathcal{O}_W^n) = \xi_i & \longrightarrow \Lambda^i(\mathcal{I}(\infty)) & \longrightarrow \Lambda^i(\mathcal{O}_W^n(\infty) \oplus \mathcal{O}_W) & = \Lambda^i(\mathcal{O}_W^n(\infty)) \oplus \Lambda^{i-1}(\mathcal{O}_W^n(\infty)) \\
\downarrow & \downarrow & & \downarrow \\
\Lambda^{i-1}(\mathcal{O}_W^n) = \xi_{i-1} & \longrightarrow \Lambda^{i-1}(\mathcal{I}(\infty)) & \longrightarrow \Lambda^{i-1}(\mathcal{O}_W^n(\infty) \oplus \mathcal{O}_W) & = \Lambda^{i-1}(\mathcal{O}_W^n(\infty)) \oplus \Lambda^{i-2}(\mathcal{O}_W^n(\infty))
\end{array}$$

$$\begin{array}{ccc} a & \longrightarrow & (a,(-1)^i d_i(a)) \\ \downarrow & & \downarrow \\ da & \longrightarrow & (d_i(a),0) \end{array}$$

(n.b. The Koszul differential induced by *minus* the projection $\mathcal{O}_W^n(\infty)\oplus \mathcal{O}_W \to \mathcal{O}_W$ is $e_1\wedge\cdots\wedge e_i + e_1\wedge\cdots\wedge e_{i-1}\wedge f \to (-1)^{i-2} f\, e_1\wedge\cdots\wedge e_{i-1}$ for $e_1,\ldots,e_n \in \mathcal{O}_W^n(\infty)$ and $f\in \mathcal{O}_W$.

We turn now to items (i), (ii), and (iii).

Observe that $\mathcal{O}_W(-P)\otimes\mathcal{O}_W(-\infty) = \mathcal{O}_W(\tilde{X})$, so that we have an exact sequence

$$0\to \mathcal{I}(\infty)\to \mathcal{O}_W^n(\infty)\oplus\mathcal{O}_W\to\mathcal{O}_W(\tilde{X})\to 0. \tag{4.8}$$

Hence if we restrict to $\tilde{X}$, the map $\mathcal{O}_{W|\tilde{X}}\subset \big(\mathcal{O}_W^n(\infty)\oplus\mathcal{O}_W\big)\big|_{\tilde{X}}$ to $\mathcal{O}_W(\tilde{X})\big|_{\tilde{X}}$ vanishes. Therefore, on restricting to $\tilde{X}$, $\mathcal{I}(\infty)$ splits as a direct sum

$$\mathcal{I}(\infty)\big|_{\tilde{X}} = \mathcal{I}_1\oplus\mathcal{O}_{\tilde{X}}\subset \mathcal{O}_W(\infty)^n\big|_{\tilde{X}}\oplus\mathcal{O}_{\tilde{X}}.$$

The Koszul complex associated to $\mathcal{I}(\infty)\to\mathcal{O}_W$ therefore restricts to the Koszul complex for minus the projection $\mathcal{I}_1\oplus\mathcal{O}_{\tilde{X}}\to\mathcal{O}_{\tilde{X}}$, which is split acyclic.

Turning to (ii) observe that on $P=\mathbf{P}(N_{Y/X}\otimes N^{-1}_{\infty/\mathbf{P}^1}\oplus\mathbf{C}), \mathcal{I}(\infty)$ restricts to the kernel of the map

$$\pi_P^*(\mathcal{N}^*_{Y/X}\otimes\mathcal{N}_{\infty/\mathbf{P}^1})\oplus\mathcal{O}_P\to\mathcal{O}_P(\tilde{X}\cap P)$$

i.e. to the analogue of the locally free sheaf $\mathcal{H}$ of Lemma 4.1; (ii) now follows from the discussion of loc.cit.

Finally, to prove (iii), we need only show that the map $\tilde{\xi}_\bullet\to j_*\mathcal{O}_{Y\times\mathbf{P}^1}$, induced by the map $\tilde{\xi}_0=\mathcal{O}_W\to j_*\mathcal{O}_{Y\times\mathbf{P}^1}$, is a quasi-isomorphism of complexes. We know that $\tilde{\xi}_\bullet|_{W-W_\infty}=p_W^*\xi_\bullet$ is a resolution of $j_*\mathcal{O}_{Y\times\mathbf{C}}$, and we know that $\xi_\bullet|_{\tilde{X}}$ is acyclic; it suffices therefore to verify (iii) in a neighbourhood U of $Y\times\{\infty\}\subset W$. If we choose a local equation $t=0$ for ∞ in $\mathbf{P}^1$, then $x_1=\ldots=x_n=t=0$ is a local system of equations for $Y\times\{\infty\}\subset X\times\mathbf{P}^1$. Hence $\mathcal{I}$ is isomorphic to the kernel of map $\mathcal{O}_W^{n+1}\to\mathcal{O}_W$ sending $(a_1,\ldots,a_n,b)$ to $\sum_{i=1}^n a_ix_i+bt$. We can choose U so that $\frac{x_1}{t},\ldots,\frac{x_n}{t}, t$ is part of a system of coordinates on U, with $Y\times\mathbf{P}^1\subset U$ given by the equations $\frac{x_1}{t}=\ldots=\frac{x_n}{t}=0$. (This is a part of the standard description of coordinates on a blow up.) The locally free sheaf $\mathcal{I}(\infty)$ is the kernel of the map

$$\mathcal{O}_W^{n+1}\to\mathcal{O}_W$$

$$(a_1,\ldots,a_n,b)\to\sum_{i=1}^n a_ix_i/t+b.$$

Therefore the map $\mathcal{O}_W^n \to \mathcal{O}_W^{n+1}$ sending $(a_1, \ldots, a_n)$ to $(a_1, \ldots, a_n, -\sum a_i x_i/t))$ is an isomorphism onto $\mathcal{I}(\infty)$. Composing with minus the projection $\mathcal{O}_W^{n+1} = \mathcal{O}_W^n \oplus \mathcal{O}_W \to \mathcal{O}_W$, we identify the Koszul complex $\Lambda^\bullet \mathcal{I}(\infty)$ with the Koszul complex $K_\bullet(\frac{x_1}{t}, \ldots, \frac{x_n}{t})$, which is a resolution of $j_* \mathcal{O}_{Y \times \mathbf{P}^1}$. □

We return now to the general case of a resolution $\xi_\bullet \to i_* \eta$. Note that there is a natural map $\tilde{\xi}_0 \to j_* \eta$.

Theorem 4.8.

(1) $\tilde{\xi}_\bullet$ *is a resolution of* $j_* p_Y^* \eta$.

(2) $\tilde{\xi}_\bullet|_{\tilde{X}}$ *is split acyclic.*

(3) *There is a natural exact sequence of complexes of locally free sheaves on* P:

$$0 \to \pi_P^* \mathcal{L}_\bullet \to f^* \tilde{\xi} \to K_\bullet(\varphi) \otimes \pi_P^* \eta \to 0 \tag{4.9}$$

where $K_\bullet(\varphi)$ *is the Koszul complex on* $P = \mathbf{P}(N_{Y/X} \otimes N_{\infty/\mathbf{P}^1}^{-1} \oplus \mathbf{C})$ *described in* a), *and* $\mathcal{L}_\bullet$ *is the split acyclic complex described in* (4.6).

Proof. (1) is a local question on X, so we may suppose that $\xi_\bullet = \zeta_\bullet \oplus (K_\bullet(\underline{x}) \otimes \mathcal{V})$ where $\zeta_\bullet$ is acyclic, $K_\bullet(\underline{x})$ is the Koszul complex associated to equations $x_1 = \ldots = x_n = 0$ of $Y \subset X$ and $\mathcal{V}$ is locally free on X such that $j^* \mathcal{V} \cong \eta$. Then $\tilde{\xi}_\bullet \cong \tilde{\zeta}_\bullet \oplus \widetilde{K_\bullet(\underline{x})} \otimes p_W^* \mathcal{V}$. By Lemma 4.5, $\tilde{\zeta}_\bullet$ is acyclic. By Lemma 4.7 $\widetilde{K_\bullet(\underline{x})}$ is a resolution of $j_* \mathcal{O}_{Y \times \mathbf{P}^1}$, and hence $\tilde{\xi}_\bullet$ is a resolution of $\mathcal{O}_{Y \times \mathbf{P}^1} \otimes p_W^* \mathcal{V} \cong j_*(p_Y^* \eta)$, thus proving part (1). Next, observe that by $\tilde{\zeta}_\bullet|_{W_\infty}$ is split acyclic, and hence $\tilde{\zeta}_\bullet|_{\tilde{X}}$ is split acyclic too, while $\widetilde{K_\bullet(x_1, \ldots, x_n)}|_{\tilde{X}}$ is split acyclic by part (i) of Lemma 4.7. Hence $\tilde{\zeta}_\bullet|_{\tilde{X}}$ is split acyclic, at least locally on X. However $\tilde{\xi}|_{\tilde{X} - \tilde{X} \cap P}$ is split acyclic by Lemma 4.5, and the local splittings described above are compatible with this splitting, and hence uniquely determined, and patch together into a global splitting.

Finally, to prove (iii), we start by observing that by Remark 4.2 (ii), there is an epimorphism of locally free sheaves on P:

$$\varepsilon_i \colon f^* \tilde{\xi}_i \to H_i(p_W^* \xi_\bullet \otimes (\mathcal{O}_W(-\infty) \to \mathcal{O}_W))(i\infty) \cong \Lambda^i(\mathcal{H}) \otimes \pi_P^* \eta. \tag{4.10}$$

Here $\mathcal{H}$ is the kernel of the map $\mathcal{N}_{Y/X}^* \otimes \mathcal{N}_{\infty/\mathbf{P}^1} \oplus \mathcal{O}_P \to \mathcal{O}_P(\tilde{X})$. In order to show that this map is compatible with differentials, we can work locally on X, and write $\xi_\bullet = \zeta_\bullet \oplus (K_\bullet(\underline{x}) \otimes \mathcal{V})$ as above. Then $f^*(\tilde{\xi}) \cong f^*(\tilde{\zeta}) \oplus f^* \widetilde{K_\bullet(\underline{x})} \otimes \pi_P^* \eta$, and since $\tilde{\zeta}$ is acyclic, the map ε factors through the projection onto the second factor in the direct sum. But by Lemma 4.7, the map $f^*\widetilde{K_\bullet(\underline{x})}\otimes\pi_P^*\eta\to\Lambda^*(\mathcal{H})\otimes\pi_P^*(\eta)$ is an isomorphism of complexes. Hence ε is a map of complexes, and its kernel is, locally on X, $\tilde{\zeta}_\bullet|_P$. There is a map of complexes $f^*\pi_P^*\eta\to f^*(\tilde{\xi}_\bullet\otimes(\mathcal{O}_W\to\mathcal{O}_W(\infty)))$ given by the inclusions

$$\pi_P^*(\mathcal{B}_{i+1}\otimes\mathcal{N}^i_{\infty/\mathbf{P}^1})\subset\pi_P^*(i^*\xi_i\otimes\mathcal{O}_{\mathbf{P}^1}(\infty))=f^*\pi^*(\xi_i(i\infty)).$$

It suffices to show that the image of this map is contained in $\tilde{\xi}_\bullet$, and equals the kernel of ε. Again this is a local question, and writing, as usual, $\xi_\bullet\cong\zeta_\bullet\oplus(K_\infty(\underline{x})\otimes\mathcal{V})$, we see that $\mathcal{L}_\bullet\cong\tilde{\zeta}|_{Y\times\{\infty\}}=\operatorname{Ker}\varepsilon$ as desired. The complex $\tilde{\xi}_\bullet$ on W constructed above coincides (if $\eta\neq 0$) with the complex obtained by MacPherson's Grassmannian graph construction, which is described in [BaFM]. In [BaFM] it is shown that the parameter space for the Grassmannian graph construction applied to $\xi_\bullet$ is *locally* isomorphic to W. It then follows from the fact that the map from the Grassman bundle to $X\times\mathbf{P}^1$ is separated, that the two constructions are globally isomorphic. □

Lemma 4.9. *The normal bundle of* $Y\times\mathbf{P}^1$, *in* W,

$$\mathcal{N}_{Y\times\mathbf{P}^1/W}\cong p_Y^*\mathcal{N}_{Y/X}\otimes q_Y^*\mathcal{O}_{\mathbf{P}^1}(-\infty).$$

Proof. (This also follows directly from description $\mathcal{N}^*$ as $\mathcal{I}/\mathcal{I}^2$; see 1.1).

$$\mathcal{N}^*_{Y\times\mathbf{P}^1/W}\cong\mathcal{T}or_1^{\mathcal{O}_W}(\mathcal{O}_Y,\mathcal{O}_Y).$$

Let $\tilde{\xi}_\bullet\to i_*\mathcal{O}_Y$ be a resolution; then

$$\mathcal{T}or_1^{\mathcal{O}_W}(\mathcal{O}_Y,\mathcal{O}_Y)\cong\mathcal{H}_1(j^*\tilde{\xi}_\infty)\cong\mathcal{H}_1(p_Y^*\xi_\bullet)\otimes q_Y^*\mathcal{O}_{\mathbf{P}^1}(\infty)$$

and by the discussion after Corollary 4.6

$$\cong p_Y^*\mathcal{N}^*_{Y/X}\otimes q_Y'^*\mathcal{O}_{\mathbf{P}^1}(\infty).\qquad\square$$

Remark 4.10. All the results of a), b), c) remain valid for arithmetic varieties over any base instead of the complex numbers.

(d) *Bott–Chern currents and deformation to the normal cone.* We keep the notations of the previous paragraph, except that we denote by $N=N_{Y/X}$ the normal bundle of Y in X, and by $N'=N_{Y\times\mathbf{P}^1/W}$ the normal bundle of $Y \times \mathbf{P}^1$ in W. We also call φ the projection $p_W : W \to X$, and $\psi = \pi_P : P \to Y$ its restriction to P.

We now put metrics on the bundles considered so far: We fix metrics g^N g^η on N, η, and h^ξ on ξ. We assume that h^ξ satisfies assumption (A) with respect to g^N, g^η.

On $\mathcal{O}_{\mathbf{P}^1}(-\infty)$ we put the standard metric, and on N' the metric $g^{N'}$ coming from the isomorphism of Lemma 4.9.

On $\tilde{\xi}$ we choose a metric $h^{\tilde{\xi}}$ which satisfies assumption (A) with respect to g^η and $g^{N'}$, whose restriction to $W_0 = X \times \{0\}$ coincides with h^ξ, and whose restriction to the blow up $\tilde{X} \subset W_\infty$ of X along Y is split acyclic (these assumptions are compatible since (ξ, h) satisfies assumption (A) and $\tilde{X}$ does not meet $Y \times \{\infty\}$).

On the bundle H on P whose sheaf of sections is $\mathcal{H}$, we put the metric g^H induced from its inclusion into $\psi^*(\mathcal{N}^*_{Y/X} \oplus \mathcal{N}^*_{\infty/\mathbf{P}^1})$ (see Lemma 4.1). Note that H restricted to Y is isometric to N^*. Let σ be the canonical section of H^* and $\sigma^*(\tilde{e}(g^{H^*}))$ the corresponding Euler-Green current on P (see Remark 3.16).

For each $j \geq 0$ consider the exact sequence of sheaves

$$A_j : 0 \to \pi_P^* \mathcal{L}_j \to f^* \tilde{\xi}_j \xrightarrow{\varepsilon} K_j(\varphi) \otimes \pi_P^* \eta \to 0$$

defined in Theorem 4.8 (3). On the bundle L_j corresponding to

$$\mathcal{L}_j = (\mathcal{B}_{j+1} \otimes \mathcal{N}^j_{\infty/\mathbf{P}^1}) \oplus (\mathcal{B}_j \otimes \mathcal{N}^{j-1}_{\infty/\mathbf{P}^1})$$

we put the orthogonal direct sum of the induced metrics, so that the complex $L_\bullet$ attached to $\mathcal{L}_\bullet$ becomes split acyclic as a complex of Hermitian holomorphic vector bundles. On $K_j(\varphi) = \Lambda^j H$ we put the metric induced from g^H. Let g^{A_j} be the corresponding metric on the complex A_j. Then $T(g^{A_j})$ is a smooth current on P.

Finally, let z be the standard coordinate on $\mathbf{P}^1$ and $\operatorname{Log}|z|^2$ the function on W obtained by pulling back $\operatorname{Log}|z|^2$ on $\mathbf{P}^1$ via $p_W : W \to \mathbf{P}^1$. Notice that $\operatorname{Log}|z|^2$ is integrable on W, since near $\mathbf{P}(N)$ we have the equations

$$z^{-1} = \frac{x_0}{x_i} y_i$$

(where $y_i = 0$ are local equations of Y in X and $x_0, \ldots, x_e$ are coordinates in $N \oplus 1$). Therefore

$$\operatorname{Log}|z|^2 = -\operatorname{Log}|y_i|^2 - \operatorname{Log}\left|\frac{x_i}{x_0}\right|^2 \tag{4.11}$$

is locally integrable.

Theorem 4.11. *Let C be the current defined by*

(4.12)
$$C = T(h^{\xi}) + \varphi_*\left[\mathrm{Log}|z|^2 ch(h^{\tilde{\xi}})\right] - \psi_*\left[Td^{-1}(g^{H^*})\,\sigma^*(\tilde{e}(g^{H^*}))\right] ch(g^{\eta})\delta_Y + \psi_*\left[\sum_{j=0}^{m}(-1)^j T(g^{A_j})\right]\delta_Y.$$

Then $C \in P_Y^0$. Furthermore the integral $\psi_*\left[Td^{-1}(g^{H^*})\sigma^*(\tilde{e}(g^{H^*}))\right]$ along the fibers of ψ is a smooth closed differential form on Y, whose cohomology class does not depend on the choice of the metric g^N.

Proof. By Theorem 1.9 the wave front set $WF(T(h^{\tilde{\xi}}))$ in $T^*_{\mathbb{R}}W$ of the current $T(h^{\tilde{\xi}})$ is included in $N'^*_{\mathbb{R}}$. Also the projection $q_W: W \to \mathbb{P}^1$ is a submersion near $Y \times \mathbb{P}^1 \subset W$. In particular we deduce from [H, Theorem 8.2.4] that

$$WF(T(h^{\tilde{\xi}})) \cap WF(-\mathrm{Log}|z|^2) = \emptyset.$$

By [H, Theorem 8.2.10], the product of currents $\mathrm{Log}|z|^2 T(h^{\tilde{\xi}})$ is well-defined, and we can use on this product the usual rules of differential calculus. In particular

$$\begin{aligned}(4.13)\qquad &\frac{1}{2i\pi}\bar{\partial}\partial\left(\mathrm{Log}|z|^2\right)T(h^{\tilde{\xi}}) - \mathrm{Log}|z|^2\frac{\bar{\partial}\partial}{2i\pi}T(h^{\tilde{\xi}}) \\ &= \frac{\bar{\partial}}{2i\pi}((\partial\,\mathrm{Log}|z|^2)T(h^{\tilde{\xi}})) + \frac{\partial}{2i\pi}(\mathrm{Log}|z|^2(\bar{\partial}T(h^{\tilde{\xi}})))\end{aligned}$$

and so

$$\frac{1}{2i\pi}\bar{\partial}\partial\left(\mathrm{Log}|z|^2\right)T(h^{\tilde{\xi}}) - \mathrm{Log}|z|^2\frac{\bar{\partial}\partial}{2i\pi}T(h^{\tilde{\xi}})$$

lies in $P^{W,0}$. Using (2.16) and (4.11) (which is of special interest where q_W is not a submersion), we find that

$$\frac{\bar{\partial}\partial}{2i\pi}\mathrm{Log}|z|^2 = \delta_{W_0} - \delta_{W_\infty}. \qquad (4.14)$$

It follows from our previous considerations on wave front sets that the restrictions of the current $T(H^{\tilde{\xi}})$ to W_0 and W_∞ are well-defined. Also, by construction $T(h^{\tilde{\xi}})\big|_{W_0} = T(h^{\xi})$. We deduce from Theorem 1.9 and

equations (4.13) and (4.14) that
(4.15)

$$T(h^\xi)\delta_{W_0} - T(h^{\tilde{\xi}})\delta_{W_\infty} - \mathrm{Log}|z|^2 (Td^{-1}(g^{N'})ch(g^\eta))\delta_{Y\times \mathbf{P}^1} - ch(\tilde{\xi}))$$
$$= \frac{\bar{\partial}}{2i\pi}((\partial \,\mathrm{Log}|z|^2)T(h^{\tilde{\xi}})) + \frac{\partial}{2i\pi}((\mathrm{Log}|z|^2)\bar{\partial}T(h^{\tilde{\xi}}))$$

lies in P_0^W. Using Lemma 3.10, we find easily that,

$$Td^{-1}(g^{N'}) = Td^{-1}(g^N) + (Td^{-1})'(g^N)\cdot\left(\frac{-R}{2i\pi}\right), \tag{4.16}$$

where R denotes the curvature of the holomorphic Hermitian connection on $\mathcal{O}(-1)$. By integrating (4.15) along the fibers of $\varphi\colon W \to X$ and using (4.16), we find that

$$\begin{aligned} &T(h^\xi) - \varphi_*\left[T(h^{\tilde{\xi}})\delta_{W_\infty}\right] + \varphi_*\left[\mathrm{Log}|z|^2 ch(h^{\tilde{\xi}})\right] \\ &\quad - \left[\int_{\mathbf{P}^1} \mathrm{Log}(|z|^2)\cdot\left(\frac{-R}{2i\pi}\right)\right](Td^{-1})'(g^N)ch(g^\eta)\delta_Y \\ &= \frac{\bar{\partial}}{2i\pi}\varphi_*\left(\partial(\mathrm{Log}|z|^2)T(h^{\tilde{\xi}})\right) + \frac{\partial}{2i\pi}\varphi_*\left((\mathrm{Log}|z|^2)\bar{\partial}T(h^{\tilde{\xi}})\right). \end{aligned} \tag{4.17}$$

We now calculate $\varphi_*\left(T(h^{\tilde{\xi}})\delta_{W_\infty}\right)$.

Let $\tilde{A}_u$ be the superconnection (1.10) associated with the chain complex of Hermitian vector bundles $(\tilde{\xi}, \tilde{v})$. Let k be the imbedding $W_\infty \to W$. Note that $\mathbf{P}(N)$ is of measure zero in W_∞. Near $\mathbf{P}(N)$, W_∞ is the union of two smooth manifolds intersecting transversally along $\mathbf{P}(N)$. Therefore, if α is a smooth form on W, the form $k^*\alpha$ is unambiguously defined on $W \setminus \mathbf{P}(N)$, and defines an integrable current on W_∞. Also, as a current on M, $k^*(\alpha)\delta_{W_\infty}$ is exactly the product of the currents α and δ_{W_∞}.

Note that Y is a submanifold of W_∞. Also, by Section 4b) and our choice of metrics, the normal bundle to Y in W_∞ coincides—as a holomorphic Hermitian vector bundle—with N.

For $0 < \mathrm{Re}(s) < 1$ let $\zeta_{\tilde{\xi}}^\infty(s)$ be the current on W_∞

$$\begin{aligned} \zeta_{\tilde{\xi}}^\infty(s) = \frac{1}{\Gamma(s)}\int_0^{+\infty} u^{s-1} &\Big[k^* Tr_s\left[N_H \exp(-\tilde{A}_u^2)\right] \\ &+ \varphi^{-1}\left[(Td^{-1})'(g^N)ch(\eta)\right]\delta_Y\Big] du. \end{aligned} \tag{4.18}$$

Note that since the forms $Tr_s\left[N_H \exp(-\tilde{A}_u^2)\right]$ decay exponentially fast as $u \to +\infty$ on compact subsets of $W_\infty \setminus Y$, and in particular near $\mathbf{P}(N)$, the arguments of [BGS4] can be used verbatim to prove the existence of $\zeta_{\tilde{\xi}}^{\infty}(s)$. Furthermore this function of s extends holomorphically at $s = 0$. Let

$$T^{\infty}(h^{\tilde{\xi}}) = \varphi\big[\zeta_{\tilde{\xi}}^{\infty\prime}(0)\big]. \tag{4.19}$$

We claim that

$$T(h^{\tilde{\xi}})\delta_{W_\infty} = T^{\infty}(h^{\tilde{\xi}})\delta_{W_\infty}. \tag{4.20}$$

In fact, we can replace in both sides of (4.20) integration in u from 0 to $+\infty$ by integration from 0 to T $(0 < T < +\infty)$, and then equality is an obvious consequence of the previous considerations. We now make $T \to +\infty$. Since by Theorem 1.9, the truncated integrals approximate $T(h^{\tilde{\xi}})$ in $\mathcal{D}'_{N'_{\bullet}}(W)$, and since multiplication by δ_{W_∞} maps continuously $\mathcal{D}'_{N'_{\bullet}}(W)$ in $\mathcal{D}'(W)$, we obtain (4.20).

We claim that on $W_\infty \setminus P$ the smooth current $T^{\infty}(h^{\tilde{\xi}})$ vanishes identically. In fact, on $W_\infty \setminus P$ the complex $(\tilde{\xi}, \tilde{v})$ splits as a holomorphic Hermitian complex. We can now use the argument of [BGS1, Corollary 1.30]. On $W_\infty \setminus P$ we have

$$\tilde{A}_u^2 = \big(\nabla^{\tilde{\xi}}\big)^2 + uI_{\tilde{\xi}}. \tag{4.21}$$

Using (4.18) and (4.21), we find that on $W_\infty \setminus P$, $\zeta_{\tilde{\xi}}^{\infty}(s)$ is the smooth form $Tr_s\big[N_H \exp(-(\nabla^{\tilde{\xi}})^2)\big]$, which does not depend on s, so that $T^{\infty}(h^{\tilde{\xi}}) = 0$ on $W_\infty \setminus P$.

Therefore the support of the current $T^{\infty}(h^{\tilde{\xi}})$ is included in the manifold P. More precisely the restriction of the current $T^{\infty}(h^{\tilde{\xi}})$ to the manifold P is exactly the singular current associated with the holomorphic complex $\tilde{\xi} \mid P$ of Hermitian vector bundles, which provides a resolution of the direct image $s_*\eta$ of η by the immersion $s: Y \to P$. We now consider $T^{\infty}(h^{\tilde{\xi}})$ as a current on P.

By (4.9) we have on P an exact sequence of complexes of sheaves

$$0 \to \pi_P^* \mathcal{L}_\bullet \to f^*\tilde{\xi} \xrightarrow{\varepsilon} K_\bullet(\varphi) \otimes \pi_P^*\eta \to 0$$

where $\mathcal{L}$ is split acyclic, and ε is a map of resolutions of $s_*\eta$. The Hermitian metrics on $K_\bullet(\varphi) \otimes \pi_P^*\eta$ and $\tilde{\xi}$ both verify assumption (A) with respect to the same couple of metrics on N and η. Finally, the holomorphic Hermitian complex $L_\bullet$ defined from $\mathcal{L}_\bullet$ splits. It now follows from Theorem 2.9 that

$$T^{\infty}(h^{\tilde{\xi}}) = T\big(K_\bullet(\varphi) \otimes \pi_P^*\eta\big) - \sum_{j=0}^{m}(-1)^j T(g^{A_j}) \text{ in } P_Y^P/P_Y^{P,0}. \tag{4.22}$$

By Theorem 3.17 we know that

$$T(K_\bullet(\varphi)\otimes\pi_P^*\eta) - Td^{-1}(g^{H^*})\sigma^*(\tilde{e}(g^{H^*}))\,ch(g^\eta) \tag{4.23}$$

lies in $\tilde{P}_Y^{P,0}$. From (4.20), (4.22), (4.23) we deduce that

$$\begin{aligned}\varphi_*\left[T(h^{\tilde{\xi}})\delta_{W_\infty}\right] &= \psi_*\left[Td^{-1}(g^{H^*})\sigma^*(\tilde{e}(g^{H^*}))\right]ch(g^\eta)\delta_Y \\ &\quad - \psi_*\left[\sum_{j=0}^{m}(-1)^j T(g^{A_j})\right]\delta_Y \text{ in } P_Y^X/P_Y^{X,0}.\end{aligned} \tag{4.24}$$

Also

$$\int_{\mathbf{P}^1}\mathrm{Log}|z|^2 R = 0$$

(map z to $1/z$).

The conormal bundle to W_∞ in W is well defined out of $\mathbf{P}(N)$. Near $\mathbf{P}(N)$, W is made of two smooth submanifolds intersecting transversally along $\mathbf{P}(N)$. On $\mathbf{P}(N)$, we define the conormal bundle to W_∞ to be the union of the conormal bundles to these two submanifolds.

Using [H, Theorem 8.2.13] and the fact that the map of φ can be expressed as the composition of an immersion and a submersion, we find that if ω is a current on W

$$WF(\varphi_*\omega)\subset\left\{p\in T_R^*Y\setminus\{0\};\ \varphi^*p\in\{0\}\cup WF(\omega)\right\}.$$

Now by [H, Theorem 8.2.9], the wave front sets of the currents $\partial(\mathrm{Log}|z|^2 T(h^{\tilde{\xi}}))$ and $\mathrm{Log}|z|^2\bar{\partial}T(h^{\tilde{\xi}})$ are included in the sum of the conormal bundles to W_∞ and to $Y\times\mathbf{P}^1$ in W. Using the local form of the map φ given in the proof of Lemma 4.7 (iii), it is clear that

$$\begin{aligned}WF(\varphi_*\partial(\mathrm{Log}|z|^2)T(h^{\xi})) &\subset N_R^* \\ WF\left(\varphi_*((\mathrm{Log}|z|^2)\bar{\partial}T(h^{\xi}))\right) &\subset N_R^*.\end{aligned}$$

From (4.17), (4.24) and from the previous considerations, we find that C lies in $P_Y^{X,0}$.

We now prove that the form $\psi_*\left[Td^{-1}(g^{H^*})\sigma^*(\tilde{e}(g^{H^*}))\right]$ is closed. Let Q be the bundle of unitary frames in the vector bundle N. Then Q is a $U(e)$-principal bundle ($e=\mathrm{codim}(Y)$) which we equip with the connection ∇^N. We equip $\mathbf{C}^e$ with its canonical Hermitian metric. Then $U(e)$ acts on $\mathbf{C}^e$ and it also acts on $\mathbf{P}(\mathbf{C}^e\oplus 1)$ as a group of holomorphic transformations. One then immediately verifies that

$$P=\mathbf{P}(N\oplus 1)=Q\times_{U(e)}\mathbf{P}(\mathbf{C}^e\oplus 1).$$

Also we can form on the single "fiber" $\mathsf{P}(\mathbf{C}^e \oplus 1)$ the holomorphic Hermitian vector bundle H_0, with a morphism $\varphi_0 : \mathcal{H}_0 \to \mathcal{O}_{\mathsf{P}(\mathbf{C}^e \oplus 1)}$ and the holomorphic Hermitian Koszul chain complex $\Lambda H_0 = K(\varphi_0)$. The group $U(e)$ acts naturally on H_0 as a group of holomorphic unitary transformations, which preserve the map φ_0. One has

$$H = Q \times_{U(e)} H_0$$

and

$$K(\varphi) = Q \times_{U(e)} K(\varphi_0).$$

The connection ∇^N induces a connection on the fibration $P \to Y$. In particular the curvature T of the fibration $P \to Y$ is obtained by lifting the action of $(\nabla^N)^2$ on the fibers N to P; it is a two form on Y with values in vector fields along the fibers.

The previous considerations show that T lifts to a two form $\tilde{T}$ on Y with values in infinitesimal unitary transformations of H along the fibers. Let $\tilde{T}_0$ be the horizontal part of $\tilde{T}$ with respect to the connection ∇^H. Then $\tilde{T}_0$ is a two-form on Y with values in skew-adjoint endomorphisms of H, so that

$$\tilde{T} = -\nabla^H_T + \tilde{T}_0.$$

The connection ∇^N induces a splitting

$$T_{\mathbf{R}} P = \psi^* T_{\mathbf{R}} Y \oplus T^V_{\mathbf{R}} P.$$

Let R be the restriction of the curvature $(\nabla^H)^2$ to vectors of $T^V_{\mathbf{R}} P$. Then we find that

$$(\nabla^H)^2 = R + \tilde{T}_0$$

i.e.:

- on vectors of $T^V_{\mathbf{R}} P$, $(\nabla^H)^2$ coincides with R.
- on horizontal vectors, $(\nabla^H)^2$ coincides with $\tilde{T}_0$.
- if $U \in \psi^*\, T_{\mathbf{R}} Y$, and $V \in T^V_{\mathbf{R}} P$, then $(\nabla^H)^2(U, V) = 0$.

In the Chern–Weil formula for $Td(g^H)$, we now replace $(\nabla^H)^2$ by $R + \tilde{T}_0$. Similarly in the formula for $\tilde{e}(g^H)$, we also replace $(\nabla^H)^2$ by $R + \tilde{T}_0$. Remember that $\tilde{T}_0$ can be canonically expressed in terms of $(\nabla^N)^2$.

Let u_0 be a unitary frame in N. We consider u_0 as a linear isometry from $\mathbf{C}^e$ into N. Let $\mathcal{U}(e)$ be the Lie algebra of $U(e)$. The previous considerations show that for $A \in \mathcal{U}(e)$ there exists a smooth form $\omega(A)$ on $\mathsf{P}(\mathbf{C}^e \oplus 1)$ with the following properties:

- the map $A \to \int_{\mathsf{P}(\mathbf{C}^e \oplus 1)} \omega(A)$ is ad-invariant.

• $\psi_*\left[Td^{-1}(g^{H^*})\sigma^*(\tilde{e}(g^{H^*}))\right] = \int_{\mathbf{P}(\mathbf{C}^e\oplus 1)} \omega\left(u_0^{-1}(\nabla^N)^2 u_0\right)$.

Using the standard Chern–Weil theory, it is now clear that

$$\psi_*\left[Td^{-1}(g^{H^*})\,\sigma^*(\tilde{e}(g^{H^*}))\right]$$

is a closed form.

The fact that the cohomology class of $\psi_*\left[Td^{-1}(g^{H^*})\sigma^*(\tilde{e}(g^{H^*}))\right]$ does not depend on the metric g^H is now obtained by the usual argument in Chern–Weil theory.

(e) **Proof of (3.95).** We use the notations from (3.95). In particular X is an arithmetic variety, (E,h) an hermitian vector bundle over X, $p: X^E \to X$ the canonical projection from its total space to X, and $\hat{e}(p^*E, p^*h) \in \widehat{CH}^r(X^E)$ the Euler class defined before Theorem 3.22. Let s be an algebraic section of E such that the corresponding holomorphic section s_∞ satisfies the hypotheses of Remark 3.5. Using [GS1] and the fact that s is transverse to $s_0(X_F)$, we get

$$\hat{e}(E,h) = s^*\left(\hat{e}(p^*E, p^*h)\right).$$

We want to show that this class is also equal to

$$s_0^*\left(\hat{e}(p^*E, p^*h)\right).$$

Let $E \oplus 1$ be the direct sum on X of E with the trivial line bundle, and $\mathbf{P}(E \oplus 1)$ the associated projective space. Consider the map

$$\psi: X \times \mathbf{A}^1 \to X^E$$

sending (x,t) to $(x, ts(x))$. On $X \times \{0\}$ ψ coincides with s_0, and with s on $X \times \{1\}$. The closure of the image of ψ in $\mathbf{P}(E \oplus 1)$ is the variety W studied in paragraph (b) above. Let H be the kernel of the map $E^* \oplus 1 \to \mathcal{O}(1)$ on $\mathbf{P}(E \oplus 1)$, endowed with the induced metric, and let $\alpha \in \widehat{CH}^r(W)$ be the restriction of $\hat{e}(H^*)$. By the argument of [GS1, Theorem 4.4.5] (see also Lemma 4.12 below), and using the fact that the function $\mathrm{Log}|z/(z-1)|^2$ vanishes at infinity, we get the following equation in $\widehat{CH}^r(X)$:

$$i_0^*\alpha - i_1^*\alpha = -a(p_{W*}((\mathrm{Log}|z/(z-1)|^2 e(H^*))) \tag{4.25}$$

where p_W is the projection from W to X, $e(H^*)$ is the Euler form of the dual of H, and $i_t: X \to \mathbf{P}(E \oplus 1)$ maps x to $(x, ts(x)) \in X^E \subset \mathbf{P}(E \oplus 1)$.

When restricted to X^E, the bundle H^* becomes isomorphic to p^*E, and its canonical section becomes the tautological section of p^*E, but its metric is not p^*h. Let $\tilde{c}_r$ be the corresponding Bott-Chern class. From (3.93) we get

$$i_0^*\alpha - i_1^*\alpha = (s_0^* - s^*)(\hat{e}(p^*E, p^*h) + \tilde{c}_r). \tag{4.26}$$

On the other hand, the restriction β of $\hat{c}_r(H^*)$ to W satisfies

$$i_0^*\beta - i_1^*\beta = -a(p_{W^*}(\mathrm{Log}|z/(z-1)|^2 e(H^*)) \tag{4.27}$$

by the same argument as above, since $e(H^*)$ is the r-th Chern form of H^*. We also have

$$i_0^*\beta - i_1^*\beta = (s_0^* - s^*)(\hat{c}_r(p^*E, p^*h) + \tilde{c}_r). \tag{4.28}$$

But the functoriality of $\hat{c}_r$ implies

$$(s_0^* - s^*)(\hat{c}_r(p^*E, p^*h)) = \hat{c}_r(E,h) - \hat{c}_r(E,h) = 0. \tag{4.29}$$

Using (4.25), (4.26), (4.27), (4.28), (4.29) we conclude that (3.95) holds. □

(f) *Immersions and the arithmetic Chern character.*

1. Let (A, Σ, F_∞) be an arithmetic ring as in 3 j) and

$$\begin{array}{ccc} Y & \xrightarrow{\;i\;} & X \\ & {\scriptstyle g}\searrow \quad \swarrow{\scriptstyle f} & \\ & B & \end{array}$$

a diagram of morphisms between arithmetic varieties over A. We assume that the morphisms f and g are smooth and projective, and that i is a closed immersion. Let η be an Hermitian vector bundle on Y and

$$\xi_\bullet \to i_*\eta$$

a resolution of its direct image on X. We choose a metric on the normal bundle N to Y in X and a metric on ξ such that hypothesis (A) is satisfied.

Let $\varphi: W \to X$ be the deformation of i to the normal cone, $j: Y \times \mathbf{P}^1 \to W$ the immersion extending i, $p_Y: Y \times \mathbf{P}^1 \to Y$ the first projection, $\tilde{\eta} = p_Y^*(\eta)$ and $\tilde{\xi}_\bullet \to j_*\tilde{\eta}$ a metrized resolution such that, on $W_0 = q_W^{-1}(0)$,

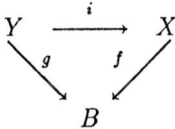

$\tilde{\xi}_\bullet$ coincides with $\xi_\bullet$ (we denote by $q_W: W \to \mathbf{P}^1$ the projection). We also assume that the metric on $\tilde{\xi}$ is compatible with the metric on the normal bundle $N(-1)$ of $Y \times \mathbf{P}^1$ in W as in 4d). As we saw in 4b) above, $W_\infty = q_W^{-1}(\infty)$ has two components, the blow up $\tilde{X}$ of X along Y and the projective space $P = \mathbf{P}(N \oplus 1)$ over Y. Let $\psi: \mathbf{P}(N \oplus 1) \to Y$ be the projection and ξ_∞ the restriction of $\tilde{\xi}_\bullet$ on P. We also assume that $\tilde{\xi}_\bullet$ restricted to $\tilde{X}$ is split acyclic as complex of hermitian vector bundles [BGS1]. We shall consider the Chern character

$$\widehat{ch}(\tilde{\xi}_\bullet) = \sum_{p \geq 0} (-1)^p \widehat{ch}(\tilde{\xi}_p)$$

in $\widehat{CH}(W)_{\mathbf{Q}}$ [GS2].

Let $a: \tilde{A}(X_{\mathbf{R}}) \to \widehat{CH}(X)$ and $\omega = \widehat{CH}(X) \to A(X_{\mathbf{R}})$ be defined as in the proof of Theorem 3.22. Finally let $ch(h^{\tilde{\xi}})$ be the Chern character of $\tilde{\xi}_\bullet$ and

$$\beta = \int_{W/X} ch(h^{\tilde{\xi}}) \mathrm{Log}|z|^2$$

its integral against the integrable function $\mathrm{Log}|z|^2$ (defined via $q_W: W \to \mathbf{P}^1$). We view β as a current in $\tilde{\mathcal{D}}(X_{\mathbf{R}})$.

We refer to [GS1] for the definition of f_* and i^* on $\widehat{CH}$.

Lemma 4.12. *For any $\alpha \in \widehat{CH}(X)$, the following identity holds:*

$$f_*\left(\widehat{ch}(\xi_\bullet)\alpha\right) = g_*\left(\psi_*(\widehat{ch}(\xi_\infty))i^*(\alpha)\right) + af_*(\beta\omega(\alpha)).$$

Proof. The proof below will imply that $dd^c f_*(\beta\omega(\alpha))$ is smooth, hence the current $f_*(\beta\omega(\alpha))$ lies in $\tilde{A}(B_{\mathbf{R}})$ and its image by a makes sense.

Let us represent $\widehat{ch}(\tilde{\xi}_\bullet)$ by an arithmetic cycle (Z, g). Since $\tilde{\xi}_\bullet$ is acyclic outside $Y \times \mathbf{P}^1$ we may assume (up to linear equivalence) that Z is supported on $Y \times \mathbf{P}^1$, and moreover that $Z = p_Y^*(T) + S \times \{1\}$, with S and T two cycles on Y.

By definition of $\widehat{CH}^*(W)$ we have $\widehat{\mathrm{div}}(z) = \left(\mathrm{div}(z), -\mathrm{Log}|z|^2\right) = 0$ in this group. Therefore

$$f_*\varphi_*\left(\widehat{ch}(\tilde{\xi}_\bullet)\widehat{\mathrm{div}}(z)\alpha\right) = 0. \tag{4.30}$$

Now $\widehat{ch}(\tilde{\xi}_\bullet)\widehat{\mathrm{div}}(z)$ is the class of

$$\begin{aligned}(Z, g)\left(\mathrm{div}(z), -\mathrm{Log}|z|^2\right) = \big(&T \times \{0\} - T \times \{\infty\}, \\ &g|_{W_0} - g|_{W_\infty} - ch(h^{\tilde{\xi}})\mathrm{Log}|z|^2\big)\end{aligned} \tag{4.31}$$

since $S \times \{1\}$ does not meet $\operatorname{div}(z) = W_0 - W_\infty$. On the blow-up $\tilde{X}$ of X along Y the restriction of $\tilde{\xi}_\bullet$ is split acyclic, therefore we may assume that $g|_{\tilde{X}} = 0$. Using (4.30) and (4.31) we get

$$\begin{aligned} 0 &= f_* \varphi_* \big(\widehat{ch}(\tilde{\xi}_\bullet) \widehat{\operatorname{div}}(z) \alpha\big) \\ &= f_* \big(\widehat{ch}(\tilde{\xi}_\bullet)\alpha\big) - g_* \big(\psi_* (\widehat{ch}(\xi_\infty))\big) i^*(\alpha) - f_* \varphi_* \big(0, ch(h^{\tilde{\xi}}) \mathrm{Log}|z|^2 \omega(\alpha)\big) \end{aligned}$$

and we get the lemma since, by definition,

$$\beta = \varphi_* \big(ch(h^{\tilde{\xi}}) \mathrm{Log}|z|^2\big). \qquad \square$$

Let now $\widehat{Td}(N) \in \widehat{CH}(Y)_{\mathbf{Q}}$ be the Todd class of the normal bundle to Y in X (defined in [GS2]; $\widehat{Td}$ is not to be confused with the arithmetic Todd genus Td^A of [GS3]).

Theorem 4.13. *For any α in $\widehat{CH}(X)$, the following identity holds in $\widehat{CH}(B)_{\mathbf{Q}}$*

$$f_* \big(\widehat{ch}(\xi_\bullet)\alpha\big) = g_* \big(\widehat{Td}(N)^{-1} \widehat{ch}(\eta) i^*(\alpha)\big) - a f_* \big(T(h^\xi)\omega(\alpha)\big).$$

Proof. From Lemma 4.12, we have

$$f_* \big(\widehat{ch}(\xi_\bullet)\alpha\big) = g_* \big(\psi_* (\widehat{ch}(\xi_\infty) i^*(\alpha))\big) + a f_* (\beta \omega(\alpha)). \tag{4.32}$$

From Theorem 4.11

$$\begin{aligned} -f_*(\beta\omega(\alpha)) = f_* &\big(T(h^\xi)\omega(\alpha)\big) \\ &- g_* \big(\psi_* \big(Td^{-1}(g^{H^*}) \sigma^* (\tilde{e}(g^{H^*})) ch(g^\eta) i^* \omega(\alpha)\big)\big) \\ &+ \sum_{j \geq 0} (-1)^j g_* \big(\psi_* (T(g^{A_j})) i^* \omega(\alpha)\big). \end{aligned} \tag{4.33}$$

From [GS2, Theorem 4.8(ii)] and the fact that $L_\bullet$ is split acyclic, we know that

$$\widehat{ch}(\xi_\infty) = \widehat{ch}\big(K(\varphi) \otimes \pi_P^* \eta\big) + \sum_{j=0} (-1)^j a(T(g^{A_j})). \tag{4.34}$$

Combining (4.32), (4.33), and (4.34) we obtain

$$\begin{aligned} f_* \big(\widehat{ch}(\xi_\bullet)\alpha\big) &= g_* \big(\psi_* \big(\widehat{ch}(K(\varphi) \otimes \pi_P^* \eta)\big) i^* \alpha\big) \\ &+ a g_* \big(\psi_* \big(Td^{-1}(g^{H^*}) \sigma^* (\tilde{e}(g^{H^*})) ch(g^\eta) i^* \omega(\alpha)\big)\big) \\ &- a f_* \big(T(h^\xi)\omega(\alpha)\big). \end{aligned} \tag{4.35}$$

Since $\tilde{\pi}_P^*\eta = \psi^*(\eta)$ we get

$$\psi_*\big(\widehat{ch}(K(\varphi)\otimes \pi_P^*\eta)\big) = \psi_*\big(\widehat{ch}(K(\varphi))\big)\widehat{ch}(\eta) \tag{4.36}$$

and by a standard formula

$$\widehat{ch}(K(\varphi)) = \hat{c}_r(H^*)\widehat{Td}(H^*)^{-1}, \tag{4.37}$$

where r is the rank of H. From Theorem 3.22 we know that $\hat{c}_r(H^*) = \hat{e}(H^*)$ is the class of $(Y\times\{\infty\}, -\sigma^*(\tilde{e}(g^{H^*}))$ in $\widehat{CH}(P)$. Since the restriction of H^* to $Y\times\{\infty\}$ coincides with N, we conclude from (4.36) and (4.37) that (4.38)

$$\begin{aligned} g_*\psi_*\big(\widehat{ch}(K(\varphi)\otimes\pi_P^*\eta)\big) &= g_*\big(\widehat{Td}(N)^{-1}\widehat{ch}(\eta)\big) \\ &\quad - a g_*\psi_*\big(Td^{-1}(g^{H^*})\sigma^*(\tilde{e}(g^{H^*}))ch(g^\eta)\big). \end{aligned}$$

From (4.35) and (4.38) we conclude that

$$f_*\big(\widehat{ch}(\xi_\bullet)\alpha\big) = g_*\big(\widehat{Td}(N)^{-1}\widehat{ch}(\eta)i^*(\alpha)\big) - af_*\big(T(h^\xi)\omega(\alpha)\big).$$

REFERENCES

[BaFM] Baum, P., Fulton, W., MacPherson, R., *Riemann–Roch for singular varieties,* Publ. Math. IHES **45**, 101–146 (1975).

[B1] Bismut, J.M., *Localisation du caractère de Chern en géométrie complexe et superconnexions,* C.R.A.S t. **307**, série I, 523–526 (1988).

[B2] Bismut, J.M., *Superconnection currents and complex immersions. Invent. Math.* **99**, 59–113 (1990).

[B3] Bismut, J.M., *Equivariant Bott–Chern currents and the Ray–Singer analytic torsion,* to appear in *Math. Annalen.*

[BGS1] Bismut, J.M., Gillet, H., Soulé, C., *Analytic torsion and holomorphic determinant bundles,* I. Comm. Math. Phys. **115**, 49–78 (1988).

[BGS2] Bismut, J.M., Gillet, H., Soulé, C., *Analytic torsion and holomorphic determinant bundles,* II. Comm. Math. Phys. **115**, 79–126 (1988).

[BGS3] Bismut, J.M., Gillet, H., Soulé, C., *Analytic torsion and holomorphic determinant bundles,* III. Comm. Math. Phys. **115**, 301–351 (1988).

[BGS4] Bismut, J.M., Gillet, H., Soulé, C., *Bott–Chern currents and complex immersions.* Duke Math Journal **60**, 255–284 (1990).

[BGS5] Bismut, J.M., Gillet, H., Soulé, C., *Classes caractéristiques secondaires et immersions en géométrie complexe,* C.R.A.S t. **307**, Série I, 565–567 (1988).

[BoC] Bott, R., Chern, S.S., *Hermitian vector bundles and the equidistribution of the zeros of their holomorphic sections,* Acta Math. **114**, 71–112 (1968).

[E] Eilenberg, S., *Homological dimension and local syzygies,* Annals of Math. **64**, 328–336 (1956).

[GS1] Gillet, H., Soulé, C., *Arithmetic Intersection Theory,* 1988, Preprint IHES.

[GS2] Gillet, H., Soulé, C., *Characteristic classes for algebraic vector bundles with Hermitian metrics,* I, *Annals of Math.* **131**, 163–243 (1990); II, to appear.

[GS3] Gillet, H., Soulé, C., *Analytic torsion and the arithmetic Todd genus,* to appear in *Topology.*

[Ha] Hartshorne, R., *Algebraic Geometry,* Graduate Texts in Math., **52**, Berlin–Heidelberg–New York, Springer (1977).

[H] Hörmander, L., *The analysis of linear partial differential operators,* Vol. I., Grundl. der Math. Wiss., Band 256, Berlin–Heidelberg–New York: Springer (1983).

[Ma] Manin, Yu. L., *New dimensions in geometry,* in Lecture Notes in Math. **1111**, 59–101, Berlin–Heidelberg–New York, Springer (1985).

[MQ] Mathai, V., Quillen D., *Superconnections, Thom classes and equivariant differential forms,* Topology **25**, 85–110 (1986).

[Q1] Quillen, D., *Superconnections and the Chern character,* Topology **24**, 89–95 (1985).

[Q2] Quillen, D., *Determinants of Cauchy–Riemann operators over a Riemann surface,* Funct. Anal. Appl. **14**, 31–34 (1985).

[Se] Serre, J.P., *Algèbre locale. Multiplicités,* Lecture Notes in Math. **11**, Berlin–Heidelberg–New York, Springer (1965).

[SGA6] Grothendieck, A. and al., *Théorie des intersections et Théorème de Riemann–Roch,* Lecture Notes in Math. **225**, Berlin–Heidelberg–New York, Springer (1971).

J.-M. Bismut
Dept. de Math. Bât 425
Univ. de Paris-Sud
91405-Orsay,
France

H. Gillet
Dept. of Math.
Univ. of Illinois
Chicago, IL 60638
USA

C. Soulé
CNRS and IHES
35, route de Chartres
91440 Bures-sur-Yvette,
France

L-Functions and Tamagawa Numbers of Motives

SPENCER BLOCH* AND KAZUYA KATO

dedicated to A. Grothendieck

0. Introduction

The notion of a motif was first defined and studied by A. Grothendieck, and this paper is an attempt to understand some of the implications of his ideas for arithmetic. We will formulate a conjecture on the values at integer points of L-functions associated to motives. Conjectures due to Deligne and Beilinson express these values "modulo $\mathbf{Q}^*$ multiples" in terms of archimedean period or regulator integrals. Our aim is to remove the $\mathbf{Q}^*$ ambiguity by defining what are in fact Tamagawa numbers for motives. The essential technical tool for this is the Fontaine-Messing theory of p-adic cohomology. As evidence for our Tamagawa number conjecture, we show that it is compatible with isogeny, and we include strong results due to one of us (Kato) for the Riemann zeta function and for elliptic curves with complex multiplication.

To recall how the Tamagawa numbers of algebraic groups are related to special values of L-functions, we consider the example $SL_{n,\mathbf{Q}}$. The Tamagawa number, which is 1 in this case, is defined to be the volume of the adelic points modulo global points

$$SL_n(\mathbf{Q})\backslash \prod_{p\leq\infty} {}'SL_n(\mathbf{Q}_p) = SL_n(\mathbf{Q})\backslash SL_n(\mathbb{A}_{\mathbf{Q}})$$

with respect to a canonical measure on $SL_n(\mathbb{A}_{\mathbf{Q}})$, the Tamagawa measure. To define this, choose an isomorphism (det=highest exterior power)

$$\det_{\mathbf{Q}}(sl_n(\mathbf{Q})) \cong \mathbf{Q}. \tag{0.1}$$

Associated to this choice, one can define Haar measures μ_p on $SL_n(\mathbf{Q}_p)$ for $p \leq \infty$, and a measure μ on $SL_n(\mathbb{A}_{\mathbf{Q}})$. The product formula implies

*Partially funded by the NSF.

that μ is independent of the choice in (0.1). To actually do the calculation, note the strong approximation theorem gives

$$\mu(SL_n(\mathbf{Q})\backslash SL_n(\mathbb{A}_{\mathbf{Q}})) = \mu_\infty(SL_n(\mathbf{Z})\backslash SL_n(\mathbf{R}) \prod_{p<\infty} \mu_p(SL_n(\mathbf{Z}_p))$$

If (0.1) is chosen so $\det_{\mathbf{Z}}(\mathrm{sl}_n(\mathbf{Z})) \cong \mathbf{Z}$, the μ_p satisfy

$$\mu_p(SL_n(\mathbf{Z}_p)) = \prod_{i=2}^{n}(1-p^{-i})$$

so the fact that the Tamagawa number is 1 is equivalent to

$$\prod_{i=2}^{n} \zeta(i) = \mu_\infty(SL_n(\mathbf{Z})\backslash SL_n(\mathbf{R}))$$

where ζ is the Riemann zeta function.

We turn now to the case of a motif. Broadly speaking, a motif M is a sutiable direct factor of $H^r(X, \mathbf{Z}(n))$ for some universal cohomlogy theory for smooth, complete varieties X. Suppose X is defined over $\mathbf{Q}$, so one has an associated L-function $L(M,s)$. Deligne and Beilinson have formulated conjectures sbout the special value $L(M,0)$ [**De2**], [**Be1**]. To fix ideas, suppose $M = H^r(X, \mathbf{Z}(n))$. We will always assume $n \geq (r+1)/2$ (i.e., the weight of $M \leq -1$), but for purposes of this introduction let us assume M has weight < -2. The intermediate jacobian

$$A(\mathbf{C}) = H^r(X(\mathbf{C}),\mathbf{C})/\{H^r(X,\mathbf{Z}(n)) + F^n H^r(X(\mathbf{C}),\mathbf{C})\}$$

has a real structure given by invariants of conjugation acting simultaneously on the topological space $X(\mathbf{C})$ and on the coefficients $\mathbf{C}$. Moreover, the canonical identification

$$H^r(X(\mathbf{C}),\mathbf{C}) \cong H^r_{DR}(X/\mathbf{Q}) \otimes \mathbf{C}$$

defines a $\mathbf{Q}$-structure on the tangent space $H^r(X(\mathbf{C}),\mathbf{C})/F^n H^r(X(\mathbf{C}),\mathbf{C})$ to $A(\mathbf{C})$. The choice of an isomorphism

(0.2) $$\det(H^r_{DR}(X/\mathbf{Q})/F^n H^r_{DR}(X/\mathbf{Q})) \xrightarrow{\sim} \mathbf{Q}$$

thus determines a Haar measure ω_∞ on $A(\mathbf{C})$. In the cases considered by Deligne, $A(\mathbf{R})$ is compact, and his conjecture reads

$$L(M,0) \in \omega_\infty(A(\mathbf{R})) \cdot \mathbf{Q},$$

i.e. $L(M, 0)$ is a rational multiple of the volume of $A(\mathbf{R})$. Deligne has also pointed out [**De5**] that Beilinson's conjectures can be formulated similarly. In the Beilinson case, $A(\mathbf{R})$ is not compact, but the image of the regulator map on a suitable motivic cohomology group defines a subgroup one might call $A(\mathbf{Z}) \subset A(\mathbf{R})$. Then Beilinson conjectures that $A(\mathbf{Z}) \subset A(\mathbf{R})$ is discrete and co-compact, and

$$L(M, 0) \in \omega_\infty(A(\mathbf{R})/A(\mathbf{Z})) \cdot \mathbf{Q}^*.$$

Notice that ω_∞ depends up to a rational number on the non-canonical choice we made in (0.2), so the $\mathbf{Q}$ or $\mathbf{Q}^*$ ambiguity appears necessary. (However, when $n > r$, so $F^n H^r(X(\mathbf{C}), \mathbf{C}) = (0)$, a polarization on $H^r_{DR}(X/\mathbf{Q})$ determines a preferred isomorphism (0.2).) We will define abelian groups $A(\mathbf{Q}_p)$ for all p, with Haar measure μ_p (depending in the expected way on the choice of (0.2)) such that for almost all p,

$$\mu_p(A(\mathbf{Q}_p)) = P_p(M, 1),$$

where (ignoring factors at bad primes)

$$L(M, s) = \prod P_p(M, p^{-s})^{-1}.$$

We also define (more precisely, we postulate) an abelian group $A(\mathbf{Q})$ mapping to all the $A(\mathbf{Q}_p), p \leq \infty$. The properties we need essentially follow from Beilinson's conjectures for motivic cohomology, however, and one can hope that $A(\mathbf{Q})$ is isomorphic to the relevant motivic cohomology (essentially $H^{r+1}_{\mathcal{M}}(X, \mathbf{Z}(n))$, although one has to be careful about bad fibres). The Tamagawa number is defined by

$$\mathrm{Tam}(M) = \mu(\prod_{p \leq \infty} A(\mathbf{Q}_p)/A(\mathbf{Q}))$$

Just as in the algebraic group case, this is canonically defined independent of any choices. Roughly speaking, the statement that $\mathrm{Tam}(M)$ is a rational number is equivalent to the Beilinson and Deligne conjectures. In fact, if the motivic cohomology term $A(\mathbf{Q})$ is defined using a regular model of X over $\mathbf{Z}$, then the Beilinson conjecture says $L(M, 0) \in \mu_\infty(A(\mathbf{R})/A(\mathbf{Q})) \cdot \mathbf{Q}^*$. Since $\mu_p(A(\mathbf{Q}_p)) = L_p(M, 0)^{-1}$ for almost all p, and since the Euler product expression for $L(m, 0)$ converges in weights ≤ -3, we get the assertion.

Clearly it would be more elegant to work with the variety over $\mathbf{Q}$ rather than over $\mathbf{Z}$. We do this, denoting the corresponding groups $B(\mathbf{Q}_p)$ and

$B(\mathbf{Q})$. One might think of $A(\mathbf{Q}_p) = B(\mathbf{Z}_p)$ and $A(\mathbf{Q}) = B(\mathbf{Z})$. The quotient $B(\mathbf{Q}_p)/A(\mathbf{Q}_p)$ must be discrete and should have rank equal to the order of local zero of $P_p(M^*(1), u)$ at $u = 1$. Unfortunately, our description of it relies on conjectural properties of motivic cohomology. We therefore prefer to emphasize the A-theory, where at least the local properties are actually proved.

We define a group $\text{Ш}(M)$, which we conjecture to be finite, and we propose a conjectural Tamagawa number formula

$$\mathrm{Tam}(M) = \#H^0(\mathrm{Gal}(\overline{\mathbf{Q}}/\mathbf{Q}), M^* \otimes \mathbf{Q}/\mathbf{Z}(1)) \cdot (\#\text{Ш}(M))^{-1}$$

which is precisely analogous to the formula of Ono [**On2**] for tori. We prove that the truth of this formula is isogeny invariant in a suitable sense.

The outline of this paper is as follows. §1 is a summary of some of the "linear algebra" associated to the rings B_{crys} and B_{DR} of Fontaine. §2 contains a comparison result between Fontaine-Messing theory and the Coates-Wiles homomorphism which is the key to results on the Riemann zeta function proved in §6. Strictly speaking this section could be posponed until after §4, but we have put it earlier because it gives some feeling for the deeper, non-formal aspects of the problem. §3 studies the cohomology of the local galois groups with values in the Tate module T associated to the motive. In particular, the subgroup $A(\mathbf{Q}_p) = H^1_f(\mathbf{Q}_p, T) \subset H^1(\mathbf{Q}_p, T)$ which generalizes the notion of the image of the $\mathbf{Q}_p$-points of an abelian variety under the boundary map of the Kummer sequence, is introduced and the exponential map, which generalizes the p-adic exponential for a p-divisible group and enables us define the local Tamagawa measure, is defined. §4 is devoted to local volume calculations. It contains the proof that $\mu_p(A(\mathbf{Q}_p)) = P_p(M, 1)$ for almost all p, as well as a much more precise result for the motif $\mathbf{Z}(n)$ associated to the Riemann zeta function. §5 contains the global conjectures. The main result here is the invariance of our conjecture under isogeny. Finally, §6 and 7 contain the arguments for the Riemann zeta function and for CM elliptic curves.

Here, then, is a picture of the land of Tamagawa numbers.

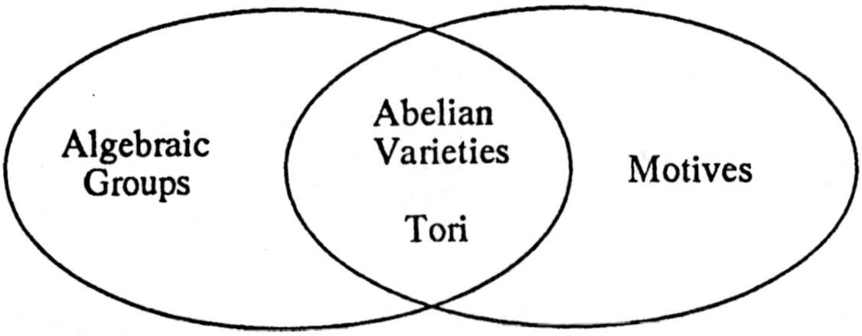

The reader should look at it closely. Anyone who perceives a larger "third oval" containing the other two and explaining their relation, must contact us immediately.

We would like to thank Professors Deligne, Fontaine, Jannsen, Messing, Perrin-Riou, and Soulé for helpful conversations on subjects related to this paper.

1. The rings B_{crys} and B_{DR}

The purpose of this section is to summarize properties of the ring B_{crys} which we will use. Much of the discussion is for the convenience of the reader and duplicates material in [**Fo**] and [**FM**].

Let K be a complete characteristic 0 discrete valuation field with valuation ring Λ and residue field $k = \Lambda/\mathfrak{m}$ perfect of characteristic $p > 0$. Let $\overline{K}$ be the algebraic closure of K, $A = \overline{\Lambda}$ the integral closure of Λ in $\overline{K}$, and write $A_n = A/p^n A$.

Let $W_n(A_1)$ be the ring of p-Witt vectors of length n on A_1. Elements in $W_n(A_1)$ are represented in Witt coordinates by n-tuples $(a_0, \dots, a_{n-1})$ of elements of A_1. There is a surjective ring homomorphism

$$(1.1)\qquad \begin{gathered}\Theta_n : W_n(A_1) \twoheadrightarrow A_n\\ \Theta_n(a_0, \dots, a_{n-1}) = \hat{a}_0^{p^n} + p\hat{a}_1^{p^{n-1}} + \dots + p^{n-1}\hat{a}_{n-1}\end{gathered}$$

where $\hat{a}$ is any lifting of a. Define

$$(1.2)\qquad \begin{gathered}\vartheta_n : W_{n+1}(A_1) \to W_n(A_1),\\ \vartheta_n(a_0, \dots, a_n) = (a_0^p, \dots, a_{n-1}^p).\end{gathered}$$

One checks that

$$(1.3)\qquad \Theta_n \circ \vartheta_n \equiv \Theta_{n+1} \mod p^n.$$

Let $\varprojlim A$ be the multiplicative monoid which is the inverse limit of copies of A under the p-th power map. Reduction mod p induces an isomorphism of multiplicative monoids, $\varprojlim A \cong R \stackrel{\text{def}}{=} \varprojlim A_1$. Note that R is a ring. One can identify $\varprojlim_{\vartheta} W_n(A_1)$ with $W(R) = \varprojlim_{v} W_n(R)$ where $v_n : W_{n+1}(R) \to W_n(R)$ is the projection $v_n(a_0, \dots, a_n) = (a_0, \dots, a_{n-1})$. The Teichmuller map $[\,] : A_1 \to W_n(A_1)$ (resp. $R \to W(R)$) is given by $[a] = (a, 0, 0, \dots, 0)$.

It is compatible with multiplication. The diagram

$$\begin{array}{ccccc} R & \xrightarrow{[\,]} & W(R) & \longrightarrow & W_n(A_1) \\ \| & & & & \downarrow{\scriptstyle \Theta_n} \\ \varprojlim A & \longrightarrow & A & \longrightarrow & A_n \end{array} \tag{1.4}$$

commutes. Finally, W and W_n have a frobenius operator f (induced by the p-th power operator on the argument ring when that ring has characteristic p). The image of $[]$ is the set of elements x in $W(R)$ such that $f(x) = x^p$. We will fix a generator ε of $\mathbf{Z}_p(1)$ with $\varepsilon \stackrel{\text{def}}{=} [1, \zeta_1, \zeta_2, \ldots] \in \varprojlim A$, where $\zeta_{i+1}^p = \zeta_i$. It follows from (1.4) above that the image e_n of ε in $W_n(A_1)$ lies in $1 + \mathrm{Ker}(\Theta_n)$, and $f(e_n) = e_n^p$.

Following [**FM**], we define

$$B_n = W_n^{PD}(A_1);\, B_\infty = \varprojlim B_n$$

(limit with respect to maps $B_{n+1} \to B_n$ induced by

$$\vartheta_n : W_{n+1}(A_1) \to W_n(A_1).)$$

to be the $W_n(A_1)$-algebras obtained by adjoining divided powers to the ideals $\mathrm{Ker}(\Theta_n)$ in a universal way compatible with the divided powers on the ideal (p). Θ_n extends to surjections

$$\begin{aligned} \Theta_n &: B_n \twoheadrightarrow A_n; \mathrm{Ker}(\Theta_n) \stackrel{\text{def}}{=} J_n \\ \Theta_\infty &: B_\infty \twoheadrightarrow \widehat{A} \stackrel{\text{def}}{=} \varprojlim A_n \end{aligned} \tag{1.6}$$

(limit with respect to $\bmod p^n : A_{n+1} \to A_n$).

The rings B_n are linked with the crystalline cohomology via the theorem of Fontaine and Messing [FM]

$$B_n \cong \Gamma((\mathrm{Spec}(A_n))_{\mathrm{crys}}, \mathcal{O}_{\mathrm{crys}}). \tag{1.7}$$

The frobenius f on $W_n(A_1)$ satisfies $f(\mathrm{Ker}(\Theta_n)) \subset \mathrm{Ker}(\Theta_n) + (p)$, so there is an induced frobenius f on the divided envelope B_n. (This also follows directly from (1.7)). In addition, the Galois group $G = \mathrm{Gal}(\overline{K}/K)$ acts on B_n.

For $x \in J_n$, the divided power structure gives us for every $N \geq 1$ an element $\gamma_N(x) \in J_n$ having all the natural properties of $x^N/N!$. In particular, $\log(1+x) \in J_n$ is defined for $n \leq \infty$. We define

(1.8) $$t_n = \log(e_n) \in J_n; t = t_\infty = \log(e_\infty) \in J_\infty.$$

We can think of t as $\log([\varepsilon])$ in a certain $W(R)$-algebra. Note that $f(t_n) = p \cdot t_n$. The assignments $\zeta_n \mapsto t_n; \varepsilon \mapsto t$ give G-equivariant inclusions

(1.9) $$\mu_{p^n} \hookrightarrow J_n; \mathbf{Z}_p(1) \hookrightarrow J_\infty.$$

Here is some further notation we will need:

(1.10) $$B^+_{\text{crys}} = B_\infty \otimes \mathbf{Q}; J_{\mathbf{Q}} = J_\infty \otimes \mathbf{Q}; B_{\text{crys}} = B^+_{\text{crys}}[t^{-1}];$$

$J_n^{[r]}$ = r-th divided power of J_n (generated by products $\prod \gamma_{N_i}(x_i)$ with $\sum N_i \geq r$.)

The completion $\mathbf{C}_p$ of $\overline{K}$ is related to B^+_{crys} via

(1.11) $$0 \rightarrow J_{\mathbf{Q}} \rightarrow B^+_{\text{crys}} \rightarrow \mathbf{C}_p \rightarrow 0.$$

Writing K_0 for the quotient field of $W(k)$, one has

(1.12) $$H^0(G, B^+_{\text{crys}}) = H^0(G, B_{\text{crys}}) \cong K_0.$$

The above constructions also lead to exact sequences for $r \geq 0$

(1.13) $$0 \rightarrow \mathbf{Q}_p(r) \rightarrow J^{[r]}_{\mathbf{Q}} \xrightarrow{1-p^{-r}f} B^+_{\text{crys}} \rightarrow 0.$$

In a sense, B_{crys} is the ring of all p-adic periods associated to varieties with good reduction at p. In order to consider also bad reduction primes, we need the larger ring

(1.14) $$B^+_{DR} = \varprojlim_r B^+_{\text{crys}}/J^{[r]}_{\mathbf{Q}}$$

B^+_{DR} is a complete discrete valuation ring with residue field $\mathbf{C}_p$ and maximal ideal generated by the image of the element t described above. We write

(1.15) $$B_{DR} = B^+_{DR}[t^{-1}]$$

for the corresponding discretely valued field. As G-module,

(1.15.1) $$(B_{DR})^i/(B_{DR})^{i+1} \cong \mathbf{C}_p(i); \quad -\infty < i < \infty.$$

Also

$$H^0(G, B^+_{DR}) \cong H^0(G, B_{DR}) \cong K. \tag{1.16}$$

Proposition 1.17. *The following sequences are exact*

$$0 \to \mathbf{Q}_p \xrightarrow{\alpha} B^{f=1}_{\mathrm{crys}} \oplus B^+_{DR} \xrightarrow{\beta} B_{DR} \longrightarrow 0 \tag{1.17.1}$$

where $\alpha(x) = (x, x)$ *and* $\beta(x, y) = x - y$;

$$0 \longrightarrow \mathbf{Q}_p \xrightarrow{\alpha} B_{\mathrm{crys}} \oplus B^+_{DR} \xrightarrow{\beta} B_{\mathrm{crys}} \oplus B_{DR} \longrightarrow 0 \tag{1.17.2}$$

where $\alpha(x) = (x, x)$ *and* $\beta(x, y) = (x - f(x), x - y)$.

Proof. We have $B^{f=1}_{\mathrm{crys}} \cap B^+_{DR} = \mathbf{Q}_p$ by ([**Fo**], (4.12) and (4.13)). Also $1 - f : B_{\mathrm{crys}} \to B_{\mathrm{crys}}$ is surjective because $B_{\mathrm{crys}} \supset J^{[r]}_{\mathbf{Q}} \cdot t^{-r}$ for $r \geq 0$, and $1 - p^{-r} f : J^{[r]}_{\mathbf{Q}} \twoheadrightarrow B^+_{\mathrm{crys}}$ by ([**FM**], 2.1). It remains to prove that $B^{f=1}_{\mathrm{crys}} + B^+_{DR} = B_{DR}$.

Lemma 1.17.3. *Fix* $r \geq 1$. *Then* $B^+_{DR} \subset t \cdot B^+_{DR} + \{a \in B_{\mathrm{crys}} \mid f(a) = p^r \cdot a\}$.

Note the lemma implies $t^{-r} \cdot B^+_{DR} \subset B^+_{DR} + B^{f=1}_{\mathrm{crys}}$ as claimed.

Proof of (1.17.3). The ring $\widehat{A} \subset \mathbf{C}_p$ is is defined in (1.6). Taking limits with respect to the p-th power map, we get $\varprojlim \widehat{A} \cong \varprojlim A \cong R$ so there is a Teichmuller map making the diagram

$$\begin{array}{ccc} \varprojlim \widehat{A} & \xrightarrow{[\;]} & B_\infty \\ \downarrow & & \downarrow \\ \widehat{A} & \longrightarrow & \mathbf{C}_p \end{array}$$

commute. If $a = (a_0, \ldots) \in \varprojlim \widehat{A}$ satisfies $a_0 \in 1 + p \cdot \widehat{A}$, then $[a] \in 1 + ((p) + J_\infty)$, so $\log([a]) \in B_\infty$ is defined and maps to $\log(a_0) \in \mathbf{C}_p$. Also $f(\log([a])) = p \cdot \log([a])$, so

$$f\{\log([a^{(1)}]) \cdot \ldots \cdot \log([a^{(r)}])\} = p^r \cdot \log([a^{(1)}]) \cdot \ldots \cdot \log([a^{(r)}]).$$

Since $\mathbf{C}_p$ is generated as a $\mathbf{Q}_p$-vector space by expressions $\log(a^{(1)}) \cdot \ldots \cdot \log(a^{(r)})$, we find $B^+_{\mathrm{crys}} \subset J_{\mathbf{Q}} + \{a \in B_{\mathrm{crys}} \mid f(a) = p^r \cdot a\}$. Since $B^+_{DR}/t \cdot B^+_{DR} \cong B^+_{\mathrm{crys}}/J_{\mathbf{Q}}$, the lemma is proved. Q.E.D.

This completes the proof of Proposition (1.17). Q.E.D.

Remarks 1.18. We will need to consider continuous Galois cohomology for sequences obtained from the sequences in (1.17) by tensoring with continuous representations on free $\mathbf{Z}_p$-modules of finite rank. For more about the topology on the rings B_{crys} and B_{DR} the reader is referred to [Fo]. Let us, however, at least sketch e.g. how one sees that the sequence (1.17.1) leads to a long exact sequence of continuous cohomology. Since $B_{DR} = \varinjlim t^{-n} B_{DR}^+$ it will suffice to show the map $\beta_n : (t^{-n} B_{\text{crys}}^+)^{f=1} \oplus B_{DR}^+ \twoheadrightarrow t^{-n} B_{DR}^+$ admits a continuous (but not G-equivariant!) splitting. For this, one reduces easily to showing $\overline{\beta} : (t^{-n} B_{\text{crys}}^+)^{f=1} \twoheadrightarrow t^{-n} B_{DR}^+ / B_{DR}^+$ admits a continuous section. The group on the right has a subgroup $\overline{V}$ consisting of the image mod B_{DR}^+ of the group $V \subset (t^{-n} B_{\text{crys}}^+)^{f=1}$ of all expressions of the form

$$\sum_{1 \le j \le n} \prod_{i \le j} \log([a^{i,j}]) \cdot t^{-j} \tag{1.18.1}$$

where $a^{i,j} = (a_k^{i,j})_{k=0,1,2,\ldots}$ with $a_0^{i,j} \in 1 + p \cdot \widehat{A}$ and $(a_k^{i,j})^p = (a_{k-1}^{i,j})$. Note that such an expression makes sense also in $B_{\text{crys}}^{f=1}$. We have, with notation as above

$$[a^{i,j}] \in 1 + pW(R) + \operatorname{Ker}(\Theta),$$

so $\log([a^{i,j}]) \in p \cdot B_\infty + J_\infty$. It follows that $V \subset t^{-n} B_\infty$. One sees easily that $\bigcap p^m \overline{V} = (0)$, and the $p^m \overline{V}$ form a basis of neighborhoods of $t^{-n} B_{DR}^+ / B_{DR}^+$. Since $\overline{V}$ is separated, it is free over $\mathbf{Z}_p$ so the surjection $V \twoheadrightarrow \overline{V}$ splits. To check that a splitting of this map induces a continuous map

$$t^{-n} B_{DR}^+ / B_{DR}^+ = \overline{V} \otimes \mathbf{Q} \to V \otimes \mathbf{Q} \subset (t^{-n} B_{\text{crys}}^+)^{f=1}$$

it suffices to show that the $p^m V \to 0$ in $(t^{-n} B_{\text{crys}}^+)^{f=1}$, but this follows because $p^m t^{-n} B_\infty \to 0$ in B_{crys} as $m \to \infty$.

2. The Coates-Wiles homomorphism and Fontaine-Messing theory

The Coates-Wiles homomorphism has an important role in the local theory of Cyclotomic fields [Wa]. The aim of of this section is to derive a formula (2.1.1) describing the relation between the Coates-Wiles homomorphism and the Fontaine-Messing theory of p-adic periods [FM]. The result will be used in §6 to study the Tamagawa number of the motif $\mathbf{Z}(r)$.

For the remainder of this section, K denotes an unramified finite extension of $\mathbf{Q}_p$ with ring of integrs $\mathcal{O}_K$. Let

$$G = \operatorname{Gal}(K(\zeta_{p^\infty})/K)$$

be the Galois group of the p-cyclotomic extension, and let

$$P = \mathrm{Gal}(K(\zeta_{p^\infty})^{\mathrm{ab}}/K(\zeta_{p^\infty}))$$
$$U = \varprojlim_n(\mathcal{O}_K[\zeta_{p^n}]^*)$$

where $K(\zeta_{p^\infty})^{\mathrm{ab}}$ is the maximal abelian extension of $K(\zeta_{p^\infty})$ and the $\varprojlim$ is taken with respect to the norms. We regard P and U as G-modules in the natural way. By the local class field theory, U is identified with the inertia subgroup of P.

The Coates-Wiles homomorphism is a canonical continuous G-homomorphism

$$\phi^r_{CW} : U \to K(r) = K \otimes \mathbf{Q}(r)$$

defined for $r \geq 1$.

Next we denote the continuous Galois cohomology $H^q(\mathrm{Gal}(\overline{K}/K),)$ by $H^q(K, \ )$. Since

$$P/U \cong \hat{\mathbf{Z}} = \varprojlim_n \mathbf{Z}/n\mathbf{Z}$$

with the trivial action of G, and since

$$H^q(G, \mathbf{Q}_p(r)) = (0) \quad \text{for } q, r \geq 1$$

we have for $r \geq 1$

$$\begin{aligned} H^1(K, K(r)) &\cong H^0(G, H^1(K(\zeta_{p^\infty}), K(r))) \\ &\cong \mathrm{Hom}_G(P, K(r)) \cong \mathrm{Hom}_G(U, K(r)). \end{aligned}$$

We regard ϕ^r_{CW} as an element of $H^1(K, K(r))$ via these isomorphisms.

The main result of this section is:

Theorem 2.1. *Let K be a finite unramified extension of $\mathbf{Q}_p$ for p an odd prime, and take $r \geq 1$. Then the boundary map from (1.13)*

$$\partial^r : K = H^0(K, B^+_{\mathrm{crys}}) \to H^1(K, \mathbf{Q}_p(r))$$

satisfies

$$\partial^r(a) = T(a \cdot \phi^r_{CW})/(r-1)!$$

where $T : H^1(K, K(r)) \to H^1(K, \mathbf{Q}_p(r))$ is the trace for $K/\mathbf{Q}_p$. In particular, if $K = \mathbf{Q}_p$, we have

(2.1.1) $$\phi^r_{CW} = (r-1)! \cdot \partial^r(1).$$

The above result will be deduced from an explicit reciprocity law for the motif $\mathbb{Z}(r)$ given in Theorem 2.6. Here we regard the classical explicit reciprocity law as the explicit reciprocity law for $\mathbb{Z}(1)$ or for $\mathbb{G}_m$. We expect there will be explicit reciprocity laws for elliptic curves with CM.

We now recall Coleman's definition of the Coates-Wiles homomorphism [Co2]. Consider the power series ring

$$\mathcal{R} = \mathcal{O}_K[[T]]; \quad z = 1 + T.$$

It will be convenient to regard $\mathcal{R}$ as the ring of functions on the formal multiplicative group $\widehat{G}_m$ over $\mathcal{O}_K$ with the canonical function z. Let for $a \in \mathbb{Z}_p$

$$\sigma_a \text{ (resp. } f, \text{ resp. } \varphi) : \mathcal{R} \to \mathcal{R}$$

be the unique ring homomorphism characterized by the property that the restriction of σ_a (resp. f, resp. φ) to $\mathcal{O}_K$ is the identity map (resp. the frobenius of $\mathcal{O}_K$, resp. the frobenius of $\mathcal{O}_K$), and

$$\sigma_a(z) = z^a \text{ (resp. } f(z) = z^p, \text{ resp. } \varphi(z) = z).$$

Here

$$z^a = \sum_{n=0}^{\infty} (n!)^{-1} \Big(\prod_{i=0}^{n-1} (a - i) \Big) (z - 1)^n \in \mathcal{R}.$$

In other words, σ_a comes from the a-th power map $\widehat{\mathbb{G}}_m \to \widehat{\mathbb{G}}_m$. Since σ_p and f are finite flat maps of degree p, there are norm and trace maps

$$N_p, T_p \text{(resp. } N_f, T_f) : \mathcal{R} \to \mathcal{R}$$

satisfying

$$\begin{aligned} &N_p \circ \sigma_p(x) = x^p; \quad T_p \circ \sigma_p(x) = px; \\ &N_f \circ f(x) = x^p; \quad T_f \circ f(x) = px; \quad \forall x \in \mathcal{R}. \end{aligned}$$

We define the action of $G = \mathrm{Gal}(\mathbb{Q}_p(\zeta_{p^\infty})/\mathbb{Q}_p)$ on $\mathcal{R}$ as follows. The action of G on $\mathbb{Z}_p(1)$ defines an isomorphism $\kappa : G \cong \mathbb{Z}_p^*$. The action of $\kappa^{-1}(a)$ on $\mathcal{R}$ is by σ_a. All the above operations on $\mathcal{R}$ commute with this G-action. Fix a generator $\nu = (\zeta_{p^n})_{n=0,1,\ldots}$ for $\mathbb{Z}_p(1)$. Define $\rho_n : \mathcal{R} \to \mathcal{O}_K[\zeta_{p^n}]$ for $n \geq 1$ to be the unique $\mathcal{O}_K$ homomorphism such that $\rho_n(z) = \zeta_{p^n}$. Then ρ_n is also compatible with the G-action,and we have commutative diagrams

(2.1.1)
$$\begin{array}{ccccccc} \mathcal{R} & \xrightarrow{\sigma_p} & \mathcal{R} & \quad & \mathcal{R}^* & \xrightarrow{N_p} & \mathcal{R}^* \\ \rho_n \downarrow & & \downarrow \rho_{n+1} & & \rho_{n+1} \downarrow & & \downarrow \rho_n \\ \mathcal{O}_K[\zeta_{p^{n+1}}] & \hookrightarrow & \mathcal{O}_K[\zeta_{p^{n+1}}] & & \mathcal{O}_K[\zeta_{p^{n+1}}]^* & \xrightarrow{\text{norm}} & \mathcal{O}_K[\zeta_{p^n}]^*. \end{array}$$

Theorem 2.2. (Coleman) *There exists an isomorphism of G-modules*

$$U \cong \{a \in \mathcal{R}^* \mid N_f(a) = a\}; \quad u = \varprojlim u_n \mapsto g_u$$

where g_u is characterized as the unique element of $\mathcal{R}^*$ such that

$$\rho_n(\varphi^{-n}(g_u)) = u_n$$

for any $n \geq 1$.

Write*

$$t = \log(z) = \sum_{i=1}^{\infty}(-1)^{i-1}T^i/i; \quad T = z - 1,$$

and consider the inclusion $\mathcal{R} \subset K[[t]] \cong K[[T]]$. Define the homomorphisms $\phi^r : U \to K$ $(r \geq 0)$ by

$$\log(g_u) = \sum_{n=1}^{\infty} \phi^r(u)t^r/r! \in K[[t]].$$

The map

$$\phi^r_{CW} : U \to K(r); \quad u \mapsto \phi^r(u) \otimes \nu^{\otimes r}$$

is compatible with the G-action and is independent of the choice of ν generating $\mathbf{Z}_p(1)$. We call these maps Coates-Wiles homomorphisms

Let

$$\widehat{\Omega}^1_{\mathcal{R}} = \varprojlim_n \Omega^1_{\mathcal{R}/\mathbf{Z}}/p^n\Omega^1_{\mathcal{R}/\mathbf{Z}},$$

a free $\mathcal{R}$ module of rank one with standard basis dz/z. Define

$$T_p : \widehat{\Omega}^1_{\mathcal{R}} \to \widehat{\Omega}^1_{\mathcal{R}}; \quad T_p(h \cdot dz/z) = p^{-1}T_p(h) \cdot dz/z.$$

(Note $T_p(\mathcal{R}) \subset p\mathcal{R}$.) We have

$$\text{(2.3)} \qquad \phi^r_{CW}(u)/(r-1)! = \text{Res}(t^{-r}dg_u/g_u) \otimes \nu^{\otimes r}; \text{ for } u \in U,$$

where Res is the residue for the field of normal power series $K((t))$. Since $T = \exp(t) - 1$, we get

$$\text{(2.4)} \qquad \text{Res}(t^{-r}\widehat{\Omega}^1_{\mathcal{R}}) \subset ((r-1)!)^{-1}.\mathcal{O}_K.$$

---

* The referee points out that there is a natural homomorphism $\nu : \mathcal{R} \to \mathcal{R}$ where R is as in §1 such that $\nu(z) = \varepsilon$. Thus t in §2 $=\log(z)$ while t in §1 $=\log([\nu(z)])$. For $u = \lim u_n$ in U, $\nu(g_u) = (a_k)$ with $a_k = \lim u_{n+k}^{p^n}$.

We rewrite theorem (2.1) in the form of the explicit reciprocity law for the motif $\mathbf{Z}(r)$. Consider the cup product pairing

$$H^1(K,\mathbf{Q}_p(r)) \times (\varprojlim_n H^1(K(\zeta_{p^n}),\mathbf{Z}/p^n(1)\otimes\mathbf{Q}) \to$$
$$\to \varprojlim_n H^2(K(\zeta_{p^n}),\mathbf{Z}/p^n(r+1))\otimes\mathbf{Q} \cong \mathbf{Q}_p(r),$$

where the inverse limits are taken with respect to the norm maps, and the last isomorphism is induced by the canonical isomorphism

$$H^2(K(\zeta_{p^n}),\mathbf{Z}/p^n(r+1)) \cong \mathbf{Z}/p^n(r).$$

Theorem (2.1) is equivalent to the assertion for $a \in K$ and $u \in U$ that the above pairing sends

$$(2.5) \qquad (\partial^r(a),u) \in H^1(K,\mathbf{Q}_p(r)) \times \varprojlim H^1(K(\zeta_{p^n}),\mathbf{Z}/p^n(1))$$
$$\mapsto \mathrm{Tr}_{K/\mathbf{Q}_p}\mathrm{Res}(at^{-r}dg_u/g_u)\otimes\nu^{\otimes r}.$$

Here we denote the image of u in $\varprojlim H^1(K(\zeta_{p^n}),\mathbf{Z}/p^n(1))$ under the canonical Kummer isomorphism $K(\zeta_{p^n})^\times/K(\zeta_{p^n})^{\times n} \cong H^1(K(\zeta_{p^n}),\mathbf{Z}/p^n(1))$ by the same letter u.

In Theorem (2.6) below, we prove the mod p^n version of this result for every $n \geq 1$. For this, we need an integral version of sequence (1.13), proved in ([**FM**] III, 1.1, and III 5.2). Write $[x]$ for the integer part of a real number x, and define

$$c = c(r) = \sum_{i=0}^{\infty}[r(p-1)^{-1}p^{-i}]; \quad J_\infty^{(r)} = \{x \in J_\infty^{[r]} \mid f(x) \in p^r B_\infty\}$$

Then the sequence

$$(2.5.1) \qquad 0 \to p^{-c}\mathbf{Z}_p(r) \to J_\infty^{(r)} \xrightarrow{1-p^{-r}f} B_\infty \to 0$$

is exact. For $m \geq c(r)$, let

$$\partial^r_{n,m} : \mathcal{O}_K \to H^1(K,\mathbf{Z}/p^n(r))$$

be the composite

$$\mathcal{O}_K \to H^0(K,B_\infty) \xrightarrow{\partial} H^1(K,p^{-c}\mathbf{Z}_p(r)) \xrightarrow{p^m}$$
$$H^1(K,\mathbf{Z}_p(r)) \to H^1(K,\mathbf{Z}/p^n(r)),$$

where ∂ is the boundary map from (2.5.1). Thus

$$\partial^r = p^{-m}(\mathbf{Q} \otimes \varprojlim_n \partial^r_{n,m})$$

The explicit reciprocity law is the following:

Theorem 2.6. *Let K be an unramified finite extension of $\mathbf{Q}_p$, and let $\mathcal{R}$, ν be as above. Then for $r, n \geq 1$ and $m \geq c(r+1)$, there is a commutative diagram*

$$\begin{array}{ccc}
\mathcal{O}_K \times \mathcal{R}^\times & \xrightarrow{(a,b)\mapsto a\cdot db/b} & \hat{\Omega}^1_{\mathcal{R}} \\
\downarrow{\scriptstyle (\partial^r_{n,m},\rho_n)} & & \downarrow{\scriptstyle \varepsilon^r_{n,m}} \\
H^1(K, \mathbf{Z}/p^n(r)) \times H^1(K(\zeta_{p^n}), \mathbf{Z}/p^n(1)) & \longrightarrow & H^2(K(\zeta_{p^n}), \mathbf{Z}/p^n(r+1)) \\
 & & \wr\| \\
 & & \mathbf{Z}/p^n(r)
\end{array}$$

where

$$\varepsilon^r_{n,m}(\omega) \equiv p^m \operatorname{Tr}_{K/\mathbf{Q}_p} \operatorname{Res}(t^{-r} T^n_p(\omega)) \otimes \nu^{\otimes r} \mod p^n.$$

Note $c(r+1) \geq c(r-1) \geq \operatorname{ord}_p((r-1)!)$ so the expression on the right is integral by (2.4). If $r < p-2$, we can take $m = 0$ since $c(p-2) = 0$. It is probable that the hypothesis on m in (2.6) can be weakened to $m \geq c(r)$.

Corollary 2.6.1. *For $u = (u_n)_n \in U$, the lower pairing in the above diagram sends $(\partial^r_{n,c(r)}(a), u_n)$ to $p^{c(r)} \cdot \operatorname{Tr}_{K/\mathbf{Q}_p} \operatorname{Res}(at^{-r} \cdot dg_u/g_u) \otimes \nu^{\otimes r}$ mod p^n.*

Note that the corollary is an easy consequence of the theorem. Indeed, (2.6) shows that the image of $(\partial^r_{n,m}(a), u_n)$ in $\mathbf{Z}/p^n(r)$ for $m \geq c(r+1)$ is $\varepsilon^r_{n,m}(a \cdot d(\varphi^{-n}(g_n))/\varphi^{-n}(g_n))$. Since $N_f(g_u) = g_u$, $T_f(dg_u/g_u) = dg_u/g_u$ and hence $T^n_p(d(\varphi^{-n}(g_n))/\varphi^{-n}(g_n)) = dg_u/g_u$. We can replace m by $c(r)$ by passing to the inverse limit.

Note (2.1) is deduced from (2.6.1) as in (2.5).

To prove (2.6), we must use some of the more difficult aspects of the theory of [FM]. We need the exact sequence of sheaves

$$(2.7) \qquad 0 \to S^r_n \to \underline{\underline{J}}^{(r)}_n \xrightarrow{1-p^{-r}f} \underline{\underline{\mathcal{O}}}^{\text{crys}}_n \longrightarrow 0$$

on the small syntomic site over $\mathrm{Spec}(\mathcal{R})$ or that over $\mathrm{Spec}(\mathcal{O}_K[\zeta_{p^n}])$. Here $\underline{\underline{\mathcal{O}}}_n^{\mathrm{crys}}$ is defined so that $\underline{\underline{\mathcal{O}}}_n^{\mathrm{crys}}(X')$ for a syntomic scheme X' over X is the ring of global sections of the structural sheaf of the crystalline site on $X' \otimes \mathbf{Z}/p^n\mathbf{Z}$. If we denote by $\underline{\underline{\mathcal{O}}}_n$ the sheaf

$$\underline{\underline{\mathcal{O}}}_n(X') = \Gamma(X' \otimes \mathbf{Z}/p^n\mathbf{Z}, \underline{\underline{\mathcal{O}}}),$$

and by $\underline{\underline{J}}_n^{[r]} \subset \underline{\underline{\mathcal{O}}}_n^{\mathrm{crys}}$ the r-th divided power of the ideal

$$\mathrm{Ker}(\underline{\underline{\mathcal{O}}}^{\mathrm{crys}} \to \underline{\underline{\mathcal{O}}}_n)$$

then $\underline{\underline{J}}^{(r)}$ is defined by

$$\underline{\underline{J}}_n^{(r)} = \mathrm{Image}\Big(\{a \in \underline{\underline{J}}_{n+r}^{[r]} \mid f(a) \in p^r \cdot \underline{\underline{\mathcal{O}}}_{n+r}^{\mathrm{crys}}\} \to \underline{\underline{J}}_n^{[r]}\Big).$$

(The sheaf $\underline{\underline{J}}_n^{(r)}$ is not introduced in [**FM**] but the authors learned its definition from Fontaine and Messing. The exactness of (2.7) follows from [**FM**] III 1.1.)

We also need the canonical homomorphism

$$\alpha = \alpha_n^r : H^q(\mathrm{Spec}(\mathcal{O}_K[\zeta_{p^n}])_{\mathrm{syn}}, S_n^r) \to H^q(K(\zeta_{p^n}), \mathbf{Z}/p^r(r)) \tag{2.9}$$

of [**FM**] III, 5. The maps α_n^r for various r have the following compatibility with the product structures, by the construction in *op. cit.*. Let $c(r)$ be as above, and define $\alpha_{n,m}^r = p^{m-c(r)}\alpha_n^r$ for $m \geq c(r)$. Then, if $m \geq c(r)$, $m' \geq c(r')$, and $m + m' \geq c(r + r')$, we have

$$\alpha_{n,m}^r(x) \cdot \alpha_{n',m'}^r(y) = \alpha_{n,m+m'}^{r+r'}(xy)$$

for the products

$$S_n^r \times S_n^{r'} \to S_n^{r+r'}; \mathbf{Z}/p^n(r) \times \mathbf{Z}/p^n(r') \to \mathbf{Z}/p^n(r+r').$$

Now consider the commutative diagram

$$
\begin{array}{ccc}
\mathcal{O}_K \times \mathcal{R}^\times & \overset{(a,b)\mapsto a\,db/b}{\longrightarrow} & \Omega^1_{\mathcal{R}} \\
\downarrow & 1 & \downarrow \\
H^0(\mathrm{Spec}(\mathcal{R})_{syn}, \underline{\underline{\mathcal{O}}}{}^{\mathrm{crys}}_n) \times H^1(\mathrm{Spec}(\mathcal{R})_{syn}, S^1_n) & \longrightarrow & H^1(\mathrm{Spec}(\mathcal{R})_{syn}, \underline{\underline{\mathcal{O}}}{}^{\mathrm{crys}}_n) \\
\downarrow{\scriptstyle \rho_n} & & \downarrow{\scriptstyle \rho_n} \\
H^0(\mathrm{Spec}(\mathcal{O}_K[\zeta_{p^n}])_{syn}, \underline{\underline{\mathcal{O}}}{}^{\mathrm{crys}}_n & \longrightarrow & H^1(\mathrm{Spec}(\mathcal{O}_K[\zeta_{p^n}])_{syn}, \underline{\underline{\mathcal{O}}}{}^{\mathrm{crys}}_n) \\
\times & & \Bigg| \\
H^1(\mathrm{Spec}(\mathcal{O}_K[\zeta_{p^n}])_{syn}, S^1_n) & 2 & \Bigg| \\
\downarrow & & \downarrow \\
H^1(\mathrm{Spec}(\mathcal{O}_K[\zeta_{p^n}])_{syn}, S^r_n) & \longrightarrow & H^2(\mathrm{Spec}(\mathcal{O}_K[\zeta_{p^n}])_{syn}, S^{r+1}_n) \\
\times & & \Bigg| \\
H^1(\mathrm{Spec}(\mathcal{O}_K[\zeta_{p^n}])_{syn}, S^1_n) & & \Bigg| \\
\downarrow{\scriptstyle (\alpha^r_{n,m}, \alpha^1_{n,0})} & & \downarrow{\scriptstyle \alpha^{r+1}_{n,m}} \\
H^1(K(\zeta_{p^n}), \mathbf{Z}/p^n(r)) \times H^1(K(\zeta_{p^n}), \mathbf{Z}/p^n(1)) & \longrightarrow & H^2(K(\zeta_{p^n}), \mathbf{Z}/p^n(r+1)) \\
 & & \downarrow \\
 & & \mathbf{Z}/p^n(r)
\end{array}
$$

Here the vertical arrows of the square (1) (resp. (2)) are defined via the isomorphism

$$H^q(\mathrm{Spec}(\mathcal{R})_{\mathrm{syn}}, \underline{\underline{\mathcal{O}}}{}^{\mathrm{crys}}_n) \cong H^q(\widehat{\Omega}^{\cdot}_{\mathcal{R}/\mathbf{Z}} \otimes \mathbf{Z}/p^n\mathbf{Z})$$

([FM]) and the first Chern class map

$$\mathcal{R}^\times \to H^1(\mathrm{Spec}(\mathcal{R})_{syn}, S^1_n)$$

[Gros] (resp. by the exact sequences

$$(2.11)\qquad 0 \to S_n^i \to \underline{\underline{J}}_n^{(i)} \xrightarrow{1-p^{-i}f} \underline{\underline{\mathcal{O}}}^{\text{crys}} \to 0 \quad (i = r, r+1)).$$

In the diagram, m is any integer such that $m \geq c(r+1)$. The composite of all vertical arrows on the left-hand side is equal to

$$(a, b) \mapsto (\partial_{n,m}^r(a), \rho_n(b)).$$

Let

$$\delta_{n,m}^r : \widehat{\Omega}_{\mathcal{R}}^1 \to \mathbf{Z}/p^n(r)$$

be the composite of all vertical arrows on the right side. Now (2.6) is reduced to the

Key Lemma 2.12. $\delta_{n,m}^r = \varepsilon_{n,m}^r$.

Proof. In the case $r = 1$ this interpretation of the usual explicit reciprocity law for Hilbert symbols via Fontaine-Messing theory was proven in [Ka]. We shall reduce (2.12) to this case. We extend $\delta_{n,m}^r$ to a homomorphism

$$\widetilde{\delta}_{n,m}^r : \widehat{\Omega}_{\mathcal{R}}^1 \to \mathbf{Z}_p(r)$$

defined as the inverse limit over i of

$$(2.12.1)\qquad \widehat{\Omega}_{\mathcal{R}}^1 \to \mathbf{Z}/p^{n+i}(r); \quad \omega \mapsto \delta_{n+i,m}^r((p^{-1}\sigma_p)^i(\omega))$$

where $p^{-1}\sigma_p(h \cdot dz/z) = \sigma_p(h) \cdot dz/z$. The compatibility of the homomorphisms (2.12.1) with the reduction maps $\mathbf{Z}/p^{n+i+1}(r) \to \mathbf{Z}/p^{n+i}(r)$ follows because by (2.1.1), the injection "p"; $\mathbf{Z}/p^{n+i}(r) \to \mathbf{Z}/p^{n+i+1}(r)$ sends $\delta_{n+i,m}^r((p^{-1}\sigma_p)^i(\omega))$ to

$$\delta_{n+i+1,m}(\sigma_p(p^{-1}\sigma_p)^i(\omega)) = p \cdot \delta_{n+i+1,m}^r((p^{-1}\sigma_p)^{i+1}(\omega))$$

Let

$$\widetilde{\delta}_n^r = p^{-m}(\mathbf{Q}_p \otimes \delta_{n,m}^r) : \widetilde{\Omega}_{\mathcal{R}}^1 \to \mathbf{Q}_p(r),$$

which is independent of m. The desired equation (2.12) follows from

$$\widetilde{\delta}_n^r(\omega) = \operatorname{Tr}_{K/\mathbf{Q}_p} \operatorname{Res}(t^{-r}T_p^n(\omega))$$

for $\omega \in \widehat{\Omega}_{\mathcal{R}}^1$. Using Lemmas 2.13 and 2.14 below, this formula in turn is reduced to the case $r = 1$ where it is proven in [Ka].

Lemma 2.13. *Let $e \gg 0$ be a sufficiently large integer. Then:*
(i) *δ_n^r annihilates $(z^{p^n}-1)^e \cdot \widehat{\Omega}^1_{\mathcal{R}} \otimes \mathbf{Q}$.*
(ii) *Let $\gamma \in \mathcal{R} \otimes \mathbf{Q}$ be such that*

$$\gamma = \sum_{i=1}^{e-1} (-1)^{i-1}(z^{p^n}-1)^i/i \mod (z^{p^n}-1)^{r+1}\mathcal{R} \otimes \mathbf{Q}$$

Then if $r \geq 2$, the following diagram commutes

$$\begin{array}{ccccc} \omega & \widehat{\Omega}^1_{\mathcal{R}} & \xrightarrow{\tilde{\delta}_n^{r-1}} & \mathbf{Q}_p(r-1) \\ \downarrow & \downarrow & & \downarrow \otimes\nu \\ \gamma\omega & \widehat{\Omega}^1_{\mathcal{R}} & \xrightarrow{\tilde{\delta}_n^{r}} & \mathbf{Q}_p(r). \end{array}$$

Lemma 2.14. *If $h : \widehat{\Omega}^1_{\mathcal{R}} \otimes \mathbf{Q} \to \mathbf{Q}_p(r)$ is a $\mathbf{Q}_p[G]$-homomorphism annihilating $(z^{p^n}-1)^{r-1}\widehat{\Omega}^1_{\mathcal{R}} \otimes \mathbf{Q}$, then $h = 0$.*

Note $\tilde{\delta}_n^r$ and $\omega \mapsto \mathrm{Tr}_{K/\mathbf{Q}_p} \mathrm{Res}(t^{-r}T_p^n(\omega))$ are $\mathbf{Q}_p[G]$-homomorphisms. By (2.13) and by induction on r, the difference of these two annihilates $(z^{p^n}-1)\widehat{\Omega}^1_{\mathcal{R}} \otimes \mathbf{Q}$. We first prove (2.13) with $e = r+2$. (It will follow from (2.12) that the lemma is true with $e = r$.) We show that the image of $(z^{p^n}-1)^e\widehat{\Omega}^1_{\mathcal{R}}$ in $H^2(\mathrm{Spec}(\mathcal{O}_K[\zeta_{p^n}])_{syn}, S_n^{r+1})$ under the composition of right-hand vertical arrows from diagram (2.10) is zero, from which (2.13))(i) follows.

Regard z^{p^n} as an element of $H^0(\mathrm{Spec}(\mathcal{O}_K[\zeta_{p^n}])_{syn}, \underline{\underline{\mathcal{O}}}_n^{syn})$. It suffices to show there exists a homomorphism of sheaves on $\mathrm{Spec}(\mathcal{O}_K[\zeta_{p^n}])_{syn}, \lambda : \underline{\underline{\mathcal{O}}}_n^{\mathrm{crys}} \to \underline{\underline{J}}_n^{(r+1)}$ such that $(1 - p^{-r-1}f) \circ \lambda$ is multiplication by $(z^{p^n}-1)^e$ on $\underline{\underline{\mathcal{O}}}_n^{\mathrm{crys}}$.

Let $h = (z^{p^n}-1)/(z^{p^{n-1}}-1)\mathcal{R}$, so $(h) = \mathrm{Ker}\,\rho_n$. We regard h as a section of $\underline{\underline{J}}_n^{(1)}$ on $\mathrm{Spec}(\mathcal{O}_K[\zeta_{p^n}])_{syn}$. We have

$$(p^{-1}f)(z^{p^n}-1) = (z^{p^n}-1) \cdot (p^{-1}f)(h).$$

This implies, for a local section x of $\mathcal{O}_n^{\mathrm{crys}}$ and for $e \geq r+1$, we have

$$(p^{-r-1}f)((z^{p^n}-1)^e x) = p^{e-r-1}(z^{p^n}-1)^e((p^{-1}f)(h))^e f(x).$$

It follows, for $e \geq r+2$, the sum

$$\lambda(x) \stackrel{\text{def}}{=} \sum_{i=0}^{\infty}(p^{-r-1}f)^i((z^{p^n}-1)^e x)$$

is a finite sum and belongs to $\underline{\underline{J}}_n^{(e)} \subset \underline{\underline{J}}_n^{(r+1)}$. Clearly,

$$(1-p^{-r-1}f)\lambda(x) = (z^{p^n}-1)^e x.$$

Next, (2.13)(ii) follows from the commutative diagram

$$\begin{array}{ccccc}
\widehat{\Omega}^1_{\mathcal{R}} & \longrightarrow & H^2(\mathrm{Spec}(\mathcal{O}_K[\zeta_{p^n}])_{syn}, S_n^r) & \xrightarrow{\alpha^r_{n,m}} & H^2(K(\zeta_{p^n}), \mathbf{Z}/p^n(r)) \\
\big\downarrow{\scriptstyle (e-1)!\gamma} & & \big\downarrow{\scriptstyle (e-1)!\cdot\log(z^{p^n})} & & \big\downarrow{\scriptstyle (e-1)!\cdot\nu} \\
\widehat{\Omega}^1_{\mathcal{R}} & \longrightarrow & H^2(\mathrm{Spec}(\mathcal{O}_K[\zeta_{p^n}])_{syn}, S_n^{r+1} & \xrightarrow{\alpha^{r+1}_{n,m}} & H^2(K(\zeta_{p^n}), \mathbf{Z}/p^n(r+1))
\end{array}$$

for $m \geq c(r+1)$. The commutativity of the left-hand square follows from the proof of (2.13)(i), and the right-hand square commutes because

$$\alpha^1_{n,0}(\log(z^{p^n})) \equiv \nu \mod p^n$$

[FM].

Finally, we prove (2.14). For $i \geq 0$, $(z^{p^n}-1)^i\widehat{\Omega}^1_{\mathcal{R}} \otimes \mathbf{Q}$ is stable under the action of G, and there exists an isomorphism of $K[G]$-modules

$$\begin{gathered}
K[\mathbf{Z}/p^n\mathbf{Z}(1)](i+1) \cong (z^{p^n}-1)^i\widehat{\Omega}^1_{\mathcal{R}}/(z^{p^n}-1)^{i+1}\widehat{\Omega}^1_{\mathcal{R}} \\
[a\nu \mod p^n] \otimes \nu^{\otimes i+1} \mapsto z^a(z^{p^n}-1)^i \cdot dz/z \quad (a \in \mathbf{Z}).
\end{gathered}$$

Since the action of G on $K[\mathbf{Z}/p^n\mathbf{Z}(1)]$ factors through a finite quotient of G, we have

$$\mathrm{Hom}_{\mathbf{Q}_p[G]}(K[\mathbf{Z}/p^n\mathbf{Z}(1)](i), \mathbf{Q}_p(r)) = (0) \text{ if } i \neq r,$$

which proves (2.14).

This completes the proof of (2.6). Q.E.D.

3. H^1 of local Galois representations

In this section K denotes a finite extension of $\mathbf{Q}_p$. Given a motif M over K, we need some p-adic space with a measure on it in order to construct the Tamagawa measure. In the example of an abelian variety A, this space is $A(K)$, the space of K-p;oints of A, and it embeds in $H^1(K,T)$ where $T = \varprojlim_n {}_nA(\overline{K})$ is the Tate module, and the map is the inverse limit of boundary maps from exact sequences

$$0 \longrightarrow {}_nA(\overline{K}) \xrightarrow{n} A(\overline{K}) \longrightarrow A(\overline{K}) \longrightarrow 0.$$

We can interpret $A(K)$ motivically as the group of extensions defined over K of the dual abelian variety A^* by $\mathbf{G}_m$, but we don't really understand the category of mixed motives well enough to say much about this group of extension when A is replaced by a more general motive M. On the other hand, the motivic interpretation of T via étale cohomology generalizes easily. Our problem becomes how to identify the subgroup $A(K) \subset H^1(K,T)$, together with theTamagawa measure on it, in some fashion which can be generalized to other motives.

By using the rings B_{crys} and B_{DR} of Fontaine as modified in [**FM**] (cf. §1) we shall define for any $\widehat{\mathbf{Z}}$-module T of finite rank, with a continuous action of $\mathrm{Gal}(\overline{K}/K)$, subgroups

$$(3.1) \qquad H^1_e(K,T) \subset H^1_f(K,T) \subset H^1_g(K,T) \subset H^1(K,T)$$

called the exponential part, the finite part, and the geometric part. In particular

$$(3.2) \qquad H^1_g(K,T) = \mathrm{Ker}(H^1(K,T) \to H^1(K, B_{DR} \otimes_{\widehat{\mathbf{Z}}} T)).$$

For the Tate module T of an abelian variety A, these three subgroups coincide and equal the image of $A(K)$ in $H^1(K,T)$. As a consequence, we remark that for an abelian variety

$$(3.3) \quad H^1(K, A(\overline{K})) \cong H^1(K, T \otimes \mathbf{Q}/\mathbf{Z})/H^1_*(K,T) \otimes \mathbf{Q}/\mathbf{Z} \quad (* = e, f, g).$$

This group is needed to define Ш(A), the Tate-Shafarevich group, whose order enters into the Tamagawa number conjecture (equivalent by [**Bl1**] to the Birch and Swinnerton-Dyer conjecture) for A. When T is the Tate module of a general motif, the groups

$$(3.4) \quad H^1_*(K,T) \quad \text{and} \quad H^1(K, T \otimes \mathbf{Q}/\mathbf{Z})/H^1_*(K,T) \otimes \mathbf{Q}/\mathbf{Z} \quad (* = e, f, g)$$

will play analogous roles.

We shall also prove a duality result (3.8) for these groups, generalizing Tate duality

$$A(K) \cong \text{ Pontryagin dual of } H^1(K, A^*(\overline{K})).$$

The notion of the motivic part of H^1 has been considered independently by a number of other authors.

Let k be the residue field of K, and let K_0 be the quotient field of the Witt vectors $W(k)$. $K_0 \subset K$ is the maximum unramified subfield of K. For a finite dimensional $\mathbf{Q}_p$-vector space V endowed with a continuous $\mathrm{Gal}(\overline{K}, K)$-action, Fontaine defines

$$\text{(3.5)} \quad \mathrm{Crys}(V) \overset{def}{=} H^0(K, B_{\mathrm{crys}} \otimes V); \quad DR(V) \overset{def}{=} H^0(K, B_{DR} \otimes V).$$

Then $\mathrm{Crys}(V)$ is a K_0-vector space with a frobenius

$$f = (f \text{ on } B_{\mathrm{crys}}) \otimes (\mathrm{id}. \text{ on } V).$$

On the other hand, $DR(V)$ is a K-vector space with a decreasing filtration

$$\text{(3.6)} \qquad DR(V)^i = H^0(K, B^i_{DR} \otimes V) \quad (i \in \mathbf{Z}),$$

and $K \otimes_{K_0} \mathrm{Crys}(V) \hookrightarrow DR(V)$. One has ([Fo], (5.1))

$$\dim_{K_0}(\mathrm{Crys}(V)) \le \dim_K(DR(V)) \le \dim_{\mathbf{Q}_p}(V).$$

If $\dim_{K_0}(\mathrm{Crys}(V)) = \dim_{\mathbf{Q}_p}(V)$ (resp. if $\dim_K(DR(V)) = \dim_{\mathbf{Q}_p}(V)$) V is said to be a crystalline (resp. a de Rham) representation of $\mathrm{Gal}(\overline{K}/K)$.

For a prime number ℓ and a $\mathbf{Q}_p$-vector space V of finite dimension endowed with a continuous action of $\mathrm{Gal}(\overline{K}/K)$ we define $\mathbf{Q}_\ell$-subspaces

$$\text{(3.7)} \qquad H^1_e(K,V) \subset H^1_f(K,V) \subset H^1_g(K,V) \subset H^1(K,V)$$

as follows. If $\ell \neq p$, let

$$\text{(3.7.1)} \qquad \begin{cases} H^1_e(K,V) = (0); H^1_g(K,V) = H^1(K,V); \\ H^1_f(K,V) = \mathrm{Ker}(H^1(K,V) \to H^1(K_{nr}, V)), \end{cases}$$

where K_{nr} is the maximal unramified extension of K. If $\ell = p$, let

$$\text{(3.7.2)} \qquad \begin{cases} H^1_e(K,V) = \mathrm{Ker}(H^1(K,V) \to H^1(K, B^{f=1}_{\mathrm{crys}} \otimes V)) \\ H^1_f(K,V) = \mathrm{Ker}(H^1(K,V) \to H^1(K, B_{\mathrm{crys}} \otimes V)) \\ H^1_g(K,V) = \mathrm{Ker}(H^1(K,V) \to H^1(K, B_{DR} \otimes V)). \end{cases}$$

For a prime number ℓ and a free $\mathbf{Z}_\ell$-module T of finite rank endowed with a continuous action of $\mathrm{Gal}(\overline{K}/K)$-action, we define

(3.7.3)

$$H^1_*(K,T) \stackrel{def}{=} \iota^{-1}(H^1_*(K,T\otimes\mathbf{Q})); \quad \iota : H^1(K,T) \to H^1(K,T\otimes\mathbf{Q})$$
$$(* = e,f,g).$$

Thus $H^1_*(K,T)$ always contains the torsion subgroup of $H^1(K,T)$. If T is a free $\widehat{\mathbf{Z}}$-module of finite rank with continuous $\mathrm{Gal}(\overline{K}/K)$-action, let $T_\ell = T\otimes_{\widehat{\mathbf{Z}}}\mathbf{Z}_\ell$. Define

$$H^1_*(K,T) \stackrel{def}{=} \prod H^1_*(K,T_\ell) \quad (* = e,f,g). \tag{3.7.4}$$

The following remark is easily proved and gives some insight into the groups H^1_*. Let V be as above, and let $\alpha\in H^1(K,V)$. α corresponds to an extension

$$0 \longrightarrow V \longrightarrow E \longrightarrow \mathbf{Q}_\ell \longrightarrow 0.$$

Assume $\ell\neq p$ (resp. $\ell = p$, resp. $\ell = p$) and that V is an unramified (resp. crystalline, resp. de Rham) representation. Then E is an unramified (resp. crystalline, resp. de Rham) representation if and only if $\alpha\in H^1_f(K,V)$ (resp. $\alpha\in H^1_f(K,V)$, resp. $\alpha\in H^1_g(K,V)$).

Proposition 3.8. *Let ℓ be a prime number, and let V (resp. T) be a finite-dimensional $\mathbf{Q}_p$-vector space (resp. free $\widehat{\mathbf{Z}}$-module of finite rank) with a continuous Galois action. If $\ell = p$, we assume V (resp. $T\otimes\mathbf{Q}$) is a de Rham representation. Let $V^* = Hom_{\mathbf{Q}_\ell}(V,\mathbf{Q}_\ell)$ (resp.$T^* = Hom_{\mathbf{Z}_\ell}(T,\mathbf{Z}_\ell)$). Then in the perfect pairing*

$$H^1(K,V)\times H^1(K,V^*(1))\times H^2(K,\mathbf{Q}_\ell(1)) \longrightarrow^{\cong} \mathbf{Q}_\ell$$

resp.

$$H^1(K,T)\times H^1(K,T^*\otimes\mathbf{Q}_\ell/\mathbf{Z}_\ell(1)) \to H^2(K,\mathbf{Q}_\ell/\mathbf{Z}_\ell(1)) \cong \mathbf{Q}_\ell/\mathbf{Z}_\ell,$$

$H^1_e(K,V)$ and $H^1_g(K,V^*(1))$ (resp. $H^1_e(K,T)$ and $H^1_g(K,T^*(1)\otimes\mathbf{Q}_\ell/\mathbf{Z}_\ell)$ are the exact annihilators of each other. The same statement holds with e replaced by g and g by e and also when e and g are both replaced by f.

Proof. The assertion for T follows easily from that for V. Furthermore,the case $\ell\neq p$ is easy, so we consider only the case of V with $\ell = p$. Recall (1.14) $B^+_{DR}\subset B_{DR}$ is the discrete valuation ring.

Lemma 3.8.1. *Let V be a de Rham representation of* $\mathrm{Gal}(\overline{K}/K)$. *Then* $H^1(K, B_{DR}^+ \otimes V) \to H^1(K, B_{DR} \otimes V)$ *is injective.*

Proof of 3.8.1. Consider the exact sequence
(3.8.2)

$$0 \to H^0(K, B_{DR}^+ \otimes V) \to H^1(K, B_{DR} \otimes V) \to H^0(K, B_{DR}/B_{DR}^+ \otimes V).$$

We have

$$\begin{aligned}\dim_K(DR(V)) &= \dim_K H^0(K, B_{DR} \otimes V)\\ &\underset{(1)}{\leq} \dim_K H^0(K, B_{DR}^+ \otimes V) + \dim_K H^0(K, B_{DR}/B_{DR}^+ \otimes V)\\ &\underset{(2)}{\leq} \sum_{i \in \mathbf{Z}} \dim_K H^0(K, \mathbf{C}_p(i) \otimes V) \underset{(3)}{\leq} \dim_{\mathbf{Q}_p}(V),\end{aligned}$$

where (2) follows from $B_{DR}^i/B_{DR}^{i+1} \cong \mathbf{C}_p(i)$ (1.15.1), and (3) is a result of Tate [**Ta2**]. Since V is assumed de Rham, all these inequalities are in fact equalities. In particular, (1) is an equality, proving the lemma. Q.E.D.

Remark 3.8.3. The same argument shows

$$H^1(K, B_{DR}^i \otimes V) \to H^1(K, B_{DR}^{i-j} \otimes V)$$

is injective for any $i \in \mathbf{Z}$ and any j with $0 \leq j \leq \infty$.

Corollary 3.8.4 *Let V be a de Rham representation as above. Then there is a commutative diagram of exact sequences:*

$$\begin{array}{ccccccccc} 0 \to & H^0(K,V) & \xrightarrow{\alpha} & \mathrm{Crys}(v)^{f=1} \oplus DR(V)^0 & \xrightarrow{\beta} & DR(V) & \to & H^1_e(K,V) & \to 0\\ & \| & & \cap & & \downarrow{\scriptstyle (0,id)} & & \cap & \\ 0 \to & H^0(K,V) & \xrightarrow{\alpha} & \mathrm{Crys}(V) \oplus DR(V)^0 & \xrightarrow{\gamma} & \mathrm{Crys}(V) \oplus DR(V) & \to & H^1_f(K,V) & \to 0\end{array}$$

Here $\mathrm{Crys}(V)^{f=1} = \{a \in \mathrm{Crys}(V) \mid f(a) = a\}$, $\alpha(x) = (x, x)$, $\beta(x, y) = x - y$, *and* $\gamma(x, y) = (x - f(x), x - y)$. *In particular,*

$$\dim_{\mathbf{Q}_p} H^1_f(K, V) = \dim_{\mathbf{Q}_p}(DR(V)/DR(V)^0) + \dim_{\mathbf{Q}_p} H^0(K, V),$$

and

$$H^1_f(K, V)/H^1_e(K, V) \cong \mathrm{Crys}(V)/(1 - f)\,\mathrm{Crys}(V).$$

Proof. The exact sequences are obtained by tensoring the sequences (1.17) with V and taking Galois cohomology, using Lemma 3.8.1.

We now consider the assertion of (3.8) for $H^1_f(K,V)$ and $H^1_f(K,V^*(1))$ in the case $\ell = p$. First we show these two groups annihilate each other. Consider the commutative diagram

$$\begin{array}{ccc}
\mathrm{Crys}(V^*(1)) \oplus DR(V^*(1)) & \xrightarrow{\delta} & H^1(K,V^*(1)) \\
\downarrow{\scriptstyle (1)} & & \downarrow{\scriptstyle (2)} \\
H^1(K, B_{\mathrm{crys}} \otimes V \otimes V^*(1)) \oplus H^1(K, B_{DR} \otimes V \otimes V^*(1)) & \xrightarrow{\varepsilon} & H^2(K, V \otimes V^*(1)) \\
 & & \downarrow \\
 & & \mathbf{Q}_p
\end{array}$$

where (1) and (2) are given by cup product with an element $\alpha \in H^1(K,V)$ and δ and ε are the connecting maps of the exact sequences (1.17.2) $\otimes V^*(1)$ and (1.17.2)$\otimes V \otimes V^*(1)$, respectively. If $\alpha \in H^1_f(K,V)$, the map (1) is zero since (1) depends only on the image of α in $H^1(K, B_{\mathrm{crys}} \otimes V)$. It follows in this case that (2) is zero on the image $H^1_f(K,V^*(1))$ of δ.

To see that $H^1_f(K,V)$ and $H^1_f(K,V^*(1))$ are exact annihilators, it is enough to show

$$\text{(3.8.5)} \qquad \dim_{\mathbf{Q}_p} H^1_f(K,V) + \dim_{\mathbf{Q}_p} H^1_f(K,V^*(1)) = \dim_{\mathbf{Q}_p} H^1(K,V).$$

Fontaine has shown in **[Fo]** that V de Rham implies $V^*(1)$ de Rham, so we can apply the dimension formula in (3.8.4) together with local duality to these two representations

$$\dim_{\mathbf{Q}_p} H^1_f(K,V) + \dim_{\mathbf{Q}_p} H^1_f(K,V^*(1)) = \dim_{\mathbf{Q}_p}(DR(V)/DR(V)^0)$$
$$+ \dim_{\mathbf{Q}_p}(DR(V^*(1))/DR(V^*(1))^0) + \dim_{\mathbf{Q}_p} H^0(K,V) + \dim_{\mathbf{Q}_p} H^2(K,V).$$

One knows

$$\sum(-1)^i \dim_{\mathbf{Q}_p} H^i(K,V) = -[K:\mathbf{Q}_p] \dim_{\mathbf{Q}_p} V \qquad \text{([Se],II(5.7))}$$
$$\dim_K DR(V)^0 + \dim_K DR(V^*(1))^0 = \dim_K DR(V) \quad \text{[Fo]}$$

The desired equation (3.8.5) follows. Finally we prove the assertion of (3.8) for the groups $H^f_e(K,V)$ and $H^1_g(K,V^*(1))$ when $\ell = p$. Consider

the commutative diagram
(3.8.6)

$$\begin{array}{ccccc}
H^1(K,V)\times DR(V^*(1)) & & \xrightarrow{(id.,\delta)} & & H^1(K,V)\times H^1(K,V^*(1)) \\
\downarrow & & & & \downarrow \\
H^1(K,B_{DR}\otimes V)\times DR(V^*(1)) & \xrightarrow{\cup} & H^1(K,B_{DR}(1)) & \xrightarrow{\varepsilon} & H^2(K,\mathbf{Q}_p(1))\cong \mathbf{Q}_p
\end{array}$$

where δ and ε are connecting homomorphisms of the exact sequences (1.17.1) $\otimes V^*(1)$ and (1.17.1)$\times\mathbf{Q}_p(1)$ respectively, and $\cup$ is the cup product. Simple manipulations reduce us to showing that the composition $\varepsilon\circ\cup$ along the bottom is a perfect pairing.

To this end, recall we have defined (1.8) an element $t\in B_{\rm crys}^{f=p}$ whose image in B_{DR}^+ generates the maximal ideal and on which Galois acts via the cyclotomic character. Multiplication by t therefore identifies the sequence obtained from (1.17.1) by tensoring with $\mathbf{Q}_p(1)$ with the top line of the following diagram, where all maps are the natural ones:
(3.8.7)

$$\begin{array}{ccccccccc}
0 & \longrightarrow & \mathbf{Q}_p(1) & \longrightarrow & B_{\rm crys}^{f=p}\oplus B_{DR}^1 & \longrightarrow & B_{DR} & \longrightarrow & 0 \\
 & & \| & & \cup & & \cup & & \\
0 & \longrightarrow & \mathbf{Q}_p(1) & \longrightarrow & (B_{\rm crys}^{f=p}\cap B_{DR}^+)\oplus B_{DR}^1 & \longrightarrow & B_{DR}^+ & \longrightarrow & 0 \\
 & & \| & & \downarrow{\scriptstyle pr_1} & & \downarrow & & \\
0 & \longrightarrow & \mathbf{Q}_p(1) & \longrightarrow & B_{\rm crys}^{f=p}\cap B_{DR}^+ & \longrightarrow & \mathbf{C}_p & \longrightarrow & 0.
\end{array}$$

We will show the analogous pairing
(3.8.8)

$$H^1(K,\mathbf{C}_p\otimes V)\times H^0(K,\mathbf{C}_p\otimes V^*)\xrightarrow{\cup} H^1(K,\mathbf{C}_p)\xrightarrow{\gamma} H^2(K,\mathbf{Q}_p(1)\cong\mathbf{Q}_p$$

is perfect, where γ is defined using the bottom row of (3.8.7). Assuming this for a moment, to see that the bottom line of (3.8.6) is perfect we tensor (3.8.7) with V and V^* and identify $DR(V^*(1))\cong DR(V^*)$ (with a shift of Hodge filtration) via multiplication by t. The pairing

$$\begin{array}{ccc}
H^1(K,B_{DR}^+\otimes V)\times DR(V^*(1))^{-1} & \longrightarrow & H^2(K,\mathbf{Q}_p(1))\cong\mathbf{Q}_p \\
\| & & \\
DR(V^*)^0 & &
\end{array}$$

then factors through (3.8.8). Since, by (3.8.1), $DR(V^*)^0\twoheadrightarrow H^0(K,\mathbf{C}_p\otimes V^*)$, it follows that any element in $H^1(K,B_{DR}^+\otimes V)$ which is orthogonal to $DR(V^*(1))^{-1}$ lies in $H^1(K, B^1_{DR} \otimes V) \cong H^1(K, B^+_{DR} \otimes V(1))$. We can replace V by $V(1)$ and iterate this argument to get the null space in $H^1(K, B^+_{DR} \otimes V)$ contained in $\cap H^1(K, B^i_{DR} \otimes V)$. Replacing V by $V(n)$ for $n > 0$, we get a similar result for the part of the null space contained in $H^1(K, B^n_{DR} \otimes V)$. But $B^i_{DR}/B^{i+1}_{DR} \cong \mathbf{C}_p(i)$ and $H^1(K, \mathbf{C}_p \otimes V(i)) = (0)$ for all but finitely many i. (Indeed, de Rham representations have a Hodge-Tate decomposition $\mathbf{C}_p \otimes V \cong \mathbf{C}_p(n_i)$ and by a result of Tate, $H^i(K, \mathbf{C}_p(j))$ has dimension 1 if $j = 0$ and $i = 0$ or 1, and is zero otherwise.) Using convergence properties of continuous cohomology, it follows that

$$H^1(K, B^i_{DR} \otimes V) = \begin{cases} (0) & i \gg 0 \\ H^1(K, B_{DR} \otimes V) & i \ll 0. \end{cases}$$

This implies that the null space in $H^1(K, B_{DR} \otimes V)$ is trivial. The proof that the null space in $DR(V^*)$ is trivial is similar, recalling that

$$H^1(K, B^n_{DR} \otimes V) \twoheadrightarrow H^1(K, \mathbf{C}_p(n) \otimes V)$$

by (1.15.1).

It remains to show the pairing (3.8.8) is perfect. Since V has a Hodge-Tate decomposition we reduce immediately to the case $V = \mathbf{Q}_p(n)$. Both cohomology groups are trivial,and there is nothing to prove unless $n = 0$. Finally therefore, everything boils down to showing the boundary map $\gamma : H^1(K, \mathbf{C}_p) \to H^2(K, \mathbf{Q}_p(1)) \cong \mathbf{Q}_p$ is non-zero. Consider the commutative diagram

$$\begin{array}{ccccccccc}
0 & \longrightarrow & \mathbf{Z}_p(1) & \longrightarrow & \varprojlim \overline{K}^* & \xrightarrow{pr_1} & \overline{K}^* & \longrightarrow & 0 \\
 & & \| & & \cup & & \cup & & \\
0 & \longrightarrow & \mathbf{Z}_p(1) & \longrightarrow & \varprojlim (1+pA)^{\times p^{-n}} & \longrightarrow & (1+pA)^{\times} & \longrightarrow & 0 \\
 & & \downarrow & & \downarrow{\scriptstyle \log[\,]} & & \downarrow{\scriptstyle \log} & & \\
0 & \longrightarrow & \mathbf{Q}_p(1) & \longrightarrow & B^{f=p}_{\mathrm{crys}} \cap B^+_{DR} & \longrightarrow & \mathbf{C}_p & \longrightarrow & 0.
\end{array}$$

where $A \subset \overline{K}$ is the ring of integers, the inverse limits are with respect to the p-th power map, and log[] denotes the log of the Teichmuller representative (§1). If we identify $H^1(K, \mathbf{Q}_p(1)) \cong (\varprojlim K^*/K^{*p^n}) \otimes \mathbf{Q}$, it follows immediately that the boundary map $\partial : K = H^0(K, \mathbf{C}_p) \to H^1(K, \mathbf{Q}_p(1))$ is the p-adic exponential. Thus, the composition

$$H^1(K, \mathbf{Q}_p) \to H^1(K, \mathbf{C}_p) \to H^2(K, \mathbf{Q}_p(1))$$

is cup product with exp(1). This is non-zero since the pairing

$$H^1(K, \mathbf{Q}_p) \otimes H^1(K, \mathbf{Q}_p(1)) \to H^2(K, \mathbf{Q}_p(1))$$

is perfect. This completes the proof of (3.8). Q.E.D.

Example 3.9. Consider the case $V = \mathbf{Q}_p(r)$. By (3.8) and (3.8.4) we have the following table for the dimensions of $H^1_*(K, \mathbf{Q}_p(r))$ for $* = e, f, g$.

r	$H^1_e(K, \mathbf{Q}_p(r))$	$H^1_f(K, \mathbf{Q}_p(r))$	$H^1_g(K, \mathbf{Q}_p(r))$	$H^1(K, \mathbf{Q}_p(r))$
$r < 0$	0	0	0	$[K : \mathbf{Q}_p]$
$r = 0$	0	1	1	$[K : \mathbf{Q}_p] + 1$
$r = 1$	$[K : \mathbf{Q}_p]$	$[K : \mathbf{Q}_p]$	$[K : \mathbf{Q}_p] + 1$	$[K : \mathbf{Q}_p] + 1$
$r > 1$	$[K : \mathbf{Q}_p]$	$[K : \mathbf{Q}_p]$	$[K : \mathbf{Q}_p]$	$[K : \mathbf{Q}_p]$

For $r = 0$, $H^1_f(K, \mathbf{Q}_p) \subset \mathrm{Hom}(\mathrm{Gal}(\overline{K}/K), \mathbf{Q}_p)$ are the unramified homomorphisms. For $r = 1$, $H^1_f(K, \mathbf{Q}_p(1))$ is the image of the exponential map

$$(\mathcal{O}_K^\times) \otimes \mathbf{Q} \to (\varprojlim K^*/K^{*p^n}) \otimes \mathbf{Q} \cong H^1(K, \mathbf{Q}_p(1)).$$

The following exponential map plays an important role in later sections.

Definition 3.10. Let V be a de Rham representation of $\mathrm{Gal}(\overline{K}/K)$. We define the exponential map

$$DR(V)/DR(V)^0 \to H^1_e(K, V)$$

to be the connecting homomorphism in (3.8.4). It is surjective with kernel $\mathrm{Crys}(V)^{f=1}/H^0(K, V)$.

Example 3.10.1. Let G be a commutative formal Lie group of finite height over $\mathcal{O}_K$. Write T for the p-adic Tate module of G, and let $V = T \otimes \mathbf{Q}$. Then V is a de Rham representation, and $DR(V)/DR(V)^0$ is identified with the tangent space of G_K ([**Fo**],§6). We claim that the expenential map (3.10) coincides with the classical one, i.e. that the following diagram commutes;

$$\begin{array}{ccc} \tan(G_K) & \xrightarrow{\exp} & G(\mathcal{O}_K) \otimes \mathbf{Q} \\ \| & & \downarrow \partial \\ DR(V)/DR(V)^0 & \xrightarrow{\exp} & H^1(K, V), \end{array}$$

where the upper (resp. lower) exp is the exponential map in the classical sense (resp. in (3.10)), and ∂ is defined by the boundary maps of the Kummer sequences (A = integral closure of $\mathcal{O}_K$)

$$0 \to T/p^nT \to G(A) \xrightarrow{p^n} G(A) \to 0.$$

To see this, for $\chi \in T^*(1)$, consider the commutative diagram of exact sequences

$$\begin{array}{ccccccccc}
0 & \longrightarrow & T & \longrightarrow & \varprojlim G(A) & \longrightarrow & G(A) & \longrightarrow & 0 \\
 & & \downarrow{\scriptstyle \chi} & & \downarrow{\scriptstyle \chi} & & \downarrow{\scriptstyle \chi} & & \\
0 & \longrightarrow & \mathbb{Z}_p(1) & \longrightarrow & \varprojlim A^\times & \longrightarrow & A^\times & \longrightarrow & 0 \\
 & & \downarrow & & \downarrow{\scriptstyle \log[\ ]} & & \downarrow{\scriptstyle \log} & & \\
0 & \longrightarrow & \mathbb{Q}_p(1) & \longrightarrow & B_{\mathrm{crys}}^{f=p} \cap B_{DR}^+ & \longrightarrow & \mathbb{C}_p & \longrightarrow & 0 \\
 & & \Vert & & \cap\downarrow & & \cap\downarrow & & \\
0 & \longrightarrow & \mathbb{Q}_p(1) & \longrightarrow & B_{\mathrm{crys}}^{f=1}(1) & \longrightarrow & (B_{DR}/B_{DR}^+)(1) & \longrightarrow & 0.
\end{array}$$

(Note χ corresponds to a homomorphism $G \to \widehat{\mathbb{G}}_m$ defined over A.) We get, therefore, a map

$$T^*(1) \to \mathrm{Hom}(\text{“top sequence”}, \text{“bottom sequence”}).$$

i.e., a map "top sequence"$\to$ "bottom sequence"$\otimes T(-1)$. Taking cohomology yields a commutative square

$$\begin{array}{ccc}
G(\mathcal{O}_K) & \xrightarrow{\partial} & H^1(K,T) \\
\downarrow & & \downarrow \\
DR(V)/DR(V)^0 & \xrightarrow{\exp} & H^1(K,V).
\end{array}$$

By the construction, the left vertical arrow of this diagram is the classical logarithm when we identify $DR(V)/DR(V)^0$ with the tangent space of G_K. This proves (3.10.2) commutative.

It follows from (3.8.4) and the duality (3.8) that in this case

$$H^1_f(K,V) = H^1_e(K,V); \ \ \mathrm{rank}\ \mathrm{Hom}(\widehat{\mathbb{G}}_m, G) = \dim_{\mathbb{Q}_p} H^1_g(K,V)/H^1_f(K,V).$$

Example 3.11. Let A be an abelian variety over K. Write $T = \prod T_\ell$ for the (total) Tate module. Then V_p is a de Rham representation [Fo] and the diagram

$$\begin{array}{ccc} \tan(A) & \xrightarrow{\exp} & A(K)\otimes\mathbf{Q} \\ \wr\| & & \wr\| \quad \partial \\ DR(V_p)/DR(V_p)^0 & \xrightarrow{\exp} & H^1_e(K,V_p) \end{array} \tag{3.11.1}$$

commutes, with $\tan(A)$ the tangent space of A at the origin, upper exp the classical one, lower exp as in (3.10), and ∂ from the Kummer sequence. Moreover,

$$H^1_e(K,T) = H^1_f(K,T) = H^1_g(K,T). \tag{3.11.2}$$

Indeed, for (3.11.1), we can replace K by a finite extension so the Neron model has semi-stable reduction and then apply the previous example. For (3.11.2), it suffices by duality to show one of the equalities, e.g. $H^1_g = H^1_f$, and we may replace T by V. The p-part follows from (3.8.4) because $\mathrm{Crys}(V)^{f=1} = (0)$. Indeed, the piece of $\mathrm{Crys}(V)$ of slope 0 can be identified with a part of the Weil cohomology of the abelian variety quotient of the special fibre, and the frobenius has no eigenvalues 1 by the Weil conjectures. For the non p-part, it suffices to note

$$H^1(\mathrm{Gal}(K^{nr}/K), H^0(\mathrm{Gal}(\overline{K}/K^{nr}), V)) = (0)$$

because $A(K)$ has no divisible elements.

4. Volumes and L-functions; the local situation

In this section, we consider a relation between local L-functions and measures on Galois cohomology groups. We denote by K a finite extension of $\mathbf{Q}_p$, by K_{nr} the maximal unramified extension of K, and by K_0 the maximal unramified subfield of K over $\mathbf{Q}_p$.

For a prime number ℓ and a $\mathbf{Q}_\ell$-vector space V of finite dimension endowed with a continuous action of $\mathrm{Gal}(\overline{K}/K)$, put

$$P(V,u) = \begin{cases} \det_{\mathbf{Q}_\ell}(1 - f_K u : H^0(K_{nr}, V)) \in \mathbf{Q}_\ell[u] & \text{if } \ell \neq p \\ \det_{K_0}(1 - f_K u : \mathrm{Crys}(V)) \in K_0[u] & \text{if } \ell = p \end{cases}$$

Here if $\ell \neq p$, f_K denotes the action of an element of $\mathrm{Gal}(\overline{K}/K)$ which acts on $\mathbf{Z}_\ell(-1)$ by $p^{[K_0:\mathbf{Q}_p]}$. If $\ell = p$, f_K denotes the K_0-linear map $f^{[K_0:\mathbf{Q}_p]}$. We call $P(V,u)^{-1}$ the local L-function attached to V. The aim of this section is to prove Theorems 4.1 and 4.2 below.

Theorem 4.1. *Let ℓ and V be as above, and assume $P(V,1) \neq 0$.*

(i) *Assume $\ell \neq p$. Then $(0) = H^1_e(K,V) = H^1_f(K,V)$. If V is unramified and T is a $\mathrm{Gal}(\overline{K}/K)$-stable $\mathbf{Z}_\ell$-sublattice in V, then*

$$\# H^1_f(K,T) = |P(V,1)|_\ell^{-1}$$

where $|\ |_\ell$ is the normalized absolute value on $\mathbf{Q}_\ell$.

(ii) *Assume $\ell = p$ and V is a de Rham representation. Then*

$$DR(V)/DR(V)^0 \xrightarrow[\simeq]{\exp} H^1_e(K,V) = H^1_f(K,V)$$

(iii) *Assume $\ell = p$, K is unramified over $\mathbf{Q}_p$, V is a crystalline representation, and the following conditions (*) holds:*
(*) There exists $i \leq 0$ and $j \geq 1$ with $j - i < p$ such that $DR(V)^i = DR(V)$ and $DR(V)^j = (0)$.
Let $D \subset \mathrm{Crys}(V) = DR(V)$ be a strongly divisible lattice (i.e., a finitely generated $\mathcal{O}_K$-submodule of $DR(V)$ such that

$$D = \sum p^{-i} f(D^i); \text{ with } D^i = D \cap DR(V)^i).$$

Let $T \subset V$ be the Galois stable sublattice constructed from D in[**FL**] *(cf. below for a review). Then*

$$\mu(H^1_f(K,T)) = |P(V,1)|_p^{-1}$$

where μ is the Haar measure of $H^1_f(K,V)$ induced from the Haar measure of D/D^0 having total measure 1 via the exponential map. Here $|\ |_p$ is the absolute value on $K = K_0$ such that $|p|_p = p^{-1}$.

Theorem 4.2. *Let K be an unramified finite extension of $\mathbf{Q}_p$, p an odd prime, and let $r \geq 2$. Consider the Haar measure μ on $H^1(K,\hat{\mathbf{Z}}(r))$ induced by the canonical Haar measure on K via the exponential map*

$$\exp : K \xrightarrow[\sim]{} H^1(K,\hat{\mathbf{Z}}(r)) \otimes \mathbf{Q}.$$

Then

$$\mu(H^1(K,\hat{\mathbf{Z}}(r))) = (1 - q^{-r})|(r-1)!|_K \cdot \# H^0(K,\mathbf{Q}_p/\mathbf{Z}_p(1-r)),$$

where q is the order of the residue field of K and $|\ |_K$ is the normalized absolute value of K.

Parts (i) and (ii) of Theorem 4.1 are straightforward, using the fact that $H^1_f(T_\ell)$ is the coinvariants of frobenius for an unramified representation when $\ell \neq p$, and $H^1_f(V_p)/H^1_e(V_p)$ has rank equal to the K_0-rank of $\mathrm{Crys}(V_p)^{f=1}$ by (3.8.4). Note this latter group is zero since $P(V,1) \neq 0$. The proof of (iii) is more delicate. It depends on the theory of Fontaine-Laffaille [**FL**] reviewed below.

A filtered Dieudonné module over $\mathcal{O}_K$ is an $\mathcal{O}_K$-module D of finite type endowed with:

(a) a decreasing filtration $(d^i)_{i\in\mathbf{Z}}$ where the D^i are direct factors of D;

(b) a family of frobenius linear maps $f_i : D^i \to D$ satisfying

$$D^i = D \text{ for } i << 0; D^i = (0) \text{ for } i >> 0,$$
$$f_i|_{D^{i+1}} = p \cdot f_{i+1} \text{ for any } i,$$
$$D = \sum_{i\in\mathbf{Z}} f_i(D^i).$$

The important and somewhat surprising fact is that the category of filtred Dieudonné modules over $\mathcal{O}_K$ is abelian.

To a filtered Dieudonné module D satisfying the conditioned

$$(*)\qquad \exists i,j \in \mathbf{Z} \text{ such that } D^i = D, D^j = (0), \text{ and } j - i < p,$$

one can associated naturally a $\mathbf{Z}_p$-module $T(D)$ of finite type endowed with a continuous action of $\mathrm{Gal}(\overline{K}/K)$, as follows. (The definition of $T(D)$ introduced here is the modified one of Fontaine-Messing [**FM**], and differs slightly from the original definition in [**FL**], but the results of that paper used below continue to hold with the same proofs.)

If $D^1 = (0)$ and $D^{2-p} = D$, let

$$T(D) = \mathrm{Ker}(1 - f : \mathrm{Fil}^0(B_\infty \otimes_{\mathcal{O}_K} D) \to B_\infty \otimes_{\mathcal{O}_K} D).$$

Here

$$\mathrm{Fil}^0(B_\infty \otimes D) = \sum J^{[i]}_\infty \otimes D^{-i} \subset B_\infty \otimes D,$$

and f is the unique homomorphism such that

$$f(x \otimes y) = p^{-i} f(x) \otimes f_{-i}(y); \quad 0 \leq i < p, x \in J^{[i]}_\infty, y \in D^{-i}.$$

(Note $f(J_\infty^{[i]}) \subset p^i B_\infty$ for $i < p$ [**FM**].) To handle shifts in the filtration, define the Tate twist $D_{(r)}$ for $r \in \mathbf{Z}$ by

$$(D_{(r)})^i = D^{i+r}; f_i \text{ for } D_{(r)} = f_{r+i} \text{ for } D.$$

Then, more generally for a filtered Dieudonné module D satisfying the condition (*), take $r \in \mathbf{Z}$ such that $(D_{(r)})^1 = (0)$ and $(D_{(r)})^{2-p} = D_{(r)}$. Define

$$T(D) = T(D_{(r)})(-r),$$

where the $(-r)$ is the Tate twist as a galois module. Then $T(D)$ is independent of the choice of r as above. Indeed, if $r, r' \in \mathbf{Z}$ satisfy the above condition and $r \leq r'$, the map from $T(D)$ defined using r to $T(D)$ defined using r' given by $x \mapsto (t^{r'-r}x) \otimes t^{\otimes(r-r')}$ with t as in (1.8) is bijective.

Theorem 4.3. (Fontaine-Laffaille) *The functor $D \mapsto T(D)$ is exact and fully faithful for D's satisfying (*) with fixed i and j. It commutes with Tate twists and preserves ranks as well as lengths for objects of finite length. Moreover, $T(D) \otimes \mathbf{Q}$ is a crystalline representation of $Gal(\overline{K}/K)$ and the canonical map $B_{\mathrm{crys}} \otimes_{\mathbf{Z}_p} T(D) \to B_{\mathrm{crys}} \otimes_{O_K} D$ is a bijection whose inverse induces and isomorphism*

$$D \otimes_{O_K} K \xrightarrow{\sim} \mathrm{Crys}(T(D) \otimes \mathbf{Q}) = DR(T(D) \otimes \mathbf{Q})$$

which preserves the frobenius filtrations. (Here the frobenius on $D \otimes K$ means $p^i f_i$ for $i \ll 0$. This is independent of i.)

Theorem 4.1(iii) follows from Lemma 4.5 below.

Lemma 4.4. *For a filtered Dieudonné module D, let*

$$h^0(D) = Ker(1 - f_0 : D^0 \to D), h^1(D) = \mathrm{Coker}(1 - f_0 : D^0 \to D)$$
$$h^i(D) = (0) \text{ for } i \geq 2.$$

Let 1_{FD} be the "unit filtered Dieudonné module" defined by

$$(1_{FD})^i = \begin{cases} O_K & \text{for } i \leq 0 \\ 0 & \text{for } i > 0 \end{cases}$$

and f_i on $(1_{FD})^i$ for $i \leq 0$ is p^{-i} times the usual frobenius. Then

$$h^i(D) \cong Ext^i(1_{FD}, D) \text{ for all } i.$$

Proof. *The family* $\{h^i\}_{i\geq 0}$ *is a cohomological functor and we see easily that* $H^0(D) \cong Hom(1_{FD}, D)$. *To prove the lemma, it therefore suffices to show that* h^1 *is effacable, i.e that for any* D *and any* $x \in h^1(D)$, *there exists an injection* $D \hookrightarrow E$ *such that the image of* x *in* $h^1(E)$ *is zero. Take a representative* $\tilde{x} \in D$ *of* $x \in D/(1-f_0)D^0$, *and define* E *by*

$$E = D \oplus \mathcal{O}_K, E^i = D^i \text{ for } i > 0; E^i = D^i \oplus \mathcal{O}_K(\tilde{x}, 1) \text{ for } i \leq 0,$$
$$f_i(\tilde{x}, 1) = p^{-i}(0,1) \text{ for } i \leq 0.$$

Then we have an exact sequence $0 \to D \to E \to 1_{FD} \to 0$, *and the boundary map* $h^0(1_{FD}) \to h^1(D)$ *sends 1 to* x. *Hence* x *dies in* $h^1(E)$, *proving the lemma.*

Lemma 4.5. *Let* D *be a filtered Dieudonné module. Assume there exist* $i \leq 0$ *and* $j \geq 1$ *such that* $j - i < p$, $D^i = D$, *and* $D^j = (0)$, *i.e. that* D *and* 1_{FD} *satisfy condition (*) above for common* i *and* j. *Let* $T = T(D)$, $V = T \otimes \mathbf{Q}$, *and consider the canonical maps*

$$\theta_k : h^k(D) \cong Ext^k(1_{FD}, D) \to Ext^k_{\text{Gal}}(\mathbf{Z}_p, T) = H^k(K, T).$$

Then
(a) θ_0 *is an isomorphism and* θ_1 *is an injection.*
(b) *We have a commutative diagram with exact rows*

$$\begin{array}{ccccccccc} 0 \to & H^0(D) \otimes \mathbf{Q} & \to & D^0 \otimes \mathbf{Q} & \xrightarrow{1-f_0} & D \otimes \mathbf{Q} & \to & h^1(D) \otimes \mathbf{Q} & \to 0 \\ & \downarrow{\scriptstyle \theta_0} & & \downarrow{\scriptstyle a} & & \downarrow{\scriptstyle b} & & \downarrow{\scriptstyle \theta_1} & \\ 0 \to & H^0(K, V) & \to & \text{Crys}(V) \oplus DR(V)^0 & \to & \text{Crys}(V) \oplus DR(V) & \to & H^1(K, V) & \end{array}$$

where $a(x) = (x, x)$, $b(x) = (x, 0)$ *with the identification*

$$D \otimes \mathbf{Q} = \text{Crys}(V) = DR(V),$$

and the lower row is that of (3.8.4). In particular, θ_1 *induces an isomorphism*

$$h^1(D) \otimes \mathbf{Q} \cong H^1_e(K, V),$$

and $\exp : DR(V)/DR(V)^0 \to H^1(K, V)$ *coincides with the composite*

$$(D/D^0) \otimes \mathbf{Q} \xrightarrow[1-f]{} (D/(1-f_0)D^0) \otimes \mathbf{Q} = h^1(D) \otimes \mathbf{Q} \xrightarrow[\theta_1]{} H^1(K, V).$$

(c) *If, moreover, D is torsion free, then θ_1 induces an isomorphism*

$$h^1(D) \cong H^1_e(K,T) = Ker(H^1(K,T) \to H^1(K,V)/H^1_e(K,V)).$$

Proof. *The assertions in (a) are a formal consequence of the full faithfulness in (4.3). Let r be such that $(D_{(r)})^1 = (0)$ and $(D_{(r)})^{2-p} = D_{(r)}$. There is a commutative diagram*

$$\begin{array}{ccccccc} 0 \to V & \to & \mathrm{Fil}^0(B_\infty \otimes D_{(r)}) \otimes \mathbf{Q}_p(-r) & \xrightarrow{c} & B_\infty \otimes D_{(r)} \otimes \mathbf{Q}_p(-r) & \to 0 \\ \| & & \downarrow a & & \downarrow b & \\ 0 \to V & \to & (B_{\mathrm{crys}} \otimes V) \oplus (B^+_{DR} \otimes V) & \to & (B_{\mathrm{crys}} \otimes V) \oplus (B_{DR} \otimes V) & \to 0, \end{array}$$

with $c = (1-f) \otimes id$. (Surjectivity of c follows from [FM] (2.3), and the lower line is (1.17.2)$\otimes V$.) The assertions in (b) are straightforward from this. Finally, (c) is equivalent to showing the cokernel of $h^1(D) \to H^1(K,T)$ is torsion free. This follows from the diagram

$$\begin{array}{ccccccc} h^0(D/pD) & \longrightarrow & h^1(D) & \xrightarrow{p} & h^1(D) & \longrightarrow & H^1(D/pD) \\ \|\wr & & \cap & & \cap & & \cap \\ H^0(K,T/pT) & \longrightarrow & H^1(K,T) & \longrightarrow & H^1(K,T) & \longrightarrow & H^1(K,T/pT). \end{array}$$

We turn now to the proof of Theorem 4.2. The non-p-part is easy, so we consider only the p-part. It is convenient to work in the derived category C of the category of abelian groups. For a morphism $h : X \to Y$ in C such that the kernel and cokernel of $H^i(h) : H^i(X) \to H^i(Y)$ are zero for almost all i, and are finite for all i, write

$$\chi(h) = \prod_{i \in \mathbf{Z}} (\#(\mathrm{Coker}\, H^i(h)) \cdot \#(Ker H^i(h))^{-1})^{(-1)^i}. \tag{4.6}$$

For an object X of C such that $H^i(X)$ is zero for almost all i and finite for all i, write

$$\chi(X) = \prod (\# H^i(X))^{(-1)^i}.$$

Let C' be the category obtained from C by inverting morphisms in C satisfying the above condition for h. For objects X, Y of C, we write $X \to Y$ for a morphism in C and $X \dashrightarrow Y$ for a morphism in C'. Note the definition *of $\chi(h)$ extends naturally to any isomorphism in C'. With these notations we have*

(4.6.1)

$$\begin{aligned}&\mu(H^1(K,\mathbb{Z}_p(r)))\cdot\#(H^0(K,\mathbb{Q}_p/\mathbb{Z}_p(1-r)))^{-1}\\&=\mu(H^1(K,\mathbb{Z}_p(r)))\cdot\prod_{i\neq 1}\#(H^i(K,\mathbb{Z}_p(r)))^{(-1)^i}\quad\textit{Tate duality}\\&=\chi(\mathcal{O}_K\overset{\exp}{\dashrightarrow}R\Gamma(K,\mathbb{Z}_p(r))[1]).\end{aligned}$$

Our aim is to prove this number is equal to

$$|1-q^{-r}|_p^{-1}\cdot|(r-1)!|_K.$$

Let $G=Gal(K(\zeta_{p^\infty})/K)$, $P=Gal(K(\zeta_{p^\infty})^{ab}/K(\zeta_{p^\infty}))$, $U=\varprojlim\mathcal{O}_K[\zeta_{p^n}]^\times$ as in §2. Consider the morphisms

$$\begin{aligned}R\Gamma(K,\mathbb{Z}_p(r))[1]&=R\Gamma(G,R\Gamma(K(\zeta_{p^\infty}),\mathbb{Z}_p(r)))[1]\\&\xrightarrow{a}R\Gamma(G,Hom(P,\mathbb{Z}_p(r)))\\&\xrightarrow{b}R\Gamma(G,Hom(U,\mathbb{Z}_p(r)))\\&\overset{c}{\dashrightarrow}R\Gamma(G,Hom(\mathcal{O}_K[[G]],\mathbb{Z}_p(r)))=Hom_{\mathbb{Z}_p}(\mathcal{O}_K,\mathbb{Z}_p(r)),\end{aligned}$$

where a is defined because

$$H^q(K(\zeta_{p^\infty}),\mathbb{Z}_p(r))=\begin{cases}\mathbb{Z}_p(r) & q=0\\ Hom(P,\mathbb{Z}_p(r)) & q=1\\ 0 & q\geq 2,\end{cases}$$

and $\chi(a)=\chi(R\Gamma(G,\mathbb{Z}_p(r)))$. Also b is defined by the inclusion $U\subset P$, so

$$\chi(b)=\chi(R\Gamma(G,Hom(P/U,\mathbb{Z}_p(r))))^{-1}=\chi(R\Gamma(G,\mathbb{Z}_p(r)))^{-1}.$$

*Finally, c is defined by an exact sequence of Coleman [***Co***]*

$$0\to\mathbb{Z}_p(1)\to U'\xrightarrow{\varphi}\mathcal{O}_K[[G]]\to\mathbb{Z}_p(1)\to 0 \tag{4.7}$$

*where U' is the pro-p part of U ($U/U'\cong\mathbb{F}_q^\times$), so $\chi(c)=1$. Here $\varphi(u)\in\mathcal{O}_K[[G]]$ for $u\in U'$ is characterized by $(1-p^{-1}f)\log(g_u)=\varphi(u)z$. (In [***Co***], Coleman considers the case $K=\mathbb{Q}_p$, but the exact sequence (4.7) holds for any unramified finite extension K of $\mathbb{Q}_p$ with the same proof.) Consequently, the number (4.6.1) equals*

$$\chi(c\circ b\circ a\circ\exp:\mathcal{O}_K\dashrightarrow Hom_{\mathbb{Z}_p}(\mathcal{O}_K,\mathbb{Z}_p(r))).$$

CLAIM 4.8: The map $c \circ b \circ a \circ \exp$ coincides with the map

$$K \to \mathrm{Hom}_{\mathbf{Q}_p}(K, \mathbf{Q}_p(r))$$
$$x \mapsto \{y \mapsto (r-1)!^{-1}\,\mathrm{Tr}_{K/\mathbf{Q}_p}((1-p^{-r}f)(x)\cdot(1-p^{r-1}f)^{-1}(y)) \otimes \nu^{\otimes r}\}$$

where ν is the fixed generator of $\mathbf{Z}_p(1)$. Note this will confirm that

$$\chi(c \circ b \circ a \circ \exp) = |1-q^{-r}|_p^{-1} \cdot |(r-1)!|_K$$

and complete the proof of (4.2). Q.E.D.

To justify the claim, we use the main result (2.1) from §2, that the boundary map $\partial : K \to H^1(K, \mathbf{Q}_p(r))$ from the Fontaine-Messing sequence

$$0 \to \mathbf{Q}_p(r) \to J_{\mathbf{Q}}^{[r]} \to B_{\mathrm{crys}}^{+} \to 0$$

is given by

(4.8.1) $$\partial(x) = \mathrm{Tr}_{K/\mathbf{Q}_p}(x\phi^r_{CW})/(r-1)!$$

It follows from the description of exp given in (4.5)(b) that

(4.8.2) $$\exp = \partial \circ (1-p^{-r}f).$$

Also, writing $\phi^r_{CW}(u) = \phi^r(u) \otimes \nu^{\otimes r}$ we have by definition $\phi^r(u) = (zd/dz)^r \log(g_u)|_{z=1}$ with g_u as in (2.2). Using

$$(zd/dz) \circ f = pf \circ (zd/dz)$$

and properties of g_u, we find

$$\begin{aligned}(1-p^{r-1}f)\phi^r(u) &= ((zd/dz)^r(1-p^{-1}f)\log(g_u))\big|_{z=1} \\ &= ((zd/dz)^r\varphi(u)z)\big|_{z=1} = \kappa_r(\varphi(u)).\end{aligned}$$

Here $\kappa_r : \mathcal{O}_K[[G]] \to \mathcal{O}_K$ is the $\mathcal{O}_K$-algebra homomorphism extending the character $G \to \mathbf{Z}_p^\times$ given by the action of G on $\mathbf{Z}_p(r)$. Thus

(4.8.3) $$\phi^r_{CW}(u) = (1-p^{r-1}f)^{-1}\kappa_r(\varphi(u)) \otimes \nu^{\otimes r}.$$

The claim follows by combining (4.8.1), (4.8.2), and (4.8.3).

This completes the proof of (4.2).

5. Global conjectures

In this section we formulate our conjecture (5.15) for special values of Hasse L-functions. After some discussion of H^1 of global Galois representations, we consider Tamagawa measures, Tamagawa numbers, and the conjectural Tamagawa number formula. Compatibility of the conjecture with isogeny is proven in (5.14), and a geometric analog is demonstrated in (5.21).

In what follows, $\mathbf{A}_f = \hat{\mathbf{Z}} \otimes \mathbf{Q}$ denotes the ring of finite adeles of $\mathbf{Q}$.

We first discuss global versions of the finite and geometric parts of H^1. K denotes a finite extension of $\mathbf{Q}$, and $\mathcal{O}_K$ denotes the ring of integers of K. For a place v of K, K_v will be the completion of K at v.

Definition 5.1. Let $\Lambda = \mathbf{Z}_\ell, \mathbf{Q}_\ell, \hat{\mathbf{Z}}$, or $\mathbf{A}_f$, and let T be a free Λ-module of finite rank endowed with a continuous Λ-linear action of $\mathrm{Gal}(\overline{K}/K)$. For a non-empty open set $U \subset \mathrm{Spec}(\mathcal{O}_K)$, we define $H^1_{f,U}(K,T) \subset H^1(K,T)$ to be the set of cohomolmogy classes whose images in $H^1(K_v,T)$ belong to $H^1_f(K_v,T)$ (resp. to $H^1_g(K_v,T)$) for any finite place $v \in U$ (resp. $v \notin U$). We define

$$H^1_g(K,T) = \varinjlim_U H^1_{f,U}(K,T).$$

If $\Lambda = \mathbf{Z}_\ell$, and if T is unramified on U and $\ell \notin U$, we regard $H^1_{f,U}(K,T)$ as a sub-$\mathbf{Z}_\ell$-module of $H^1(U,T) = \varprojlim H^1(U_{et}, T/\ell^n T)$.

To give some feeling for these groups, we formulate a conjecture relating $H^1_{f,U}$, H^1_g, and K-theory. Let X be a smooth proper scheme over K. Fix $m, r \in \mathbf{Z}$, and let

$$\Psi = \begin{cases} gr^r(K_{2r-m-1}(X) \otimes \mathbf{Q}) & \text{if } m \neq 2r-1 \\ (CH^r(X) \otimes \mathbf{Q})_{\text{homologically}\sim 0} & \text{if } m = 2r-1 \end{cases}$$

where gr is taken with respect to the γ-filtration **[Be1]**, and "hom ~ 0" denotes the subgroup of cycles homologically equivalent to 0.

Beilinson [op. cit.] has conjectured that if $m \neq 2r-1$, $2r-2$, and if X has proper regular model $\mathcal{X}$ over $\mathbf{Z}$ and $\Phi \subset \Psi$ denotes the image $gr^r(K_{2r-m-1}(\mathcal{X}) \otimes \mathbf{Q})$ in Ψ, then his regulator map induces an isomorphism

$$\Phi \otimes \mathbf{R} \cong H^m_{DR}(X_{\mathbf{R}}/\mathbf{R})/(\mathrm{Fil}^r\, H^m_{DR}(X_{\mathbf{R}}/\mathbf{R}) + H^m(X(\mathbf{C}), \mathbf{R}(2\pi i)^r))^+.$$

Here $X_{\mathbf{R}} = X \times_{\mathbf{Q}} \mathbf{R}$, Fil is the Hodge filtration on the de Rham cohomology, $X(\mathbf{C})$ is the set of $\mathbf{C}$-points of X as a $\mathbf{Q}$-scheme, endowed with the natural topology, and $(\ )^+$ is the $\mathrm{Gal}(\mathbf{C}/\mathbf{R})$-fixed part, with $\mathrm{Gal}(\mathbf{C}/\mathbf{R})$ acting on both $X(\mathbf{C})$ and $\mathbf{R}(2\pi i)^r$.

We want to formulate a conjecture which will be analogous to Beilinson's conjectures and also to Tate's conjecture

$$CH^r(X)/CH^r(X)_{\mathrm{hom}\sim 0} \otimes \mathbb{A}_f \cong H^0(K, H^{2r}(X_{\overline{K}}, \mathbb{A}_f(r))).$$

Note there is a canonical homomorphism

$$\Psi \to H^1(K, H^m(X_{\overline{K}}, \mathbb{A}_f(r))). \tag{5.2}$$

Indeed, Soulé [So] has constructed chern class maps

$$K_{2r-m-1}(X) \to H^{m+1}(X, \mathbb{Z}/n\mathbb{Z}(r))$$

for the étale topology. A weight argument using the Weil conjectures shows $K_{2m-m-1}(X) \to H^{m+1}(X_{\overline{K}}, \mathbb{A}_f(r))$ is zero if $m \neq 2r-1$, and the construction of (5.2) follows using the spectral sequence

$$E_2^{i,j} = H^i(K, H^j(X_{\overline{K}}, \mathbb{Z}/n\mathbb{Z}(r))) \Rightarrow H^{i+j}(X, \mathbb{Z}/n\mathbb{Z}(r)).$$

The following conjecture was independently formulated in a slightly different form by Jannsen.

Conjecture 5.3. (i) *The above homomorphism induces an isomorphism*

$$\Psi \otimes \mathbb{A}_f \cong H^1_g(K, H^m(X_{\overline{K}}, \mathbb{A}_f(r))).$$

(ii) *For each non-empty open set U of Spec($\mathcal{O}_K$), there exists a sub-$\mathbb{Q}$-vector space $\Phi_U \subset \Psi$ characterized by the property*

$$\Phi_U \otimes \mathbb{A}_f \cong H^1_{f,U}(K, H^m(X_{\overline{K}}, \mathbb{A}_f(r)))$$

via the isomorphism (i).
(iii) *If $m \neq 2r-1$ and X has a proper regular model $\mathcal{X}$ over U, then Φ_U coincides with the image of $gr^r(K_{2r-m-1}(\mathcal{X}) \otimes \mathbb{Q})$ in Ψ. If $m = 2r-1$, then $\Phi_U = \Psi$.*

This conjecture is true if $X = \mathrm{Spec}(K)$ and $m = 0$, $r \geq 0$ by Soulé [So2]. In the case $m = r = 1$, the conjecture is equivalent to (5.4.1) or (5.4.2) below, applied to the Picard variety A of X.

Proposition 5.4. *Let A be an abelian variety or an algebraic torus over K, and let T be its Tate module. The Kummer sequences induce an injection $A(K) \otimes \hat{\mathbb{Z}} \hookrightarrow H^1_g(K, T)$. This is an isomorphism when A*

is a torus. If A is an abelian variety, the following two conditions are equivalent:

(5.4.1) $$A(K) \otimes \hat{\mathbf{Z}} \cong H^1_g(K,T);$$

(5.4.2) *The ℓ-primary part* $Ш(A)\{\ell\}$ *of*

$$(A) = Ker(H^1(K,A) \to \oplus H^1(K_v,A))$$

is finite for all ℓ.

The proof is omitted.

We next discuss a Galois representation with a de Rham structure a *motivic pair* which provides a good axiomatic foundation for our Tamagawa measures. From now on it will be convenient to take our base field to be $\mathbf{Q}$. This simplifies definitions, and as usual one areduces to this case by Weil restriction of scalars, anyway. We write p for either a prime number or the real place, ∞, of $\mathbf{Q}$. For each p, we fix an embedding $\overline{\mathbf{Q}} \hookrightarrow \overline{\mathbf{Q}}_p$. We write $V_p = V \otimes \mathbf{Q}_p$.

Definition 5.5. A motivic pair (V, D) is a pair of finite dimensional $\mathbf{Q}$-vector spaces with the following extra structure (i)–(iii) satisfying axioms (P1)–(P4).

(i) $V \otimes \mathbf{A}_f$ has a continuous $\mathbf{A}_f$-linear galois action such that $V \subset V \otimes \mathbf{A}_f$ is stable under Gal$(\mathbf{C}/\mathbf{R}) \subset$ Gal$(\overline{\mathbf{Q}}/\mathbf{Q})$.

(ii) D has a decreasing filtration $(D^i)_{i\in\mathbf{Z}}$ by $\mathbf{Q}$-subspaces such that $D^i = (0)$ for $i \gg 0$ and $D^i = D$ for $i \ll 0$.

(iii) For $p < \infty$ we are given an isomorphism of $\mathbf{Q}_p$-vector spaces

$$\theta_p : D_p \cong DR(V_p)$$

preserving filtrations. For $p = \infty$, we are given an isomorphism of $\mathbf{R}$-vector spaces

$$\theta_\infty : D_\infty \cong (V_\infty \otimes_{\mathbf{R}} \mathbf{C})^+$$

Here $D_p = D \otimes \mathbf{Q}_p$, and $DR(\ )$ is defined with respect to the action of Gal$(\overline{\mathbf{Q}}_p/\mathbf{Q}_q)$. We denote the Gal$(\mathbf{C}/\mathbf{R})$ fixed part by $(\ )^+$, where the action of $\sigma \in$ Gal$(\mathbf{C}/\mathbf{R})$ on $V_\infty \otimes \mathbf{C}$ is $\sigma \otimes \sigma$. We shall regard the maps θ_p as being identifications.

These data are subject to the following axioms:

(P1) There exists a non-empty open set U of Spec$(\mathbf{Z})$ such tht for any $p \in U$, V_ℓ is unramified at p for $\ell \neq p$ and V_p is crystalline.

(P2) Let M be a $\mathbf{Z}$-lattice in V and let L be a $\mathbf{Z}$-lattice in D. Then there exists a finite set S of primes of $\mathbf{Q}$ ("bad primes") with $\infty \in S$ and such that for all $p \neq S$, V_p is a crystalline representaton of $\mathrm{Gal}(\overline{\mathbf{Q}}_p/\mathbf{Q}_p)$, condition (*) in §4 is satisfied by the filtration on $DR(V)$, $L \otimes \mathbf{Z}_p$ is a strongly divisible lattice in $D_p = \mathrm{Crys}(V_p)$,and $M \otimes \mathbf{Z}_p$ coincides with the $\mathrm{Gal}(\overline{\mathbf{Q}}_p/\mathbf{Q}_p)$-stable lattice in V_p corresponding to $L \otimes \mathbf{Z}_p$ via (5.5)(iii).

(P3) Let $p < \infty$, and let $P_p(V_\ell, u)$ be the polynomial $P(V_\ell, u)$ of §4, defined for the $\mathrm{Gal}(\overline{\mathbf{Q}}/\mathbf{Q})$-module V_ℓ. Then $P_p(V_\ell, u) \in \mathbf{Q}[u]$ for all ℓ and these polynomials are independent of ℓ.

(P4) If $p < \infty$, there exists a $\mathrm{Gal}(\overline{\mathbf{Q}}_p/\mathbf{Q}_p)$-stable $\mathbf{Z}$-lattice $T \subset V \otimes \mathbb{A}_f$ such that $H^0(\mathbf{Q}_{p,nr}, T \otimes \mathbf{Q}_\ell/\mathbf{Z}_\ell)$ is divisible for almost all ℓ. (This condition is easily seen to be independent of the choice of T.)

Definition 5.5.1. A motivic pair (V, D) has *weights* $\leq w$ if for each $p < \infty$, the polynomial $P_p(V, u)$ has the form $\prod(1 - \alpha_i u)$ with $|\alpha_i| \leq p^{w/2}$ in $\mathbf{C}[u]$, and if $D_\infty^i \cap V_\infty^+ = (0)$ for $i > w/2$.

Suppose (V, D) has weights $\leq w$, and S is a finite set of places of $\mathbf{Q}$ containing ∞. The L-function $L_S(V, s)$ is defined by

$$L_S(V,s) = \prod_{p \notin S} P_p(V, p^{-s})^{-1}.$$

This product converges absolutely for $\mathbf{Re}(s) > w/2 + 1$.

Let (V, D) be a motivic pair with weights ≤ -1. Fix a $\mathbf{Z}$-lattice M in V such that $M \otimes \hat{\mathbf{Z}}$ is $\mathrm{Gal}(\overline{\mathbf{Q}}/\mathbf{Q})$-stable in $V \otimes \mathbb{A}_f$. (Note $\mathbf{Z}$-lattices in V are in one-to-one correspondence with $\hat{\mathbf{Z}}$-lattices in $V \otimes \mathbb{A}_f$, so there are many such M.) We define groups $A(\mathbf{Q}_p)$ for $p \leq \infty$ associated to M, which are analogous to groups of $\mathbf{Q}_p$-rational points of a commutative algebraic group. Here the letter A is simply a notation. When (V, D) has weights ≤ -3, we define the Tamagawa measure on $\prod_{p \leq \infty} A(\mathbf{Q}_p)$.

First, let

$$(5.6) \qquad A(\mathbf{Q}_p) = \begin{cases} H_f^1(\mathbf{Q}_p, M \otimes \hat{\mathbf{Z}}) & \text{if } p < \infty \\ ((D_\infty \otimes_{\mathbf{R}} \mathbf{C})/((D_\infty^0 \otimes_{\mathbf{R}} \mathbf{C}) + M))^+ & \text{if } p = \infty \end{cases}$$

(The inclusion $M \hookrightarrow D_\infty \otimes_{\mathbf{R}} \mathbf{C}$ is given by the identification $D_\infty \otimes_{\mathbf{R}} \mathbf{C} = V_\infty \otimes \mathbf{C}$.) We regard $A(\mathbf{Q}_p)$ for $p < \infty$ as a compact group with the natural topology, and $A(\mathbf{R})$ as a locally compact group.

For $p \leq \infty$, we have the exponential homomorphism

$$(5.7) \qquad \exp : D_p/D_p^0 \dashrightarrow A(\mathbf{Q}_p)$$

which is a local isomorphism defined on a neighborhood of zero in D_p/D_p^0. Indeed, for $p < \infty$, our hypothesis $w \leq -1$ implies $P_p(V,1) \neq 0$, so we know from (4.1)(ii) that

$$\exp : D_p/D_p^0 \cong H_f^1(\mathbf{Q}_p, V_p)$$

From (P3) and (P4), we easily see that

$$A(\mathbf{Q}_p)/H_f^1(\mathbf{Q}_p, M \otimes \mathbf{Z}_p) \cong \prod_{\ell \neq p} H^0(\mathbf{Q}_p, M \otimes \mathbf{Q}_\ell/\mathbf{Z}_\ell)$$

is finite, so (5.7) is a local isomorphism as claimed. For $p = \infty$, we define the exponential homomorphism to be the evident canonical map, which is defined on the total space D_∞/D_∞^0.

Now assume the weights are ≤ -3. We define the Tamagawa measure on $\prod A(\mathbf{Q}_p)$ as follows. Fix an isomorphism

$$\omega : \det_{\mathbf{Q}}(D/D^0) \cong \mathbf{Q}.$$

For each $p \leq \infty$ this gives

$$\det_{\mathbf{Q}_p}(D_p/D_p^0) \cong \mathbf{Q}_p$$

This trivialization of the determinant gives a Haar measure on the p-adic space D_p/D_p^0 and hence a Haar measure $\mu_{p,\omega}$ on $A(\mathbf{Q}_p)$ via the exponential map. By (4.1), for a sufficiently large finite set S of places of $\mathbf{Q}$ containing ∞, we have for $p \neq S$

$$\mu_{p,\omega}(A(\mathbf{Q}_p)) = P_p(V,1). \tag{5.8}$$

Since the weights are ≤ -3, the product

$$L_S(V,0)^{-1} = \prod_{p \notin S} \mu_{p,\omega}(A(\mathbf{Q}_p))$$

converges, so the product measure $\mu = \prod_{p \leq \infty} \mu_{p,\omega}$ on $\prod_{p \leq \infty} A(\mathbf{Q}_p)$ is defined. Since for $a \in \mathbf{Q}^\times$, $\mu_{p,a\omega} = |a|_p \mu_{p,\omega}$, the product formula implies that μ is independent of the choice of ω.

Definition 5.9. μ is the Tamagawa measure for the motivic pair (D,V).

If we only assume the weights of (D,V) are ≤ -1, we can define the Tamagawa measure μ on $\prod A(\mathbf{Q}_p)$ if we assume $L_S(V,s)$ can be analytically continued to $\mathbf{Re}(s) > -\varepsilon$ for some $\varepsilon > 0$. Indeed, in this case, let $r = \mathrm{ord}_{s=0} L_S(V, s)$, and define

$$(5.9.1) \qquad \mu = |\lim_{s\to 0} s^r L_S(V,s)^{-1}| \cdot \prod_{p\in S} \mu_{p,\omega} \cdot \prod_{s\notin S} (P_p(V,1)^{-1} \mu_{p,\omega}).$$

We hope that a motif over $\mathbf{Q}$ gives a motivic pair. Consider, for example, the pure motif $H^m(X)(r)$ where X is a smooth, proper, (not necessarily geometrically connected) scheme over $\mathbf{Q}$. Define

$$V = H^m(X(\mathbf{C}), \mathbf{Q}((2\pi i)^r); \quad D = H^m_{DR}(X/\mathbf{Q}).$$

By Artin's theorem, $V \otimes \mathbb{A}_f \cong H^m_{et}(X_{\overline{\mathbf{Q}}}, \mathbb{A}_f)(r)$ whence the galois acton. The filtration on D is deduced fromt he Hodge filtration on H_{DR} by

$$D^i \stackrel{\text{def}}{=} \mathrm{Fil}^{r+i} H^m_{DR}(X/\mathbf{Q}).$$

The isomorphism

$$\theta_\infty : D_\infty \cong (\mathbf{C} \otimes_{\mathbf{R}} V_\infty)^+$$

is standard. The de Rham conjecture of Fontaine [**Fo**] for the scheme $X_{\mathbf{Q}_p}$ for $p < \infty$ proved recently by Faltings [**Fa**] says that there exists a canonical isomorphism

$$\theta_p : D_p \cong DR(V_p).$$

The work of Fontaine and Messing shows that (V, D) has the properties (P1), (P2). Moreover, (P3) holds for almost all p ([**Del**], [**FM**]), and (P3) and (P4) hold if $m = 1$. Thus (V, D) is a motivic pair if $m = 1$. Although (P1)–(P4) involve a number of unproven properties of motives, we will see in (5.15.2) below that it is possible to formulate a conjecture about the ℓ-part of the Tamagawa number under much weaker assumptions.

We now introduce the global points $A(\mathbf{Q})$. Let (V, D) be a motivic pair of weights ≤ -3, and assume that we are given a finite dimensional $\mathbf{Q}$-vector space Φ endowed with an isomorphism of $\mathbf{R}$-vector spaces

$$R_\infty : \Phi \otimes \mathbf{R} \cong D_\infty/(D^0_\infty + V^+_\infty)$$

and an isomorphism of $\mathbb{A}_f$-modules

$$R_{\mathrm{Gal}} : \Phi \otimes \mathbb{A}_f \cong H^1_{f,\mathrm{Spec}(\mathbf{Z})}(\mathbf{Q}, V \otimes \mathbb{A}_f).$$

Fix a $\mathbf{Z}$-lattice M in V such that $M \otimes \hat{\mathbf{Z}}$ is galois stable in $V \otimes \mathbb{A}_f$. Define $A(\mathbf{Q}) \subset H^1_{f,\mathrm{Spec}(\mathbf{Z})}(\mathbf{Q}, M \otimes \hat{\mathbf{Z}})$ to be the inverse image of $R_{\mathrm{gal}}(\Phi)$. Note that $A(\mathbf{Q})$ is a finitely generated abelilan group such that

$$A(\mathbf{Q}) \otimes \hat{\mathbf{Z}} = H^1_{f,\mathrm{Spec}(\mathbf{Z})}(\mathbf{Q}, M \otimes \hat{\mathbf{Z}}) \text{ and } A(\mathbf{Q}) \otimes \mathbf{Q} = \Phi.$$

Lemma 5.10. (i) $A(\mathbf{Q})_{\rm tor} \cong H^0(\mathbf{Q}, M \otimes \mathbf{Q}/\mathbf{Z})$.
(ii) $A(\mathbf{Q}_p)_{\rm tor} \cong H^0(\mathbf{Q}_p, M \otimes \mathbf{Q}/\mathbf{Z})$ for $p \leq \infty$.

Proof. Exercise

Of course there are natural homomorphisms $A(\mathbf{Q}) \to A(\mathbf{Q}_p)$ for $p < \infty$, as well as $A(\mathbf{Q}) \to A(\mathbf{R})/A(\mathbf{R})_{\rm cpt} = D_\infty/(D^0_\infty + V^+_\infty)$. Here $A(\mathbf{R})_{\rm cpt}$ denotes the maximal compact subgroup of $A(\mathbf{R})$. (5.10)(ii) above for $p = \infty$ implies that $A(\mathbf{Q})_{\rm tor} \subset A(\mathbf{R})_{\rm tor}$, so we can choose $h : A(\mathbf{Q}) \to A(\mathbf{R})$ lifting the above map. (We expect, of course, that if (V, D, Φ) comes from a motif, a canonical such lifting is given; but this is not necessary to define the Tamagawa number.) We define

$$(5.11) \qquad \mathrm{Tam}(M) = \mu((\prod A(\mathbf{Q}_p))/A(\mathbf{Q})).$$

Our hypotheses imply that the image of $A(\mathbf{Q})$ in $A(\mathbf{R})/A(\mathbf{R})_{\rm cpt}$ is discrete and co-compact, so $\mathrm{Tam}(M)$ is defined.

When (V, D) is associated to the motive $H^m(X)(r)$, the conjectures of Beilinson suggest that

$$\Phi = \mathrm{Image}(gr^r(K_{2r-m-1}(\mathcal{X}) \otimes \mathbf{Q}) \to gr^r(K_{2dr-m-1}(X) \otimes \mathbf{Q})$$

with $R_{\rm gal}$ given by the chern class map and R_∞ by the Beilinson regulator, has the required properties. If so, we say that the triple (V, D, Φ) comes from the motive $H^m(X)(r)$.

Let (V, D, Φ, M) be as above, and let $A(\mathbf{Q})$ and $A(\mathbf{Q}_p)$ be the corresponding groups. In order to formulate our Tamagawa number conjecture, we consider the map

$$(5.12) \qquad \alpha_M : \frac{H^1(\mathbf{Q}, M \otimes \mathbf{Q}/\mathbf{Z})}{A(\mathbf{Q}) \otimes \mathbf{Q}/\mathbf{Z}} \longrightarrow \bigoplus_{p \leq \infty} \frac{H^1(\mathbf{Q}_p, M \otimes \mathbf{Q}/\mathbf{Z})}{A(\mathbf{Q}_p) \otimes \mathbf{Q}/\mathbf{Z}}$$

Define

$$(5.13) \qquad Ш(M) = \mathrm{Ker}(\alpha_M).$$

Proposition 5.14 (i) *For any prime number ℓ, the ℓ-primary part $Ш(M)\{\ell\}$ is finite.*
(ii) *$\mathrm{Coker}(\alpha_M)$ is finite and is isomorphic to the Pontryagin dual of the finite group $H^0(\mathbf{Q}, M^* \otimes \mathbf{Q}/\mathbf{Z}(1))$, with $M^* = Hom(M, \mathbf{Z})$.*
(iii) *Assume $Ш(M)$ is finite, and define $\chi(M) \in \mathbf{R}^\times$ by*

$$\chi(M) = \mathrm{Tam}(M) \cdot \#(Ш(M)) \cdot \#(H^0(\mathbf{Q}, M^* \otimes \mathbf{Q}/\mathbf{Z}(1)))^{-1}.$$

Then for any $\mathbf{Z}$-lattice $M' \subset V$ such that $M' \otimes \hat{\mathbf{Z}}$ is galois stable in $V \otimes \mathbb{A}_f$, we have $Ш(M')$ also finite and $\chi(M) = \chi(M')$.

Conjecture 5.15.(Tamagawa number conjecture) *Assume the triple (V, D, Φ) comes from a motif. Let M be a $\mathbf{Z}$-lattice in V such that $M \otimes \hat{\mathbf{Z}}$ is galois stable in $V \otimes \mathbb{A}_f$. Then $Ш(M)$ is finite, and*

$$\mathrm{Tam}(M) = \frac{\#(H^0(\mathbf{Q}, M^* \otimes \mathbf{Q}/\mathbf{Z}(1)))}{\#(Ш(M))}.$$

By (5.14), the validity of this conjecture is independent of the choice of M. If (V, D, Φ) comes from $H^m(X)(r)$, we can take

$$M = H^m(X(\mathbf{C}), \mathbf{Z}(2\pi i)^r)/\,\mathrm{tors}.$$

However, when the motive is only the image of a projector, there is frequently no canonical choice for M so it is nice to have a conjecture which is "isogeny invariant".

The conjecture can be rewritten to emphasize the role of the L-function (5.15.1)

$$L_S(V, 0) = \frac{\#(Ш(M))}{\#(H^0(\mathbf{Q}, M^* \otimes \mathbf{Q}/\mathbf{Z}(1)))}\mu_{\infty,\omega}(A(\mathbf{R})/A(\mathbf{Q}))\cdot \prod_{p \in S - \infty} \mu_{p,\omega}(A(\mathbf{Q}_p))$$

where S is sufficiently large (depending on ω as well as (V, D, Φ, M)).

Remark 5.15.2. Suppose (V, D, Φ, M) corresponds to $H^m(X)(r)$ with $m - 2r \le -3$ and X smooth and proper over $\mathbf{Q}$. The problem of even defining the two sides of (5.15) involves difficult unsolved questions relating to global behavior at all primes. Such questions as finiteness of $Ш(M)$ and axiom (P4) can be avoided if we are content to work modulo $\mathbf{Z}_{(\ell)}^\times = \{a/b | a, b \in \mathbf{Z}; \ell \nmid ab\}$. Indeed, we can forget $V_{\ell'}$ for $\ell' \neq \ell$, and we only need assume $P_\ell(V_\ell, 1) \neq 0$ and find $\Phi \subset gr^r(K_{2r-m-1}(X) \otimes \mathbf{Q})$ such that

$$\Phi \otimes \mathbf{R} \cong D_\infty/(D_\infty^0 + V_\infty^+);$$
$$\Phi \otimes \mathbf{Q}_\ell \cong \mathrm{Ker}(H^1(\mathbf{Q}, V_\ell) \to H^1(\mathbf{Q}_\ell, V_\ell)/H^1_f(\mathbf{Q}_\ell, V_\ell) \oplus \prod_{p \neq \ell} H^1(\mathbf{Q}_p, V_\ell)).$$

(Note $P_\ell(V_\ell, 1) \neq 0$ holds at least if X is projective and has good reduction at ℓ by **[FM]**, **[Fa]** and the Weil conjectures.) To define the ℓ-Tamagawa number

$$\mathrm{Tam}^{(\ell)}(M) \in \mathbf{R}^\times/\mathbf{Z}_{(\ell)}^\times$$

we use the groups

$$A^{(\ell)}(\mathbf{Q}_p) \stackrel{\text{def}}{=} \begin{cases} H^1_f(\mathbf{Q}_\ell, M \otimes \mathbf{Z}_\ell) & \text{if } p = \ell \\ H^1(\mathbf{Q}_\ell, M \otimes \mathbf{Z}_\ell)_{\text{tor}} & \text{if } p \neq \ell, p < \infty \end{cases}$$

One *can* show finiteness for

$$Ш^{(\ell)}(M) \stackrel{\text{def}}{=}$$
$$\operatorname{Ker}\Big(\frac{H^1(\mathbf{Q}, M \otimes \mathbf{Q}_\ell/\mathbf{Z}_\ell)}{\text{image}(\Phi)} \to \frac{H^1(\mathbf{Q}_\ell, M \otimes \mathbf{Q}_\ell/\mathbf{Z}_\ell)}{H^1_f(\mathbf{Q}_\ell, M \otimes \mathbf{Z}_\ell) \otimes \mathbf{Q}/\mathbf{Z}} \bigoplus$$
$$\bigoplus_{p \neq \ell} H^1(\mathbf{Q}_p, M \otimes \mathbf{Q}_\ell/\mathbf{Z}_\ell)\Big)$$

and we can actually ask if

$$\operatorname{Tam}^{(\ell)}(M) \cdot \#Ш^{(\ell)}(M) \cdot \#H^0(\mathbf{Q}, M^* \otimes \mathbf{Q}_\ell/\mathbf{Z}_\ell(1)) = 1 \text{ in } R^\times/\mathbf{Z}^\times_{(\ell)}.$$

We turn now to the proof of (5.14). The finitenes of $Ш\{\ell\}$ and $\operatorname{Coker}(\alpha_M)\{\ell\}$ follows from

Lemma 5.16. *Let ℓ be a prime number, U a non-empty open set of Spec($\mathbf{Z}$) not containing ℓ, T a free $\mathbf{Z}_\ell$-module of finite rank with a continuous action of Gal($\overline{\mathbf{Q}}/\mathbf{Q}$), and $V = T \otimes \mathbf{Q}_\ell$. Assume conditions (a)–(d) below hold:*

(a) *V is unramified on U.*
(b) *V is a de Rham representation of Gal($\overline{\mathbf{Q}}_\ell/\mathbf{Q}_\ell$).*
(c) *$P_p(V, 1) \neq 0$ for any $p \notin U$, $p \neq \infty$.*
(d) *$P_p(V(-1), 1) \neq 0$ for any $p \in U$.*

Let

$$\alpha_T : \frac{H^1(\mathbf{Q}, T \otimes \mathbf{Q}/\mathbf{Z})}{H^1_{f,\operatorname{Spec}(\mathbf{Z})}(\mathbf{Q}, T) \otimes \mathbf{Q}/\mathbf{Z}} \to \bigoplus_{p \leq \infty} \frac{H^1(\mathbf{Q}_p, T \otimes \mathbf{Q}/\mathbf{Z})}{H^1_f(\mathbf{Q}_p, T) \otimes \mathbf{Q}/\mathbf{Z}}$$
(5.16.1)
$$\beta_T : H^2(\mathbf{Q}, T \otimes \mathbf{Q}/\mathbf{Z}) \to \bigoplus_{p \leq \infty} H^2(\mathbf{Q}_p, T \otimes \mathbf{Q}/\mathbf{Z}).$$

(For $p = \infty$, $H^1_f(\mathbf{Q}_p, T) \otimes \mathbf{Q}/\mathbf{Z}$ is understood to be 0.) Then $\ker(\alpha_T)$ *is finite, β_T is surjectrive,* $\operatorname{Coker}(\alpha_T)$ *and Ker(β_T) are of co-finite type (i.e., $\cong (\mathbf{Q}_\ell/\mathbf{Z}_\ell)^r \oplus$ (finite) for some $r < \infty$), and*

$$H^1_{f,\operatorname{Spec}(\mathbf{Z})}(\mathbf{Q}, V) = \dim(V) - \dim(DR(V)^0) - \dim(V^+)$$
$$+ \operatorname{corank}(\operatorname{Coker}(\alpha_T)) + \operatorname{corank}(\operatorname{Ker}(\beta_T)).$$

In particular, if $H^1_{f,\mathrm{Spec}(\mathbf{Z})}(\mathbf{Q},V)=\dim V-\dim(DR(V)^0)-\dim V^+$ then $\mathrm{Coker}(\alpha_T)$ and $\ker(\beta_T)$ are finite.

Remark 5.16.2. The bijectivity of β_T is proven in ([Ja],Th. 3) in a general setting.

Proof of 5.16. Let

$$\alpha_{T,U}:\frac{H^1(U,T\otimes\mathbf{Q}/\mathbf{Z})}{H^1_{f,\mathrm{Spec}(\mathbf{Z})}(\mathbf{Q},T)\otimes\mathbf{Q}/\mathbf{Z}}\to\bigoplus_{p\notin U}\frac{H^1(\mathbf{Q}_p,T\otimes\mathbf{Q}/\mathbf{Z})}{H^1_f(\mathbf{Q}_p,T)\otimes\mathbf{Q}/\mathbf{Z}}$$

$$\beta_{T,U}:H^2(U,T\otimes\mathbf{Q}/\mathbf{Z})\to\bigoplus_{p\notin U}H^2(\mathbf{Q}_p,T\otimes\mathbf{Q}/\mathbf{Z}).$$

$\beta_{T,U}$ is surjective by Tate duality, and localization for étale cohomology yields

$$\mathrm{Ker}(\alpha_{T,U})\cong\mathrm{Ker}(\alpha_T)$$

$$0\to\mathrm{Coker}(\alpha_{T,U})\to\mathrm{Coker}(\alpha_T)\to\mathrm{Ker}(\beta_{T,U})\to\mathrm{Ker}(\beta_T)\to 0.$$

By the definition of $H^1_{f,\mathrm{Spec}(\mathbf{Z})}$ and the finite generation of $H^*(U,T)$, it follows that $\mathrm{Ker}(\alpha_{T,U})$, and hence $\mathrm{Ker}(\alpha_T)$, is finite. Also $\mathrm{Coker}(\alpha_T)$ and $\mathrm{Ker}(\beta_T)$ are of co-finite type, and

$$\begin{aligned}&\mathrm{corank}(\mathrm{Coker}(\alpha_T))+\mathrm{corank}(\mathrm{Ker}(\beta_T))=\mathrm{corank}(\mathrm{Coker}(\alpha_{T,U}))\\&\quad+\mathrm{corank}(\mathrm{Ker}(\beta_{T,U}))=\\&\sum_{i=0}^{2}(-1)^i\,\mathrm{corank}\,H^i(U,T\otimes\mathbf{Q}/\mathbf{Z})-\sum_{\substack{p\notin U\\ p\neq\infty}}\sum_{i=0}^{2}(-1)^i\,\mathrm{corank}(H^i(\mathbf{Q}_p,T\otimes\mathbf{Q}/\mathbf{Z})\\&\quad+\dim H^1_{f,\mathrm{Spec}(\mathbf{Z})}(\mathbf{Q},V)-\dim H^1_f(\mathbf{Q}_\ell,V).\end{aligned}$$

We have

$$\sum(-1)^i\,\mathrm{corank}\,H^i(U,T\otimes\mathbf{Q}/\mathbf{Z})=\dim(V^+)-\dim(V)\qquad[\mathrm{Ta1}]$$

$$\sum(-1)\,\mathrm{corank}\,H^i(\mathbf{Q}_p,T\otimes\mathbf{Q}/\mathbf{Z})=\begin{cases}-\dim(V) & \text{if } p=\ell\\ & \text{(cf [Se])}\\ 0 & \text{if } p\neq\ell,p\neq\infty\end{cases}$$

$$\dim H^1_f(\mathbf{Q}_\ell,V)=\dim(D/D^0)\qquad(4.1).$$

These prove the formula in (5.14).

Proof of the remainder of (5.14) (ii). Let ℓ be a given prime number. By (3.8), we have

$$\frac{H^1(\mathbf{Q}_p, M \otimes \mathbf{Q}_\ell/\mathbf{Z}_\ell)}{A(\mathbf{Q}_p) \otimes \mathbf{Q}_\ell/\mathbf{Z}_\ell} \cong \mathrm{Hom}(H^1_f(\mathbf{Q}_p, M^* \otimes \mathbf{Z}_\ell(1)), \mathbf{Q}_\ell/\mathbf{Z}_\ell)$$

Let $U \subset \mathrm{Spec}(\mathbf{Z})$ be non-empty open such that $M \otimes \mathbf{Q}_\ell/\mathbf{Z}_\ell$ is unramified over U, and consider the diagram

$$\begin{array}{ccccccc} H^1(U, M \otimes \mathbf{Q}_\ell/\mathbf{Z}_\ell) & \to & \bigoplus_{p \notin U} H^1(\mathbf{Q}_p, M \otimes \mathbf{Q}_\ell/\mathbf{Z}_\ell) & \to & H^1(U, M^* \otimes \mathbf{Z}_\ell(1))^* & \to & 0 \\ \downarrow & & \downarrow & & & & \\ \frac{H^1(U, M \otimes \mathbf{Q}_\ell/\mathbf{Z}_\ell)}{A(\mathbf{Q}) \otimes \mathbf{Q}/\mathbf{Z}} & \xrightarrow{\alpha_{U,\ell}} & \bigoplus_{p \notin U} \frac{H^1(\mathbf{Q}_p, M \otimes \mathbf{Q}_\ell/\mathbf{Z}_\ell)}{A(\mathbf{Q}_p) \otimes \mathbf{Q}_\ell/\mathbf{Z}_\ell} & & & & \end{array}$$

(The top row is exact by Tate duality.) Since $\mathrm{Coker}(\alpha_{U,\ell})$ is finite, we see that

$$\begin{aligned} \mathrm{Coker}(\alpha_{U,\ell}) &\cong H^1(U, M^* \otimes \mathbf{Z}_\ell(1))^*/(\mathrm{div}) \\ &\cong \mathrm{Hom}(H^1(U, M^* \otimes \mathbf{Z}_\ell(1))_{\mathrm{tor}}, \mathbf{Q}_\ell/\mathbf{Z}_\ell) \\ &\cong \mathrm{Hom}(H^0(U, M^* \otimes \mathbf{Q}_\ell/\mathbf{Z}_\ell(1)), \mathbf{Q}_\ell/\mathbf{Z}_\ell) \\ &\cong \mathrm{Hom}(H^0(\mathbf{Q}, M^* \otimes Q_\ell/\mathbf{Z}_\ell(1)), \mathbf{Q}_\ell/\mathbf{Z}_\ell). \end{aligned}$$

This last expression is independent of U, so

$$\mathrm{Coker}(\alpha_{U,\ell}) \cong \mathrm{Coker}(\alpha_M)\{\ell\}.$$

Further, our weight hypothesis implies $\mathrm{wt}(M^*(1)) \geq 1$, so

$$\mathrm{Coker}(\alpha_M) \cong \mathrm{Hom}(H^0(\mathbf{Q}, M^* \otimes \mathbf{Q}/\mathbf{Z}(1)), \mathbf{Q}/\mathbf{Z})$$

is finite. This proves (5.14)(ii).

Proof of 5.14(iii). We may assume $M' \subset M$ and $\#(M/M')$ is a power of ℓ. Let $A'(\mathbf{Q})$ and $A'(\mathbf{Q}_p)$ be groups corresponding to M'. For U a sufficiently small open set in $\mathrm{Spec}(\mathbf{Z})$, we have by (5.8)

$$\frac{\mathrm{Tam}(M)}{\mathrm{Tam}(M')} = \frac{\prod_{p \notin U} \chi(A'(\mathbf{Q}_p) \to A(\mathbf{Q}_p))}{\chi(A'(\mathbf{Q}) \to A(\mathbf{Q}))}. \tag{5.17}$$

(Here χ is as in (4.6).) As in the proof of (5.16), we have

$$(5.18)\qquad \frac{\chi(\alpha_M)}{\chi(\alpha_{M'})} = \frac{\prod_{p\notin U}\chi\left[\frac{H^1(\mathbf{Q}_p,M'\otimes\mathbf{Q}_\ell/\mathbf{Z}_\ell)}{A'(\mathbf{Q}_p)\otimes\mathbf{Q}_\ell/\mathbf{Z}_\ell}\to\frac{H^1(\mathbf{Q}_p,M\otimes\mathbf{Q}_\ell/\mathbf{Z}_\ell)}{A(\mathbf{Q}_p)\otimes\mathbf{Q}_\ell/\mathbf{Z}_\ell}\right]}{\chi\left[\frac{H^1(U,M'\otimes\mathbf{Q}_\ell/\mathbf{Z}_\ell)}{A'(\mathbf{Q})\otimes\mathbf{Q}_\ell/\mathbf{Z}_\ell}\to\frac{H^1(U,M\otimes\mathbf{Q}_\ell/\mathbf{Z}_\ell)}{A(\mathbf{Q})\otimes\mathbf{Q}_\ell/\mathbf{Z}_\ell}\right]}.$$

A theorem of Tate ([**Ta3**]) gives

$$(5.19)\qquad \begin{gathered} H^q(U,M/M')\cong\bigoplus H^q(\mathbf{Q}_p,M/M') \text{ for } q\geq 3;\\ \prod_{0\leq q\leq 2}\#H^q(U,M/M')^{(-1)^q}=\prod_{p\notin U}\prod_{0\leq q\leq 2}H^q(\mathbf{Q}_p,M/M')^{(-1)^q}\end{gathered}$$

Lemma 5.20. (i) *For $K=\mathbf{Q}$ or $K=\mathbf{Q}_p$, there is an exact sequence*
(5.20.1)

$$\begin{aligned} &0\to H^0(K,M/M')\to A'(K)\to A(K)\to\\ &H^1(K,M/M')\to\frac{H^1(K,M'\otimes\mathbf{Q}/\mathbf{Z})}{A'(K)\otimes\mathbf{Q}/\mathbf{Z}}\to\frac{H^1(K,M\otimes\mathbf{Q}/\mathbf{Z})}{A(K)\otimes\mathbf{Q}/\mathbf{Z}}\\ &\to H^2(K,M/M')\to H^2(K,M'\otimes\mathbf{Q}/\mathbf{Z})\to H^2(K,M\otimes\mathbf{Q}/\mathbf{Z}).\end{aligned}$$

(ii) *For U as above, there is an exact sequence*

$$(5.20.2)\qquad \begin{aligned} &0\to H^0(K,M/M')\to A'(\mathbf{Q})\to A(\mathbf{Q})\to\\ &H^2(U,M/M')\to\frac{H^1(U,M'\otimes\mathbf{Q}_\ell/\mathbf{Z}_\ell)}{A'(K)\otimes\mathbf{Q}_\ell/\mathbf{Z}_\ell}\to\frac{H^1(U,M\otimes\mathbf{Q}_\ell/\mathbf{Z}_\ell)}{A(K)\otimes\mathbf{Q}_\ell/\mathbf{Z}_\ell}\\ &\to H^2(U,M/M')\to H^2(U,M'\otimes\mathbf{Q}_\ell/\mathbf{Z}_\ell)\\ &\to H^2(U,M\otimes\mathbf{Q}_\ell/\mathbf{Z}_\ell).\end{aligned}$$

The proof of the lemma is easy, and we omit it. It is now straightforward if somewhat tedious to combine formulas (5.17)–(5.19) with the above exact sequences and deduce $\chi(M)=\chi(M')$. This finishes the proof of (5.14).

We now show that an analogoue of (5.15) for function fields in one variable over finite fields is true for the non p part ($p=$ characteristic) with no reference to a motif. We do not know how to treat the p-primary part.

Let C be a smooth proper connected curve over a finite field k of characteristic p, with function field K. Let $\ell\neq p$ be a prime number and let T be a free $\mathbf{Z}_\ell$-module of finite rank endowed with a continuous action of $\mathrm{Gal}(K_{\mathrm{sep}}/k)$. Assume that T is unramified on some non-empty open set of C.

For a place v of K, let K_v be the completion of K at v, and define $H^1_f(K_v,T) \subset H^1(K_v,T)$ and $P_v(T,u) \in \mathbf{Q}_\ell[u]$ just as in the number field case. For a non-empty open set U of C, define $H^1_{f,U}(K,T) \subset H^1(K,T)$ again as in the number field case, and let

$$Z(U,T,u) = \prod P_v(T, u^{\deg(v)}) \in \mathbf{Q}_\ell[[u]].$$

Then by Grothendieck, $Z(U,T,u) \in \mathbf{Q}_\ell(u)$, and we have also

$$H^1_{f,U}(K,T) = H^1(U, j_*T); \quad j : U' \hookrightarrow U, \ T \text{ unramified on } U'.$$

Proposition 5.21. *With notations as above, assume that for some isomorphism* $\iota : \overline{\mathbf{Q}}_\ell \cong \mathbf{C}$, *the sheaf* T *is* ι*-mixed [De3] on some open set of* C.

(i) *Assume* T *has* ι*-weights* ≤ -1. *Then for any place* v *of* K, $H^1_f(K_v,T)$ *is finite and is isomorphic to* $H^0(K_v, T \otimes \mathbf{Q}/\mathbf{Z})$. *If* T *is unramified at* v, *the order of this group is equal to* $|P_v(T,1)|_\ell$,

(ii) *Assume* T *has* ι*-weight* ≤ -2. *Then* $H^1_{f,C}(K,T)$ *is finite and is isomorphic to* $H^0(K, T \otimes \mathbf{Q}/\mathbf{Z})$.

(iii) *Assume* T *has* ι*-weights* ≤ -3. *Then the kernel and cokernel of* α : $H^1(K, T \otimes \mathbf{Q}/\mathbf{Z}) \to \bigoplus_{\text{all } v} H^1(K_v, T \otimes \mathbf{Q}/\mathbf{Z})$ *are finite.*

(iv) *Assume* T *has* ι*-weights* ≤ -3. *If* U *is a non-empty open set of* C *on which* T *is unramified, we have*

$$\frac{\prod \# H^0(K_v, T \otimes \mathbf{Q}/\mathbf{Z})}{\# H^0(K, T \otimes \mathbf{Q}/\mathbf{Z})} \cdot |Z(U,T,1)|_\ell = \frac{\# \operatorname{Coker}(\alpha)}{\# \operatorname{Ker}(\alpha)}$$

(v) *Assume* T *has* ι*-weights* ≤ -3. *Then* Coker(α) *is isomorphic to the Pontryagin dual of the finite group* $H^0(K, T^* \otimes \mathbf{Q}/\mathbf{Z}(1))$.

Proof. The proof used Deligne's theory of weights. We prove here only (iii) and (iv), leaving the rest for the reader. Assume T unramified over U. By [**Gr**],

$$Z(U,T,u) = \det(1 - fu : R\Gamma_c(U_{\bar{k}}, T \otimes \mathbf{Q}))^{-1}. \tag{5.22}$$

The eigenvalues of f on $H^q_c(U_{\bar{k}}, T \otimes \mathbf{Q})$ are not 1 by [**De3**], and hence the kernel and cokernel of $1 - f$ on $H^q_c(U_{\bar{k}}, T \otimes \mathbf{Q}/\mathbf{Z})$ are finite. This yields finiteness for $H^q_c(U, T \otimes \mathbf{Q}/\mathbf{Z})$, and hence by (5.22)

$$|Z(U,T,1)|_\ell = \chi(1 - f : R\Gamma_c(U_{\bar{k}}, T \otimes \mathbf{Q}/\mathbf{Z})) = \chi(R\Gamma_c(U, T \otimes \mathbf{Q}/\mathbf{Z})).$$

On the other hand, by Jannsen's theorem [**Ja**]

$$H^2(K, T\otimes \mathbf{Q}/\mathbf{Z}) \cong \bigoplus_{\text{all } v} H^2(K_v, T\otimes \mathbf{Q}/\mathbf{Z}).$$

Tate duality yields, therefore

$$0 \to H_c^0(U) \to H^0(K) \to \oplus_{v\in C-U} H^0(K_v) \to H_c^1(U) \to H^1(K) \to \\ \oplus_{\text{all } v} H^1(K_v) \to H_c^2(U) \to 0$$

$$H_c^q(U) = (0) \text{ for } q \geq 3, \text{ (cohomology with coefficients in } T\otimes \mathbf{Q}/\mathbf{Z}).$$

This proves (5.21)(iii) and (iv).

For a motif of weights ≤ -1 which is not necessarily of weights ≤ -3, we formulate the conjecture on the special value of L-functions at $s=0$ as follows. We consider here two cases

(i) motives of pure weight -1

(ii) anisotropic motives of pure weight -2, where "anisotropic" means that $H^0(\mathbf{Q}, V^* \otimes \mathbf{A}_f(1)) = (0)$.

It is probable that (considered modulo torsion) a motif of weights ≤ -1 is a successive extension of motives of types (i) and (ii), together with

(iii) motives of the form (Artin motif)(1)

(iv) motives of weights ≤ -3.

Motives of type (iii) are treated by the classical Tamagawa number formula, and we have already treated motives of type (iv), so it suffices in some sense to consider cases (i) and (ii). (Taking such pure pieces makes the statement of the conjecture simpler.) In these cases the Tamagawa number conjecture again has the form

$$\text{Tam}(M) = \#H^0(\mathbf{Q}, M^*\otimes \mathbf{Q}/\mathbf{Z})/\#\text{Ш}(M).$$

The right hand side is defined as before. To define $\text{Tam}(M)$, we assume analytic continuation of the L-function and define the Tamagawa measure μ on $\prod A(\mathbf{Q}_p)$ by (5.9.1). In case (ii) we assume

$$\Phi\otimes\mathbf{R} \cong D_\infty/(D_\infty^0 + V_\infty^+)$$

and hence $\text{Tam}(M) = \mu(\prod A(\mathbf{Q}_p)/A(\mathbf{Q}))$ is defined. In case (i), we define

$$\text{Tam}(M) = \mu(\prod A(\mathbf{Q}_p)) \cdot H/\#A(\mathbf{Q})_{\text{tor}}$$

where H is the descriminant of the height pairing

$$A(\mathbf{Q}) \times A^*(\mathbf{Q}) \to \mathbf{R} \quad ([\mathbf{Be}], [\mathbf{B}])$$

Here $A^*(\mathbb{Q}) \subset H^1_{f,\mathrm{Spec}(\mathbb{Z})}(\mathbb{Q}, M^* \otimes \hat{\mathbb{Z}}(1))$ is defined to be the inverse image of $\mathrm{Image}(\Phi^*) \subset H^1_{f,\mathrm{Spec}(\mathbb{Z})}(\mathbb{Q}, V^* \otimes \mathbb{A}_f(1))$ with Φ^* as follows. In the case of a motif $H^{2r-1}(X)(r)$ with X smooth and proper of dim n,

$$\Phi^* = (CH^{n+1-r}(X) \otimes \mathbb{Q})_{\mathrm{hom}\sim 0}.$$

In general, motives of type (i) should be direct summands of such defined by correspondence projectors, and one takes for Φ^* the image of the cycles under this correspondence.

Finally, the reader may be concerned that our groups $A(\mathbb{Q}_p)$ ar always compact for $p < \infty$, while $A(\mathbb{R})$ need not be compact. This break in the parallelism between $\mathbb{R}$ and $\mathbb{Q}_p$ arises because in defining A we chose to work with Φ rather than Ψ. It can be remedied as follows. Consider the groups V, D, and Ψ corresponding to $H^m(X)(r)$ with $m - 2r \leq -1$. Let $M \subset V$ be a lattice with $M \otimes \hat{\mathbb{Z}}$ galois stable. Define groups $B(\mathbb{Q}_p) \supset A(\mathbb{Q}_p)$ by taking $B(\mathbb{R}) = A(\mathbb{R})$ and $B(\mathbb{Q}_p) =$ inverse image in $H^1_g(\mathbb{Q}_p, M \otimes \hat{\mathbb{Z}})$ of

$$\mathrm{Im}(\Psi \to H^1_g(\mathbb{Q}_p, V \otimes \mathbb{A}_f)/H^1_f(\mathbb{Q}_p, V \otimes \mathbb{A}_f)).$$

(We assume (5.3).) The topology on $A(\mathbb{Q}_p)$ is extended to $B(\mathbb{Q}_p)$ by taking $B(\mathbb{Q}_p)/A(\mathbb{Q}_p)$ to be discrete. We define $B(\mathbb{Q}) \subset A(\mathbb{Q})$ using Ψ in place of Φ.

Assume $m - 2r \leq -3$. Then $B(\mathbb{Q}_p) = A(\mathbb{Q}_p)$ for almost all p. Define

$$\mathrm{Tam}(M)^\sim = \mu(\prod B(\mathbb{Q}_p)/B(\mathbb{Q}))$$

$$Ш(M)^\sim = \mathrm{Ker}\left(\frac{H^1(\mathbb{Q}, M \otimes \mathbb{Q}/\mathbb{Z})}{B(\mathbb{Q}) \otimes \mathbb{Q}/\mathbb{Z}} \to \bigoplus_{p \leq \infty} \frac{H^1(\mathbb{Q}_p, M \otimes \mathbb{Q}/\mathbb{Z})}{B(\mathbb{Q}_p) \otimes \mathbb{Q}/\mathbb{Z}}\right)$$

Then conjecture (5.15) is equivalent to

$$\mathrm{Tam}(M)^\sim = \frac{\#H^0(\mathbb{Q}, M^* \otimes \mathbb{Q}/\mathbb{Z}(1))}{\#Ш(M)^\sim}.$$

6. The Riemann zeta function

Theorem 6.1. *Let $r \geq 2$ be given.*

(i) *If r is even, the Tamagawa number conjecture is true modulo a power of 2 for the motif $\mathbb{Q}(r)$.*

(ii) *Let r be odd. Then the Tamagawa number conjecture is true modulo a power of 2 for the motif $\mathbb{Q}(r)$ if the conjecture (6.2) below is true in the case $\alpha = 1$.*

The unfortunate power of 2 ambiguity in the theory is due to technical problems the Fontaine-Messing theory has at primes dividing 2.

The compatibility conjecture necessary to to prove the theorem for odd twists of $\mathbf{Q}$ concerns cyclotomic elements

$$c_r(\alpha) \in gr^r_\gamma(K_{2r-1}(\mathbf{Q}(\alpha)) \otimes \mathbf{Q})$$

defined by Beilinson [**Be1**]. Here α is a root of 1, and $c_r(\alpha)$ is characterized by the property that for any $\iota : \mathbf{Q}(\alpha) \to \mathbf{C}$, the image of $c_r(\alpha)$ in $\mathbf{C}/\mathbf{Q}(2\pi i)^r$ under the regulator map associated to ι is equal to the class of $-(r-1)!\sum_{i=1}^{\infty} \iota(\alpha)^n/n^r$.

Deligne and Soulé defined cyclotomic elements in $H^1(\mathbf{Q}, \hat{\mathbf{Z}}(r))$. For an n-th root of 1, there exists an element

$$c_{r,n}(\alpha) \in H^1(\mathbf{Q}(\alpha), \hat{\mathbf{Z}}(r))$$

which is characterized moodulo torsion elements by the following property. For any $m \geq 1$ and any field F over $\mathbf{Q}(\alpha)$ containing all m-th roots of α, the image of $c_{r,n}(\alpha)$ in $H^1(F, \mathbf{Z}/m\mathbf{Z}(r)) \cong (F^\times/F^{\times m}) \otimes \mathbf{Z}/m\mathbf{Z}(r-1)$ is equal to

$$\sum_{\zeta}\{1-\zeta\} \otimes [\zeta^n]^{\otimes(r-1)}.$$

where if $\alpha \neq 1$ (resp. $\alpha = 1$) ζ ranges over all m-th roots of α (resp. all m-th roots of α except 1). (cf. [**De4**] [**Ih**] [**So2**] [**So5**].)

Conjecture 6.2. *The images of $c_r(\alpha)$ and $n^{1-r}c_{r,n}(\alpha)$ coincide up to sign in $H^1(\mathbf{Q}(\alpha), \mathbb{A}_f(r))$.*

This conjecture is proved in Soulé [**So5**] in the case $r = 2$.

In [**De4**], Deligne constructed an extension of motives in his sense

$$0 \to \mathbf{Q}(r) \to E \to \mathbf{Q} \to 0$$

whose class has the same image as $c_r(1)$ in $\mathbf{C}/\mathbf{Q}(2\pi i)^r$ and the same image as $c_{r,1}(1)$ in $H^1(\mathbf{Q}, \hat{\mathbf{Z}}(r))$. However, the authors do not know if Deligne's extension corresponds to an element of $gr^r_\lambda(K_{2r-1}(\mathbf{Q}(\alpha)) \otimes \mathbf{Q})$.

The rest of this section is devoted to the proof of (6.1). We do not assume (6.2) until the end.

Lemma 6.3. *Assume $r \geq 2$.*

(i) $\mathrm{Tam}(\mathbf{Z}(r)) = \pm \frac{\# H^0(\mathbf{Q}, \mathbf{Q}/\mathbf{Z}(1-r))}{\# H^0(\mathbf{Q}, \mathbf{Q}/\mathbf{Z}(r))} \cdot \frac{2}{\zeta(1-r)}$.

(ii) *If r is odd,* $\mathrm{Tam}(\mathbf{Z}(r)) = \pm \frac{\# H^0(\mathbf{Q}, \mathbf{Q}/\mathbf{Z}(1-r))}{\chi(A(\mathbf{Q}) : \mathbf{Z} \cdot c_r(1))}$.
Here $\mathbf{Z} \cdot c_r(1) \subset A(\mathbf{Q}) \otimes \mathbf{Q}$ is the free abelian group generated by $c_r(1)$, and by definition

$$\chi(A(\mathbf{Q}) : \mathbf{Z} \cdot c_r(1)) = [A(\mathbf{Q}) : L]/[\mathbf{Z} \cdot c_r(1) : L]$$

for any free subgroup L of $A(\mathbb{Q})$ of finite index whose image in $A(\mathbb{Q})\otimes\mathbb{Q}$ is contained in $\mathbb{Z}\cdot c_r(1)$.

Proof. Take the canonical base $1\in\mathbb{Q}=H^0_{DR}(\mathrm{Spec}(\mathbb{Q}))$, and let μ_p be the corresponding measure on $A(\mathbb{Q}_p)$ for $p\leq\infty$. Then, if $p\neq\infty$, the table (3.9) gives $A(\mathbb{Q}_p)=H^1(\mathbb{Q}_p,\widehat{\mathbb{Z}}(r))$ and by (4.2) we have

$$\begin{aligned}\mu_p(A(\mathbb{Q}_p)) &= |(r-1)!|_p(1-p^{-r})\#H^0(\mathbb{Q}_p,\mathbb{Q}_p/\mathbb{Z}_p(1-r))\\ &= |(r-1)!|_p(1-p^{-r})\#H^0(\mathbb{Q},\mathbb{Q}_p/\mathbb{Z}_p(1-r)).\end{aligned}$$

Hence

$$\prod_{p<\infty}\mu_p(A(\mathbb{Q}_p))=\frac{\#H^0(\mathbb{Q},\mathbb{Q}/\mathbb{Z}(1-r))}{(r-1)!\cdot\zeta(r)}$$

If r is even, $A(\mathbb{R})=\mathbb{R}/(2\pi)^r\mathbb{Z}$ with the Lebesgue measure μ_∞, and we have

(6.4)

$$\begin{aligned}\mathrm{Tam}(\mathbb{Z}(r)) &= (\#A(\mathbb{Q}))^{-1}\prod_{p\leq\infty}\mu_p(A(\mathbb{Q}_p))\\ &= \frac{1}{\#H^0(\mathbb{Q},\mathbb{Q}/\mathbb{Z}(r))}\cdot(2\pi)^r\cdot\frac{\#H^0(\mathbb{Q},\mathbb{Q}/\mathbb{Z}(1-r))}{(r-1)!\zeta(r)}\\ &= \pm\frac{\#H^0(\mathbb{Q},\mathbb{Q}/\mathbb{Z}(1-r))}{\#H^0(\mathbb{Q},\mathbb{Q}/\mathbb{Z}(r))}\cdot\frac{2}{\zeta(1-r)}.\end{aligned}$$

If r is odd, $A(\mathbb{R})=A(\mathbb{C})^+=(\mathbb{C}/(2\pi i)^r\mathbb{Z})^+\cong\mathbb{R}\oplus\mathbb{Z}/2\mathbb{Z}$, $A(\mathbb{Q})\hookrightarrow A(\mathbb{R})$ and the calculation reads

$$\begin{aligned}\mathrm{Tam}(\mathbb{Z}(r)) &= \mu\Big(\prod_{p\leq\infty}A(\mathbb{Q}_p)/A(\mathbb{Q})\Big)\\ &= \mu_\infty(A(\mathbb{R})/A(\mathbb{Q}))\#H^0(\mathbb{Q},\mathbb{Q}/\mathbb{Z}(1-r))((r-1)!\zeta(r))^{-1}\\ &= \#H^0(\mathbb{Q},\mathbb{Q}/\mathbb{Z}(1-r))\chi(A(\mathbb{Q}):\mathbb{Z}\cdot c_r(1))^{-1}.\end{aligned}$$

Note that here we did not use conjecture (6.2). This proves the lemma.

To finish the proof of Theorem 6.1, we must show

(6.5) $\zeta(1-r)=2\cdot\#(\text{Ш})\cdot\#H^0(\mathbb{Q},\mathbb{Q}/\mathbb{Z}(r))^{-1}$ for r even

(6.6) $\#(\text{Ш})=\chi(A(\mathbb{Q}):\mathbb{Z}\cdot c_r(1))$ for r odd.

Lemma 6.7 *If $p\neq 2$, the p-primary part of Ш is isomorphic to $H^2(\mathbb{Z}[1/p],\mathbb{Z}_p(r))$. If r is even, it is also isomorphic to $H^1(\mathbb{Z}[1/p],\mathbb{Q}_p/\mathbb{Z}_p(r))$.*

Proof. Consider the commutative diagram of exact sequences:

$$\begin{array}{ccccccc}
0 \longrightarrow & \dfrac{H^1(\mathbf{Z}[1/p], \mathbf{Q}_p/\mathbf{Z}_p(r))}{A(\mathbf{Q})\otimes\mathbf{Q}_p/\mathbf{Z}_p} & \longrightarrow & \dfrac{H^1(\mathbf{Q}, \mathbf{Q}_p/\mathbf{Z}_p(r))}{A(\mathbf{Q})\otimes\mathbf{Q}_p/\mathbf{Z}_p} & \longrightarrow & \bigoplus_{\ell\neq p} H^0(\mathbf{F}_\ell, \mathbf{Q}_p/\mathbf{Z}_p(r-1)) & \longrightarrow 0 \\
& \downarrow s & & \downarrow t & & \| & \\
0 \longrightarrow & \dfrac{H^1(\mathbf{Q}_p, \mathbf{Q}_p/\mathbf{Z}_p(r))}{A(\mathbf{Q}_p)\otimes\mathbf{Q}_p/\mathbf{Z}_p} & \longrightarrow & S & \longrightarrow & \bigoplus_{\ell\neq p} H^0(\mathbf{F}_\ell, \mathbf{Q}_p/\mathbf{Z}_p(r-1)) & \longrightarrow 0
\end{array}$$

where

$$S = \bigoplus_{\ell} \frac{H^1(\mathbf{Q}_\ell, \mathbf{Q}_p/\mathbf{Z}_p(r))}{A(\mathbf{Q}_\ell)\otimes\mathbf{Q}_p/\mathbf{Z}_p} = \frac{H^1(\mathbf{Q}_p, \mathbf{Q}_p/\mathbf{Z}_r(r))}{A(\mathbf{Q}_p)\otimes\mathbf{Q}_p/\mathbf{Z}_p} \oplus \bigoplus_{\ell\neq p} H^0(\mathbf{F}_\ell, \mathbf{Q}_p/\mathbf{Z}_p(r-1)).$$

We have

$$\text{Ш}\{p\} = \operatorname{Ker}(t) \cong \operatorname{Ker}(s) \cong \operatorname{Ker}(H^2(\mathbf{Z}[1/p], \mathbf{Z}_p(r)) \to H^2(\mathbf{Q}_p, \mathbf{Z}_p(r)).$$

The map on the right is zero, as follows from the Tate duality sequence

$$H^2(\mathbf{Z}[1/p], \mathbf{Z}_p(r)) \to \quad H^2(\mathbf{Q}_p, \mathbf{Z}_p(r)) \quad \to H^0(\mathbf{Z}[1/p], \mathbf{Q}_p/\mathbf{Z}_p(1-r))^* \to 0$$

$$H^0(\mathbf{Q}_p, \mathbf{Q}_p/\mathbf{Z}_p(1-r))^*$$

This proves Lemma (6.7)

The proof of (6.5) now follows from the result of Mazur-Wiles [MW]

$$\zeta(1-r) = \pm \prod_{p<\infty} \frac{\# H^1(\mathbf{Z}[1/p], \mathbf{Q}_p/\mathbf{Z}_p(r))}{\# H^0(\mathbf{Z}[1/p], \mathbf{Q}_p/\mathbf{Z}_p(r))} \qquad (r \geq 2 \text{ even}).$$

We consider (6.6). Let

$G = \operatorname{Gal}(\mathbf{Q}_p(\zeta_{p^\infty})/\mathbf{Q}_p = \operatorname{Gal}(\mathbf{Q}(\zeta_{p^\infty})/\mathbf{Q})$;
$P = \operatorname{Gal}(\mathbf{Q}_p(\zeta_{p^\infty})^{ab}/\mathbf{Q}_p(\zeta_{p^\infty}))$; $\mathcal{X} = \pi_1(\operatorname{Spec}(\mathbf{Z}[1/p][\zeta^{p^\infty}])^{ab}$;
$C =$ the $\widehat{\mathbf{Z}}[[G]]$ – submodule of P generated by $(1-\zeta_{p^n})_n \in \varprojlim \mathbf{Q}_p(\zeta_{p^n})^\times \subset P$.

We have a canonical G-homomorphism $P/C \to \mathcal{X}$. Consider the following commutative diagram of exact sequences

$$\begin{array}{ccc}
0 & & \\
\downarrow & & \\
H^2(\mathbf{Z}[1/p],\mathbf{Z}_p(r))^* & & 0 \\
\downarrow & & \downarrow \\
H^1(\mathbf{Z}[1/p],\mathbf{Q}_p/\mathbf{Z}_p(1-r)) & \xrightarrow{a} & H^0(G,(P/C)^*(1-r)) \\
\downarrow & & \downarrow \\
H^1(\mathbf{Q}_p,\mathbf{Q}_p/\mathbf{Z}_p(1-r)) & \xrightarrow{b} & H^0(G,P^*(1-r)) \\
\downarrow & & \downarrow \\
H^1(\mathbf{Z}[1/p],\mathbf{Z}_p(r))^* & \xrightarrow{c} & H^0(G,C^*(1-r)) \\
\downarrow & & \downarrow \\
H^2(\mathbf{Z}[1/p],\mathbf{Q}_p/\mathbf{Z}_p(1-r)) & \xrightarrow{d} & H^1(G,(P/C)^*(1-r)) \\
\downarrow & & \downarrow \\
0 & & 0
\end{array}$$

Here $*$ denotes $\mathrm{Hom}_{\mathrm{cont}}(\ ,\mathbf{Q}_p/\mathbf{Z}_p)$, so $P^* = H^1(\mathbf{Q}_p(\zeta_{p^\infty}),\mathbf{Q}_p/\mathbf{Z}_p)$. The map a factors as $a = a_1 \circ a_2$:

$$H^1(\mathbf{Z}[1/p],\mathbf{Q}_p/\mathbf{Z}_p(1-r)) \overset{a_2}{\cong} H^0(G,\mathcal{X}^*(1-r)) \xrightarrow{a_1} H^0(G,(P/C)^*(1-r))$$

with a_2 bijective. (Note $\mathcal{X}^* = H^1(\mathbf{Z}[1/p][\zeta_{p^\infty}],\mathbf{Q}_p/\mathbf{Z}_p)$.) Moreover, the map b is bijective, and c is the surjection induced by

$$\begin{aligned}\varprojlim \mathbf{Z}[1/p][\zeta_{p^n}]^\times &\longrightarrow \varprojlim H^1(\mathbf{Z}[1/p][\zeta_{p^n}],\mathbf{Z}_p(r))(1-r) \\ &\overset{\mathrm{Norm}}{\longrightarrow} H^1(\mathbf{Z}[1/p][\zeta_{p^n}],\mathbf{Z}_p(r))(1-r).\end{aligned}$$

Finally, d factors as

$$H^2(\mathbf{Z}[1/p],\mathbf{Q}_p/\mathbf{Z}_p(1-r)) \overset{d_2}{\cong} H^1(G,\mathcal{X}^*(1-r)) \xrightarrow{d_1} H^1(G,(P/C)^*(1-r))$$

with d_2 bijective

By Mazur-Wiles [**MW**], the kernel and cokernel of

$$(P/C)^+(p) \to \mathcal{X}^+(p)$$

(where $(\ )^+$ denotes the Gal($\mathbf{C}/\mathbf{R}$)-invariants and (p) the pro-p-part) have the same images in the Grothendieck group of

$$\{\text{finitely generated torsion } \mathbf{Z}_p[[G]]\text{-modules}\}\Big/\{\text{finite } \mathbf{Z}_p[[G]]\text{-modules}\}.$$

This shows

$$\#\operatorname{Coker}(a_1)\cdot\#\operatorname{Ker}(a_1)^{-1} = \#\operatorname{Coker}(d_1)\cdot\#\operatorname{Ker}(d_1)^{-1}.$$

Consequently we have

(6.8) $$\#H^2(\mathbf{Z}[1/p],\mathbf{Z}_p(r)) = \#\operatorname{Ker}(c).$$

On the other hand, by the definition of c, we see that

(6.9) $$\#\operatorname{Ker}(c) = [H^1(\mathbf{Z}[1/p],\mathbf{Z}_p(r)) : \mathbf{Z}_p\cdot c_{r,1}(1)].$$

If we assume the case $\alpha = 1$ of (6.2), we have

$$\chi(A(\mathbf{Q}),\mathbf{Z}\cdot c_r(1)) = \prod[H^1(\mathbf{Z}[1/p],\mathbf{Z}_p(r)) : \mathbf{Z}_p\cdot c_{r,1}(1)],$$

and hence (6.8)and (6.9) prove (6.6). This completes the argument.

7. Complex multiplication

In this section, for an elliptic curve over $\mathbf{Q}$ with complex multiplication, we reduce the ℓ-primary part of our Tamagawa number conjecture for $H^1(E)(2)$ (i.e., for $L(H^1(E),2)$, where E is such an elliptic curve) to the question $\dim gr^2(K_2(E)\otimes\mathbf{Q}) \overset{?}{=} 1$ and to a certain problem on the Galois cohomology of the twist of the Tate module $T_\ell E(1)$, for good primes ℓ (cf. (7.3)). The latter problem is solve in the case ℓ is regular for E (cf. (7.4)).

7.1. In this section, we review the "cyclotomic elements" in K_2 of elliptic curves (cf. [**B14**] [**DW**]). Generally let k be a field and E an elliptic curve over k. For non-zero integers a, n such that $(a,n) = 1$ and such tht all points in ${}_aE \cup {}_nE$ are k-rational (here ${}_bE = \operatorname{Ker}(b : E \to E)$ for $b \in \mathbf{Z}$), and for $\beta \in {}_nE - \{0\}$, we define

$$c_n^a(\beta) \in \Gamma(E_{\text{zar}}, K_2)/K_2(k)$$

as follows. Take functions g, s, t_γ ($\gamma \in {}_aE - \{0\}$) on E such that $\mathrm{div}(g) = a^2(0) - {}_aE$, $\mathrm{div}(s) = N(\beta - n(0)$, $\mathrm{div}(t_\gamma) = a(\gamma) - a(0)$. Then,

$$c_n^a(\beta) \stackrel{\mathrm{def}}{=} a\{g(\beta)^{-1}g, s\} - \sum_{\substack{a\gamma=0 \\ \gamma\neq 0}} \{s(\gamma), t_\gamma\} \in K_2(k(E))/K_2(k).$$

It is easily checked that $c_n^a(\beta)$ belongs to the kernel of the tame symbol map

$$K_2(k(E))/K_2(k) \to \bigoplus_{\substack{x\in E \\ \text{closed}}} k(x)^\times,$$

which is $\Gamma(E_{\mathrm{zar}}, K_2)/K_2(k)$, and that $c_n^a(\beta)$ is independent of the choices of g, s, t_γ. Note

$$\text{(7.1.2)} \qquad c^a(\beta) \stackrel{\mathrm{def}}{=} (an)^{-1}c_n^a(\beta) \in \Gamma(E_{\mathrm{zar}}, K_2) \otimes \mathbf{Q}$$

is independent of the chocie of $n \neq 0$ such that $n\beta = 0$

In the rest of this section and §§(7-2)–(7.4), let E be an elliptic curve over $\mathbf{Q}$ with complex multiplication by the ring of integers $\mathcal{O}_K$ of a quadratic imaginary field K. Since we will be interested in the motif $H^1(E)(2)$, we denote by D and V the corresponding $\mathbf{Q}$-vector spaces $H^1_{DR}(E/\mathbf{Q})$ and $H^1(E(\mathbf{C}), \mathbf{Q}(2\pi i)^2)$, respectively, and we identify V_ℓ with $T_\ell E(1) \otimes \mathbf{Q}$. Fix $K \xrightarrow{\subset} \mathbf{C}$, let ψ be the Hecke character of K of A_0-type of type $(1,0)$ associated to E, and let f be the conductor of ψ. We fix a generator δ of $f^{-1}H_1(E(\mathbf{C}), \mathbf{Z})$ for the $\mathcal{O}_K$-module structure by complex multiplication. By the identification

$${}_fE(\mathbf{C}) = f^{-1}H_1(E(\mathbf{C}), \mathbf{Z})/H_1(E(\mathbf{C}), \mathbf{Z}),$$

$\delta \bmod H_1(E(\mathbf{C}), \mathbf{Z})$ is a generator of $E(\mathbf{C})$ as an $\mathcal{O}_K/f$-module. Take $v \in \mathcal{O}_K - \{0\}$ such that $v\delta \in H_1(E(\mathbf{C}), \mathbf{Z})^+$ and for a non-zero integer a which is prime to f, let

$$\begin{aligned} \text{(7.1.3)} \qquad c_v^a(E) &= N_v N_{K(f)/K}(c^a(\delta)) \\ &\in (\Gamma((E \otimes K)_{\mathrm{zar}}, K_2) \otimes \mathbf{Q})^{\mathrm{Gal}(K/\mathbf{Q})} \\ &= \Gamma(E_{\mathrm{zar}}, K_2) \otimes \mathbf{Q} = gr^2(K_2(E) \otimes \mathbf{Q}) \end{aligned}$$

where $K(f)$ denotes the ray class field of modulus f over K, we regard $c^a(\delta)$ as an element of $\Gamma((E \otimes K(f))_{zar}, K_2) \otimes \mathbf{Q}$ (it is seen easily that $c_n^a(\delta)$ is invariant under $\mathrm{Gal}(\overline{Q}/K(f))$; note that K_2 of an algebraic number field is a torsion group) $N_{K(f)/K}$ is the norm map

$$\Gamma((E \otimes K(f))_{\mathrm{zar}}, K_2) \otimes \mathbf{Q} \to \Gamma((E \otimes K)_{\mathrm{zar}}, K_2) \otimes \mathbf{Q},$$

and N_v is the norm map

$$\Gamma((E \otimes K)_{\text{zar}}, K_2) \otimes \mathbf{Q} \to \Gamma((E \otimes K)_{\text{zar}}, K_2) \otimes \mathbf{Q}$$

associated to the finite flat morphism $v : E \to E$ (complex multiplication). Using the fact that multiplication by $m \neq 0$ on E induces multiplication by m on $\Gamma(E, K_2)/K_2(k)$ we can show that

$$c_v(E) \underset{dfn.}{=} (a^2 - \psi(a)^{-1})^{-1} c_v^a(E) \tag{7.1.4}$$

is independent of a and $c_{nv}^a(E) = nc_v^a(E)$ for $n \in \mathbf{Z} - \{0\}$. Note that $\psi(a) = \pm a$. It is conjectured that $\dim_{\mathbf{Q}}(gr^2(K_2(E) \otimes \mathbf{Q})) = 1$.

It is known ([B14] [DW]) that

(7.1.5) the regulator map $gr^2(K_2(E) \otimes \mathbf{Q}) \to D_\infty / V_\infty^+$ sends $c_v(E)$ to $\pm 2^{-1}(\lim_{s \to 0} s^{-1} L(H^1(E), s)) \cdot v\delta$

where the L-function $L(H^1(E), s)$ includes the Euler factors at all bad places, and we regard $v\delta$ as an element of $H^1(E(\mathbf{C}), \mathbf{Z} \cdot 2\pi i)^+ \subset D_\infty$ via the Poincaré duality $H_1(E(\mathbf{C}), \mathbf{Z}) \cong H^1(E(\mathbf{C}), \mathbf{Z}\ 2\pi i)$ which preserves the action of Gal$(\mathbf{Z}/\mathbf{R})$.

In (7.2)and (7.5) below, we relate the above cyclotomic elements to the following cyclotomic elements in $H^1(\mathbf{Q}, T_\ell E(1))$ considered by Soulé [So6] (Soulé treats $H^1(\mathbf{Q}, T_\ell E(r))$ for all $f \geq 1$). This implies that an analogue of the conjecture (6.2) for K_2 of elliptic curves is true.

In the rest of this section, we fix a prime $\ell \neq 2, 3$ such that E has a good reduction at ℓ. Let the fixed generator δ of $f^{-1}H_1(E(\mathbf{C}), \mathbf{Z})$ and v, a be as above, and let

$$\begin{aligned} e_v^a(E) &\overset{\text{def}}{=} (N_{K_i/K}(\{g(\ell^{-i}\delta) g(\gamma_i)^{-1}\} \otimes [v\gamma_i]))_{i \geq 1} \\ &\in (\lim_i H^1(K, {}_{\ell^i}E(1))^{\text{Gal}(K/\mathbf{Q})} = H^1(\mathbf{Q}, T_\ell E(1)) \end{aligned}$$

where g is a function on E such that $\text{div}(g) = a^2(0) - {}_aE$, γ_i is the image of $\ell^{-i}\delta$ in ${}_{\ell^i}E$ under ${}_{\ell^i f}E(\mathbf{C}) = {}_{\ell^i}E(\mathbf{C}) \oplus {}_fE(\mathbf{C})$, K_i is the ray class field of modulus $\ell^i f$ over K, and $N_{K_i/K}$ is the norm map

$$K_i^\times / K_i^{\times \ell^i} \otimes {}_{\ell^i}E \cong H^1(K_i, {}_{\ell^i}E(1)) \to H^1(K, {}_{\ell^i}E(1)).$$

It is easily see that

$$e_v(E) \overset{\text{def}}{=} (a^2 - \psi(a)^{-1})^{-1} e_v^a(E) \tag{7.1.6}$$

is independent of a, and $e_v(E) \in H^1(\mathbf{Q}, T_\ell E(1))$ in $H^1(\mathbf{Q}, V_\ell)$ (the last fact is seen by choosing suitable a). Note the groups

$$H^1(\mathbf{Q}, T_\ell E(1)), \quad H^1_{f,\mathrm{Spec}(\mathbf{Z})}(\mathbf{Q}, T_\ell E(1)), \quad H^1(\mathbf{Z}[1/\ell], j_* T_\ell E(1))$$

($j : \mathrm{Spec}(\mathbf{Q}) \to \mathrm{Spec}(\mathbf{Z}[1/\ell])$) coincide, and have no torsion.
In the following (7.1)–(7.4), let E, ℓ, δ, a, v be as above.

Proposition 7.2. *The regulator map sends $c_v(E)$ to $\kappa e_v(E)$ where κ is the value at $s = 0$ of the Euler factor at ℓ of the L-function $L(H^1(E), s)$.*

Proposition 7.3. *Let Φ be the one-dimensional $\mathbf{Q}$-vector subspace of $\mathrm{gr}^2(K_2(E) \otimes \mathbf{Q})$ generated by $c_v(E)$ with v as above (so Φ is independent of the choice of v). Then the ℓ-Tamagawa number of $M = H^1(E(\mathbf{C}), \mathbf{Z}(2\pi i)^2)$ with respect to Φ (cf. (5.15.2)) is defined if $e_v(E)$ generates $H^1(\mathbf{Q}, V_\ell)$. If this is the case, $H^2(\mathbf{Z}[1/\ell], j_* E(1))$ is finite and the ℓ-Tamagawa number conjecture (5.15.3) is equivalent to*

$$\#H^1(\mathbf{Z}[1/\ell], j_* T_\ell E(1))/\mathbf{Z}_\ell e_v(E) = \#H^2(\mathbf{Z}[1/\ell], j_* T_\ell E(1)). \tag{7.3.1}$$

for v prime to ℓ.

Remark 7.3.2. Kolyvagin and Rubin have recently proved the Iwasawa main conjecture ([dS] III §1). Using this and an argument similar to (6.6), we can prove (7.3.1) in the case ℓ splits in K and $e_v(E)$ generates $H^1(\mathbf{Q}, V_\ell)$.

By using the method of Soulé in [So6], we can show

Proposition 7.4. *If ℓ is "regular" for E ([So6], 3.3.1), $H^1(\mathbf{Q}, V_\ell)$ is generated by $e_v(E)$ and the ℓ-Tamagawa number conjecture is true.*

Now we prove (7.2)–(7.4). We deduce (7.2) from

Proposition 7.5. *Let k be a field, E an elliptic curve over k, a and n non-zero integers, and ℓ a prime number. Assume $(a, n) = 1$, $\ell \nmid a$, $\ell \neq \mathrm{char}(k)$, and all points in ${}_{\ell^\infty}E \cup {}_aE \cup {}_{n^2}E$ are k-rational. Then, for $\beta \in E_n - \{0\}$, the image of $c_n^a(\beta)$ in $H^1(k, E_{\ell^s}(1)) \cong k^\times/(k^\times)^{\ell^s} \otimes E_{\ell^s}$ is equal to*

$$-\sum_{\ell^s\gamma=\beta} a\{g(\gamma)\} \otimes [n\gamma] + \sum_{\substack{\ell^s\gamma=0 \\ \gamma\neq 0}} a\{g(\gamma)\} \otimes [n\gamma].$$

where g is a function on E such that $\mathrm{div}(g) = {}_aE - a^2(0)$.

In the situation of (7.2), let k be the minimal extension of $\mathbf{Q}$ for which all points of ${}_{\ell^\infty}E \cup {}_aE \cup {}_{N(f)^2}E$ are k-rational. Then,

$$H^1(\mathbf{Q}, T_\ell(E(1)) \to H^1(k, T_\ell E(1))$$

is injective. By comparing the image of $e_v^a(E)$ in $H^1(k, T_\ell E(1))$ and that of $c_v^a(E)$ described by (7.5), we obtain (7.2) (assuming (7.5)).

For the proof of (7.5), we need the following two lemmas. Let the assumptions and notations be as in (7.5).

Lemma 7.6. *For any integer b which is prime to a, $N_b(g)g^{-1}$ is a constant where N_b is the norm map associated to the finite flat morphism $b : E \to E$. If we denote this constant by λ_b, we have $c_n^a(\beta) \to -a\{\lambda_n g(\beta)\} \otimes [\beta]$ in $H^1(k, {}_nE(1))$.*

Lemma 7.7. *Let $m = \ell^i$ for some $i \geq 1$. Then, the image of $c_n^a(\beta)$ in $H^1(k, T_\ell E(1))$ is equal to that of*

$$\sum_{m\gamma=\beta} c_{mn}^a(\gamma) - n \sum_{\substack{m\gamma=0 \\ \gamma\neq 0}} c_m^a(\gamma).$$

vskip 1pc

We prove (7.5) assuming (7.6), (7.7). Let $m = \ell^i$. If $m\gamma = \beta$, the image of $c_{mn}^a(\gamma)$ in $H^1(k, {}_{mn}E(1))$ is equal to $-a\{\lambda_{mn} g(\gamma)\} \otimes [\gamma]$ by (7.6) (which we apply by replacing n by mn) and hence its image in $H^1(k, {}_mE(1))$ is equal to $-a\{\lambda_{mn} g(\gamma)\} \otimes [n\gamma]$. On the other hand, if $m\gamma = 0$ and $\gamma \neq 0$, the image of $c_m^a(\gamma)$ in $H^1(k, {}_mE(1))$ is equal to $-a\{\lambda_m g(\lambda)\} \otimes [\gamma]$ by (7.6) (with n replaced by m). Hence by (7.7) and by using $\lambda_{mn} = \lambda_m \lambda_n^{m^2}$, we see that the image of $c_n^a(\gamma)$ in $H^1(k, {}_mE(1))$ is equal to (7.5.1).

Proof of (7.6). We prove first a weaker version of (7.6):

(7.8) let $U = E - ({}_aE \cup {}_nE)$. Then, $c_n^a(\beta)$ and $-\{\lambda_n g(\beta)\} \otimes \beta$ have the same image in

$$H^1(k, H^1(\overline{U}, \mathbf{Z}/n\mathbf{Z}(2)) \subset H^2(U, \mathbf{Z}/n\mathbf{Z}(2)) / \operatorname{Image}(H^2(k, \mathbf{Z}/n\mathbf{Z}(2))).$$

Proof. Let s be a function on E such that $\operatorname{div}(s) = n(\beta) - n(0)$. Then

$$\operatorname{div}(s \circ [n]) = n\Big(\sum_{n\gamma=\beta} (\gamma) - \sum_{n\gamma=0} (\gamma)\Big),$$

where $[n]$ denotes the multiplication $n : E \to E$. Since $\sum_{nx=\beta}(\gamma) - \sum_{n\gamma=0}(\gamma)$ is a principal divisor, by replacing s by cs for some $c \in k^\times$ if necessary, we have

$$s \circ [n] = (s')^n \quad \text{for some } s' \in \mathcal{O}(U')^\times$$

where $U' = [n]^{-1}(U)$. By $N_n(g) = \lambda_n g$, we have in $H^2(U, \mathbf{Z}/n\mathbf{Z}(2))$,

$$\{\lambda_n g, s\} = \{N_n(g), s\} = N_n\{g, s \circ [n]\} = nN_n\{g, s'\} = 0$$

where the last two N_n are the norm map $H^2(U', \mathbf{Z}/n\mathbf{Z}) \to H^2(U, \mathbf{Z}/n\mathbf{Z}(2))$. Furtherfore $s(\gamma) \in (k^\times)^n$ for any $\gamma \in {}_aE - \{0\}$ by $s(\gamma) = (s \circ [n])(n^{-1}\gamma)$. These prove

$$c_n^a(\beta) \to -a\{\lambda_n g(\beta), s\} \quad \text{in } H^2(U, \mathbf{Z}/n\mathbf{Z}(2)).$$

Since $[\beta] \to s$ under

$$\begin{aligned} {}_nE \cong H^1(E, \mathbf{Z}/n\mathbf{Z}(1))/(H^1(k, \mathbf{Z}/n\mathbf{Z}(1)) \to \\ H^1(U, \mathbf{Z}/n\mathbf{Z}(1))/H^1(k, \mathbf{Z}/n\mathbf{Z}(1)) \cong \mathcal{O}(U)^\times/(k^\times \cdot \mathcal{O}(U)^{\times n}), \end{aligned}$$

this proves (7.8).

For the proof of (7.6), it remains to show thast

$$H^1(k, H^1(\overline{E}, \mathbf{Z}/n\mathbf{Z})(2))) \to H^1(k, H^1(\overline{U}, \mathbf{Z}/n\mathbf{Z}(2)))$$

is injective. Twisting by (-1) (note k contains all n-th roots of 1) and taking the Galois cohomology of the exact sequence

$$0 \to H^1(\overline{E}, \mathbf{Z}/n\mathbf{Z}(1)) \to H^1(\overline{U}, \mathbf{Z}/n\mathbf{Z}(1)) \to \left(\oplus_{x \in E-U} \mathbf{Z}/n\mathbf{Z}\right)^0 \to 0,$$

where $(\ )^0$ denotes the kernel of the summation $\oplus \mathbf{Z}/n\mathbf{Z} \to \mathbf{Z}/n\mathbf{Z}$, we are reduced to proving the surjectivity of

$$H^0(k, H^1(\overline{U}, \mathbf{Z}/n\mathbf{Z}(1))) \to (\oplus_{x \in E-U} \mathbf{Z}/n\mathbf{Z})^0$$

Let $x \in E - U$. Since all points in ${}_{n^2}E$ are k-rational by the assumption, we have $x = ny$ for some $y \in E(k)$. Hence

$$(x) - (0) = n((y) - (0)) + \mathrm{div}(h)$$

for some $h \in k(E)^\times$. Then h defines an element of

$$H^0(k, H^1(\overline{U}, \mathbf{Z}/n\mathbf{Z}(1)) = \operatorname{Ker}(H^0(k, \overline{k}(E)^\times/\overline{k}(E)^{\times n}) \to \bigoplus_{\substack{x\in\overline{U} \\ \text{closed}}} \mathbf{Z}/n\mathbf{Z})$$

whose image in $(\bigoplus_{x\in E-U} \mathbf{Z}/n\mathbf{Z})^0$ equals $(x) - (0)$.

Proof of (7.7). Since $H^1(k, T_\ell E(1))$ is torsion free and N_m acts on $H^1(k, T_\ell E(1))$ as the multiplication by m, it is enough to prove

$$N_m\Big(\sum_{m\gamma=\beta} c^a_{mn}(\gamma) - n \sum_{\substack{m\gamma=0 \\ \gamma\neq 0}} c^a_m(\gamma)) = mc^a_n(\beta) \tag{7.9}$$

in $\Gamma(E_{\mathrm{zar}}, K_2)/K_2(k)$.

The proof of (7.9) is a straightforward computation on symbols and is omitted.

Proposition (7.3) follows from (7.2), (7.1.4) and the following lemma.

Lemma 7.10. *Let X be a smooth projective scheme over $\mathbf{Q}$ having potentially good reductions at all finite places of $\mathbf{Q}$. Consider the motif $H^m(X)(r)$ with $m, r \in \mathbf{Z}$ such that $r > \sup(m, 1)$, and let V and D be the associated $\mathbf{Q}$-vector spaces. Fix a prime number ℓ such that X has good reduction at ℓ and such that $\ell > r+1$. Assume that $P_p(H^m(\overline{X}, \mathbf{Q}_\ell), T) \in \mathbf{Q}[T]$ for all $p < \infty$ and $L(H^m(X), s) = \prod_{p<\infty} P_p(H^m(\overline{X}, \mathbf{Q}_\ell), p^{-s})^{-1}$ has a meromorphic analytic continuation to the whole s-plane satisfying the conjectural functional equation. Assume further that we find a $\mathbf{Q}$-vector subspace Φ in the image of*

$$gr^r_\gamma(K_{2r-m-1}(X) \otimes \mathbf{Q})$$

such that

$$\Phi \otimes R \xrightarrow[\cong]{} D_R/V_R^+, \quad \Phi \otimes Q_\ell \xrightarrow[\cong]{} H^1_{f,\mathrm{Spec}(\mathbf{Z})}(\mathbf{Q}, V_\ell).$$

Let S be a finite set of places of $\mathbf{Q}$ containing ∞, ℓ and all finite places at which X has bad reduction, and let $U = \mathrm{Spec}(\mathbf{Z}) - S$. Then for any $\mathbf{Z}$-lattice M in V such that $M \otimes \mathbf{Z}$ is $\mathrm{Gal}(\overline{\mathbf{Q}}/\mathbf{Q})$-stable, $H^2(U, M \otimes \mathbf{Z}_\ell)$ is finite and we have an equation in $\mathbf{R}^\times/\mathbf{Z}^\times_{\ell)}$

$$\begin{aligned}
&\mathrm{Tam}^{(\ell))}(M)\#Ш^{(\ell)}(M)(\#H^0(\mathbf{Q}, M^* \otimes \mathbf{Q}_\ell/\mathbf{Z}_\ell(1))^{-1} \\
&\quad = \lim_{s\to m+1-r} (L_S(H^m(X), s)^{-1}(s-m-1+r)^n) \\
&\quad \cdot \mathrm{vol}((D_\infty/V_\infty^+)/\mathrm{Image}(A^{(\ell)}(\mathbf{Q})))\#(A^{(\ell)}(\mathbf{Q})_{\mathrm{tor}})^{-1}\#H^2(U, M \otimes \mathbf{Z}_\ell)
\end{aligned}$$

where n is the order of $L_S(H^m(X), s)$ at $s = m+1-r$ and other notations are as in (5.15.2).

Proof. Let $L_p(s)$ be the local Euler factor of $L(H^m(X), s)$ for $p < \infty$ and let $L_\infty(s)$ be the gamma factor. We may assume that there exists a $\mathbf{Z}$-lattice L in D such that the Hodge filtration and the frobenius of D_ℓ induce on $L \otimes \mathbf{Z}_\ell$ a structure of a strongly divisible lattice and the $\mathrm{Gal}(\overline{\mathbf{Q}}_\ell/\mathbf{Q}_\ell)$-module $M \otimes \mathbf{Z}_\ell$ corresponds to $L \otimes \mathbf{Z}_\ell$. Take $\omega : \det_{\mathbf{Q}}(D) \xrightarrow{\cong} \mathbf{Q}$ which induces $\det_{\mathbf{Z}}(L) \xrightarrow{\cong} \mathbf{Z}$. By the functional equation

$$L_S(H^m(X), r) \prod_{p \in S} L_p(r) = \lim_{s \to m+1-r} L_S(H^m(X), s) \prod_{p \in S} L_p(s)$$

in $\mathbf{R}^x/\mathbf{Z}^x_{(\ell)}$, (7.10) is reduced to

Lemma 7.11. *We have $\mathbf{R}^x/\mathbf{Z}^x_{(\ell)}$:*

(1) *For $p < \infty$,*

$$L_p(r)^{-1} L_p(m+1-r) = \mu_{p,\omega}(A^{(\ell)}(Q_p)) \#(H^1(\mathbf{Q}_p, M \otimes \mathbf{Q}_\ell/\mathbf{Z}_\ell)/(A^{(\ell)}(\mathbf{Q}_p) \otimes \mathbf{Q}_\ell/\mathbf{Z}_\ell))^{-1}$$

(2) *On the other hand we have*

$$L_\infty(r)^{-1} \lim_{s \to m+1-r} L_\infty(s)(s-m-1+r)^n$$
$$= \mu_{\infty,\omega}(A(\mathbf{R})/A^{(\ell)}(\mathbf{Q}))\, \mathrm{vol}((D_\infty/V_\infty^+)/\,\mathrm{Image}(A^{(\ell)}(\mathbf{Q})))^{-1} \#(A^{(\ell)}(\mathbf{Q})_{\mathrm{tor}}).$$

Proof. The case $p = \ell$ of (1) follows from (4.1) and the duality (3.8). The case $p \neq \ell$ of (1) is an easy exercise and left to the reader. Finally (2) is a problem of the ratio of two volumes on D_∞/V_∞^+; one is the measure in (7.10) and the other is induced via the isomorphism $A(\mathbf{R})/A(\mathbf{R})_{cpt} \cong D_\infty/V_\infty^+$ by $\mu_{\infty,\omega}$ on $A(\mathbf{R})$ and by the Haar measure on $A(\mathbf{R})_{cpt}$ with the total volume one. We are reduced to

7.12. *With the assumption and notations as above, the composite*

$$\det_{\mathbf{Z}}(M) \otimes_{\mathbf{Z}} \mathbf{C} \cong \det_{\mathbf{Q}}(D) \otimes_{\mathbf{Q}} \mathbf{C} \xrightarrow{\omega} \mathbf{C}$$

sends $\det_{\mathbf{Z}}(M) \otimes_{\mathbf{Z}} \mathbf{Z}_{(\ell)}$ onto $\mathbf{Z}_{(\ell)}(2\pi i)^{(r-\frac{1}{2}m)\dim(V)}$.

Proof. By Poincaré duality and by the hard Lefschetz, we have an isomorphism of motives (with $\mathbf{Q}$-coefficients)

$$\det(H^m(X)(r)) \otimes \det(H^m(X)((r)) \xrightarrow{\cong} \mathbf{Q}((2r-m)\dim(V)).$$

The de Rham, the ℓ-adic and the Hodge realization of this isomorphism satisfy in $\mathbf{R}^x/\mathbf{Z}^x_{(\ell)}$

$$\begin{aligned}
&\chi(\det_{\mathbf{Z}}(L) \otimes \det_{\mathbf{Z}}(L) \dashrightarrow \mathbf{Z}) \\
&= \chi(\det_{\mathbf{Z}_\ell}(M \otimes \mathbf{Z}_\ell) \otimes \det_{\mathbf{Z}_\ell}(M \otimes \mathbf{Z}_\ell) \dashrightarrow \mathbf{Z}_\ell((2r-m)\dim(V))) \\
&= \chi(\det_{\mathbf{Z}}(M) \otimes \det_{\mathbf{Z}}(M) \dashrightarrow \mathbf{Z}(2\pi i)^{(2r-m)\dim(V)})
\end{aligned}$$

in $\mathbf{R}^x/\mathbf{Z}^x_{(\ell)}$. Hence the commutative diagram

$$\begin{array}{ccc}
(\det_{\mathbf{Z}}(L) \otimes_{\mathbf{Z}} \mathbf{C}) \otimes_{\mathbf{C}} (\det_{\mathbf{Z}}(L) \otimes_{\mathbf{Z}} C) & \xrightarrow{\cong} & \mathbf{Z} \otimes_{\mathbf{Z}} \mathbf{C} \\
\downarrow & & \downarrow \\
(\det_{\mathbf{Z}}(M) \otimes_{\mathbf{Z}} \mathbf{C}) \otimes_{\mathbf{C}} (\det_{\mathbf{Z}}(M) \otimes_{\mathbf{Z}} \mathbf{C}) & \xrightarrow{\cong} & \mathbf{Z}(2\pi i)^{(2r-m)\dim(V)} \otimes_{\mathbf{Z}} \mathbf{C}
\end{array}$$

shows that the left vertical arrow sends $\det(M) \otimes \det(M) \otimes \mathbf{Z}_{(\ell)}$ onto $\det_{\mathbf{Z}}(L) \otimes \det_{\mathbf{Z}}(L) \otimes \mathbf{Z}_{(\ell)}(2\pi i)^{(2r-m)\dim(V)}$. This proves (7.12)

7.13. Here we review the theory of Hurwitz numbers from the point of view of Fontaine-Messing theory. This will be used for the proof of (7.4).

For a non-zero integer n, let θ_a be the function on E characterized by the properties $\operatorname{div}(\theta_a) = 12((a^2(0) - {}_{a^2}E)$ and $N_b(\theta_a) = \theta_a$ for any non-zero element $b \in \mathcal{O}_K$ which is prime to a (cf. [dS], II 2.3). Assume ℓ splits in K, let $\ell\mathcal{O}_K = \mathfrak{m}\overline{\mathfrak{m}}$ with $\mathfrak{m}$ a maximal ideal of $\mathcal{O}_K$, and put $\tau = \psi(\mathfrak{m})$ (so τ is a generator of $\mathfrak{m}$). Let T be the component of $T_\ell E$ on which $\mathcal{O}_K$ acts via $\mathcal{O}_K \to \mathcal{O}_K$. T is a free $\mathbf{Z}_\ell$-module of rank one. For $r \in \mathbf{Z}$, we define $e_{r,E} \in H^1(K, T^{\otimes r})$ as

$$e_{r,E} = 12^{-1}(a^2 - \psi(a)^{-r})^{-1}(N_{K_i'/K}(\{\theta_a(\tau^{-i}\delta)\} \otimes [\gamma_i']^{\otimes r})_i \in H^1(K, T^{\otimes r})$$

where γ' denotes the ${}_{\mathfrak{m}^i}E$-component of the image of $\ell^{-i}\delta$ in ${}_{\ell^i f}E(\mathbf{C})$ and K_i' denotes the ray class field of modulus $\mathfrak{m}^i f$ over K. Then, $e_{r,E}$ is independent of the choice of a. On the other hand, let L be the filtered Dieudonné module corresponding to the $\operatorname{Gal}(\overline{K}_{\mathfrak{m}}/K_{\mathfrak{m}})$- module T in the sense of Fontaine-Laffaille. Then $L^0 = 0$, $L^{-1} = L$ and $L \otimes \mathbf{Q} \xrightarrow{\cong} H^1_{DR}(E/\mathbf{Q})/\operatorname{Fil}^1 H^1_{DR}(E/\mathbf{Q}) \otimes K_{\mathfrak{m}}$. For $r \geq 2$, let

$$\partial^r : L^{\otimes r} \otimes \mathbf{Q} \xrightarrow{\cong} H^1(K_{\mathfrak{m}}, T^{\otimes r}) \otimes \mathbf{Q}$$

be the boundary map obtained from (1.13) $\otimes T^{\otimes r}(-r)$ and from the isomorphism $L^{\otimes r} \cong H^0(K_{\mathfrak{m}}, B^+_{\mathrm{crys}} \otimes T^{\otimes r}(-r))$. Then, if $r < p-1$, ∂^r gives an isomorphism without $\otimes\mathbf{Q}$. The ℓ-adic theory of Hurwitz numbers (cf. [dS] II§4) is interpreted as:

(7.13.1) Let $r \geq 3$. Then, if ω is a differential $H^1_{DR}(E/\mathbf{Q})/\operatorname{Fil}^1 \xrightarrow{\cong} \mathbf{Q}$ and (ω, δ) denotes the integration of ω on δ, $(\omega,\delta)^{-r}L(\psi^{-r},0)$ belongs to K and the cup product with $c_{-r,E}$

$$L^{\otimes r} \otimes \mathbf{Q} \underset{\cong}{\xrightarrow{\partial^r}} H^1(K_m, T^{\otimes r}) \otimes \mathbf{Q} \xrightarrow{\cup c_{-r,E}} H^2(K_{\mathfrak{m}}, \mathbf{Q}_\ell(1)) = \mathbf{Q}_\ell$$

coincides with

$$\pm(1-\tau^r p^{-1})((\omega,\delta)^{-r}L(\psi^{-r},0)) \cdot \omega^{\otimes r}.$$

In fact, this theory is described usually by using the Coates-Wiles homomorphisms applied to the norm compatible system $(\theta_a(\tau^{-i}\delta))_i$ via an isomorphism of formal groups $\widehat{E} \cong \widehat{G}_m$ over $\widehat{K}_{\mathfrak{m},nr}$ (cf. [dS], II §4), but the relation between the Coates-Wiles homomorphism and Fontaine-Messing theory proved in §2 enables us to rewrite the theory as above.

As a relation with cyclotomic elements in (7.1), the formula

$$(7.13.2) \qquad e_v(E) = (1-\overline{\tau})ve_{1,E} \quad \text{in } H^1(K, T(1))$$

is proved easily. By (7.2), if $\widehat{e}_v(E) \in H^1(\mathbf{Z}[1/\ell], T_\ell E(1))$ denotes the image of $c_v(E)$, we have

$$(7.13.3) \qquad ve_{1,E} = (1-\tau)\widehat{e}_v(E) \quad \text{in } H^1(K, T(1)).$$

(Note $\kappa^{-1} = (1-\tau)(1-\overline{\tau})$ with κ as in (7.2).)

(7.14). Now we prove (7.4). In [So6], Soulé proved that $e_v(E)$ generates $H^1(\mathbf{Q}, V_\ell)$ and $H^2(\mathbf{Z}[1/\ell], j_*V_\ell) = 0$ in the case ℓ is regular for E. The following argument is inspired by [So6]. If ℓ is regular for E, the following conditions are satisfied ([Y]).

(i) ℓ splits in K.

(ii) $(\omega,\delta)^{2-\ell}L(\psi^{2-\ell},0)$ is an ℓ-unit if the differential ω is a generator of $\Gamma(\mathfrak{F}, \Omega_{\mathfrak{F}/\mathbf{Z}_{(\ell)}})$ for the proper smooth model $\mathfrak{F}$ of E over $\mathbf{Z}_{(\ell)}$.

(iii) Write $\ell O_K = \mathfrak{m}\overline{\mathfrak{m}}$ as in (7.13). Then, if F denotes the field $K({}_{\mathfrak{m}}E)$, $\operatorname{Pic}(O_F)/\ell\operatorname{Pic}(O_F)$ has no nontrivial element on which the action of $\operatorname{Gal}(F/K)$ is the same as that on ${}_{\mathfrak{m}}E(1)$.

Under the conditions (i)–(iii) we prove

$$\#(H^1(\mathbf{Z}[1/\ell], j_*T_\ell E(1))/\mathbf{Z}_\ell c_v(E)) = \#H^2(\mathbf{Z}[1/\ell], j_*T_\ell E(1)) = \kappa^{-1}$$

in $\mathbf{Q}^\times/\mathbf{Z}^\times_{(\ell)}$. Here and in the following, v is taken to be prime to ℓ.

By (ii) and (7.13.1), $e_{2-\ell,E}$ generates $H^1(K_{\mathfrak{m}}, T^{\otimes(2-\ell)}(1))$. By using $(T/\ell)^{\otimes(2-\ell)} \cong T/\ell$, we see that $e_{1,E}$ generates $H^1(K_{\mathfrak{m}}, T(1))$. Hence by (7.13.3), $\widehat{e}_v(E)$ generates $H^1(\mathbf{Z}[1/\ell], j_*T_\ell E(1))$ and we obtain

$$\#(H^1(\mathbf{Z}[1/\ell], j_*T_\ell E(1))/\mathbf{Z}_\ell e_v(E)) = \kappa^{-1}$$

in $\mathbf{Q}^\times/\mathbf{Z}_{(\ell)}^\times$.

Next we prove $\#H^2(\mathbf{Z}[1/\ell], j_*T_\ell E(1)) = \kappa^{-1}$ in $\mathbf{Q}^\times/\mathbf{Z}_{(\ell)}^\times$. It is seen easily that $\#H^2(\mathbf{Q}_\ell, T_\ell E(1)) = \kappa^{-1}$ in $\mathbf{Q}^\times/\mathbf{Z}_{(\ell)}^\times$ and hence it suffices to prove that the canonical map $H^2(\mathbf{Z}[1/\ell], j_*T_\ell E(1)) \to H^2(\mathbf{Q}_\ell, T_\ell E(1))$ is bijective. By Tate duality, the kernel (resp. cokernel) of this map is isomorphic to the Pontrjagin dual of

$$\ker(H^1(\mathbf{Z}[1/\ell], j_*H) \xrightarrow{i} H^1(\mathbf{Q}_\ell, H)) \quad (\text{resp. } H^0(\mathbf{Q}, H))$$

where $H = H^1(E \otimes \overline{\mathbf{Q}}, \mathbf{Q}_\ell/\mathbf{Z}_\ell)$. The injectivity of i is proved by using the conditions (iii) and the fact $H^0(\mathbf{Q}, H) = 0$ is seen easily.

Remark 7.15. (7.3) suggests that an Iwasawa main conjecture should exist even in the case ℓ does not split in K.

BIBLIOGRAPHY

[Be1] Beilinson, A., *Higher regulators and values of L-functions*, J. Soviet Math. **30** (1985), 2036–2070.

[Be2] Beilinson, A., *Height pairings between algebraic cycles*, in *Current Trends in Arithmetical Algebraic Geometry*, Contemporary Math. **67** (1987), 1–24.

[Bl1] Bloch, S., *A note on height pairings, Tamagawa numbers, and the Birch and Swinnerton-Dyer conjecture*, Inv. Math. **58** (1980), 65–76.

[Bl2] Bloch, S., *Algebraic K-theory and zeta functions and elliptic curves*, Proc. Int. Congress Helsinki (1978).

[Bl3] Bloch, S., *Height pairings for algebraic cycles*, J. Pure and Appl. Algebra **34** (1984).

[Bl4] Bloch, S., *Lectures on Algebraic Cycles*, Duke Univ. Math Series, 1981.

[Co1] Coleman, R., *Division values in local fields*, Inv. Math. **53** (1979), 91–116.

[Co2] Coleman, R., *Local units modulo circular units*, Proc. Amer. Math. Soc. **89** (1983), 1–7.

[De1] Deligne, P., *La conjecture de Weil I*, Publ. Math. IHES **43** (1981), 273–307.

[De2] Deligne, P., *Valeurs de functions L et périods d'intégrales*, Proc. Symp. Pure Math. **33** (1979) AMS, 313–346.

[De3] Deligne, P., *La conjecture de Weil II*, Publ. Math. IHES **52** (1981).

[De4] Deligne, P., in preparation.

[De5] Deligne, P., Letter to C. Soulé.

[DW] Deninger, C., *On the Belinson conjecture for elliptic curves with complex multiplication*, in *Belinson's Conjectures on Special Values of L-functions*, Perspectives in Math., Academic Press 1988.

[Fa] Faltings, G., *Crystalline cohomology and p-adic Galois representations*, (preprint).

[Fo] Fontaine, J.M., *Sur certains types de représentations p-adiques du groupe de Galois d'un corps local: construction d'un anneau de Barsotti-Tate*, Ann. of Math. **115** (1982), 529–577.

[FL] Fontaine, J.M. and Laffaille, G., *Construction de représentations p-adiques*, Ann. Sci. ENS. **15** (1982), 547–608.

[FM] Fontaine, J.M. and Messing, W., *P-adic periods and p-adic étale cohomology*, in *Current Trends in Arithmetical Algebraic Geometry*, Cont. Math. **67** (1987) AMS, 179–207.

[Gros] Gros, M., *Régulateurs syntomiques*, preprint (1987).

[Grot] Grothendieck, A. *Formule de Lefchetz et rationalité des fonctions L*, Sém. Bourbaki, exposé 279 (1964).

[Ih] Ihara, Y. *Profinite braid groups, Galois representations, and complex multiplications*, Ann. of Math. **123** (1986), 43–106.

[Ja] Jannsen, U., *On the ℓ-adic cohomology of varieties over number fields and its Galois cohomology*, preprint (1987).

[Ka] Kato, K., *The explicit reciprocity law and the cohomology of Fontaine-Messing*, to appear in Bull. Math. Soc. France.

[MW] Mazur, B. and Wiles, A., *Class fields of abelian extensions of* $\mathbb{Q}$, Inv. Math. **76** (1984), 179–330.

[On1] Ono, T., *Arithmetic of algebraic tori*, Ann. of Mth. **74** (1961), 101–139.

[On2] Ono, T., *On the Tamagawa number of algebraic tori*, Ann. of Math. **78** (1963), 47–73.

[Se] Serre, J.-P., *Cohomologie Galoisienne*, LMN No. 5, Springer Verlag, 1973.

[dS] de Shalit, E., *Iwasawa Theory of Elliptic Curves with Complex Multiplication*, Academic Press, 1987.

[So1] Soulé, C., *k-théorie des anneaux d'entiers de corps de nombres et cohomologie étale*, Inv. Math. **55** (1979), 251–295.

[So2] C. Soulé, "On higher p-adic regulators", in *Algebraic K-theory*, Evanston, 1980, LNM 854, 371–401, Springer Verlag (1981).

[So3] Soulé, C., *The rank of étale cohomology of varieties over p-adic number fields*, Comp. Math. **53** (1984), 113–131.

[So4] Soulé, C., *Opérations en k-théorie algébrique*, Canadian J. Math., Vol. XXXVII, (1985).

[So5] Soulé, C., *Eléments cyclotomiques en K-théorie*, IHES preprint, 1985.

[So6] Soulé, C., *p-adic K-theory of elliptic curves*, Duke Math. J. **54** (1987), 249–269.

[Ta1] Tate, J., *Duality theorems in Galois cohomology over number fields*, Proc. ICM Stockholm 1962, 288–295 Institute Mittag-Leffler (1963).

[Ta2] Tate, J., *p-divisible groups*, in Proc. of a Conf. On Local Fields, Driebergen 1966, 153–183, Springer Verlag (1967).

[Ta3] Tate, J. .*On the conjectures of Birch and Swinnerton-Dyer and a geometric analog*, Sem. Bourbaki exposé 306, 1965–66, in *Dix Exposés sur la Cohomologie des Schémas*, North Holland (1968).

[Wa] Washington, L., *Introduction to cyclotomic fields*, Graduate Texts in Math., Vol. 83, Springer (1982).

[We] Weil, A. *Adeles and algebraic groups*, Progress in Math. **23**, Birkhauser, (1982).

[Ya] Yager, R., *A Kummer criterion for imaginary quadratic fields*, Comp. Math. **47** (1982), 31–42.

Spencer Bloch
Department of Mathematics
University of Chicago
Chicago, Illinois 60637
and
Kazuya Kato
Department of Mathematics
University of Tokyo
Hongo 113
Tokyo, Japan

Bitorseurs et Cohomologie Non Abélienne

LAWRENCE BREEN*

1. Introduction

L'ubiquité de la notion de G-torseur (ou G-espace principal homogène) en mathématiques provient de sa double nature : nature géométrique d'une part, qu'il n'y a pas lieu d'illustrer ici tant les exemples abondent dans les domaines les plus divers, et nature cohomologique d'autre part, liée à l'interprétation de l'ensemble des classes d'isomorphisme de G-torseurs comme premier groupe de cohomologie à valeurs dans G. C'est, en définitive, la flexibilité apportée par cette seconde interprétation, avec son cortège de suites exactes longues, qui rend aisée l'étude dans de nombreuse situations de l'objet géométrique de départ.

Or il existe une autre notion, assez proche de la précédente, mais qui fait à l'heure actuelle figure de parent pauvre : c'est celle de (G, H)-bitorseur (ou espace (G, H)-principal). Introduite dans [53] dans un cas particulier, elle a ensuite été étudiée dans le cadre général des topos par J. Giraud [26] et A. Grothendieck [31]. Elle n'a cependant pas eu, et de loin, la popularité de sa proche parente, bien qu'elle possède des propriétés agréables qui font défaut à cette dernière, notamment l'existence, lorsque $G = H$, d'une loi de groupe sur l'ensemble de ses classes à isomorphisme près. La raison principale, me semble-t-il, de ce manque de faveur, provient de l'absence d'une interprétation cohomologique de l'ensemble des classes de (G, H)-bitorseurs, analogue à celle mentionnée plus haut dans le cas des G-torseurs, et qui permette d'en appréhender la fonctorialité et le comportement relatif aux suites exactes. On en était donc réduit, jusqu'ici, à travailler avec l'objet géométrique lui-même, avec pour conséquence une lourdeur dans le discussion peu propice à une large diffusion du concept en question.

(*)U.A. 742 du C.N.R.S.

Le premier but de ce travail est de remédier à cet état de fait en expliquant quelle est, pour une paire de groupes G, H d'un topos T, l'interprétation cohomologique de l'ensemble des classes de (G,H)-bitorseurs. Le résultat, d'ailleurs facile à obtenir, ne semble pas figurer dans la littérature si ce n'est, dans un cas particulier, sous la forme d'un calcul de cocycles proposé en exercice par J.-P. Serre dans [53], p. I.6.3. Il s'énonce, dans le cadre général, en termes d'algèbre homotopique plutôt qu'homologique, c'est à dire en termes d'objets simpliciaux de T plutôt que de complexes de groupes abéliens. Ceci n'est certes guère surprenant dans la mesure où l'on considère des groupes G et H non abéliens, mais il y a là un élément de complication qui en a certainement retardé la découverte. Néanmoins, les objets simpliciaux qui apparaissent ici sont, du point de vue des topologues, extrêmement simples et ont été très étudiés ces dernières années, sous le nom de modules croisés (voir par exemple [10]). J'espère convaincre ici le lecteur de leur innocuité et du fait que, même dans les situations les plus courantes, telle l'étude des extensions de groupes, on ne peut guère éviter d'y avoir recours. De manière plus générale, le texte qui suit se veut une illustration du fait que la topologie algébrique dans le cadre des topos, telle qu'elle a été développée par K. Brown [8], L. Illusie [35], J.F. Jardine [36] et quelques autres, est une généralisation de la topologie algébrique usuelle qui n'est pas introduite par simple goût du formalisme. C'est plutôt une extension très naturelle de l'algèbre homologique, et qui doit nécessairement être prise en compte dès qu'on s'éloigne des groupes abéliens.

De l'interprétation cohomologique des (G,H)-bitorseurs découlent divers autres résultats, notamment la fonctorialité en (G,H) des objets en question, qui n'est pas celle à laquelle on s'attendrait de prime abord, mais qui s'exprime néanmoins agréablement en termes d'algèbre simpliciale. D'autre part, la description qui vient d'être trouvée suggère, dans le cas particuliers où $G = H$, une variante de la notion de (G,G)-bitorseur que j'appelle (G,π)-torseur et qui, si elle peut paraître quelque peu artificielle du point de vue géométrique, s'impose naturellement lorsque l'on se place dans une perspective cohomologique. Ceci permet alors, sous certaines hypothèses, d'associer à des suites exactes courtes de coefficients des suites exactes de (G,π)-torseurs.

La loi de groupe sur l'ensemble des classes d'isomorphisme de (G,G)-bitorseurs provient en fait d'une loi de composition $\wedge\colon \mathcal{C} \times \mathcal{C} \to \mathcal{C}$ qui fait de la catégorie $\mathcal{C}$ des (G,G)-bitorseurs une *gr*-catégorie, dans la terminologie de A. Grothendieck, c'est à dire que l'on dispose notamment d'isomorphismes naturels d'associativité $X \wedge (Y \wedge Z) \simeq (X \wedge Y) \wedge Z$ pour tous objets X, Y, Z, satisfaisant aux conditions de cohérence dites "du pentagone" ([44]). Ceci suggère la possibilité de considérer les torseurs (en un sens convenable) sous cette *gr*-catégorie (ou plutôt sous le *gr*-champ obtenu par faisceautisation lorsque l'on se sera placé dans une situation relative). Les classes d'équivalence des objets en question forment alors un ensemble pointé qui mérite, plus encore que l'ensemble défini en [26], d'être appelé l'ensemble de cohomologie non abélienne $H^2(G)$. A la différence de l'objet de loc.cit., il s'agit ici d'un ensemble qui a de bonnes propriétés de fonctorialité en G, analogues à celles mentionnées ci-dessus pour les (G, G)-bitorseurs. A la description géométrique des objets qui vient d'être esquissée correspond par ailleurs une description cohomologique, et cette dernière peut même s'expliciter de manière concrète en termes de cocycles, une fois choisi un hyperrecouvrement approprié de l'objet final. On retrouve alors des conditions de cocycles considérées notamment, dans un cas très particuliers, par P. Dedecker dans [15] (2.2).

Lorsque l'on teste la définition proposée de $H^2(G)$ dans la situation correspondant aux extensions de groupes dans un topos, on retrouve par voie cohomologique la description des extensions en question obtenue par A. Grothendieck de manière géométrique dans [31] exposé VII. §1.1. En outre, l'interprétation cohomologique mentionnée permet de comprendre sans difficulté quelle est la fonctorialité en P de l'ensemble $\mathrm{Ext}(P, G)$. Celle-ci était certainement connue de Dedecker (voir ci-dessous 8.6, note en bas de page) mais celui-ci préfère discuter l'énoncé de fonctorialité plus général concernant ce qu'il appelle les (G, π)-extensions (et qui sont aux extensions ce que les (G, π)-torseurs sont aux (G, G)-bitorseurs), sans s'attarder à l'assertion correspondante, pourtant très frappante, dans le cas des extensions ordinaires.

L'étude du H^2 non abélien a déjà été entreprise un certain nombre de fois, dans des cadres plus ou moins généraux, mais la comparaison entre les différentes théories (outre les auteurs déjà cités, il convient de mentionner ici J. Duskin [23], [24] et J.-C. Douai [20]) ne l'a guère été jusqu'ici, si ce n'est, récemment, dans [11]. Je me suis donc attaché à préciser la relation entre mon approche et celles de P. Dedecker et J. Giraud. Les diverses notions introduites par ce dernier, telle celle du lien attaché à une gerbe, et les différentes suites exactes rattachant le H^2 non abélien à des objets plus accessibles, s'interprètent de manière très simple, et à vrai dire automatique, dès lors que l'on dispose d'une formulation cohomologique, plutôt que géométrique, des ensembles considérés. En ce qui concerne l'approche de Duskin et de Douai, il semble qu'elles soient l'une et l'autre très proches de la mienne, à ceci près qu'ils se voient contraints, la notion de bitorseur leur faisant défaut, de recourir à des descriptions très peu explicites des objets considérés. Cependant, d'après [11], Duskin a récemment décrit son $H^2(G)$ non abélien comme l'ensemble, dans sa terminologie, de classes de 2-torseurs sous un 2-hypergroupoïde associé au groupe G, ce qui devrait en rendre aisée la comparaison avec l'objet proposé ici. Mentionnons enfin le fait que l'idée selon laquelle le coefficient d'un H^2 non abélien devra être constitué d'un module croisé qui en déterminera la fonctorialité, et non pas du seul groupe G, figure de manière très explicite chez Dedecker (voir notamment l'introduction de [15]), ainsi que chez Douai.

Pour en terminer avec ces comparaisons, il convient de signaler que la théorie du H^2 non abélien présentée ici est un peu trop limitée, par rapport à celles de [26], [23], dans la mesure où seuls apparaissent les ensembles de gerbes localement de la forme TORS(G) (pour un groupe donné G du topos), que j'appelle ici les G-gerbes. La raison en est que l'on a considéré comme seuls coefficients des modules croisés, c'est à dire en définitive des groupes simpliciaux de T. Le passage de tels groupes à des groupoïdes aurait permis d'obtenir des gerbes générales et une description de ces gerbes compatible à celle de [19] proposition 3.3, mais ce au prix d'un facteur de complication supplémentaire. C'est pourquoi il m'a paru judicieux de me restreindre ici au cas des G-gerbes (déjà suffisant pour analyser les extensions de groupes), la discussion générale étant néanmoins préparée en 2.10–2.15 par la description cohomologique du champ des (G, H)-bitorseurs, et de ses propriétés de fonctorialité.

J'ai profité, au cours de la préparation de ce travail, de discussions avec J. Benabou, D. Conduché, P. Deligne et J. Lannes. Les commentaires qui m'ont été fait par K.H. Ulbrich et le referee à la lecture d'une première version du texte m'ont permis d'y apporter des améliorations. Je remercie également Madame C. Simon, pour le soin avec lequel elle s'est chargée de la dactylographie du manuscrit. Enfin, le lecteur le plus superficiel de ce texte y reconnaitra l'influence d'Alexandre Grothendieck. Les questions de cohomologie non abélienne et d'algèbre simpliciale dans le cadre des topos l'ont constamment intéressé, de [27]–[31] à [33]. Plus encore que dans ces textes, c'est dans [32] que je me suis familiarisé avec son point de vue sur les n-champs, et sur la relation entre la topologie algébrique usuelle et sa variante dans le cadre faisceautique, ainsi que sur la manière dont il convient d'interpréter en termes géométriques des énoncés cohomologiques ou homotopiques. C'est bien sa façon d'aborder ces problèmes qui sous-tend l'approche adoptée dans ce texte. Qu'il trouve ici l'expression de mon admiration et de ma gratitude.

2. Bitorseurs

Nous commencerons par rappeler quelques énoncés de base concernant les bitorseurs ([48], [31], [26]). On a tout d'abord :

Définition 2.1. Soient G et H deux groupes d'un topos T. On appelle

(G, H)-bitorseur de T la donnée d'un objet P de T, sur lequel G (resp. H) agit à gauche (resp. à droite), ces deux opérations commutant entre elles et satisfaisant aux conditions suivants :

(i) P est une G-torseur à gauche

(ii) Le morphisme $H \to \mathrm{Aut}_G(P)$ qui décrit l'action de H sur P est un (anti)-isomorphisme.

La seconde condition peut d'ailleurs, par *loc. cit.* III 1.5.3, être remplacée par:

(ii′) P est un H-torseur à droite.

Lorsque $G = H$, on dira simplement que P est un G-bitorseur.

Exemple 2.2. (i) On appelle G lui-même, muni de l'action de G à gauche et à droite par translation, le G-bitorseur trivial, noté G_{triv}.

(ii) Soit P un torseur à gauche sous un groupe abélien A. On en fait un A-bitorseur par l'action à droite de A sur P définie par *

$$pa = ap. \tag{2.2.1}$$

(iii) Soit u un automorphisme de G. Le G-torseur à gauche trivial G est muni d'une structure de G-bitorseur pour la multiplication à droite par un élément g' de G définie par :

$$g.g' =: gu(g') \tag{2.2.2}$$

pour tout $g \in G$. Ce bitorseur sera noté (G, u).

(iv) Soit $1 \to G \to E \to P \to 1$ une suite exacte de groupes de T. Alors l'objet en ensembles sous-jacent à E est un G_P-bitorseur de $T_{|P}$, pour l'action à gauche et à droite de G_P sur E déduite de la loi de groupe de E.

2.3. Soient P, Q une paire de (G, H)-bitorseurs. On appelle morphisme de (G, H)-bitorseurs tout morphisme $f: P \to Q$ qui est (G, H)-équivariant. En particulier, une trivialisation d'un G-bitorseur P consiste en la donnée d'une telle flèche $f: G \to P$ de source le bitorseur trivial, c'est à dire une section $s = f(1)$ de P qui est centrale, en ce sens que $gs = sg$ pour tout $g \in G$. Les (G, H)-bitorseurs d'un topos T forment une catégorie, que l'on notera $\mathrm{BITORS}(T; G, H)$ (ou $\mathrm{BITORS}(G, H)$, ou même $\mathrm{BITORS}(G)$ lorsque $G = H$). Soit $\mathcal{C}$ la catégorie fibrée au-dessus de T dont la fibre en

---

* Pour ne pas alourdir le texte on écrira constamment en notation ensembliste des énoncés ayant cours dans T.

un objet X de T est la catégorie $\mathcal{C}_X = \mathrm{B\scriptstyle ITORS}(T_{|X}, G_{|X}, H_{|X})$ des (G,H)-bitorseurs au-dessus de X. Pour X variable, les objets et les morphismes de la catégorie $\mathcal{C}_X$ se recollent, tout comme dans le cas des catégories de torseurs. On dit alors que $\mathcal{C}$ est un champ, qui sera noté BITORS(G,H) (resp. BITORS(G)).

Soient maintenant des objets P et Q de $\mathrm{B\scriptstyle ITORS}(G,H)$ et $\mathrm{B\scriptstyle ITORS}(H,K)$ respectivement. Leur produit contracté $P \overset{H}{\wedge} Q$ est l'objet de T, quotient de $P \times Q$ par la relation $(ph,q) = (p,hq)$. L'action à gauche de G sur P (resp. à droite de K sur Q) en fait un élément de $\mathrm{B\scriptstyle ITORS}(G,K)$. Ce produit contracté est associatif, à un isomorphisme naturel près, et cet isomorphisme d'associativité est cohérent au sens de [44]. En outre, à un (G,H)-bitorseur P est associé le (H,G)-bitorseur opposé P^0, de même objet en ensemble que P, mais sur lequel H (resp. G) agit à gauche (resp. à droite) par la règle

$$h * p = ph^{-1} \qquad (\text{resp. } p * g = g^{-1}p).$$

En particulier, lorsque $G = H$, le produit contracté munit la catégorie $\mathrm{B\scriptstyle ITORS}(G)$ d'une structure de gr-catégorie par laquelle G est l'objet neutre et P^0 un inverse de P. Cette structure étant définie de manière compatible sur chacune des categories $\mathrm{B\scriptstyle ITORS}(G_{|X})$ qui constituent les fibres du champ BITORS(G), on dira que la champ en question est un gr-champ. La discussion précédente est donc résumée par l'énoncé suivant :

Proposition 2.3. *Le produit contracté, l'objet trivial et la loi d'inverse mentionnée munissent le champ* BITORS(G) *d'une structure de gr-champ.*

En particulier le faisceau $\pi_0(\text{BITORS}(G))$ des classes d'isomorphisme de G-bitorseurs est un groupe de T, tandis que le faisceau $\pi_1(\text{BITORS}(G))$ des automorphismes du G-bitorseur trivial G est un groupe abélien de T.

2.4. Le principal résultat de ce paragraphe est la proposition suivante. C'est sur lui que reposera l'interprétation cohomologique des bitorseurs. On note $i: G \to \mathrm{Aut}(G)$, (resp. $i': G \to \mathrm{Aut}^o(G)$) l'homomorphisme de conjugaison intérieure qui envoie g vers i_g (resp. i'_g) défini par :

$$i_g(g_1) = gg_1g^{-1} \qquad (\text{resp. } i'_g(g_1) = g^{-1}g_1g).$$

Proposition 2.4. *La donnée d'un G-bitorseur de T équivaut à celle d'un G-torseur à gauche P de T et d'une section s_P de l'* $\mathrm{Aut}^o(G)$*-torseur à* *gauche $i'_*(P)$ qui s'en déduit par extension du groupe structural par $i': G \to \mathrm{Aut}^o(G)$.*

En effet, par définition de $i'_* P$, l'ensemble sous-jacent à ce torseur est l'objet $\mathrm{Aut}^o(G) \overset{G}{\wedge} P$ de T, G agissant à droite sur $\mathrm{Aut}^o(G)$ par $(u, g) \to i'_g \circ u$.* Par [26] III.1.6.1.ii), il existe, pour tour G-torseur à gauche P, un isomorphisme canonique :

$$\mathrm{Aut}^o(G) \overset{G}{\wedge} P \longrightarrow \mathrm{Hom}_G(P, \mathrm{Aut}(G)) \tag{2.4.1}$$

de but le faisceau des applications G-équivariantes de P dans $\mathrm{Aut}(G)$ (sur lequel G agit maintenant à gauche à travers i, c'est à dire par $(g, u) \longrightarrow i_g \circ u$). L'isomorphisme (2.4.1) est défini en associant à toute section locale (u, p) de $\mathrm{Aut}^o(G) \overset{G}{\wedge} P$ le morphisme local de P vers $\mathrm{Aut}(G)$ qui envoie un élément gp de P vers $i_g \circ u$. La donnée d'une section de $i'_* P = \mathrm{Aut}^o(G) \overset{G}{\wedge} P$ équivaut donc à celle d'une section de $\mathrm{Hom}_G(P, \mathrm{Aut}(G))$, c'est à dire à une application G-équivariante globalement définie de P vers $\mathrm{Aut}(G)$. La proposition équivaut donc au lemme suivant :

Lemme 2.5. *La donnée d'un G-bitorseur équivaut à celle d'un G-torseur à gauche P et d'un morphisme : $j: P \to \mathrm{Aut}(G)$ G-équivariant pour l'action à gauche de G sur $\mathrm{Aut}(G)$ définie par $(g, u) \to i_g \circ u$.*

Ceci se démontre en observant que si P est un G-bitorseur, le G-torseur à gauche sous-jacent est muni d'un tel morphisme j, défini par :

$$pg = j(p)(g)p. \tag{2.5.1}$$

Puisque cette action à droite commute avec l'action à gauche, on a :

$$(pg_1)g_2 = j(p)(g_1)pg_2 = j(p)(g_1)j(p)(g_2)p$$

c'est à dire que $j(p)(g_1 g_2) = j(p)(g_1)j(p)(g_2)$. Ainsi $j(p)$ est un endomorphisme de G, et même un automorphisme de G puisque l'action à droite ainsi définie de G sur P fait de P un G-torseur à droite.

---

* Pour toute paire d'éléments, $u, v \in \mathrm{Aut}(G)$, $u \circ v$ désignera toujours leur produit dans $\mathrm{Aut}(G)$, obtenu par composition des automorphismes. On notera $u * v$ $(= v \circ u)$ le composé de u et v dans le groupe opposé $\mathrm{Aut}^o(G)$.

Vérifions la propriété d'équivariance de la flèche j : on considère l'équation :

$$(2.5.2) \qquad (g_1p)g = j(g_1p)(g)g_1p$$

de type (2.5.1) qui caractérise l'élément $j(g_1p)(g)$ de G, et on la compare à l'équation :

$$(2.5.3) \qquad g_1(pg) = g_1j(p)(g)p$$

obtenu à partir de (2.5.1) par multiplication à gauche par g_1. De la comparaison des termes de droite des équations (2.5.2) et (2.5.3) résulte que $j(g_1p) = i_{g_1} \circ j(p)$, ce qui est l'assertion d'équivariance souhaitée. Inversement, si l'application $j: P \to \mathrm{Aut}(G)$ est donnée, la formule (2.5.1) définit une action à droite de G sur P et l'équivariance de j implique que cette action commute à celle de gauche. Elle fait de P un G-torseur à droite, puisque j est à valeurs dans le faisceau des automorphismes de G, ce qui termine la démonstration.

Exemples 2.6. (i) Si P est le torseur trivial de G, la donnée d'un morphisme G-équivariant $j: G \to \mathrm{Aut}(G)$ équivaut à celle d'une section $u = j(1)$ de $\mathrm{Aut}(G)$. La structure de bitorseur sur G est alors celle de l'exemple 2.2. iii).

(ii) Si $G = A$ est un groupe abélien de T, l'application constante $j: P \to \mathrm{Aut}(A)$, d'image l'élément neutre de $\mathrm{Aut}(A)$, est A-équivariante, et définit sur P la structure de bitorseur donnée par (2.2.1). Toute autre application constante j (il y en a autant que de section de $\mathrm{Aut}(A)$) définit sur P une structure de bitorseur, qui diffère de la précédente.

Remarque 2.7. Soient, dans la terminologie du lemme 2.5, (P_1, j_1) et (P_2, j_2) deux G-bitorseurs et $f: P_1 \to P_2$ un morphisme de G-torseurs à gauche. Alors f est un morphisme de G-bitorseurs si et seulement si le diagramme :

$$(2.7.1) \qquad \begin{array}{ccc} P_1 & \xrightarrow{f} & P_2 \\ & {\scriptstyle j_1}\searrow \quad \swarrow{\scriptstyle j_2} & \\ & \mathrm{Aut}(G) & \end{array}$$

est commutatif.

La proposition 2.4 permet de comprendre quelle est la fonctorialité de $\mathrm{BITORS}(G)$. On prendra garde que, malgré la notation adoptée,

BITORS(G) n'est pas fonctorielle en le groupe G ! L'énoncé minimal de fonctorialité est le suivant :

Lemme 2.8. *Soient $u: G \to G'$ un homomorphisme de T et $v: \mathrm{Aut}(G) \to \mathrm{Aut}(G')$ un morphisme u-équivariant de T, où G (resp. G') agit à droite sur $\mathrm{Aut}^o(G)$ (resp. $\mathrm{Aut}^o(G')$) à travers i, c'est à dire par $(f,g) \to f * i'_g = i'_g \circ f$. Alors, la paire (u,v) définit un morphisme de champs ("extension du groupe structural") :*

$$(u,v)_*: \mathrm{BITORS}(G) \longrightarrow \mathrm{BITORS}(G')$$

qui est compatible, par oubli de structure, avec le morphisme $u_*: \mathrm{TORS}(G) \to \mathrm{TORS}(G')$ d'extension du groupe structural pour les G-torseurs à gauche.

En effet, la flèche canonique extension du groupe structural $\lambda: P \to u_*P = G' \overset{G}{\wedge} P$ est u-équivariante à gauche, par construction. Puisque v l'est à droite, la paire (v, λ) induit ([26], III.1.3.2) un morphisme :

$$(v,\lambda)_*: \mathrm{Aut}^o(G) \overset{G}{\wedge} P \to \mathrm{Aut}^o(G') \overset{G'}{\wedge} u_*P.$$

La section donnée de $\mathrm{Aut}^o(G) \overset{G}{\wedge} P$ s'envoie alors sur une section de $\mathrm{Aut}^o(G) \overset{G'}{\wedge} u_*P$, qui définit une structure de G'-bitorseur sur u_*P. En terme plus concrets, $(u,v)_*P$ est donc le G-torseur à gauche u_*P, sur lequel G' agit à droite par la règle :

$$(g',p)g'_1 = (g'v(j(p))(g'_1),p) \tag{2.8.1}$$

pour $g', g'_1 \in G'$, $p \in P$, où $j: P \to \mathrm{Aut}(G)$ est définie par (2.5.1).

Il importe de remarquer que, sous les hypothèses précédentes, le morphisme $(u,v)_*$ ne respecte en général pas le produit contracté. Pour obtenir cette propriété, il convient de renforcer nos hypothèses sur la paire (u,v) :

Proposition 2.9. *Soient $u: G \to G'$, et $v: \mathrm{Aut}(G) \to \mathrm{Aut}(G')$ une paire d'homomorphismes de T tels que :*

(i) *v soit u-équivariant à travers i', c'est à dire que le diagramme :*

$$\begin{array}{ccc} G & \xrightarrow{u} & G' \\ {\scriptstyle i'}\downarrow & & \downarrow{\scriptstyle i'} \\ \mathrm{Aut}^o(G) & \xrightarrow{v} & \mathrm{Aut}^o(G') \end{array} \tag{2.9.1}$$

ou encore,

$$\begin{array}{ccc} G & \xrightarrow{u} & G' \\ {\scriptstyle i}\downarrow & & \downarrow{\scriptstyle i} \\ \mathrm{Aut}(G) & \xrightarrow{v} & \mathrm{Aut}(G') \end{array} \tag{2.9.2}$$

soit commutatif.

(ii) *La flèche u est v-équivariante, lorsque* $\mathrm{Aut}^o(G)$ (*resp.* $\mathrm{Aut}^o(G')$) *agit à droite sur G (resp. G') de la manière habituelle (c'est à dire par $(g,f) \to f(g)$).*

Alors le morphisme $(u,v)_*$ est un foncteur additif au sens de [16] *pour le produit contracté des bitorseurs et définit donc un morphisme de Gr-champs :*

$$(u,v)_*\colon \mathrm{BITORS}(G) \longrightarrow \mathrm{BITORS}(G'). \tag{2.9.2}$$

Pour le démontrer, on vérifie tout d'abord que le morphisme canonique $\lambda\colon P \to u_*P$, qui est, comme on l'a déjà dit, u-équivariant pour l'action à gauche de G et G', l'est également pour l'action à droite. En effet :

$$\lambda(pg) = \lambda(j(p)(g)p) = u(j(p)(g))\lambda(p) = (u(j(p)(g)), p)$$

tandis que :

$$\lambda(p)u(g) = (1,p)u(g) = (v(j(p))(u(g)), p)$$

la dernière égalité provenant de (2.8.1). La condition (ii) montre que ces deux expressions sont égales. Ainsi, pour toute paire P_1, P_2 de G-bitorseurs, on dispose de flèches canoniques $\lambda_1\colon P_1 \to u_*P_1$ et $\lambda_2 : P_2 \to u_*P_2$ qui sont respectivement u-équivariante à droite et à gauche, et qui induisent donc une flèche :

$$\lambda_1 \wedge \lambda_2\colon P_1 \overset{G}{\wedge} P_2 \longrightarrow u_*P_1 \overset{G'}{\wedge} u_*P_2 \tag{2.9.3}$$

qui est en outre u-équivariante à gauche puisque λ_1 l'est. La flèche (2.9.3) induit donc un morphisme (c'est à dire un isomorphisme) de G'-torseurs à gauche :

$$u_*(P_1 \overset{G}{\wedge} P_2) \xrightarrow{\sim} u_*P_1 \overset{G'}{\wedge} u_*P_2. \tag{2.9.4}$$

Il reste à démontrer que cette flèche respecte la G'-structure à droite. Pour cela, on observe que si $j_1 : P_1 \to \mathrm{Aut}(G)$ et $j_2 : P_2 \to \mathrm{Aut}(G)$ sont deux morphismes qui décrivent par (2.5.1) des structures de bitorseurs sur les torseurs à droite P_1 et P_2, alors la flèche :

$$j = \text{“}j_1 \circ j_2\text{”} : P_1 \overset{G}{\wedge} P_2 \longrightarrow \mathrm{Aut}(G) \tag{2.9.5}$$

qui envoie (p_1, p_2) sur $j_1(p_1) \circ j_2(p_2)$ est bien définie et décrit la structure de bitorseur du produit contracté de P_1 et de P_2. En effet, pour tout $p_1 \in P_1$, $p_2 \in P_2$, $g \in G$, on a :

$$\begin{aligned} (p_1, p_2)g &= (p_1, j_2(p_2)(g)p_2) = (p_1 j_2(p_2)(g), p_2) \\ &= (j_1(p_1)(j_2(p_2)(g))p_1, p_2) = (j_1(p_1) \circ j_2(p_2))(g)(p_1, p_2) \end{aligned} \tag{2.9.6}$$

ce qui décrit de la manière dite l'action à droite de G sur $P_1 \overset{G}{\wedge} P_2$. Si on applique cette remarque aux paires de torseurs (P_1, P_2) et $(u_* P_1,\ u_* P_2)$, on constate, compte tenu de la remarque 2.7, que l'isomorphisme (2.9.4) préserve la G'-structure à gauche si et seulement si le diagramme :

$$\begin{array}{ccc} P_1 \overset{G}{\wedge} P_2 & \xrightarrow{\text{“}j_1 \circ j_2\text{”}} & \mathrm{Aut}(G) \\ {\scriptstyle \lambda_1 \wedge \lambda_2}\downarrow & & \downarrow{\scriptstyle v} \\ u_* P_1 \overset{G'}{\wedge} u_* P_2 & \xrightarrow[\text{“}(v \circ j_1) \circ (v \circ j_2)\text{”}]{} & \mathrm{Aut}(G') \end{array}$$

commute, ce qui est bien le cas si v est un homomorphisme.

2.10. Les résultats précédents s'étendent au cas des (G, H)-bitorseurs. Ainsi, si P est le G-torseur à gauche sous-jacent à un (G, H)-bitorseur, la multiplication à droite par G est décrite par un morphisme :

$$j : P \longrightarrow \mathrm{Isom}(H, G) \tag{2.10.1}$$

à valeurs dans le faisceau des isomorphismes de groupe de H vers G, défini par la règle :

$$ph = j(p)(h)p \tag{2.10.2}$$

qui généralise (2.5.1). Le raisonnement précédent montre que j est G-équivariant, lorsque G agit à gauche sur $\mathrm{Isom}(H, G)$ par la règle $(g, u) \to i_g \circ u$. On invoque à nouveau [26] III.1.6.1 (ii) pour conclure que la donnée d'un morphisme j équivaut à celle d'une section de l'espace fibré $\mathrm{Isom}(H,G) \overset{G}{\wedge} P$ de fibre $\mathrm{Isom}(H,G)$ associé à P, pour l'action à droite de G sur $\mathrm{Isom}(H,G)$ définie par $(u,g) \to i'_g \circ u$. En résumé :

Proposition 2.11. *Soient G, H deux groupes de T. La donnée d'un (G,H)-bitorseur de T équivaut à celle d'un G-torseur à gauche P, et d'une section de l'espace fibré $\mathrm{Isom}(H,G) \overset{G}{\wedge} P$ associé à P, pour l'action à droite de G sur $\mathrm{Isom}(H,G)$ définie par $(u,g) \to i'_g \circ u$.*

On en déduit l'énoncé suivant, d'apparence plus générale :

Proposition 2.12. *Soit G un groupe de T, et X un $\mathrm{Aut}^o(G)$-torseur à droite (sur lequel on fait agir G à droite à travers i'). Alors la donnée d'un G-torseur à gauche P, et d'une section de l'espace fibré $X \overset{G}{\wedge} P$ de fibre X associé à P, équivaut à celle d'une structure de $(G, {}^X G)$-bitorseur sur P, où ${}^X G = X \overset{\mathrm{Aut}^o(G)}{\wedge} G$ est le groupe obtenu en tordant par X le groupe G (sur lequel $\mathrm{Aut}^o(G)$ agit à gauche par $(g,u) \to u^{-1}(g)$).*

C'est en effet une simple reformulation de la proposition précédente : par définition du groupe tordu ${}^X G$ (voir [26] 1.6.7, 2.3.6, 2.3.2) on dispose d'une flèche $\mathrm{Aut}^o(G)$-équivariante $m_X : X \to \mathrm{Bij}(G, {}^X G)$* où $\mathrm{Aut}^o(G)$ agit à droite sur $\mathrm{Bij}(G, {}^X G)$ par $(u,f) \to f o u^{-1}$. Par *loc. cit.*, puisque $\mathrm{Aut}^o(G)$ agit sur G par automorphismes de groupes, m_X se factorise par un morphisme (donc un isomorphisme) d' $\mathrm{Aut}^o(G)$-torseurs :

$$m_X : X \longrightarrow \mathrm{Isom}(G, {}^X G).$$

En composant avec la flèche qui associe à un isomorphisme son inverse, on trouve donc un isomorphisme d' $\mathrm{Aut}^o(G)$-torseurs à droite :

$$n_X : X \xrightarrow{\sim} \mathrm{Isom}({}^X G, G)$$

où $\mathrm{Aut}^o(G)$ agit à droite sur $\mathrm{Isom}({}^X G, G)$ par $(f,u) \to u \circ f$. On retrouve la proposition précédente en posant $H = {}^X G$.

Les propositions 2.11 et 2.12 permettent de comprendre quelle est la fonctorialité du champ $\mathrm{BITORS}(G,H)$ par rapport aux groupes G et H. Commençons par énoncer cette fonctorialité au moyen de la proposition

* On note $\mathrm{Bij}(G_1, G_2)$ plutôt que $\mathrm{Isom}(G_1, G_2)$ le faisceau des bijections entre les ensembles sous-jacents à G_1 et G_2, pour ne pas le confondre avec le faisceau des isomorphismes de groupes.

2.12. Soit donc X un $\mathrm{Aut}^o(G)$-torseur à droite (resp. X' un $\mathrm{Aut}^o(G')$-torseur à droite). On se donne une paire de flèches (u,w), où $u: G \to G'$ est un homomorphisme et $w: X \to X'$ est une application u-équivariante (pour l'action à droite de G (resp. G') sur X (resp. X') à travers i'). On en déduit alors, pour tout G-torseur à gauche P, un morphisme $(u,w)_*: X \overset{G}{\wedge} P \longrightarrow X' \overset{G'}{\wedge} P'$ où $P' = u_*(P)$ s'obtient à partir de P par extension du groupe structural. Cette flèche envoie une section donnée de $X \overset{G}{\wedge} P$ vers une section de $X' \overset{G'}{\wedge} P'$. On en déduit, par la proposition 2.12 :

Proposition 2.13. *Soient (G,X) et (G',X') deux paires d'objets de T du type qui vient d'être indiqué. Un homomorphisme $u: G \to G'$ et une flèche u-équivariante $w: X \to X'$ définissent un morphisme de champs*

$$(u,w)_*: \mathrm{BITORS}(G, {}^X G) \longrightarrow \mathrm{BITORS}(G', {}^{X'} G')$$

qui induit, par oubli de structure, le morphisme d'extension du groupe structural

$$u_*: \mathrm{TORS}(G) \longrightarrow \mathrm{TORS}(G') \tag{2.13.1}$$

sur les torseurs à gauche.

Si l'on traduit cet énoncé dans le langage des (G,H)-bitorseurs en posant $H = {}^X G$ (resp. $X = \mathrm{Isom}(H,G)$), on obtient la généralisation suivante du lemme 2.8 :

Proposition 2.14. *Soient (G,H) et (G',H') deux paires de groupes de T. On suppose données un homomorphisme $u: G \to G'$ et une application $w: \mathrm{Isom}(H,G) \to \mathrm{Isom}(H',G')$ qui soit u-équivariante (pour l'action à droite de G (resp. G') sur $\mathrm{Isom}(H,G)$ (resp. $\mathrm{Isom}(H',G')$) à travers i', définie par $(f,g) \to f * i'_g = i'_g \circ f$. Alors la paire (u,w) induit un morphisme de champs.*

$$(u,w)_*: \mathrm{BITORS}(G,H) \longrightarrow \mathrm{BITORS}(G',H')$$

dont on déduit le morphisme (2.13.1) par oubli de structure.

On remarquera que pour tout (G,H)-bitorseur P, la structure de H'-torseur à droite ainsi obtenue sur u_*P est donnée explicitement par la formule suivante, qui généralise (2.8.1)

$$(g',p)h' = (g'(w \circ j)(p)(h'),p). \tag{2.14.1}$$

Remarque 2.15. (i) Cette discussion mériterait d'être complétée par l'examen des conditions à imposer à (u, w) pour que le foncteur $(u, w)_*$ soit compatible au produit contracté.

(ii) On renvoie à (4.5) pour une dernière variante des propositions 2.4 et 2.12.

(iii) Pour ne pas interrompre le cours de la discussion, on a omis de signaler en 2.10 le fait que, s'il existe un (G, H)-bitorseur P de T, alors les groupes G et H sont localement isomorphes, un tel isomorphisme étant défini par la section $f(p)$ de $\mathrm{Isom}(H, G)$ associée à une section locale donnée p de P. Par (2.10.2), $f(p)$ peut être considéré comme l'opérateur de "conjugaison par p", qui envoie $h \in H$ vers l'élément "php^{-1}" de G.

2.16. La proposition (2.4) a diverses conséquences, qui s'énoncent très aisément en termes cohomologiques. Considérons tout d'abord un homomorphisme quelconque $\partial: G_1 \to G_0$ de T, et notons $H^0(T, G_1 \to G_0)$ (ou simplement $H^0(G_1 \to G_0)$) l'ensemble des classes d'isomorphismes de paire (P, s), où P est un G_1-torseur à gauche et s une section du G_0-torseur induit $\partial_* P$. Cet ensemble est pointé par la paire triviale (G_1, e) et on note de même $H^{-1}(G_1 \to G_0)$ le groupe des automorphismes de cette objet trivial.* On sait (voir par exemple [17] 2.4.3.1) que ces ensembles de cohomologie vivent dans une suite exacte (dont les trois premiers termes sont en fait des groupes et des ensembles pointés)

$$(2.16.1) \qquad \begin{aligned} &0 \to H^{-1}(T, G_1 \to G_0) \to H^0(T, G_1) \xrightarrow{\partial_*} H^0(T, G_0) \\ &\xrightarrow{\alpha} H^0(T, G_1 \to G_0) \xrightarrow{\beta} H^1(T, G_1) \xrightarrow{\partial_*} H^1(T, G_0). \end{aligned}$$

Rappelons que la flèche α envoie une section s de G_0 vers le G_1-torseur trivial, muni de la trivialisation de $\partial_*(G_1) = G_0$ définie par s, tandis que β est définie par oubli de structure. En particulier, à l'homomorphisme $i': G \to \mathrm{Aut}^o(G)$ de T associé à un groupe G correspond, compte tenu de la proposition 2.4, une suite exacte:

$$(2.16.2) \qquad \begin{aligned} &0 \to \pi_1(\mathrm{BITORS}(G)) \to H^0(G) \xrightarrow{H^0(i')} H^0(\mathrm{Aut}^o(G)) \\ &\xrightarrow{\alpha} \mathrm{Bitors}(G) \to \mathrm{Tors}(G) \xrightarrow{H^1(i')} \mathrm{Tors}(\mathrm{Aut}^o(G)) \end{aligned}$$

où $\mathrm{Bitors}(G)$ (resp. $\mathrm{Tors}(G)$) désigne l'ensemble $\pi_0(\mathrm{BITORS}(G))$ (resp. $\pi_0(\mathrm{TORS}(G))$) des classes d'isomorphismes de G-bitorseurs (resp. de G-torseurs) de T. On a déjà observé (2.3) que le premier terme de cette suite

* Les objets similaires définis à partir de torseurs à droite sont respectivement notés $H^1(G_1 \to G_0)$ et $H^0(G_1 \to G_0)$ dans [17] 2.4.3.

est un groupe abélien, et on constate maintenant, par exactitude de la suite en question, que ce n'est autre que le groupe $H^0(ZG)$ des section du centre $ZG = \ker i'$ du groupe G. Ici la flèche α envoie une section u de $\mathrm{Aut}(G)$ vers le bitorseur noté (G, u) en (2.2.iii) et l'exactitude en $\mathrm{Bitors}(G)$ de la suite (2.16.2) montre que le procédé de construction indiqué en (2.2.2) pour les G-bitorseurs à G-torseur à gauche sous-jacent trivial est le seul possible. L'exactitude en $H^0(\mathrm{Aut}^o(G))$ nous indique enfin qu'un tel bitorseur (G, u) n'est isomorphe au bitorseur trivial G_{triv} que si u est intérieur, de la forme $u = i'_{g_1}$ et on constate que, dans ce cas, le morphisme de translation à droite par l'élément g_1 de G définit bien un isomorphisme $G_{triv} \to (G, u)$ de G-bitorseurs.

3. Topologie générale

L'interprétation de $\mathrm{BITORS}(G)$ qui vient d'être donnée, et la suite exacte (2.16.2) qui en découle, comportent deux lacunes. Tout d'abord, elles ne prennent pas en compte la loi de groupe de $\mathrm{Bitors}(G)$ introduite précédemment. D'autre part, on souhaite prolonger la suite exacte en question d'un cran vers la droite, mais ceci n'est pas réalisable dans le cadre étroit proposé ci-dessus, et pour y aboutir de la manière la plus satisfaisante, il nous faut introduire différentes considérations d'algèbre simpliciale qui sont en principe bien connues. Il a néanmoins paru nécessaire d'en rappeler l'essentiel ici, à la fois pour la commodité du lecteur peu familiarisé avec la topologie simpliciale dans le cadre des topos, et parce qu'il n'existe pas de références détaillées auxquelles on pourrait le renvoyer. Signalons cependant d'ores et déjà que le but recherché ici est la construction de la suite exacte homotopique (3.9.1) : le lecteur pour qui cette suite est familière (ou qui en connait l'existence dans le cadre de la topologie usuelle et qui est prêt à admettre qu'elle existe également dans le cadre des topos) est invité à se reporter directement au prochain paragraphe.

3.1. La suite exacte (2.16.2) a été construite à partir de l'homomorphisme de groupe $i' : G \to \mathrm{Aut}^o(G)$, mais ne prenait pas en compte l'action à droite naturelle de $\mathrm{Aut}^o(G)$ sur G. Lorsque l'on introduit celle-ci, on est en présence d'un module croisé (au sens de J.H.C. Whitehead [60]), dont voici la définition :

Définition 3.2. On appelle module croisé à droite (ou simplement module croisé) de T la donnée d'un homomorphisme $\partial : G_1 \to G_0$ de T et d'une action à droite de G_0 sur G_1, notée $(g_1, g_0) \to g_1^{g_0}$, telle que :

$$\partial(g_1^{g_0}) = \partial(g_1)^{g_0} \tag{3.2.1}$$

$$(g_1')^{\partial g_1} = (g_1')^{g_1} \tag{3.2.2}$$

où pour tous éléments x, y d'un groupe G, on pose $y^x = i'_x(y) = x^{-1}yx$.

De la même manière, un morphisme $(f_1, f_0): (G_1 \xrightarrow{\partial} G_0) \to (G'_1 \xrightarrow{\partial'} G'_0)$ de modules croisés consiste en la donnée d'une paire d'homomorphismes $(f_1: G_1 \to G'_1,\ f_0: G_0 \to G'_0)$ tels que le diagramme :

$$\begin{array}{ccc} G_1 & \xrightarrow{f_1} & G'_1 \\ {\scriptstyle\partial}\big\downarrow & & \big\downarrow{\scriptstyle\partial} \\ G_0 & \xrightarrow[f_0]{} & G'_0 \end{array}$$

commute, et que l'homomorphisme f_1 soit f_0-équivariant. Enfin (voir [9]) on appellera homotopie entre les paires (f_1, f_0) et $(\bar{f}_1, \bar{f}_0)$ la donnée d'un application $h: G_0 \to G'_1$ telle que :

$$h(g_1 g_2) = (h(g_1)^{\bar{f}_0(g_2)})h(g_2) \tag{3.2.3}$$

$$f_0(g) = \bar{f}_0(g)\partial h(g) \tag{3.2.4}$$

$$f_1(g) = \bar{f}_1(g)h(\partial g) \tag{3.2.5}$$

On peut, si on préfère, considérer des modules croisés gauches, définis par un homomorphisme $\partial_1: H_1 \to H_0$ et une action à gauche de H_0 sur H_1, notée $(h_0, h_1) \to {}^{h_0}h_1$, telle que les conditions :

$$\partial_1({}^{h_0}h_1) = i_{h_0}(h_1) \tag{3.2.6}$$

$${}^{\partial_1 h_1}(h'_1) = i_{h_1}(h'_1) \tag{3.2.7}$$

similaires à (3.2.1)–(3.2.2) soient satisfaites. A chaque module croisé droit $\partial: G_1 \to G_0$ correspond un module croisé gauche $\partial_1: G_1 \xrightarrow{\partial} G_0 \xrightarrow{-1} G_0^o$ (le groupe $H_0 = G_0^o$ agissant à gauche sur $H_1 = G_1$ par ${}^{g_0}g_1 =: g_1^{g_0}$) et réciproquement. Sauf mention du contraire, nous considérerons toujours les modules croisés droits.

3.3. Soit $\mathcal{G}$ un groupe simplicial et T, et $N\mathcal{G}$ son complexe normalisé (ou complexe de Moore). Rappelons qu'il s'agit là du complexe de groupe défini par :

$$(N\mathcal{G})_i = \bigcap_{j<i} \ker(d_j: \mathcal{G}_i \to \mathcal{G}_{i-1}),$$

la différentielle $\partial\colon (N\mathcal{G})_i \rightarrow (N\mathcal{G})_{i-1}$ s'obtenant par restriction de $d_i\colon \mathcal{G}_i \rightarrow \mathcal{G}_{i-1}$ à $(N\mathcal{G})_i$. Les relations entre les opérateurs face impliquent que $N\mathcal{G}$ est un complexe, et l'on dispose maintenant d'une définition très élémentaire des groupes d'homotopie du groupe simplicial $\mathcal{G}$. On pose en effet (voir [45] proposition 17.4) :

$$\pi_i(\mathcal{G}) = H_i(N\mathcal{G}) \qquad i \geq 0.$$

Considérons maintenant un groupe simplicial $\mathcal{G}$ de T pour lequel $\pi_i(\mathcal{G}) = 0$ lorsque $i \geq 2$. Il lui correspond donc un complexe normalisé $N\mathcal{G}$, qui est exact sauf en degrés 0 et 1. Puisque l'image de $\partial\colon (N\mathcal{G})_2 \rightarrow (N\mathcal{G})_1$ est un sous-groupe distingué de $(N\mathcal{G})_1$, on peut alors associer à $N\mathcal{G}$ le complexe tronqué de longueur 1 $N'\mathcal{G}$ défini par :

$$(N\mathcal{G})_1/\operatorname{Im}\partial \xrightarrow{\partial} (N\mathcal{G})_0 = \mathcal{G}_0$$

de noyau (resp. de conoyau) $\pi_1\mathcal{G}$ (resp. $\pi_0\mathcal{G}$). Le complexe $N'\mathcal{G}$ est muni d'une structure naturelle de module croisé : si l'on fait agir $\mathcal{G}_0$ sur $\mathcal{G}_1$, et notamment sur $(N\mathcal{G})_1$ (resp. $(N'\mathcal{G})_1$), par :

(3.3.1) $$y^x = s_0(x)^{-1} y\, s_0(x)$$

la condition (3.2.1) est immédiatement vérifiée, tandis que (3.2.2) provient, par restriction à $(N\mathcal{G})_1$, de l'identité suivante, satisfaite par toute paire d'éléments u, v de $\mathcal{G}_1$:

(3.3.2) $$u^{-1}vu = (v^{d_0 u})d_0(z)$$

où z est l'élément de $\mathcal{G}_2$ défini par $z = (s_0(v)^{s_1(u)})^{-1} s_0(v^u)$. L'identité (3.3.2) résulte en effet directement des identités simpliciales entre opérateurs face et dégénérescence (un élément tel que z étant parfois appelé le relèvement de Peiffer de u et v, *cf.* [12] définition 2.1).

Si $f\colon \mathcal{G} \rightarrow \mathcal{G}'$ est un morphisme de groupe simpliciaux, alors la flèche induite $N'f\colon N'\mathcal{G} \rightarrow N'\mathcal{G}'$ est un morphisme de modules croisés, et on vérifie que si $h\colon f' \xrightarrow{\sim} f'$ est une homotopie entre deux morphismes $f, f'\colon \mathcal{G} \rightarrow \mathcal{G}'$ au sens de [45] définition 5.1, alors il lui correspond une homotopie entre les morphismes $N'f$ et $N'f'$.

En fait, et c'est là l'intérêt de la notion de module croisé, toute l'information homotopique concernant un groupe simplicial $\mathcal{G}$ satisfaisant à la condition

(3.3.3) $$\pi_i(\mathcal{G}) = 0 \quad \text{pour} \quad i \geq 2$$

s'exprime en termes de l'objet bien plus simple qu'est le module croisé associé $N'\mathcal{G}$. On peut notamment, à partir d'un module croisé quelconque $\partial: G_1 \to G_0$ de T, reconstruire un groupe simplicial, obtenant ainsi un foncteur quasi-inverse à N' dans la catégorie homotopique. Pour cela, on commence par associer à $\partial: G_1 \to G_0$ le groupe simplicial tronqué au niveau 1 :

$$G_0 \ltimes G_1 \overset{\longrightarrow}{\underset{\longrightarrow}{\longleftarrow}} G_0 \tag{3.3.4}$$

le produit semi-direct étant défini au moyen de l'action à droite donnée de G_0 sur G_1. Ici, l'homomorphisme d_1 (resp. s_0) est la projection (resp. l'injection) canonique, tandis que l'on pose :

$$d_0(g_0, g_1) = g_0(\partial g_1). \tag{3.3.5}$$

C'est une conséquence de (3.2.2) que les éléments des sous-groupes $\ker(d_0)$ et $\ker(d_1)$ de $G_0 \times G_1$ commutent entre eux (dans la terminologie de [42], $(G_0 \times G_1, d_0, d_1)$ est un 1-cat groupe). On sait alors (voir *loc.cit.* lemme 2.2 ou [12] théorème 1.5) qu'il existe un unique groupe simplicial $\mathcal{G}$, de tronqué le groupe (3.3.4), et dont le normalisé est le complexe $\partial: G_1 \to G_0$ de départ. Le groupe simplicial en question s'explicite en définissant, par récurrence sur i, un groupe $\mathcal{G}_i$, muni d'un homomorphisme $u_i: \mathcal{G}_i \to \mathcal{G}_0 = G_0$, de la manière suivante. On pose :

$$\mathcal{G}_i = \mathcal{G}_{i-1} \ltimes G_1 \tag{3.3.6}$$

(muni de la loi de groupe du produit semi-direct, pour l'action à droite de $\mathcal{G}_{i-1}$ à travers u_{i-1}) et l'on définit u_i par $u_0 = Id_{G_0}$ et, pour $y \in \mathcal{G}_{i-1}$, $g_1 \in G_1$, par :

$$u_i(y, g_1) = u_{i-1}(y)\partial(g_1).$$

Le i-ème composante de $\mathcal{G}$ est, en définitive, un produit semi-direct itéré

$$\mathcal{G}_i = (((G_0 \ltimes G_1) \ltimes G_1) \ldots \ltimes G_1) \tag{3.3.7}$$

d'ensemble sous-jacent $G_0 \times G_1^i$, u_i étant définie par

$$u_i(g_0, g_1, \ldots, g_i) = g_0\partial(g_1 \ldots g_i). \tag{3.3.8}$$

Quand aux opérateurs face et dégénérescence, ce sont les homomorphismes $d_j: \mathcal{G}_j \to \mathcal{G}_{j-1}$ et $s_j: \mathcal{G}_j \to \mathcal{G}_{j+1}$ donnés pour $g_0 \in G_0$, $g_1, \ldots, g_i \in G_1$ par les formules

$$\begin{aligned} d_j(g_0, g_1, \ldots, g_i) &= (g_0\partial g_1, \ldots, g_i) && j = 0 \\ &= (g_0, \ldots, g_j g_{j+1}, \ldots, g_i) && 1 \leq j < i \\ &= (g_0, g_1, \ldots, g_{i-1}) && j = i \end{aligned} \tag{3.3.9}$$

$$s_j(g_0, g_1, \ldots, g_i) = (g_0, g_1, \ldots, g_{j-1}, e, g_{j+1}, \ldots, g_i) \quad 0 \leq j \leq i.$$

Une autre manière de comprendre la construction précédente est la suivante : au module croisé $\partial: G_1 \to G_0$ est associé un groupoïde $\mathcal{C}$, telle que $\mathrm{ob}(\mathcal{C}) = G_0$, tandis qu'une flèche $(g_0, g_1): g_0 \to g_0(\partial g_1)$ est définie par un élément (g_0, g_1) du produit semi-direct introduit en (3.3.4). La composition des flèches est définie en prenant pour flèche composée $g_0 \xrightarrow{(g_0,g_1)} g_0(\partial(g_1)) \xrightarrow{g_0(\partial g_1), g_2} g_0 \partial g_1 \partial g_2$, la flèche $(g_0, g_1 g_2)$. $\mathcal{C}$ est alors une gr-catégorie (stricte en ce sens que les lois d'associativité sont strictement satisfaites), pour la loi de groupe définie sur les objets par celle de G_0 et sur les flèches par la loi de groupe du produit semi-direct $G_0 \times G_1$. Le nerf du groupoïde $\mathcal{C}$ est alors le groupe simplicial $\mathcal{G}$ décrit ci-dessus.

Dans le cas où $\partial: G_1 \to G_0$ est un homomorphisme de groupes abéliens, la décomposition (3.3.7) de $\mathcal{G}_i$ en produits est directe et l'on retrouve ici le transformé de Dold–Puppe du complexe de longueur 1 défini par ∂, tandis que la gr-catégorie associée est la gr-catégorie sous-jacente à la catégorie de Picard considérée en [16] 1.4.10. En particulier, si le module croisé est concentré en degré -1, (c'est à dire qu'il est de la forme $\partial: G_1 \to (e)$ où G_1 est un groupe nécéssairement abélien), alors le groupe simplicial associé $\mathcal{G}$ n'est autre que le classifiant BG_1 de G_1, muni de sa structure de groupe usuelle. Par ailleurs, au module croisé discret $e \to G_0$ correspond le groupe simplicial constant défini par le groupe G_0.

3.4. Une autre description peut être donné de l'ensemble simplicial sous-jacent au groupe simplicial $\mathcal{G}$ associé à un module croisé $\partial: G_1 \to G_0$. On rappelle tout d'abord qu'à tout groupe G de T correspond un groupe simplicial $EG = \mathrm{cosq}^o G$, de $i$ème composante $(EG)_i = \mathrm{Hom}(\Delta_i, G) = G^{i+1}$ (Δ_i désignant l'ensemble fini totalement ordonné non vide $[0, i]$). Les applications croissantes $\Delta_j \to \Delta_i$ définissent par composition les opérateurs face et dégénérescence. Le groupe simplicial constant G agit sur EG par $g(g_0, \ldots, g_i) = (gg_0, \ldots, gg_i)$. Soit alors X un objet de T sur lequel le groupe G opère à droite. On associe à la donnée (X, G) le produit contracté $X \overset{G}{\wedge} EG$, muni de la structure d'ensemble simplicial induite par celle de EG. Il s'agit là d'un objet bien connu des topologues, et dont la cohomologie à valeurs dans un ensemble (resp. groupe, resp. groupe abélien) de T mérite le nom de "cohomologie G-équivariante de X" (voir [47] §1). Notons par ailleurs que la construction de cet objet est fonctorielle en (X, G) : à tout homomorphisme $u: G \to G'$ et à toute application $w: X \to X'$ u-équivariante est associé un morphisme d'objets simpliciaux de T :

$$(u, w)_*: X \overset{G}{\times} EG \to X' \overset{G'}{\times} EG'.$$

Une seconde description de l'objet $X \overset{G}{\times} EG$ s'obtient en observant que sa composante de degré i peut être identifiée à l'ensemble $X \times G^i$ par la flèche :

$$X \overset{G}{\times} G^{i+1} \to X \times G^i \tag{3.4.1}$$

définie par $(x, g_1, \ldots, g_{i+1}) \to (xg_1, g_1^{-1}g_2, \ldots, g_i^{-1}g_{i+1})$, l'application inverse étant donnée par

$$(x, \gamma_1, \ldots, \gamma_i) \to (x, 1, \gamma_1, \gamma_1\gamma_2, \ldots, \gamma_1 \ldots \gamma_i).$$

Par transport de structure, on obtient ainsi un objet simplicial, de $i$ème composante $X \times G^i$, pour lequel la structure simpliciale est définie par les formules :

$$\begin{aligned} d_j(x, g_1, \ldots, g_i) &= (xg_1, \ldots, g_i) & j = 0 \\ &= (x, g_1, \ldots, g_jg_{j+1}, \ldots, g_i) & 1 \leq j \leq i \\ &= (x, g_1, \ldots, g_{i-1}) & j = i \end{aligned} \tag{3.4.2}$$

$$s_j(x, g_1, \ldots, g_i) = (x, g_1, \ldots, g_{j-1}, e, g_{j+1}, \ldots, g_i) \qquad 0 \leq j \leq i.$$

En particulier, lorsque X est l'objet final e de T, la flèche (3.4.1) identifie $e \overset{G}{\times} EG$ avec le classifiant BG de G et la projection canonique $p: EG \to BG$, définie par :

$$p(g_1, \ldots, g_{i+1}) = (g_1^{-1}g_2, \ldots, g_i^{-1}g_{i+1}) \tag{3.4.3}$$

fait de EG le G-torseur gauche universel sur BG. Plus généralement, la flèche G-équivariante qui envoie un G-objet X vers le G-objet final e induit une projection $q: X \overset{G}{\wedge} EG \to BG$ par laquelle $X \overset{G}{\wedge} EG$ peut être considéré comme l'espace fibré de fibre X associé au G-torseur gauche universel. Le lemme suivant résulte de la comparaison de formules (3.3.9) et (3.4.2) :

Lemme 3.5. *Soit $\partial: G_1 \to G_0$ un module croisé, de groupe simplicial associé $\mathcal{G}$. Alors les flèches* (3.4.1) *identifient l'objet simplicial $G_0 \overset{G_1}{\times} EG_1$ de T (pour l'action à droite de G_1 sur G_0 à travers ∂) avec l'ensemble simplicial sous-jacent à $\mathcal{G}$.*

3.6. Nous abordons, après ces rappels, la description promise des termes de la suite exacte (2.16.1). A un homomorphisme $\partial: G_1 \to G_0$ de groupes de T correspond un morphisme d'espaces classificants

$B\partial\colon BG_1 \to BG_0$. Ceci est un morphisme d'objet simpliciaux pointés de T. On observera d'autre part que les objets simpliciaux en question satisfont à la condition de Kan (définie, dans le cas d'un topos dans [35] 2.3.8, voir également [36] où on les appelle les faisceaux simpliciaux fibrants). En effet, la variante faisceautique de [45] théorème 17.1 montre que le groupe simplicial EG satisfait à la condition de Kan. Par ailleurs, le lemme 18.2 de *loc.cit.*, implique que p est une fibration au sens de [35] 2.3.8, [36] §1 (ou encore, dans la terminologie de K. Brown, une fibration locale, voir [8] chapitre II.§1.). Il résulte alors de [45] proposition 7.5 faisceautisée que BG est de Kan, puisque EG l'est et que la projection canonique $p\colon EG \to BG$ est un épimorphisme.

Au morphisme $f := B\partial\colon BG_1 \to BG_0$ est associé (voir [25] VI.1.3) une suite exacte au sens homotopique :

$$(3.6.1)\quad \ldots \to \Omega^2 BG_0 \to \Omega\Gamma f \to \Omega BG_1 \to \Omega BG_0 \to \Gamma f \to BG_1 \to BG_0.$$

L'exactitude signifie ici que pour tout objet simplicial X de T, la suite induite :

$$\ldots \to [X, \Omega BG_1] \to [X, \Omega BG_0] \to [X, \Gamma f] \to [X, BG_1] \to [X, BG_0]$$

(où $[X,Y]$ désigne les classes d'homotopie de morphismes de X dans Y) est une suite exacte d'ensembles pointés (et même de groupes pour les termes situés à gauche de $[X, \Gamma f]$ puisque, fonctoriellement en Y, l'espace des lacets ΩY d'un ensemble simplicial Y est muni d'une loi de groupe à homotopie près).

En fait, les termes de la suite exacte (3.6.1) s'interprètent aisément : en effet, par définition, l'objet Γf est la fibre homotopique de f, définie par le carré cartésien suivant (voir [25] VI.1.2) :

$$(3.6.2)\qquad \begin{array}{ccc} \Gamma f & \longrightarrow & \mathrm{Hom.}(\Delta[1], BG_0) \\ \big\downarrow & & \big\downarrow{\scriptstyle e} \\ BG_1 & \longrightarrow & BG_0 \end{array}$$

où $\mathrm{Hom.}(X,Y)$ désigne l'objet simplicial des application pointées de X vers Y, et e l'évaluation d'un chemin pointé en son extrémité libre. Or $\mathrm{Hom.}(\Delta(1), BG_0)$ s'identifie à EG_0, muni de sa projection canonique vers BG_0. En effet, un terme de la composante de degré n de $\mathrm{Hom.}(\Delta(1), BG_0)$ est décrit par une application simpliciale de $\Delta(n) \times \Delta(1)$ vers BG_0, et la décomposition simpliciale bien connue du prisme $\Delta(n) \times \Delta(1)$ (voir par exemple [39] I 4.6) signifie que cette dernière est décrite par $n+1$ éléments $c_0, \dots, c_n$ de $(BG_0)_{n+1}$ satisfaisant aux relations indiquées en *loc.cit.* La bijection de $(EG_0)_{n+1}$ vers Hom.$(\Delta(1), BG_0)$ envoie $(g_0, \dots, g_n)$ vers les éléments c_i définis par :

$$\begin{aligned} c_0 &= (g_0, g_0^{-1}g_1, g_1^{-1}g_2, \dots, g_{n-1}^{-1}g_n) \\ c_1 &= (e, g_1, g_1^{-1}g_2, \dots, g_{n-1}^{-1}g_n) \\ c_2 &= (e, e, g_2, g_2^{-1}g_3, \dots, g_{n-1}^{-1}g_n) \\ &\vdots \\ c_n &= (e, e, \dots, e, g_n) \end{aligned} \tag{3.6.3}$$

Le lemme suivant en résulte :

Lemme 3.7. *L'objet Γf associé à $f = B\delta\colon BG_1 \to BG_0$ est caractérisé, à isomorphisme près, par le carré cartésien*

$$\begin{array}{ccc} \Gamma f & \longrightarrow & EG_0 \\ \downarrow & & \downarrow{\scriptstyle p} \\ BG_1 & \xrightarrow[f]{} & BG_0 \end{array} \tag{3.7.1}$$

On constate alors que Γf est ici une vieille connaissance : en effet, par fonctorialité, on dispose d'un carré commutatif

$$\begin{array}{ccc} EG_1 & \xrightarrow{\bar{f}} & EG_0 \\ {\scriptstyle p}\downarrow & & \downarrow{\scriptstyle p} \\ BG_1 & \xrightarrow[f]{} & BG_0 \end{array} \tag{3.7.2}$$

Puisque $(f, \bar{f})$ est un δ-morphisme de torseurs, il induit par extension du groupe structural de G_1 à G_0, un morphisme de G_0-torseurs au dessus de f, c'est à dire un carré commutatif:

$$\begin{array}{ccc} G_0 \overset{G_1}{\wedge} EG_1 & \longrightarrow & EG_0 \\ \downarrow & & \downarrow \\ BG_1 & \xrightarrow[f]{} & BG_0 \end{array} \tag{3.7.3}$$

soit encore, par (3.7.1), un morphisme (et donc un isomorphisme) de G_0-torseurs sur $BG_1 \colon G_0 \overset{G_1}{\wedge} EG_1 \xrightarrow{\sim} \Gamma f$. En définitive, en tenant compte du lemme 3.5, on a démontré le :

Lemme 3.8. *Soit $f = B\delta \colon BG_1 \to BG_0$ le morphisme d'objets simpliciaux pointés associé à un homomorphisme $\partial \colon G_1 \to G_0$ de groupes de T. Alors la flèche $\Gamma f \to BG_1$ s'identifie à la projection sur BG_1 du G_0-torseur $G_0 \overset{G_1}{\wedge} EG_1$ obtenu à partir du G_1-torseur universel sur BG_1 par extension du groupe structural par ∂. En particulier, si $\partial \colon G_1 \to G_0$ est un module croisé, alors Γf s'identifie à l'objet simplicial, sous-jacent au groupe simplicial $\mathcal{G}$ associé au module croisé en question.*

En termes de cette identification, les carrés cartésien (3.7.2) et (3.7.3) deviennent :

(3.8.1)
$$\begin{array}{ccc} EG_1 & \xrightarrow{b} & EG_0 \\ {\scriptstyle a}\downarrow & & \downarrow{\scriptstyle p} \\ BG_1 & \xrightarrow[B(\partial)]{} & BG_0 \end{array}$$

où la $i$ème composante a_i (resp. b_i) de a (resp. b) est définie, pour $z = (g_0, x) \in G_0 \times G_1^i$, par :

(3.8.2)
$$a_i(g_0, \ldots, x) = x$$

(3.8.3)
$$b_i(z) = (u_0(z), \ldots, u_i(z)).$$

En particulier, on constate que $\mathcal{G}$, image inverse par $B(\partial)$ du G_0-torseur universel, est un G_0-torseur sur BG_1, pour l'action à gauche de G_0 sur $\mathcal{G}$, définie sur la composante de degré i par :

$$g_0'(g_0, x) = (g_0' g_0, x).$$

D'autre part, si l'on considère le diagramme (3.8.1) dans le cas où le groupe G_1 est trivial, et donc BG_1 est l'ensemble simplicial trivial, on retrouve le fait bien connu que l'espace des lacets ΩBG_0, (défini comme étant Γf pour $f \colon e \to BG_0$ l'application triviale, voir [25] V 3.2.1) s'identifie au groupe simplicial constant G_0 (plongé diagonalement dans chaque composante de

EG_0. Enfin, la flèche $\Omega BG_0 \to \Gamma f$ de la suite exacte (3.6.1) peut maintenant être explicitée. Par définition (voir *loc.cit.* V 3.4.4) c'est la flèche $\Gamma f_1 \to \Gamma f$ déduite par fonctorialité de Γ à partir du diagramme commutatif

$$\begin{array}{ccc} e & \xrightarrow{f_1} & BG_0 \\ \downarrow & & \| \\ BG_1 & \xrightarrow[f]{} & BG_0 \end{array}$$

c'est à dire, compte tenu des précédentes identifications, l'injection canonique c du groupe simplicial constant G_0 dans $\mathcal{G}$. La discussion précédente est résumé par l'assertion suivante :

Proposition 3.9. *Soit $\partial: G_1 \to G_0$ un homomorphisme de groupes de T. Les identifications précédentes font alors correspondre à la suite* (3.6.1) *la suite exacte (au sens homotopique) suivante :*

$$e \to \Omega \mathcal{G} \to G_1 \xrightarrow{\partial} G_0 \xrightarrow{c} \mathcal{G} \to BG_1 \xrightarrow{B(\partial)} BG_0 \tag{3.9.1}$$

où $\mathcal{G}$ est l'ensemble simplicial associé à ∂.

En effet, il ne reste plus qu'à observer que ΩG_0 est réduit à l'objet final e, puisque tout chemin à valeurs dans l'objets simplicial constant G_0 est constant.

Exemple 3.10. Soit $\partial: G_1 \to G_0$ un monomorphisme de T, et

$$G_1 \xrightarrow{\partial} G_0 \xrightarrow{\rho} K \tag{3.10.1}$$

la "suite exacte" associée, en ce sens que K est un objet des classes à droite de G_0 modulo G_1, quotient de G_0 par la relation d'équivalence

$$G_0 \times G_1 \overset{p_1}{\underset{m}{\rightrightarrows}} G_0$$

où $m(g_0, g_1) = g_0(\partial g_1)$. La fibre homotopique Γf de $f = B\partial$ s'identifie, comme on vient de le voir, à l'ensemble simplicial $\mathcal{G}$, qu'il est plus commode de représenter ici par $G_0 \overset{G_1}{\wedge} EG_1$. La projection de cet ensemble simplicial sur l'ensemble simplicial constant K, définie sur la composante de degré i par $r(g_0, g_1, \ldots, g_i) = g_0 G_1$ est un quasi-isomorphisme. Pour s'en convaincre, il suffit de remarquer que la suite exacte d'homotopie induite par (3.9.1) se réduit à $1 \to \pi_1(\mathcal{G}) \to G_1 \to G_0 \to \pi_0(\mathcal{G}) \to 1$, c'est à dire, en définitive, à (3.10.1).

3.11. Revenons maintenant à la situation qui nous intéresse, c'est à dire celle où l'homomorphisme $\partial: G_1 \to G_0$ est un module croisé de T. La flèche $c: G_0 \to \mathcal{G}$, (dont chaque composante est la composée de sections canoniques de produits semi-directs) est un homomorphisme. Ceci nous permet de prolonger d'un cran vers la droite la suite exacte (3.9.1). En effet le foncteur "classifiant" B, appliqué à un groupe simplicial G, défini un ensemble bisimplicial BG dont la $i$ème composante, pour la première structure simpliciale, est l'ensemble simplicial $B(G_i)$, les premiers opérateurs face et dégénérescence étant déduits par fonctorialité des opérateurs correspondants dans G. On peut également considérer BG comme ensemble simplicial, si l'on préfère, en appliquant le foncteur diagonal Δ qui transforme les objets bisimpliciaux en objets simpliciaux ([35] I 1.1). Une autre variante de l'ensemble classifiant d'un groupe simplicial G est fournie par l'ensemble simplicial $\overline{W}(G)$ (voir [45] §2.1) défini par $\overline{W}(G)_0 = e$ et $\overline{W}(G)_n = G_{n-1} \times \ldots \times G_0$ pour tout $n > 0$, les opérateurs face et dégénérescence étant définis par :

$$(3.11.1) \qquad d_i(g) = e \quad \text{pour} \quad g \in \overline{W}(G)_1 = G_0 \qquad (i = 0, 1)$$

$$\begin{aligned} d_i(g_{n-1}, \ldots, g_0) &= (g_{n-2}, \ldots, g_0) && i = 0 \\ &= (d_{i-1}g_{n-1}, \ldots, d_1 g_{n-i+1}, (d_0 g_{n-1}) g_{n-i-1}, \\ &\qquad g_{n-i-2}, \ldots, g_0) && 1 \leq i \leq n \\ s_0(g_{n-1}, \ldots, g_0) &= (e, g_{n-1}, \ldots, g_0) \\ s_i(g_{n-1}, \ldots, g_0) &= (s_{i-1}g_{n-1}, \ldots, s_0 g_{n-i}, e, g_{n-i-1}, \ldots, g_0) && 1 \leq i \leq n. \end{aligned}$$

Ces deux constructions sont équivalentes, car l'application $\bar{f}: \Delta BG \to \bar{W}G$ définie pour, $(g_1, \ldots, g_n) \in (\Delta BG)_n = G_n \times \ldots \times G_n$, par :

$$(3.11.2) \qquad \bar{f}(g_1, \ldots, g_n) = (d_0 g_1, \ldots, d_0^n g_n)$$

est par [62] une équivalence d'homotopie. Si on applique le foncteur classifiant à l'homomorphisme $c: G_0 \to \mathcal{G}$, on obtient un morphisme d'ensemble simpliciaux $Bc: BG_0 \to B\mathcal{G}$, et donc une suite :

$$(3.11.2) \qquad e \to \Omega\mathcal{G} \to G_1 \xrightarrow{\partial} G_0 \xrightarrow{c} \mathcal{G} \to BG_1 \to BG_0 \xrightarrow{B(c)} B\mathcal{G}$$

qui prolonge (3.9.1). Cette suite est exacte au sens homotopique en BG_0 : en effet, la $i$ème composante de $B(c)$ est la flèche $BG_0 \to B\mathcal{G}_i$ induite par l'injection canonique de G_0 dans $\mathcal{G}_i$ et la $i$ème composante de la fibre homotopique $\Gamma B(c)$ de $B(c)$ est donc définie par le carré cartésien :

$$\begin{array}{ccc} \Gamma B(c)_i & \longrightarrow & E\mathcal{G}_i \\ \downarrow & & \downarrow \\ BG_0 & \xrightarrow[f]{} & B\mathcal{G}_i \end{array}$$

L'exemple 3.10 montre alors que $\Gamma(Bc)_i$ s'identifie à l'ensemble simplicial constant G_1^i et $\Gamma(Bc)$ n'est autre que BG_1, ce qui est l'assertion souhaitée. En définitive, on a démontré la proposition suivante :

Proposition 3.12. *Soit $\partial: G_1 \to G_0$ un module croisé de T. Alors la suite* (3.11.2) *est exacte au sens homotopique.*

4. Cohomologie des Modules Croisés

4.1. Nous sommes maintenant en mesure de définir, de manière plus satisfaisante qu'en 2.16, des groupes de cohomologie $H^i(T, G_1 \to G_0)$. On associe en effet à tout homomorphisme de groupes $\partial: G_1 \to G_0$ un ensemble simplicial $\mathcal{G}$, défini par (3.3.7), ou encore par (3.5), en remarquant que dans un cas comme dans l'autre, la structure de module croisé de $\mathcal{G}$ n'intervient pas dans la définition de $\mathcal{G}$ comme ensemble simplicial. Il suffit alors de poser :

$$\text{(4.1.1)} \qquad \begin{aligned} H^0(T, G_1 \to G_0) &= \mathrm{Hom}_{D.(T)}(e, \mathcal{G}) \\ H^{-1}(T, G_1 \to G_0) &= \mathrm{Hom}_{D.(T)}(e, \Omega\mathcal{G}) \end{aligned}$$

où e est l'objet final de la catégorie $D(T)$, localisée de la catégorie des objets simpliciaux de T par rapport aux quasi-isomorphismes ([35] définition I 2.3.5). $H^0(T, G_1 \to G_0)$ est donc un ensemble pointé, et $H^{-1}(T, G_1 \to G_0)$ un groupe, puisque ΩG en est un à homotopie près.

De la même manière, si $\partial: G_1 \to G_0$ est un module croisé, on définit $H^i(T, G_1 \to G_0)$ comme ci-dessus, pour $i = 0, -1$, et on pose :

$$\text{(4.1.2)} \qquad H^1(T, G_1 \to G_0) = \mathrm{Hom}_{D.(T)}(e, B\mathcal{G}).$$

Dans ce cas $H^i(T, G_1 \to G_0)$ est un ensemble pointé (resp. un groupe, resp. un groupe abélien pour $i = 1$ (resp. $i = 0$, resp. $i = -1$) puisque $B\mathcal{G}$ (resp. $\mathcal{G}$, resp. $\Omega\mathcal{G}$) possède la même propriété dans la catégorie homotopique.

On remarquera que les définitions (4.1.1) et (4.1.2) nous fournissent notamment, pour $i = 0, 1$, en posant $G_0 = G$, $G_1 = e$, des objets $H^i(T, G)$ de cohomologie à valeurs dans G, mais que ceux-ci coïncident avec les objets de cohomologie usuels. En effet, on sait, par [35] I théorème 2.3.12 que la catégorie $D.(T)$ est équivalente à la catégorie localisée par rapport aux quasi-isomorphismes de la catégorie $\mathrm{Hot.}(T)_{Kan}$ (dont les objets sont les objets simpliciaux de Kan de T, et les flèches des classes d'homotopie d'applications simpliciales). En outre, les quasi-isomorphismes permettent dans $\mathrm{Hot.}(T)_{Kan}$ un calcul de fractions à droite (l'hypothèse, faite dans *loc.cit.* que T possède assez de points se révélant inutile par [36] théorème 1.10). On a donc :

$$\mathrm{Hom}_{D.(T)}(e, Y) = \varinjlim_{X' \to e} [X', Y] \tag{4.1.3}$$

où $X' \to e$ parcourt les classes d'homotopie de quasi-isomorphismes (que l'on peut même supposer fibrants, voir [35] I proposition 2.3.11, [37] §1). En particulier lorsque Y est un groupe simplicial constant G, on a :

$$\mathrm{Hom}_{D.(T)}(e, G) = [X', G].$$

Or le groupe $\mathrm{Hom}(X'_0, G)$ des morphismes de X' dans G s'identifie à $\ker(H^0(X'_0, G) \rightrightarrows H(X'_1, G))$ c'est à dire, puisque $\pi_0(X') = e$ est le conoyau des flèches $X'_1 \rightrightarrows X'_0$, au groupe $H^0(e, G)$ des sections de e à valeurs dans G. En outre deux applications de l'objet simplicial X' dans l'objet constant G ne peuvent être homotopes que si elles sont égales. On a donc une identification

$$\mathrm{Hom}_{D.(T)}(e, G) = H^0(e, G)$$

du terme de gauche avec le groupe de cohomologie usuel de T à valeurs dans G.

L'identification de $\mathrm{Hom}_{D.(T)}(e, BG)$ avec l'ensemble usuel $H^1(T, G)$ des classes d'isomorphisme de G-torseurs est démontrée en [37] Cor. 1.4. Elle s'obtient en observant que, par (4.1.3), l'ensemble $\mathrm{Hom}_{D.(T)}(e, BG)$ est décrit par une classe d'homotopie de flèches $X' \to BG$, pour X' un objet simplicial acyclique, c'est à dire par un G-torseur trivial sur X'_0 muni de données de recollement sur X'_1 qui satisfait à la condition de cocycle usuelle au dessus de X'_2; il revient au même de dire qu'on est en présence d'une classe de 1-cohomologie à valeurs dans G calculée au moyen de l'hyper-recouvrement défini par X'. L'identification à une classe de G-torseurs de T équivaut à l'assertion que ce H^1 peut être calculé (à la Cech) à partir d'un recouvrement ordinaire, ce qui se démontre comme dans *loc.cit.* ou en invoquant [26] III proposition 3.6.3.

Remarquons enfin, en revenant à un complexe $G_1 \to G_0$, d'objet simplicial associé $\mathcal{G}$, que la définition de $H^0(T, G_1 \to G_0)$ proposée en (4.1.1) coïncide avec celle donnée de $H^0(T, G_1 \to G_0)$ en (2.16). Ceci résulte maintenant de la discussion précédente et du caractère cartésien du diagramme (3.8.1), une fois faite la remarque que la donnée d'une section du "G_0-torseur universel" EG_0 équivaut à celle d'un G_0-torseur P et d'une section s de P.

4.2. Soit $\partial: G_1 \to G_0$ un homomorphisme de groupes de T (resp. un module croisé) de T. La proposition 3.9 (resp. 3.12) nous fournit une suite exacte à 6 termes (resp. 7 termes)

$$\begin{aligned} 0 \to H^{-1}(G_1 \to G_0) \to H^0(G_1) \to H^0(G_0) \to H^0(G_1 \to G_0) \\ \to H^1(G_1) \to H^1(G_0) \end{aligned} \tag{4.2.1}$$

(resp.

$$\begin{aligned} 0 \to H^{-1}(G_1 \to G_0) \to H^0(G_1) \to H^0(G_0) \to H^0(G_1 \to G_0) \\ \to H^1(G_1) \to H^1(G_0) \to H^1(G_1 \to G_0)), \end{aligned} \tag{4.2.2}$$

et les identifications précédentes montrent que (4.2.1) coïncide avec (2.16.1). Si l'on part du module croisé $i': G \to \mathrm{Aut}^o(G)$, la suite exacte (4.1.4) devient (2.16.2) prolongée d'un terme vers la droite, c'est à dire :

$$\begin{aligned} 0 \to H^0(ZG) \to H^0(G) \to H^0(\mathrm{Aut}^o(G)) \to H^0(G \to \mathrm{Aut}^o(G)) \\ \to H^1(G) \to H^1(\mathrm{Aut}^o(G)) \to H^1(G \to \mathrm{Aut}^o(G)), \end{aligned} \tag{4.2.3}$$

à ceci près que l'on dispose maintenant d'une loi de groupe sur $H^0(G \to \mathrm{Aut}^o(G))$.

Proposition 4.3. *La loi de groupe définie sur* $H^0(G \to \mathrm{Aut}^o(G))$ *par la structure de module croisé de* $G \to \mathrm{Aut}^o(G)$ *est l'opposée de la loi de groupe induite sur* BITORS(G) *par le produit contracté des bitorseurs.*

Ceci se démontre par un calcul de cocycle. Pour en rendre le calcul plus familier, on supposera, bien que cela ne joue en fait aucun rôle, que l'on est en présence d'éléments de $H^0(G \to \mathrm{Aut}^o(G))$ décrits par des classes d'applications $X' \to \mathcal{G}$, où l'"hyperrecouvrement" X' de l'objet final e de T est construit à partir d'un recouvrement $(U_i)_{i \in I}$ de e, c'est à dire que $X'_0: \bigcup U_i \to e$ est un épimorphisme, et que $X' = \mathrm{cosq}^0 X'_0$ est l'objet simplicial de $i$ème composante $(X'_0)^{i+1}$, les opérateurs face étant définis par les projections canoniques et les dégénérescences via le choix d'un point u_i dans chaque ouvert U_i ([51] §4). Si E est un G-bitorseur de T trivialisé par le recouvrement (U_i), le G-torseur gauche sous-jacent est donc muni de sections locales $s_i \in H^0(U_i, E)$, et il est donc décrit par des cochaines $f_{ij} \in H^0(U_{ij}, G)$ telles que :

$$s_i = f_{ij} s_j \tag{4.3.1}$$

et qui satisfont à la condition de cocycle

$$f_{ij} \cdot f_{jk} = f_{ik}. \tag{4.3.2}$$

Une section t de l'objet induit $\mathrm{Aut}^o(G) \overset{G}{\wedge} E$ est alors décrite localement par $t_i = t_{|U_i} = g_i(1, s_i)$, où $g_i \in H^0(U_i, \mathrm{Aut}^o(G))$ satisfait, sur U_{ij}, à la condition

$$g_i * i'_{f_{ij}} = g_j \quad \text{dans} \quad \mathrm{Aut}^o(G) \tag{4.3.3}$$

c'est à dire

$$i'_{f_{ij}} \circ g_i = g_j \quad \text{dans} \quad \mathrm{Aut}(G). \tag{4.3.4}$$

A ces données correspond le morphisme d'objets simpliciaux

$$p: X' \longrightarrow \mathcal{G} \tag{4.3.5}$$

défini en degré 0 (resp. 1) par

$$\begin{aligned} \varphi_{0|U_i} &= g_i : U_i \to \mathrm{Aut}^o(G) \\ \varphi_{1|U_i} &= (g_i, f_{ij}) : U_{ij} \to \mathrm{Aut}^o(G) \times G \end{aligned} \tag{4.3.6}$$

et dont, plus généralement, la restriction à $U_{i_0 \ldots i_s}$ de la s ième composante est

$$(g_{i_0}, f_{i_0 i_1}, \ldots, f_{i_{s-1} i_s}) : U_{i_0 \ldots i_s} \to \mathrm{Aut}^o(G) \times G^s. \tag{4.3.7}$$

Soit $\bar{g}\, s_i$ une section de E au dessus de U_i. La multiplication à gauche par $g \in H^0(U_i, G)$ est définie par :

$$g * (\bar{g}\, s_i) = g \bar{g}\, s_i \tag{4.3.8}$$

et, par (2.5.1), celle de droite est définie par :

$$s_i * g = g_i(g) s_i \tag{4.3.9}$$

et donc, en toute généralité, par :

$$(\bar{g}\, s_i) * g = \bar{g} g_i(g) s_i. \tag{4.3.10}$$

Soient donc $E^{(1)}$ et $E^{(2)}$ deux G-bitorseurs trivialisés localement sur un recouvrement $(U_i)_{i \in I}$ par des sections $s_i^{(1)}$ et $s_i^{(2)}$ en termes desquels ils sont décrits par les cocyles $(g_i^{(1)}, f_{ij}^{(1)})$ et $(g_i^{(2)}, f_{ij}^{(2)})$. La multiplication à droite dans $E^{(1)} \overset{G}{\wedge} E^{(2)}$, trivialisé localement par $s_i^{(1)} \wedge s_i^{(2)}$, nous est connue par le calcul effectué en (2.9.6), que nous pouvons reprendre dans le contexte présent : on a, par (4.3.8) et (4.3.9),

$$\begin{aligned}(s_i^{(1)} \wedge s_i^{(2)}) g &= s_i^{(1)} \wedge g_i^{(2)}(g) s_i^{(2)} = s_i^{(1)} g_i^{(2)}(g) \wedge s_i^{(2)}, \\ &= (g_i^{(1)} \circ g_i^{(2)}(g))(s_i^{(1)} \wedge s_i^{(2)})\end{aligned} \tag{4.3.11}$$

ce qui nous montre, par comparaison avec 4.3.9, que les premières composantes des paires de cocycles correspondant au produit contracté de $(g_i^{(1)}, f_{ij}^{(1)})$ et $(g_i^{(2)}, f_{ij}^{(2)})$ sont les $(g_i^{(1)} \circ g_i^{(2)})$, la loi de composition étant calculée dans $\mathrm{Aut}(G)$. Par ailleurs, le cocycle qui décrit la donnée de recollement de $p^{(1)} \wedge p^{(2)}$ par rapport aux trivialisation $s_i^{(1)} \wedge s_i^{(2)}$ s'obtient en comparant $s_i^{(1)} \wedge s_i^{(2)}$ à $s_j^{(1)} \wedge s_j^{(2)}$. Par (4.3.1), on a :

$$\begin{aligned}s_i^{(1)} \wedge s_i^{(2)} &= (f_{ij}^{(1)} s_j^{(1)}) \wedge (f_{ij}^{(2)} s_j^{(2)}) \\ &= f_{ij}^{(1)} (s_j^{(1)} \cdot f_{ij}^{(2)}) \wedge s_j^{(2)} \\ &= (f_{ij}^{(1)} g_j^{(1)}(f_{ij}^{(2)})) s_j^{(1)} \wedge s_j^{(2)}.\end{aligned} \tag{4.3.12}$$

Mais la relation (4.3.4) permet de reécrire ceci sous la forme :

$$s_i^{(1)} \wedge s_i^{(2)} = (g_i^{(1)}(f_{ij}^{(2)}) f_{ij}^{(1)}) s_j^{(1)} \wedge s_j^{(2)}. \tag{4.3.13}$$

En définitive, les formules (4.3.11) et (4.3.13) nous montrent que la loi de composition de bitorseurs s'énonce en termes de cocyles de la manière suivante :

$$(g_i^{(1)}, f_{ij}^{(1)}) \times (g_i^{(2)}, f_{ij}^{(2)}) = (g_i^{(2)} * g_i^{(1)}, g_i^{(1)}(f_{ij}^{(2)}) f_{ij}^{(1)}). \tag{4.3.14}$$

C'est donc l'opposée de la loi

$$(g_i^{(1)}, f_{ij}^{(1)}) \times (g_i^{(2)}, f_{ij}^{(2)}) = (g_i^{(1)} * g_i^{(2)}, g_i^{(2)}(f_{ij}^{(1)}) f_{ij}^{(2)}). \tag{4.3.15}$$

Or cette dernière n'est autre que la loi du produit semi direct $\mathrm{Aut}^o(G) \times G$, lorsque $\mathrm{Aut}^o(G)$ agit à droite sur G par $g \times u = u(g)$, c'est à dire la loi de groupe définie par la structure de groupe simplicial de $\mathcal{G}$. Ceci termine la démonstration.

On retiendra de la proposition que la suite exacte (4.2.3), dont les premiers termes s'identifient à ceux de (2.16.2) n'est pas compatible à la structure de groupe usuelle de Bitors(G). Si l'on veut préserver celle-ci, il convient de remplacer les premiers termes de (4.2.3) par les groupes opposés, sans changer les flèches. On obtient alors une suite exacte à sept termes (dont les quatre premiers sont des groupes) :

$$(4.3.16) \qquad \begin{aligned} &0 \to H^0(ZG) \to H^0(G^0) \xrightarrow{H^0(i')} H^0(\mathrm{Aut}(G)) \to \mathrm{Bitors}(G) \\ &\to \mathrm{Tors}(G) \to \mathrm{Tors}(\mathrm{Aut}^o(G)) \to H^1(\mathcal{G}) \end{aligned}$$

Quitte à remplacer $H^0(i')$ par la flèche correspondante induite par $i: G \to \mathrm{Aut}(G)$ (sans changer, toutefois, la flèche $H^1(i')$) on peut remplacer cette suite par une suite exacte :

$$(4.3.17) \qquad \begin{aligned} &0 \to H^0(ZG) \to H^0(G) \xrightarrow{H^0(i)} H^0(\mathrm{Aut}(G)) \to \mathrm{Bitors}(G) \\ &\to \mathrm{Tors}(G) \to \mathrm{Tors}(\mathrm{Aut}^o(G)) \to H^1(\mathcal{G}). \end{aligned}$$

La signification de l'exactitude de ces suites en $H^0(\mathrm{Aut}(G))$ a été analysée en 2.16. Quant à l'exactitude en Bitors(G) de (4.3.17), elle équivaut à une partie de la proposition IV 5.2.9 ii de [26].

4.4. La fin de ce paragraphe sera consacrée à l'examen de variantes de la proposition 4.3. Remarquons tout d'abord que la discussion précédente s'est déroulée au niveau des objets, et non de leurs classes à isomorphisme près (ou si l'on veut : au niveau des cocycles plutôt que des classes de cohomologie). On a donc obtenu une description de la gr-catégorie BITORS(G), et même du gr-champ BITORS(G) qui lui a été associé en 2.3. Cette description peut s'expliciter de la même manière qu'en [16] 1.4.21 en associant à un module croisé $G_1 \to G_0$ de T la catégorie fibrée dont la fibre au-dessus d'un objet X de T est la gr-catégorie associée comme il a été dit en 3.3 au module croisé (usuel) $G_1(X) \to G_0(X)$. Cette catégorie fibrée est un gr-préchamp, en ce sens que les morphismes se recollent lorsque X varie, et il lui correspond donc par faisceautisation un gr-champ, qu'on appelera le gr-champ associé au module croisé $G_1 \to G_0$. On obtient alors, en reprenant la discussion précédente, la version renforcée suivante de la proposition 4.4 :

Théorème 4.5. *Le* gr-*champ associé au module croisé* $i'\colon G \to \mathrm{Aut}^o(G)$ *de* T *est équivalent au* gr-*champ* $\mathrm{B{\scriptstyle ITORS}}^0(T;G)$ *opposé au* gr-*champ* $\mathrm{B{\scriptstyle ITORS}}(T;G)$, *cette équivalence étant fonctorielle en le module croisé.*

Par ailleurs, la discussion précédente mériterait d'être reprise dans le cas où i' est remplacé par un module croisé quelconque $\partial\colon G_1 \to G_0$. Nous avons déjà explicité en (4.2.2) la suite exacte correspondante, et seule manque l'interprétation géométrique du groupe $H^0(G_1 \to G_0)$. Il s'agit là d'un objet intermédiaire entre le groupe des G-bitorseurs et celui des (G,H)-bitorseurs, et qui peut paraître assez artificiel du point de vue géométrique. Posons, pour des raisons de commodité, $G_1 = G$ et désignons par π le groupe opposé à G_0. On dispose donc d'un module croisé (droit) $\partial\colon G \to \pi^0$. L'ensemble sous-jacent à $H^0(G \to \pi^0)$ coïncide, comme on l'a dit en 4.1, avec celui des classes d'isomorphie de G-torseurs E munis d'une section s de $\partial_*(E)$, ou encore, en faisant appel comme en 2.4 à [26] III 1.6.1. ii, avec celui des classes paires (E,t), où E est un G-torseur et $t\colon E \to \pi$ est une flèche G-équivariante pour l'action à gauche de G sur π à travers l'homomorphisme $\partial_1 = i \circ \partial\colon G \to \pi^0 \to \pi$. La structure de module croisé définit une loi de composition sur les paires (E,t). En effet E est muni d'une structure de G-bitorseur, pour laquelle la multiplication à droite par G est définie par la règle :

$$eg =: (g^{t(e)})e \tag{4.5.1}$$

pour toute section $e \in E$. On fait alors correspondre aux paires (E_1,t_1) et (E_2,t_2) la paire :

$$(E_1,t_1) \wedge (E_2,t_2) = (E_1 \overset{G}{\wedge} E_2, t) \tag{4.5.2}$$

définie par le torseur obtenu par produit contracté des bitorseurs E_1 et E_2, muni de la flèche équivariante $t =$ "$t_1 \wedge t_2$" : $E_1 \overset{G}{\wedge} E_2 \to \pi$ définie, comme en (2.9.5) par :

$$t(e_1,e_2) = t_1(e_1)t_2(e_2). \tag{4.5.3}$$

Appelons (G,π)-torseur (à ne pas confondre avec les (G,H)-bitorseurs, qui sont plutôt, par la proposition 2.11, apparentés à des "$(G, \mathrm{Isom}(H,G)^0)$-torseurs") une paire (E,t), munie de la loi de composition qui vient d'être introduite. Les mêmes arguments que précédemment démontrent la généralisation suivante du théorème 4.5 :

Théorème 4.6. *Le gr-champ associé à un module croisé $\mathcal{M}: G \xrightarrow{\partial} \pi^0$ de T est équivalent au gr-champ opposé à celui des (G,π)-torseurs de T, cette équivalence étant fonctorielle en le module croisé $\mathcal{M}$.*

En particulier, la suite exacte (4.2.2) prend maintenant la forme :

$$(4.6.1)\qquad \begin{aligned} &0 \to H^{-1}(G \to \pi^0) \to H^0(G) \to H^0(\pi^0) \to ((G,\pi) - \mathrm{Tors})^0 \\ &\to \mathrm{Tors}(G) \to \mathrm{Tors}(\pi^0) \to H^1(\mathcal{M}), \end{aligned}$$

ce qui généralise (4.3.16).

Remarque 4.7. (i) Il n'est pas inintéressant d'expliciter rapidement la description à la Cech d'un (G,π)-torseur trivialisé sur un recouvrement $(U_i)_{i\in I}$ de e : un tel objet est décrit par des paires (g_i, f_{ij}) (où $g_i \in \Gamma(U_i, \pi^0)$ et $f_{ij} \in \Gamma(u_{ij}, G)$ satisfont à la condition (4.3.2) et à la règle suivante dans π^0 :

$$(4.7.1)\qquad g_i\partial(f_{ij}) = g_j,$$

qui est analogue à (4.3.3). Un morphisme $E \to E'$ entre des (G,π)-torseurs (trivialisés respectivement sur U_i par des sections s_i et s'_i, en termes desquelles ils sont décrits par des paires (g_i, f_{ij}) et (g'_i, f'_{ij})) s'exprime au moyen de sections $f_i \in \Gamma(U_i, G)$ telles que l'on ait :

$$(4.7.2)\qquad g'_i = g_i\partial(f_i)$$

dans π^0, et également

$$(4.7.3)\qquad f'_{ij} = f_i^{-1} f_{ij} f_j.$$

Ces identités expriment notamment la manière dont une paire (g_i, f_{ij}) est transformée, lorsque la famille de section locales (s_i) de E donnée est remplacée par la famille $s'_i = f_i s_i$. En particulier, une paire (g_i, f_{ij}) décrit un G-bitorseur trivialisé si et seulement si elle est de la forme $(\partial(f_i), f_i^{-1}f_j)$ pour une famille $f_i \in H^0(U_i, G)$ de sections de G. Les formules (4.3.6)–(4.3.7) définissent maintenant l'application $\varphi: X' \to \mathcal{G}$ correspondant au (G,π)-torseur E, $\mathcal{G}$ désignant ici le groupe simplicial associé en 3.3 au module croisé $\partial: G \to \pi^0$, et l'on vérifie aisément que la donnée de sections (f_i) de G satisfaisant à (4.7.2) et (4.7.3) équivaut à celle d'une homotopie $\varphi' \simeq \varphi: X' \to \mathcal{G}$ au sens simplicial (voir [45] définition 5.1) entre les flèches φ et φ' associées aux paires (g_i, f_{ij}) et (g'_i, f'_{ij}). Enfin la loi de composition est donnée par le règle analogue à (4.3.14).

(ii) On renvoie à [7] proposition 2.1 (i) et [35] I proposition 3.2.2.10 pour l'étude de suites exactes longues associées à des fibrations dans T, dans des situations plus commutatives que celle considérérée ci-dessus.

5. Suites Exactes

5.1. Nous pouvons maintenant aller plus loin dans l'étude de la fonctorialité. Soit en effet $f: G \to G'$ un homomorphisme de groupes simpliciaux de T. Lorsque chaque composante $f_i: G_i \to G'_i$ de f est un épimorphisme, l'argument de [45] lemme 18.2, convenablement faisceautisé, montre que f est un morphisme fibrant. On pose $H = \ker f$.

Les considérations détaillées en 3.6, étendues au cas où l'on part d'un homomorphisme de groupes simpliciaux tel que f, plutôt que de groupes, et un raisonnement analogue à celui de 3.10, montrent que l'on dispose alors d'une suite exacte analogue à (3.9.1) :

$$\ldots \Omega G \to \Omega G' \to H \to G \to G' \to BH \to BG \to BG'. \tag{5.1.1}$$

Il en résulte, en appliquant le foncteur $\mathrm{Hom}_{D\ (T)}(e, -)$, une suite exacte de cohomologie associée

$$\begin{aligned} &\ldots H^{-1}(G) \to H^{-1}(G') \to H^0(H) \to H^0(G) \to H^0(G') \\ &\qquad \to H^1(H) \to H^1(G) \to H^1(G'). \end{aligned} \tag{5.1.2}$$

Considérons le cas où $\pi_i(G) = \pi_i(G') = 0$ pour $i \geq 2$, c'est à dire où l'on part d'un épimorphisme de modules croisés $(f_1, f_0): (G_1 \xrightarrow{\partial} G_0) \to (G'_1 \xrightarrow{\partial} G'_0)$, soit encore de la donnée d'une paire d'épimorphismes $f_1: G_1 \to G'_1$, $f_0: G_0 \to G'_0$ satisfaisant aux conditions expliquées en 3.2. Dans ce cas $H = (\ker f_1 \xrightarrow{\partial_H} \ker f_0)$ est encore un module croisé, d'où une suite exacte (5.1.2). Un dévissage permet, pour tout module croisé $G = (G_1 \xrightarrow{\partial} G_0)$ d'identifier $H^{-1}(G)$ au groupe de cohomologie usuel $H^0(\ker \partial)$ (voir 5.2.2). Ainsi la suite (5.1.2) prend maintenant la forme suivante :*

$$\begin{aligned} &0 \to H^0(\ker \partial_H) \to H^0(\ker \partial) \to H^0(\ker \partial') \to H^0(H) \\ &\qquad \to H^0(G) \to H^0(G') \to H^1(H) \to H^1(G) \to H^1(G'). \end{aligned} \tag{5.1.3}$$

En particulier, partons d'un épimorphisme $(G \xrightarrow{i'} \mathrm{Aut}^o(G)) \to (G' \xrightarrow{i'} \mathrm{Aut}^o(G'))$ c'est à dire de la donnée d'une paire d'épimorphisme $v_1: G \to G'$ et $v_0: \mathrm{Aut}(G) \to \mathrm{Aut}(G')$ tels que $i' \circ v_1 = v_0 \circ i'$ et que, pour tout

* Pour un énoncé de ce type dans le cas particulier du topos $T = (G-\mathrm{mod})$ des G-modules associé à un groupe abstrait G, voir notamment [14] théorème 6.

$g \in G$, $u \in \mathrm{Aut}(G)$, $v_0(u)(v_1(g)) = v_1(u(g))$. On pose $K = \ker v_1$, (resp. $\Phi = \ker v_0$). A la suite exacte de modules croisés

$$\begin{array}{ccccccccc} 1 & \longrightarrow & K & \xrightarrow{j_1} & G & \xrightarrow{v_1} & G' & \longrightarrow & 1 \\ & & {\scriptstyle i'_{|K}}\big\downarrow & & \big\downarrow{\scriptstyle i'} & & \big\downarrow{\scriptstyle i'} & & \\ 1 & \longrightarrow & \Phi & \xrightarrow[j_0]{} & \mathrm{Aut}^o(G) & \xrightarrow[v_0]{} & \mathrm{Aut}^o(G') & \longrightarrow & 1 \end{array} \tag{5.1.4}$$

correspond donc une suite exacte :

$$\begin{aligned} &0 \to H^0(K \cap ZG) \to H^0(ZG) \to H^0(ZG') \to H^0(K \to \Phi) \\ &\to \mathrm{Bitors}(G)^0 \to \mathrm{Bitors}(G')^0 \to H^1(K \to \Phi) \\ &\to H^1(G \to \mathrm{Aut}^0(G)) \to H^1(G' \to \mathrm{Aut}^0(G')) \end{aligned} \tag{5.1.5}$$

dont les termes $H^0(K \cap ZG)$ et $H^0(K \to \Phi)$ ne possèdent pas en général d'interprétation géométrique agréable. Tout au plus résulte-t-il formellement de la suite exacte (5.1.1) que $H^0(K \to \Phi)$ (resp. $H^{-1}(K \to \Phi) = H^0(K \cap ZG)$) est le groupe $\pi_0(\mathcal{D})$ des classes à isomorphisme près (resp. le groupe $\pi_1(\mathcal{D})$ des automorphismes de l'objet neutre) de la *gr*-catégorie $\mathcal{D}$ opposée à celle des G-bitorseurs P munis d'une trivialisation du G'-bitorseur induit $(v_1, v_0)_*P$. On peut également décrire $\mathcal{D}$ comme étant la catégorie des K-torseurs à gauche munis d'une trivialisation de $\Phi^0 \overset{K}{\wedge} P$, avec une loi de composition appropée, ou encore à la catégorie opposée à celle des K-torseurs P munis d'une structure de G-bitorseurs sur $(j_1)_*P$ induisant sur $(v_1)_*(j_1)_*P$ la structure de G'-bitorseur trivial, mais aucune de ces descriptions n'est très satisfaisante. Remarquons cependant que l'action de Φ sur K définit un homomorphime $\lambda\colon \Phi \to \mathrm{Aut}^o(K)$ qui n'est autre que la flèche obtenue en restreignant à K un élément de Φ, vu comme automorphisme de G. Le noyau de λ consiste donc en les automorphismes u de G qui s'insèrent dans le diagramme :

$$\begin{array}{ccccccccc} 1 & \longrightarrow & K & \xrightarrow{j_1} & G & \xrightarrow{u_1} & G' & \longrightarrow & 1 \\ & & {\scriptstyle 1_K}\big\downarrow & & {\scriptstyle u}\big\downarrow & & {\scriptstyle 1_{G'}}\big\downarrow & & \\ 1 & \longrightarrow & K & \xrightarrow[j_1]{} & G & \xrightarrow[u_1]{} & G' & \longrightarrow & 1 \end{array}$$

et ce noyau est donc égal au groupe $\mathrm{Crhom}(G', ZK)$ des homomorphismes croisés de G' vers ZK. Si l'on suppose que $\mathrm{Crhom}(G', ZK) = 0$ et que l'on fait l'hypothèse (à vrai dire très rectrictive) que λ est un épimorphisme, c'est à dire que tout automorphisme de K s'obtient, après changement de base, par restriction à K d'un automorphisme de G, alors Φ peut être remplacé par $\mathrm{Aut}^o(K)$ dans le diagramme (5.1.4), et on aura dans ce cas une suite exacte (5.1.3) dont l'opposée sera la suite :

$$(5.1.6)\qquad \begin{aligned} 0 \to H^0(ZK) \to H^0(ZG) \to H^0(ZG') \to \mathrm{Bitors}\, K \\ \to \mathrm{Bitors}\, G \to \mathrm{Bitors}\, G' \to \ldots \end{aligned}$$

5.2. Il existe un autre cas où les termes de la suite de cohomologie associée à la suite exacte (5.1.4) sont faciles à interpréter. C'est celui où le morphisme de modules croisés de départ est $(v_1, 1): (G \to \mathrm{Aut}^o(G)) \to (\mathrm{Int}(G) \to \mathrm{Aut}^o(G))$ où $v_1: G \to \mathrm{Int}(G)$ est la projection de G sur le groupe $G/ZG = \mathrm{Int}(G)$ des automorphismes intérieurs de G. La suite exacte de modules croisés correspondante est alors :

$$(5.2.1)\qquad \begin{array}{ccccccccc} 1 & \longrightarrow & ZG & \longrightarrow & G & \xrightarrow{v_1} & \mathrm{Int}(G) & \longrightarrow & 1 \\ & & \big\downarrow & & \big\downarrow{\scriptstyle i} & & \big\downarrow{\scriptstyle i'} & & \\ 1 & \longrightarrow & 1 & \longrightarrow & \mathrm{Aut}^o G & = & \mathrm{Aut}^o(G) & \longrightarrow & 1 \end{array}$$

Par ailleurs la projection π de $\mathrm{Aut}^o(G)$ sur $\mathrm{Aut}^o(G)/\mathrm{Int}(G) = \mathrm{Aut}^o(G)/i'(G) = \mathrm{Out}^o(G)$* définit un quasi-isomorphisme entre le module croisé $(\mathrm{Int}(G) \xrightarrow{i'} \mathrm{Aut}^o(G))$ et le module croisé $(0 \to \mathrm{Out}^o(G))$. On en déduit une suite exacte de cohomologie, dépendant fonctoriellement de $i': G \to \mathrm{Aut}^o(G)$, et dont les termes s'interprètent au moyen de la proposition 4.3. On trouve, en désignant à nouveau par $\mathcal{G}$ le groupe simplicial associé au module croisé défini par i', l'isomorphisme :

$$(5.2.2)\qquad H^0(ZG) = H^{-1}(ZG[1]) \xrightarrow{\sim} H^{-1}(\mathcal{G})$$

déjà obtenu, et la suite exacte à six termes (dont les deux derniers sont des ensembles pointés)**

$$(5.2.3)\qquad \begin{aligned} 0 \to H^1(ZG) \xrightarrow{\alpha} H^0(\mathcal{G}) \xrightarrow{\beta} H^0(\mathrm{Out}^0(G)) \xrightarrow{\gamma} H^2(ZG) \\ \xrightarrow{H^1(\alpha)} H^1(\mathcal{G}) \xrightarrow{H^1(\beta)} H^1(\mathrm{Out}^0(G)) \end{aligned}$$

* Nous adoptons la notation anglo-saxonne $\mathrm{Out}(G)$, moins encombrante que le $\mathrm{Autex}(G)$ de [26].

** On prendra garde que γ n'est en général pas un homomorphisme, voir [26] IV 5.2.7 (5).

soit encore, en passant aux groupes opposés

$$\begin{aligned} (5.2.4) \qquad & 0 \to \mathrm{Tors}(ZG) \xrightarrow{\alpha} \mathrm{Bitors}(G) \xrightarrow{\beta} H^0(\mathrm{Out}(G)) \to H^2(ZG) \\ & \to H^1(\mathcal{G}) \to \mathrm{Tors}(\mathrm{Out}^0(G)) \end{aligned}$$

Il s'agit là du dévissage "intelligent" des groupes $H^*(\mathcal{G})$, qui apporte un complément d'information à celles obtenues grâce à (4.3.17). Les quatre premiers termes de cette suite figurent dans [26] IV Proposition 5.2.7 (i), 5.2.4 (i) (l'exactitude de la suite en question en la source et le but de β est également examinée, sous une perspective qui diffère quelque peu de la nôtre, dans le récent article [41]). Nous dirons plus bas comment s'interprètent les deux derniers termes. Bornons nous pour l'instant à décrire géométriquement les flèches qui induisent α et β par passage aux classes d'isomorphisme ; en ce qui concerne β, il suffit d'observer que si on se donne un G-bitorseur, défini comme il a été expliqué en 2.5 par un G-torseur P et une flèche $j: P \to \mathrm{Aut}(G)$, alors l'application composée

$$(5.2.5) \qquad P \xrightarrow{j} \mathrm{Aut}(G) \xrightarrow{\pi} \mathrm{Out}(G)$$

est, par G-équivariance de j, constante sur les G-orbites de P. Elle se factorise donc par la section souhaitée de $\mathrm{Out}(G)$, et la commutativité de (2.7.1) implique qu'à deux bitorseurs isomorphes correspond la même section. Quant à la flèche α du diagramme 5.2.4, elle associe à un ZG-torseur P, le G-torseur à gauche $P_1 = G \overset{ZG}{\wedge} P$ (induit par extension du groupe structural), muni de la trivialisation canonique de $i'_* P_1 \in \mathrm{Tors}(\mathrm{Aut}^o(G))$, c'est à dire de la multiplication à droite sur P_1 définie par :

$$(5.2.6) \qquad (g_1, p)g = (g_1, gp).$$

5.3. La discussion prédédente gagne à être étendue au cas de la cohomologie à valeur dans un module croisé droit quelconque $\mathcal{M}: (G \xrightarrow{\partial'} \pi^0)$ (voir notamment [15] théorème 2.2, pour un énoncé de type (5.1.2) dans ce cas). On se limitera ici à expliciter l'analogue de la suite exacte (5.2.3) dans ce contexte général : on a défini, pour un tel $\mathcal{M}$, des groupes d'homotopie $\pi_i(\mathcal{M})$ $(i = 0, 1)$ par $\pi_0(\mathcal{M}) = \mathrm{coker}\,\partial'$ (resp. $\pi_1(\mathcal{M}) = \ker \partial'$). Les axiomes imposés en (3.2) font de $\pi_0(\mathcal{M})$ (resp. $\pi_1(\mathcal{M})$ un groupe (resp. un groupe abélien) de T. La filtration intelligente (ou "décomposition de Postnikoff") de $\mathcal{M}$ est donnée par un triangle distingué :

$$\pi_1(\mathcal{M})[1] \to \mathcal{M} \to \pi_0(\mathcal{M})[0]$$

qui induit un isomorphisme $H^{-1}(\mathcal{M}) \simeq H^0(\pi_1(\mathcal{M}))$ et une suite exacte:

$$
(5.2.7)\qquad \begin{array}{l} 0 \to H^1(\pi_1(\mathcal{M})) \to ((G,\pi)\text{-Tors})^0 \to H^0(\pi_0\mathcal{M})) \\ \quad \to H^2(\pi_1(\mathcal{M})) \to H^1(\mathcal{M}) \to H^1(\pi_0(\mathcal{M})) \end{array}
$$

qui généralise (5.2.3) et dont les termes s'interprètent de manière tout à fait analogue.

6. Interpretation de $H^1(\mathcal{M})$

Ce paragraphe et le suivant sont consacrés à l'interprétation géométrique du terme $H^1(\mathcal{M})$ qui apparaît dans les suites exactes (4.6.1) et (5.2.7) (et donc notamment dans les cas particuliers (4.3.16) et (5.2.4)). On commence pour cela par introduire la définition suivante, calquée sur celle d'un torseur sous un groupe.

Définition 6.1. Soit $\mathcal{C}$ un *gr*-champ sur un site E. On dit qu'un E-champ en groupoïdes $\mathcal{D}$ est un pseudo-torseur à gauche sous le *gr*-champ $\mathcal{C}$ si $\mathcal{D}$ est muni d'un morphisme de E-champs

$$
(6.1.1)\qquad m : \mathcal{C} \times \mathcal{D} \to \mathcal{D}
$$

et d'une transformation naturelle μ* entre les deux flèches du diagramme:

$$
(6.1.2)\qquad \begin{array}{ccc} \mathcal{C} \times \mathcal{C} \times \mathcal{D} & \xrightarrow{m \times 1} & \mathcal{C} \times \mathcal{D} \\ {\scriptstyle 1 \times m}\downarrow & \overset{\mu}{\Swarrow} & \downarrow{\scriptstyle m} \\ \mathcal{C} \times \mathcal{D} & \xrightarrow[m]{} & \mathcal{D} \end{array}
$$

(notée $\mu_{C_1,C_2;\mathcal{D}} : (C_1 \cdot C_2) \cdot D \xrightarrow{\sim} C_1(C_2 \cdot D)$ pour le triple d'objets $(C_1, C_2, D)) \in \mathcal{C} \times \mathcal{C} \times \mathcal{D}$ satisfaisant aux conditions suivantes:

(i) La transformation μ est astreinte à la condition de cohérence (de style condition du pentagone, voir [43]) exprimée par la commutativité du diagramme suivant, pour tout objet (C_1, C_2, C_3, D) de $\mathcal{C}^3 \times D$:

$$
(6.1.3)\qquad \begin{array}{ccccc} & ((C_1C_2)C_3)D & \longrightarrow & (C_1(C_2C_3))D & \\ \swarrow & & & & \searrow \\ (C_1C_2)(C_3D) & & & & C_1((C_2C_3)D) \\ & \searrow & & \swarrow & \\ & & C_1(C_2(C_3D)) & & \end{array}
$$

* Un morphisme de morphismes de E-champs dans la terminologie de [26] II définition 1.2.1.

(ii) La transformation μ induit sur la sous-catégorie $\{0\} \times \mathcal{D}$ de $\mathcal{C} \times \mathcal{D}$ (où $\{0\}$ désigne la sous catégorie de $\mathcal{C}$ réduite à l'objet unité 0 de $\mathcal{C}$), un morphisme équivalent à l'identité, c'est à dire que l'on se donne, pour tout objet D de $\mathcal{D}$, un isomorphisme fonctoriel $C_D : OD \xrightarrow{\sim} D$, et on requiert la compatibilité de cet isomorphisme à μ décrite par la commutativité des diagrammes (de style condition AO de [48] I. 2.2.1) suivants :

(6.1.4)
$$\begin{array}{ccccccc} C(OD) & \xrightarrow{\mu} & (CO)D & \quad & O(CD) & \xrightarrow{\mu} & (OC)D \\ & \searrow & \swarrow & & & \searrow & \swarrow \\ & CD & & & & CD & \end{array}$$

Enfin on exige que le morphisme de E-champs induit $(m, pr_2) : \mathcal{C} \times \mathcal{D} \to \mathcal{D} \times \mathcal{D}$ soit une équivalence.

Lorsque l'on impose en outre l'existence d'un raffinement R de E tel que, pour tout objet S de R, la catégorie fibre $\mathcal{D}_S$ soit non vide, alors on dit que $\mathcal{D}$ est un torseur à gauche sous $\mathcal{C}$ (ou $\mathcal{C}$-torseur lorqu'il n'y a pas d'ambiguité).

Soient d'autre part $\mathcal{D}$ et $\mathcal{D}'$ deux $\mathcal{C}$-torseurs. Un morphisme $(p, q) : \mathcal{D} \to \mathcal{D}'$ consiste en la donnée d'un morphisme de champs $p : \mathcal{D} \to \mathcal{D}'$ et d'une transformation naturelle q :

(6.1.5)
$$\begin{array}{ccc} \mathcal{C} \times \mathcal{D} & \xrightarrow{1 \times p} & \mathcal{C} \times \mathcal{D}' \\ \downarrow & \overset{q}{\Swarrow} & \downarrow \\ \mathcal{D} & \xrightarrow[p]{} & \mathcal{D}' \end{array}$$

exprimant l'équivariance de p par rapport à l'action de $\mathcal{C}$ sur $\mathcal{D}$:

$$Cp(D) \overset{q_{C,D}}{\underset{\sim}{\longrightarrow}} p(CD).$$

On exige également la condition de compatibilité entre les flèches μ et q exprimée par la commutativité du diagramme suivant

(6.1.6)
$$\begin{array}{ccc} CC'p(D) & \xrightarrow{q_{C'D}} & C(p(C'D)) \\ {\scriptstyle q_{CC',D}} \downarrow & & \downarrow {\scriptstyle q_{C,p(C'D)}} \\ p((CC')D) & \xrightarrow[p(\mu)]{} & p(C(C'D)) \end{array}$$

Un tel morphisme $(p,q):\mathcal{D} \to \mathcal{D}'$ est toujours une équivalence. On appellera enfin transformation naturelle entre deux morphismes (p,q) et $(p',q'):\mathcal{D} \to \mathcal{D}'$, la donnée d'une transformation naturelle $\varphi: p \Rightarrow p'$ compatible à l'action de $\mathcal{C}$, en ce sens que les diagramme

$$\text{(6.1.7)} \qquad \begin{array}{ccc} Cp(D) & \xrightarrow{q} & p(CD) \\ {\scriptstyle 1\times\varphi}\downarrow & & \downarrow{\scriptstyle \varphi} \\ Cp'(D) & \xrightarrow[q']{} & p'(CD) \end{array}$$

commutent.

Le champ $\mathcal{C}$ lui-même est muni d'une structure de $\mathcal{C}$-torseur définie par la loi de groupe de $\mathcal{C}$, et pour laquelle la 2-flèche μ (6.1.2) est la donnée d'associativité de $\mathcal{C}$ et (6.1.3) est la condition du pentagone usuel. On dit que $\mathcal{C}$ est alors le $\mathcal{C}$-torseur trivial.

Proposition 6.2. *Soit $\mathcal{M} = (G \xrightarrow{\partial} \pi^0)$ un module croisé de T, et $\mathcal{C} = ((G,\pi)\text{-Tors})^0)$ le* gr-*champ auquel il correspond par le théorème* 4.6. *Alors l'ensemble pointé $H^1(\mathcal{M})$ est en bijection avec celui des classes d'équivalence de $\mathcal{C}$-torseurs de T.*

En particulier, pour $\mathcal{G} = (G \xrightarrow{i} \mathrm{Aut}^o(G))$, on interprête ainsi l'ensemble $H^1(\mathcal{G})$, comme l'ensemble des classes d'équivalence de torseurs sous la *gr*-catégorie opposée à celle des G-bitorseurs. La suite de ce paragraphe sera principalement consacrée à la démonstration de cette proposition, c'est à dire à vérifier, compte tenu de la définition (4.1.2), que la donnée d'une classe de $\mathcal{C}$-torseurs de T équivaut à celle d'une flèche, élément de $[e, B\mathcal{M}]_{D.(T)}$ dans la catégorie dérivée de T.

La proposition se démontre de la même manière qu'est obtenue la description d'un torseur ordinaire en termes de cocycles (voir [36] §1), mais avec un degré supplémentaire de complication. On se bornera à l'expliciter pour $\mathcal{G} = (G \to \mathrm{Aut}^o(G))$ et donc $\mathcal{C} = (\mathrm{BITORS}(G))^0$, la démonstration générale étant identique en tous points. Soit donc $\mathcal{D}$ un $\mathcal{C}$-torseur de T muni d'une section au dessus d'un raffinement donné S de l'objet final e de T, c'est à dire d'un objet X de la catégorie fibre $\mathcal{D}_S$. Celui-ci définit un morphisme de $\mathcal{C}_{|S}$-torseurs de source le $\mathcal{C}_{|S}$-torseur trivial:

$$\text{(6.2.1)} \qquad \begin{array}{rcl} \chi: \mathcal{C}_{|S} & \longrightarrow & \mathcal{D}_{|S} \\ C & \longrightarrow & CX \end{array}$$

pour lequel la transformation naturelle (6.1.5)

$$\text{(6.2.2)} \qquad q_{C,C'}: C(C'(X)) \xrightarrow{\sim} (CC')X$$

n'est autre que $\mu_{C,C',X}$ (voir 6.1.2). Soient $p_1, p_2 = S \times_e S \rightrightarrows S$ les deux projections. A l'isomorphisme de $\mathcal{C}_{|S\times_e S}$-torseurs canonique $f: p_1^*(\mathcal{D}_{|S}) \xrightarrow{\sim} p_2^*(\mathcal{D}_{|S})$ correspond donc, une fois choisi un foncteur quasi-inverse χ^{-1} de χ, un automorphisme

(6.2.3) $$\varphi = \chi^{-1} \circ f \circ \chi: \mathcal{C}_{|S\times S} \to \mathcal{C}_{|S\times S}$$

du $\mathcal{C}_{|S\times S}$-torseur trivial. On peut, en outre, supposer que la restriction à la diagonale est le foncteur identité

(6.2.4) $$1 = \Delta\varphi: \mathcal{C}_{|S} \to \mathcal{C}_{|S}.$$

La condition de compatibilité

(6.2.5) $$f_{23} \circ f_{12} \simeq f_{13}$$

entre les images inverse $f_{ij} = p_{ij}(f)$ de f par les projections $p_{ij}: S^3 \to S^2$ sur les (i,j)-ème facteurs induit une transformation naturelle

(6.2.6)

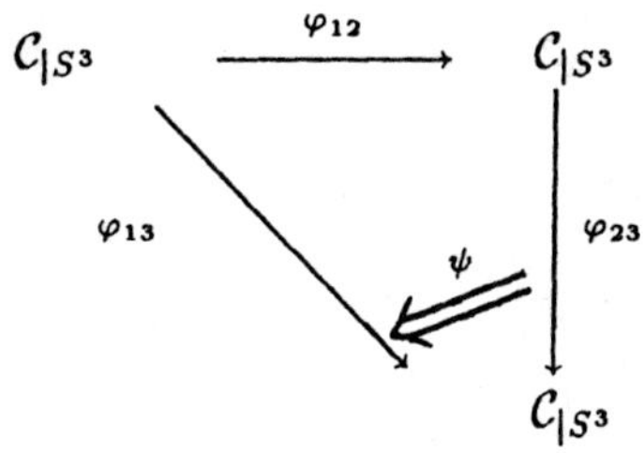

entre les automorphismes correspondants $\varphi_{ij} = p_{ij}^*(\varphi)$ du $\mathcal{C}_{|S^3}$-torseur trivial. On prendra garde que ψ n'est en général pas l'identité, puisque l'identification (6.2.1) de $\mathcal{C}_{|S}$ à $\mathcal{D}_{|S}$ ne respecte l'action à gauche de $\mathcal{C}$ qu'au morphisme (6.2.2) près. La condition de compatibilité (6.1.5) pour le morphisme (6.2.1) (c'est à dire, en définitive, la condition de cohérence (6.1.3)) implique que les images de la transformation ψ par les projections $p_{ijk}: S^4 \to S^3$ sur les (i,j,k)-ième facteurs sont reliées par la relation suivante (ou l'on désigne pour tout diagramme $A \xrightarrow{f} B \underset{g}{\overset{h}{\rightrightarrows}} C \xrightarrow{k} D$ (avec $\Uparrow u$) dans une 2-catégorie $\mathcal{C}$, par $k * u: kg \Rightarrow kh$ (resp. par $u * f: gf \Rightarrow hf$) la 2-flèche induite par u) :

(6.2.7) $$p_{134}^*\psi \circ [p_{34}^*\varphi * p_{123}^*\psi] = p_{124}^*\psi \circ [p_{234}^*\psi * p_{12}^*\varphi].$$

Cette relation se comprend en examinant le diagramme suivant, dont chacun des sommets est la catégorie $\mathcal{C}_{|S^4}$ (notée $\mathcal{C}$ pour ne pas alourdir le diagramme) et chacune des quatres 2-flèches l'une de celles qui figurent dans la relation (6.2.7), et qui représente le défaut de commutativité d'un triangle de type (6.2.6) restreint à S^4 :

(6.2.8)

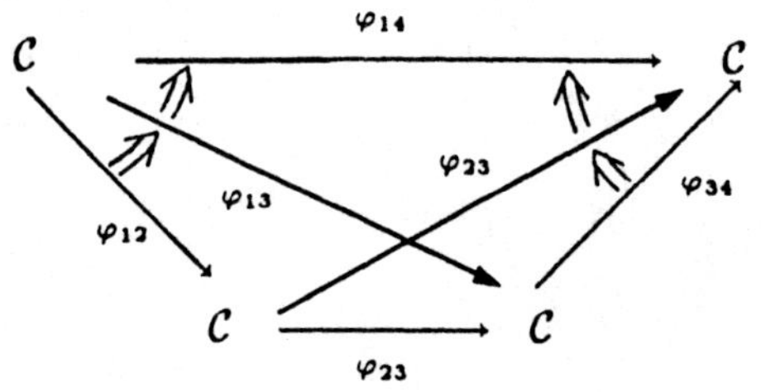

A la description géométrique en termes de "conditions de descente" du $\mathcal{C}$-torseur $\mathcal{D}$ qui vient d'être proposée correspond une description en termes de cocycles. Le procédé indiqué en (6.2.1) définit notamment une équivalence de gr-catégories

(6.2.9)
$$\mathcal{C}^0 \longrightarrow \mathrm{Aut}_{\mathcal{C}-tor}(\mathcal{C})$$

entre la gr-catégorie $\mathcal{C}^0$ opposée à $\mathcal{C}$ et celle des automorphismes du $\mathcal{C}$-torseur trivial. Ainsi la donnée de recollement φ (6.2.3) et le défaut de condition de cocycle ψ (6.2.6) sont décrits respectivement par un objet g de $\mathcal{C}^0_{S\times S}$ et par une flèche

(6.2.10)
$$f: g_{23} \cdot g_{12} \longrightarrow g_{13}$$

dans $\mathcal{C}^0_{|S^3}$, qui satisfait aux conditions de compatibilité qui découlent de (6.2.8).

6.3. Choisissons un morphisme couvrant $S' \xrightarrow{p} S\times S$ au-dessus duquel le torseur sous-jacent au bitorseur g est trivialisé. La condition 6.2.4 montre que l'on peut même supposer que S' est muni d'une application $s_0: S \to S'$ telle que l'application ps_0 soit la diagonale de S. On dispose donc d'un objet simplicial tronqué $S' \rightleftarrows S$ de T dont les opérateurs face d_0 et d_1 sont définis respectivement en composant p avec la projection sur le second et le premier facteur de $S \times S$. Un tel objet tronqué est canoniquement prolongeable en son cosquelette $\mathrm{cosq}^1(S' \rightrightarrows S)$. Rappelons (voir par exemple [21]) qu'il s'agit là de l'objet simplicial de T dont les éléments de la composante de degré n (que nous noterons de manière abrégée $(S'/S)_n$) sont en correspondance avec les applications du 1-squelette du n-simplexe type $\Delta(n)$ dans l'objet tronqué $S' \rightrightarrows\!\!\!\leftarrow S$. Ainsi $(S'/S)_2$ s'identifie à l'objet des triplets (x,y,z) de $S' \times S' \times S'$ tels que :

$$d_0x = d_0y, \quad d_1x = d_0z, \quad d_1y = d_1z, \tag{6.3.1}$$

muni des opérateurs face vers S' définis par :

$$d_0(x,y,z) = x, \quad d_1(x,y,z) = y, \quad d_2(x,y,z) = z \tag{6.3.2}$$

et des opérateurs de dégénérescence définis pour tout $x \in S'$ par

$$s_0(x) = (x, x, s_0d_1x), \quad s_1(x) = (s_0d_0x, x, x). \tag{6.3.3}$$

De même, $(S'/S)_3$ consiste en les sextuplets (x,y,z,u,v,w) de S' tels que :

$$\begin{aligned} d_0x &= d_0y = d_0u \\ d_1x &= d_0z = d_0v \\ d_1y &= d_1z = d_0w \\ d_1u &= d_1v = d_1w. \end{aligned} \tag{6.3.4}$$

Les opérateurs face $d_i\colon (S'/S)_3 \to (S'/S)_2$ sont données par les formules :

$$\begin{aligned} d_0(x,y,z,u,v,w) &= (x,y,z) \\ d_1(x,y,z,u,v,w) &= (x,u,v) \\ d_2(x,y,z,u,v,w) &= (y,u,v) \\ d_3(x,y,z,u,v,w) &= (z,v,w) \end{aligned} \tag{6.3.5}$$

tandis que les $s_i\colon (S'/S)_2 \to (S'/S)_3$ le sont par :

$$\begin{aligned} s_0(x,y,z) &= (x,y,z,y,z,s_0d_1y) \\ s_1(x,y,z) &= (x,x,s_0d_0z,y,z,z) \\ s_2(x,y,z) &= (s_0d_0x,x,x,y,y,z). \end{aligned}$$

En outre, on dispose d'une application de $\operatorname{cosq}^1(S' \rightrightarrows\!\!\!\leftarrow S)$ vers son $0-$*ième* cosquelette $\operatorname{cosq}^0(S' \rightrightarrows\!\!\!\leftarrow S)$, qui n'est autre que le nerf, au sens usuel, du recouvrement $S \to e$ de l'objet final de T, c'est-à-dire l'objet simplicial

$$S \times S \times S \times S \begin{smallmatrix}\to\\\to\\\to\\\to\end{smallmatrix} S \times S \times S \begin{smallmatrix}\to\\\to\\\to\end{smallmatrix} S \times S \rightrightarrows S$$

ou d_i est l'opérateur de projection qui omet le $(i+1)$-ième terme, tandis que s_i est la diagonale qui répète le $(i-1)$-ième terme. La projection en question de $\operatorname{cosq}^1(S' \rightrightarrows S)$ vers le nerf de S

$$(6.3.6)\qquad \begin{array}{ccccccc} (S'/S)_3 & \begin{smallmatrix}\to\\\to\\\to\\\to\end{smallmatrix} & (S'/S)_2 & \begin{smallmatrix}\to\\\to\\\to\end{smallmatrix} & S' & \rightrightarrows & S \\ \downarrow{\scriptstyle p_3} & & \downarrow{\scriptstyle p_2} & & \downarrow{\scriptstyle p_1} & & \| \\ S \times S \times S \times S & \begin{smallmatrix}\to\\\to\\\to\\\to\end{smallmatrix} & S \times S \times S & \begin{smallmatrix}\to\\\to\\\to\end{smallmatrix} & S \times S & \rightrightarrows & S \end{array}$$

est décrite explicitement en bas degré par les formules

$$(6.3.7)\qquad \begin{array}{rll} p_1(x) = (d_1x, d_0x) & & (x \in S') \\ p_2(x,y,z) = (d_1y, d_1x, d_0x) & & (x,y,z) \in (S'/S)_2 \\ p_3(x,y,z,u,v,w) = (d_1u, d_1y, d_1x, d_0x) & & (x,y,z,u,v,w) \in (S'/S)_3. \end{array}$$

Après ces rappels de la théorie simpliciale, nous sommes en mesure de revenir à notre démonstration. Rappelons donc que S' a été choisi de manière que la bitorseur g qui décrit la donnée de recollement (6.2.3) possède une section au dessus de S'. Sa restriction à S' est donc, par (2.6), décrite par un élément, que nous noterons à nouveau g, de $H^0(S', \mathrm{Aut}^o(G))$, et dont on peut supposer que le composé avec $s_0 \colon S \to S'$ est l'identité. Le changement de base p_2 fait alors correspondre à la flèche f (6.2.10) une flèche

$$(6.3.8)\qquad f \colon d_0^*g \cdot d_2^*g \to d_1^*g$$

de $C^0_{|(S'/S)_2}$ et la discussion de 4.7 montre que celle-ci correspond à un élément, que nous noterons encore f, de $H^0(((S'/S)_2, G)$ tel que, pour (6.2.10), on ait la relation

$$(6.3.9)\qquad d_0^*g * d_2^*g * i'_f = d_1^*g$$

dans $H^0((S'/S)_2, \mathrm{Aut}^o(G))$. On vérifie alors, en utilisant notamment la description donnée en (4.3), (4.7) du produit contracté de deux (G,π)-torseurs (resp. de deux bitorseurs), que la condition de compatibilité (6.2.8) correspond à l'identité suivante entre éléments de $H^0((S'/S)_3, G)$:

$$(6.3.10)\qquad {}^{((d_2d_3)^*g)}d_0^*f \cdot d_2^*f = d_3^*f \cdot d_1^*f.$$

On peut enfin supposer que la restriction f de S' au moyen de chacune des deux flèches s_i (6.3.3) est triviale.

Une telle paire de sections (g,f), satisfaisant aux identités (6.3.9) et (6.3.10), définit un morphisme d'objets simpliciaux

(6.3.11) $$\lambda\colon \mathrm{cosq}^1(S' \rightrightarrows S) \to \overline{W}(\mathcal{G}^0)$$

à valeurs dans l'ensemble simplicial associé en (3.11) au groupe simplicial $\mathcal{G}^0$ opposé à $\mathcal{G}$. Les conditions de cocycle (6.3.9)–(6.3.10) montrent en effet que les flèches

(6.3.12) $$\begin{gathered}\lambda_0 \colon S \to e\\ \lambda_1 = g\colon S' \to \mathcal{G}'_0 = \mathrm{Aut}(G)\end{gathered}$$

et les flèches

$$\begin{aligned}\lambda_2 &: (S'/S)_2 \to \mathcal{G}_1^0 \times \mathcal{G}_0^0 = (\mathrm{Aut}^o(G) \times G)^0 \times \mathrm{Aut}(G)\\ \lambda_3 &: (S'/S)_3 \to \mathcal{G}_2^0 \times \mathcal{G}_1^0 \times \mathcal{G}_0^0\end{aligned}$$

définies respectivement par $\lambda_2 = ((d_2^* g, f), d_0^* g)$ et par

$$\lambda_3 = [((d_2 d_3)^* g, d_3^* f, (d_3^* f)^{-1} d_2^* f), \quad ((d_0 d_3)^* g, d_0^* f), (d_0 d_1)^* g]$$

déterminent de manière unique un morphisme

(6.3.13) $$\mathrm{tr}^3(\mathrm{cosq}^1(S' \rightrightarrows S) \longrightarrow \mathrm{tr}^3\, \overline{W}\mathcal{G}^0$$

entre les tronqués à l'ordre trois des ensembles simpliciaux considérés.

Or $\overline{W}(\mathcal{G}^0)$ est isomorphe à son troisième cosquelette (puisque $\mathcal{G}^0$ l'est à son deuxième cosquelette), comme on le vérifie par exemple en utilisant la propriété d'adjonction du foncteur cosquelette (voir *loc.cit.* 0.7), ainsi que celle du foncteur $\overline{W}$ ([45] théorème 27.1). Il en résulte que (6.3.13) détermine de manière unique le morphisme (6.3.11).

6.4. En résumé, la discussion précédente nous a permis d'associer à un $\mathcal{C}$-torseur $\mathcal{D}$ une application λ (6.3.11) dont la source est l'hyper-recouvrement $\mathrm{cosq}^1(S' \rightrightarrows S)$ de e. Soient maintenant $\mathcal{D}$ et $\mathcal{D}'$ deux tels torseurs. On peut les supposer tous deux munis d'objets au dessus d'un raffinement commun suffisamment fin S de e, et également choisir un raffinement S' de $S \times S$ au-dessus duquel les deux bitorseurs g et g' qu'ils définissent possèdent tous deux des sections. Il en résulte une paire d'applications λ, λ' (6.3.11) de source le même hyper-recouvrement $\mathrm{cosq}^1(S' \rightrightarrows S)$

de e. Soit maintenant $S_1 \longrightarrow S$ un raffinement quelconque de S. Le diagramme cartésien :

$$(6.4.1) \qquad \begin{array}{ccc} S'_1 & \longrightarrow & S_1 \times S_1 \\ \big\downarrow & & \big\downarrow \\ S' & \longrightarrow & S \times S \end{array}$$

définit un objet simplicial tronqué $S'_1 \rightrightarrows S_1$ et donc un raffinement $\mathrm{cosq}^1(S'_1 \rightrightarrows S_1)$ de l'hyperrecouvrement $\mathrm{cosq}^1(S' \rightrightarrows S)$ considéré ci-dessus.

Montrons maintenant que la donnée d'un morphisme $\chi: \mathcal{D} \to \mathcal{D}'$ définit, pour un choix d'un raffinement approprié S_1 de S, une homotopie entre les applications

$$(6.4.2) \qquad \tilde{\lambda}, \tilde{\lambda}': \mathrm{cosq}^1(S'_1 \rightrightarrows S_1) \to \overline{W}(\mathcal{G})$$

déduites des applications λ, λ'. Ceci terminera la première partie de cette démonstration, puisqu'on aura ainsi construit une application de l'ensemble pointé des classes de $\mathcal{C}$-bitorseurs vers $H^1(\mathcal{G}) = [e, \overline{W}\mathcal{G}]_{D(T)}$.

Soient donc $\mathcal{D}$ et $\mathcal{D}'$ deux $\mathcal{D}$-torseurs décrits, au moyen d'identifications locales $\mathcal{C}_{|S} \simeq \mathcal{D}_{|S}$ et $\mathcal{C}_{|S} \simeq \mathcal{D}'_{|S}$ par des paires (g, f) et (g', f') satisfaisant aux conditions (6.2.10) et par les conditions de cocycle qui découlent de (6.2.8). Le morphisme $\chi: \mathcal{D} \to \mathcal{D}'$ s'exprime localement au moyen d'un morphisme du $\mathcal{C}_{|S}$-Torseurs

$$(6.4.3) \qquad x: \mathcal{C}_{|S} \longrightarrow \mathcal{C}_{|S}$$

et la compatibilité de x aux données de recollement de $\mathcal{D}$ et de $\mathcal{D}'$ s'exprime par une 2-flèche

$$(6.4.4) \qquad \begin{array}{ccc} \mathcal{C}_{|S\times S} & \xrightarrow{\ \varphi\ } & \mathcal{C}_{|S\times S} \\ {\scriptstyle p_1^* x}\big\downarrow & \overset{\varepsilon}{\swArrow} & \big\downarrow{\scriptstyle p_2^* x} \\ \mathcal{C}_{|S\times S} & \xrightarrow[\ \varphi'\ ]{} & \mathcal{C}_{|S\times S} \end{array}$$

qui doit être compatible aux 2-flèches de type (6.2.6) exprimant les défauts de transitivité de φ et de φ'. Ainsi, le diagramme de 2-flèches constitué par le prisme suivant (dont la face antérieure verticale est la 2-flèche ε_{13}) est commutatif :

(6.4.5)

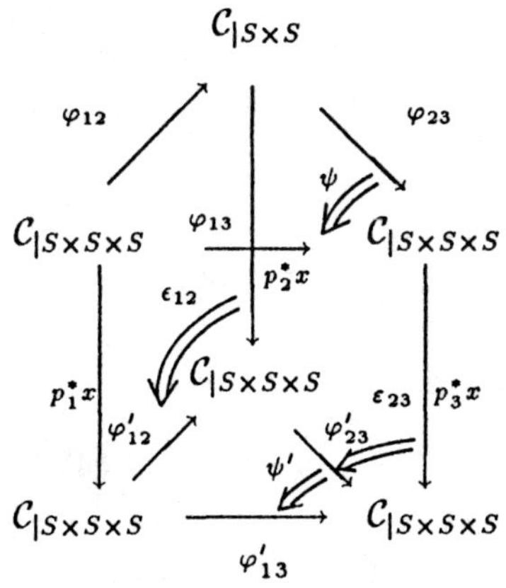

La 1-flèche x est décrite, via (6.2.9), par un objet $\overline{X}$ de $\mathcal{C}_S^0$. On choisit le raffinement S_1 de S de manière que $\overline{X}$ possède une section au-dessus de S_1 et soit donc décrit par une section $h \in H^0(S_1, \mathrm{Aut}^o(G))$. La donnée de la 2-flèche ε (6.4.4) équivaut à celle d'une section de $H^0(S_1 \times S_1, G)$ et celle-ci induit une section (également notée ε) de $H^0(S_1', G)$, telle que l'on ait la relation

(6.4.6) $$g' * d_1^* h = (d_0^* h) * g * i'_\varepsilon$$

dans $H^0(S_1', \mathrm{Aut}^o(G))$ (les éléments g et g' de $H^0(S', \mathrm{Aut}^o(G))$ définis par les bitorseurs du même nom étant ici assimilés aux éléments de $H^0(S_1',\ \mathrm{Aut}^o(G))$ qu'il définissent). La condition de compatibilité (6.4.5) s'exprime de la manière suivante dans $\mathrm{cosq}^1(S_1' \rightrightarrows S_1)_2$, les sections f et f' de $H^0((S_1'/S_1)_2, G)$ étant ici identifiées, comme il a été expliqué plus haut, aux restrictions au-dessus du raffinement $(S_1'/S_1)_2$ de S^3 des 2-flèches ψ et ψ' :

(6.4.7) $$f \cdot d_1^* \varepsilon = ((d_2^* g)(d_0^* \varepsilon)) \cdot (d_2^* \varepsilon) \cdot ((d_1 d_2)^* h)(f').$$

La donnée d'une telle paire (h, ε) définit une homotopie (au sens simplicial, voir [45] définition 5.1) $x\colon d' \simeq d$ entre les flèches $\tilde{\lambda}$, $\tilde{\lambda}'$ (6.4.2) associées aux $\mathcal{C}$-torseurs $\mathcal{D}$ et $\mathcal{D}'$. En effet, telle homotopie est déterminée par les flèches qui la définissent en bas degré, à savoir :

(i) $$h_0 = h\colon S_1 \to (\overline{W}\mathcal{G}^0)_1 = \mathrm{Aut}(G)$$

(*ii*) $$\tilde{h}_0, \tilde{h}_1 : S'_1 \to (\overline{W}\mathcal{G}^0)_2 = (\mathrm{Aut}^\circ(G) \times G)^0 \times \mathrm{Aut}(G)$$

définies par :

$$\tilde{h}_0 = (d_1^* h, 1, g') \quad \text{et} \quad \tilde{h}_1 = (g, \varepsilon, d_0^* h).$$

(*iii*) $$(h_0, h_1, h_2) : (S'_1/S_1)_2 \to (\overline{W}\mathcal{G}^0)_3 = \mathcal{G}_2^0 \times \mathcal{G}_1^0 \times \mathcal{G}_0^0$$

défini par

$$\begin{aligned} h_0 &= ((d_1 d_2)^* h, 1, 1,)\ (d_2^* g', f'),\ d_0^* g') \\ h_1 &= ((d_2^* g, d_2^* \varepsilon, (d_1^* h)(f')),\ (d_0^* h, 1),\ d_0^* g') \\ h_2 &= (d_0^* g, f, ((i'_{f^\circ} d_2^* g)(d_0^* \varepsilon)^{-1}) d_1^* \varepsilon,\ d_0^*(g(\varepsilon)), d_0 d_0 h). \end{aligned}$$

On vérifie que ceci définit bien, compte tenu des conditions de cocycle (6.4.6) et (6.4.7), une homotopie entre les morphismes simpliciaux d et d'.

6.5. Inversement, partons d'un élément de $[e, \overline{W}(\mathcal{G}^0)]_{D.(T)}$, représenté par un morphisme $\lambda : \mathcal{S} \to \overline{W}(\mathcal{G}^0)$ d'objets simpliciaux de T, où

$$\mathcal{S}. = \dots S^{(3)} \ \substack{\rightrightarrows \\ \rightrightarrows}\ S^{(2)} \ \substack{\to \\ \rightrightarrows}\ S' \rightrightarrows S$$

est un hyper-recouvrement quelconque de l'objet final e de T. $\mathcal{S}$ peut être identifié, en ce qui nous concerne, à son troisième cosquelette, et on a des projections naturelles

$$S' = \mathrm{cosq}^3(\mathcal{S}.) \to \mathrm{cosq}^2(\mathcal{S}.) \to \mathrm{cosq}^1(\mathcal{S}.) \to \mathrm{cosq}^0(\mathcal{S}.).$$

Le fait que $\mathcal{S}^{\cdot}$ soit un hyper-recouvrement implique que les composantes de chacune de ces projections sont des épimorphismes. Les objets $\mathrm{cosq}^1(\mathcal{S}.)$ et $\mathrm{cosq}^0(\mathcal{S}.)$ ont été décrits ci-dessus. Quant à $\mathrm{cosq}^2(\mathcal{S}.)$, ses composantes de degré ≤ 2 coïncident avec celles de $\mathcal{S}^{\cdot}$, tandis que $\mathrm{cosq}^2(\mathcal{S})_3$ est l'ensemble des quadruplets $(a, b, c, d) \in S^{(2)}$ tels que :

(6.5.1) $$\begin{matrix} d_0 a = d_0 b & d_0 c = d_1 a & d_0 d = d_2 a \\ d_1 c = d_1 b & d_1 d = d_2 b & d_2 d = d_2 c. \end{matrix}$$

La projection

$$\begin{array}{ccccccc} \mathrm{cosq}^2(\mathcal{S})_3 & \substack{\rightrightarrows \\ \rightrightarrows} & S^{(2)} & \substack{\to \\ \rightrightarrows} & S' & \rightrightarrows & S \\ \big\downarrow{\scriptstyle q_3} & & \big\downarrow{\scriptstyle q_2} & & \| & & \| \\ (S'/S)_3 & \substack{\rightrightarrows \\ \rightrightarrows} & (S'/S)_2 & \substack{\to \\ \rightrightarrows} & S' & \rightrightarrows & S \end{array}$$

est définie sur les composantes qui nous concernent par les flèches q_2 et q_3 données par

(6.5.2) $$q_2(\alpha) = (d_0\alpha,\ d_1\alpha,\ d_2\alpha)$$

et

(6.5.3) $$q_3(a,b,c,d) = (d_0a, d_1a, d_2a, d_1b, d_2b, d_2c).$$

Enfin la projection de $\mathcal{S}$ vers $\operatorname{cosq}^2(\mathcal{S})$ est définie en degré 3 par l'application $r_3\colon \mathcal{S}^{(3)} \to \operatorname{cosq}^2(\mathcal{S})_3$ définie par

(6.5.4) $$r_3(\beta) = (d_0\beta, d_1\beta, d_2\beta, d_3\beta).$$

L'application $\mathcal{S}. \to \overline{W}(\mathcal{G}^0)$ considérée est déterminée, comme on l'a vu, par une paire de sections $g \in \Gamma(S', \mathrm{Aut}^o(G))$ et $f \in \Gamma(S^{(2)}, G)$ qui satisfont aux identités (6.3.9), (6.3.10) et dont les composées avec les opérateurs de dégénérescence s_i sont triviales. Puisque l'application r_3 (6.5.4) est un épimorphisme, la relation (6.3.10), a priori satisfaite entre des sections au dessus de $S^{(3)}$, l'est en fait déjà sur $\operatorname{cosq}^2(\mathcal{S})_3$. En d'autre termes l'application $\mathcal{S}. \to \overline{W}(\mathcal{G}^0)$ considérée se factorise par $\lambda\colon \operatorname{cosq}^2(\mathcal{S}.) \to \overline{W}(\mathcal{G}^0)$. Nous allons maintenant démontrer, en utilisant d'autres propriétés similaires de l'hyper-recouvrement, que l'on a même une factorisation par $\operatorname{cosq}^1(\mathcal{S})$. Pour cela on observe que la formule :

(6.5.5) $$\theta_1(\lambda, \mu) = (s_0d_0\lambda, \lambda, \mu, s_1d_2\mu)$$

définit une application

(6.5.6) $$\theta_1\colon S^{(2)} \times_{(S'/S)_2} S^{(2)} \to \operatorname{cosq}^2(\mathcal{S})_3.$$

On en déduit, en prenant l'image inverse par θ_1 des membres de gauche et de droite de l'identité satisfaite par f, que $pr_1^*(f) = pr_2^*(f)$, où pr_i désigne la projection de la source de θ_1 sur son i ème facteur. Ainsi f est compatible à la descente par l'épimorphisme q_2, et définit un élément que l'on notera à nouveau f, de $H^0((S'/S)_2, G)$. Puisque q_3 est également un épimorphisme, la relation (6.3.10), *a priori* satisfaite au dessus de $\operatorname{cosq}^2(\mathcal{S})_3$, l'est encore sur $(S'/S)_3$, ce qui termine la vérification que l'application associée à la paire initiale (g, f) se factorise par un morphisme $\lambda\colon \operatorname{cosq}^1(\mathcal{S}) \to \overline{W}(\mathcal{G}^0)$.

Nous abordons maintenant la dernière partie de la démonstration. Il s'agit d'effectuer une descente de $\operatorname{cosq}^1(\mathcal{S})$ à $\operatorname{cosq}^0(\mathcal{S})$, mais on aurait tort d'imaginer qu'elle s'exprime à nouveau par la factorisation de l'application

λ à travers $\operatorname{cosq}^0(\mathcal{S})$, et c'est dans le sens plus sophistiqué des champs que nous allons effectuer la descente en question. Soit donc $p_1: S' \to S \times S$ la composante de degré un de cette projection, définie en (6.3.7). Considérons les applications :

$$\theta_2: S' \times_{S \times S} S' \to (S'/S)_2$$
$$\theta_3: S' \times_{S \times S} S' \times_{S \times S} S' \to (S'/S)_3$$

définies par :

(6.5.7) $$\theta_2(\lambda, \mu) = (s_0 d_0 \lambda, \lambda, \mu)$$

(6.5.8) $$\theta_3(\lambda, \mu, \nu) = (s_0 d_0 \lambda, s_0 d_0 \lambda, s_0 d_0 \lambda, \lambda, \mu, \nu).$$

Par la relation (6.3.9) la section $f \in H^0(S'/S)_2, G)$ décrit un isomorphisme

(6.5.9) $$f: d_0^* g \cdot d_2^* g \simeq d_1^* g$$

entre les bitorseurs sur $(S'/S)_2$ images inverse par les d_i du bitorseur sur S (à torseur sous-jacent trivial) défini par g. On en déduit, par image inverse par θ_2 un isomorphisme

(6.5.10) $$\theta_2^*(f): pr_2^* g \xrightarrow{\sim} pr_1^* g$$

entre les images inverses du bitorseur en question par les deux projections $pr_i: S' \times_{S \times S} S' \rightrightarrows S'$. L'image inverse par θ_3 de la condition (6.3.10) implique alors que les isomorphismes satisfont au-dessus de $S' \times_{S \times S} S' \times_{S \times S} S'$ à la condition de recollement. Il en résulte que le bitorseur sur S' défini par la section $g \in H^0(S', \operatorname{Aut}^o(G))$ descend, non pas en une section de $\operatorname{Aut}^o(G)$ sur $S \times S$, mais en un objet sur $S \times S$ du champ associé à $G \to \operatorname{Aut}^o(G)$, c'est à dire en un objet $\mathbf{g}$ sur $S \times S$ du champ $\mathcal{C}$ des G-bitorseurs. Comparons maintenant les bitorseurs $d_0^* \mathbf{g} \cdot d_2^* \mathbf{g}$ et $d_1^* \mathbf{g}$ déduits de $\mathbf{g}$ par les projections $d_i: S^3 \to S^2$. Le faisceau $\operatorname{Isom}(d_0^* \mathbf{g} \cdot d_2^* \mathbf{g}, d_1^* \mathbf{g})$ sur S^3 est muni, par (6.3.9), d'une section au dessus de $(S'/S)_2$ et on sait que ses images inverses dans $(S'/S)_3$ satisfont à la relation (6.3.10). Nous allons montrer que la section en question peut être descendue relativement à l'épimorphisme $p_2: (S'/S)_2 \to S^3$, en introduisant les applications

(6.5.11) $$\theta_4, \theta_5: (S'/S)_2 \times_{S^3} (S'/S)_2 \to (S'/S)_3$$

définies par :

$$\theta_4(x, y, z, x', y', z') = (s_0 d_0 x', x', x, y', y, z)$$
$$\theta_5(x, y, z, x', y', z') = (x', x, s_0 d_1 x', y', z', z).$$

Considérons le diagramme

$$(6.5.12)\qquad \begin{array}{ccc} pr_1^* d_0^* g \cdot pr_1^* d_2^* g & \xrightarrow{pr_1^* f} & pr_1^* d_1^* g \\ \wr\big| & & \big|\wr \\ pr_2^* d_0^* g \cdot pr_1^* d_2^* g & \xrightarrow{p^* f} & pr_2^* d_1^* g \\ \wr\big| & & \big\| \\ pr_2^* d_0^* g \cdot pr_2^* d_2^* g & \xrightarrow{pr_2^* f} & pr_2^* d_1^* g \end{array}$$

où pr_i (resp. p) : $(S'/S)_2 \times_{S^3} (S'/S)_2 \to (S'/S)_2$ désigne la projection sur le i ème facteur (resp. l'application définie par $p(x,y,z,x',y',z') = (x',y',z)$), et les isomorphismes verticaux sont les isomorphismes canoniques exprimant la descente par rapport à p_2. La restriction à $(S'/S)_2 \times_{S^3} (S'/S)_2$ par θ_4 (resp. θ_5) de la relation (6.3.10) exprime la commutativité du carré supérieur (resp. inférieur) de ce diagramme. La commutativité du carré extérieur exprime alors la compatibilité souhaitée de f à la descente, d'où l'isomorphisme souhaité (6.2.10) sur S^3. Puisque cet isomorphisme satisfait à la relation (6.3.10) sur $(S'/S)_3$, et que le morphisme p_3 (6.3.7) est un épimorphisme, la relation en question est satisfaite sur S^3, ce qui signifie que la flèche (6.2.10) satisfait bien aux conditions de compatibilité qui découlent de (6.2.8). On est donc dans la situation examinée au début de cette démonstration, c'est à dire qu'on est bien en présence d'un $\mathcal{C}$-torseur de T muni d'une section au dessus de S, ce qui termine la démonstration de la proposition 6.2.

Remarques 6.6. (i) On peut également exprimer une 2-flèche $\mathcal{D} \underset{x'}{\overset{x}{\Downarrow}} \mathcal{D}'$ entre deux morphismes de $\mathcal{C}$-torseurs en termes de cocyles. L'énoncé auquel on aboutit en considérant la description donnée des $\mathcal{C}$-torseurs, de leur 1-flèches et 2-flèches, exprime que le 2-champ de T dont la fibre S est la 2-catégorie fibrée TORS$(\mathcal{C}_{|S})$ équivaut au 2-champ associé, en un sens approprié, à celui défini par les sections de l'objet simplicial $\overline{W}(\mathcal{G}^0)$ de T. Malheureusement, la notion de 2-champ n'est pas formalisée dans la littérature (voir cependant [34] I.4.3 pour quelques indications à ce propos) et il nous paraît hors de notre propos de considérer ici cette question.

(ii) Des conditions de cocycle analogue à (6.3.9), (6.3.10) figurent notamment dans [24] II définition 2.0 3°c, ou encore, dans le cas où T est le topos classifiant d'un groupe, dans [15] (voir également dans ce cas (8.10.3)–(8.10.4) ci-dessous).

6.7. Il nous faut maintenant examiner la fonctorialité en $\mathcal{C}$ de la 2-catégorie TORS($\mathcal{C}$) associée à une gr-catégorie $\mathcal{C}$, et pour cela esquisser une théorie du produit contracté dans les catégories. Soient $\mathcal{D}$ et $\mathcal{D}'$ deux groupoïdes. On suppose donnée une action à gauche de $\mathcal{C}$ sur $\mathcal{D}'$ satisfaisant aux conditions exprimées par la commutativité des diagrammes (6.1.3) et (6.1.4), mais on ne suppose pas pour l'instant que $\mathcal{D}'$ soit $\mathcal{C}$-pseudo torseur, (c'est à dire que $\mathcal{C} \times \mathcal{D}' \to \mathcal{D}' \times \mathcal{D}'$ soit une équivalence). On suppose également que $\mathcal{C}$ agit à droite, avec des conditions de commutativité analogues, sur $\mathcal{D}$. On définit alors* un groupoïde produit contracté $\mathcal{D} \overset{\mathcal{C}}{\wedge} \mathcal{D}'$, dont les objets sont ceux de la catégorie produit $\mathcal{D} \times \mathcal{D}'$, mais pour lequel une flèche $(D_1, D_1') \to (D_2, D_2')$ consiste en une classe d'équivalence de triples $(\varphi_1, G, \varphi_1')$, où G est un objet de $\mathcal{C}$, $\varphi_1 \colon D_1 G \to D_2$ (resp. $\varphi_1' \colon D_1' \to G D_2'$) une flèche de $\mathcal{D}$ (resp. $\mathcal{D}'$), deux tels triples $(\varphi_1, G, \varphi_1')$ et (ψ_1, G', ψ_1') étant équivalents s'il existe un isomorphisme $\lambda \colon G \to G'$ dans $\mathcal{C}$ tel que les diagrammes

(6.7.1) 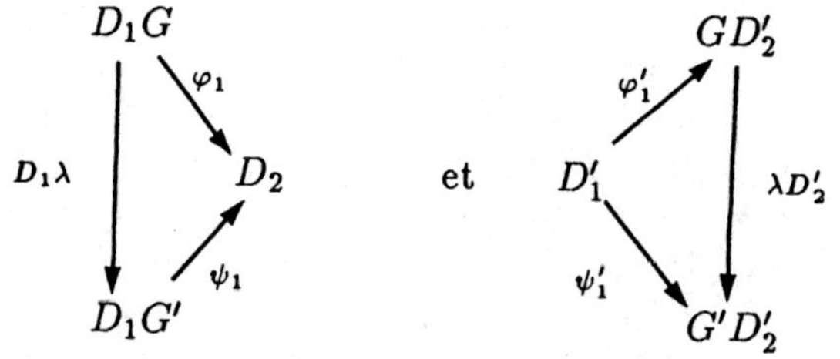

commutent. La composée de deux flèches
$(\varphi_1, G, \varphi_1') \colon (D_1, D_1') \to (D_2, D_2')$ et $(\varphi_2, H, \varphi_2') \colon (D_2, D_2') \to (D_3, D_3')$
est la flèche définie par l'objet GH de $\mathcal{C}$ et les classes des deux flèches

$$
(6.7.2) \qquad
\begin{array}{l}
D_1(GH) \xrightarrow{\mu} (D_1 G)H \xrightarrow{\varphi_1 H} D_2 H \xrightarrow{\varphi_2} D_3 \\[2ex]
D_1' \xrightarrow{\varphi_1'} G D_2' \xrightarrow{G\varphi_2'} G(H D_3') \xrightarrow{\mu} (GH) D_3'
\end{array}
$$

μ désignant l'isomorphisme canonique d'associativité (6.1.2) dans chacun des cas. Enfin le morphisme identique associé à l'objet (D, D') de $\mathcal{D}_1 \overset{\mathcal{C}}{\wedge} \mathcal{D}'$ est défini par l'objet unité O de $\mathcal{C}$ et les flèches canoniques $DO \to D$ et $D' \to OD'$.

Ce produit contracté possède les propriétés auxquelles on s'attend. Ainsi, il est associatif en ce sens que l'on dispose d'une famille cohérente d'équivalences $(\mathcal{D}_1 \overset{\mathcal{C}}{\wedge} \mathcal{D}_2) \overset{\mathcal{C}'}{\wedge} D_3 \simeq \mathcal{D}_1 \overset{\mathcal{C}}{\wedge} (\mathcal{D}_2 \overset{\mathcal{C}'}{\wedge} \mathcal{D}_3)$ et unitaire, l'objet unité

* Je dois cette définition à J. Bénabou.

$\mathcal{O}$ définissant des équivalences

$$\mathcal{D}_1 \to \mathcal{D}_1 \overset{\mathcal{C}}{\wedge} \mathcal{C}, \qquad \mathcal{D}_2 \to \mathcal{C} \overset{\mathcal{C}}{\wedge} \mathcal{D}_2. \tag{6.7.3}$$

On observera par ailleurs que la définition du produit contracté s'étend des torseurs sous une gr-catégorie aux torseurs sous un gr-champ $\mathcal{C}$. En effet si $\mathcal{D}_1$ et $\mathcal{D}_2$ sont des torseurs sous un tel $\mathcal{C}$, alors la construction précédente nous fournit une catégorie fibrée, et on désigne ici par $\mathcal{D}_1 \overset{\mathcal{C}}{\wedge} \mathcal{D}_2$ le champ associé à celle-ci ([26] II §2).

La définition qui vient d'être esquissée du produit contracté nous permet maintenant de comprendre quelle est la fonctorialité de TORS($\mathcal{C}$) : soit en effet $\varphi: \mathcal{C} \to \mathcal{C}'$ un morphisme de gr-champs. Le morphisme φ définit une action à droite de $\mathcal{C}$ sur $\mathcal{C}'$ à travers φ, qui satisfait aux axiomes (6.1.3)–(6.1.4). Ainsi φ définit un foncteur

$$\varphi_*: \mathrm{TORS}(\mathcal{C}) \to \mathrm{TORS}(\mathcal{C}') \tag{6.7.4}$$

qui associe à un $\mathcal{C}$-torseur $\mathcal{D}$ le produit contracté $\varphi_*(\mathcal{D}) = \mathcal{C}' \overset{\mathcal{C}}{\wedge} \mathcal{D}$, muni de l'action à gauche de $\mathcal{C}'$ induite par la structure de gr-champ donnée de $\mathcal{C}'$. Il en résulte notamment, par passage aux classes d'équivalence, un morphisme d'ensembles pointés

$$H^1(\varphi): H^1(\mathcal{C}) \to H^1(\mathcal{C}'). \tag{6.7.5}$$

On observera par ailleurs que pour tout objet $\mathcal{D}$ de TORS($\mathcal{C}$), il existe un foncteur $\varphi_{\mathcal{D}}: \mathcal{D} \to \varphi_*(\mathcal{D})$ induit par φ, qui est défini par

$$\mathcal{D} \xrightarrow{\sim} \mathcal{C} \overset{\mathcal{C}}{\wedge} \mathcal{D} \xrightarrow{\varphi \wedge 1} \mathcal{C}' \overset{\mathcal{C}}{\wedge} \mathcal{D} = \varphi_*(\mathcal{D}). \tag{6.7.6}$$

dont la première flèche est celle de (6.7.3), et que $\varphi_{\mathcal{D}}$ est φ-équivariant, en ce sens qu'il existe une transformation naturelle $\varphi_{\mathcal{D}}(CD) \sim \varphi(C)\varphi_{\mathcal{D}}(D)$ compatible en un sens évident aux isomorphismes d'associativité μ et d'unicité C_D définis en 6.1.

Remarque 6.8. Le morphisme (6.7.5) s'interprète, en termes topologiques, de la manière suivante : soit G (resp. G') le nerf de $\mathcal{C}$ (resp. $\mathcal{C}'$). Ces objets, à défaut d'être des groupes simpliciaux de T, sont des ensembles simpliciaux, munis d'un loi de composition associative à homotopie près, et qui en fait des "A_∞-espaces" au sens de Stasheff [56], puisqu'on c'est donné la condition d'associativité optimale, décrite par la condition du pentagone (les conditions d'associativité supérieures n'apparaissent pas ici puisque $\pi_i G = 0$ pour $i \geq 2$ lorsque $\mathcal{C}$ est un gr-champ en groupoïdes). Or à tout A_∞-espace G est associé, comme dans le cas où G est un groupe, son espace classifiant BG, et cette construction est fonctorielle en le A_∞-espace G (voir [57] §8, [1] p. 34, [50] ; le point est fait dans l'introduction de [58] sur les différentes constructions possibles d'un tel classifiant, ce texte étant par ailleurs consacré à la comparaison entre celles-ci). Un morphisme de gr-champs $\varphi: \mathcal{C} \to \mathcal{C}'$ induit un morphisme de A_∞-espaces $\varphi: G \to G'$ et donc une application $BG \to BG'$ d'ensembles pointés, qui en vertu de la proposition 6.2, induit (6.7.4)–(6.7.5). Remarquons enfin que le foncteur classifiant B induit, par passage aux classes d'homotopie, une application

$$B: [G, G']_{A_\infty} \longrightarrow [BG, BG']_{pt} \tag{6.8.1}$$

des classes d'homotopie de morphismes de A_∞-espaces à celles des objets simpliciaux pointés, et cette application est même une bijection d'inverse l'application induit par le foncteurs "espace" de lacets Ω :

$$\Omega: [X, Y]_{pt} \longrightarrow [\Omega X, \Omega Y]_{A_\infty} \tag{6.8.2}$$

(pour une démonstration de la bijectivité de (6.8.1) dans le cas où G et G' sont des groupes, voir [57] théorème 8.12 ; il est simple de se ramener à ce cas dans la situation qui nous concerne, en remplaçant, par [46] proposition 4.2, les gr-champs $\mathcal{C}$ et $\mathcal{C}'$ par des gr-champs équivalents dans lesquels la multiplication est strictement associative).

7. Gerbes Associées à un Groupe

Nous venons d'identifier le terme $H^1(\mathcal{M})$ à un ensemble de classes à équivalence près de torseurs sous le gr-champ défini par $\mathcal{M}$. Ici, nous comparons cette ensemble à celui des classes des gerbes introduites par Giraud dans [26] (voir également [18] Appendice). Pour une comparaison des gerbes à des torseurs sous des objets appropriés, voir également [23], [24] I §5,6 (et notamment le théorème 5.9), ainsi que [20] chapitres II et III (notamment le corollaire III.4.3).

Rappelons brièvement qu'on appelle gerbe sur un site E la donnée d'un champ en groupoïde $\mathcal{D}$ sur le site tel que :

(i) Il existe un raffinement R de E tel que pour tout $S \in \mathrm{Ob}(R)$ la fibre $\mathcal{D}_S$ de $\mathcal{D}$ en S soit non vide.

(ii) Pour tout $S \in \mathrm{Ob}(E)$, deux objets quelconques de D_S sont localement isomorphes.

On introduit ici une notion plus restrictive :

Définition 7.1. Soit G un faisceau de groupes sur E. On appelle G-gerbe un champ en groupoïdes $\mathcal{D}$ pour lequel il existe une famille d'équivalences

$$\gamma_{(S)}: \mathcal{D}_{(S)} \to \mathrm{TORS}(S, G_S) \tag{7.1.1}$$

entre les restrictions $\mathcal{D}_{(S)}$ de $\mathcal{D}$ au-dessus des objets S d'un raffinement R de E et les champs de G-torseurs au-dessus des objets S. Un morphisme de G-gerbes est un morphisme de gerbes $\mathcal{D} \to \mathcal{D}'$ qui est compatible, au-dessus d'un raffinement commun $\tilde{R}$ de E, aux équivalences $\gamma_{(S)}$ et $\gamma'_{(S)}: \mathcal{D}'_{(S)} \to \mathrm{TORS}(S, G_S)$.

Une G-gerbe est une gerbe au sens usuel : en effet, la catégorie $\mathrm{TORS}(S, G_S)$ n'est jamais vide, puisqu'elle contient le G_S-torseur trivial, et d'autre part les objets de $\mathrm{TORS}(S, G_S)$ sont localement isomorphes, donc la condition (ii) est également satisfaite. Par ailleurs, il est clair que la gerbe $\mathrm{TORS}(G)$ est munie d'une structure de G-gerbe. On dira que c'est la G-gerbe triviale. Inversement, si $\mathcal{D}$ est une gerbe au sens usuel, alors la donnée d'une famille d'équivalences (7.1.1) équivaut à celle d'une famille d'objets X_S dans les fibres $\mathcal{D}_S$ de $\mathcal{D}$ au-dessus des objets S du raffinement R de E, et d'une famille d'isomorphismes

$$\gamma_S: G_S \to \mathrm{Aut}(X_S). \tag{7.1.2}$$

En effet, toute gerbe est localement une gerbe de torseurs. Plus précisément, par [26] III corollaire 2.2.6, la donnée d'objets X_S dans les fibres $\mathcal{D}_S$ d'une gerbe quelconque $\mathcal{D}$ détermine une famille d'équivalences

$$\mathcal{D}_{(S)} \to \mathrm{TORS}(S, \mathrm{Aut}(X_S)). \tag{7.1.3}$$

Si l'on dispose en outre d'une famille d'isomorphismes γ_S (7.1.2), pour un groupe G défini sur E tout entier, alors on en déduit, en composant (7.1.3) et l'équivalence déduite de (7.1.2), une famille d'équivalences (7.1.1).

Désignons par $\mathrm{GER}(G)$ la 2-catégorie des G-gerbes de E. Nous allons maintenant examiner quel est le lien entre cette notion et celle de torseur sous $\mathrm{BITORS}^0(G)$. Avant d'en arriver là, rappelons quel est l'énoncé analogue à celui auquel nous serons intéressés, dans le cas plus élémentaire des torseurs sous un groupe H, plutôt que sous un gr-champ tel que $\mathrm{BITORS}^0(G)$. Soit donc F, et un faisceau d'ensemble sur E, muni d'une action à droite d'un groupe H, et P un H-torseur. Le faisceau $F^P = F \overset{H}{\wedge} P$, est dit l'objet tordu de F par P ([26] III, définition 2.3.2, exemple 2.3.6). Remarquons par ailleurs que le passage de F à F^P est la construction, bien connue notamment en topologie, par laquelle on associe à un H-torseur P l'espace fibré sur e de fibre F correspondant. L'objet F^P ainsi obtenu est localement isomorphe à F, le choix d'une trivialisation locale s de P détermine un isomorphisme local $1 \wedge s \colon F \to F^P$. Inversement, à un faisceau $\mathcal{F}$ sur E, localement isomorphe à F, est associé le faisceau $\mathrm{Isom}(F, \mathcal{F})$, qui est un $\mathrm{Aut}^o(F)$-torseur (et même un $(\mathrm{Aut}^o(F), \mathrm{Aut}^o(\mathcal{F}))$-bitorseur. Si l'on part d'une H-torseur P, alors la flèche $P \to \mathrm{Isom}(F, F^p)$ qui envoie une section s de P vers $1 \wedge s$ induit un isomorphisme d'$\mathrm{Aut}^o(F)$-torseurs :

$$\mathrm{Aut}^o(F) \overset{H}{\wedge} P \longrightarrow \mathrm{Isom}(F, F^P) \tag{7.1.4}$$

(voir [26] III.2.3.2.1). En particulier, lorsque l'homomorphisme $H \to \mathrm{Aut}^o(F)$ qui décrit l'action de H sur F est un isomorphisme, on obtient un isomorphisme :

$$P \longrightarrow \mathrm{Isom}(F, F^P). \tag{7.1.5}$$

Inversement, le morphisme d'évaluation $F \times \mathrm{Isom}(F, \mathcal{F}) \longrightarrow \mathcal{F}$ induit par passage au quotient un morphisme de $\mathrm{Aut}(\mathcal{F})$-torseurs à droite $F \overset{\mathrm{Aut}^o(F)}{\wedge} \mathrm{Isom}(F, \mathcal{F}) \longrightarrow \mathcal{F}$. En particulier, lorsque $H = \mathrm{Aut}^o(\mathcal{F})$, on a donc un isomorphisme :

$$F^{\mathrm{Isom}(F,\mathcal{F})} \simeq \mathcal{F}. \tag{7.1.6}$$

On retiendra de cette discussion le fait que les foncteurs $F \overset{H}{\wedge} -$ et $\mathrm{Isom}(F, -)$ définissent, lorsque $H = \mathrm{Aut}^o(F)$, une équivalence entre la catégorie des H-torseurs et celle des objets localement isomorphes à F. Dans le cas des Gl_n-torseurs et des fibrés vectoriels de rang n associés, l'identification correspondante est d'ailleurs bien familière.

7.2. Dans le cas qui nous intéresse, le rôle du groupe H est joué par le gr-champ $\mathcal{C} = \mathrm{BITORS}^0(G)$. Ce gr-champ agit à droite sur le champ $\mathrm{TORS}(G)$ par la flèche induite par l'action à gauche de $\mathrm{BITORS}(G)$ sur $\mathrm{TORS}(G)$ définie par le produit contracté :

$$\begin{aligned} \mathrm{TORS}(G) \wedge \mathrm{BITORS}^0(G) &\longrightarrow \mathrm{TORS}(G) \\ (Q, P) &\longrightarrow P \overset{G}{\wedge} Q \end{aligned} \tag{7.2.1}$$

En outre, l'important énoncé [26] IV. 5.2.5.3 (conséquence peu mise en valeur dans *loc. cit.* de la proposition III, 2.6.1) montre que la flèche :

$$\begin{aligned} \mathrm{BITORS}^0(G) &\longrightarrow \mathrm{Equ}^0(\mathrm{TORS}(G)) \\ P &\longrightarrow (Q \to P \overset{G}{\wedge} Q) \end{aligned} \tag{7.2.2}$$

induite par 7.2.1 est une équivalence du gr-champ BITORS$^0(G)$ avec le gr-champ opposé à celui des auto-équivalences du champ TORS(G).

Un raisonnement analogue à celui indiqué ci-dessus pour les torseurs sous les groupes donne maintenant l'énoncé suivant :

Proposition 7.3. *Les foncteurs quasi-inverses*

(7.3.1) $$\Phi : \mathcal{D} \longrightarrow \mathrm{TORS}(G) \overset{\mathrm{BITORS}^0(G)}{\wedge} \mathcal{D}$$

et

(7.3.2) $$\Psi : \mathcal{E} \longrightarrow \mathrm{Equ}(\mathrm{TORS}(G), \mathcal{E})$$

définissent une équivalence entre le 2-champ des BITORS$^0(G)$-Torseurs et celui des G-gerbes. (Equ *désignant ici le champs des équivalences, voir* [26] IV *proposition* 5.2.5).

Pour démontrer cette proposition, il suffit de réécrire dans ce contexte la démonstration qui vient d'être rappelée de la relation entre H-torseurs et objets localement isomorphe au H-objet F. Contentons nous de vérifier les deux assertions suivantes. D'une part $\Phi(\mathcal{D})$ est bien, pour tout BITORS$^0(G)$-torseur $\mathcal{D}$, une G-gerbe. En effet, la fibre $\mathcal{D}_{(S)}$ de $\mathcal{D}$ en S est par hypothèse localement non vide. Le choix d'un objet d de $\mathcal{D}_{(S)}$ définit une équivalence d: BITORS$^0(G_S) \xrightarrow{\sim} \mathcal{D}_S$, d'où une équivalence induite

$$\mathrm{TORS}(G_S) \longrightarrow \mathrm{TORS}(G_S) \overset{\mathrm{BITORS}^0(G_S)}{\wedge} \mathrm{BITORS}^0(G_S)$$
$$\xrightarrow{1\wedge d} \mathrm{TORS}(G_S) \wedge \mathcal{D}_S$$

(où la première flèche est définie par (6.7.3)), qui munit $\Phi(\mathcal{D})$ d'une structure de G-gerbe. Soit d'autre part $\mathcal{E}$ une G-gerbe, alors l'identification déjà signalée de Equ^0(TORS(G), TORS(G)) avec BITORS$^0(G)$ montre bien que $\Psi(\mathcal{E})$ est muni d'une action à gauche de BITORS$^0(G)$ et deux objets de φ, ψ de $\Psi(\mathcal{E})$, différent par une auto-équivalence $\psi \circ \varphi^{-1}$ de TORS(G), ce qui confère bien à $\Psi(\mathcal{E})$ une structure de pseudo-torseur sous BITORS$^0(G)$. Enfin $\Psi(\mathcal{E})$ est bien localement non vide, puisqu'une équivalence locale de $\mathcal{E}$ avec TORS(G) identifie localement, comme on l'a vu, Equ(TORS$(G), \mathcal{E}$) à BITORS(G).

7.4. La proposition précédente nous permet d'interpréter en termes de gerbes les constructions cohomologiques faisant intervenir l'ensemble $H^1(\mathcal{G})$

examinées aux paragraphes précédents. Ainsi, on dispose maintenant, à la différence de [26], d'une énoncé de fonctorialité pour les G-gerbes. Soit en effet $\gamma:(G \to \mathrm{Aut}^o(G)) \to (G' \to \mathrm{Aut}^o(G'))$ un morphisme de modules croisés (voir la définition 3.2). Celui-ci définit par la proposition 2.9 un morphisme $\Gamma:\mathcal{G} \to \mathcal{G}'$ entre les gr-champ opposés aux gr-champs des bitorseurs correspondants, induisant par (6.7.4) un morphisme

$$\Gamma_*:\mathrm{TORS}(\mathcal{G}) \longrightarrow \mathrm{TORS}(\mathcal{G}'). \tag{7.4.1}$$

A Γ_* correspond un morphisme de 2-champs

$$\Gamma_{**}:\mathrm{GER}(G) \longrightarrow \mathrm{GER}(G') \tag{7.4.2}$$

tel que le carré de 2-champs

$$\begin{array}{ccc} \mathrm{TORS}(G) & \underset{\psi}{\overset{\Phi}{\rightleftarrows}} & \mathrm{GER}(G) \\ \Gamma_*\big\downarrow & & \big\downarrow\Gamma_{**} \\ \mathrm{TORS}(\mathcal{G}') & \underset{\psi}{\overset{\Phi}{\rightleftarrows}} & \mathrm{GER}(G') \end{array} \tag{7.4.3}$$

soit essentiellement commutatif, et dont la flèche Γ_{**} s'explicite de la manière suivante : un objet $\mathcal{E}$ de $\mathrm{GER}(G)$ est localement équivalent à $\mathrm{TORS}(G)_S$, la donnée de recollement entre deux telles trivialisations de $\mathcal{E}$ étant décrite par une auto-équivalence de $\mathrm{TORS}(G)_{|S\times_e S}$, c'est à dire compte tenu de (7.2.2), par une section s de $\mathrm{BITORS}(G)$ au dessus de $S \times_e S$. L'objet $\mathcal{E}' = \Gamma_{**}(\mathcal{E})$ correspondant de $\mathrm{GER}(G')$ est défini localement par $\mathcal{E}'_{|S} = \mathrm{TORS}(G')_S$, la donnée de recollement permettant de construire $\mathcal{E}'$ à partir de sa définition locale étant fournie par l'auto-équivalence de $\mathrm{TORS}(G')_{|S\times_e S}$ définie par la section $\Gamma(s)$ de $\mathrm{BITORS}(G')$ au dessus de $S \times S$. Enfin on dispose maintenant d'un morphisme de champs $\mathcal{E} \to \mathcal{E}'$ défini localement par le foncteur canonique $\mathrm{TORS}(G)_{|S} \to \mathrm{TORS}(G')_{|S}$ de changement de groupe structural : en effet, l'équivariance de ce foncteur par rapport à Γ implique qu'il est compatible aux données de recollement, et il est donc globalement défini.

7.5. Il est maintenant possible d'interpréter au niveau des gerbes les suites exactes (4.3.17) et (5.2.3). De ce point de vue, la plus intéressante des deux suites est la seconde (en ce qui concerne la première, voir 7.8). Les flèches α et β situées à la gauche de 5.2.4, qui ont été décrites en termes géométriques en 5.2, induisent par faisceautisation des morphismes de gr-champs (le terme de droite désignant le gr-champ discret de groupe d'objets $\mathrm{Out}^0(G)$) :

$$\mathrm{TORS}(ZG) \xrightarrow{\alpha} \mathrm{BITORS}^0(G) \xrightarrow{\beta} \mathrm{Out}^0(G)$$

et donc des flèches

$$H^1(\mathrm{TORS}(ZG)) \xrightarrow{H^1\alpha} H^1(\mathrm{BITORS}^0(G)) \xrightarrow{H^1\beta} H^1(\mathrm{Out}^0(G))$$

entre les classes de torseurs sous les gr-champs en question. Laissons de côté ici l'interprétation de $H^1(\alpha)$, pour laquelle il faudrait identifier $H^1(\mathrm{TORS}(ZG))$ au groupe de cohomologie abélienne usuel $H^2(ZG)$ (voir à ce propos [22], [5], et également, pourvu que l'on identifie $H^1(\mathrm{TORS}(ZG))$ aux classes de gerbes abéliennes correspondantes, [26] IV, 3.4), pour en venir à l'interprétation de $H^1(\beta)$, qui permet de relier les groupes de cohomologie introduits ici à ceux de Giraud. Commençons tout d'abord par observer que la variante élémentaire du foncteur (7.3.1) rappelée en 7.1, associe à un torseur P sous le groupe $\mathrm{Out}^0(G)$ le lien $\varphi(P) = \mathrm{lien}(G) \overset{\mathrm{Out}^0(G)}{\wedge} P$ où l'isomorphisme

$$\mathrm{Hom}_{\mathrm{lien}}(\mathrm{lien}(G), \mathrm{lien}(G)) \simeq \mathrm{Out}(G) \tag{7.5.1}$$

de [26] IV Corollaire 1.1.7.3 définit l'action à droite de $\mathrm{Out}^0(G)$ sur $\mathrm{lien}(G)$. Le lien $\varphi(P)$ est, par construction, localement isomorphe à $\mathrm{lien}(G)$, et l'isomorphisme (7.5.1) permet d'associer à tout lien L localement isomorphe à $\mathrm{lien}(G)$ un G-torseur $\Psi(L)$ défini par

$$\Psi(L) = \mathrm{Hom}_{\mathrm{lien}}(\mathrm{lien}(G), L), \tag{7.5.2}$$

le foncteur Ψ ainsi défini étant quasi-inverse de Φ. Désignons par $\mathrm{LIEN}(G)$ le champ des liens localement isomorphes à $\mathrm{lien}(G)$. Puisque un morphisme de gerbes induit, par loc. cit. IV corollaire 2.2.3, un morphisme de liens, on dispose d'un foncteur lien :$\mathrm{GER}(G) \to \mathrm{LIEN}(G)$.

Proposition 7.6. *Le carré de 2-champs*

$$\begin{array}{ccc} \mathrm{TORS}(\mathrm{BITORS}^0(G)) & \xrightarrow{\Phi} & \mathrm{GER}(G) \\ {\scriptstyle \mathrm{TORS}(\beta)}\downarrow & & \downarrow{\scriptstyle \mathrm{lien}} \\ \mathrm{TORS}(\mathrm{Out}^0(G)) & \xrightarrow[p]{} & \mathrm{LIEN}(G) \end{array} \tag{7.6.1}$$

(dont la ligne inférieure est constituée par un morphisme de 1-champs, vus comme 2-champs dépourvus de 2-flèches non triviales) est essentiellement commutatif.

Preuve. Si P est un G-bitorseur, et

$$j_P\colon \mathrm{TORS}(G) \longrightarrow \mathrm{TORS}(G) \tag{7.6.2}$$

l'auto-équivalence de la gerbe triviale qui lui est associée par (7.2.2), alors l'isomorphisme induit

$$\mathrm{lien}(j_p)\colon \mathrm{lien}(G) \longrightarrow \mathrm{lien}(G) \tag{7.6.3}$$

est, par *loc. cit.* IV lemme 3.2.5.(i), 3.2.5.1, l'automorphisme de lien(G) décrit, via (7.5.1), par la section $\beta(P)$ de $\mathrm{Out}^0(G)$. Si $\mathcal{D}$ est un objet de $\mathrm{TORS}(\mathrm{BITORS}^0(G))$, de G-gerbe associée $\Phi(\mathcal{D})$, alors les données de recollement qui permettent de reconstruire $\mathcal{D}$ à partir de sa description locale sont fournies, comme on l'a vu, par une section locale P de BITORS(G) sur un hyperrecouvrement approprié S' de l'objet final e de T, tandis que les données de recollement correspondantes pour $\Phi(\mathcal{D})$ sont fournies par l'auto-équivalence locale (7.6.2) correspondante. Ainsi, le lien de $\Phi(\mathcal{D})$ n'est autre, localement, que lien(G), recollé au moyen de l'isomorphisme (7.6.3) déduit de j_P, c'est à dire, comme on vient de le voir, au moyen de la section correspondante $\beta(P)$ de $\mathrm{Out}^0(G)$ sur l'hyperrecouvrement. Mais ceci revient à dire que lien($\Phi(\mathcal{D})$), est l'objet de fibre lien(G) associé à l'$\mathrm{Out}^0(G)$-torseur $\beta_*(\mathcal{D})$, ce qui démontre la commutativité du diagramme (7.6.1).

7.7. Soit G un groupe de T et $\mathbf{G}$ une gerbe dont le lien L est localement isomorphe à lien(G), et définit donc par *loc. cit.* IV corollaire 1.1.7.3 un élément $[L] \in H^1(\mathrm{Out}^0(G))$. Pour toute section locale x de $\mathbf{G}, \mathrm{Aut}(x)$ est un représentant de L, d'où, quitte à restreindre encore, un isomorphisme de groupes $\mathrm{Aut}(x) \xrightarrow{\sim} G$, défini à conjugaison intérieure dans G près (voir *loc. cit.* IV proposition 1.1.7). Ainsi la gerbe $\mathbf{G}$, localement équivalente à TORS(Aut(x)), est localement équivalente à TORS(G), et c'est donc, dans notre terminologie, une G-gerbe. L'ensemble noté $H^2(L)$ par Giraud, des classes d'équivalences de gerbes liées par L s'envoie surjectivement, par oubli de l'isomorphisme $L \xrightarrow{\sim} \mathrm{lien}(\mathbf{G})$, pour L localement isomorphe à lien(G), vers la fibre en $[L]$ de la flèche $H^1(\beta)\colon H^1(\mathcal{G}) \to H^1(\mathrm{Out}^0(G))$, que nous noterons $\mathsf{H}^2(L)$ pour la distinguer de $H^2(L)$. Un élément de $\mathsf{H}^2(L)$ est donc une classe d'équivalence de gerbes $\mathbf{G}$ dont le lien est isomorphe (par un isomorphisme non spécifié), à L. Du point de vue cohomologique qui a été exposé ici, l'ensemble $H^1(\mathcal{G})$ des G-gerbes semble un objet plus intéressant que le sous-ensemble $\mathsf{H}^2(L)$ ou que $H^2(L)$. Le seul argument qui puisse faire pencher la balance en faveur de ce dernier est le fait que c'est lui qui pour G abélien s'identifiera au groupe de cohomologie usuel $H^2(G)$).* On a notamment vu que la fonctorialité par rapport au module croisé $\mathcal{G}$ s'exprime le plus naturellement sur l'ensemble $H^1(\mathcal{G})$ tout entier. Néanmoins, comme la suite exacte (5.2.3) est tout entière fonctorielle en $\mathcal{G}$, on en déduit un énoncé de fonctorialité pour $\mathsf{H}^2(L)$ dont nous n'aurons pas besoin par la suite, mais qu'il paraît intéressant de décrire brièvement. Soit $\gamma = (u, v)\colon \mathcal{G} \to \mathcal{G}'$ un morphisme au sens de (3.2) entre les modules croisés associés à des groupes G et G', et $\overline{v}\colon \mathrm{Out}^0(G) \to \mathrm{Out}^0(G')$ la flèche induite par v. A tout L localement isomorphe à lien(G) correspond un $\mathrm{Out}^0(G)$-torseur P défini par (7.5.2), et donc, par extension du groupe structural, un $\mathrm{Out}^0(G')$-torseur $\bar{v}_*(P)$ lui-même associé à un lien $\bar{v}_*(L)$ localement isomorphe à lien(G'). Mais on a vu en 7.4 qu'à γ était associé une paire de foncteurs compatibles (7.4.1) et (7.4.2), et le caractère fonctoriel de la proposition (7.6) implique notamment que les foncteurs en question induisent au niveau des liens (resp. des $\mathrm{Out}^0(G)$-torseurs) les applications $\bar{v}_*$ qui viennent d'être mentionnées. Ainsi $\gamma = (u, v)$ induit une flèche $\gamma_*\colon H^2(X, L) \to H^2(X, \bar{v}_*(L))$, restriction au sous-ensemble $\mathsf{H}^2(X, L)$ de l'application $\gamma_*\colon H^1(\mathcal{G}) \to H^1(\mathcal{G}')$ définie par γ.

7.8. Il est également question dans *loc. cit.* IV, 3.1.1 du sous-ensemble $H^2(L)'$ de $H^2(L)$, défini comme l'ensemble des classes d'équivalence, de gerbes représentées par des gerbes neutres, c'est à dire qui admettent des sections, ou encore qui sont de la forme TORS(G) où $G = \mathrm{Aut}(x)$ est le groupe des automorphismes de la section x donnée de la gerbe considérées (*loc. cit.* III corollaire 2.2.6). Ce sous-ensemble s'interprète aisément une fois décrite en termes géométriques la flèche $H^1(\mathrm{Aut}^0(G)) \to H^1(\mathcal{G})$ de (4.3.16). On obtient une première interprétation, assez formelle, du sous-ensemble analogue $\mathsf{H}^2(L)'$ de $\mathsf{H}^2(L)$ en remarquant que la construction 2.2 (iii) définit un morphisme de *gr*-champs

$$\text{(7.8.1)} \qquad \mathrm{Aut}^0(G) \longrightarrow \mathrm{BITORS}^0(G)$$

du champ discret d'objets les sections de $\mathrm{Aut}^0(G)$ vers le champ opposé à

---

* La distinction qu'il y a lieu de faire entre la fibre $\mathsf{H}^2(L)$ de $[L]$ en $H^1(\beta)$ et l'ensemble $H^2(L)$ a été signalée par Ulbrich, dans le cas des extensions, dans l'introduction de [60]. On prendra garde que l'assertion, dans [23] théorème 5, que ces deux ensembles peuvent être mis en bijection semble incorrecte.

celui des bitorseurs, d'où par (6.7.4) une flèche induite :

$$\text{(7.8.2)} \qquad \mathrm{TORS}(\mathrm{Aut}^0(G)) \longrightarrow \mathrm{TORS}(\mathrm{BITORS}^0(G)).$$

Une meilleure interprétation de la flèche en question, d'ailleurs plus proche de l'esprit de *loc. cit.*, s'obtient en observant que la flèche (7.8.1) s'identifie à la flèche

$$\text{(7.8.3)} \qquad \mathrm{Aut}^0(G) \longrightarrow \mathrm{Equ}^0(\mathrm{TORS}(G), \mathrm{TORS}(G))$$

définie par la fonctorialité en G du champ $\mathrm{TORS}(G)$, compte tenu de l'identification (7.2.2). Mais $H^1(\mathrm{Aut}^0(G))$ s'identifie à l'ensemble des classes d'isomorphisme de groupes localement isomorphes à G, et la flèche induite par (7.8.3) envoie un tel groupe G' vers la classe de la G-gerbe $\mathrm{TORS}(G')$. L'exactitude en $H^1(\mathrm{Aut}^o(G))$ de la suite (4.3.17) s'interprète de la manière suivante. Soit G' un groupe de T localement isomorphe à G. Alors la G-gerbe $\mathrm{TORS}(G')$ est équivalente à la G-gerbe triviale $\mathrm{TORS}(G)$ si et seulement si il existe un G-torseur P tel que le groupe adjoint $P^{ad} = \mathrm{Aut}_G(P)$ de P soit isomorphe à G'. En d'autres termes une condition nécessaire et suffisante pour que $\mathrm{TORS}(G')$ soit équivalente à $\mathrm{TORS}(G)$ est qu'il existe un (G, G')-bitorseur P. Sous cette forme, cette assertion de style Morita s'obtient d'ailleurs immédiatement en se référant, comme on l'a déjà fait en (7.2.2), à [26] IV 5.2.5.3.

Considérons maintenant le triangle :

$$\text{(7.8.4)} \qquad \begin{array}{ccc} H^1(\mathrm{Aut}^o(G)) & \longrightarrow & H^1(\mathcal{G}) \\ & \searrow & \downarrow \\ & & H^1(\mathrm{Out}^o(G)) \end{array}$$

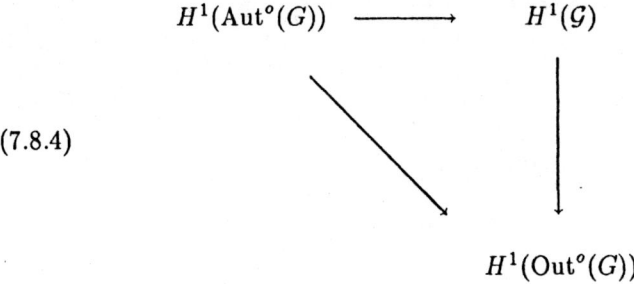

dont la flèche diagonale est induite par l'épimorphisme canonique $\mathrm{Aut}(G) \longrightarrow \mathrm{Out}(G)$ et envoie donc G' sur le lien $L = \mathrm{lien}(G')$. La commutativité de ce triangle implique que la flèche horizontale induit une application entre les fibres des deux autres flèches du triangle au dessus la classe $[L]$ d'un lien L localement isomorphe à G. L'interprétation qui vient d'être donnée de la flèche horizontale implique notamment que l'application sur les fibres a pour image le sous-ensemble $\mathsf{H}^2(L)'$ de $\mathsf{H}^2(L)$. Il conviendrait donc, pour adopter une notation consistente avec celle de *loc. cit.* de désigner par $H^1(\mathcal{G})'$ l'image de l'application $H^1(\mathrm{Aut}^o(G)) \longrightarrow H^1(\mathcal{G})$ mais nous n'aurons guère à considérer ici cet ensemble.

Remarque 7.9. Terminons cette discussion des suites exactes (4.3.17) et (5.2.3) et de leurs interprétations géométriques en observant qu'il conviendrait de prolonger (5.2.3) d'un cran vers la droite par un groupe $H^3(ZG)$ (voir *loc. cit.* VI théorème 2.3). Dans le cas des extensions de groupe, qui seront examinées au prochain paragraphe, il s'agit là de l'application classique qui associe à un "noyau abstrait" une classe d'obstruction à se réalisation par une extension de groupe (voir *loc. cit.* IV proposition 3.6.3 ou [4] proposition 1.1, et également [11]). Au vu des références citées, on doit s'attendre à ce qu'une section $x \in H^1(\mathrm{Out}^0(G))$ soit envoyée vers une classe de cohomologie $H^3((ZG)_x)$ à valeurs dans ZG, tordu de manière appropriée par x, plutôt qu'à une simple application cobord $\partial: H^1(\mathrm{Out}^0(G)) \longrightarrow H^3(ZG)$.

8. Théorie de Schreier–Grothendieck

8.1. Un lieu de passage obligé, pour qui s'intéresse à la cohomologie non abélienne, est l'étude du cas où le topos T considéré est un topos classifiant BP associé à un groupe G (abstrait ou lui-même objet d'un topos), et de la relation entre la cohomologie de BP et les extensions du groupe P (voir [26] VIII, [24] III notamment). Dans le cadre du topos ponctuel, c'est là la théorie classique de Schreier [49] (des références plus modernes sont [38] §2.4, [43] IV) permettant de décrire une extension E d'un groupe P par un groupe G, c'est à dire un groupe qui vit dans une suite exacte :

$$1 \longrightarrow G \longrightarrow E \longrightarrow P \longrightarrow 1 \tag{8.1.1}$$

en termes des groupes de cohomologie de P à valeurs dans G. Dans le cadre d'un topos quelconque T, l'étude des extensions a été menée à bien par A. Grothendieck [31] §1.1. Résumons en les points principaux : on observe tout d'abord que la structure de groupe E fait notamment de E un de G-bitorseur au dessus de P, comme il a été dit en 2.2 iv. La donnée d'une loi de composition sur E équivaut alors à celle d'un isomorphisme de G-bitorseurs ψ sur $P \times P$:

$$\psi: p_1^*E \overset{G}{\wedge} p_2^*E \longrightarrow m^*E, \tag{8.1.2}$$

où m (resp. p_i): $P \times P \to P$ désigne la loi de groupe de P (resp. la i-ème projection). Enfin l'associativité de la loi de composition équivaut à la commutativité du diagramme de torseurs sur P^3 suivant :

$$\begin{array}{ccc} p_1^*E \wedge p_2^*E \wedge p_3^*E & \longrightarrow & m_{12}^*E \wedge p_3^*E \\ \downarrow & & \downarrow \\ p_1^*E \wedge m_{23}^*E & \longrightarrow & m_{123}^*E \end{array} \tag{8.1.3}$$

où $m_{ij}: P^3 \to P^2$ (resp. $m_{123}: P^3 \to P$) désigne la flèche définie par la loi de groupe partielle associée aux facteurs i, j correspondants.

La description de l'extension E en termes de G-bitorseur sous-jacent, et de flèches compatibles entre des bitorseurs induits peut maintenant être reformulée d'une manière extrêmement compacte, et qui permet de donner un sens précis à l'affirmation de Dedecker ([13] note en bas de page 228) suivant laquelle une extension correspond à une gerbe (en fait, compte tenu de la proposition 7.3, à une G-gerbe sur BP, voir également à ce sujet, outre les références déjà mentionnées, [55], [35] §2, [60] théorème 6.1).

Proposition 8.2.* *Il existe un isomorphisme canonique d'ensembles pointés, fonctoriel en le groupe P et le module croisé $\mathcal{G}: G \xrightarrow{i'} \mathrm{Aut}^o(G)$ de T :*

$$\mathrm{Ext}(P, G) \xrightarrow{\sim} \tilde{H}^1(BP, \mathcal{G}^0).$$

Preuve. La discussion précédente peut être résumé par l'assertion que la donnée de l'extension E de P par G équivaut à celle du morphisme de gr-champs

$$\varphi_E: P \longrightarrow \mathrm{BITORS}(G) \tag{8.2.1}$$

(de source le gr-champ discret d'objets les sections de P, munies de la loi de composition définie par la loi de groupe de P) défini par :

$$\varphi_E(p) = p^*E, \tag{8.2.2}$$

c'est à dire qui associe à une section p de P la fibre de E en p, munie de la structure de bitorseur induite par celle de E. De même, à tout morphisme

---

* On rappelle que pour tout objet X de T pointé par une section $s: e \longrightarrow X$, la cohomologie réduite de X est définie par $\tilde{H}_i(X) = \ker(H^i(X) \xrightarrow{s^*} H^i(e))$.

d'extensions

$$\begin{array}{ccccccccc} 1 & \longrightarrow & G & \longrightarrow & E & \longrightarrow & P & \longrightarrow & 1 \\ & & \| & & \downarrow & & \| & & \\ 1 & \longrightarrow & G & \longrightarrow & E' & \longrightarrow & P & \longrightarrow & 1 \end{array}$$

correspond un morphisme $\varphi_\lambda : \varphi_E \to \varphi_{E'}$ (au sens de [16] 1.4.6) entre les foncteurs correspondant. Ceci démontre l'analogue non commutatif suivant de la proposition 1.4.23 de *loc. cit.*

Lemme 8.3. *La construction précédente définit une équivalence de champs*

(8.3.1) $$\varphi : \mathrm{EXT}(P, G) \longrightarrow \mathrm{HOM}(P, \mathrm{BITORS}(G))$$

entre le champ des extensions de P par G et celui des foncteurs additifs du gr-champ associé à P vers celui des G-bitorseurs.

Remarque 8.4. Le donnée du morphisme (8.2.1) permet de résoudre une question d'algèbre homologique générale laissée de côté dans le paragraphe précédent, à savoir l'interprétation, dans le contexte des G-torseurs, de la notion de "gerbe des relèvements d'un torseur" qui permet, dans la théorie de Giraud, d'associer à une extension de groupes (8.1.1) un opérateur cobord : $H^1(P) \longrightarrow H^2(G)$. Ceci s'obtient facilement dans notre contexte, en observant que le morphisme de *gr*-champs (8.2.1) induit par fonctorialité un 2-foncteur

(8.4.1) $$(\varphi_E)_* : \mathrm{TORS}(P) \longrightarrow \mathrm{TORS}(\mathrm{BITORS}(G))$$

et donc par composition avec la loi d'inverse un foncteur $\mathrm{TORS}(P) \longrightarrow \mathrm{TORS}(\mathrm{BITORS}^0(G))$ qui définit par passage aux classes d'isomorphisme de cobord en question. Cette construction a l'avantage, par rapport à celle de [26] IV, 2.5.8, de ne faire intervenir qu'implicitement l'extension E de P par G dans la définition du cobord, alors qu'elle intervient explicitement, par le biais de la "gerbe des relèvements à E" dans *loc. cit.* Finissons-en enfin avec ces remarques en observant que nous avons déjà utilisé un tel cobord en (7.8.2) pour interpréter la flèche de droite de (4.3.16). Dans ce cas, l'extension considérée était l'extension de $\mathrm{Aut}(G)$ par G définie par le produit semi-direct, pour l'action à gauche canonique de $\mathrm{Aut}(G)$ sur G.

8.5. Les considérations précédentes rendent immédiate la fin de la démonstration de la proposition 8.2 : il suffit en effet de considérer l'application composée

$$\text{Ext}(P, G) \xrightarrow{\varphi} [P, \mathcal{G}^0]_{A_\infty} \xrightarrow{B} [BP, B\mathcal{G}^0]_{pt} \tag{8.5.1}$$

où φ s'obtient, compte tenu de la proposition 4.3, en passant aux classes à isomorphisme près, et où B est la bijection classifiante (6.8.1). Ceci termine la démonstration, puisque, par définition, $[BP, B\mathcal{G}^0] = H^1(BP, \mathcal{G}^0)$, la cohomologie réduite correspondant alors aux classes d'homotopie d'applications pointées.

8.6. La proposition précédente affirme notamment que tout morphisme $(u,v)\colon (G \to \text{Aut}^o(G)) \to (G' \to \text{Aut}^o(G'))$ de modules croisés (au sens de la définition 3.2) induit une application pointée

$$(u,v)_*\colon \text{Ext}(P,G) \longrightarrow \text{Ext}(P,G') \tag{8.6.1}$$

qui ne semble pas figurer de manière explicite dans la littérature,* même dans le cas ponctuel pourtant très étudié depuis Schreier, et qu'il est instructif d'expliciter : soit donc une extension E de P par G, de classe $[E] \in \text{Ext}(P,G)$. L'extension induite $(u,v)_*[E]$ est représentée par un groupe $(u,v)_*(E)$, d'ensemble sous-jacent $G' \overset{G}{\times} E$, muni de la loi de groupe définie par :

$$(g_1, e_1)(g_2, e_2) = (g_1 v(i_{e_1})(g_2), e_1 e_2) \tag{8.6.2}$$

où $i_e \in \text{Aut}(G)$ est l'élément défini par la conjugaison intérieure dans E, c'est à dire :

$$i_e(g) = ege^{-1}. \tag{8.6.3}$$

En d'autres termes $(u,v)_* E$ s'obtient en passant au quotient à partir du produit semi-direct $G' \times E$ (pour l'action à gauche de E sur G' définie par ${}^e g' = v(i_e)(g')$), par la relation $(g'u(g), e) = (g', ge)$.

---

* On en trouve une variante, dans le cadre du topos ponctuel, dans [14] 4.3, (voir également la proposition 8.12 ci-dessous), dans le cadre plus général des extensions à valeurs dans un module croisé quelconque, le cas des extensions de groupes étant peu mis en valeur (voir cependant [13] remarque 11.2). Le cas particulier d'un groupe abstrait commutatif est traité dans [3] I définition 1.5 et celui d'un homomorphisme surjectif $G \to G/H$ (où H est un sous-groupe caractéristique de G) dans [4].

8.7. L'interprétation de l'ensemble Ext(P,G) donnée par la proposition 8.2 nous permet de lui appliquer les dévissages de style (4.3.16) et (5.2.3). Le premier définit une suite exacte d'ensemble pointés

$$0 \to \tilde{H}^0(BP, \mathcal{G}^0) \to \tilde{H}^1(BP, G^0) \to \tilde{H}^1(BP, \mathrm{Aut}(G)) \to \mathrm{Ext}(P, G)$$

soit encore

(8.7.1)
$$0 \to \tilde{H}^0(BP, \mathcal{G}^0) \to \mathrm{Hom}(P, G^0) \to \mathrm{Hom}(P, \mathrm{Aut}(G)) \to \mathrm{Ext}(P, G)$$

tandis que l'autre dévissage permet d'une part d'identifier $\tilde{H}^0(BP, \mathcal{G}^0)$ à Hom(P, ZG) (ce que l'exactitude de (8.7.1) rend d'ailleurs évident) et d'autre part d'obtenir une suite exacte d'ensembles pointés

(8.7.2) $$0 \to \tilde{H}^2(BP, ZG) \to \mathrm{Ext}(P, G^0) \xrightarrow{\beta'} \tilde{H}^1(BP, \mathrm{Out}(G)).$$

Comme le terme de droite de cette suite exacte n'est autre que Hom$(P$, Out$(G))$, on reconnait en (8.7.2) une partie de la classique discussion, étendue ici au cas d'un topos quelconque, concernant le noyau abstrait associé à une extension de groupe et la classification des extensions à noyau abstrait $\psi \in$ Hom$(P, \mathrm{Out}(G))$ donné (voir [43] IV théorème 8.7, 8.8, [26] VIII 7.4). Il y manque des considérations sur l'obstruction à réaliser un tel ψ par une extension, considérations pour lesquelles il conviendrait de prolonger d'un cran vers la droite la suite (8.7.2) (voir à ce sujet la remarque 7.9).

La suite (8.7.2) peut se comparer à la suite de type 5.2.3 correspondante du topos $T_{|P}$: en effet (6.8.2) définit une application inverse de l'application classifiante B introduite en (8.5.1). De la flèche $[P, \mathcal{G}^0]_{A_\infty} \longrightarrow [P, \mathcal{G}^0]$ obtenue par oubli de la structure A_∞ on déduit donc une application (dite de suspension)

(8.7.3) $$S_{\mathcal{G}^0} \colon \tilde{H}^1(BP, \mathcal{G}^0) \longrightarrow \tilde{H}^0(P, \mathcal{G}^0)$$

qui associe à une extension $[E]$ de P par G le bitorseur sous-jacent. En faisant de même pour les modules croisés $ZG[1]$ et Out(G) qui dévissent $\mathcal{G}^0$, on en déduit un diagramme commutatif

(8.7.4)
$$\begin{array}{ccccccc}
0 & \longrightarrow & \tilde{H}^2(BP, ZG) & \longrightarrow & \mathrm{Ext}(P,G) & \xrightarrow{\beta'} & \mathrm{Hom}(P, \mathrm{Out}(G)) \\
 & & \Big\downarrow{\scriptstyle S_{ZG[1]}} & & \Big\downarrow{\scriptstyle S_{\mathcal{G}^0}} & & \Big\downarrow{\scriptstyle S_{\mathrm{Out}(G)}} \\
0 & \longrightarrow & \tilde{H}^1(P, ZG) & \longrightarrow & \mathrm{Bitors}(P,G) & \xrightarrow{\beta} & \tilde{H}^0(P, \mathrm{Out}(G))
\end{array}$$

dont les lignes horizontales sont (8.7.2) et (5.2.4). Le carré de droite précise notamment quelle est la compatibilité du noyau abstrait d'une extension avec l'application β décrite géométriquement en 5.2 (voir également 8.12).

8.8. Dans le cas de groupes P, G d'un topos quelconque, il existe un autre dévissage de l'ensemble $\mathrm{Ext}(P, G)$, qui s'obtient à partir de la description géométrique rappelée en 8.1 d'une extension E de P par G, et qui permet d'analyser le noyau de la flèche "oubli de structure" (8.7.3). Un élément de la fibre distinguée de (8.7.3), c'est à dire une extension à bitorseur sous-jacent trivial, est décrit par une flèche (8.1.2), c'est à dire ici par un automorphisme du G-bitorseur trivial sur $P \times P$, soit encore, par (5.2.2), par un élément de $H^0(P \times P, ZG)$. La condition d'associativité (8.1.3) dit que c'est un élément du noyau de l'application cobord usuelle $H^0(P\times P, ZG) \to H^0(P\times P\times P, ZG)$ c'est à dire, dans la terminologie classique, "un système de facteurs de P à valeurs dans le groupe abélien ZG." Deux tels éléments décrivent des extensions équivalentes s'ils diffèrent, en tant que systèmes de facteurs par un élément de la forme $f = \partial g$ avec $g \in H^0(P, ZG)$. On obtient donc ainsi une suite exacte

$$(8.8.1) \qquad 0 \longrightarrow H^2\{P; ZG\} \longrightarrow \mathrm{Ext}(P, G) \longrightarrow H^0(P, \mathcal{G}^0)$$

(où $H^2\{\ ,\ \}$ désigne ici le groupe des classes de facteurs de P à valeurs dans ZG mentionné) qui est à vrai dire implicite dans [31] et qui est l'analogue non-commutatif de la suite exacte bien connue ([51] VII proposition 4 a)

$$(8.2.2) \qquad 0 \longrightarrow H^2_s\{A, B\} \longrightarrow \mathrm{Ext}^1(A; B) \longrightarrow H^1(A; B)$$

(où A et B sont supposés commutatifs, et $H^2_s\{\ \ \}$ désigne les systèmes de facteurs symétriques).

Remarque 8.9. (i) Les dévissages de type (8.8.1) ont souvent été examinés sous des hypothèses plus restrictives, qui donnent lieu à des situations intermédiaires entre (8.8.1) et (8.8.2), notamment lorsque G est abélien (voir par exemple [2], [59]). Signalons d'autre part le fait bien connu que, quitte à rigidifier la discussion précédente, on peut calculer le terme de gauche de (8.8.1) au moyen de facteurs normalisés.

(ii) De la suite spectrale de [7] (5.1), on déduit en bas degrés que la suite exacte (8.8.2) peut être prolongée de deux crans, une fois le terme de droite remplacé par le sous-groupe $H^1(A, B)_{\mathrm{prim}}$, des classes de cohomologie primitives. Les méthodes géométriques devraient permettre de prolonger (8.8.1), une fois le terme de droite restreint, d'au moins un cran par la groupe de systèmes de facteurs supérieur $H^3\{P, ZG\}$ (mais il y aurait lieu d'examiner si ici le coefficient ZG doit être tordu ou non).

8.10. La suite exacte (8.8.1) décrit des extensions de P par G à bitorseur sous-jacent trivial en termes de systèmes de facteurs de P à valeurs dans ZG. La théorie de Schreier classique, bien qu'elle ne vive que dans le cadre plus restrictif du topos ponctuel, est en un sens plus ambitieuse, puisqu'elle fournit une description de toutes les extensions de P par G. On peut la comprendre dans notre contexte en examinant le diagramme

(8.10.1)

$$\begin{array}{ccccccc} & & & & \mathrm{Ext}(P,G) & & \\ & & & & \downarrow & & \\ H^0(P,G^0)) & \longrightarrow & H^0(P,\mathrm{Aut}(G)) & \longrightarrow & \mathrm{Bitors}(P,G) & \longrightarrow & H^1(P,G) \\ & & & \searrow & \downarrow & & \\ & & & & H^0(P,\mathrm{Out}(G)) & & \end{array}$$

fabriqué à partir de portions des suites exactes (4.3.17), (5.2.4) et de la flèche (8.7.3). Il convient alors d'analyser, dans le cadre d'un topos T, la classification des extensions de P par G munis d'une trivialisation s du G-torseurs à gauche sous-jacent (il en existe toujours lorsque T est le topos ponctuel). Il s'agit donc là, on l'a vu, d'extension décrites (par exactitude de la suite horizontale du diagramme (8.10.1)) par une section $\varphi \in H^0(P,\mathrm{Aut}(G))$, la multiplication à droite par G dans l'existence E étant décrite (voir 2.2.2, 2.16) par

$$s(p)g = (\varphi(p)(g))s(p)$$

ce qu'il est plus naturel d'écrire sous la forme

$$s(p)gs(p)^{-1} = \varphi(p)(g). \tag{8.10.2}$$

Ainsi φ associe, comme il est d'usage, à tout $p \in P$, l'action du relevé $s(p)$ de p sur G par conjugaison intérieure dans E.

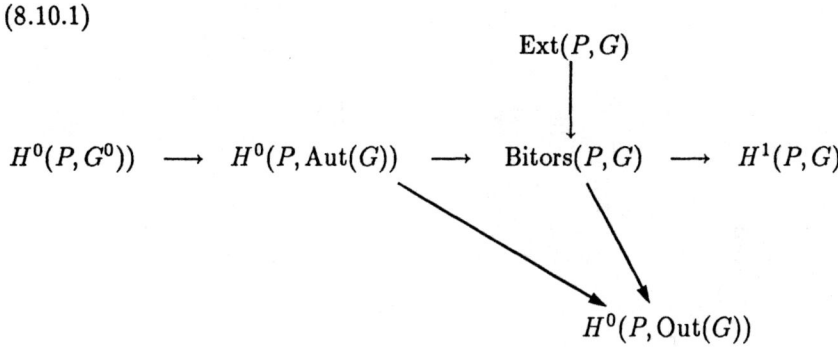

Le reste de l'analyse de la structure de E est en réalité une paraphrase de la discussion classique, mais qu'il est néanmoins instructif d'expliciter : commençons par remarquer que si (G,φ_1) et (G,φ_2) sont deux bitorseurs à torseur sous-jacent trivialisés, définis, par le procédé décrit en (2.2.2), par une paire de sections $\varphi_1,\varphi_2 \in H^0(P,\mathrm{Aut}(G))$, alors la donnée d'un automorphisme du bitorseurs $\lambda:(G,\varphi_1) \to (G,\varphi_2)$ équivaut à celle d'un automorphisme du G-torseur trivial, décrit par une section f du groupe G^0 opposé à G et dont la compatibilité à la multiplication à droite par G s'exprime par la relation

$$\varphi_1 = i_f \circ \varphi_2 \quad (\text{soit encore } i'_f \circ \varphi_1 = \varphi_2)$$

dans Aut(G). En particulier, la donnée d'une loi de composition sur le bitorseur (G,φ), au moyen d'un automorphisme (8.1.2)

$$(G, p_1^*\varphi \circ p_2^*\varphi) = p_1^*(G,\varphi) \wedge p_2^*(G,\varphi) \longrightarrow m^*(G,\varphi) = (G, m^*\varphi),$$

équivaut à celle d'une section $f(x,y) \in H^0(P^2,G)$ telle que, en notation ensembliste,

$$(8.10.3) \qquad i_{f(x,y)} \circ \varphi(xy) = \varphi(x) \circ \varphi(y) \qquad \forall\, x,y \in P.$$

D'autre part, si l'on considère une paire de tels morphismes

$$\lambda:(G,\varphi_1) \longrightarrow (G,\varphi_2)$$
$$\lambda':(G,\varphi'_1) \longrightarrow (G,\varphi'_2)$$

décrits respectivement par les sections f et f' de G, alors le morphisme induit par produit contracté

$$\lambda \wedge \lambda':(G,\varphi_1 \circ \varphi'_1) \longrightarrow (G,\varphi_2 \circ \varphi'_2)$$

est décrit par l'élément $(\lambda \overset{G}{\wedge} \lambda')(1 \wedge 1) = \lambda(1) \overset{G}{\wedge} \lambda'(1) = g \overset{G}{\wedge} g' = g\varphi_1(g')$ de G.

Il résulte de ces remarques que le diagramme (8.1.3) exprimant l'associativité de la loi de groupe s'explicite ici de la manière suivante (les flèches étant décrites en notation ensembliste par les sections de G qui les définissent) :

$$\begin{array}{ccc}
(G,\varphi(x) \circ \varphi(y) \circ \gamma(z)) & \xrightarrow{f(x,y)} & (G,\varphi(xy) \circ \varphi(z)) \\
{\scriptstyle \varphi(x)f(y,z)} \big\downarrow & & \big\downarrow {\scriptstyle f(xy,z)} \\
(G,\varphi(x) \circ \varphi(yz)) & \xrightarrow[f(x,yz)]{} & (G,\varphi(xyz))
\end{array}$$

La commutativité du diagramme s'exprime donc bien (en prenant garde que le groupe des automorphismes de G-torseur trivial est G^0), par la formule

$$(8.10.4) \qquad (\varphi(x)f(y,z))f(x,yz) = f(x,y)f(xy,z).$$

Les formules (8.10.3) et (8.10.4) sont bien les classiques relations de Schreier (voir [43] IV (8.5)–(8.6), [55] 1.14(5)).

8.11. Toute la discussion précédente peut maintenant être étendue du cas où l'on a affaire au module croisé $(G \to \mathrm{Aut}^o(G))$ à celui d'un module croisé droit $\mathcal{M} = (G \to \pi^0)$ de T. On est alors en présence, par le théorème 4.6, de ce qu'on a appelé des $(G-\pi)$-torseurs, et les suites exactes (4.3.16) et (5.2.3) sont remplacées par (4.6.1) et (5.2.7). Il s'agit alors d'interpréter, comme en (8.2), l'ensemble $\tilde{H}^1(BP, \mathcal{M}^0)$, P étant à nouveau un groupe de T. La flèche de suspension

$$(8.11.1) \qquad S_{\mathcal{M}^0}: \tilde{H}^1(BP, \mathcal{M}^0) \longrightarrow \tilde{H}^0(P, \mathcal{M}^0)$$

définie comme en (8.7.3) envoie un élément de $\tilde{H}^1(BP, \mathcal{M}^0)$ sur la classe d'un (G,π)-torseur de $T_{|P}$, c'est à dire, rappelons-le, un G-bitorseur E au dessus de P, muni d'une flèche $f: E \to \pi$ équivariante pour l'action à gauche de G sur π à travers $\partial_1 = i \circ \partial: G \to \pi^0 \to \pi$, et telle que la multiplication à droite par G soit définie par (4.5.1). Lorsque l'on remonte de $\tilde{H}^0(P, \mathcal{M}^0)$ à $\tilde{H}^1(BP, \mathcal{M}^0)$, en se donnant, comme en 8.1, un morphisme de (G,π)-torseurs (8.1.2) puis en exigeant que la condition d'associativité analogue à (8.1.3) soit satisfaite, on constate que, non seulement on a muni E d'une structure de groupe qui en fait, comme précédemment, une extension de P par G, mais qu'on a de plus exigé que la flèche $j: E \to \pi$ soit un homomorphisme. La condition d'équivariance mentionnée précédemment s'exprime ici par l'assertion que $j \circ k = \partial$, k désignant l'injection de G dans E définie par la rigidification $1 \in E$.

Ainsi donc en définitive un élément de $\tilde{H}^1(BP, \mathcal{M}^0)$ est une classe d'équivalence d'extensions de groupes E, munies d'un homomorphisme de groupes $j: E \to \pi$ tel que le diagramme

$$\begin{array}{ccccccccc} 1 & \longrightarrow & G & \xrightarrow{k} & E & \longrightarrow & P & \longrightarrow & 1 \\ & & {\scriptstyle \partial_1}\downarrow & \swarrow{\scriptstyle j} & & & & & \\ & & \pi & & & & & & \end{array}$$

commute, l'homomorphisme $j: E \to \pi$ étant en outre astreint par (4.5.1) à satisfaire à la condition

$$(8.11.2) \qquad {}^{j(e)}g = ege^{-1},$$

(pour la structure de module croisé gauche déduite sur ∂_1, comme il a été expliqué en 3.2, de la structure de module croisé droit de $\mathcal{M}$). Une telle paire (E,j) est précisément, dans le cas du topos ponctuel, ce que Dedecker appelle un π-extension de P par le module croisé gauche $\partial_1 : G \longrightarrow \pi$ ([14] définition 1, [13] définition 9.1). Nous dirons plutôt qu'il s'agit d'une (G,π)-extension de P, ou bien d'une $\mathcal{M}$-extension de P. La discussion précédente est donc résumée par la généralisation suivante de la proposition 8.2.

Proposition 8.12. *Soit $\mathcal{M} = (G \xrightarrow{\partial} \pi^0)$ un module croisé droit de T et $G \xrightarrow{\partial_1} \pi$ le module croisé gauche correspondant. Il existe alors un isomorphisme canonique d'ensembles pointés.*

(8.12.1) $$\mathrm{Ext}_\pi(P,G) \simeq \tilde{H}^1(BP,\mathcal{M}^0),$$

où le terme de gauche désigne, tout comme dans loc. cit., *l'ensemble des classes d'isomorphisme de (G,π)-extensions de P associées au module croisés gauche $\partial_1 : G \longrightarrow \pi$.*

On trouvera une variante de ce résultat, dans le cas ensembliste dans [13] théorème 3. Il rend évident la fonctorialité en $\mathcal{M}$ de l'ensemble $\mathrm{Ext}_\pi(P,G)$ d'ailleurs signalée dans [15] 3.2. Ce résultat englobe notamment, comme cela a été expliqué ci-dessus pour les extensions ordinaires, une théorie de Schreier décrivant les (G,π)- extensions E de P, dans le cas où le G-torseur sur P défini par E possède une section, en termes de cocycles $f : P \times P \longrightarrow G$, $\psi : P \longrightarrow \pi$ qui satisfont aux formules analogues à (8.10.3)–(8.10.4), avec $G \xrightarrow{i} \mathrm{Aut}(G)$, remplacé par $\partial_1 : G \longrightarrow \pi$, (voir [13] (3.1)–(3.2)). En définitive, la proposition 8.12 et les conséquences qui en découlent sont à la théorie mentionnée de Dedecker ce que la discussion de 8.1–8.10 est à la théorie de Schreier classique. Ainsi, l'isomorphisme (8.12.1) permet, dans le cas d'un topos T quelconque, d'insérer l'ensemble $\mathrm{Ext}_\pi(P,G)$ dans la suite exacte de type (5.2.7) au dessus de BP, et même dans le diagramme suivant similaire à (8.7.4) :

(8.12.2)
$$\begin{array}{ccccccc}
0 \longrightarrow & \tilde{H}^2(B,P,\pi_1(\mathcal{M}^0)) & \longrightarrow & \mathrm{Ext}_\pi(P,G) & \xrightarrow{\beta_1} & \mathrm{Hom}(P,\pi_0(\mathcal{M}^0)) \\
& \downarrow & & \downarrow & & \downarrow \\
0 \longrightarrow & \tilde{H}^1(P,\pi_1(\mathcal{M}^0)) & \longrightarrow & ((G,\pi)\text{-TORS})_{|P} & \xrightarrow{\beta_2} & \tilde{H}^0(P,\pi_0(\mathcal{M}^0))
\end{array}$$

Ici la ligne inférieure est suite analogue à (5.2.7), au dessus de P, et les deux premières flèches verticales sont les morphismes de suspension définis comme en 8.7.3. La flèche β_1 est appelée "application crête" dans *loc. cit.*

(2.3) et la commutativité du carré de droite rend explicite la compatibilité entre la crête d'un (G,π)-extension et la flèche β_2, qui associe au (G,π)-torseur $(E,j) \mid P$ sous-jacent à l'extension en question la flèche $\bar{j}: P \to \pi_0(\mathcal{M}^0)$ induite, comme en (5.2), par j.

BIBLIOGRAPHIE

[1] Adams, J.F., *Infinite loop space,* Annals of Mathematical Studies **90**, Princeton University Press, 1978.

[2] André, M., *Cohomologie des groupes topologiques,* Batelle Institute Report **22** (1969).

[3] Beyl, F.R. et Tappe, J., *Group extensions, representations and the Schur multiplier,* LNM **958**, Springer-Verlag, 1982.

[4] Berrick, A.J., *Group extensions and their trivialisations,* L'enseignement Mathématique **31** (1985), 151–172.

[5] Bourn, M., *Higher cohomology groups as classes of principal group actions,* Université de Picardie (1985).

[6] Bousfield, A.K. and Friedlander, E.M., *Homotopy theory of Γ-spaces, spectra and bisimplicial sets.* Dans *Geometric applications of homotopy theory II: proceedings Evanston 1977* (M.G. Barratt et M.E. Mahowald, ed.) LNM **658**, Springer-Verlag, 1978.

[7] Breen, L., *Extensions of abelian sheaves and Eilenberg–MacLane algebras,* Invent. Math. **9** (1969), 15–44.

[8] Brown, K.S., *Abstract homotopy theory and generalised sheaf cohomology,* Trans. A.M.S. **186** (1973), 419–458.

[9] Brown, R. et Spencer, C.B., *G-groupoids, crossed modules and the fundamental groupoid of a topological group,* Proc. Kon. Nederl. Akad. Wet. **79** (1976), 296–302.

[10] Brown, R., *From groups to groupoids, a brief survey,* Bull. London Math. Soc. **19** (1987), 113–134.

[11] Cegarra, A.M., Bullejos, M., and Garzon, A.R., *Higher dimensional obstruction theory in algebraic categories,* Journal of Pure and Applied Algebra **49** (1987), 43–102.

[12] Conduché, D., *Modules croisés généralisés de longueur* 2. J. Pure and Appl. Algebra **34** (1984), 155–178.

[13] Dedecker, P., *Algèbre Homologique non abélienne,* Colloque de Topologie algébrique, Bruxelles, 1964.

[14] Dedecker, P., *Les foncteurs* Ext_π, H^2_π *et* H^2_π *non abéliens,* CRAS **258** (1964), 4891–4894. *Premier dérivé du foncteur Hom non abélien* CRAS **259** (1964), 2054–2057.

[15] Dedecker, P., *Three dimensional non abelian cohomology for groups,* dans "Category theory, homology theory and their applications II,"

LNM **92**, 1969, 32–64.

[16] Deligne, P., *La formule de dualité globale,* exposé XVIII de SGA 4, LNM **305**, Springer-Verlag, 1973.

[17] Deligne, P., *Interprétation modulaire et techniques de construction de modèles canoniques,* dans "Automorphic forms, Representations and L-functions," Proc. Symp. Pure Math. **33** (1979), II, 247–289.

[18] Deligne, P. et Milne, J.S., *Tannakian catégories,* dans "Hodge Cycles, Motives and Shimura varieties", LNM **900**, Springer-Verlag, 1982.

[19] Deligne, P., *Catégories Tannakiennes,* ce volume.

[20] Douai, J-C., *2-Cohomologie Galoisienne des groupes semi-simples,* thèse de Doctorat, Université de Lille I (juin 1976).

[21] Duskin, J., *Simplicial methods and the interpretation of "triple" cohomology,* Memoirs of the AMS **163** (1975).

[22] Duskin, J., *Higher dimensional torsors and cohomology of topos: the abelian theory,* dans "Applications of Sheaves (Durham 1977)," LNM **753**, 255–279, Springer-Verlag, 1979.

[23] Duskin, J., *An outline of non-abelian cohomology in a topos* (1)*: the theory of bouquets and gerbes,* Cahiers de topologie et géométrie différentielle XXIII (1982).

[24] Duskin, J., *Non abelian cohomology in a topos,* to appear in Memoirs of the A.M.S.

[25] Gabriel, P. et Zisman, M., *Calculus of fractions and homotopy theory,* Ergebnisse der Math. und ihrer Grenzgebiete, n.s. **35**, Springer-Verlag, 1967.

[26] Giraud, J., *Cohomologie non abélienne,* Grundlehren der mathematischen Wissenchaften in Einzeldarstellungen **179**, Springer-Verlag, 1971.

[27] Grothendieck, A., *A general theory of fibre spaces with structure,* University of Kansas (1959).

[28] Grothendieck, A., *Sur quelques points d'algèbre homologique,* Tohoku Math. J. **9** (1957), 119–221.

[29] Grothendieck, A., *Classes de Chern et représentations linéaires de groupes discrets* dans Giraud et al. "Dix exposés sur la cohomologie des schémas," Masson-North Holland Publ. Co., 1968.

[30] Grothendieck, A., *Catégories cofibrés additives et complexe cotangent relatif,* LNM **79**, Springer-Verlag, 1968.

[31] Grothendieck, A., *Séminaire de Géométrie Algébrique du Bois-Marie,* 1967–69 (SGA 7) I LNM **288**, Springer-Verlag, 1972.

[32] Grothendieck, A., Lettres à l'auteur du 5/02/75 et du 17/02/75.

[33] Grothendieck, A., *A la poursuite des champs,* 1983.

[34] Hakim, M., *Topos annelé et schémas relatifs,* Ergebnisse der Math. und ihrer Grenzgebiete **64**, Springer-Verlag, 1972.

[35] Illusie, L., *Complexe cotangent et déformations* I, II, LNM **239**, **283**, Springer-Verlag, 1971–1972.

[36] Jardine, J.F., *Simplicial objets in a Grothendieck topos,* Contemporary Math. **55** (1989), 193–239.

[37] Jardine, J.F., *Universal Hasse-Witt classes,* Contemporary Math. **83** (1989), 83–100.

[38] Kirillov, A., *Eléments de la théorie des représentations,* MIR, 1974.

[39] Lamotke, K., *Semisimpliziale algebraische Topologie,* Grundlehren der mathematischen Wissenschaften in Einzeldarstellungen **147**, Springer-Verlag, 1968.

[40] Langlands, R.P. et Rapoport, M., *Shimura Varietäten und Gerben,* J. Reine Angew. Math. **378** (1987), 113–220.

[41] Mackenzie, K., *Classification of principal bundles and Lie groupoids with prescribed gauge group bundle,* J. Pure Appl. Alg. **58** (1989), 181–208.

[42] Loday, J.L., *Spaces with finitely many non trivial homotopy groups,* J. Pure Appl. Alg. **24** (1982), 179–202.

[43] MacLane, S., *Homology,* Grundlehren der mathematischen Wissenschaften in Einzeldarstellungen **114**, Springer-Verlag, 1963.

[44] MacLane, S., *Categories for the working mathematician,* Graduate Texts in Mathematics **5**, Springer-Verlag, 1972.

[45] May, J.P., *Simplicial objects in algebraic topology,* Van Nostrand, 1967.

[46] May, J.P., *E_∞-spaces, group completions and permutative categories,* dans "New developments in topology", ed. G. Segal, L.M.S. Lecture Note Series **11**, Cambridge University Press, 1974.

[47] Quillen, D., *The spectrum of an equivariant cohomology ring* I, Ann. of Math. **94** (1971), 549–572.

[48] Saavedra Rivano, N., *Catégories Tannakiennes,* LNM **265**, Springer-Verlag, 1972.

[49] Schreier, O., *Uber die Erweiterung von Gruppen* I. Monatsh. Math. Phys. **34** (1926), 165–180, II. Abh. Math. Sem. Hamburg **4** (1926), 321–346.

[50] Segal, G., *Homotopy everything H-spaces,* non publié.

[51] Segal, G., *Classifying spaces and spectral sequences,* Publ. Math. I.H.E.S. **34** (1968), 105–112.

[52] Serre, J-P., *Groupes algébriques et Corps de Classes,* Actualités Scientifiques et Industrielles, Hermann, 1959.

[53] Serre, J-P., *Cohomologie Galoisienne,* LNM **5**, Springer-Verlag, 1964.

[54] Sinh, Hoang Xuan, Gr-*catégories,* thèse de l'Université de Paris VII, 1975.

[55] Springer, T.A., *Non abelian H^2 in Galois cohomology,* Proc. Symp. Pure Math. **9** (1966), 164–182.

[56] Stasheff, J., *Homotopy associativity of H-spaces* I, T.A.M.S. **108** (1963), 275–292.

[57] Stasheff, J., *H-spaces from a homotopy point of view,* LNM **161**, Springer-Verlag, 1970.

[58] Thomason, R.W., *Uniqueness of delooping machines,* Duke Math. J. **46** (1979), 217–252.

[59] Ulbrich, K.H., *Sur les extensions de groupes à noyau abélien,* CRAS **295** (1982), 715–718.

[60] Ulbrich, K.H., *On non abelian H^2 for profinite groups,* preprint.

[61] Whitehead, J.H.C., *Note on the previous paper: on adding relations to homotopy groups,* Ann. of Math. **47** (1946), 806–810.

[62] Zisman, M., *Suite spectrale d'homotopie et ensemble bisimpliciaux,* Université scientifique et médicale de Grenoble (mai 1975).

Université Paris-Nord
C.S.P.
Département de Mathématiques
Avenue Jean-Baptiste Clément,
93430 Villetaneuse
France

Non-commutative Ruelle-Sullivan type currents

JEAN-LUC BRYLINSKI*

1. Introduction

We present here a simple direct construction which, under certain circumstances, produces a map $H_*(\mathfrak{g}, \mathbf{C}) \xrightarrow{\chi} H_*^{c\ell}(M, \mathbf{C})$ if $\mathfrak{g}$ is a Lie algebra which acts infinitesimally on the manifold M. Here $H_*(\mathfrak{g}, \mathbf{C})$ means Lie algebra homology (with trivial coefficients), and $H_*^{c\ell}(M, \mathbf{C})$ denotes homology with closed supports, i.e., Borel-Moore homology [2]. Such a map may be defined in the following two cases.

(a) M is a C^∞-manifold, and there is a volume form on M preserved by the $\mathfrak{g}$-action.

(b) M is a complex manifold, for which the "Hodge to de Rham" spectral sequence degenerates at E_1, and $\mathfrak{g}$ acts on M by holomorphic vector fields.

For example, in case (a), if $\mathfrak{g}$ is one-dimensional, a generator of its degree one homology maps to the class of a current on M, which is identical with the Ruelle-Sullivan current defined in [13]. For general non-commutative Lie algebras, we obtain currents on M which we would like to view as non-commutative analogs of the Ruelle-Sullivan current.

The formula for χ is quite simple in case (a): let ω be the preserved volume form; to a class in $\bigwedge^k(\mathfrak{g})$, say $X_1 \wedge \cdots \wedge X_k$, we associate the cohomology class of $(-1)^k i(X_1) \ldots i(X_k)\omega$, which we view as a homology class, using Poincaré duality. The construction therefore really depends on the choice of ω. This should be contrasted with the case of an action of a topological group G on a connected topological space X, where the map $H_*(G) \to H_*(X)$ is defined simply by functoriality, choosing a base point in X. It should be pointed out that for $G = \mathbf{R}$ acting on $X = S^1$ by

* Supported in part by the N.S.F.

translation, the map $H_1(G,\mathbf{C}) \to H_1(X,\mathbf{C})$ is 0, whereas $\chi: H_1(\mathfrak{g},\mathbf{C}) \to H_1(X,\mathbf{C})$ is an isomorphism. This simple example shows that even if the Lie algebra action extends to a Lie group action, the Lie algebra homology may be related in a more interesting way with the homology of the manifold than the homology of the Lie group would be (this has in part to do with the group being non-compact).

A similar construction produces a class in $H^2(\mathfrak{g},\mathbf{C})$ if $\mathfrak{g}$ acts on a symplectic manifold, preserving the symplectic form (hence this gives a central extension of $\mathfrak{g}$). This is the well-known class constructed by Kostant [8] and independently Souriau [14], which is of central importance in geometric quantization. The case of symplectic Lie algebra actions is connected with the *canonical homology* of the manifold, as developed in [3]. In particular, for Poisson actions on symplectic manifolds, there is a natural map $\chi: H_*(\mathfrak{g}, C^\infty(\mathfrak{g}^*)) \to H_*^{cl}(M,\mathbf{C})$ where $\mathfrak{g}$ acts on $\mathfrak{g}^*$ by the coadjoint action. This map deserves further study.

I have given, in §6.7, some speculative ideas on how the constructions of this article might be related with the map from the 2-dimensional cohomology of the Lie algebra of vector fields on the circle to that of the moduli space of curves of given genus due to Arbarello, De Concini, Kač and Procesi [1]. These ideas are however still in primitive form.

It is a pleasure to dedicate this article to Prof. Alexander Grothendieck whose colossal work in so many branches of mathematics had an enormous and extremely beneficial influence on mathematicians in my generation. We learned his modern algebraic geometry not so much from the E.G.A. volumes but from his crystal clear exposition of his new ideas in so many places, for instance, his Séminaire Bourbaki talks.

The research in this article has been influenced by discussions with Armand Borel, Pierre Deligne, Mikhail Gromov and Victor Kač, and I thank them warmly. I am very grateful to Pierre Cartier for carefully reading this article; if the signs are all correct, it is certainly due to him.

2. The main construction

Let M be a finite-dimensional smooth manifold. Let $C^\infty(M)$ be the algebra of complex-valued smooth functions on M, $C_c^\infty(M) \subset C^\infty(M)$ the ideal of compactly-supported smooth functions. The Lie algebra $\mathrm{Vect}(M)$ of smooth vector fields on M with complex coefficients acts on $C^\infty(M)$ and on $C_c^\infty(M)$.

Recall the definition of the standard complexes $C_*(\mathfrak{g},V)$ and $C^*(\mathfrak{g},V)$ for a Lie algebra $\mathfrak{g}$ over a field k and a $\mathfrak{g}$-module V, [4], [5].

We have: $C_k(\mathfrak{g},V) = \bigwedge^k(\mathfrak{g}) \otimes V$, d maps $C_k(\mathfrak{g},V)$ to $C_{k-1}(\mathfrak{g},V)$ and is given by:

$$d(X_1 \cdots X_k \otimes v) = \sum_{1 \leq i < j \leq k} (-1)^{i+j} [X_i, X_j] \wedge \cdots \wedge \widehat{X_i} \wedge \cdots \wedge \widehat{X_j} \cdots \otimes v + \sum_{1 \leq i \leq k} (-1)^i X_1 \wedge \cdots \wedge \widehat{X_i} \wedge \cdots X_k \otimes X_i \cdot v.$$

The groups $H_k(\mathfrak{g}, V) = H_k(C_*(\mathfrak{g}, V))$ are the *homology groups* of $\mathfrak{g}$ with coefficients in V.

We have: $C^k(\mathfrak{g}, V) = \mathrm{Hom}_k(\bigwedge^k \mathfrak{g}, V)$, d increases degrees by one:

$$(d\omega)(X_1, \ldots, X_{k+1}) = \sum_{1 \leq i < j \leq k+1} (-1)^{i+j} \omega([X_i, X_j], \ldots, \widehat{X_i}, \ldots, \widehat{X_j}, \ldots) + \sum_{1 \leq i \leq k+1} (-1)^{i+1} X_i \cdot \omega(X_1, \ldots, \widehat{X_i}, \ldots, X_{k+1}).$$

The groups $H^k(\mathfrak{g}, V) = H^k(C^*(\mathfrak{g}, V))$ are the *cohomology groups* of $\mathfrak{g}$ with coefficients in V.

Notice that the dual of a $\mathfrak{g}$-module is also a $\mathfrak{g}$-module in a natural way. The dual of the complex $C_*(\mathfrak{g}, V)$ is isomorphic to $C^*(\mathfrak{g}, V^*)$, hence the dual of $H_*(\mathfrak{g}, V)$ is isomorphic to $H^*(\mathfrak{g}, V^*)$.

Let $A^k(M)$ denote the space of smooth differential forms of degree k on M (with complex coefficients). Let $A^*(M)$ be the de Rham complex and $A^*_c(M)$ the sub-complex of compactly supported differential forms.

The following construction is well-known [6].

Construction 2.1. There is a morphism of complexes

$$\chi^*: A^*(M) \to C^*(\mathrm{Vect}(M), C^\infty(M))$$

defined by $\chi^*(\omega)(X_1, \ldots, X_k) = i(X_k) \ldots i(X_1)\omega$ for ω in $A^k(M)$, and $X_1, \ldots, X_k$ in $\mathrm{Vect}(M)$, where i denotes interior product. This can also be written as

$$\chi^*(\omega)(X_1, \ldots, X_k) = \langle X_1 \wedge \cdots \wedge X_k, \omega \rangle,$$

where $\langle \cdot, \cdot \rangle$ denotes the pairing of k-vector fields and k-forms. χ^* also induces a morphism of complexes

$$\chi^*: A^*_c(M) \to C^*(\mathrm{Vect}(M), C^\infty_c(M)).$$

This also suggests a dual construction. Let $A^m_{cl}(M)$ denote the space of *closed* m-forms.

Construction 2.2. There is a morphism of complexes

$$\chi^*: C_*(\mathrm{Vect}(M), A^m_{cl}(M)) \to A^{m-*}(M)$$

such that

$$\chi(X_1 \wedge \cdots \wedge X_k \otimes \omega) = (-1)^k i(X_1)\dots i(X_k)\omega.$$

χ also induces a morphism of complexes

$$\chi: C_*(\mathrm{Vect}(M), A^m_{cl,c}(M)) \to A^{m-*}_c(M).$$

Verification. Note that $\mathrm{Vect}(M)$ acts on $A^m(M)$ using Lie derivatives. We have

$$\begin{aligned}
&\chi \circ d(X_1 \wedge \cdots \wedge X_k \otimes \omega)\\
&\quad = (-1)^{k-1} \sum_{1\le i<j\le k} (-1)^{i+j} i([X_i, X_j])\dots \widehat{i(X_i)}\dots \widehat{i(X_j)}\dots i(X_k)\omega\\
&\qquad + (-1)^{k-1} \sum_{1\le i\le k} (-1)^i i(X_1)\dots \widehat{i(X_i)}\dots i(X_k)\theta(X_i)\omega.
\end{aligned}$$

Now recall the formulas: $i([X,Y]) = [\theta(X), i(Y)]$ and $\theta(X) = di(X) + i(X)d$, where X and Y are vector fields. It follows that for i fixed we have:

$$\begin{aligned}
&\sum_{i<j\le k} (-1)^{i+j} i([X_i, X_j])\dots \widehat{i(X_i)}\dots \widehat{i(X_j)}\dots i(X_k)\omega\\
&\quad = \sum_{i<j\le k} (-1)^i i(X_1)\dots \widehat{i(X_i)}\dots i(X_{j-1}) i([X_i, X_j])\dots i(X_k)\omega\\
&\quad = \sum_{i<j\le k} (-1)^i i(X_1)\dots \widehat{i(X_i)}\dots i(X_{j-1}) \cdot [\theta(X_i), i(X_j)]\dots i(X_k) \cdot \omega\\
&\quad = (-1)^{i+1} \cdot \big(i(X_1)\dots \widehat{i(X_i)}\dots i(X_k)\theta(X_i)\omega\\
&\qquad - i(X_1)\dots i(X_{i-1})\theta(X_i) i(X_{i+1})\dots i(X_k)\omega\big).
\end{aligned}$$

So we find

$$\chi \circ d(X_1 \wedge \cdots \wedge X_k \otimes \omega) = (-1)^{k-1} \sum_{1 \leq i \leq k} (-1)^i i(X_1) \dots i(X_{i-1}) \theta(X_i) \dots i(X_k) \omega.$$

On the other hand,

$$\begin{aligned} \chi \circ d(X_1 \wedge \dots \wedge X_k \otimes \omega) &= (-1)^k d i(X_1) i(X_2) \dots i(X_k) \omega \\ &= (-1)^k \theta(X_1) i(X_2) \dots i(X_k) \omega \\ &\quad + (-1)^{k+1} i(X_1) d i(X_2) \dots i(X_k) \omega \\ &= (-1)^k \theta(X_1) i(X_2) \dots i(X_k) \omega \\ &\quad + (-1)^{k+1} i(X_1) \theta(X_2) \dots i(X_k) \omega \\ &\quad + (-1)^{k+2} i(X_1) i(X_2) \theta(X_3) \dots \omega \\ &\quad + \cdots + (-1)^{2k-1} i(X_1) \dots \theta(X_k) \omega \\ &\quad + (-1)^{2k} i(X_1) \dots i(X_k) d\omega. \end{aligned}$$

Noticing that $d\omega = 0$ we obtain the equality of this sum and of $\chi \circ d(X_1 \wedge \cdots \wedge X_k \otimes \omega)$.

This construction admits the following variant. Suppose that M is an oriented manifold of dimension n. For any integer i, let us denote by $A_c^i(M)$ the space of smooth differentiable forms on M of degree i with compact support and endow this space with the C^∞-topology. The topological dual of $A_c^i(M)$ shall be denoted by $S_i(M)$, and its elements are called *currents* of dimension i (or degree $n-i$). We identify any form ω in $A^{n-i}(M)$ with the current $\alpha \mapsto \int_M \alpha \wedge \omega$ of degree $n-i$ (for α running over $A_c^i(M)$). Hence $A^{n-i}(M)$ is considered as a subspace of $S_i(M)$. The operators d, $i(X)$, and $\theta(X)$ admit of the following extensions to the space of currents:

$$\begin{aligned} d: S_i(M) &\to S_{i-1}(M) & \langle \alpha, dC \rangle &= (-1)^i \langle d\alpha, C \rangle \\ i(x): S_i(M) &\to S_{i+1}(M) & \langle \alpha, i(X)C \rangle &= (-1)^i \langle i(X)\alpha, C \rangle \\ \theta(X): S_i(M) &\to S_i(M) & \langle \alpha, \theta(X)C \rangle &= -\langle \theta(X)\alpha, C \rangle. \end{aligned}$$

Using the differential d, we get a complex $S_*(M)$, and $A^{n-*}(M)$ is a subcomplex of $S_*(M)$.

Variant 2.3. There is a morphism of complexes:

$$\chi: C_*(\mathrm{Vect}(M), S_0(M)) \to S_*(M)$$

such that

$$\chi(X_1 \wedge \cdots \wedge X_k \otimes \omega) = (-1)^k i(X_1) \ldots i(X_k)\omega.$$

Now let the complex Lie algebra $\mathfrak{g}$ act on M, i.e., we consider a Lie algebra homomorphism $\rho: \mathfrak{g} \to \mathrm{Vect}(M)$. In cases of interest to us ρ is injective, so we identify $\mathfrak{g}$ with its image under ρ. Construction 2.2 gives a morphism of complexes

$$\chi: C_*(\mathfrak{g}, A^m_{c\ell}(M)) \to A^{m-*}(M).$$

Proposition 2.4. *Let $\omega \in A^m_{c\ell}(M)$ be preserved by $\mathfrak{g}$, i.e., $\theta(X)\cdot\omega = 0$ for all X in $\mathfrak{g}$. Then there is a morphism of complexes*

$$\chi: C_*(\mathfrak{g}, \mathbf{C}) \to A^{m-*}(M)$$

such that

$$\chi(X_1 \wedge \cdots \wedge X_k) = (-1)^k i(X_1) \ldots i(X_k)\omega.$$

Hence χ defines a map $\chi: H_k(\mathfrak{g}, \mathbf{C}) \to H^{m-k}(M, \mathbf{C})$.

Here $\mathbf{C}$ is the trivial, one-dimensional $\mathfrak{g}$-module.

Proof. We have a $\mathfrak{g}$-linear map $\mathbf{C} \to A^m_{c\ell}(M)$ sending 1 to ω, hence a morphism of complexes $C_*(\mathfrak{g}, \mathbf{C}) \to C_*(\mathfrak{g}, A^m_{c\ell}(M))$. Compose this with $C_*(\mathfrak{g}, A^m_{c\ell}(M)) \to A^{m-*}(M)$. Q.E.D.

Variant 2.5. Let $\omega \in S_0(M)$ be preserved by $\mathfrak{g}$. Then there is a morphism of complexes $\chi: C_*(\mathfrak{g}, \mathbf{C}) \to S_*(M)$ such that $\chi(X_1 \wedge \cdots \wedge X_k) = (-1)^k i(X_1) \ldots i(X_k)\omega$. Hence χ defines a map $\chi: H_k(\mathfrak{g}, \mathbf{C}) \to H_k^{c\ell}(M, \mathbf{C})$.

Remark 2.6. If M is orientable, the maps of 2.4 and 2.5 agree if we use Poincaré duality to identify $H_k^{c\ell}(M, \mathbf{C})$ with $H^{n-k}(M, \mathbf{C})$.

Remark 2.7. It is also useful to write the morphism of complexes χ of 2.5 in the following way: $\chi(X_1 \wedge \cdots \wedge X_k)$ is the linear form on $A_c^k(M)$ which maps α to $(-1)^k \omega(i(X_1)i(X_2) \ldots i(X_k)\alpha) = (-1)^{k(k+1)/2} \int_M <X_1 \wedge \cdots \wedge X_k, \alpha> \cdot \omega$, where $<\cdot, \cdot>$ denotes the pairing between k-vector fields and k-forms.

Remark 2.8. If the n-form ω (of top degree) nowhere vanishes, a vector field X on M is called *unimodular* (with respect to ω) if it preserves ω, that is, if $\theta(X) \cdot \omega = 0$. The action of $\mathfrak{g}$ on M is called unimodular if each X in $\mathfrak{g}$ acts as a unimodular vector field.

The Construction 2.4 for unimodular Lie algebra actions may be generalized to the case where $\theta(X) \cdot \omega = \lambda(X) \cdot \omega$ for $X \in \mathfrak{g}$, where $\lambda: \mathfrak{g} \to \mathbf{C}$ is a Lie algebra character. Such vector fields are called *conformal unimodular* (see [9]).

Proposition 2.9. *Let $\omega \in A^n(M)$ be preserved by $\mathfrak{g}$ up to scalars, i.e., assume $\theta(X) \cdot \omega = \lambda(X) \cdot \omega$, where λ is a Lie algebra character of $\mathfrak{g}$. Then the construction of 2.4 defines a morphism of complexes*

$$\chi: C_*(\mathfrak{g}, \mathbf{C}_\lambda) \to A^{n-*}(M),$$

where $\mathbf{C}_\lambda$ is any one-dimensional $\mathfrak{g}$-module on which $\mathfrak{g}$ acts by λ. Hence we have $\chi: H_*(\mathfrak{g}, \mathbf{C}_\lambda) \to H^{n-*}(M)$.

We will not write down explicitly the version of Proposition 2.9 relating to currents.

Remark 2.10. Let $\pi: Y^n \to X^n$ be an étale map, and let $\mathfrak{g}$ be a complex Lie algebra which acts on X, preserving an n-form ω. The canonical lift of this action to Y then preserves the pull-back form $\pi^*\omega$, and we have a commutative diagram

$$\begin{array}{ccc} H_i(\mathfrak{g}, \mathbf{C}) & \xrightarrow{\chi_X} & H^{n-1}(X, \mathbf{C}) \\ & \searrow^{\chi_Y} & \downarrow \pi^* \\ & & H^{n-i}(Y, \mathbf{C}). \end{array}$$

3. Some Examples

3.1. Let $\mathfrak{g}$ be the abelian Lie algebra $\mathbf{C}^m$; assume that $\mathfrak{g}$ acts on M and preserves the n-form ω of maximum degree. Let $(X_1, \ldots, X_m)$ denote the canonical basis of C^m. Then the map constructed in 2.5,

$$\chi: H_m(\mathbf{C}^m, \mathbf{C}) \to H_m(M, \mathbf{C}),$$

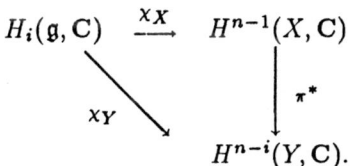

maps $X_1\wedge\cdots\wedge X_m$ to the current $\alpha\mapsto(-1)^{m(m+1)/2}\int<X_1\wedge\cdots\wedge X_m,\alpha>\cdot\omega$.

This is easily identified with the *Ruelle-Sullivan current* [13] in case the action of $\mathbf{C}^m$ is the complexification of a free action of $\mathbf{R}^m$ which then defines a foliation of M with m-dimensional leaves. Notice that the n-form ω, preserved by the action, gives rise to a transverse volume form $\alpha=i(X_1)\cdots i(X_m)\omega$, also preserved by the action. Conversely, such a preserved transverse volume form α corresponds to a preserved volume form ω.

This example is meant to justify the title of this paper: for $\mathfrak{g}$ non-abelian, classes in $H_k(\mathfrak{g},\mathbf{C})$ map to (classes of) currents on M, which we view as non-commutative analogs of Ruelle-Sullivan currents.

3.2 Let G be a real Lie group of dimension m, and $\mathfrak{g}$ its (complexified) Lie algebra. A $\mathfrak{g}$-invariant m-form on G determines an element of $\bigwedge^m(\mathfrak{g})^*$, its value at the origin; of course, any invariant m-form is determined by its value at the origin. The Construction 2.5 gives a map $\chi\colon H_k(\mathfrak{g},\mathbf{C})\to H_k(G,\mathbf{C})$ (for a given choice of $\omega\in\bigwedge^m(\mathfrak{g})^*$).

Let ${}^t\chi\colon H_c^k(G,\mathbf{C})\to H^k(\mathfrak{g},\mathbf{C})$ be the transpose map, and recall that $H^k(\mathfrak{g},\mathbf{C})$ is the cohomology of the complex of left-invariant differential forms on G.

Proposition 3.3. *${}^t\chi$ is up to sign the averaging map: ${}^t\chi(\alpha)=(-1)^{k(k+1)/2}\int_G g^*(\alpha)\cdot\omega(g)$, where $g^*(\alpha)$ is the inverse image of α under left-translation by g.*

Proof. Let $X_1\wedge\cdots\wedge X_k$ be an element of $\bigwedge^k(\mathfrak{g})$. Then we have:

$$\begin{aligned}<{}^t\chi(\alpha),X_1\wedge\cdots\wedge X_k> &= <\alpha,\chi(X_1\wedge\cdots\wedge X_k)>\\ &=(-1)^{k(k+1)/2}\int_G <X_1\wedge\cdots\wedge X_k,\alpha>\cdot\omega(g)\\ &=(-1)^{k(k+1)/2}\int_G <X_1\wedge\cdots\wedge X_k,g^*(\alpha)>\cdot\omega(g)\end{aligned}$$

(using the G-invariance of the X_i). Q.E.D.

Corollary 3.4. *If G is compact and connected, then $\chi\colon H_k(\mathfrak{g},\mathbf{C})\to H_k(G,\mathbf{C})$ is an isomorphism.*

3.5 Assume that the real Lie group G acts differentiably on the manifold M. Then the (complexified) Lie algebra $\mathfrak{g}$ of G acts on M. If M

is connected, there is a well-defined map $H_k(G, \mathbf{C}) \to H_k(M, \mathbf{C})$ induced from the map $g \mapsto gx_0$ from G into M, where x_0 is a base point.

On the other hand, if ω is a preserved n-form (under the action of $\mathfrak{g}$), we have the map $H_k(\mathfrak{g}, \mathbf{C}) \to H_k^{cl}(M, \mathbf{C})$. And if we choose a G-invariant m-form β on G, we have a map $H_k(\mathfrak{g}, \mathbf{C}) \to H_k^{cl}(G, \mathbf{C})$ described in 3.3.

Proposition 3.6. *Assume G is compact. Choose volume forms β and ω as before, and assume that the volume* $\mathrm{vol}(M) = \int_M \omega$ *is finite. Then the map $\chi: H_k(g, M) \to H_k^{cl}(M, \mathbf{C})$ is equal to* $\mathrm{vol}(M)/\mathrm{vol}(G)$ *times the composite map*

$$H_k(\mathfrak{g}, \mathbf{C}) \to H_k^{cl}(G, \mathbf{C}) = H_k(G, \mathbf{C}) \to H_k(M, \mathbf{C}) \to H_k^{cl}(M, \mathbf{C}).$$

Proof. Let $X_1 \wedge \cdots \wedge X_k \in \bigwedge^k \mathfrak{g}$; let us evaluate its image in $H_k(M, \mathbf{C})$ under the composite map above against a k-form α on M. We obtain $(-1)^{k(k+1)/2} \int_G <X_1 \wedge \cdots \wedge X_k, m_x^* \alpha> \cdot \beta$, where $m_x: G \to M$ is given by $m_x(\mathfrak{g}) = \mathfrak{g} \cdot x$. This is true for any choice of $x \in M$, so multiplied by $\mathrm{vol}(M)$ it is equal to $(-1)^{k(k+1)/2} \int_M \left(\int_G <X_1 \wedge \cdots \wedge X_k, m_x^* \alpha> \cdot \beta\right) \omega$, which may be written in more detail as $(-1)^{k(k+1)/2} \int_M \int_G <X_1 \wedge \cdots \wedge X_k, \alpha_{g \cdot x}> \beta(g) \omega_x$. Using the $\mathfrak{g}$-invariance of ω, this is simply $\mathrm{vol}(G)$ times $<\chi(X_1 \wedge \cdots \wedge X_k), \alpha>$. Q.E.D.

Corollary 3.7. *Let M be a* non-compact *connected manifold, countable at infinity. Let G be a connected compact Lie group which acts smoothly on M. The composite map $H_k(G, \mathbf{C}) \to H_k(M, \mathbf{C}) \to H_k^{cl}(M, \mathbf{C})$ is zero.*

Proof. One may deduce it from Proposition 3.6 by choosing ω of infinite volume. More rigorously, one notices that the one-point compactification $\widehat{M} = M \cup \mathrm{pt}$ is connected, and admits a continuous G-action. The map $H_k(G, \mathbf{C}) \to H_k(\widehat{M}, \mathbf{C})$ factors through $H_k(\mathrm{pt}, \mathbf{C})$, so its composition with the restriction map $H_k(\widehat{M}, \mathbf{C}) \to H_k^{cl}(M, \mathbf{C})$ is zero. Q.E.D.

3.8 Let G be a real Lie group of dimension m and $\mathfrak{g}$ its (complexified) Lie algebra. Let ω be a left-invariant form of maximal degree m on G. Then G acts on $\bigwedge^m(\mathfrak{g}^*)$ by the m-th exterior power of the co-adjoint action. Let Γ be a discrete subgroup of G, which fixes $\omega \in \bigwedge^m(\mathfrak{g}^*)$; then ω descends to an m-form on G/Γ, which is preserved by $\mathfrak{g}$; hence we get a map $\chi: H_k(\mathfrak{g}, \mathbf{C}) \to H_k^{cl}(G/\Gamma, \mathbf{C})$.

If G is nilpotent and simply-connected, one may choose Γ such that G/Γ is compact. If $\omega \neq 0$, then the map $H_k(\mathfrak{g}, \mathbf{C}) \to H_k(G/\Gamma, \mathbf{C})$ is the transpose of the map constructed by Nomizu [11], which he proved to be an isomorphism.

It follows from 2.10 that in any case, we have a commutative diagram

$$\begin{array}{ccc} H_k(\mathfrak{g}, \mathbf{C}) & \longrightarrow & H_k^{cl}(G/\Gamma, \mathbf{C}) \\ & \searrow & \downarrow \\ & & H_k^{cl}(Y, G, \mathbf{C}). \end{array}$$

4. Abelian extensions of Lie algebras

In [12], Pressley and Segal construct a central extension of a Lie group G by the circle group if G acts smoothly on a simply-connected manifold, preserving a 2-form Ω such that the cohomology class of $\Omega/2\pi$ is integral. We present here, as a consequence of the general construction in §2, an analogous additive construction. This allows us to recover a famous construction of Kostant [8] and Souriau [14]. A better understanding of the analogy would probably involve Chern-Simons classes (in particular, cohomology with coefficients in $\mathbf{R}/\mathbf{Z}$) and, in the complex-analytic case, Beilinson-Deligne cohomology.

We consider a connected symplectic manifold M (it may be infinite-dimensional), with 2-form Ω. Let the Lie algebra $\mathfrak{g}$ act on M in such a way that $\theta(X)\cdot\Omega = \mu(X)\cdot\Omega$, where μ is a Lie algebra character (such a Lie algebra of vector fields is called "conformally symplectic").

Proposition 4.1. (i) *For any $x \in M$, the element β_x of $\bigwedge^2(\mathfrak{g})^*$ defined by $\beta_x(X,Y) = \Omega(X,Y)(x)$ is a 2-cocycle of $\mathfrak{g}$ with values in the one-dimensional $\mathfrak{g}$-module $\mathbf{C}_{-\mu}$. Its cohomology class in $H^2(\mathfrak{g}, \mathbf{C}_{-\mu})$ is independent of x.* (ii) *Let x_1, x_2 be two points of M, and let $\gamma\colon [0,1] \to M$ be a smooth path from x_1 to x_2. Then the 2-cocycle $\beta_{x_2}-\beta_{x_1}$ is the boundary of the 1-cochain $\alpha\colon \mathfrak{g} \to \mathbf{C}_{-\mu}$ defined by $\alpha(X) = \int_0^1 \Omega(X, d\gamma/dt)\cdot dt$.*

Proof. (i) This is a generalization of Proposition 2.4 and is proved in the same way; we have a $\mathfrak{g}$-linear map $\mathbf{C}_\mu \to A^2_{c\ell}(M)$, sending 1 to ω, hence a morphism of complexes from $C_*(\mathfrak{g}, \mathbf{C}_\mu)$ to $C_*(\mathfrak{g}, A^2_{c\ell}(M))$; this latter complex maps to $A^{2-*}(M)$. This induces a map $H_k(\mathfrak{g}, \mathbf{C}_\mu) \to H^{2-k}(M, \mathbf{C})$; in particular, for $k=2$, this gives a linear form on $H_2(\mathfrak{g}, \mathbf{C}_\mu)$, i.e., an element of $H^2(\mathfrak{g}, \mathbf{C}_{-\mu})$. This is clearly, by construction, the class of β_x, for any $x \in M$.

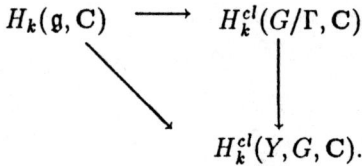

(ii) For X and Y in $\mathfrak{g}$ fixed, $\beta_x(X,Y)$ is a function of $x \in M$ with derivative

$$\begin{aligned} di(Y)i(X)\Omega &= \theta(Y)i(X)\Omega - i(Y)di(X)\Omega \\ &= i([Y,X])\Omega + i(X)\theta(Y)\Omega - i(Y)\theta(X)\Omega \\ &= i([Y,X])\Omega + \mu(Y)\cdot i(X)\Omega - \mu(X)\cdot i(Y)\Omega. \end{aligned}$$

Evaluate this against $d\gamma/dt$ and integrate from 0 to 1 to obtain

$$\begin{aligned} \beta_{x_2}(X,Y) - \beta_{x_1}(X,Y) &= \int_0^1 \Omega\left([Y,X], \frac{d\gamma}{dt}\right)\, dt \\ &\quad + \mu(Y)\int_0^1 \Omega\left(X, \frac{d\gamma}{dt}\right)\, dt \\ &\quad - \mu(X)\int_0^1 \Omega\left(Y, \frac{d\gamma}{dt}\right)\, dt \\ &= -\alpha([X,Y]) + \mu(Y)\cdot\alpha(X) - \mu(X)\cdot(Y) \\ &= (d\alpha)(X,Y). \end{aligned}$$

Q.E.D.

Corollary 4.2. *Since the cohomology class $\beta = [\beta_x]$ in $H^2(\mathfrak{g}, \mathbf{C}_{-\mu})$ of the 2-cocycle β_x is independent of $x \in M$, there is a well-defined abelian extension*

$$0 \to \mathbf{C}_{-\mu} \to \tilde{\mathfrak{g}} \to \mathfrak{g} \to 0.$$

If $\mu = 0$, this is a central extension.

4.3 Let (M,Ω) be a finite-dimensional simply-connected manifold. Let $\mathfrak{g}$ be the Lie algebra of all smooth vector fields on M which preserve Ω; then the central extension $\tilde{\mathfrak{g}}$ is the Lie algebra of C^∞-functions on M under Poisson bracket, and the projection map $\tilde{\mathfrak{g}} \to \mathfrak{g}$ associates to a smooth function F the hamiltonian vector field H_F. This central extension was discovered by Kostant [8] and Souriau [14].

Proposition 4.4. *Let $M = \mathbf{R}^2$ with coordinates (x_1, x_2), let $\Omega = dx_1 \wedge dx_2$, and let $\mathfrak{g}$ be the Lie algebra of smooth vector fields on M which are conformal symplectic (i.e., of type $u_1\frac{\partial}{\partial x_1} + u_2\frac{\partial}{\partial x_2}$ with the divergence $\frac{\partial u_1}{\partial x_1} +$*

$\frac{\partial u_2}{\partial x_2}$ constant). Then the cohomology class β in $H^2(\mathfrak{g}, \mathbf{C}_{-\mu})$ is non-trivial (hence the extension $\tilde{\mathfrak{g}}$ of $\mathfrak{g}$ by $\mathbf{C}_{-\mu}$ does not split).

Proof. Choosing the point $x = 0$ in $\mathbf{R}^2$, the 2-cocycle β_x is given by:

$$\beta_x\left(u_1\frac{\partial}{\partial x_1} + u_2\frac{\partial}{\partial x_2}, v_1\frac{\partial}{\partial x_1} + v_2\frac{\partial}{\partial x_2}\right) = \begin{vmatrix} u_1 & u_2 \\ v_1 & v_2 \end{vmatrix}(0,0).$$

Let $\mathcal{A} = \mathbf{R}^2 \subset \mathfrak{g}$ be the Lie algebra of translations spanned by $\frac{\partial}{\partial x_1}$ and $\frac{\partial}{\partial x_2}$. Then $\beta_x\left(\frac{\partial}{\partial x_1}, \frac{\partial}{\partial x_2}\right) = 1$, hence $[\beta_x]$ restricts to a non-zero element of $H^2(\mathcal{A}, \mathbf{C}_{-\mu}) = H^2(\mathcal{A}, \mathbf{C})$. Hence the class $[\beta_x] \in H^2(\mathfrak{g}, \mathbf{C}_{-\mu})$ cannot be trivial. Q.E.D.

Remark 4.5. The same proof shows that β restricts to a non-zero cohomology class for the Cremona Lie algebra of vector fields on $\mathbf{C}^2$ with polynomial coefficients and constant divergence. The Lie algebra is in some loose sense a Lie algebra analogue of the group of biregular transformations of $\mathbf{C}^2$.

5. Poisson actions on symplectic manifolds

5.1 Let M be a Poisson C^∞-manifold with Poisson bracket $\{,\}$. Recall [10], [15] that $\{,\}$ gives a Lie algebra structure to $C^\infty(M)$ and that $\{,\}$ satisfies the Leibnitz identity $\{f, gh\} = \{f,g\}h + \{f,h\}g$. There is a tensor $G \in \bigwedge^2(T_M) = \bigwedge^2(T_M^*)^*$ given by $G(df, dg) = \{f,g\}$: this tensor defines a map $I: T_M^* \to T_M$ such that $<I(\alpha_1), \alpha_2> = G(\alpha_1, \alpha_2)$ for 1-forms α_1 and α_2. For $f \in C^\infty(M)$, $H_f = I(df)$ is the *hamiltonian vector field* of f; it acts on functions by $H_f(g) = \{f, g\}$.

We recall that the *canonical homology* $H_k^{\mathrm{can}}(M)$ is the homology of the complex vector spaces $C_*(M) = A^*(M)$

$$\cdots \to A^{k+1}(M) \xrightarrow{\delta} A^k(M) \xrightarrow{\delta} A^{k-1}(M) \to \cdots.$$

where

$$\begin{aligned} &\delta(f_0 df_1 \wedge \cdots \wedge df_k) \\ &\quad = \sum_{1\le i\le k} (-1)^{i+1}\{f_0, f_i\} df_1 \wedge \cdots \wedge \widehat{df_i} \wedge \cdots \wedge df_k \\ &\qquad + \sum_{1\le i<j\le k} (-1)^{i+j} f_0 d\{f_i, f_j\} \wedge df_1 \wedge \cdots \wedge \widehat{df_i} \wedge \cdots \wedge \widehat{df_j} \wedge \cdots \wedge df_k. \end{aligned}$$

This was introduced in [3].

If M is a symplectic manifold, so that I is bijective and the 2-form Ω on M is such that: $<\alpha,\xi> = \Omega(\xi, I(\alpha))$, for any 1-form α and vector field ξ, then we proved in [3, Corollary 2.2.2] that $H_k^{\mathrm{can}}(M)$ is canonically isomorphic to $H^{2n-k}(M, \mathbf{C})$, where $\dim(M) = 2n$. Using Poincaré duality on M, this also means we have an isomorphism $H_k^{\mathrm{can}}(M) \xrightarrow{\sim} H_k^{c\ell}(M)$.

Proposition 5.2. *The isomorphism* $H_k^{\mathrm{can}}(M) \underset{\varphi}{\xrightarrow{\sim}} H^{2n-k}(M, \mathbf{C})$ *of loc. cit. is given up to sign by the map on differential forms*

$$\varphi: f_0 df_1 \wedge \cdots \wedge df_k \mapsto f_0 i(H_{f_1}) i(H_{f_2}) \ldots i(H_{f_k})\omega,$$

where the $2n$*-form* ω *is* $\omega = \Omega^n/n!$.

Proof. The isomorphism of *loc. cit.* is given by the duality operator $*: A^k(M) \to A^{2n-k}(M)$. Let α be a k-form; then $*\alpha$ is characterized by the fact that $\beta \wedge (*\alpha) = \bigwedge^k G(\beta, \alpha) \cdot \omega$ for every k-form β. If β has compact support and $\alpha = f_0 df_1 \wedge \cdots \wedge df_k$, we get:

$$\begin{aligned}
\int_M \beta \wedge (*\alpha) &= \int_M \bigwedge^k G(\beta, \alpha) \cdot \omega \\
&= (-1)^{k(k+1)/2} \int_M f_0 [i(H_{f_1}) \ldots i(H_{f_k})\beta]\omega \\
&= (-1)^{k(k-1)/2} \int_M f_0 \beta \wedge [i(H_{f_1}) \ldots i(H_{f_k})\omega] \\
&= (-1)^{k(k-1)/2} \int_M \beta \wedge \varphi(\alpha).
\end{aligned}$$

It follows that $\varphi(\alpha)$ is equal to $*\alpha$, up to sign. Q.E.D.

Remark 5.3. Up to sign, the isomorphism $H_k^{\mathrm{can}}(M) \xrightarrow{\sim} H_k(M, \mathbf{C})$ may be described as follows: to the k-form α associate the linear form on $H_c^k(M, \mathbf{C})$: $\beta \to \int_M \bigwedge^k G(\beta, \alpha) \cdot \omega$. This follows directly from the definition of $*\alpha$ and has been used in the proof of 5.2.

5.4 Let the complex Lie algebra $\mathfrak{g}$ act smoothly on the Poisson manifold M. We say that the action of $\mathfrak{g}$ *preserves the Poisson structure* on M if $X \cdot \{f, g\} = \{X \cdot f, g\} + \{f, X \cdot g\}$, that is, if any $X \in \mathfrak{g}$ induces a derivation of the Lie algebra $C^\infty(M)$ (under Poisson bracket).

Proposition 5.5. *Assume M is a* symplectic *manifold. Then there is a morphism of complexes $\psi: C_*(\mathfrak{g}, \mathbf{C}) \to C_*(M)$ such that*

$$\psi(X_1 \wedge \cdots \wedge X_k) = I^{-1}(X_1) \wedge \cdots \wedge I^{-1}(X_k).$$

Proof. One could show this by an easy computation. However, we will instead deduce this as a corollary of a construction in 5.9. Q.E.D.

5.6 Let the action of $\mathfrak{g}$ on M preserve the Poisson structure. This action is called a *Poisson action* if for every $X \in \mathfrak{g}$ there is given a smooth function μ_X on M such that:

(a) $X = I(d\mu_X)$
(b) $X \mapsto \mu_X$ is a Lie algebra homomorphism.

Equivalently, the action of $\mathfrak{g}$ is a Poisson action if there is a smooth map $\mu: M \to \mathfrak{g}^*$ (called the *moment map*) such that (identifying $\mathfrak{g}$ to the space of linear functions on $\mathfrak{g}^*$)

(a′) for any $X \in \mathfrak{g}$, $X = I(d(X \circ \mu))$
(b′) μ is a morphism of Poisson manifolds, if $\mathfrak{g}^*$ is endowed with its standard Poisson structure, for which $\{f, g\}(\ell) = <\ell, [df(\ell), dg(\ell)]>$.

5.7 The determination of the canonical homology of the Poisson manifold $\mathfrak{g}^*$ is essentially done in [7,§7] (except that we view it as a C^∞-manifold, not as an algebraic manifold). We have the equality $H_k^{\text{can}}(\mathfrak{g}^*) = H_k(\mathfrak{g}_{\mathbf{R}}, C^\infty(\mathfrak{g}^*))$, where $\mathfrak{g}_{\mathbf{R}}$ is $\mathfrak{g}$ viewed as a *real Lie algebra*, which acts on $C^\infty(\mathfrak{g}^*)$ via its co-adjoint action on $\mathfrak{g}^*$. Indeed $H_k(\mathfrak{g}_{\mathbf{R}}, C^\infty(\mathfrak{g}^*))$ is equal to $H_k(\mathfrak{g} \oplus \mathfrak{g}, C^\infty(\mathfrak{g}^*))$, and the canonical complex of the Poisson manifold $\mathfrak{g}^*$ identifies with the standard complex $H_k(\mathfrak{g} \oplus \mathfrak{g}, C^\infty(\mathfrak{g}^*))$.

5.8 Let's assume instead that $\mathfrak{g}$ is a *real Lie algebra*. Then the considerations of 5.7 simplify to $H_k^{\text{can}}(\mathfrak{g}^*) = H_k(\mathfrak{g}, C^\infty(\mathfrak{g}^*))$.

5.9 Let the real Lie algebra $\mathfrak{g}$ act on the Poisson manifold M. Let the action be Poisson, and let $\mu: M \to \mathfrak{g}^*$ be the moment map. Then $\mu^*: H_k^{\text{can}}(\mathfrak{g}^*) \to H_k^{\text{can}}(M)$ is defined; using 5.8, this gives a linear map $\mu: H_k(\mathfrak{g}, C^\infty(\mathfrak{g}^*)) \to H_k^{\text{can}}(M)$. Concretely, we have $\mu^*(X_1 \wedge \cdots \wedge X_k \otimes f) = (f \circ \mu)(d\mu_{X_1}) \wedge \cdots \wedge (d\mu_{X_k})$ using the notation $\mu_X = X \circ \mu$ for $X \in \mathfrak{g}$ as in 5.6.

5.10 Assume that (M, Ω) is a symplectic manifold and that the Lie algebra $\mathfrak{g}$ acts on (M, Ω). If M is simply-connected, there is a central extension $\tilde{\mathfrak{g}}$ of $\mathfrak{g}$ such that the action of $\tilde{\mathfrak{g}}$ (which is induced from the $\mathfrak{g}$-action) is a Poisson action. We express this by saying that the action of $\mathfrak{g}$ is *almost Poisson.* The map of complexes $C_*(\tilde{\mathfrak{g}}) \hookrightarrow C_*(\tilde{\mathfrak{g}}, C^\infty(\tilde{\mathfrak{g}}^*)) \xrightarrow{\mu^*} C_*(M)$ clearly factors through the quotient complex $C_*(\mathfrak{g})$ of $C_*(\tilde{\mathfrak{g}})$. Furthermore, since μ is well-defined up to the addition of a constant element of $\tilde{\mathfrak{g}}^*$, this map of complexes is independent of the choice of μ.

Now if $\mathfrak{g}$ acts on any symplectic manifold (M, Ω), the action of $\mathfrak{g}$ is locally "almost Poisson," hence the morphism of complexes of Proposition 5.5 exists globally on M.

Proposition 5.11. *Let (M, Ω) be a symplectic manifold of finite dimension $2n$. Let the Lie algebra $\mathfrak{g}$ act on M, preserving Ω. Then we have a commutative diagram*

$$\begin{array}{ccc} & & H_k^{\mathrm{can}}(M, \mathbf{C}) \\ & \overset{\psi}{\nearrow} & \\ H_k(\mathfrak{g}, \mathbf{C}) & & \downarrow \varphi \\ & \underset{(-1)^k \chi}{\searrow} & \\ & & H^{2n-k}(M, \mathbf{C}), \end{array}$$

where χ is as in 2.4 (or 2.9), ψ as in 5.5, and φ as in 5.2.

Proof. We will show that we have a commutative diagram on the level of complexes. The question becomes local on M, so that we may assume the action of $\mathfrak{g}$ is almost Poisson in the sense of 5.10, or that $X = I(d\mu_X) = H_{\mu_X}$ for all X in $\mathfrak{g}$, with the notation of 5.6. For $X_1, \ldots, X_k$ in $\mathfrak{g}$, we have

$$\begin{aligned} \psi(X_1 \wedge \cdots \wedge X_k) &= I^{-1}(X_1) \wedge \cdots \wedge I^{-1}(X_k) \\ &= d\mu_{X_1} \wedge \cdots \wedge d\mu_{X_k} \end{aligned}$$

and

$$\begin{aligned} (\varphi \circ \psi)(X_1 \wedge \cdots \wedge X_k) &= i(H_{\mu_{X_1}}) \ldots i(H_{\mu_{X_k}})\omega \\ &= i(X_1) \ldots i(X_k)\omega \\ &= (-1)^k \chi(\omega). \end{aligned}$$

Q.E.D.

Example 5.12. Consider again a *Poisson action* of a real Lie algebra $\mathfrak{g}$ on the symplectic manifold M. Then according to 5.9 we obtain a linear map $\mu^*: H_k(\mathfrak{g}, C^\infty(\mathfrak{g}^*)) \to H_k^{\mathrm{can}}(M)$. First consider the abelian Lie algebra $\mathbf{R}$ and the induced map $H_1(\mathbf{R}, C^\infty(\mathbf{R})) \to H_1^{\mathrm{can}}(M) = H_1(M, \mathbf{C})$. We claim this map is trivial; since the class $X \otimes f$ in $H_1(\mathbf{R}, C^\infty(\mathbf{R})) = \mathbf{R} \otimes C^\infty(\mathbf{R})$ (for X the generator of the Lie algebra $\mathbf{R}$) maps to the class in $H_1^{\mathrm{can}}(M)$ of the differential form $(f \circ \mu) d\mu$, where $\mu = \mu_X$, and since $f = g'$ for some $g \in C^\infty(\mathbf{R})$, it is enough to show that the class of dh in $H_1^{\mathrm{can}}(M)$ is always trivial for any $h \in C^\infty(M)$. This is easily done using a partition of unity and the fact that the compactly-supported canonical homology of a ball in $\mathbf{R}^{2n}$ is trivial except in degree 0.

However, the exact currents obtained as $(f \circ \mu) d\mu$, as above, appear to have geometric interest. They may be thought of as "weighted Ruelle-Sullivan currents," the function f giving a weight to each hypersurface $\mu =$ constant (a fibre of the moment map).

Example 5.13. We give an example of a Poisson action of the abelian Lie algebra $\mathfrak{g} = \mathbf{R}^2$ on the 2-torus $\mathbf{T} = \mathbf{R}^2/\mathbf{Z}^2$ such that the mapping $\mu^*: H_1(\mathfrak{g}, C^\infty(\mathfrak{g}^*)) \to H_1^{\mathrm{can}}(\mathbf{T}^2, \mathbf{C})$ is not trivial. Let (x, y) be the cartesian coordinates on $\mathbf{R}^2$. Then $e^{2\pi i x}$ and $e^{-2\pi i x}$ are functions on $\mathbf{T}^2$ which Poisson commute. Let $(\vec{\imath}, \vec{\jmath})$ be the standard basis of $\mathfrak{g}$, and let the moment map for the Poisson action be defined by $\mu(\vec{\imath}) = e^{2\pi i x}$, $\mu(\vec{\jmath}) = e^{-2\pi i x}$.

Let u and v be the coordinate functions on $\mathfrak{g}^*$ relative to the basis $(\vec{\imath}, \vec{\jmath})$. Consider the class $\vec{\imath} \otimes v$ in $\mathfrak{g} \otimes C^\infty(\mathfrak{g}^*)$; it maps to the class of the 1-form $e^{-2\pi i x} d(e^{2\pi i x}) = (2\pi i) dx$ in the canonical homology group; this class is non-trivial because its image in $H_1(\mathbf{T}^2, \mathbf{C})$ (de Rham cohomology) is $* (2\pi i dx) = 2\pi i dx$, which is not exact.

6. Lie algebras of holomorphic vector fields

We consider a *complex manifold* M of (complex) dimension n and denote by $\mathcal{V}(M)$ the Lie algebra of holomorphic vector fields on M; it is a sub-Lie algebra of $\mathrm{Vect}(M)$. A *holomorphic* action of a Lie algebra $\mathfrak{g}$ on M is simply a Lie algebra homomorphism $\mathfrak{g} \to \mathcal{V}(M)$.

For $X \in \mathcal{V}(M)$, we have operations $i(X)$ and $\theta(X)$ defined on the sheaves Ω_M^k of germs of holomorphic differential forms.

We generalize (the holomorphic version of) Construction 2.2 as follows.

Construction 6.1. Let F be an additive functor from the category of sheaves of complex vector spaces on M to the category of complex vector spaces. For $m \geq 0$, let $F(\Omega_M^m)_{c\ell}$ denote the kernel of $d: F(\Omega_M^m) \to$

$F(\Omega_M^{m+1})$. Then there is a morphism of complexes

$$\chi: C_*(\mathcal{V}(M), F(\Omega_M^m)_{cl}) \to F(\Omega_M^{m-*})$$

such that

$$\chi(X_1 \wedge \cdots \wedge X_k \otimes \omega) = (-1)^k i(X_1) \dots i(X_k)\omega.$$

Remark 6.2. $i(X)$ and $\theta(X)$ induce maps on the $F(\Omega_M^k)$ since they define morphisms of sheaves and F is a functor. The verification that χ is a morphism of complexes is the same as in 2.2 since relations like $i([X,Y]) = [\theta(X), i(Y)]$ hold on the $F(\Omega_M^k)$.

Now consider a Lie algebra $\mathfrak{g}$ acting holomorphically on M.

We next turn to a variant of Proposition 2.4.

Proposition 6.3. *Let $\omega \in F(\Omega_M^m)_{cl}$ be preserved by $\mathfrak{g}$, i.e., $\theta(X) \cdot \omega = 0$ for all $X \in \mathfrak{g}$. Then there is a morphism of complexes*

$$\chi: C_*(\mathfrak{g}, \mathbf{C}) \to F(\Omega_M^{m-*})$$

such that $\chi(X_1 \wedge \cdots \wedge X_k) = (-1)^k i(X_1) \dots i(X_k)\omega$. Hence χ defines a map $\chi: H_k(\mathfrak{g}, \mathbf{C}) \to H^{m-k}(F(\Omega_M^*))$.

Proof. As in 2.4, there is a $\mathfrak{g}$-linear map $\mathbf{C} \to F(\Omega_M^m)_{cl}$ sending 1 to ω, hence a morphism of complexes $C_*(\mathfrak{g}, \mathbf{C}) \to C_*(\mathfrak{g}, F(\Omega_M^m)_{cl})$. Compose this with the morphism $C_*(\mathfrak{g}, F(\Omega_M^m)_{cl})$ of 6.1. Q.E.D.

Rather than continuing to analyze the above constructions for the case of a general functor F, we now assume that F is the functor $\mathcal{A} \mapsto H^q(M, \mathcal{A})$ for some $q \geq 0$. Then $F(\Omega_M^m)_{cl}$ is the kernel of $d: H^q(M, \Omega_M^m) \to H^q(M, \Omega_M^{m+1})$, therefore to apply Proposition 6.3, one has to find criteria for such a (d-closed) class in $H^q(M, \Omega_M^m)$ to be preserved by $\mathfrak{g}$.

Lemma 6.4. *If the "Hodge to de Rham" spectral sequence $E_1^{p,q} = H^q(M, \Omega_M^p) \Rightarrow H_{\mathrm{DR}}^*(M) = H^*(M, \mathbf{C})$ degenerates at E_1 (for instance, if M is compact Kähler), then each class in $H^q(M, \Omega_M^p)$ is preserved by $\mathcal{V}(M)$.*

Proof. For $X \in \mathcal{V}(M)$, $\theta(X)$ defines an endomorphism of the complex of sheaves Ω_M^*, hence of any reasonable double complex used to resolve it (for instance, the Čech double complex for a covering of M by Stein open sets or the Godement canonical resolution of Ω_M^* or the Dolbeault double complex). Hence $\theta(X)$ defines an endomorphism of the whole spectral sequence. Now $\theta(X)$ annihilates the E_∞-term since this is de Rham cohomology (using C^∞-differential forms), and the relation $\theta(X) = d\circ i(X) + i(X)\circ d$ holds in the C^∞-de Rham complex. Under the assumption that the spectral sequence degenerates, E_1 is equal to E_∞, so is annihilated by $\theta(X)$.

Q.E.D.

Proposition 6.5. *Under the assumption of 6.4, there is a well-defined map* $\chi: H_k(\mathcal{V}(M), \mathbf{C}) \otimes H^q(M, \Omega_M^p) \to H^q(M, \Omega_M^{p-k})$.

Proof. It suffices to combine 6.3 and 6.4 noting that every class in $H^q(M, \Omega_M^p)$ is closed. Q.E.D.

Corollary 6.6. *Let M be a compact Kähler manifold. Then there is a well-defined map* $\chi: H_k(\mathcal{V}(M), \mathbf{C}) \otimes H^m(M, \mathbf{C}) \to H^{m-k}(M, \mathbf{C})$ *compatible with the map in* 6.5.

Proof. Simply use the Hodge decomposition of $H^m(M, \mathbf{C})$.

Q.E.D.

Example 6.7. Under the assumption of 6.6, any holomorphic vector field X defines a map $H^m(M, \mathbf{C}) \to H^{m-1}(M, \mathbf{C})$. (Use X to construct a class in $H_1(\mathbf{C}, \mathbf{C})$ where $\mathbf{C} \subset \mathcal{V}(M)$ is the one-dimensional Lie algebra spanned by X, and apply 6.6.) This map has square zero. It would not exist, in general, if M was not Kähler.

For $m = 2n$, a top-dimensional cohomology class $[\omega] \in H^{2n}(M, \mathbf{C})$ maps to a class in $H^{2n-1}(M, \mathbf{C}) = H_1(M, \mathbf{C})$ which is described by the following linear form on $H_1(M, \mathbf{C})$: $\alpha \mapsto \int_M <\alpha, X> \omega$. This is a holomorphic version of the Ruelle-Sullivan current.

6.8 In [1], Arbarello, De Concini, Kač and Procesi construct a canonical homomorphism $H^2(\underline{d}, \mathbf{C}) \to H^2(M_g, \mathbf{C})$ where $\underline{d}$ is the Lie algebra of vector fields on the circle, and M_g is the moduli space of smooth curves of genus g.

From our point of view, it is natural to wonder how to construct a map on the homology (rather than cohomology) $H_k(\underline{d}, \mathbf{C}) \to H_k^{c\ell}(M_g'', \mathbf{C})$, where as in *loc. cit.*, M_g'' is the moduli space of triples (C, p, v), where C is a smooth curve of genus g, p a point of C,and v a non-zero tangent vector to C at p.

The construction of [1] involves the moduli space $\widehat{M}_g$ of triples (C, p, z), where z is a local parameter around the point p. The projection $\widehat{M}_g \to M_g''$ is a homotopy equivalence; there is a holomorphic action of $\underline{d}$ on $\widehat{M}_g$, constructed in [1].

M_g'' is a smooth complex manifold of dimension $n = 3g-1$, for all $g \geq 1$. We speculate the existence of a class ω in $H^{n-1}(\widehat{M}_g, \Omega^{n+1})$ which is preserved by the action of $\underline{d}$. We note that $n-1$ is the holomorphic homological dimension of $\widehat{M}_g$, and we assume that ω is transgressive, so as to give a class in $H^{2n}(\widehat{M}_g, \mathbf{C})$. Then the construction in 6.3 would produce a map:

$$\begin{aligned} H_k(\underline{d}, \mathbf{C}) &\to H^{n-1}(\widehat{M}_g, \Omega^{n-k+1})_{c\ell} \\ &= \ker[d\colon H^{n-1}(\widehat{M}_g, \Omega^{n-k+1}) \to H^{n-1}(\widehat{M}_g, \Omega^{n-k+2})]. \end{aligned}$$

Then maybe an analysis of the Hodge to de Rham spectral sequence for the (infinite-dimensional) complex manifold $\widehat{M}_g$ would show that the classes in $H^{n-1}(\widehat{M}_g, \Omega^{n-k+1})_{c\ell}$ would be transgressive, giving rise to corresponding classes in $H^{2n-k}(\widehat{M}_g, \mathbf{C})$, i.e., in $H^{2n-k}(M_g'', \mathbf{C})$ (the projection map $\widehat{M}_g \to M_g''$ is a homotopy equivalence). Then using Poincaré duality we would identify $H^{2n-k}(M_g'', \mathbf{C})$ with $H_k^{c\ell}(M_g'', \mathbf{C})$. For $k = 2$ we would obtain a class in $H^{n-1}(M_g'', \Omega^{n-1})_{c\ell}$, that is, a class of pure Hodge type.

7. A double complex

7.1 The main construction of §2 may be explained in terms of a double complex K^{**} naturally associated with an action of a Lie algebra $\mathfrak{g}$ on the smooth manifold M. The (p,q)-term $K^{p,q}$ of this double complex is $K^{p,q} = \bigwedge^{-p}(\mathfrak{g}) \otimes_{\mathbf{C}} A^q(M)$.

The vertical differential is $(-1)^p 1 \otimes d$, where $d\colon A^q(M) \to A^{q+1}(M)$ is the exterior differential; the horizontal derivative

$$\delta\colon \textstyle\bigwedge^{-p}(\mathfrak{g}) \otimes A^q(M) \to \bigwedge^{-p-1}(\mathfrak{g}) \otimes A^q(M)$$

is that of the standard complex $C_*(\mathfrak{g}, A^q(M))$, where $A^q(M)$ is viewed as a $\mathfrak{g}$-module (action by Lie derivatives).

In other words, K^{**} is the bi-complex which computes the hyperhomology of the Lie algebra $\mathfrak{g}$ with coefficients in the de Rham complex $A^*(M)$.

7.2 Let K^* denote the simple complex $K^n = \bigoplus_{p+q=n} K^{p,q}$. Then there is a morphism of complexes $\varphi_k: K^* \to A^*(M)$ such that $\varphi(X_1 \wedge \cdots \wedge X_k \otimes \omega) = (-1)^k i(X_1) \ldots i(X_k)\omega$. The verification is a beefed-up version of the one given in 2.2 and therefore need not be given here.

7.3 To connect this with Construction 2.2, notice the obvious morphism of complexes, for a given integer m

$$C_p(\mathrm{Vect}(M), A^m_{c\ell}(M)) \to K^{p,m}$$
$$\bigwedge^{-p}(\mathrm{Vect}(M)) \otimes A^m_{c\ell}(M) \hookrightarrow \bigwedge^{-p}(\mathrm{Vect}(M)) \otimes A^m(M).$$

Composing this with the morphism φ of 7.2, we obtain again Construction 2.2.

7.4 The cohomology of K^* may be computed from the spectral sequence $E_2^{-k,q} = H_k(\mathfrak{g}, H^q(M, \mathbf{C})) = H_k(\mathfrak{g}) \otimes H^q(M, \mathbf{C})$; the d_2-differential maps $H_k(\mathfrak{g}) \otimes H^q(M) \to H_{k-2}(\mathfrak{g}) \otimes H^{q-1}(M)$.

Proposition 7.5. *The above spectral sequence degenerates at E_2 in the following cases:*

(a) *$\mathfrak{g}$ is abelian*
(b) $\mathfrak{g} = sl(2, \mathbf{C})$.

Proof. (a) We show that any class in $\bigwedge^k(\mathfrak{g}) \otimes H^q(M) = H_k(\mathfrak{g}) \otimes H^q(M)$ may be lifted to a cohomology class in the double complex K^{**}. This is enough to ensure the degeneracy of the spectral sequence. We describe the lift of the class of $(X_1 \wedge \cdots \wedge X_k) \otimes \omega$, where ω is a closed q-form. The component in $\bigwedge^{k-j}(\mathfrak{g}) \otimes A^{q-j}(M)$ of this lift is equal to

$$(\pm 1) \cdot \sum_{1 \leq i_1 < i_2 < \cdots < i_j \leq k} \Big((-1)^{\sum i_\ell} (X_1 \wedge \cdots \wedge \widehat{X_{i_1}} \cdots \wedge \widehat{X_{i_j}} \wedge \cdots \wedge X_k) \otimes i(X_{i_1}) \ldots i(X_{i_j})\omega \Big).$$

(b) Let (H, X, Y) be a basis of $\mathfrak{g}$ such that $[H, X] = 2X$, $[H, Y] = -2Y$, $[X, Y] = H$. Then the 3-cycle $H \wedge X \wedge Y$ generates $H_3(\mathfrak{g}, \mathbf{C})$. It suffices to show that for a closed q-form ω the class of $(H \wedge X \wedge Y) \otimes \omega$ in $H_3(\mathfrak{g}) \otimes A^q(M)$ may be lifted to a cohomology class in K^{**} (recall that $H_3(\mathfrak{g})$ is one-dimensional and $H_i(\mathfrak{g}) = 0$ for $i \neq 0, 3$). This lift has the following components.

$$(H \wedge X \wedge Y) \otimes \omega \quad \text{in} \quad \bigwedge^3(\mathfrak{g}) \otimes A^q(M)$$
$$(X \wedge Y) \otimes i(H) \cdot \omega - (H \wedge Y) \otimes i(X) \cdot \omega \quad \text{in} \quad \bigwedge^2(\mathfrak{g}) \otimes A^{q-1}(M)$$
$$+H \otimes i(X)i(Y)\omega + X \otimes i(H)i(Y)\omega - Y \otimes i(H)i(X) \cdot \omega \quad \text{in} \quad \mathfrak{g} \otimes A^{q-2}(M)$$
$$i(H)i(Y)i(X)\omega \quad \text{in} \quad A^{q-3}(M).$$

Q.E.D.

Remark 7.6. The proof of Proposition 7.5 produces in case (a) a quasi-isomorphism of K^{**} with the direct sum of the single complexes $\bigwedge^k(\mathfrak{g}) \otimes A^*(M)$ (translated by k). In particular, this gives a left-inverse to the morphism φ of 7.2.

In case (b), it may be checked that the composition of the inclusion map $H_3(\mathfrak{g}) \otimes H^q(M, \mathbf{C}) \to H^{q-3}(K^{**})$ with $\varphi: H^{q-3}(K^{**}) \to H^{q-3}(M, \mathbf{C})$ is 0. In other words, no new interesting homological construction, in the spirit of this paper, is found this way.

Remark 7.7. To an action of a Lie algebra $\mathfrak{g}$ by derivatives on an associative algebra A with unit, one associates the *algebraic crossed product* algebra B which is isomorphic to $A \otimes_{\mathbf{C}} U(\mathfrak{g})$ as a vector space whose product is characterized by the act that A and $U(\mathfrak{g})$ are subalgebras of B and that $[X, a] = X(a)$ for $X \in \mathfrak{g}, a \in A$.

In the case $A = C^\infty(M)$, the cyclic homology of B is probably related to the double complex K^{**} of 7.1 or more precisely to a version of it involving $\bigwedge^*(\mathfrak{g}) \otimes_{\mathbf{C}} U(\mathfrak{g}) \otimes_{\mathbf{C}} A^*(M)$ by an extension of Kassel's methods [7].

REFERENCES

[1] Arbarello, E., De Concini, C., Kač, V., and Procesi, C., *Moduli spaces of curves and representation theory*, Comm. Math. Phys. **17**(1988), 1–36.

[2] Borel, A. and Moore, J-C., *Homology theory for locally compact spaces*, Michigan Math. J. **7**(1960), 137–159.

[3] Brylinski, J-L, *A differential complex for Poisson manifolds*, J. Diff. Geom. **28**(1960), 93–114.

[4] Cartan, H., and Eilenberg, S., *Homological algebra*, Princeton University Press (1956).

[5] Chevalley, C and Eilenberg, S., *Cohomology theory of Lie groups and Lie algebras*, Trans. A.M.S. **63**(1948), 85–124.

[6] Dubrovin, B., Novikov, S.P. and Fomenko, A.T., *Modern geometry—methods and applications*, Graduate text in math. **93**(1984).

[7] Kassel, C., *Homologie cyclique des algèbres enveloppantes*, Invent. Math. **91**(1988), 221–251.

[8] Kostant, B., *Quantization and Unitary Representations. Part I: Prequantization*, Lecture Notes in Math., **170**, Springer-Verlag (1970), 87–208.

[9] Lichnérowicz, A., *Algèbre de Lie des automorphismes infinitésimaux d'une structure unimodulaire*, Ann. Inst. Fourier **24**(1974), 219–226.

[10] Lichnérowicz, A., *Les variétés de Poisson et leurs algèbres de Lie associées*, J. of Diff. Geom. **12**(1977), 253–300.

[11] Nomizu, K., *On the cohomology of compact homogeneous spaces of nilpotent Lie groups*, Ann. of Math. **59**(1954), 531–538.

[12] Pressley, A. and Segal, G., *Loop groups*, Clarendon Press (1986).

[13] Ruelle, P. and Sullivan, D. *Currents, flows and diffeomorphisms*, Topology **14**(1975), 319–327.

[14] Souriau, J.-M., *Structure des systèmes dynamiques*, Dunod, Paris, 1970.

[15] Weinstein, A., *The local structure of Poisson manifolds*, J. Diff. Geom. **18**(1983), 523–557.

Jean-Luc Brylinski
Department of Mathematics
McAllister 305
The Pennsylvania State University
University Park, PA 16801